BEHAVIORAL ECOLOGY AND CONSERVATION BIOLOGY

Edited by
Tim Caro

New York Oxford

Oxford University Press

1998

Oxford University Press

Oxford New York
Athens Auckland Bangkok Bogotá Buenos Aires Calcutta
Cape Town Chennai Dar es Salaam Delhi Florence Hong Kong Istanbul
Karachi Kuala Lumpur Madrid Melbourne Mexico City Mumbai
Nairobi Paris São Paulo Singapore Taipei Tokyo Toronto Warsaw

and associated companies in
Berlin Ibadan

Copyright © 1998 by Oxford University Press, Inc.

Published by Oxford University Press, Inc.
198 Madison Avenue, New York, New York 10016

Oxford is a registered trademark of Oxford University Press

All rights reserved. no part of this publication may be reproduced,
stored in a retrieval system, or transmitted, in any form or by any means,
electronic, mechanical, photocopying, recording, or otherwise,
without the prior permission of Oxford University Press.

Library of Congress Cataloging-in-Publication Data
Behavioral ecology and conservation biology / edited by Tim Caro.
p. cm.
Includes bibliographical references and index.
ISBN 0-19-510489-7; 0-19-510490-0 (pbk.)
1. Animal behavior. 2. Animal ecology. 3. Conservation biology.
I. Caro, T. M. (Timothy M.)
QL751.B3425 1998
591.5—dc21 97-18005

9 8 7 6 5 4 3 2

Printed in the United States of America
on recycled acid-free paper

For my parents,
Anthony Caro and Sheila Girling,
who see the fun in everything.

Preface

Behavioral ecology is concerned with the strategies individuals use to maximize their genetic representation in future generations. In contrast, conservation biology focuses on small populations and the means by which extinctions can be prevented and habitats can be conserved. The first discipline involves basic research on individuals, whereas the second focuses on the fate of populations. Aside from the fact that small populations are composed of a handful of individuals, the disciplines seem far apart, and until very recently there was little connection between them.

Within the last few years, however, a handful of behavioral ecologists, increasingly concerned about species losses, have begun to address issues in conservation biology. During the 1970s and 1980s, the goal of most students heading to field sites was to determine whether particular behavioral or life-history traits were adaptive. Since the mid-1980s, however, many have returned from the field with a different agenda. The species on which they were working were under threat from poaching or habitat fragmentation, and former study sites had been converted to agricultural land. Increasingly, behavioral ecologists began to associate with colleagues from applied disciplines such as forest managers, park wardens, and wildlife and conservation biologists. Using data collected in the course of their fieldwork on mating systems, foraging behavior, or habitat preferences, or simply through working on an endangered species, behavioral ecologists tried to apply their findings to developing management plans or change existing ones. As yet, however, they have made little impact on conservation beiology because they have yet to generate general principles by which behavioral ecology advances understanding in conservation. Nor have they influenced their parent field because, in regard to conservation, they have been working in isolation, lacking a common forum for expressing their results and ideas.

In addition, there is now a cohort of incoming graduate students fascinated by the major advances in behavioral ecology since the 1970s but keen to slow the losses of species and habitats they see going on around them. While they recognize that the time has come to make behavioral ecology more relevant to a world rapidly becoming dominated by conservation issues, they do not have the expertise to put this into practice. Unfortunately, many of us are not yet able to provide a lead in bringing these disciplines together either.

To provide a forum in which to link these disciplines and to generate new avenues of research, I invited a team of international behavioral ecologists to show how their discipline relates to conservation biology. I selected those behavioral ecologists who I knew were tackling theoretical or practical conservation issues and challenged each of them to formulate principles that extend beyond their own case studies. In every chapter except the first, the afterword, and the epilogue, contributors were asked first to identify a problem in conservation biology; second, to review the various ways in which it has been tackled; third, to review pertinent behavioral ecological data and present their own data that bear on the issue; fourth, to assess the strengths and weaknesses of the behavioral ecological approach as it relates to their chosen topic in conservation biology; and fifth, to suggest conservation recommendations in the light of behavioral ecological information. In juxtaposing different studies using a common framework, I hoped to generate preliminary general principles, or at least themes, by which behavioral ecological information might bear on conservation problems.

As a result of this exercise, six themes emerged and are represented by the sections of this book: applying baseline behavioral ecological data to conservation problems and to conservation intervention programs; the way in which animal mating systems affect both the conservation of species and management decisions; the importance of dispersal and inbreeding avoidance for conservation; and the behavioral ecological study of humans. The book does not aim to be comprehensive, however, and there are other important links not covered here. Nevertheless, the contributors have identified many ways in which environmentally active behavioral ecologists can use their expertise to move the conservation agenda forward. While the conservation success of individual studies will be measured by their ability to alter policy, the success of the book will hinge on its ability to shed new light on conservation problems, generate new avenues of interdisciplinary research, and encourage behavioral ecologists to be effective conservation biologists.

Many people reviewed the chapters in this book, and I would like to thank Steve Albon, Michael Alvard, Joel Berger, Marc Bekoff, Monique Borgerhoff Mulder (2 chapters), Rachel Brock, Scott Carroll, Alice Clarke, Scott Creel, Eberhard Curio, Clare FitzGibbon, Mike Fogarty, Matt Gommper, Paule Gros (2), Mart Gross, Phil Hedrick, Charles Janson, Astrid Kodric-Brown, Walt Koenig, John Lazarus, Mike Lombardo, Marc Mangel, Peter Marler, Manfred Milinski, Peter Moyle, Craig Packer, Chris Ray, John Robinson, Paul Siri, Andrew Smith, Tracey Spoon, Judy Stamps (2), Mark Stanley Price, Cathy Toft, James Umbanhowar, Dirk Van Vuren (2), Nadja Wielebnowski, and 17 additional anonymous reviewers. I thank Eric Paulovich for retyping many of the manuscripts, Kirk Jensen, Lisa Stallings, and Karla Pace for assistance at Oxford University Press, and Monique Borgerhoff Mulder and Barnabas Caro for support in the final stages of editing.

T.C.

Davis, California
July 1997

Contents

Contributors

Michael Alvard
Department of Anthropology
State University of New York at Buffalo
380 MFAC
Buffalo, NY 14261
USA
Alvard@acsu.buffalo.edu

Joel Berger
Program in Ecology, Evolution,
and Conservation Biology
University of Nevada
1000 Valley Road
Reno, NH 89512
USA
berger@ers.unr.edu

Tim Caro
Department of Wildlife, Fish,
and Conservation Biology
and Center for Population Biology
University of California, Davis
Davis, CA 95616
USA
tmcaro@ucdavis.edu

Scott Creel
Department of Biology
Montana State University
Bozeman, Montana 59717
USA
screel@gemini.oscs.montana.edu

Eberhard Curio
Arbeitsgruppe für Verhaltensforschung
Fakultät für Biologie
Ruhr-Universität Bochum
D-44780 Bochum
Germany
eberhard.curio@ruhr-uni-bochum.de

Martin Daly
Department of Psychology
McMaster University
Hamilton, Ontario L8S 4K1
Canada
daly@mcmaster.ca

Andy Dobson
Department of Ecology
and Evolutionary Biology
Princeton University
Princeton, NJ 08544
USA
andy@eno.Princeton.edu

Sarah Durant
Institute of Zoology

Zoological Society of London
Regent's Park
London NW1 4RY
UK
s.durant@ucl.ac.uk

John Eadie
Department of Wildlife, Fish,
and Conservation Biology
University of California, Davis
Davis, CA 95616
USA
jmeadie@ucdavis.edu

Clare FitzGibbon
Large Animal Research Group
Department of Zoology
University of Cambridge
Cambridge CB2 3EJ
UK
sokoke@users.africaonline.co.ke

Stephen Gordon
Department of Psychology
McMaster University
Hamilton, Ontario L8S 4K1
Canada

Mats Grahn
Molecular Population Biology Laboratory
Department of Animal Ecology
Ecology Building
University of Lund
S-223 63 Lund
Sweden
Mats.Grahn@zooekol.lu.se

Correigh Greene
Animal Behavior Graduate Group
University of California, Davis
Davis, CA 95616
USA
cmgreene@ucdavis.edu

Alexander Harcourt
Department of Anthropology
University of California, Davis
Davis, CA 95616
USA
ahharcourt@ucdavis.edu

Åsa Langefors
Molecular Population Biology Laboratory
Department of Animal Ecology
Ecology Building
University of Lund
S-223 63 Lund
Sweden
Asa.Langefors@zooekol.lu.se

Marc Mangel
Department of Environmental Studies
University of California, Santa Cruz
Santa Cruz, CA 95064
USA
msmangel@cats.ucsc.edu

Peter McGregor
Behaviour Group
Zoological Institute
University of Copenhagen
Tagensvej 16, DK-2200
Copenhagen N
Denmark
pkmcgregor@zi.ku.dk

Tom Peake
Behaviour Group
Zoological Institute
University of Copenhagen
Tagensvej 16, DK-2200
Copenhagen N
Denmark
tmpeake@zi.ku.dk

Joyce Poole
P.O. Box 24467
Nairobi
Kenya

Daniel Rubenstein
Department of Ecology
and Evolutionary Biology
Princeton University
Princeton, NJ 08544
USA
dir@princeton.edu

Yvonne Sadovy
Department of Ecology and Biodiversity
The University of Hong Kong

Pokfulam Road
Hong Kong
yjsadovy@hkusua.hku.hk

Brad Semel
Illinois Department of Natural Resources
Division of Natural Heritage
110 James Road
Spring Grove, IL 60081
USA
bsemel@dnrmail.state.il.us

Paul Sherman
Section of Neurobiology and Behavior
Cornell University
Ithaca, NY 14853
USA
tmn3@cornell.edu

Mandy Tocher
Science and Research Division
P.O. Box 5244
Dunedin
New Zealand
mtocher@doc.govt.nz

James Umbanhowar
Center for Population Biology
University of California, Davis
Davis, CA 95616
USA
jaumbanhowar@ucdavis.edu

Dirk Van Vuren
Department of Wildlife, Fish,
and Conservation Biology
University of California, Davis
Davis, CA 95616
USA
dhvanvuren@ucdavis.edu

Amanda Vincent
Department of Biology
McGill University
1205 Avenue Dr. Penfield
Montréal, Québec H3A 1B1
Canada
amanda_Vincent@maclan.mcgill.ca

Torbjörn von Schantz
Molecular Population Biology Laboratory
Department of Animal Ecology
Ecology Building
University of Lund
S-223 63 Lund
Sweden
zoo_TVC@luecology.ecol.lu.se

Bruce Waldman
Department of Zoology
University of Canterbury
Private Bag 4800
Christchurch
New Zealand
bw@zool.canterbury.ac.nz

Nadja Wielebnowski
Department of Wildlife, Fish,
and Conservation Biology
University of California
Davis, CA 95616
USA
nwielebn@sivm.si.edu

Margo Wilson
Department of Psychology
McMaster University
Hamilton, Ontario L8S 4K1
Canada
wilson@mcmaster.ca

BEHAVIORAL ECOLOGY AND CONSERVATION BIOLOGY

1

The Significance of Behavioral Ecology for Conservation Biology

Tim Caro

Two Disciplines

Imagine that you are a beginning graduate student or professor interested in topics in both behavioral ecology and conservation biology. You want to apply the principles and methods in behavioral ecology to the conservation of species and habitats. What should you study? How can you make a meaningful contribution that will last for 5, perhaps 10 years? More specifically, is it really sensible to ask if a species' mating system can affect its survival chances, or whether patterns of sex allocation might influence the success of a reintroduction program, or if territorial behavior affects plans to bolster wild populations?

These are difficult questions to answer because the two disciplines are so different. Conservation biology is a theoretical and applied discipline aimed at preventing population extinction, whereas behavioral ecology attempts to understand the way in which behavioral and morphological traits contribute to the survival and reproduction of individual animals and plants under different ecological circumstances. Although it is a truism that populations are composed of individuals, it has proved relatively difficult to understand how the behavior of individuals affects population dynamics. Nevertheless, advances are being made (see Lomnicki, 1988; Sutherland, 1996; Clutton-Brock et al., 1997; Goss-Custard and Sutherland, 1997). For example, Sutherland and Dolman (1994) have applied models of interference competition and resource depletion to determine density-dependent mortality within habitat patches; Ives and Dobson (1987) have modeled how antipredator behavior affects population dynamics (see also FitzGibbon and Lazarus, 1995); Eadie and Fryxell (1992) have shown how female reproductive strategies can affect population dynamics; and Lima and Zollner (1996) are beginning to apply patterns of animal dispersal and habitat selection to landscape ecology. Of more direct relevance to conservation biology, Caughley and Gunn (1996) have collated case studies to highlight how a species' natural history predisposes it to changes in habitat, to predation or competition from introduced animals, or to hunting. These developments are still limited in scope, however, because they have yet to demonstrate how the swathe of concepts in behavioral ecology can be applied to conservation problems. The purpose of this book is to show how this can be accomplished. In this first chapter, I review the significance for conservation biology of various components of behavioral ecology, broadly divided into quantitative natural history, concepts, and methods, by collating a number of disparate studies including those in this book (see also Caro and Durant, 1995; Clemmons and Buchholz, 1997; Strier, 1997). First, however, I sketch the field of conservation biology in order to flag, in advance, those areas likely to benefit from behavioral ecology and those that have little to gain from it.

Conservation Biology: A Sketch

Conservation biology is a multidisciplinary science that can be divided into three general areas: documenting the extent of biodiversity; understanding the nature, causes, and consequences of loss of genes, populations, species, and habitats; and attempting to develop practical methods to prevent species' extinctions and to allow for continuation of ecosystem processes (Soulé, 1985; Wilson, 1992; Meffe and Carroll, 1997). It uses principles from ecology, population genetics, and systematics to describe the breadth of biological diversity and ways to conserve it (Simberloff, 1988). Increasingly, conservation biology employs economic and philosophical concepts as well as anthropological and sociological data to understand the effects of human activity on species and habitats.

Loss of Biodiversity

It is difficult to appreciate current rates of species loss unless we have some estimate of both the number of species alive today and their rates of decline. Unfortunately, only a small (but unknown) proportion of taxonomic diversity has been documented (May, 1988, 1995), so indirect methods of estimating the number of extant species have been devised. These include the use of environmental variables, indicator groups, or higher taxa as measures of species diversity (Gaston, 1996). Current and future rates of extinction are estimated using species-area relationships, extrapolating from known rates of loss in historical times, and collating additions to Red Data books which list threatened and endangered species (Smith et al., 1993; Whitmore and Sayer, 1992; Mace, 1994; Bibby, 1995; Steadman, 1997).

To maximize conservation effort, we also need to identify areas of greatest biodiversity. At a gross level these include tropical rainforests, coral reefs, and the ocean floor. At a more local scale, centers of diversity are pinpointed using geographic ranges of species in well-known classes such as butterflies and mammals (World Conservation and Monitoring Centre, 1992). To determine whether taxonomic, specific "hot spots" predict areas of high biodiversity in other taxa, researchers examine the extent to which hot spots overlap (e.g., Balmford and Long, 1995; Williams et al., 1996; Robbins and Opler, 1997). These analyses are facilitated by computer-aided superimposition of geographic ranges on each other (GAP analysis; Caicco et al., 1995). It is difficult to envisage how behavioral ecology can enlarge our understanding of the extent and loss and location of biodiversity other than in the trivial sense of encouraging fieldwork in many areas of the world.

Factors Affecting Small Populations

Populations may be subject to sustained pressures (for instance, from hunting) and suffer deterministic decline, or they may be subject to stochastic events such as occasional droughts or floods. A central focus of conservation biology since its inception in the mid

1980s has been to describe and predict the responses of small populations to genetic, demographic, and environmental stochasticity (Caughley, 1994). Thus, when a local population declines to low levels, genetic variation is lost as a result of genetic drift, genetic bottlenecks, and inbreeding. Loss of genetic variation is usually measured as a reduction in average heterozygosity per individual per locus. Reduction in heterozygosity results in the expression of deleterious recessive alleles, which can be manifested in reduced fecundity, higher juvenile mortality, and increased susceptibility to disease, each of which may pose a threat to a population's persistence (Allendorf and Leary, 1986; Frankham, 1995). In the long term, loss of rare alleles from a population may result in a reduced ability to adapt to changing environmental conditions. The expected rate of loss of genetic variation depends on the genetic effective population size, which is lowered by a skewed sex ratio of breeding animals, high variance in family size, and fluctuations in the number of breeding individuals (Franklin, 1980). The study of mating systems therefore has a direct bearing on effective population size (N_e).

Demographic stochasticity refers to the effects of random alterations in population age-structure and sex ratio on the survival and reproduction of individuals. It is manifested only at very small population sizes of around 50 breeding individuals or less. Demographic information is clearly relevant in predicting small population persistence.

Environmental stochasticity refers to unpredictable events such as changes in weather or biotic factors, such as food supply, predation, or disease (Scott, 1988), which alter the population's mean rate of reproduction or mortality. Environmental factors are viewed as key to population persistence because they affect large and small populations alike, and nowadays many are human in origin. Such anthropogenic factors include habitat fragmentation (Harris, 1984), competition or predation by exotic species (Atkinson, 1989), and hybridization with either native or introduced species (see Diamond, 1984). In a small population, genetic, demographic, and environmental factors all combine to affect the probability of the population persisting. These interactions are modeled using population viability analyses (PVAs), which predict the probability of a population going extinct over a specified time period (Shaffer, 1981; Gilpin and Soulé, 1986; Lande, 1988; Boyce, 1992). They can be used to generate genetic and demographic minimum viable population sizes (MVPs) for a variety of species. PVAs require demographic data collected in the course of long-term field studies.

When habitats become fragmented, the total area is reduced in size and what is left becomes restricted to smaller, more isolated patches. Rapid fragmentation may result in rapid loss of species between remaining fragments, crowding inside fragments, and changes in community structure (Lovejoy et al., 1986; Bierregaard et al., 1992). Subsequently, populations living in habitat fragments may be affected by changes in microclimate around the edges of patches (Murcia, 1995) and by increased rates of predation and parasitism from outside, which are often facilitated by human settlement between fragments (Brittingham and Temple, 1983). Individuals may also experience difficulties in moving between fragments (Dunning et al., 1992). Habitat fragmentation is therefore seen as one of the greatest threats to biodiversity (Harris, 1984; Wilcove et al., 1986). Individuals' dispersal and habitat selection decisions will have a strong influence on population responses to habitat fragmentation.

Fragmentation may lead to population subdivision. In some instances, subpopulations in habitat patches may go extinct independently but with equal probability, and these vacant habitat patches may subsequently be recolonized; these are known as metapopulations (Levins, 1970; Gilpin and Hanski, 1991; Hanski and Gilpin, 1996). Metapopulations have been used as models for examining effects of habitat fragmentation on the viability of once-continuous populations and for examining population viability of species living in naturally patchy environments. In partially isolated subpopulations, genetic variation is subject to different rates of drift and loss compared to a single, contiguous population (Lande and Barrowclough, 1987; Wade and McCaughley, 1988; Hedrick and Gilpin, 1996). Despite the prominent position of metapopulations as a heuristic tool in conservation biology, it remains an open question as to whether naturally or anthropogenically fragmented populations conform to the assumptions of metapopulation theory (Harrison, 1994). The study of dispersal and its extent, rate, and consequences therefore has direct relevance to metapopulation theory.

Practical Aspects of Conservation Biology

Conservation biology employs a wide variety of methods in seeking solutions to loss of biodiversity. These include setting priorities as to which species, populations, or habitats require conservation protection, designing reserves, assessing the effects of different forms of protection, managing reserves, limiting trade in wildlife, and captive breeding and reintroductions.

It is now recognized that we will be unable to save all species; systematists have therefore devised objective methods for ranking species according to taxonomic uniqueness and representation in ecological communities (Vane-Wright et al., 1991; Faith, 1995). In addition, management plans for subpopulations of endangered species are formulated by a combination of ranking both local population sizes and imminent threats to them (e.g., IUCN/SSC Asian Rhino Specialist Group, 1989). Behavioral ecology has little to contribute here.

The size and shape of reserves once occupied a central place in conservation biology because island biogeography theory predicts that large areas hold more species than small areas (MacArthur and Wilson, 1967; Diamond, 1975). This prompted a debate as to whether a single large reserve would hold more species than several small ones of the same total area. The issue was never resolved and is now largely forgotten (Soulé and Simberloff, 1986), except that all agree that a large reserve will protect a greater number of species for longer than a small reserve (Newmark, 1987). Interest now focuses on connections between protected reserves or unprotected habitat fragments (Hansson et al., 1995), their efficacy in allowing organisms to move between them (Harrison, 1992; Rosenberg et al., 1997), and their limitations, as they may facilitate the spread of disease and fire (Noss, 1987; Simberloff and Cox, 1987; Hess, 1994). Animal dispersal is perhaps the key element in corridor design.

In addition, the concept of a reserve has broadened from that of an area free from human activity (other than photographic tourism) to include multiple-use areas consisting of a central, well-protected core area surrounded by buffer zones that allow for human use, as well as modified landscapes that include human settlements and roads (Forman, 1993). In the tropics, reserves in which resource exploitation occurs (extractive reserves) are seen as an important alternative to full protection (Fearnside, 1989). There, economic benefits that local people derive from removing nontimber forest products, from subsistence hunting, or even from selective logging, are increasingly viewed as ensuring the long-term future of wilderness areas (Redford and Padoch, 1992; Western et al., 1994; Redford and Mansour, 1996; Freese, 1997). The study of human foraging and resource use is important in understanding the likely success of extractive reserves.

Finally, conservation biologists are attempting to formulate general management principles for protected areas. As illustrations, these include preventing illegal hunting by concentrating antipoaching efforts in small areas (Leader-Williams and Albon, 1988), practicing disturbance regimes that promote habitat mosaics (Warren, 1990), encouraging periodic fires to stimulate regeneration (Knight and Wallace, 1989), and monitoring populations within protected areas (Goldsmith, 1991). Behavioral ecology has only marginal relevance for reserve maintenance, principally through devising monitoring strategies.

In some cases, trade in wildlife and wildlife products is one of the greatest drains on biodiversity (Fitzgerald, 1989; Dobson, 1996), and international laws such as the Convention on International Trade in Endangered Species (CITES) are in place to regulate this trade. The system used to classify species according to risk of extinction has recently been made more quantitative and objective and incorporates extinction probabilities over specified time periods (Mace, 1995). Behavioral ecologists have little to offer in this area.

As natural habitats rapidly disappear, the importance of zoological institutions in conserving endangered species has grown (Soulé et al., 1986; Conway, 1989). Zoo breeding programs manage their populations genetically so as to maximize the number of founders, equalize founder representation, and minimize inbreeding (Foose and Ballou, 1988; Ballou and Foose, 1996). Pedigree records have been constructed that enable zoos to identify distantly related breeding partners; these can then be moved between institutions. Zoos increasingly use artificial insemination, oocyte and semen collection, *in vitro* fertilization, and cryopreservation techniques in breeding (Holt, 1994; Asa et al., 1996; Wildt, 1996). Hybridization between species is actively discouraged (see Wayne and Jenks, 1991, for a controversial example). Because zoos have a limited capacity to hold large numbers of animals (Maguire and Lacey, 1990; Snyder et al., 1996), new priorities for breeding endangered species are under discussion (Balmford et al., 1996). Captive breeding programs require information from natural populations to maintain and breed endangered species effectively.

One of the goals of captive breeding is to reintroduce species into the wild (Wilson and Stanley Price, 1994). Reintroduction is composed of four phases: a feasibility study, a preparation phase, a release phase, and postrelease monitoring. Despite this protocol, the number of attempts has been limited (Stuart, 1991), and successes are few (Beck et al., 1994). Currently, this makes it difficult to generalize about the factors that promote success (Stanley Price, 1989), but some evidence suggests that specific techniques used in reintroductions may be more important than life-history variables of the species involved (Kleiman et al., 1994; Veltman et al., 1996). Naturalistic studies may be critical in devising prerelease training and postrelease monitoring strategies in reintroduction attempts.

Finally, conservation biology is explicitly a value-laden discipline (Soulé, 1985; Barry and Oelschlaeger, 1996), and there is active debate about the relative importance of reasons to conserve biodiversity (e.g., Ehrenfeld, 1976; Oldfield, 1984). These can be dichotomized as utilitarian reasons and reasons of intrinsic value. The former includes conserving biodiversity to provide genetic material to improve commercial crops (Vietmeyer, 1986), or for use as new drugs (Farnsworth, 1988), new foods, or commercial products (Plotkin, 1988) and because of its importance in maintaining ecosystem stability (Naeem et al., 1994; Johnson et al., 1996; Hooper and Vitousek, 1997; Tilman et al., 1997). The latter set of reasons center on the intrinsic worth of nonhuman species (Norton, 1986), and a considerable interest in conservation ethics has developed among philosophers (e.g., Taylor, 1986; Rolston, 1988).

The scope of behavioral ecology's potential impact on conservation biology is summarized in table 1-1.

Table 1-1. Areas of conservation likely or unlikely to benefit from behavioral ecology.

Area	Likely to benefit from behavioral ecology? Yes	No
Extent of biodiversity		×
Loss of biodiversity		×
Genetic stochasticity and N_e	×	
Demographic stochasticity and population viability analyses	×	
Environmental stochasticity	×	
Habitat fragmentation and metapopulations	×	
Prioritizing management plans		×
Reserve connectivity	×	
Extractive reserves	×	
Managing reserves		×
Trade in wildlife		×
Captive breeding	×	
Species reintroductions	×	
Philosophical issues		×

Links Between Behavioral Ecology and Conservation Biology

Quantitative Natural History

General Measures of Behavior

Behavioral ecological studies collect routine data on activity patterns, residential group size, home range size, and territorial behavior; such information is utilized widely by conservation biologists. For instance, activity patterns (whether a species is diurnal, crepuscular, or nocturnal) affect monitoring schedules and strategies.

Patterns of grouping affect population subdivision and hence metapopulation structure, which in turn affects population persistence (Hanski and Gilpin, 1991, 1996). In chapter 5, Durant explores the effects of population subdivision and dispersal between subgroups on population persistence time under different management intervention regimes. Group size is also positively correlated with the prevalence and intensity of contagious parasites across taxa (Cote and Poulin, 1995; Dobson and Poole, chapter 8, this volume). Composition of subgroups can also affect persistence: where residential groups are small in size or unusual in composition, reproductive output may be disproportionately lowered (the Allee effect; Saether et al., 1996). Conversely, individuals may be attracted to each other and only breed in the presence of others, as noted for several bird species (Muller et al., 1997). Stamps (1988) has shown experimentally that juvenile *Anolis aeneus* lizards prefer to settle in sites next to territory owners rather than in adjacent empty habitat patches. In chapter 8, Dobson and Poole explore other means by which conspecific attraction influences population viability, as in the way roving solitary male and groups of female elephants (*Loxondata africana*) find each other in exploited populations (see also Smith and Peacock, 1990; Reed and Dobson, 1993).

Home range size has a direct bearing on reserve size and shape: some national parks such as Serengeti were delineated to encompass the annual home range of individuals of the keystone species they were trying to protect (Grzimek and Grzimek, 1959).

Territorial behavior has the potential to reduce the number of individuals that a given area can support; thus the ecological and social circumstances that promote territoriality will affect population persistence. In chapter 12, Eadie, Sherman, and Semel highlight a counterintuitive consequence of territoriality in cavity-nesting birds. When artificial nest sites are provided to increase population size, species lacking territorial behavior suffer from high rates of conspecific brood parasitism, which lowers reproductive success, whereas territorial defense of nest-boxes prevents parasitism, enabling the population to benefit from this management procedure.

Baseline behavioral and ecological data collected in the course of field studies have the potential to predict a population's response to habitat disturbance. For example, frequency of intraspecific interactions may affect spread of disease within populations (Anderson and May, 1979). In chapter 3, Harcourt attempts to determine which ecological and behavioral variables successfully predict primate species' ability to survive in selectively logged forest. Similarly, it has been argued that information about a species' behavior under different ecological and social conditions is the key to success of reintroductions (Stanley Price, 1989). Indeed, Curio (see chapter 7) suggests that a better understanding of antipredator behavior and its development would lead to greater survival and reproduction of captive birds released into the wild. Finally, many problems in *ex situ* conservation have been solved or ameliorated using knowledge of activity patterns, diet, and social behavior (Kleiman, 1994; Carlstead, 1997). In chapter 6, Wielebnowski systematically reviews important areas of behavioral research that are crucial to captive propagation of mammals, including space requirements, social organization, parental care, and sexual behavior.

Foraging Behavior

Foraging behavior affects population persistence in the general sense that carnivores are at greater risk of extinction than herbivores, and that, within a dietary class, specialist foragers are more extinction prone than generalists (Brown, 1971). Thus a knowledge of the food requirements of rare species is useful. At a general level, habitat preferences contingent on feeding requirements will influence species' responses to habitat disturbance and fragmentation. More specifically, foraging theory has shown how patterns of interference between competitors and depletion of feeding patches predict the distribution of individuals and hence the number that a site can hold (see Goss-Custard and Sutherland, 1997, for a review). Moreover, models have been developed to predict the distribution of animals in patches of differing profitability when individuals are equally competitive (Milinski, 1979) or when some dominate others at resources (Parker and Sutherland, 1986). In theory, these tools might allow conservation biologists to predict consequences of habitat fragmentation if factors such as food abundance, food distribution, and individuals' competitive abilities were known. Monaghan (1996) has shown how patterns of feeding in seabirds such as foraging trip length and number of dives can be used to monitor marine fish populations.

Antipredator Behavior

Predation is an important source of mortality in the wild (Edmunds, 1974). Many species of mammals and flightless birds have gone extinct on islands as a result of introduced domestic cats, rats, and mongooses (Caughley and Gunn, 1996). These endemics lost certain aspects of predator avoidance behavior as they evolved on islands without terrestrial predators; preliminary evidence suggests that some continental prey species are also losing their ability to respond to cues of predators in places where large carnivores have been extirpated (see Berger, chapter 4, this volume). Understanding antipredator behavior may allow us to make predictions about the fate of populations subjected to introduced predators or conservation manipulation. Controversial data (Berger et al., 1993; Berger and Cunningham, 1994b; Loutit and Montgomery, 1994) suggest that black rhinoceros *Diceros bicornis* calves suffer higher mortality in the presence of large carnivores if their mothers are dehorned as part of a conservation procedure, presumably as a result of mothers' inability to fend off predatory attack. Prior knowledge of maternal antipredator behavior and the function of horns in females would have spoken against dehorning. Since grouping confers advantages on individuals through early detection of predators (FitzGibbon, 1990), warning conspecifics of danger (Sherman, 1981), reducing vigilance time (Elgar, 1989), and diluting predation risk (Hamilton, 1971), variation in patterns of grouping make species and age–sex classes differentially vulnerable to hunting by humans. In chapter 16, FitzGibbon discusses how antipredator strategies affect the probability of being selected by subsistence hunters. She notes that individuals respond to predation risk by reducing time spent in other activities such as feeding (Lima and Dill, 1990), but the extent to which human hunting pressure indirectly affects reproductive rates in this way remains unexplored.

Whether antipredator behavior is principally innate or is modified by experience is critical to captive release programs (Curio, chapter 7), and some attempts have been made to train captive-bred individuals to recognize and avoid predators (Maloney and McLean, 1995; McLean et al., 1995). Although we know that predator recognition is innate in some species such as motmots *Eumomota superciliosa* (Smith, 1975) but must be learned in others such as vervet monkeys *Cercopithecus aethiops* (Cheney and Seyfarth, 1990) and Belding's ground squirrels *Spermophilus beldingi* (Mateo, 1996), the ecological circumstances under which different sorts of predator template development occurs are still poorly understood. Innate recognition may facilitate rapid antipredator responses, but it may limit our ability to transfer local stocks between areas with different predators. Learned predator recognition, in contrast, may result in high mortality initially, but it does allow reintroduced prey to respond to predators or free-living prey to cope with exotic or reintroduced predators.

Demographic and Life History Variables

Behavioral ecological field studies usually have records on the age and sex of individuals, so it is relatively easy to construct the age structure of the population and the sex ratios of juvenile and adult segments. Estimates of N_e are greatly influenced by population demographics such as these. As an illustration, in chapter 10, Creel uses long-term demographic records from social carnivore species to investigate the demographic factors affecting N_e.

Because behavioral ecologists are interested in the consequences of traits on individual reproduction and survival, studies usually strive to collect detailed records on individuals' age at first reproduction, interbirth intervals, litter sizes, offspring sex ratios, and longevity, although information on age-specific mortality is more difficult to collect. Life history variables are necessary for the construction of PVAs used to calculate both demographic and genetic MVPs (Burgman et al., 1993). Although influential and early PVAs used data collected by field biologists to make predictions about population persistence (grizzly bear

Ursus arctos: Shaffer, 1983; spotted owl *Strix occidentalis:* Thomas et al., 1990), they were based on relatively crude empirical data. In the context of politically charged controversies where models are scrutinized closely, resulting recommendations may be called into question (Chase, 1996). Detailed information on individual life histories can remedy this because they provide an exact measure of variance, which greatly increases the resolution of PVA and N_e calculations. In chapter 5, Durant highlights the importance of life history variables in affecting population persistence using three mammal species with different social structures. Her models show that skewed sex ratios are a powerful indicator of extinction and that extending adult survival is effective in increasing population growth rates.

Concepts in Behavioral Ecology

Competition between Males

Certain behavioral and morphological traits are solely used in competition for access to members of the opposite sex (Darwin, 1871; Andersson, 1994). Contests between males have led to the evolution of weaponry such as beetle and deer antlers (Eberhard, 1979; Clutton-Brock, 1982), to infanticidal behavior, in which males kill offspring fathered by competitors (Hausfater and Hrdy, 1984), and to mate guarding (Carroll, 1993). In some species, high costs of fighting have selected for polymorphic fighting tactics in which some males compete severely over females, whereas others avoid confrontation and mate surreptitiously (Gross, 1985, 1996).

The outcome of contests between males will affect who breeds and hence variance in male reproductive success, which affects N_e. Another example of how variance in male reproductive success affects genetic components of populations comes from bison *Bison bison* in the Badlands National Park. There, one lineage is currently being lost because males of that lineage are more timid and less able to defend females than those of the more successful line (Berger and Cunningham, 1994a). Knowledge of patterns of mating and the cues that individuals use to select mates can also help identify the potential for hybridization between species (Grant and Grant, 1997); because one of the goals of conservation is species integrity and persistence, hybridization is normally viewed with alarm (see O'Brien and Mayr, 1991).

In chapter 11, Greene, Umbanhowar, Mangel, and Caro show that in infanticidal species where males kill offspring fathered by other males, population growth rates are reduced, and it may be difficult for such species to withstand exploitation. In theory, N_e will be lowered in species in which infanticide or mate guarding is prevalent.

Different fighting tactics are usually associated with different mortality schedules; thus the frequency with which males display different behavioral and morphological tactics will affect population growth rate and N_e. In chapter 19, Rubenstein draws attention to the fact that those Atlantic salmon *Salmo salar* migrating to the sea are more likely to be caught by fishermen than conspecifics that remain behind and mature in their natal streams and that this differential offtake selects for philopatric phenotypes.

In a number of reef-dwelling fishes, intersexual competition for mating opportunities has led to the evolution of protandry, in which individuals change from male to female, or protygyny, in which they change from female to male. Sex change may be precipitated by reaching a certain age or size or by social circumstances when it pays individuals to become the currently rarer sex. In chapter 9, Vincent and Sadovy argue that exploitation of larger fish for the tropical fish trade may remove males of certain species, for example, from a population. Thus the speed with which females are able to change to being male will be crucial in determining a population's response to exploitation.

Mate Choice by Females

Among females, mate choice is increasingly seen as a means to enhance offspring fitness through choice of males with heritable, condition-enhancing traits such as resistance to parasites (Andersson, 1994). In the wild, loss of males from small populations may theoretically limit the range of males from which to choose to the extent of slowing the rate of population increase (see Dobson and Poole, chapter 8). In captive populations, some degree of female choice may be advisable because, as managers, we are poor at picking out healthy stock, at identifying traits that enhance fitness, and we have little to no knowledge of trait heritability. In chapter 13, Grahn, Langefors, and von Schantz argue that failing to allow for mate choice in captive salmon stocks has resulted in disease-induced mortality. Mate choice has possible implications for captive breeding of endangered birds, too. Zebra finches *Poephila guttata* show preferences for mating with individuals wearing particular color bands (Burley et al., 1982) and will adjust the amount of parental investment and sex ratio of their offspring according to the color of their partner's leg bands (Burley 1981, 1986). This may upset population projections of *in situ* breeding programs.

Recently, considerable attention has focused on the relationship between mate choice and the degree of symmetry in male phenotypic traits. This is because symmetrical traits are thought to reflect developmental stability and hence the ability of the genotype to control development under a range of environmental conditions (Watson and Thornhill, 1994). There is evidence that symmetrical ornaments correlate with fitness (Møller et al., 1996) and that females choose males on the basis of ornament symmetry (Møller, 1992). Although behavioral ecologists are interested in trait asymmetry as a marker of genetic quality, asymmetry also reflects the extent of environmental insult during development and, measured repeatedly, can provide an index of a population's response to environmental change over time (Clarke, 1995; Tracy et al., 1995). Furthermore, asymmetric leg bands placed on captive male birds may modify female choice and could result in lower reproductive output in endangered species (Swaddle, 1996).

Mating Systems

The distribution of resources crucial to females, and hence the economic defendability of females by males (Emlen and Oring, 1977) interact with patterns of parental care to produce the mating system. Because N_e is lowered by increasing skew in the ratio of breeding females to breeding males, whether a population is polygynous or monogamous, or more specifically, the extent of extrapair paternity (Birkhead and Møller, 1982), therefore has a strong bearing on loss of genetic diversity. In addition, N_e is lowered by increasing variance in family size, which is marked in reproductively suppressed species (Creel and Waser, 1991, 1994). In chapter 10, Creel systematically explores aspects of the mating system that affect N_e using demographic data on wild carnivores. He reports that unequal reproductive success causes the greatest reduction in N_e, with skewed sex ratios and variation in population size being less important.

Differences in mating system will also affect a population's ability to sustain commercial exploitation or harvesting (Ginsberg and Milner-Gulland, 1994). As an example, population growth rates of monogamous species are likely to be less resilient to removal of individuals than are those of polygynous species because in monogamous species there are fewer males available for remating and parental care (see Greene et al., chapter 11).

Parental Care and Sex Allocation

Theory underlying the evolution of parental care, including intraspecific variation in care in relation to costs and benefits, parent–offspring conflict, and differential investment in sons and daughters, is well understood (Clutton-Brock, 1991). The sex that cares for offspring and the period of time over which care occurs will affect a species' ability to withstand offtake of a given magnitude and selectivity. In chapter 9, Vincent and Sadovy suggest that loss of parental males could reduce the reproductive output of an exploited population in taxa such as seahorses in which male care limits female reproductive rate.

Adaptive alterations in offspring production can affect conservation agendas in at least two ways. First, in populations that are selectively exploited, parents may bias offspring production in favor of the rarer sex. For example, Creel and Creel (1997) have shown that in lions birth sex ratios are male biased in populations hunted for male trophies, and this has ramifications for sustainable offtake. In chapter 19, Rubenstein shows how failure to account for sex ratio adjustment in onagers *Equus hemionus* jeopardized a reintroduction program. Second, bird recovery programs make use of a female's disposition to lay replacement clutches if her first brood is removed, and they also capitalize on hatching asynchrony to cross-foster the first hatchling to a conspecific or heterospecific (Curio, 1996; Stoleson and Beissinger, 1997). Both management strategies can improve reproductive output of endangered species. In addition, knowledge of siblicide in birds (Mock, 1984) is useful in preventing reproductive loss when trying to breed endangered birds.

Dispersal and Inbreeding Avoidance

In theory, one might expect the rate and distance over which animals disperse and its associated costs to have greater ramifications for conservation strategies than any other behavioral ecological variable. Rates of dispersal between habitat patches affect population persistence (Durant, chapter 5), gene flow between subgroups influences heterozygosity and genetic drift (Gilpin, 1991), and the length and direction dispersers travel and their habitat preferences during movement speak to reserve design and connectivity (Kooyman et a., 1996). Although dispersal is often studied in animal (Chepko-Sade et al., 1987) and plant populations (Stacy et al., 1996) high-quality data are, unfortunately, difficult to obtain (Johnson and Gaines, 1990); for example, maximal dispersal distances are usually underestimated in behavioral studies (Koenig et al., 1996). Outside of PVAs, there have been few attempts to use empirical data on dispersal in conservation theory or practice (but see Lidicker and Koenig, 1996). To remedy this, in chapter 14, Van Vuren outlines the way in which dispersal distance, dispersal direction, and survival of dispersers might influence design of reserves and managing areas between them.

As populations become increasingly isolated, the threat of inbreeding increases (Mills and Smouse, 1994). Knowledge of the extent of philopatry and strength of inbreeding avoidance mechanisms across taxa can identify which species are most likely to be subject to inbreeding depression (see Waldman and Tocher, chapter 15). This would then prioritize which species or populations require translocations and on which species noninvasive genetic screening might be carried out, as advocated by Waldman and Tocher. In addition, the extent to which different species tolerate inbreeding in the wild informs captive programs about the relative importance of avoiding pairing relatives in zoo breeding programs.

Cooperation and Helping

Many examples of cooperative breeding in birds and mammals are a consequence of habitat saturation (Koenig et al., 1992), a situation increasingly likely to occur as habitats become isolated and smaller in area through fragmentation. Populations of cooperative breeders are likely to have lower population growth rates than populations where all individuals pair up. For instance, Seychelles warblers *Acrocephalus sechellensis* transferred from densely populated Cousin Island to empty Aride Island exhibited faster rates of reproduction than their founding population (Komdeur et al., 1991). On the other hand, the presence of helpers or floating nonresident individuals can act as a buffer to small decreases in population size, but the dynamics of these conflicting effects are unexplored. Knowledge of the circumstances under which cooperative breeding will appear in different species may therefore inform intervention strategies, while reproductive benefits that parents receive from offspring can help zoo breeding programs set up appropriate social groups.

Methods in Behavioral Ecology

Individual Recognition

Behavioral ecology focuses on the adaptive significance of variation in individual behavior, morphology, and physiology which has forced scientists to learn to recognize individual animals. Consequently, all the individuals within a subpopulation are often known to researchers (see Kelly et al., in press), fortuitously providing an accurate estimate of local population size. Individually recognizing all members of a population is a far more accurate method of estimating population size than using classical ecological techniques such as mark–recapture methods or stratified sampling (Gros et al., 1996). In addition, long-term field studies have unusually accurate figures of changes in population size (see Clutton-Brock et al., 1982; Packer et al., 1988, for examples); in turn, this may influence effective population size. In chapter 2, McGregor and Peake review methods used to recognize animals individually and additionally show that populations can be monitored through individual call recognition rather than observing animals themselves (see also Baptista and Gaunt, 1997).

Molecular Techniques

In studies of known individuals in the wild, data are regularly collected on paternity, and sometimes maternity, using genetic markers (Schierwater et al., 1994). These data generate information on the extent to which particular individuals contribute to the next generation and hence to N_e. Knowledge of the number of lineages, or breeding groups, for a given N_e will alter estimates of the time frame over which allelic diversity is maintained (Pope, 1996). Allozymes, DNA fingerprinting, mitochondrial DNA (mtDNA), minisatellite, and microsatellite techniques shed light on population differentiation and genetic variability. These methods have identified subpopulations of humpback whales *Megaptera novaeangliae*

(Baker et al., 1993), migration routes and extent of philopatry in marine turtles (Bowen and Avise, 1996), and the degree of hybridization between canid species (Lehman et al., 1991), information almost impossible to obtain from observation alone. In addition, analyses of mtDNA sequences determine phylogenetic relationships between species, which help identify evolutionarily significant units (ESUs) for conservation (Moritz, 1994). For instance, genetic analyses of orangutan *Pongo pygmaeus* populations indicate sufficient divergence between Sumatran and Bornean populations to put a halt to breeding individuals from the two regions (Dobson, 1996). Although within- and between-species molecular comparisons are normally outside the purview of behavioral ecology, analyses that attempt to uncover associations between behavioral and morphological variables without the confounding effects of shared ancestry (Harvey and Pagel, 1991) nonetheless require phylogenetic trees derived using molecular techniques.

Human Behavioral Ecology

Because anthropogenic factors are the greatest challenge to biodiversity, many conservation biologists would argue that conservation solutions will eventually emerge from understanding human behavior. Patterns of human exploitation are better understood now than they were 15 years ago. The myth of the "noble savage" living in harmony with the environment has been exploded (Diamond, 1989; FitzGibbon, chapter 16). In a detailed exploration of this issue, in chapter 17, Alvard uses empirical data to show that Piro hunters in Peru pursue an optimal foraging strategy rather than any conservation agenda. As a second illustration, the tragedy of the commons, which describes individuals overexploiting a common property, is better understood in terms of a problem of unrestricted access rather than a problem of ownership. This focuses attention on the best ways to police resources rather than on who owns them (Feeney et al., 1990). These findings are informative in that they show which conservation strategies are likely to succeed or fail.

In spite of these advances, the rapid growth of evolutionary psychology, a subdiscipline in which contemporary human behavior is interpreted in the light of past selection pressures, has so far had little to say about patterns of current exploitation (but see Low and Heinen, 1993). Similarly, economics does not consider the underlying biological motivation in decision making. Nevertheless, it seems likely that we will be able to understand human resource use better by paying closer attention to evolved behavioral dispositions, as argued in chapter 18 by Wilson, Daly and Gordon. Specifically, they draw attention to differences in attitudes toward conservation shown by different segments of society as predicted by evolutionary theory. Certainly, the two pivotal problems in conservation biology, human population growth and resource accrual (Ehrlich and Ehrlich, 1990), are central to human behavioral ecology's focus on the social and ecological factors that affect reproductive success in different cultures.

Conclusion

This chapter and the others in this book illustrate some of the ways in which behavioral ecology contributes to solving conservation problems. Some of the principal links between topics in the study of the behavioral ecology of nonhumans and topics in conservation biology are summarized in fig. 1-1. Demographic records and information on dispersal are obviously of great importance, while population growth rate, a population's response to ex-

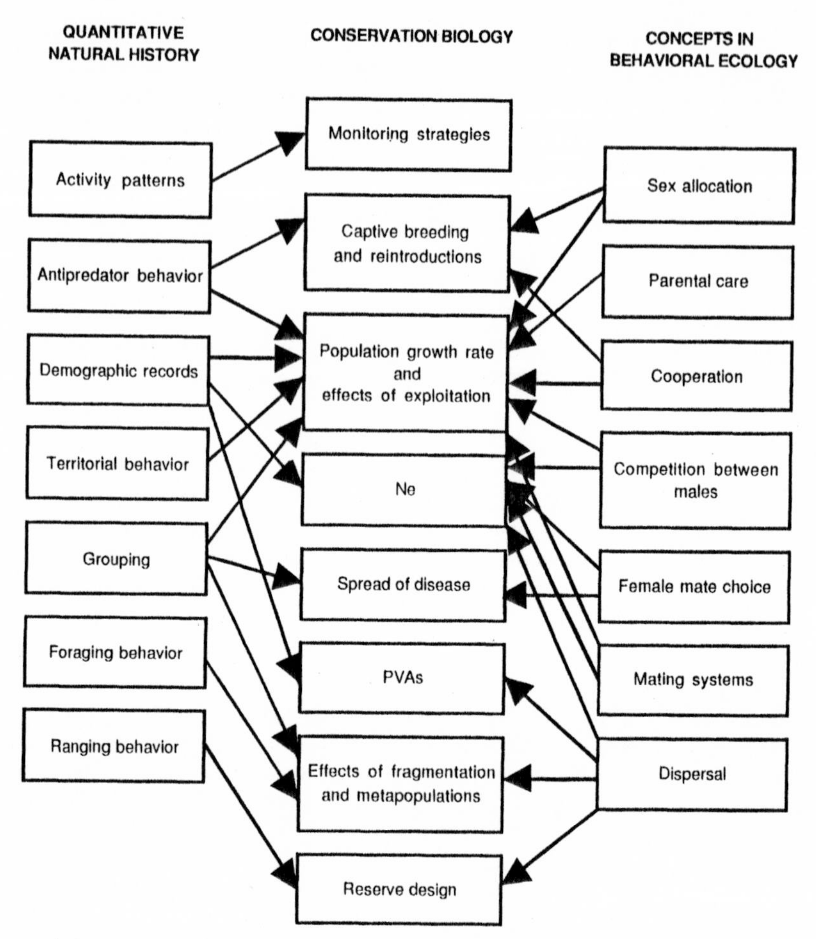

Figure 1-1 Some of the important connections between topics in the behavioral ecology of nonhumans and areas of conservation biology.

ploitation, and N_e are influenced by many factors in behavioral ecology. There are other bridges between these disciplines that have not been included, however, and many others that doubtless will appear as our knowledge of behavioral adaptations improves. Moreover, I think it fair to say that any behavioral ecologist with a conscience must be interested in the disappearance of biological diversity and, with a little thought, should be able to recognize the conservation significance of his or her research and make some contribution to solving the crisis at hand.

Summary

Conservation biology is a value-laden science that attempts to understand the causes of loss of biodiversity and ameliorate their effects. A central theme has been the dynamics of small populations focusing on loss of genetic diversity, demographic stochasticity, and environmental causes of extinction. Anthropogenically caused habitat loss and fragmentation are key problems affecting contemporary populations, and metapopulation theory has been used to model population persistence in small, remnant pieces of habitat.

In practical terms, conservation biology tries to prioritize which species and habitats demand protection. Early debate turned on the size and shape of reserves; more recently, attention has focused on limited resource exploitation as an incentive to protect habitats. Protocols for breeding endangered species in captivity have paid close attention to inbreeding avoidance, and limited attempts at reintroductions have occurred. Only some of these areas can benefit from a behavioral ecological approach.

A broad range of behavioral measures including information on grouping and territoriality are routinely used in captive breeding programs, monitoring strategies, and model construction. Competent antipredator behavior may affect the success of reintroduction attempts. Information on individual life histories is vital for constructing population viability estimates.

Competition between males affects variance in reproductive success and N_e. Mate choice influences the extent to which species hybridize and may affect disease resistance and progeny vigor. Variations in mating systems influence the rate at which heterozygosity is lost from a population. Both mating system and parental manipulation of offspring sex ratio affect a population's response to exploitation. Patterns of dispersal influence metapopulation persistence and gene flow between subpopulations.

Techniques used in behavioral ecology are useful in conservation biology. These include individual recognition and molecular techniques.

As human population growth and resource use are driving forces behind the biodiversity crisis, understanding the strategies by which people produce and limit offspring and the circumstances under which they under- or overexploit resources is critical to the conservation agenda.

Acknowledgments I thank John Eadie for discussions and Steve Albon, Joel Berger, John Eadie, Tracey Spoon, Judy Stamps, and an anonymous reviewer for critical comments.

References

Allendorf FW, Leary RF, 1986. Heterozygosity and fitness in natural populations of animals. In: Conservation biology: the science of scarcity and diversity (Soulé ME, ed). Sunderland, Massachusetts: Sinauer Associates; 57–76.

Anderson RM, May RM, 1979. Population biology of infectious diseases. Nature 280:361–367.

Andersson M, 1994. Sexual selection. Princeton, New Jersey: Princeton University Press.

Asa CS, Porton I, Baker AM, Plotka ED, 1996. Contraception as a management tool for controlling surplus animals. In: Wild mammals in captivity: principles and techniques (Kleiman DG, Allen ME, Thompson KV, Lumpkin S, eds). Chicago: Chicago University Press; 451–467.

Atkinson I, 1989. Introduced animals and extinctions. In: Conservation for the twenty-first century (Western D, Pearl MC, eds). New York: Oxford University Press; 54–75.

Baker CS, Perry A, Bannister JL, Weinrich MT, Abernethy RB, Calambokidis J, Lien J, Lamberstein RH, Urban-Ramirez J, Vasquez O, Clapham PJ, Alling A, O'Brien SJ,

Palumbi SR, 1993. Abundant mitochondrial DNA variation and world-wide population structure in humpback whales. Proc Natl Acad Sci USA 90:8239–8243.

Ballou JD, Foose TJ, 1996. Demographic and genetic management of captive populations. In: Wild mammals in captivity: principles and techniques (Kleiman DG, Allen ME, Thompson KV, Lumpkin S, eds). Chicago: Chicago University Press; 263–283.

Balmford A, Long A, 1995. Across-country analyses of biodiversity congruence and current conservation effort in the tropics. Conserv Biol 9:1539–1547.

Balmford A, Mace GM, Leader-Williams N, 1996. Designing the ark: setting priorities for captive breeding. Conserv Biol 10:719–727.

Baptista LF, Gaunt SLL, 1997. Bioacoustics as a tool in conservation studies. In: Behavioral approaches to conservation in the wild (Clemmons JR, Buchholz R, eds). Cambridge: Cambridge University Press; 212–242.

Barry D, Oelschlaeger M, 1996. A science for survival: values and conservation biology. Conserv Biol 10:905–911.

Beck BB, Rapaport LG, Price MS, Wilson A, 1994. Reintroduction of captive-born animals. In: Creative conservation: interactive management of wild and captive populations (Olney PJS, Mace GM, Feistner ATC, eds). London: Chapman and Hall; 265–284.

Berger J, Cunningham C, 1994a. Bison: mating and conservation in small populations. New York: Columbia University Press.

Berger J, Cunningham C, 1994b. Phenotypic alterations, evolutionarily significant structures, and rhino conservation. Conserv Biol 8:833–840.

Berger JC, Cunningham C, Gawuseb A, Lindeque M, 1993. "Costs" and short-term survivorship of hornless black rhinos. Conserv Biol. 7:920–924.

Bibby CJ, 1995. Recent past and future extinctions in birds. In: Extinction rates (Lawton JH, May RM, eds). Oxford: Oxford University Press; 98–110.

Bierregaard RO, Lovejoy TE, Kapos V, Dos Santos AA, Hutchins RW, 1992. The biological dynamics of tropical rainforest fragments. Bioscience 42:859–866.

Birkhead TR, Møller AP, 1992. Sperm competition in birds: evolutionary causes and consequences. London: Academic Press.

Bowen BW, Avise JC, 1996. Conservation genetics of marine turtles: In: Conservation genetics: case histories from nature (Avise JC, Hamrick JL, eds). New York: Chapman and Hall; 190–237.

Boyce MS, 1992. Population viability analyses. Annu Rev Ecol Syst 23:481–506.

Brittingham MC, Temple SA, 1983. Have cowbirds caused forest songbirds to decline? Bioscience 33:31–35.

Brown JH, 1971. Mammals on mountaintops: nonequilibrium insular biogeography. Am Nat 105:467–478.

Burgman MA, Ferson S, Akcakaya HR, 1993. Risk assessment in conservation biology. London: Chapman and Hall.

Burley N, 1981. Sex-ratio manipulation and selection for attractiveness. Science 211: 721–722.

Burley N, 1986. Sexual selection for aesthetic traits in species with biparental care. Am Nat 127:415–445.

Burley N, Krantzberg G, Radman P, 1982. Influence of colour-banding on the conspecific preferences of zebra finches. Anim Behav 30:444–455.

Caicco SL, Scott JM, Butterfield B, Csuti B, 1995. A gap analysis of management status of the vegetation of Idaho (U.S.A.). Conserv Biol 9:498–511.

Carlstead K, 1997. Effects of captivity on the behavior of wild mammals. In: Wild mammals in captivity: principles and techniques (Kleiman DG, Allen ME, Thompson KV, Lumpkin S, eds). Chicago: Chicago University Press; 317–333.

Caro TM, Durant SM, 1995. The importance of behavioral ecology for conservation biology: examples from Serengeti carnivores. In: Serengeti II: dynamics, management, and conservation of an ecosystem (Sinclair ARE, Arcese P, eds). Chicago: Chicago University Press; 451–472.

Carroll SP, 1993. Divergence in male mating tactics between two populations of the soapberry bug: I. Guarding versus nonguarding. Behav Ecol 4:156–164.

Caughley G, 1994. Directions in conservation biology. J Anim Ecol 63:215–244.

Caughley G, Gunn A, 1996. Conservation biology in theory and practice. Cambridge, Massachusetts: Blackwell Scientific.

Chase A, 1996. In a dark wood: the fight over forests and the rising tyranny of ecology. Boston: Houghton Mifflin.

Cheney DL, Seyfarth RM, 1990. How monkeys see the world. Chicago: Chicago University Press.

Chepko-Sade BD, Shields WM, Berger J, Halpin ZT, Jones WT, Rogers LL, Rood JP, Smith AT, 1987. The effects of dispersal and social structure on effective population size. In: Mammalian dispersal patterns: the effects of social structure on population genetics (Chepko-Sade BD, Halpin ZT, eds). Chicago: Chicago University Press; 287–321.

Clarke GM, 1995. Relationships between developmental stability and fitness: application for conservation biology. Conserv Biol 9:18–24.

Clemmons JR, Buchholz R (eds), 1997. Behavioral approaches to conservation in the wild. Cambridge: Cambridge University Press.

Clutton-Brock TH, 1982. The functions of antlers. Behaviour 70:108–125.

Clutton-Brock TH, 1991. The evolution of parental care. Princeton, New Jersey: Princeton University Press.

Clutton-Brock TH, Illius AW, Wilson K, Grenfell BT, MacCoil ADC, Albon SD, 1997. Stability and instability in ungulate populations: an empirical analysis. Am Nat 149:195–219.

Conway WG, 1989. The prospects for sustaining species and their evolution. In: Conservation for the twenty-first century (Western D, Pearl M, eds). Oxford: Oxford University Press; 199–209.

Cote IM, Poluin R, 1995. Parasitism and group size in social animals: a meta-analysis. Behav Ecol 6:159–165.

Creel S, Creel NM, 1997. Lion density and population structure in the Selous Game Reserve: evaluation of hunting quotas and offtake. Afr J Ecol 35:83–93.

Creel SR, Waser PM, 1991. Failures of reproductive suppression in dwarf mongooses (*Helogale parvula*): accident or adaptation? Behav Ecol 2:7–15.

Creel SR, Waser PM, 1994. Inclusive fitness and reproductive strategies in dwarf mongooses. Behav Ecol 5:339–348.

Curio E, 1996. Conservation needs ethology. Trends Ecol Evol 11:260–263.

Darwin C, 1871. The descent of man and selection in relation to sex. London: Murray.

Diamond JM, 1975. The island dilemma: lessons of modern biogeographic studies for the design of natural preserves. Biol Conserv 7:129–146.

Diamond JM, 1984. 'Normal' extinction of isolated populations. In: Extinctions (Nitecki MH, ed). Chicago: Chicago University Press; 191–246.

Diamond JM, 1989. The present, past and future human-caused extinctions. Phil Trans R Soc Land B 325:469–477.

Dobson AP, 1996. Conservation and biodiversity. New York: Scientific American Library.

Dunning JB, Danielson BJ, Pulliam HR, 1992. Ecological processes that affect populations in complex landscapes. Oikos 65:169–175.

Eadie JM, Fryxell JM, 1992. Density dependence, frequency dependence, and alternative nesting strategies in goldeneyes. Amer Nat 140:621–641.

Eberhard WG, 1979. The function of horns in *Podischnus agenor* (*Dynastinae*) and other beetles. In: Sexual selection and reproductive competition in insects (Blum MS, Blum NA, eds). New York: Academic Press; 231–258.

Edmunds M, 1974. Defense in animals: a survey of antipredator defenses. London: Longmans.

Ehrenfeld DW, 1976. The conservation of non-resources. Am Sci 64:660–668.

Ehrlich PR, Ehrlich AH, 1990. The population explosion. New York: Simon & Schuster.

Elgar MA, 1989. Predator vigilance and group size in mammals and birds: a critical review of the empirical evidence. Biol Rev 64:13–33.

Emlen ST, Oring LW, 1977. Ecology, sexual selection and the evolution of mating systems. Science 197:215–223.

Faith DP, 1995. Phylogenetic pattern and the quantification of organismal biodiversity. Phil Trans Roy Soc B 345:45–58.

Farnsworth NR, 1988. Screening plants for new medicines. In: Biodiversity (Wilson EO, ed). Washington, DC: National Academy Press; 83–97.

Fearnside PM, 1989. Extractive reserves in Brazilian amazonia. Bioscience 39:387–393.

Feeney D, Berkes F, McCay BJ, Acheson JM, 1990. The tragedy of the commons: twenty-two years later. Hum Ecol 18:1–19.

Fitzgerald S, 1989. International wildlife trade: whose business is it? Washington, DC: World Wildlife Fund.

FitzGibbon CD, 1990. Mixed-species grouping in Thomson's gazelles: the antipredator benefits. Anim Behav 39:1116–1126.

FitzGibbon CD, Lazarus J, 1995. Antipredator behavior of Serengeti ungulates: individual differences and population consequences. In: Serengeti II: dynamics, management, and conservation of an ecosystem (Sinclair ARE, Arcese P, eds). Chicago: Chicago University Press; 274–296.

Foose TJ, Ballou JD, 1988. Management of small populations. Int Zoo Yrbk 27:26–41.

Forman RRT, 1993. Landscape and regional ecology. Cambridge: Cambridge University Press.

Frankham R, 1995. Conservation genetics. Annu Rev Genet 29:305–327.

Franklin IR, 1980. Evolutionary change in small populations. In: Conservation biology: an evolutionary-ecological perspective (Soulé ME, Wilcox BA, eds). Sunderland, Massachusetts: Sinauer Associates: 135–149.

Freese CH, 1997. Harvesting wild species: implications for biodiversity conservation. Baltimore: John Hopkins University Press.

Gaston KJ, 1996. Species richness: measure and measurement. In: Biodiversity: a biology of numbers and difference (Gaston KJ, ed). Oxford: Blackwell Scientific; 77–113.

Gilpin ME, 1991. The genetic effective size of a metapopulation. Biol J Linn Soc 42:165–175.

Gilpin M, Hanski I (eds), 1991. Metapopulation dynamics: empirical and theoretical investigations. London: Academic Press.

Gilpin ME, Soulé ME, 1986. Minimum viable populations: processes of species extinction. In: Conservation biology: the science of scarcity and diversity (Soulé ME, ed). Sunderland, Massachusetts: Sinauer Associates; 19–34.

Ginsberg JR, Milner-Gulland EJ, 1994. Sex-biased harvesting and population dynamics in ungulates: implications for conservation and sustainable use. Conserv Biol 8:157–166.

Goldsmith B (ed), 1991. Monitoring for conservation and ecology. New York: Chapman and Hall.

Goss-Custard JD, Sutherland WJ, 1997. Individual behaviour, populations and conservation. In: Behavioural ecology: an evolutionary approach, 4th ed. (Krebs JR, Davies NB, eds). Oxford: Blackwell Scientific; 373–395.

Grant PR, Grant BR, 1997. Hybridization, sexual imprinting, and mate choice. Am Nat 149:1–28.

Gros PM, Kelly MJ, Caro TM, 1996. Estimating carnivore densities for conservation purposes: indirect methods compared to baseline demographic data. Oikos 77: 197–206.

Gross MR, 1985. Disruptive selection for alternative life histories in salmon. Nature 313:47–48.

Gross MR, 1996. Alternative reproductive strategies and tactics: diversity within sexes. Trends Ecol Evol 11:92–98.

Grzimek B, Grzimek M, 1959. Serengeti shall not die. Berlin: Ullstein AG.

Hamilton WD, 1971. Geometry for the selfish herd. J Theor Biol 31:295–311.

Hanski I, Gilpin M, 1991. Metapopulation dynamics: brief history and conceptual domain. Biol J Linn Soc 42:3–16.

Hanski IA, Gilpin M, 1996. Metapopulation biology: ecology, genetics and evolution. San Diego: Academic Press.

Hansson L, Fahrig L, Merriam G (eds), 1995. Mosaic landscapes and ecological processes. London: Chapman and Hall.

Harris LD, 1984. The fragmented forest: island biogeography theory and the preservation of biotic diversity. Chicago: Chicago University Press.

Harrison RL, 1992. Toward a theory of inter-refuge corridor design. Conserv Biol 6:293–295.

Harrison S, 1994. Metapopulations and conservation. In: Large-scale ecology and conservation biology (Edwards PJ, May RM, Webb NR, eds). Oxford: Blackwell Scientific; 111–128.

Harvey PH, Pagel MD, 1991. The comparative method in evolutionary biology. Oxford: Oxford University Press.

Hausfater G, Hrdy SB (eds), 1994. Infanticide: comparative and evolutionary perspectives. New York: Aldine.

Hedrick PW, Gilpin ME, 1996. Genetic effective population size of a metapopulation. In: Metapopulation biology: ecology, genetics and evolution (Hanski IA, Gilpin M, eds). San Diego: Academic Press; 166–181.

Hess GR, 1994. Conservation corridors and contagious disease: a cautionary note. Conserv Biol 8:256–262.

Holt WV, 1994. Reproductive technologies. In: Creative conservation: integrative management of wild and captive animals (Olney PJS, Mace GM, Feistner ATC, eds). London: Chapman and Hall; 144–166.

Hooper DV, Vitousek PM, 1997. The effects of plant composition and diversity on ecosystem processes. Science 277:1302–1305.

IUCN/SSC Asian Rhino Specialist Group, 1989. Asian rhinos: an action plan for their conservation. Gland: International Union for the Conservation of Nature and National Resources.

Ives AR, Dobson AP, 1987. Antipredator behavior and the population dynamics of simple predator-prey systems. Am Nat 130:431–447.

Johnson KH, Vogt KA, Clark HJ, Schmidtz OJ, Vogt DJ, 1996. Biodiversity and the productivity and stability of ecosystems. Trends Ecol Evol 11:372–377.

Johnson ML, Gaines MS, 1990. Evolution of dispersal: theoretical models and empirical tests using birds and mammals. Annu Rev Ecol Syst 21:449–480.

Kleiman DG, 1994. Mammalian sociobiology and zoo breeding programs. Zoo Biol 13:423–432.

Kleiman DG, Stanley Price MR, Beck BB, 1994. Criteria for reintroductions. In: Creative conservation: integrative management of wild and captive animals (Olney PJS, Mace GM, Feistner ATC, eds). London: Chapman and Hall; 287–303.

Kelly MJ, Laurenson MK, FitzGibbon CD, Collins DA, Durant SM, Frame GW, Bertram BCR, Caro TM, (in press). Demography of the Serengeti cheetah population: the first twenty-five years. J Zool.

Knight DH, Wallace LL, 1989. The Yellowstone fires: issues in landscape ecology. Bioscience 39:700–706.

Koenig WD, Pitelka FA, Carmen WJ, Mumme RL, Stanback MT, 1992. The evolution of delayed dispersal in cooperative breeders. Q Rev Biol 67:111–150.

Koenig WD, Van Vuren D, Hooge PN, 1996. Detectability, philopatry, and the distribution of dispersal distances in vertebrates. Trends Ecol Evol 11:514–517.

Komdeur J, Bullock ID, Rands MRW, 1991. Conserving the Seychelles warbler *Acrocephalus sechellensis* by translocation: a transfer from Cousin Island to Aride Island. Bird Conserv Int 1:177–185.

Kooyman GL, Kooyman TG, Horning M, Kooyman CA, 1996. Penguin dispersal after fledging. Nature 383:397.

Lande R, 1988. Genetics and demography in biological conservation. Science 241:1455–1460.

Lande R, Barrowclough GF, 1987. Effective population size, genetic variation, and their use in population management. In: Viable populations for conservation (Soulé ME, ed). Cambridge: Cambridge University Press; 87–124.

Leader-Williams N, Albon SD, 1988. Allocation of resources for conservation. Nature 336:533–535.

Lehman N, Eisehawer A, Hansen K, Mech LD, Peterson RO, Gogan PJP, Wayne RK, 1991. Introgression of coyote mitochondrial DNA into sympatric North American gray wolf populations. Evolution 45:104–119.

Levins R, 1970. Extinction. Lect Math Life Sci 2:75–107.

Lidicker WZ Jr, Koenig WD, 1996. Responses of terrestrial vertebrates to habitat edges and corridors. In: Metapopulations and wildlife conservation (McCullough DR, ed). Covelo, California: Island Press; 85–109.

Lima SL, Dill LM, 1990. Behavioral decisions made under risk of predation: a review and prospectus. Can J Zool 68:619–640.

Lima SL, Zollner PA, 1996. Towards a behavioral ecology of ecological landscapes. Trends Ecol Evol 11:131–135.

Lomnicki A, 1988. Population ecology of individuals. Princeton, New Jersey: Princeton University Press.

Loutit B, Montgomery S, 1994. The efficacy of rhino dehorning: too early to tell!! Conserv Biol 8:923–924.

Lovejoy TE, Bierregaard RO Jr, Rylands AB, Malcolm JR, Quintela CE, Harper LH, Brown KS Jr, Powell AH, Powell GVN, Schubart HOR, Hays MB, 1986. Edge and other effects of isolation on Amazon forest fragments. In: Conservation biology: the study of scarcity and diversity (Soulé ME, ed). Sunderland, Massachusetts: Sinauer Associates; 257–285.

Low BS, Heinen JT, 1993. Population, resources, and environment: implications of human behavioral ecology for conservation. Popul Environ 15:7–41.

MacArthur RH, Wilson EO, 1967. The theory of island biogeography. Princeton, New Jersey: Princeton University Press.

Mace GM, 1994. An investigation into methods for categorizing the conservation status of species. In: Large scale ecology and conservation biology (Edwards PJ, May RM, Webb NR, eds). Oxford: Blackwell Scientific; 295–314.

Mace GM, 1995. Classification of threatened species and its role in conservation planning. In: Extinction rates (Lawton JH, May RM, eds). Oxford: Oxford University Press; 197–213.

McLean IG, Lundie-Jenkins G, Jarman PJ, 1995. Teaching an endangered mammal to recognize predators. Biol Conserv 75:51–62.

Maguire LA, Lacey RC, 1990. Allocating scarce resources for conservation of endangered species: partitioning space for tigers. Conserv Biol 4:157–170.

Maloney RF, McLean IG, 1995. Historical and experimental learned predator recognition in free-living New Zealand robins. Anim Behav 50:1193–1201.

Mateo J, 1996. Development of alarm call recognition by juvenile Belding's ground squirrels. Anim Behav 52:489–505.

May RM, 1988. How many species are there on earth? Science 241:1441–1449.

May RM, 1995. Conceptual aspects of the quantification of the extent of biological diversity. Phil Trans R Soc B 345:13–20.

Meffe GK, Carroll RC, 1997. Principles of conservation biology, 2nd ed. Sunderland, Massachusetts: Sinauer Associates.

Milinski M, 1979. An evolutionary stable feeding strategy in sticklebacks. Z Tierpsychol 51:36–40.

Mills LS, Smouse PE, 1994. Demographic consequences of inbreeding in remnant populations. Am Nat 144:412–431.

Mock DW, 1984. Siblicidal aggression and resource monopolization in birds. Science 225:731–733.

Møller AP, 1992. Female swallow preference for symmetrical male sexual ornaments. Nature 357:238–240.

Møller AP, Cuervo JJ, Soler JJ, Zamora-Munoz C, 1996. Horn asymmetry and fitness in gemsbok, *Oryx g. gazella.* Behav Ecol 7:247–253.

Monaghan P, 1996. Relevance of the behaviour of seabirds to the conservation of marine environments. Oikos 77:227–237.

Moritz C, 1994. Defining 'evolutionary significant units' for conservation. Trends Ecol Evol 9:373–375.

Muller KL, Stamps JA, Krishnan VV, Willits NH, 1997. The effects of conspecific attraction and habitat quality on habitat selection in territorial birds (*Troglodytes aedon*). Am Nat 150:650–661.

Murcia C, 1995. Edge effects in fragmented forest: implications for conservation. Trends Ecol Evol 10:58–62.

Naeem S, Thompson LJ, Lawler SP, Lawton JH, Woodfin RM, 1994. Declining biodiversity can alter the performance of ecosystems. Nature 368:734–737.

Newmark WD, 1987. A land-bridge island perspective on mammalian extinctions in western North American parks. Nature 325:430–432.

Norton BG, 1986. The preservation of species: the value of biological diversity. Princeton, New Jersey: Princeton University Press.

Noss R, 1987. Corridors in real landscapes: a reply to Simberloff and Cox. Conserv Biol 1:159–164.

O'Brien SJ, Mayr E, 1991. Bureaucratic mischief: recognizing endangered species and subspecies. Science 251:1187–1188.

Oldfield ML, 1984. The value of conserving genetic resources. Sunderland, Massachusetts: Sinauer Associates.

Packer C, Herbst L, Pusey AE, Bygott JD, Hanby JP, Cairns EJ, Borgerhoff Mulder M, 1988. Reproductive success of lions. In: Reproductive success: studies of individual variation in contrasting breeding systems (Clutton-Brock TH, ed). Chicago: Chicago University Press; 363–383.

Parker GA, Sutherland WJ, 1986. Ideal free distributions when individuals differ in competitive ability: phenotype-limited ideal free models. Anim Behav 34:1222–1242.

Plotkin MJ, 1988. The outlook for new agricultural and industrial products from the tropics. In: Biodiversity (Wilson EO, ed). Washington, DC: National Academy Press; 106–116.

Pope TR, 1996. Socioecology, population fragmentation, and patterns of genetic loss in endangered primates. In: Conservation genetics: case histories from nature (Avise JR, Hamrick JL, eds). New York: Chapman and Hall; 119–159.

Redford KH, Mansour JA, 1996. Traditional peoples and biodiversity conservation in large tropical landscapes. Covelo, California: Island Press.

Redford KH, Padoch C (eds), 1992. Conservation of neotropical forests: working from traditional resource use. New York: Columbia University press.

Reed JM, Dobson AP, 1993. Behavioural constraints and conservation biology: conspecific attraction and recruitment. Trends Ecol Evol 8:253–256.

Robbins RK, Opler PA, 1997. Butterfly diversity and a preliminary comparison with bird and mammal diversity. In: Biodiversity II: understanding and protecting our biological resources (Reaka-Kudla ML, Wilson DE, Wilson EO, eds). Washington, DC: Joseph Henry Press; 69–82.

Rolston H, 1988. Environmental ethics: duties to and values in the natural world. Philadelphia: Temple University Press.

Rosenberg DK, Noon BR, Meslow EC, 1997. Biological corridors: form, function, and efficacy. BioScience 47:677–687.

Saether B-E, Ringsby TH, Roskaft E, 1996. Life history variation, population processes and priorities in species conservation: towards a reunion of research paradigms. Oikos 77:217–226.

Schierwater B, Streit P, Wagner GP, DeSalle (eds), 1994. Molecular ecology and evolution: approaches and applications. Basel: Birhauser Verlag.

Scott ME, 1988. The impact of infection and disease on animal populations: implications for conservation biology. Conserv Biol 2:40–56.

Shaffer ML, 1981. Minimum population sizes for species conservation. Bioscience 31:131–134.

Shaffer ML, 1983. Determining minimum viable population sizes for the grizzly bear. International Conference on Bear Research and Management 5:133–139.

Sherman PW, 1981. Kinship, demography and Belding's ground squirrel nepotism. Behav Ecol Sociobiol 8:251–259.

Simberloff D, 1988. The contribution of population and community biology to conservation science. Annu Rev Ecol Syst 19:473–511.

Simberloff D, Cox J, 1987. Consequences and costs of conservation corridors. Conserv Biol 1:63–71.

Smith AT, Peacock MM, 1990. Conspecific attraction and the determination of metapopulation colonization rates. Conserv Biol 4:320–327.

Smith FDM, May RM, Pellew R, Johnson TH, Walker KS, 1993. Estimating extinction rates. Nature 364:494–496.

Smith SM, 1975. Innate recognition of coral snake patterns by a possible avian predator. Science 87:759–780.

Snyder NFR, Derrickson SR, Beissenger SR, Wiley JW, Smith TB, Toone WD, Miller B,

1996. Limitations of captive breeding in endangered species recovery. Conserv Biol 10:338–348.

Soulé ME, 1985. What is conservation biology? Bioscience 35:727–734.

Soulé ME, Gilpin M, Conway W, Foose T, 1986. The millenium ark: how long the voyage, how many staterooms, how many passengers? Zoo Biol 5:127–138.

Soulé ME, Simberloff D, 1986. What do genetics and ecology tell us about the design of nature reserves? Biol Conserv 35:19–40.

Stacy EA, Hamrick JL, Nason JD, Hubbell SP, Foster RB, Condit R, 1996. Pollen dispersal in low-density populations of three neotropical tree species. Am Nat 148: 275–298.

Stamps JA, 1988. Conspecific attraction and aggregation in territorial species. Am Nat 131: 329–347.

Stanley Price MR, 1989. Animal re-introductions: the Arabian oryx in Oman. Cambridge: Cambridge University Press.

Steadman DW, 1997. Human-caused extinction of birds. In: Biodiversity II: understanding and protecting our biological resources (Reaka-Kudla ML, Wilson DE, Wilson EO, eds). Washington, DC: Joseph Henry Press; 139–161.

Stoleson SH, Beissinger SR, 1997. Hatching asychrony in parrots: boon or bane for sustainable use? In: Behavioral approaches to conservation in the wild (Clemmons JR, Buchholz R, eds) Cambridge: Cambridge University Press; 157–180.

Strier KB, 1997. Behavioral ecology and conservation biology of primates and other animals. Adv Study Behav 26:101–158.

Stuart SN, 1991. Reintroductions: to what extent are they needed? Symp Zool Soc Lond 62:27–37.

Sutherland WJ, 1996. From individual behaviour to population ecology. Oxford: Oxford University Press.

Sutherland WJ, Dolman PM, 1994. Combining behaviour and population dynamics with applications for predicting consequences of habitat loss. Proc R Soc Lond B 255:133–138.

Swaddle JP, 1996. Reproductive success and symmetry in zebra finches. Anim Behav 51:203–210.

Taylor PW, 1986. Respect for nature: a theory of environmental ethics. Princeton, New Jersey: Princeton University Press.

Thomas JW, Forsman ED, Lint JB, Meslow EC, Noon BR, Verner J, 1990. A conservation strategy for the Northern spotted owl. U.S. GPO 1990-791-171/20026. Washington, DC: U.S. Government Printing Office.

Tilman D, Knops J, Wedin D, Reich P, Ritchie M, Siemann E, 1997. The influence of functional diversity and composition on ecosystem processes. Science 277:1300–1302.

Tracy M, Freeman DC, Emlen JM, Graham JH, Hough RA, 1995. Developmental instability as a biomonitor of environmental stress: an illustration using aquatic plants and macroalgae. In: Biomonitors and biomarkers as indicators of environmental change: a handbook (Butterworth FM, ed). New York: Plenum: 313–337.

Vane-Wright RI, Humphries CJ, Williams PH, 1991. What to protect—systematics and the agony of choice. Biol Conserv 55:235–254.

Veltman CJ, Nee S, Crawley MJ, 1996. Correlates of introduction success in exotic New Zealand birds. Am Nat 147:542–557.

Vietmeyer ND, 1986. Lesser-known plants of potential use in agriculture and forestry. Science 232:1379–1384.

Wade MJ, McCaughley DE, 1988. Extinction and colonization: their effects on genetic differentiation of local populations. Evolution 42:995–1005.

Warren MS, 1990. The successful conservation of an endangered species, the heath fritillary butterfly *Mellicta athalia,* in Britain. Biol Conserv 55:37–56.

Watson PJ, Thornhill R, 1994. Fluctuating asymmetry and sexual selection. Trends Ecol Evol 9:21–25.

Wayne RK, Jenks SM, 1991. Mitochondrial DNA analysis implying extensive hybridization of the endangered red wolf *Canis rufus.* Nature 351:565–568.

Western D, Wright RM, Strum C (eds), 1994. Natural connections: perspectives in community-based conservation. Washington, DC: Island Press.

Whitmore TC, Sayer JA (eds), 1992. Tropical deforestation and extinction rates. London: Chapman and Hall.

Wilcove DS, McLellan CH, Dobson AP, 1986. Habitat fragmentation in the temporate zone. In: Conservation biology: the science of scarcity and diversity (Soulé ME, ed). Sunderland, Massachusetts: Sinauer Associates; 237–256.

Wildt DE, 1996. Male reproduction: assessment, management, and control of fertility. In: Wild mammals in captivity: principles and techniques (Kleiman DG, Allen ME, Thompson KV, Lumpkin S (eds). Chicago: Chicago University Press; 429–450.

Williams P, Gibbons D, Margules C, Rebelo A, Humphries C, Pessey R, 1996. A comparison of richness hotspots, rarity hotspots, and complementary areas for conserving diversity of British birds. Conserv Biol 10:155–174.

Wilson AC, Stanley Price MR, 1994. Reintroduction as a reason for captive breeding. In: Creative conservation: interactive management of wild and captive populations (Olney PJS, Mace GM, Feistner ATC, eds). London: Chapman and Hall; 243–264.

Wilson EO, 1992. The diversity of life. Cambridge, Massachusetts: Belknap Press.

World Conservation and Monitoring Centre, 1992. Global biodiversity: state of the earth's living resources. London: Chapman and Hall.

Part I

Baseline Behavioral Ecological Data and Conservation Problems

The reasons that a population declines, especially on islands, can often be understood once it is known how anthropogenic forces such as habitat modification, direct exploitation, and introduction of exotic species interact with aspects of a species' natural history (see Caughley and Gunn, 1996, for case studies). Similarly, solutions to many problems in conservation biology rely on a detailed understanding of species' natural history. For example, to protect a species effectively, we need to know its habitat requirements including home range, food and nesting sites, its main predators, and whether it migrates. In another sphere of conservation, population viability analyses, we require information on life-history variables, such as litter size and age-specific reproductive rates. Behavioral ecological study has verified and quantified these variables and has thus given important resolution to conservation models (Goss-Custard and Sutherland, 1997) and species' recovery plans. Instead of arguing over whether such quantitative data belong in the realm of behavioral ecology or sophisticated natural history (as those skeptical of the relevance of behavioral ecology to conservation biology often do), it is more instructive to examine how new methods and findings from the behavioral and ecological sciences open up avenues for monitoring populations and predicting species' responses to anthropogenic pressures.

In the first of three chapters that take up this challenge, McGregor and Peake investigate how one of the most basic variables in conservation biology, population size, is measured. They argue that the most accurate census technique is to know every individual in a given population by recognizing it individually (see Gros et al., 1996). As behavioral ecological studies in the field usually rely on knowing the behavior, morphology, or reproductive careers of individual animals (Clutton-Brock, 1988; Stacey and Koenig, 1990), researchers have been forced to devise many different methods of recognizing individuals, ranging from invasive marking to observing and recording natural variation in sometimes subtle phenotypic characters. Chapter 2 reviews the arsenal of methods that can be employed for estimating population sizes accurately and then assesses the conservation costs and benefits of each one. This is, perhaps, the most obvious way that behavioral ecological methods can serve conservation.

Ecological and behavioral variables can be used in more sophisticated ways: to predict which species are particularly prone to extinction. It has long been known that rare species, with narrow geographic ranges or small population sizes (Willis, 1974; Terborgh and Winter, 1980), or of large body size (Terborgh, 1974), or poor dispersers (Laurance, 1990), or those occupying specialist feeding niches (Brown, 1971) are all prone to extinction. In chapter 3, Harcourt reexamines these variables systematically and extends these analyses to other variables commonly collected in field studies. Controlling for phylogeny, he compares nine ecological characteristics of primate species that are known to fare poorly in selectively logged forests to those that fare well, as a proxy of risk of extinction. Analyses show that taxa with low maximum latitudes and large home ranges are at risk, and there is an indication that species living at low density or of large body size are similarly extinction prone. Nonetheless, other ecological (diet type, diet breadth, and arboreality), behavioral (territoriality), and biogeographic variables (altitudinal range and size of geographic range) do not correlate with the ability to survive well in logged forest. These findings show which variables can and cannot be used to predict susceptibility to habitat fragmentation in primates. These systematic comparisons need to be extended to other mammalian orders to determine the generality of these conclusions.

It is well known that large carnivores are particularly susceptible to habitat loss and fragmentation because they have such large home ranges compared to other taxa (Gittleman and Harvey, 1982); moreover, they are persecuted because they pose a direct danger to humans and their domestic stock (Schaller, 1996). Their loss is cause for great concern, and much conservation research has centered around the preservation and restitution of large carnivores (Clark et al., 1996), although little attention has been paid to the ramifications of loss or reintroduction of carnivores on lower trophic levels (but see Terborgh, 1992). In the third chapter in this section, Berger begins to explore this problem, but rather than focusing on the conservation of prey species, populations, or genetic material, he examines the preservation of behavior patterns in prey. Although concern has been raised about the loss of natural behaviors in zoo-bred animals (Carlstead, 1996), there has been little attempt to investigate which behavior patterns might be lost and the speed at which this might occur. By comparing moose *Alces alces* living in areas still supporting large carnivores with those where carnivores have been extirpated and in a third area where humans are the main predator, Berger uses simple field experiments involving odor and auditory playbacks to determine whether moose have lost the ability to recognize natural predators. His results suggest that responses to predators can be lost in a remarkably short period of time (50–75 years) but that responses wane at different rates according to the predator involved and according to the sensory modality.

These findings are preliminary, but they do provide the first systematic attempt to discern how behavior patterns are lost in the wild and raise important questions at two levels. First, should we be concerned about conserving animals with a different behavioral repertoire from that of their recent ancestors? In turn, this raises questions about the relative value we place on morphological, physiological, or behavioral diversity (see

Wilson, 1992) as opposed to genetic diversity, the usual focus of attention (Mallet, 1996). Second, how do various antipredator behavior patterns develop (Curio, 1993)? This is crucial because those that principally depend on experience may be reinstituted in future environments where predators will be reintroduced, whereas innate antipredator behaviors may be lost, and future prey will suffer in the face of predator reintroductions.

References

Brown JH, 1971. Mammals on mountaintops: nonequilibrium insular biogeography. Am Nat 105:467–478.

Carlstead K, 1996. Effects of captivity on the behavior of wild mammals. In: Wild mammals in captivity: principles and techniques (Kleiman DG, Allen ME, Thompson KV, Lumpkin S, eds). Chicago: Chicago University Press; 317–333.

Caughley G, Gunn A, 1996. Conservation biology in theory and practice. Cambridge, Massachusetts: Blackwell Scientific.

Clark TW, Curlee AP, Reading RP, 1996. Crafting effective solutions to the large carnivore conservation problem. Conserv Biol 10:940–948.

Clutton-Brock TH (ed), 1988. Reproductive success: studies of individual variation in contrasting breeding systems. Chicago: Chicago University Press.

Curio E, 1993. Proximate and developmental aspects of antipredator behavior. Adv Study Behav 22:135–238.

Gittleman JL, Harvey PH, 1982. Carnivore home-range size, metabolic needs and ecology. Behav Ecol Sociobiol 10:57–63.

Goss-Custard JD, Sutherland WJ, 1997. Individual behaviour, populations and conservation. In: Behavioral ecology: an evolutionary approach, 4th ed (Krebs JR, Davies NB, eds). Oxford: Blackwell Scientific; 373–395.

Gros PM, Kelly MJ, Caro TM, 1996. Estimating carnivore densities for conservation purposes: indirect methods compared to baseline demographic data. Oikos 77:197–206.

Laurance WF, 1990. Comparative responses of five arboreal marsupials to tropical forest fragmentation. J Mammal 71:641–653.

Mallet J, 1996. The genetics of biological diversity: from varieties to species. In: Biodiversity: a biology of numbers and difference (Gaston KJ, ed). Oxford: Blackwell Scientific; 13–53.

Schaller GB, 1996. Introduction: carnivores and conservation biology. In: Carnivore behavior, ecology, and evolution, vol 2 (Gittleman JL, ed). Ithaca, New York: Cornell University Press; 1–10.

Stacey PB, Koenig WD (eds), 1990. Cooperative breeding in birds: long-term studies of ecology and behavior: Cambridge: Cambridge University Press.

Terborgh J, 1974. Preservation of natural diversity: the problem of extinction prone species. Bioscience 24:715–722.

Terborgh J, 1992. Maintenance of diversity in tropical forests. Biotropica 24: 283–292.

Terborgh J, Winter B, 1980. Some causes of extinction. In: Conservation biology: an evolutionary-ecological perspective (Soulé ME, Wilcox BA, eds). Sunderland, Massachusetts: Sinauer Associates; 119–134.

Willis EO, 1974. Populations and local extinctions of birds on Barro Colorado Island, Panama. Ecol Monogr 44: 153–169.

Wilson EO, 1992. The diversity of life. Cambridge, Massachusetts: Harvard University Press.

2

The Role of Individual Identification in Conservation Biology

Peter McGregor
Tom Peake

Individual Identification

The purpose of this chapter is to discuss the role that a knowledge of the identity of individuals can play in conservation. Much of the impetus for such a discussion comes from the field of behavioral ecology. This is partly because behavioral ecologists have developed a number of techniques for identifying individuals, many of which are designed to have minimal effects on behavior. It is also partly because such techniques have uncovered differences between individuals in features relevant to conservation.

There are three main reasons an ability to individually identify animals is important to conservation. First, while the importance of appropriate methods for counting and monitoring threatened populations is recognized (Thiollay, 1989; Perrins et al., 1991; Jefferies and Mitchell-Jones, 1993; Pollard and Yates, 1993; Green, 1995b; Stewart and Hutchings, 1996), some of the assumptions that underpin such methods can only be validated by studies of individually identifiable animals. Furthermore, the extra precision of methods involving individual identification may be particularly important when assessing or predicting the results of changes in management practice or land use (e.g., Goss-Custard and Durell, 1990). This is particularly so for rare species where a small change in numbers may reflect a relatively large change in population size (e.g., many threatened bird populations number fewer than 100 individuals; Collar and Andrew, 1988). Second, individual identification is necessary to provide the detailed knowledge of the life histories of individual animals (e.g., Hammond 1990) used in a new generation of predictive models (e.g., Kenward, 1993; Bart, 1995; Sutherland, 1995). Such models may prove vital to the conservation of those species that are especially vulnerable to human disturbance and for which predictive measures may be of more importance than reactive measures. Third, individuals in the same population can differ in features relevant to their conservation. For example, they can use habitat differently (e.g., Peake, 1997), employ different behavioral strategies (e.g., Rohner, 1996), and have very different reproductive success (Newton, 1995). Such differences can be so great that the standard assumption that individuals in a population are equal in a conservation sense is no longer tenable; rather, individuals should be assumed to have different conservation values unless there is evidence to the contrary.

In the next section we outline the census and monitoring techniques commonly used in conservation, most of which do not involve individual identification. We discuss their limitations, many of which have been discovered by studies of individually identifiable animals. Next, we discuss the role of individual identification in relation to conservation. We also review the variety of techniques available to identify individuals, some of which are routinely used by behavioral ecologists. We emphasize the less invasive, and in our view, more elegant, techniques and present an example of the use of individually distinctive vocalizations in a conservation context. The costs and benefits of individual identification techniques are then discussed. We conclude with some of the issues that should be considered when deciding whether a particular conservation problem requires individual identification of the study species and, if it does, which technique should be used.

Techniques Used by Conservation Biologists and Their Limitations

Conservation efforts are typified by limited resources. Thus, estimates of the abundance of species of conservation concern need to be carried out as quickly, as cheaply, and with as little effort as possible while achieving results that are as precise and as relevant as possible. The conflict between gathering detailed information on the causes of population decline and the speed of implementing possible remedies is discussed by Green (1995b).

A wide variety of techniques has been devised for counting animals, most of which do not require that individuals be recognized. These are all capable of providing valuable information if care is taken in choice of method, sampling design, field method, data collection and analysis, and finally interpretation of results. The importance of good practice at every step of the process cannot be overstated (Krebs, 1989; Bibby et al., 1992).

These techniques vary widely in the types of questions they are able to answer as well as the various costs (time, effort, and money) required to implement them. Although our focus is the value of individual identification, we realize that other methods can provide useful results. However, we suggest that as the quality of information is inversely related to the cost of obtaining it, this trade-off needs careful assessment.

All techniques make a number of assumptions about the study population. It is important when using any technique to be aware of these assumptions and to attempt to ensure that either they are not violated or that the biases they generate are understood and accounted for. This is particularly important because endangered populations may be especially susceptible to violating assumptions due to their small size, limited dispersal opportunities, and the complications of human exploitation, interference, and habitat changes or loss.

The aim of this section is not to provide a comprehensive review of common ecological techniques, nor to recommend one over another, as many excellent texts are available for this purpose (Southwood, 1978; Krebs, 1989; Bibby et al., 1992; Pollard and Yates, 1993; Sutherland, 1996). We aim to outline some of the problems associated with common techniques, especially where these problems can be overcome using individually identifiable animals.

The best estimate of abundance equals the actual number of animals present at any one time; such census accuracy is not usually achievable and may not be desirable, as the cost of obtaining the information can outweigh its value. Although in a number of cases attempts are made to count every individual in a threatened population, the majority of census designs involve counting a sample of the population. Although it is generally the case that the greater the sampling effort, the more precise the estimate (due to the minimization of random sampling errors), this is only true if sampling design is good and effort is equally spread among sampling areas and periods. Nonrandom sampling is used in some circumstances, but the results of such studies cannot be generalized to all situations.

In many situations it may be enough to know that populations in one area are more dense than in another area, or that population density is increasing or decreasing with time in a particular area. These types of questions can be addressed by using relative methods and without knowing absolute densities or numbers. Relative density estimates are generally the simplest and cheapest type of data to collect; however, they do not allow detailed study of population density in relation to demographic parameters of survival and reproduction.

Methods of Quantifying Populations

Survey Techniques

Estimates of population numbers are generally carried out by human observers who count animals within defined areas (quadrats), count animals detected from fixed points (point counts), or note encounters while traveling through areas (transects). Field methods vary according to the amount and type of area that needs to be covered and the characteristics of the species of interest and its habitat, but a number of assumptions apply to most methods.

Observers Should be Equally Likely to Encounter/Detect Animals. Van der Meer and Camphuysen (1996) found large differences in the estimates of seabirds from different observers on a ship-based transect survey, to the extent that some observers reported 10 times more kittiwakes *Rissa tridactyla* than others. Some effects are less obvious; for example, observers have been shown to differ in their ability to detect the sounds of birds, some species being heard by some observers but not by others (Emlen and DeJong, 1992). These sorts of effect may be ameliorated by training or by calibrating observer performance against a known sample, but they should not be ignored. Where observer differences are great, an alternative, objective method may be used—for example, photographic aerial census (Krebs, 1989).

Individual Animals Should be Counted Only Once. This assumption is most likely to be violated when study animals are highly mobile or locally abundant or when counting takes place over a long time period. These effects can be avoided by increasing sampling rate or by having identifiable individuals.

Individuals Have an Equal Chance of Being Encountered. This assumption is violated in many studies of territorial songbirds in which singing males are generally much easier to detect during the breeding season than either females or nonterritorial males that do not sing. In some cases the opposite may be true: Gibbs and Wenny (1993) found that unpaired males of two species of bird were three to five times more likely to be encountered than paired males. The difference that Gibbs and Wenny (1993) found could easily lead to inappropriate action if the number of males detected was used as an indication of the conservation value of different habitats: with these two species the habitat where most males were detected would be the one with the most unpaired males and therefore probably the lower quality habitat. Differences in detectability may be exacerbated by differences between the breeding and nonbreeding season.

Indirect Estimates

A significant number of endangered species are difficult to count directly and are counted by their products such as nests, spoors, or tracks. These estimates can only produce absolute estimates when calibrated against a known pattern of production as discovered by intensive study.

In some cases estimates are produced by questioning local people (e.g., Gros et al., 1996). However, these estimates can be unreliable if those questioned have a vested interest in the outcome (e.g., hunters; Rabinowitz et al., 1995).

Mark–Recapture

The mark–recapture method involves capturing a sample of animals, marking them in some way (not necessarily individually), and releasing them back into the study population. A second sample is then captured some time later (timing depends on the technique used), and the estimate is calculated from the ratio of marked to unmarked animals in the second sample. It is important to note that any estimate derived by these methods is only representative of that fraction of a population that can be caught. Members of a population that cannot be sampled by a particular method (e.g., females of some moth species cannot be caught by light traps) do not take any part in the estimate. A variety of models is available to cover a wide range of situations, and choice of an applicable model is important if biases are to be reduced (Burnham et al., 1995).

The simplest methods involve a single capture and marking session and a single recapture session. These types of model assume that population size remains constant between capture sessions (i.e., there is no birth, death, or migration). While this will not be true for any population, the effect can be minimized by making the interval between capture sessions short. More complex models do not make this assumption because animals are individually marked, hence these techniques can also provide information on survival and recruitment rate. All mark-recapture techniques have a number of assumptions in common.

Animals have the Same Chance of Getting Caught. This assumption is commonly violated in many animal groups (Young et al., 1952). In small mammals, urine marking of traps affects trapability of voles (Stoddart, 1982) and mice (Hurst and Berreen, 1985). There are many other reasons why the assumption of equal catchability is violated for a given population; tests are available to check the assumption (Krebs, 1989) and models which are robust to its violation can then be used (e.g., Hwang and Chao, 1995). In addition to heterogeneity of capture probability, low capture rates are also a feature of many studies of endangered populations and may need to be taken into account (Rosenberg et al., 1995).

Marking does not Affect Catchability: Capture does not Affect Subsequent Recapture. Singer and Wedlake (1981) found a recapture rate of only 2% of papilionid butterflies *Graphium sarpedon* that had been previously captured and marked. When they marked individuals without capturing them, the "recapture" rate rose to 21%, indicating that capture had affected the probabilities of recapture. Similarly, *Heliconius* butterflies avoid the specific site where they have been handled (Mallet et al., 1987). Trapability of house mice has also been shown to be altered by previous capture (Hurst and Berreen, 1985). In cases where capture is difficult or requires much effort, visible marks allow individual animals to be resighted without the need for further capture (e.g., Hindell, 1991); however, unverified sightings will bias population estimates (Sheaffer and Jarvis, 1995). In some cases capture is so difficult that camera traps have been used; these photograph animals rather than capturing them, and identification may be carried out using natural markings (e.g., Karanth, 1995).

Marks Are not Lost; Marked Individuals Survive as Well as Unmarked Individuals. These assumptions are intrinsically difficult to assess (see discussion in below), but there are examples of marking affecting survivorship (e.g., Singer and Wedlake, 1981; Calvo and Furness, 1992; Daly et al., 1992).

Life-History Parameters

Many life-history parameters such as survival, reproduction, and dispersal can determine population numbers and are hence important in constructing population models to understand or predict declines. Unless the age structure of a given population can be determined easily (e.g., insect age classes are readily distinguishable as instars), these parameters cannot be measured without study of identifiable individuals (Krebs, 1989; Baker et al., 1992). Life-history parameters are generally obtained in one of two ways: by regular sampling or by following cohorts of individuals of the same age throughout their lives. The second method is unlikely to be feasible for long-lived animals.

Conservation Value of Identifying Individuals

When individual members of a study species can be identified, a wealth of information becomes available. We have divided these potential benefits into three broad categories: census accuracy and monitoring, life-history parameters, and behavioral effects.

Census and Monitoring Accuracy

All census techniques have inherent biases because of various assumptions each method makes. When a species is in danger of extinction, it is important to be aware of these inaccuracies and to minimize their effects. One way of doing this is to adjust census results from a large-scale survey with more detailed studies (e.g., aerial transects of duck populations were adjusted by detailed ground counts; Smith, 1995). Other studies have assessed census accuracy using a number of different measures, for example, four indirect estimates of cheetah *Acinonyx jubatus* densities were compared with population sizes based on individually identified cheetahs (Gros et al., 1996). Interviewing people locally was the most accurate method, but this study highlights the importance of calibrating indirect measures. Similarly, observers counting whales by spotting them from a research vessel detected fewer individuals and made less certain species identification than observers carrying out a simultaneous acoustic census based on recordings made from a hydrophone array towed by the same vessel (Kiernan, 1995; Christopher Clark, personal communication). Relatively few census techniques are assessed directly using a second method; often repeatability of the current method is considered sufficient validation of accuracy. Where a second method is used, this will often have its own suite of inaccuracies that may derive from similar or different assumptions. Census accuracy becomes particularly important when habitat changes (whether for better or for worse) are involved, as these may mean that sources of inaccuracy differ between censuses. For example, animals may move differently, or be encountered at different frequencies, in different habitats (e.g., Peake, 1997). There may also be differences in reproductive success and parental attributes in animals occupying different habitats (e.g., Riddington and Gosler, 1995). Having a more complete knowledge of individuals from small, intensive studies can allow this type of variation to be more fully understood (e.g., Gros et al., 1996). This can then be related to more cheaply obtained and extensive data (e.g., Shaugnessy, 1993). Conway et al. (1993) assessed the usefulness of playback for providing census and habitat usage data in the Yuma clapper rail *Rallus longirostris* based on a study of radio-tagged individuals. They found that about 20% of individuals responded to playback, but this varied from 7% to 40% over the course of the season. This sort of information can allow the timing of the census to be optimized, and correction factors can be applied to data collected at other times. Similarly, estimates of the rails' use of habitat obtained from the results of playback were very different from habitat use determined by radio tracking, as birds preferred to call from certain types of vegetation. Thus potential biases in data collection need to be addressed before any technique, or results derived from it, becomes useful.

Life-History Parameters

Marked, or otherwise identifiable, individuals are invaluable in the investigation of survival, long- and short-term movements, competition, behavioral strategies, and reproductive success, all of which are valuable conservation data (Newton, 1995). Such data often come from intensive studies of a relatively small sample of a population, but they can provide information that is applicable on a wider scale. For example, a study of local dialects in the song of individually marked corn buntings *Miliaria calandra* showed that an apparently continuous population of about 60 males was in essence three reproductively isolated units of 10–30 males (McGregor et al., 1997). Local dialects are a distinctive feature of corn buntings in the United Kingdom and their effect in fragmenting already small populations may play a role in the rapid decline of the species (Holland et al., 1996; Holland and McGregor, 1997).

Behavioral Effects

Although conservation biologists are undoubtedly aware of the need for accurate census and monitoring data, they are perhaps less aware of the importance of behavioral effects. In addition to showing large differences in individual animals' abilities to survive, compete, disperse, and reproduce, behavioral ecologists have shown how the behavior of individual animals and interactions between individuals determine both population structure and the response of any population to disturbance (e.g., Sutherland, 1995).

A striking illustration of the value of such knowledge to conservation involves differences in reproductive success between individuals. Many studies of population size make the implicit assumption that all individuals are equivalent in a conservation sense. Studies of individually identifiable animals have shown just how commonly and severely this assumption is transgressed. Extreme departures from the assumption are found in eusocial groups in which the queen of the colony is the only reproductive female and nonreproducing females help the queen (e.g., social Hymenoptera: Seger, 1991; mole rats *Heterocephalus glaber:* Jarvis, 1981; and *Cryptomys damarensis:* Jarvis et al., 1994). However, reproductive success can also differ greatly in apparently monogamous species. For example, a survey of lifetime reproductive success in seven species of birds found that a mere 3–9% of fledglings in one generation produced 50% of the young in the next generation (Newton, 1989, 1995). Studies of mammals have produced comparable results (Clutton-Brock, 1988), and variation in lifetime reproductive success of adult mammals can be high. For example, although 50% of male grizzly bears *Ursus arctos horribilis* of breeding age sired offspring, one male fathered more than 10% of cubs (Craighead et al., 1995). Lifetime reproductive success can also be strongly affected by territorial status; in red squirrels *Sciurius vulgaris,* 30% of resident females never reproduced, but no nonterritorial female ever reproduced (Wauters and Dhondt, 1995). An intriguing possibility for conservation is that if the individuals that make large contributions to the gene pool can be identified, then subsequent efforts could be directed toward conserving these individuals or discovering why other individuals do not achieve their reproductive potential.

Studies of individually identifiable nonbreeders have also shown their important role in population changes. For example, nonbreeding great skuas *Catharacta skua* act as a buffer against change in the size of the breeding population (Klomp and Furness, 1992). Similarly, nonterritorial "floaters" in great horned owls *Bubo virginianus* can mask population declines from traditional censuses of territorial birds, resulting in serious underestimates of the impact of these predators on their mammalian prey populations (Rohner, 1996). These examples support the suggestion that studies of the changes in numbers of nonbreeders could provide a sensitive indication of adverse environmental effects (Porter and Coulson, 1987).

The existence and scale of such differences between individuals make a strong case for future studies assuming that individuals will have different conservation value unless there is evidence to the contrary. Information of this kind can only be obtained by intensive study of the behavior of known individuals. Of particular interest may be comparison of the behavior of individuals in declining populations with those in more stable populations.

Methods of Identifying Individuals

There are two main methods of identifying individual animals. Marking techniques that render animals individually distinctive to an observer, either by adding markers or by modifying an animal's appearance (table 2-1), and the use of naturally occurring markers, based on variation in phenotype or genotype (table 2-2).

Until recently, most marking methods used external markers and relied on reflected light. The main factors to be considered when using such marks were their permanence (e.g., the retention of fish tags; Niva, 1995; Timmons and Howell, 1995) and the range at which the mark could be read (e.g., numbered leg rings used on small birds can usually only be read with the animal in the hand, whereas dye marks and collars on large vertebrates can be read from several kilometers). Radio- and satellite-tracking techniques avoid reliance on reflected light and extend the identification range to several thousand kilometers. Often the data collected in this way would not otherwise be obtainable (e.g., Walker et al., 1995; Higuchi et al., 1996). Various implantable devices, such as readable subcutaneous mi-

Table 2-1 A selection of artificial techniques used for individual identification.

Technique	Group	Example
Added markers		
Numbered leg rings	Birds	Spencer (1976)
Numbered tags	Insects	Gilbert et al. (1991)
	Fish	Adkison et al. (1995)
		Templeton (1995)
		Timmons and Howell (1995)
	Birds	Stonehouse (1978)
	Bats	Barclay and Bell (1988)
Numbered collars	Mammals	Clutton-Brock et al. (1982)
Color-coded markers	Fish	Hendry et al. (1995)
	Birds	Burley (1986)
Radio transmitters	Fish	Young (1994)
	Reptiles	Charland and Gregory (1995)
	Birds	Priede and Swift (1992)
	Mammals	Amlaner and MacDonald (1980)
Satellite transmitters	Fish	Eiler (1995)
	Birds	Meyburg et al. (1995)
Ultrasonic transmitters	Fish	Bradbury et al. (1995)
		Osborne and Bettoli (1995)
Beta lights	Mammals	Kalcounis and Brigham (1995)
Transponders	Fish	Glass et al. (1992)
Subcutaneous chips	Amphibians	Sinsch (1992), Elbin and Berger (1994)
	Birds	Hindell et al. (1996)
	Mammals	Poole (1994), Elbin and Berger (1994)
Modified appearance		
Tattoos, dye injection	Fish	Templeton (1995)
	Mammals	Poole (1994)
Fur dyes/bleaches	Mammals	Hurst et al. (1996)
Freeze-branding	Mammals	Rood and Nellis (1980)
Fin/toe clipping	Fish/amphibians	Stonehouse (1978)
Ear notching	Mammals	Stonehouse (1978)
Fur/scale clipping	Mammals/reptiles	Stonehouse (1978)
Elytra notching	Insects	Goldwasser et al. (1993)
Heat branding	Snakes	Scribner and Weatherhead (1995)

crochips and transponders (e.g., Sinsch, 1992; Elbin and Burger, 1994; Poole, 1994; Mrozek et al., 1995), are "hi-tech" marks that seem to offer novel solutions to some of the problems of external marks discussed below.

Naturally occurring variation in phenotype has been used to identify individuals for many years. This is particularly true of appearance (e.g., Caldwell, 1955; van Lawick Goodall, 1971), but various signals produced by animals also have been used (e.g., sound; Gilbert et al., 1994; see table 2-2). Humans find some types of signals difficult to identify individually even with the aid of sophisticated analysis equipment. In the case of chemical signals it is possible to train other animals to identify individuals; German Shephard dogs have been trained to identify individual Siberian tigers *Panthera tigris* by odor (Jones,

Table 2-2 A selection of natural markings used for individual identification.

Technique	Group	Example
In phenotype		
Facial characteristics	Bewick's swans	Scott (1978)
	Grey seals	Hiby and Lovell (1990)
	Chimpanzees	van Lawick Goodall (1971)
Appearance/markings	Anemone fish	Nelson et al. (1994)
	Coronella snake	Sauer (1994)
	Natterjack toads	Arak (1988)
	Ospreys	Bretagnolle et al. (1994)
	Hunting dogs	Creel and Creel (1995)
	Feral cattle	Lazo (1995)
	Cheetahs	Caro and Durant (1991)
	Sperm whales	Dufault and Whitehead (1995)
Scars	Dolphins	Lockyer and Morris (1990)
Footprints	Mountain lions	Smallwood and Fitzhugh (1993)
Weight	Hyenas	East and Hofner (personal communication)
Acoustic signals	Bitterns	Gilbert et al. (1994)
	Blue monkeys	Butynski et al. (1992)
	Fur seals	Trillmich (personal communication)
	(Bats)	Masters et al. (1995)
	(Dolphins)	Janik et al. (1994)
Electrical signals	(Knife fish)	McGregor and Westby (1992)
Chemical signals	Siberian tigers	Jones (1997)
	(House mice)	J. L. Hurst (personal communication)
In genotype	Cetaceans	Hammond et al. (1990); Palsboll et al. (1997)
	Marine turtles	Bowen (1995)

Note: Groups in parentheses indicate studies showing individually distinctive signals, but these signals were not used to identify individuals in the study cited.

1997). Molecular biology techniques such as DNA profiling now allow variation in the genotype to be used as a natural marker (e.g., Hammond et al., 1990; Bowen et al., 1995). All these techniques rely on assessments of similarity, for example, between a registration (e.g., photograph) and a catalogue of known individuals (Scott, 1978), or between the banding patterns of DNA profiles. Advances in computer technology have allowed objective and quantitative measurement of similarity. For example, photographs of pelage patterns on the head and neck of grey seals *Halichoerus grypus* can be matched automatically while making allowance for variation in the animals' orientation and posture when the photograph was taken (Hiby and Lovell, 1990). Objective and quantitative techniques can also be applied to measurement of signal similarity, for example, using cross-correlation of sound spectrograms of bird vocalizations (McGregor et al., 1994; Lessells et al., 1995).

An Example of Individual Identification Using Natural Variation: Corncrake Calls

The corncrake *Crex crex* is a globally threatened, migrant land rail which breeds throughout northern Europe (Cramp and Simmons, 1980; Collar et al., 1994; Tucker and Heath,

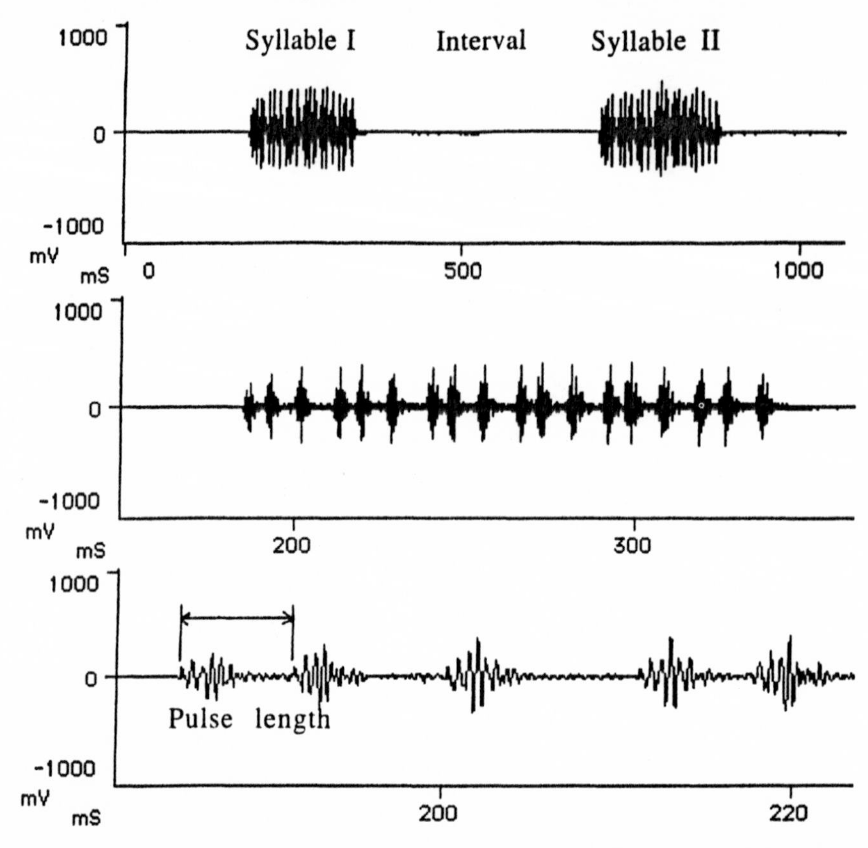

Figure 2-1 A simplified illustration of male corncrake call individuality. Waveform displays of a corncrake call at three expansions of the time scale. Corncrake calls are made up of two syllables, each of which consists of a series of irregularly spaced pulses.

1994). Male corncrakes are secretive and nocturnal, precluding the use of visual marks to identify individuals. Radio tracking has provided much useful information on the movements, habitat preferences, and reproductive behavior of corncrakes (Stowe and Hudson, 1988, 1991; Tyler and Green, 1996; Tyler et al., 1996). However, radio tracking is not feasible for long-term, wide-scale individual identification due to expense, problems of limited battery life, and the effort required to implement it. Corncrakes are readily located by their loud, distinctive "crex crex" calls, and these form the basis of many surveys of the species (e.g., Hudson et al., 1990). Male corncrakes call continuously from around 2300 to 0300 h, producing some 10,000 calls in a single night (Fangrath, 1994). This means that a large number of calls can be recorded from many males in one night (cf. Gilbert et al., 1994). Peak et al. (1998) analyzed calls from 59 males recorded in Ireland in 1993 and 1994. In 1993 all males were radio tagged, therefore, calls could be recorded from known individual males throughout the season. The timing of the pulses that make up the call of males (fig. 2-1) is individually distinctive. The distributions of pulse lengths of calls recorded from the same male are very similar, but different males have different distributions (fig. 2-2).

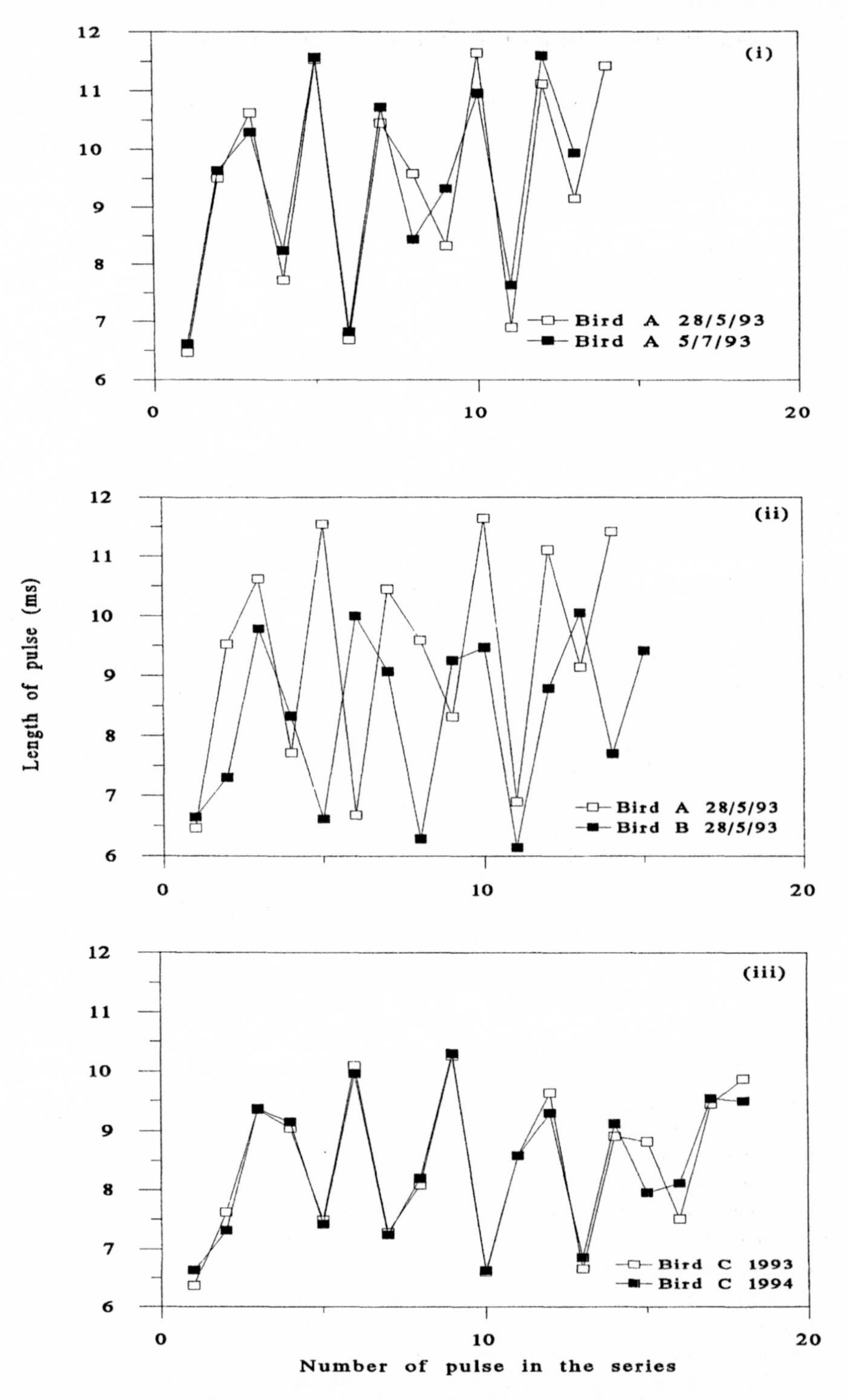

Figure 2-2 Distributions of pulse length. The distribution of pulse lengths within these syllables is individually distinctive. Each distribution is derived from the first syllable (n = 10 calls per male per night). Each graph plots two different distributions: (i) the same bird on two different nights; (ii) two different birds on the same night; and (iii) the same bird in different years.

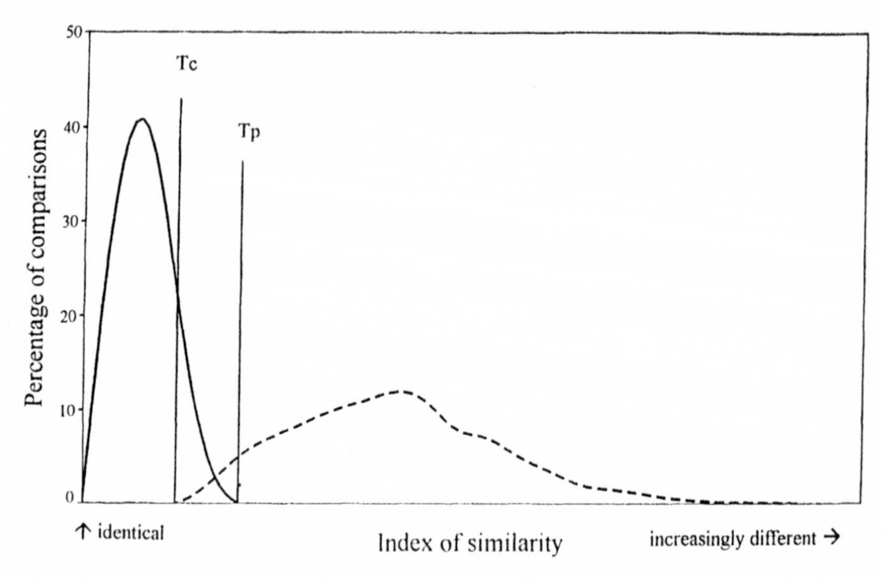

Figure 2-3 An illustration of the effects of setting threshold criteria for identification. Frequency distributions of similarity measures between two registrations (e.g., photographs, sound recordings) from the same individual (solid curve) and from two different individuals (dashed curve). The threshold T_p (all possible matches) is set at a similarity value that will correctly identify all cases of registrations from the same individual (matches), but this threshold will also include some mismatches. The threshold T_c (only correct matches) is set at a similarity value that will exclude mismatches, but it will also exclude some matches. The size of the zone between the thresholds (T_{diff}) gives an indication of the scale of identification errors related to T_c and T_p. The index of similarity between two registrations (e.g., a cross-correlation coefficient for two spectrograms) is represented on the x-axis, with identical registrations at the origin of the x-axis and similarity decreasing to the right. The curves are generated from recordings of known individual male corncrakes.

Calls recorded from the same male on the same night vary little, to the extent that we have difficulty in measuring these differences (Peake et al., 1998). The differences in calls of the same male within a season are also small, and three males recorded in both 1993 and 1994 had virtually identical call structures in both years (fig. 2-2 shows one of these males). The ability to identify individual calling corncrakes has been used in a number of ways: (1) to assess the accuracy of the traditional corncrake census technique (Peake, 1997). The census involved mapping calling males in each area on two nights within a specified period, combining the maps and using criteria derived from radio-tracking surveys to determine the number of individuals present (Hudson et al., 1990; Green, 1995a). The addition of information derived from recording calls increased census estimates by 20–30%; (2) to gather information on individual movements. These were related to habitat features within the study site (Peake, 1997); and (3) to monitor putative breeding attempts using cessation of nocturnal calling as an indicator that a mate had been attracted (Tyler and Green, 1996). Above all, this information was gathered with minimal disturbance to this protected species.

This study also illustrates a general feature of any quantitative comparison of similarity, namely, the signal detection trade-off encountered when setting threshold criteria for re-identification. These criteria are used to judge whether two registrations (e.g., spectrograms of calls or photographs) are sufficiently similar to have come from the same individual (i.e., whether they match). The problem common to all threshold criteria is that is is impossible to minimize simultaneously both incorrect matches and failure to detect correct matches (Wiley and Richards, 1983). This can be represented graphically (fig. 2-3). This figure also suggests a way of characterizing the extent to which individual distinctiveness in a population can be detected by a particular method. In the unlikely event that the distribution of similarity values for the same individual registrations does not overlap with the distribution of different individual registrations, then $T_c = T_p$, and defining the threshold criterion is simple. When the distributions overlap, the greater the area of each distribution between T_c and T_p, the less valuable the technique will be in identifying individuals.

Strengths and Weaknesses of Identifying Individuals

Costs of Methods of Individual Identification

In the previous section we distinguished methods of individual identification that marked animals (by adding marks or modifying appearance) from those methods based on naturally occurring variation. In many respects, these two groups of methods have opposite strengths and weaknesses. For example, the main advantage of individually marking animals is that the marks can potentially identify an individual unequivocally, but marking usually requires that the animal be captured. The main advantage of using naturally occurring variation is that it does not require capture and involves minimum disturbance; however, identification is generally less unequivocal than if animals are marked. This distinction between the two groups of methods recurs throughout the following discussion of the costs of individual identification methods. Costs influence the quality of the data collected, and there are costs to the study animals and to biologists. We deal with these costs at the stages of capture, handling, marking, and identification.

Capture

Any capture technique has the potential to produce biased data because it will preferentially catch particular animals (Young et al., 1952; Stoddart, 1982; Hurst and Berreen, 1985; see above). Some catching biases can be subtle; for example, most corncrakes are caught for radio tagging after being attracted by playback (Hudson et al., 1990). Indirect comparisons with males identified from their individually distinctive vocalizations strongly suggest that radio-tagged corncrakes represent the more vocally active or responsive males. The consequence is that converting a count of calling males into a population estimate on the basis of the calling activity of radio-tagged males probably underestimates numbers by 20–30% (Peake, 1997).

Capture can impose costs on the animal, generally in the form of direct physical injury from the catching and holding equipment (Mowat et al., 1994), although increased risk from predators, parasites, and pathogens are also likely (e.g., Singer and Wedlake, 1981).

Capture can also be dangerous for biologists; with large animals this is self-evident, but animals of all sizes can transmit diseases [e.g., *Salmonella* in black bears *Ursus americanus floridanus* (Dunbar et al., 1995), rabies in mammals, psittacosis in birds] and transfer parasites to humans (e.g., nematodes and ectoparasitic arthropods) and stimulate allergic reactions (e.g., mammalian hair, insect cuticle). The risks of capture for both animals and biologists are often difficult to distinguish from handling and marking costs and are dealt with in more detail in those sections below.

Handling

The effects of handling are difficult to separate from catching, but *Heliconius* butterflies avoid the site of handling, a bias that is independent of increased dispersal and mortality (Mallet et al., 1987). Handling generally involves removing an animal from the population, and this can disrupt its activities and relationships with other individuals (Cuthill, 1991). For example, when female great tits *Parus major* were caught and held for about 2 h, two males attracted new mates (Krebs et al., 1981). Similarly, when territorial male great tits were captured and held for a number of hours, their territories were reoccupied, and upon release some of the original owners (generally those held captive longest) were unable to regain their territories (Krebs, 1982).

Most marking procedures involve handling (but see Singer and Wedlake, 1981; Adkinson et al., 1995) and therefore have the potential to cause injury to animals and to transfer parasites and pathogens, both between animals and from humans. It is difficult to judge the level of such effects, and it is particularly difficult to collect data that are capable of resolving disagreements on such matters. This is illustrated by the controversy over the relationship between handling and disease transmission in wild dogs *Lycaon pictus* (Burrows et al., 1995; Ginsberg et al., 1995b; Kat et al., 1995; Morell, 1995; Villiers et al., 1995). In this case it seems most reasonable to conclude that the decimation of one population of wild dogs through disease was incidentally correlated with handling. Handling was unlikely to have been a cause of disease transmission or increased susceptibility because populations in four other ecosystems showed no effects of handling on survivorship (Ginsberg et al., 1995a).

Many adverse reactions of animals to handling are labelled as stress. (The term "stress" has many different meanings; see Broom and Johnson, 1993). Hurst and Agren (1994) discuss stress in relation to handling in a number of animal groups and offer three general conclusions: 1) trapping and handling procedures should take into account the natural behavior and needs of the species; 2) refuges can be useful for restraining and moving animals; 3) each species has its own problems. One aspect of their final point is that some species seem particularly stressed by capture. For example, greenfinches *Carduelis chloris* and bullfinches *Pyrrhulla pyrrhulla* caught by mist net are more likely to show stress effects than other species, although instances are still rare (Spencer, 1976). It is also worth pointing out that human perception of stress may be very different from the animals' experience and that interpreting physiological indications of stress may be problematic (Nimon et al., 1995). For example, a study of the heart rates of Adélie penguins *Pygoscelis adeliae* found dramatic responses to humans approaching on foot (Culik and Wilson, 1991). However, the process of surgically implanting, or externally fitting, heart rate monitors used in this study probably predisposed the birds to extreme reactions on subsequent exposure to humans. When heart rates were monitored without such procedures (artificial eggs were used instead), the effect was abolished (Nimon et al., 1995). Bearing all these points in mind, the obvious aim should be to minimize the handling period, particularly since the scale of social disruption is probably related to the time for which an animal is removed (Krebs, 1982; Cuthill, 1991).

The dangers for biologists during handling are the same as for capture (see above), except that there is a longer exposure to risk during handling than during capture.

LIBRARY, UNIVERSITY OF CHESTER

Marking

Any marking technique has the potential to produce biased data because it can modify the normal behavior and physiology of animals (Hindell et al., 1996) and affect survivorship (Singer and Wedlake, 1981; Calvo and Furness, 1992; Daly et al., 1992). Adverse effects of marking techniques are inherently difficult to assess. As Daly et al. (1992) point out, the conclusion that radio transmitters have no adverse effects (White and Garrott, 1990) could be a consequence of small sample size used to look for such effects (e.g., Pouliquen et al., 1990). In particular, increased susceptibility to predators will be difficult to detect because predation is infrequent. For example, the demonstration of increased predation risk to kangaroo rats *Dipodomys merriani* carrying transmitters required a 12-year study (Daly et al., 1992).

However, adding marks can cause direct injury to the animal (e.g., Calvo and Furness, 1992). Indeed, some marks are based on changes that would otherwise be regarded as damage, such as cutting notches in beetle elytra (Goldwasser et al., 1993), removal of scales and sections of fins in fish, toe clipping in amphibians and reptiles, and "ear punching" in small mammals (Stonehouse, 1978). Such damage is known to increase susceptibility to disease in some cases (e.g., fish in aquaria are likely to develop fin rot after fin clipping for marking; A.E. Magurran, personal communication).

There have been relatively few studies that document the effects of less severe marks. For example, cheetahs seemed unaffected by radio collars (Laurenson and Caro, 1994), and implanted radio transmitters did not change thermoregulatory patterns in the lizard *Sceloporus occidentalis* (Wang and Adolph, 1995). There have been even fewer studies comparing different marking techniques, and many of these studies compared rather obtrusive marks (e.g., the survival and breeding success of red grouse *Lagopus lagopus* fitted with necklace radio collars do not significantly differ from wing-tagged "controls"; Thirgood et al., 1995).

In general, it seems fair to conclude that the larger and heavier the mark, the more likely it is to have adverse effects. Wildlife biologists tend to follow an informal standard of a transmitter mass of 10–13% of the mass of small mammals (Madison et al., 1985). A study of meadow voles *Microtus pennsylvanicus* showed that radio collars $> 10\%$ of live body mass significantly lowered dominance status, whereas collars of $< 10\%$ did not (Berteaux et al., 1994).

Although marking by modifying appearance avoids such problems, it can incur other difficulties, such as the potentially lethal consequences of ingesting toxic fur or feather dyes or bleaches during grooming or preening. It cannot be assumed that dyes and bleaches formulated for human use will be safe for use on other species; such chemicals should be tested before use for adverse reactions in the study animal. In one instance, a human hair dye which was marketed as "not tested on animals" caused such severe hair loss and skin irritation in laboratory mice that the animals had to be euthanized (C.J. Barnard, personal communication).

The availability of internal marks such as microchips implants (e.g., Elbin and Burger, 1994) and the use of natural marks such as individually distinctive vocalizations offer the prospect of quantifying the effects of marks. As far as we are aware the only study to attempt such a comparison is that by Hindell et al. (1996), in which the effect of flipper bands on survival rate and reproductive success of royal penguins *Eudyptes schlegeli* was assessed by comparing banded birds with those implanted with subcutaneous transponders. The study found no effect of flipper bands on either survival or reproduction, despite the earlier finding that Adélie penguins with flipper bands expended 24% more energy swimming in a tank than unbanded conspecifics (Culik et al., 1993).

There may be more subtle effects of marking individuals. Marking may change behavior (e.g., Mallet et al., 1987; Daly et al., 1992) or cause differences in the behavior of others; for example, marked animals may experience more investigation from conspecifics and predators on return to the population, or they may lose social status (Adkison et al., 1995). It should be remembered that differences in appearance that are subtle or invisible to human eyes (e.g., ultraviolet reflectance patterns) may contain information on status (e.g., badges of status; see Roper, 1987; Maynard Smith and Harper, 1988), and these may be altered by marking. In zebra finches *Taeniopygia guttata,* the color of leg ring added by experimenters can affect breeding success (Burley, 1986), so the influence of marks on behavior can have important consequences for fitness.

A further cost of marking has become apparent in recent years. Many people who observe animals for pleasure, or who photograph them, find marks aesthetically unacceptable, and some may even find marks ethically questionable (Cuthill, 1991; Farnsworth and Rosovsky, 1993). As many species of conservation interest are also important subjects for ecotourism and wildlife photography, we suspect that the aesthetics of marking will become an increasingly important issue.

The costs of marks to biologists are mainly financial; marks based on electronics such as radio transmitters or transponders are the most expensive. As a result of these high costs, most studies can only afford to mark a few individuals, thereby significantly limiting the extent to which general conclusions can be drawn from the data. For example, Koubeck (1995) radio tracked five roe deer *Capreolus capreolus;* Falk and Moller (1995) tracked three northern fulmars *Fulmaris gracilis* by satellite; and Poonswad and Tsuji (1994) based conservation recommendations for three species of hornbill on radio tracking data collected from two males of each species. However, in many cases the costs of electronic marks will still be considerably less than the salaries of the trained staff needed to use them. Parker et al. (1996) reported a cautionary tale of 3 months of fieldwork radio tracking small mammals; the resulting data were unusable due to undetected and unacceptably large errors in location, possibly caused by electromagnetic interference.

Identification

Regardless of whether the method of individual identification is based on added marks or on natural marks, the process of reading marks can lead to biased data. This is most obvious when variation in a long-range signal, such as a vocalization, is used as a natural mark because such signals are often produced only by one sex or other subgroup of the population (e.g., territorial males). For example, the individually distinctive boom vocalization of bitterns *Botaurus stellaris* (McGregor and Byle, 1992; Gilbert et al., 1994) can only be used to identify booming males, and the same is true for calling corncrakes. Such systematic biases are perhaps better termed limitations of the technique, as the effects are well understood. The two main sources of bias stem from the assessments of similarity used to individually identify animals by natural variation in phenotype. First, similarity assessments are known to be affected by experience and discrimination abilities of the observer (e.g., sound spectrograms: Gilbert, 1993; cetacean photo-identification: Dufault and Whitehead, 1995).

Second, there is a tendency for particularly distinctive individuals to be overrepresented in samples because of the ease of identifying them. While these biases are more obvious in methods based on natural marks (and to some extent can be overcome by computer estimates of similarity, see above), they also occur with added marks. For example, considerable experience and skill are needed to identify reliably the color rings of a small bird moving about rapidly in the poorly lit forest canopy, and varying degrees of color-blindness are an important source of bias in male color-ring spotters. Similarity assessments of natural marks are also affected by the quality of the representation of the phenotypic feature, whether this is a photograph, sound recording, or a sighting. Once again, some of the same effects can be encountered with added marks, as transmitter batteries lose power through age or low temperatures, color marks fade, fur dyes grow out, and numbered metal rings can fall off or become unreadable through abrasion.

The costs to the animal of identification can be the same as for initial marking in cases where capture and handling are involved. By contrast, the use of natural phenotypic marks is much less intrusive, often requiring less close approach and hence less disturbance than an attempt to read color rings. Some animals are particularly sensitive to being approached; for example, the deep sea fish *Hoplostethus atlanticus* dispersed rapidly when an underwater camera was lowered toward them (Koslow et al., 1995). Initial verification that the phenotypic feature (e.g., physiognomy, markings, vocalizations, scars) is individually distinctive may also involve marking and its associated costs.

At first sight it would seem that the use of genetic markers must require capture and handling. However, the elegant technique of collecting the sloughed skin left behind after a cetacean has surfaced and extracting DNA for profiling from the skin (Whitehead and Richard, 1995) avoids the need for capture or more invasive noncapture procedures such as the use of biopsy darts (e.g., O'Corry-Crowe and Dizon, 1995).

The costs of identification to biologists are that they may require considerable time and skill (Dufault and Whitehead, 1995) and access to sophisticated equipment. In most instances these costs will be highest for identification based on natural marks because of the similarity assessments involved.

In conclusion, there are many ways an individual can be identified, but every technique has associated costs and benefits. Cost–benefit analysis and optimization are common tools in behavioral ecology, so perhaps it is fitting that this field delivers a plethora of trade-off decisions in relation to individual identification.

Summary

The ability to identify individuals can play a major role in conservation. This is because unambiguous monitoring and census data are an essential component of any management or recovery program, because good design and methods are essential, and because monitoring programs should have specific targets.

In addition, predictive models in conservation biology require detailed knowledge of population structure, and, for most species, long-term monitoring of known individuals is essential for obtaining data on life-history traits. Furthermore, in the absence of data to the contrary, it should be assumed that individuals will differ in conservation value.

When choosing a technique for individual identification, the following should be considered. Mutilation techniques should be methods of last resort, for ethical reasons if for no other. Adding marks is likely to generate biased data (either through catching bias or altered behavior by, or toward, marked individuals). The speed of identification is generally inversely related to the degree of technical sophistication employed. The lower certainty of identification sometimes associated with the use of natural marks or signals is offset by the logistic and welfare advantages of avoiding the need to capture the animal.

Acknowledgments Many colleagues provided us with examples of techniques, discussed relative merits, and commented on earlier drafts. We particularly thank Mark Avery, Jacqui Clarke, Sarah Durant, Gillian Gilbert, Lex Hiby, Jane Hurst, Vincent Janik, Gareth Jones, Anders Lund, Karen McComb, Anne Magurran, Jim Reader, Ken Smith, Fritz Trillmich, Glen Tyler, and Hal Whitehead. We also thank Paule Gros, Leonie McGregor, Dirk Van Vuren, and an anonymous referee for comments on the manuscript. Tim Caro urged us to cast our net wider. T.M.P. was supported by Biotechnology and Biological Sciences Research Council - Royal Society For the Protection of Birds Case studentship.

References

Adkison MD, Quinn TP, Rutten OC, 1995. An inexpensive, nondisruptive method of in situ dart tagging for visual recognition of fish underwater. N Am J Fish Manage 15:507–511.

Amlaner CJ, Macdonald DW, 1980. A handbook on biotelemetry and radio tracking. Oxford: Pergamon Press.

Arak A, 1988. Callers and satellites in the natterjack toad, evolutionarily stable decision rules. Anim Behav 36:416–432.

Baker CS, Straley JM, Perry A, 1992. Population characteristics of individually identified humpback whales in southeastern Alaska: Summer and fall 1986. US National Marine Fisheries Service Fishery Bulletin 90:429–437.

Barclay RMR, Bell GP, 1988. Marking and observation techniques. In: Ecological and behavioral methods for the study of bats (Kunz TH, ed). Washington, DC: Smithsonian Institute Press 59–76.

Bart J, 1995. Acceptance criteria for using individual-based models to make management decisions. Ecol Appl 5:411–420.

Berteaux D, Duhamel R, Bergeron J-M, 1994. Can radio collars affect dominance relationships in *Microtus?* Can J Zool 72:785–789.

Bibby CJ, Burgess ND, Hill DA, 1992. Bird census techniques. New York: Academic Press.

Bowen BW, 1995. Tracking marine turtles with genetic markers-voyages of the ancient mariners. Bioscience 45:528–534.

Bradbury C, Green JM, Bruce-Lockhart M, 1995. Home ranges of female cunner, *Tautogolabrus adspersus* (Labridae), as determined by ultrasonic telemetry. Can J Zool 73:1268–1279.

Bretagnolle V, Thibault JC, Dominici JM, 1994. Field identification of individual ospreys using head marking pattern. J Wildl Manage 58:175–178.

Broom DM, Johnson KG, 1993. Stress and animal welfare. London: Chapman and Hall.

Burley N, 1986. Sexual selection for aesthetic traits in species with biparental care. Am Nat 127:415–445.

Burnham KP, White GC, Anderson DR, 1995. Model selection strategy in the analysis of capture-recapture data. Biometrics 51:888–898.

Burrows R, Hofer H, East ML, 1995. Population dynamics, intervention and survival in African wild dogs (*Lycaon pictus*). Proc R Soc Lond B 262:235–245.

Butynski TM, Chapman CA, Chapman LJ, Weary DM, 1992. Use of male blue monkey "pyow" calls for long-term individual identification. Am J Primatol 28:183–189.

Caldwell DK, 1955. Notes on the spotted dolphin. J Mammal 6:467–475.

Calvo B, Furness RW, 1992. A review of the use and the effects of marks and devices on birds. Ring Migrat 13:129–151.

Caro TM, Durant SM, 1991. Use of quantitative analyses of pelage characteristics to reveal family resemblances in genetically monomorphic cheetahs. J Hered 82:8–14.

Charland MB, Gregory PT, 1995. Movements and habitat use in gravid and nongravid female garter snakes (*Colubridae, Thamnophis*). J Zool 236:543–561.

Clutton-Brock TH, 1988. Reproductive success. Chicago: Chicago University Press.

Clutton-Brock TH, Guinness FE, Albon SD, 1982. Red deer: the behaviour and ecology of two sexes. Chicago: Chicago University Press.

Collar NJ, Andrew P, 1988. Birds to watch: the ICBP World checklist of threatened birds. London: ICBP.

Collar, NJ, Crosby MJ, Stattersfield AJ, 1994. Birds to watch 2: the world list of threatened birds. London: International Council for Bird Preservation.

Conway CJ, Eddleman WR, Anderson SH, Hanebury LR, 1993. Seasonal changes in Yuma clapper rail vocalisation rate and habitat use. J Wildl Manage 57:282–290.

Craighead L, Paetkau D, Reynolds HV, Vyse ER, Strobeck C, 1995. Microsatellite analysis of paternity and reproduction in Arctic grizzly bears. J Hered 86:255–261.

Cramp S, Simmons KEL, 1980. Handbook of the birds of Europe, the Middle East and North Africa, vol 2. Oxford: Oxford University Press.

Creel S, Creel NM, 1995. Communal hunting and pack size in African wild dogs *Lycaon pictus*. Anim Behav 50:1325–1339.

Culik BM, Wilson R, 1991. Penguins crowded out? Nature 351:340.

Culik BM, Wilson RP, Bannasch R, 1993. Flipper-bands on penguins, what is the cost of a lifetime commitment? Mar Ecol Prog Ser 98:209–214.

Cuthill I, 1991. Field experiments in animal behaviour, methods and ethics. Anim Behav 42:1007–1014.

Daly M, Wilson MI, Behrends PR, Jacobs LF, 1992. Sexually differentiated effects of radio transmitters on predation risk and behaviour in kangaroo rats *Dipodomys merriami*. Can J Zool 70:1851–1855.

Dufault S, Whitehead H, 1995. An assessment of changes with time in the marking patterns used for photo-identification of individual sperm whales, *Physeter macrocephalus*. Mar Mammal Sci 11:335–343.

Dunbar MR, Wooding JB, Thomas LA, 1995. Salmonella Hartford infection in a Florida black bear (*Ursus americanus floridanus*). Fl Sci 58:252–254.

Eiler JH, 1995. A remote satellite-linked tracking system for studying Pacific salmon with radiotelemetry. Trans Am Fish Soc 124:184–193.

Elbin SB, Burger J, 1994. Implantable microchips for individual identification in wild and captive populations. Wildl Soc Bull 22:677–683.

Emlen JT, DeJong MJ, 1992. Counting birds, the problems of variable hearing abilities. J Field Ornithol 63:26–31.

Falk K, Moller S, 1995. Satellite tracking of high-arctic northern fulmars. Polar Biol 15:495–502.

Fangrath M, 1994. Analyse von Wachtelkönigrufen (*Crex crex*) (MSc thesis). Osnabrück: Universität Osnabrück.

Farnsworth EJ, Rosovsky J, 1993. The ethics of ecological field experimentation. Conserv Biol 7:463–472.

Gibbs JP, Wenny DG, 1993. Song output as a population estimator, effect of male pairing status. J Field Ornithol 64:316–322.

Gilbert FS, Haines N, Dickson K, 1991. Empty flowers. Funct Ecol 5:29–29.

Gilbert G, 1993. Vocal individuality as a census and monitoring tool, practical considerations illustrated by a study of the bittern *Botaurus stellaris* and the black-throated diver *Gavia artctica* (PhD thesis). Nottingham: University of Nottingham.

Gilbert G, McGregor PK, Tyler G, 1994. Vocal individuality as a census tool: practical considerations illustrated by a study of two rare species. J Field Ornithol 65:335–348.

Ginsberg JR, Alexander KA, Creel S, Kat PW, McNutt JW, Mills MGL, 1995a. Handling and survivorship of African wild dog (*Lycaon pictus*) in five ecosystems. Conserv Biol 9:665–674.

Ginsberg JR, Mace GM, Albon S, 1995b. Local extinction in a small and declining population, wild dogs in the Serengeti. Proc R Soc Lond B 262:221–228.

Glass CW, Johnstone ADF, Smith GW, Mojsiewicz WR, 1992. The movements of saithe (*Pollachius virens* L) in the vicinity of an underwater reef. In: Wildlife telemetry, remote monitoring and tracking of animals (Priede IG, Swift SM, eds). New York: Ellis Horwood: 328–341.

Goldwasser L, Schatz GE, Young HJ, 1993. A new method for marking Scarabaeidae and other Coleoptera. Coleopt Bull 47:21–26.

Goss-Custard JD, Durell SEA, 1990. Bird behaviour and environmental planning, approaches in the study of wader populations. Ibis 132:273–289.

Green RE, 1995a. The decline of the corncrake *Crex crex* in Britain continues. Bird Study 42:66–75.

Green RE, 1995b. Diagnosing causes of bird population declines. Ibis 137:S47–S55.

Gros PM, Kelly MJ, Caro TM, 1996. Estimating carnivore densities for conservation purposes: indirect methods compared to baseline demographic data. Oikos 77:197–206.

Hammond PS, 1990. Capturing whales on film-estimating cetacean population parameters from individual recognition data. Mammal Review 20:17–22.

Hammond PS, Mizroch SA, Donovan GP, 1990. Individual recognition of cetaceans, use of photo-identification and other techniques to estimate population parameters. Report of IWC special issue 12: International Whaling Commission 1–440.

Hendry AP, Leonetti FE, Quinn TP, 1995. Spatial and temporal isolating mechanisms, the formation of discrete breeding aggregations of sockeye salmon (*Oncorhynchus nerka*). Can J Zool 73:339–352.

Hiby L, Lovell P, 1990. Computer aided matching of natural markings, a prototype system for grey seals. Report of IWC special issue 12: 57–61.

Higuchi H, Ozaki K, Fujita G, Minton J, Mutsuyuki U, Soma M, Mita N, 1996. Satellite tracking of white-naped crane migration and the importance of the Korean demilitarised zone. Conser Biol 10:806–812.

Hindell MA, 1991. Some life-history parameters of a declining population of southern elephant seals, *Mirounga leonina*. J Anim Ecol 60:119–134.

Hindell MA, Lea M-A, Hull CL, 1996. The effects of flipper bands on adult survival rate and reproduction in the royal penguin *Eudyptes schlegeli*. Ibis 138:557–560.

Holland J, McGregor PK, 1997. Disappearing song dialects? The case of Cornish corn buntings. In: The ecology and conservation of corn buntings *Miliaria calandra* (Aebischer NJ, Donald PF, eds). UK Nature Conservation, no. 13. Peterborough: Joint Nature Conservation Committee 181–185.

Holland J, McGregor PK, Rowe CL, 1996. Microgeographic variation in the song of the corn bunting, *Miliaria calandra*, changes with time. J Avian Biol 27:47–55.

Hudson AV, Stowe TJ, Aspinall SJ, 1990. Status and distribution of corncrakes in Britain in 1988. Br Birds 83:173–187.

Hurst JL, Agren G, 1994. Stress and handling. ASAB Neslett 22:15–16.

Hurst JL, Barnard CJ, Hare R, Wheeldon EB, West CD, 1996. Housing and welfare in laboratory rats, time-budgeting and pathophysiology in single-sex groups. Anim Behav 52:335–360.

Hurst JL, Berreen JM, 1985. Observation of the trap-response of wild house mice *Mus domesticus* Rutty, in poultry houses. J Zool 207:619–622.

Hwang WD, Chao A, 1995. Quantifying the effect of unequal catchabilities on Jolly-Seber estimators via sample coverage. Biometrics 51:128–141.

Janik VM, Dehnhardt G, Todt D, 1994. Signature whistle variations in a bottlenosed dolphin, *Tursiops truncatus*. Behav Ecol Sociobiol 35:243–248.

Jarvis JUM, 1981. Eusociality in a mammal: cooperative breeding in naked mole-rat colonies. Science 212:571–573.

Jarvis JUM, O'Riain MJ, Bennett NC, Sherman PW, 1994. Mammalian eusociality, a family affair. Trends Ecol Evol 9:47–51.

Jeffries DJ, Mitchell-Jones AJ, 1993. Recovery plans for British mammals of conservation importance, their design and value. Mamm Rev 23:155–166.

Jones L, 1997. The scent of a tiger. New Scientist 155 (no. 2090):18.

Kalcounis MC, Brigham RM, 1995. Interspecific variation in wing loading affects habitat use of little brown bats (*Myotis lucifugus*). Can J Zool 73:89–95.

Karanth KU, 1995. Estimating tiger *Panthera tigris* populations from camera-trap data using capture-recapture models. Biol Conserv 71:333–338.

Kat PW, Alexander KA, Smith JS, Munson L, 1995. Rabies and African wild dogs in Kenya. Proc R Soc Lond B 262:229–233.

Kenward RE, 1993. Modelling raptor populations: to ring or to radio-tag? In: Marked individuals in the study of bird populations (LeBreton JD, North PM, eds). Basel: Birkhäser Verlag 157–167.

Kiernan V, 1995. Ocean ear beats whale watchers. New Scientist 148 (no. 2007):7.

Klomp NI, Furness RW, 1992. Non-breeders as a buffer against environmental stress, declines in numbers of great skuas on Foula, Shetland, and prediction of future recruitment. J Appl Ecol 29:341–348.

Koslow JA, Kloser R, Stanlet CA, 1995. Avoidance of a camera system by a deepwater fish, the orange roughy (*Hoplostethus atlanticus*). Deep Sea Res Ocean Res Pap 42:233–244.

Koubek P, 1995. Home range dynamics and movements of roe deer (*Capreolus capreolus*) in a floodplain forest. Folia Zool 44:215–226.

Krebs CJ, 1989. Ecological Methodology. New York: Harper and Row.

Krebs JR, 1982. Territorial defense in the great tit, *Parus major:* do residents always win? Behav Ecol Sociobiol 11:185–194.

Krebs JR, Avery MI, Cowie RJ, 1981. Effect of removal of mate on the singing behaviour of great tits. Anim Behav 29:635–637.

Laurenson MK, Caro TM, 1994. Monitoring the effects of non-trivial handling in free-living cheetahs. Anim Behav 47:547–557.

Lazo A, 1995. Ranging behaviour of feral cattle (*Bos taurus*) in Doñana National Park, SW Spain. J Zool 236:359–369.

Lessells CM, Rowe CL, McGregor PK, 1995. Individual and sex differences in the provisioning calls of European bee-eaters. Anim Behav 49:244–247.

Lockyer CH, Morris RJ, 1990. Some observations on wound healing and persistence of scars in *Tursiops truncatus*. Report of IWC special issue 10:113–118.

McGregor PK, Anderson CM, Harris J, Seal JR, Soul JM, 1994. Individual differences in songs of fan-tailed warblers *Cisticola juncidis* in Portugal. Airo 5:17–21.

McGregor PK, Byle P, 1992. Individually distinctive bittern booms: potential as a census tool. Bioacoustics 4:93–109.

McGregor PK, Holland J, Shepherd M, 1997. The ecology of corn bunting *Miliaria calandra* song dialects and their potential use in conservation. In: The ecology and conservation of corn buntings *Miliaria calandra,* (Donald PF, Aebischer NJ, eds). UK Nature Conservation, no. 13. Peterborough: Joint Nature Conservation Committee 76–87.

McGregor PK, Westby GWM, 1992. Discrimination of individually characteristic electric organ discharges by a weakly electric fish. Anim Behav 43:977–986.

Madison DM, Fitzgerald RW, McShea WJ, 1985. A user's guide to the successful radio-tracking of small mammals in the field. Laramie: University of Wyoming.

Mallet J, Longino JT, Murawski D, Murawski A, Simpson de Gamboa A, 1987. Handling effects in *Heliconius,* where do all the butterflies go? J Anim Ecol 56:377–386.

Masters WM, Raver KAS, Kazial KA, 1995. Sonar signals of big brown bats, *Eptesicus fuscus,* contain information about individual identity, age and family affiliation. Anim Behav 50:143–160.

Maynard Smith J, Harper DGC, 1988. The evolution of aggression: can selection generate variability? Phil Trans R Soc Lond B 319:557–570.

Meyburg B-U, Eichaker X, Meyburg C, Paillat P, 1995. Migrations of an adult spotted eagle tracked by satellite. Bri Birds 88:357–361.

Morell V, 1995. Dogfight erupts over animal studies in the Serengeti. Science 270:1302–1303.

Mowat G, Slough BG, Rivard R, 1994. A comparison of three live capturing devices for lynx, capture efficiency and injuries. Wildl Soc Bull 22:644–650.

Mrozek M, Fischer R, Trendelenburg M, Zillman U, 1995. Microchip implant system used for animal identification in laboratory rabbits, guineapigs, woodchucks and in amphibians. Lab Anim 29:339–344.

Nelson JS, Chou LM, Phang VPE, 1994. Pigmentation variation in the anemonefish *Amphprion ocellaris* (Teleostei, Pomacentridae): type, stability and its usefulness for individual identification. Raffles Bull Zool 42:927–930.

Newton I, 1989. Lifetime reproduction in birds. London: Academic Press.

Newton I, 1995. The contribution of some recent research on birds to ecological understanding. J Anim Ecol 64:675–696.

Nimon AJ, Schroter RC, Stonehouse B, 1995. Heart rate of disturbed penguins. Nature 374:415.

Niva T, 1995. Retention of visible implant tags by juvenile brown trout. J Fish Biol 46:997–1002.

O'Corry-Crowe GM, Dizon AE, 1995. Molecular approaches to the study of population structure and social organization in beluga whales (*Delphinapterus leucas*). Int Ethol Conf Abstracts 24:21.

Osborne R, Bettoli PW, 1995. A reusable ultrasonic tag and float assembly for use with large pelagic fish. N Am J Fish Manage 15:512–514.

Palsboll PJ, Allen J, Berube M, Clapham PJ, Feddersen TP, Hammond PS, Hudson RR, Jorgensen H, Katona S, Larsen AH, Larsen F, Lien J, Maltila DK, Sigurjonsson J, Sears R, Smith T, Sponar R, Strevick P, Oien N, 1997. Genetic tagging of humpback whales. Nature 388:767–769.

Parker N, Pascoe A, Moller H, Maloney R, 1996. Inaccuracy of a radio-tracking system for small mammals: the effect of electromagnetic interference. J Zool Lond 239:401–406.

Peake TM, 1997. Variation in the vocal behaviour of the corncrake *Crex crex*: potential for conservation. (Ph.D. thesis) Nottingham: University of Nottingham.

Peake TM, McGregor PK, Smith KS, Gilbert G, Tyler GA, Green RE, 1998. Individuality in corncrake *Crex crex* vocalizations: role in conservation of a globally threatened species. Ibis 140:121–127.

Perrins CM, Lebreton JD, Hirons GJM, 1991. Bird population studies, relevance to conservation and management. Oxford: Oxford University Press.

Pollard E, Yates TJ, 1993. Monitoring butterflies for ecology and conservation. London: Chapman and Hall.

Poole TB, 1994. Alternatives to "toe clipping" for identifying small vertebrates. ASAB Newslett 20:7–8.

Poonswad P, Tsuji A, 1994. Ranges of males of the great hornbill *Buceros bicornis,* brown hornbill *Ptilolaemus tickelli* and wreathed hornbill *Rhyticeros undulatus* in Khao Yai National Park, Thailand. Ibis 136:79–86.

Porter JM, Coulson JC, 1987. Long-term changes in recruitment to the breeding group, and the quality of recruits at a kittiwake *Rissa tridactyla* colony. J Anim Ecol 56:675–689.

Pouliquen O, Leishman M, Redhead TD, 1990. Effects of radio collars on wild mice, *Mus domesticus.* Can J Zool 68:1607–1609.

Priede, IG, Swift SM, 1992. Wildlife telemetry: remote monitoring and tracking of animals. New York: Ellis Horwood.

Rabinowitz A, Schaller GB, Uga U, 1995. A survey to assess the status of Sumatran rhinoceros and other large mammal species in Tamanthi Wildlife Sanctuary, Myanmar. Oryx 29:123–128.

Riddington R, Gosler AG, 1995. Differences in reproductive success and parental qualities between habitats in the great tit *Parus major.* Ibis 137:371–378.

Rohner C, 1996. The numerical response of great horned owls to the snowshoe hare cycle, consequences of non-territorial "floaters" on demography. J Anim Ecol 65:359–370.

Rood JP, Nellis D, 1980. Freeze marking mongooses. J Wildl Manage 44:500–502.

Roper TJ, 1987. Badges of status in avian societies. New Sci 109(1494):38–40.

Rosenberg DK, Overton WS, Anthony RG, 1995. Estimation of animal abundance when capture probabilities are low and heterogeneous. J Wildl Manage 59:252–261.

Sauer A, 1994. Individual identification of live *Coronella austriaca* (Laurenti, 1768). Salamandra 30:43–47.

Scott DK, 1978. Identification of individual Bewick's swans by bill patterns. In: Animal marking: recognition marking of animals in research (Stonehouse B, ed). London: Macmillan; 160–168.

Scribner SJ, Weatherhead PJ, 1995. Locomotion and antipredator behaviour in three species of semi-aquatic snakes. Can J Zool 73:339–352.

Seger J, 1991. Cooperation and conflict in social insects. In: Behavioural ecology, 3rd ed. (Davies NB, Krebs JR, eds). Oxford: Blackwell Scientific; 338–373.

Shaughnessy PD, 1993. Population size of the Cape fur seal *Arctocephalus pusillus,* from tagging and recapturing. S Afr Fish Res Inst Invest Rep 134:1–70.

Sheaffer SE, Jarvis RL, 1995. Bias in Canada goose population size estimates from sighting data. J Wildl Manage 59:464–473.

Singer MC, Wedlake P, 1981. Capture does affect probability of recapture in a butterfly species. Ecol Entomol 6:215–216.

Sinsch U, 1992. Two new tagging methods for individual identification of amphibians in long-term field studies: first experiences with natterjacks. Salamandra 28:116–128.

Smallwood KS, Fitzhugh EL, 1993. A rigorous technique for identifying individual mountain lions *Felis concolor* by their tracks. Biol Conserv 65:51–59.

Smith GW, 1995. A critical review of the aerial and ground surveys of breeding waterfowl in North America. Biological Science Report 5. USDI. Washington, DC: National Bureau of Statistics.

Southwood TRE, 1978. Ecological methods. London: Chapman and Hall.

Spencer R, 1976. The ringer's manual. Tring, Herts: British Trust for Ornithology.

Stewart AJA, Hutchings MJ, 1996. Conservation of populations. In: Conservation biology (Spellerberg IF, ed). Harlow: Longman.

Stonehouse B, 1978. Animal marking: recognition marking of animals in research. London: Macmillan.

Stoddart DM, 1982. Does trap odour influence estimation of population size of the short tailed vole *Microtus agrestis?* J Anim Ecol 51:375–386.

Stowe TJ, Hudson AV, 1988. Corncrake studies in the Western Isles. RSPB Conserv Rev 2:38–42.

Stowe TJ, Hudson AV, 1991. Radio-telemetry studies of corncrake in Great Britain. Die Vogelwelt 112:10–16.

Sutherland WJ, 1995. From individual behaviour to population ecology. Oxford: Oxford University Press.

Sutherland WJ, 1996. Ecological census techniques. Cambridge: Cambridge University Press.

Templeton RG (ed), 1995. Freshwater fisheries management. Oxford: Blackwell Scientific.

Thiollay J-M, 1989. Censusing of diurnal raptors in a primary rain forest, comparative methods and species detectability. J Raptor Res 23:72–84.

Thirgood SJ, Redpath SM, Hudson PJ, Hurley MM, Aebischer NJ, 1995. Effects of necklace radio transmitters on survival and breeding success of red grouse *Lagopus lagopus scoticus* Wildl Biol 1:121–126.

Timmons TJ, Howell MH, 1995. Retention of anchor and spaghetti tags by paddlefish, catfishes and buffalo fishes. N Am J Fish Manage 15:504–506.

Tucker GM, Heath MF, 1994. Birds in Europe: their conservation status. Cambridge: Birdlife International.

Tyler G, Green RE, 1996. The incidence of nocturnal song by male corncrakes *Crex crex* is reduced during pairing. Bird Study 43:214–219.

Tyler G, Green RE, Stowe TJ, Newton AV, 1996. Sex differences in the behaviour and measurements of corncrakes *Crex crex* in Scotland. Ringing Migrat 17:15–19.

Van der Meer J, Camphuysen CJ, 1996. Effect of observer differences on abundance estimates of seabirds from ship-based strip transect surveys. Ibis 138:433–437.

van Lawick Goodall J, 1971. In the shadow of man. Glasgow: Collins.

Villiers MS, Meltzer DGA, van Heerden J, Mills MGL, Richardson PRK, van Jaarsveld AS, 1995. Handling-induced stress and mortalities in African wild dogs (*Lycaon pictus*). Proc R Soc Lond B 262:215–220.

Walker K, Elliott G, Nicholls D, Murray D, Dilks P, 1995. Satellite tracking of wandering albatross (*Diomedea exulans*) from the Auckland Islands: Preliminary results. Notornis 42:127–137.

Wang JP, Adolph SC, 1995. Thermoregulatory consequences of transmitter implant surgery in the lizard *Sceloporus occidentalis*. J Herpetol 29:489–493.

Wauters LA, Dhondt AA, 1995. Lifetime reproductive success and its correlates in female Eurasian red squirrels. Oikos 72:402–410.

White GC, Garrott RA, 1990. Analysis of wildlife radio-tracking data. Riverside, California: Academic Press.

Whitehead H, Richard KR, 1995. Social organization of sperm whales *Physeter macrocephalus*. Int Ethol Conf Abstracts 24:21.

Wiley RH, Richards DG, 1983. Adaptations for acoustic communication in birds: sound transmission and signal detection. In: Ecology and evolution of acoustic communication in birds (Kroodsma DE, Miller EH, eds). New York: Academic Press; 131–181.

Young H, Neess J, Emlen JT, 1952. Heterogeneity of trap responses in a population of house mice. J Wildl Manage 16:169–180.

Young MK, 1994. Mobility of brown trout in south-central Wyoming streams. Can J Zool 72:2078–2083.

3

Ecological Indicators of Risk for Primates, as Judged By Species' Susceptibility to Logging

Alexander Harcourt

Predicting Extinction

Most primate species live in tropical forests. Tropical forests are not only disappearing fast, but disappearing at an increasing rate, at an average of more than 2.5% annually in some countries, thus giving a time to total forest disappearance in some countries of well under a century (Myers, 1984; Barnes, 1990; Skole and Tucker, 1993; Harcourt, 1995a; World Resources Institute, 1996). Primates are therefore disappearing fast: some species number fewer than 1000 individuals (Mittermeier and Cheney, 1987). The question addressed here is how might we, as biologists, identify those species that are most at risk? Some conservationists have suggested that measures of human impact on a region are a better indicator of the future than are biological measures of what is happening (Brown, 1981; Western, 1982; Weber, 1987; Leader-Williams and Albon, 1988; Peres and Terborgh, 1995; see also Harrison, 1987). However, while analysis of human impact can inform us of the fate of a whole ecosystem, it might be less applicable to predicting the future of individual species within a region.

Strong theory and a considerable body of evidence demonstrate that small populations are more likely to go extinct than large ones (Diamond, 1984; Caughley and Gunn, 1996) because they are more vulnerable to demographic and environmental stochastic effects (Lande, 1993) and to adverse genetic effects (Frankham, 1995). Consequently, the demographic analysis of small populations has become a major concern in conservation biology (Burgman et al., 1993; Caughley, 1994; Caughley and Gunn, 1996). Another reason for the "small population paradigm" (Caughley, 1994) seems to be that demographic parameters are, thanks to strong generalizable theory, very amenable to modeling. This makes their use attractive to managers and, indeed, to population ecologists in general. The influence of the small population paradigm in biological conservation is indicated by the fact that demographic data are the basis of four of the five criteria of risk adopted by the International Union for the Conservation of Nature (IUCN) in its Red Data Book categorization of levels of threat facing individual species (IUCN, 1994; Mace and Stuart, 1994). Where population viability analyses (PVAs) take due account of the process of extinction or persistence, they can be powerful and have had important successes (Wahlberg et al., 1996). However, their application can be problematical if their limitations are not realized (Caughley, 1994; Wennergren et al., 1995; Caughley and Gunn, 1996).

Both because of these problems and in order to broaden the biological base from which we might make management decisions, I review here the possibility of identifying ecological, as opposed to demographic, indicators of risk for a fairly well-known mammalian taxon, the primates. To highlight the need for such ecological indicators, I begin by describing in some detail the problems of the small population paradigm. Then in the rest of the chapter I test how well population size predicts risk, search for ecological characteristics that might distinguish species at risk from successful species, and ask whether population size or the other ecological characteristics were better indicators of risk by comparing species that do poorly in logged forest with those that do well.

The Need for Ecological Indicators of Risk

Although small populations are usually at greater risk of extinction than large ones, we have long known that certain types of species are more prone to extinction than others, to some extent independently of population size (Brown, 1971; Diamond, 1984; Harris, 1984; Rabinowitz et al., 1986; Johns, 1992; Gaston, 1994; Newmark, 1995; Leach and Givnish, 1996). For example, an important part of the species–area relationship is immigration. Immigration saves populations, both by increasing numbers and by increasing genetic diversity (MacArthur and Wilson, 1967; Simberloff and Wilson, 1969; MacArthur, 1972; Brown and Kodric-Brown, 1977; Gaston, 1994). Thus, species, including terrestrial species, whose individuals are poor dispersers are more likely to go extinct than those that are good dispersers, independently of standing numbers (MacArthur, 1972; Diamond, 1984; Burbidge and McKenzie, 1989; Laurance, 1991). Consequently, the most abundant species in undisturbed habitats are not necessarily the ones that will best survive fragmentation or disturbance (Lovejoy et al., 1984; Skorupa, 1986; Laurance, 1994). The Carolina parakeet and the passenger pigeon might be extreme examples (Bibby, 1995). Furthermore, there is theoretical reason to believe that species at the current highest densities (i.e., the best competitors and supposedly the safest species) might be at more risk than others, if being a good competitor is associated with being a poor disperser and being a poor disperser is associated with vulnerability to extinction (Tilman et al., 1994); although we still know far too little about the connection between dispersal ability and competitive superiority (Wennergren et al., 1995).

The use of demographic models to assess viability of populations or species is further complicated by lack of data. The simplist viability model, if it is to be useful, requires data on the proportion of demographic variance due to environmental variance, carrying capacity, population growth rate, and a density-dependent function (Goodman, 1987; Burgman et al., 1993; Wennergren et al., 1995). Such information is effectively unavailable and unobtainable to any usable degree of accuracy for most species (Goodman, 1987; Dobson and Lyles, 1989; Gaston, 1994; Oates, 1994a; Harcourt, 1995a, 1996). Realizing the paucity of data demanded by demographic models, Dobson and Lyles (1989) suggested a simple index of risk for primates, arguing that the results from their survey of field studies indicated that populations will be in danger of collapse when adult female survival per inter-birth interval is less than 70%. However, even these two sets of data might be practicably unobtainable for almost all species (Harcourt, 1995a).

Furthermore, even when we have large amounts of data, predictions from some of the models can be so variable that they might not be very useful. Taylor (1995) demonstrated with a voluminous data set for the Steller sea lion *Eumatomias jubatus* that predictions of probabilities of extinction within 100 years could vary from 0.12 to 0.89—in effect from very likely to go extinct to fairly unlikely to go extinct. As Taylor put it; "The results indicate that we are not ready to use PVAs, as they are currently done, to classify species" (Taylor, 1995:554; see also Wennergren et al., 1995). Nevertheless, variable as the prediction was, it still allowed the species to be classified as "Vulnerable" or "Endangered," but probably not yet "Critically Endangered," according to current IUCN criteria and nomenclature (Mace and Stuart, 1994).

Finally, at present one of the main limitations of demographic modeling might be that it is ahead of its time (Caughley, 1994; Harcourt, 1995a,b). Before we model a process, we should understand the process: "any PVA venture must start with an identification of factors affecting the species and estimation of parameters related to these factors. Modeling and risk analysis comes as a third step" (Akçakaya and Burgman, 1995:706). However, in most cases, we do not understand the process; instead, we are jumping straight to Akçakaya and Burgman's third step from knowledge (or guesswork) of numbers alone (Clutton-Brock and Albon, 1985; Hassell and May, 1985; Sinclair, 1989, 1992; Caughley, 1994; Harcourt, 1995b; Wennergren et al., 1995; Caughley and Gunn, 1996; Sutherland, 1996). Therefore, demographic modeling could, without the most explicit provisos, provide prediction in ignorance. Furthermore, insofar as much of the modeling concentrates on outcome, rather than on process, it will have problems in reliably predicting populations' responses under changing conditions, which is the issue that largely concerns conservationists (Caughley, 1994; Caughley and Gunn, 1996; Sutherland, 1996). In summary, the main use of many of the demographic models should perhaps not be to predict risk. Instead, in conjunction with sensitivity analysis, demographic models should be used to identify those parameters that strongly influence outcome. By pinpointing such parameters, we could concentrate data collection and study on the sensitive parameters, and we could concentrate management on ameliorating adverse influences on those particularly sensitive parameters (Caughley, 1977; Crouse et al., 1987; Caughley, 1994; Wennergren et al., 1995). If the demographic modeling does not provide as much predictive power as we want, the search for what I call "ecological correlates of risk" and the development of ecological models is particularly important (Caughley, 1994; Sutherland, 1996).

The Existence of Ecological Indicators of Risk

After searching qualitatively for correlates of risk for some primate species, Pearl (1992) concluded that conservation strategies would work best on a case-by-case basis. That seems unduly pessimistic. The fact that certain types of species are sometimes more at risk than others, independently of population size, implies the existence of ecological indicators of risk, although some searches have found none (Angermeier, 1995). In addition to the poor dispersers mentioned earlier, another category appears to be large-bodied, canopy frugivores, at least among birds (Diamond, 1984; Kattan et al., 1994). For primates, the extinctions of lemurs on Madagascar subsequent to the arrival of humans on the continent indicate that in this taxon too, large-bodied species were at risk (Jolly, 1986). Habitat fragmentation studies in South America indicate that nonfolivorous primates, large-bodied primates, and primates with large home-ranges might be at risk (Bernstein et al., 1976;

Lovejoy et al., 1986; Ferrari and Diego, 1995). Large body size was also a risk factor in a study of correlates of variation in population density of mammals (including primates) in South American forests, the assumption being that species whose population density varied greatly from site to site were more at risk than those with more evenly distributed densities (Robinson and Redford, 1989). However, within primates (the single order with the most species in the sample), no correlation was evident. In a quantitative review, Happel et al. (1987) found that primates listed in the IUCN Red Data Book tended to have gestation periods shorter or longer than average and to have small geographic ranges. However, a small geographic range has often been a criterion for the listing in the first place, so the finding is to some extent circular.

Response to Logging as a Means of Obtaining Indicators of Risk

Some of the most detailed studies investigating ecological indicators of risk concern primates' responses to logging, particularly selective logging. It would be surprising if the types of species that could not survive well in changed or degraded habitats did not differ from those that could survive well, given that different species have different habitat requirements and preferences, almost by definition (Harris, 1984). Some of the most detailed primate studies are Skorupa's (1986) on the effects of selective logging on the diurnal primates at one site in Africa, Johns's (1992; Johns and Johns, 1995) continuing analyses of such effects in Asian forests, and their joint review of the field (Johns and Skorupa, 1987). These studies also found that large body size and frugivory were risk factors (Skorupa, 1986; Johns and Skorupa, 1987). Skorupa (1986) found also that a large group-spread and large home-range area were risk factors, and concluded that reliance on widely dispersed, relatively rare foods made species vulnerable to disturbance by logging.

Since Johns and Skorupa's (1987) important review of responses to logging, only a few other studies have appeared on the subject (Oates, 1996; White and Tutin, in press; and other references in table 3-1). Two studies in Madagascar at different sites show 7 species (none congeneric) to range from vulnerable to safe (Ganzhorn 1995; White et al., 1995). No species was recorded from both sites. Thus, there is in effect only one record per species, and therefore these studies are not considered further. So far, only Johns has consistently studied changes over many years in the same sample plots. Others mainly use one-time samples of plots logged at different periods in the past, although two studies in Kibale forest in Uganda had samples 20 years apart (Olupot et al., 1994; Weisenseel et al., 1993). Struhsaker (1997) has collated all the Kibale Forest studies, including those on non-primates. His review for this one study site emphasises that data on vulnerability to disturbance are available for very few sites, and species (table 3-1). The main addition that I make here to Johns and Skorupa's (1987) review, besides addition of a few more studies, and of more potential correlates of vulnerability, is to incorporate modern methods of phylogenetic analysis into the statistical analysis of correlates of risk.

Methods

To estimate risk, I use Skorupa's index of response to logging, namely the density in previously logged forest over the density in primary forest at the same site. I use only comparisons at the same site in order to minimize the confounding effects of differences be-

Table 3-1 Species for which there are data on response to disturbance by selective logging, listed by continent, nature of response, and number of study sites (forests) at which the response has been measured.

Africa ($n = 9/22$)[a]		Asia[d] ($n = 6/16$)		S. America[e] ($n = 10/20$)	
Species[b]	Response[c]	Species	Response	Species	Response
Cercopithecus ascanius[1,2,3]	ssv	*Macaca fascicularis*	ssss	*Callithrix argentata*	ss
C. campbelli[4,5]	s	*M. nemestrina*	ssvv	*C. humeralifer*	s
C. diana[3,4,5]	v(v)			*C. flaviceps*	
C. erythrotis[6]		*Nasalis larvatus*	vs	*Saguinus fuscicollis*	ssss
C. lhoesti[3]	v	*Presbytis comata*	vv	*S. midas*	sss
C. mitis[1,2,3]	sss	*P. cristata*	s	*S. mystax*	sv
C. mona[3]	(s)	*P. melalophos*	ssss		
C. petaurista[4,5]	s	*P. obscura*	sssv	*Cebus albifrons*	svv
C. pogonias[2]	s	*P. rubicunda*	sv	*C. apella*	ssssv
		P. thomasi	sv	*Saimiri sciureus*	sss
Cercocebus albigena[2,3]	sv	*Simias concolor*	s	*Callicebus moloch*	sss
C. torquatys[3,4,5]	s(s)	*Hylobates agilis*	v	*C. torquatus*	svvv
		H.lar	sssv		
Colobus angolensis[2]	v	*H. muelleri*	ssv	*Alouatta belzebul*	v
C. guereza[1,2,3]	sss	*H. syndactylus*	vv	*A. fusca*	

C. polykomos[3,4,5]	v(s)			*A. seniculus*	vv
Procolobus badius[2,3,4,5]	vvv(v)	*Pongo pygmaeus*	vvv	*Ateles paniscus*	svv
Procolobus verus[3,4,5]	v(s)			*Lagothrix lagothricha*	vv
Papio anubis[1]	s	*Nycticebus coucang*	v	*Pithecia albicans*	sv
				P. hirsuta	s
Gorilla gorilla[3,7]	ssss			*Chiropotes albinasus*	sv
Pan troglodytes[1,3,7]	vvvs			*C. satanas*	v
Galago alleni[6]					
G. demidoff or *inustus*[8]	v				
P. potto[8]	v				

[a]Numbers in parentheses are *n* for genus/species.

[b]Superscripted numbers indicate references as follows: [1]Plumptre and Reynolds (1994); [2]Thomas (1991); [3]Johns and Skorupa (1987); [4]Fimbel (1994); [5]Dates et al. (1990); [6]Butynski and Koster (1994); [7]Tutin and Fernandez (1984); [8]Weisenseel et al. (1993).

[c]Response = density in disturbed forest/density in original forest at the same study site expressed as a ratio. Species with ratios of 0.5 or less are "vulnerable" (indicated by "v"); species with ratios of more than 0.5 are not vulnerable (indicated by "s," for "safe"); number of letters per species = number of study sites for which there are data; letters in parentheses are those from sites where the secondary forest <3 years old; species that have no data in the response column have had only qualitative estimates of relative abundance made. For species used in the analysis (those for which we have data from more than one site and the response is consistent; i.e., most of the sites show the same response), the response is underlined. *Nasalis* was used and categorized as vulnerable because the vulnerable response was so severe compared to the "safe" response.

[d]Data for Asia from Johns and Skorupa (1987), except for *Simias concolor,* from Watanabe (1981).

[e]Data for South America from Johns and Skorupa (1987).

tween forests at different sites, even those within the same country (e.g., Johns and Skorupa, 1987; Fimbel, 1994). Given the considerable sampling error in the estimates, I use only two categories of response: taxa are vulnerable if the index is 0.5 or less, i.e. their density in the disturbed forest is half or less than it is in primary forest at the same site; taxa with indices of more than 0.5 are not vulnerable. Table 3-1 lists all species for which studies have been conducted, but in my analysis I use only (1) diurnal species, (2) those species for which there are data from more than one site, and (3) those species for which their reaction to disturbance (as defined here) is consistently either vulnerable or not vulnerable. Although the data are fairly evenly distributed across the three major continents (table 3-1), studies differ in the number of years between disturbance and the measurement of density; in the nature, extent, and intensity of logging (and associated disturbance such as hunting); in the nature of original and secondary forest; and in the quality of census methodology. Furthermore, similar inadequacies beset the data related to the factors that I correlate with vulnerability to disturbance (table 3-2). For instance, while species are classified as either territorial or not, some species differ from site to site; and home-range size can differ enormously across study sites (Smuts et al., 1987). I have made no attempt to account for any of these problems, other than as described above. In other words, 10 years after Johns and Skorupa's review, this is still a preliminary analysis. A description and explanation of the factors that I correlate with vulnerability to logging follows.

1. *Population density.* I use population density at the site as an indirect measure of population size. Of course, the correlation is not exact, and the persistence of a population depends on far more than simply number of individuals (Goodman, 1987). However, lack of data on other demographic measures, even with this well-studied taxon, necessitates such a method. If density, the simplest demographic measure, and the one most commonly available from censuses, is not sufficient, then we will have no data to use for any but the most studied species. Consequently, density is being used by conservationists both as a criterion of risk, and as a means of obtaining population sizes. Its predictive power needs to be tested, therefore.
2. *Geographic density.* I use geographic density as an indicator of the species' overall relative competitive ability. If density correlates with population size or competitiveness, species at low density should be vulnerable.
3. *Body size.* Body size correlates with the size of the home range as well as population density (McNab, 1963; Schoener, 1968; Milton and May, 1976; Clutton-Brock and Harvey, 1977a), and thus species of large body size, which exist in smaller populations and require large areas for existence, should be at more risk. In Johns and Skorupa's (1987) review, body size correlated weakly with the response index, once diet was accounted for in the analysis ($p = .08$). They argued that large species tend to do less well than small species in secondary forest compared to primary forest.
4. *Diet.* With respect to diet, species feeding on widely dispersed, rare foods (such as fruits) should be more at risk than others if such foods are more likely to disappear with fragmentation or disturbance of the habitat (Skorupa, 1986). The same should apply to specialists (see, for example, the rarity of ant-following bird species in fragmented Central American forests; Karr, 1977), unless specialists can occupy smaller home ranges than generalists (Harris, 1984). For primates, diet was the single significant correlate of vulnerability to disturbance in Johns and Skorupa's review (1987): highly frugivorous species did poorly ($r_s = -.56, p < .01$). In Skorupa's (1986) study of correlates of vulnerability to selective logging at one site in Africa, species that had high proportions of fruit, seeds, and flowers in their diet also did significantly worse than

taxa with lower proportions ($r_s = -.83$, $n = 7$ species, $p < .05$; one tailed). However, dietary diversity (measured by the percentage of diet accounted for by the most eaten species) did not obviously correlate with vulnerability (Johns and Skorupa, 1987). Other studies have found contrary results, however, and in Johns and Skorupa's analysis, some obvious exceptions existed to the generalization that frugivores fared worse. The response of primates to logging will depend in part on the precise nature of the differences between the primary and secondary forests at the different sites; for instance, some secondary forest can contain a more consistent and abundant supply of fruit than primary forest (Johns and Skorupa, 1987; Johns, 1988; Fimbel, 1994; Plumptre and Reynolds, 1994; Hashimoto, 1995).

5. *Arboreality.* I use arboreality as a crude indication for primates' ability or propensity to cross open ground: obligately arboreal species should be at greater risk when disturbed forest is more open, given studies that have shown that inability to cross open spaces or disturbed habitat is a predictor of vulnerability (Burbidge and McKenzie, 1989; Laurance, 1990; Johns, 1992).
6. *Home ranges.* Species that require large home ranges are presumably more likely to run out of resources than are species with smaller home ranges, other things being equal; hence, large carnivores may disappear before herbivores do as fragmentation proceeds (Brown, 1971; Diamond, 1984). In Skorupa's (1986) analysis of a single site, home-range size was one of only three significant associations ($p_s = -.8$, $n = 7$ species, $p < .05$; one tailed) among the 13 factors analyzed; species with large home range did poorly.
7. *Territoriality.* If nonterritorial species usually have larger home ranges than territorial species (because they are less likely to feed on dispersed, high-quality foods), then they should be more at risk (Skorupa, 1986). Nevertheless, Skorupa (1986) found no significant effect of territoriality. White and Tutin (in press) have argued, in attempting to explain the greater vulnerability of chimpanzees than gorillas to disturbance, that territorial species might be more vulnerable to disturbance because territoriality prevents displaced animals from settling. Whether this is so depends on how compressible territories are, and indeed on how labile the territoriality is in the face of increased competition (Watson and Moss, 1970; Davies and Houston, 1984). Thus, at present, it is difficult to predict how territoriality might correlate with vulnerability to logging.
8. *Altitudinal range.* J. Skorupa (personal communication) suggested that altitudinal range might correlate with susceptibility to disturbance because the characteristics of high altitude are also those of disturbed forest. Happel et al. (1987) also tested maximum altitude as a correlate of threatened status, but did not explain their reasoning. For most species in this sample, the two measures are the same, with *Presbytis melalophos* being an obvious exception.
9. *Latitudinal limits.* Low latitudes should have lower environmental seasonality, and therefore species with low latitudinal limits would have low tolerance for changes in habitat (Stevens, 1989). (J. Skorupa suggested to me the relevance of Stevens's argument, although Stevens used it in relation to the correlation of latitudinal range with absolute latitude.)
10. *Geographic range.* Finally, species with small geographic ranges might be more vulnerable to disturbance if size of geographic range correlates with specificity of habitat requirements (see Rabinowitz et al., 1986): weeds can cope with disturbance, endemics cannot.

In any comparative analysis, the possible confounding influence of "phylogenetic inertia" needs to be accounted for, especially in statistical comparisons where inflation of the

Table 3-2 Characteristics of species for which there are comparative data on densities in original and disturbed forest at the same site.[a]

Species	Response	Response code	Site density /km²	Geog. density /km²	Female body weight (kg)	Frugivory (%)	Arboreality	Home range (km²)	Territoriality	Altitude range (m)	Latitude (°)	Geographic range (km² × 10⁶)
Africa												
Cercopithecus diana	0.4	0	2.25	54	3.9	41	1	1.09	1	110	9	0.32
Cercopithecus mitis	2.2	1	3.75	80	3.9	60	1	0.23	1	3300	34	4.17
Cercopithecus ascanius	2.7	1	6	94	3.0	47	1	0.27	1	2000	12	2.82
Procolobus badius	0.2	0	1.9	220	8.2	25	1	0.65	0	2640	14	2.24
Colobus guereza	3.6	1	1	58	8.4	16.7	1	0.28	1	4500	13	2.19
Cercocebus torquatus	1.4	1	0.7	38.5	6.2	91	1	2.5	0	350	13	1.50
Pan troglodytes	0.4	0	1.3	1	31.0	60	1	30	1	1000	14	2.45
Gorilla gorilla	1.5	1	0.5	0.5	80.0	25	0	15	0	3500	7	0.81
Asia												
Macaca fascicularis	1.7	1	1	35	3.1	95	1	0.75	0	2000	19	2.63
Presbytis comata	0.2	0	2.15	20	6.7	14	1	0.375	0	600	8	0.05
Nasalis larvatus	0.4	0	—	4.25	9.86	35	1	5.15	0	300	7	0.72
Presbytis melalophos	0.8	1	2.6	65	5.8	53	1	0.225	1	500	13	0.66
Presbytis obscura	1.0	1	4.7	72	6.6	42	1	0.33	1	1830	15	0.24
Hylobates syndactylus	0.5	0	—	3.15	10.6	41	1	0.34	1	3800	6	0.20
Hylobates lar	1.0	1	2.9	4.3	5.3	58	1	0.44	1	2400	24	1.10
H. muelleri	0.7	1	2.6	6.7	5.6	62	1	0.36	1	1500	7	0.69
Pongo pygmaeus	0.1	0	—	2.5	37.0	72	1	3.5	0	2000	6	0.32

S. America												
Callithrix argentata	2.6	1	—	5.5	0.34	75	1	0.3	1	900	20	1.23
Saguinus fuscicollis	2	1	1.8	15	0.38	37	1	0.3	1	1300	19	2.00
Saguinus midas	7.9	1	—	23.5	0.45	47	1	0.09	1	50	8	1.45
Cebus albifrons	0.5	0	1	24	2.2	25	1	1.5	0	2000	16	1.41
Cebus apella	1.5	1	1.3	40	2.5	18	1	1.5	0	2700	30	3.89
Callicebus torquatus	0.4	0	1	26	1.3	67	1	0.17	1	450	8	1.91
Callicebus moloch	2.6	1	0	38.5	1.1	54	1	0.04	1	850	23	5.13
Saimiri sciureus	1.9	1	0.5	65	0.8	10	1	0.65	0	2000	19	5.75
Lagothrix lagotricha	0.2	0	0.2	10.5	5.65	80	1	4	0	3000	16	3.80
Alouatta seniculus	0.2	0	0.5	37	5.8	42	1	0.17	0	3200	20	1.58
Ateles paniscus	0	0	0.1	14.5	8.4	83	1	2	1	2500	24	6.49
Median vulnerable	0.4	0	1.0	20.0	6.7	41.0	1	1.1	0	2000	14	1.58
Median survivor	2.0	1	1.6	39.3	3.1	47.0	1	0.3	1	2000	19	2.10
Species pairs[b] (n = 8)			3	6	4	4	1	6	3	6	6	5
p value[c] (pairs)			ns	ns	ns	ns		0.014		0.025	0.035	ns
p value (CAIC)			ns	ns	ns	ns		0.006		ns	0.04	ns

[a]Data are medians from values in the literature. Response = the ratio of density in disturbed or secondary forest/density in primary forest at the same site; response code = classification as either "vulnerable" (0) or not (1); site density = density in original forest at the site at which the vulnerability index was calculated, used as a measure of relative population size (data are groups/km^2, except for the apes which are individuals/km^2); geographic density = median density from all available records across the species' geographic range (individual/km^2), a measure of relative competitiveness; frugivory = percentage of diet that is fruit, flowers, or seed; arboreality = predominantly arboreal (1) or terrestrial (0); territoriality = territorial (1) or not (0); latitude = species maximum latitude to nearest degree. References for data on response to logging are as in table 3-1. Otherwise, data are mainly from Wolfheim (1983), especially for altitudinal range and geographical limits, and from Smuts et al. (1987), Davies and Oates (1994), Rylands (1993), and Clutton-Brock and Harvey (1977b); for geographic density from Thomas (1991); for body weight from Ford and Davis (1992), Fuentes (1994), and Gevaerts (1992); for frugivory from Mitani (1991), Mitchell et al. (1991), Peres (1994), Williamson et al. (1990), and Yamagiwa et al. (1992); for arboreality from Fuentes (1994), Happel et al. (1987), and Rodman (1978); for home range from Caldecott (1986), Mitani (1991), and Oates and Whitesides (1990); and for territoriality from Hill (1994).

[b]Species pairs = number of comparisons between closely related species that differed in response that were in the predicted direction.

[c]p values: pairs = Wilcoxon matched-pairs signed-ranks probability values for comparisons of 8 intrageneric pairs; CAIC = Wilcoxon matched-pairs signed-ranks probability value for the 7–10 CAIC comparisons.

sample size is so easy with closely related species (Harvey and Pagel, 1991). A long-standing way of obviating confounding effects of taxonomic relatedness is to compare similar species that differ only in the variable of interest (Lack, 1968). To avoid counting several similar species as independent data points in any statistical analysis, we need single comparisons among similar taxa. Purvis and Rambaut's (1995) comparative analysis by independent contrasts (CAIC), based on Felsenstein's (1985) original methodology, is designed to do precisely this. With an estimated phylogeny (i.e., evolutionary tree of the taxon), it asks whether the members of phylogenetically independent pairs of taxa differ in the direction expected: if two variables are correlated (e.g., body size with risk), most differences (contrasts) should be in the same direction. Correlation between the independent and dependent variables can thus be tested with a Wilcoxon matched-pairs signed-ranks test (Siegel, 1956). All such statistical tests in this chapter were performed with Statview (Abacus Concepts, 1990–91). For the comparisons and variables used here, an extremely useful characteristic of the CAIC is that continent is also removed as a confounding variable because the members of each pair of compared taxa are almost necessarily on the same continent. In this chapter, the significance level is set at .05, and the tests are one tailed because predictions of the direction of difference were made beforehand on the basis of other studies or general biology. Data were log transformed so that the comparison of "vulnerable" and "nonvulnerable" taxa would fall equally on either side of the zero line. The primate phylogeny used is that of Purvis (1995).

Results

With the available data set of 27 species, CAIC produced up to 10 independent contrasts (depending on the parameter under consideration; fig. 3-1). Obviously, many of these parameters are potentially interdependent: in other comparisons, body size correlates with diet, for instance, which correlates with home range (Clutton-Brock and Harvey, 1977a). I tested for interdependence of the contrasts (not the measures) by testing for correlations among the contrasts (table 3-3; Spearman rank correlation test). Probability levels are one-tailed.

In the following sections, I give two sample sizes when referring to the results of CAIC: the first is the number of comparisons in the expected direction, the second is the number in the unexpected direction.

Site Density

No relation of vulnerability with site density appears to exist (table 3-2, fig. 3-1). In other words, density in the primary forest does not predict relative density in the degraded or secondary forest. First, within sites, correlation coefficients of the density of primate species in the original forest with their vulnerability to disturbance were not only low, but also were not in any consistent direction (only forests with at least five species of diurnal primates were counted). Thus, across sites within continents, Fisher combined probability tests produced no significant results. Second, the seven independent contrasts from the data set of 23 species with the requisite data showed no consistent direction of difference (Wilcoxon $T = 13$, $n = 34, 3$, $p = .5$). The most obvious example of site density not predicting response to logging is the comparison of the African colonies, *Procolobus badius* and *Colobus guereza*. *Procolobus badius* was the species with the highest density of individuals in nat-

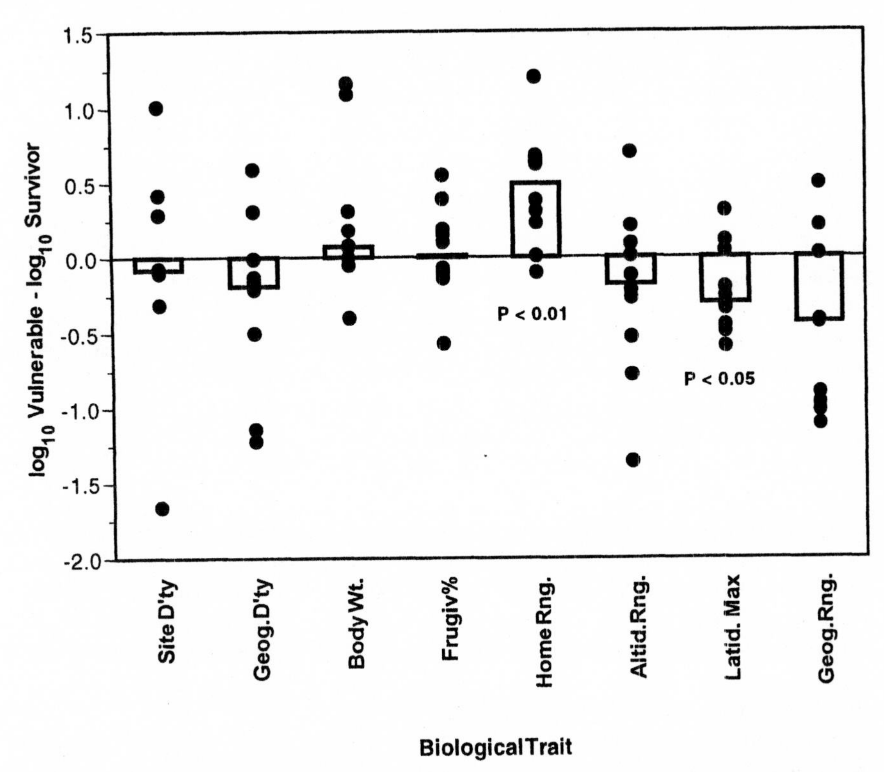

Figure 3-1 Direction of differences in various measures (logged values) between independent pairs of vulnerable and survivor species (i.e., those that did poorly or well in disturbed forest). Filled circles = taxonomically independent contrasts; histogram = median contrast. Independent pairs were detected by CAIC (Purvis and Rambaut, 1995), using the phylogeny of Purvis (Purvis, 1995). (For presentation purposes, values were log transformed before subtraction to ensure equal distance either side of zero in the figure, and columns with original values of less than 1 were multiplied by 10 to ensure positive log values). Measures are in order of presentation in the text, and are explained in the text and in table 3-2. Significant correlations were found for home range (Wilcoxon signed ranks, $T = 1$, $n = 8, 1$, $p = .006$) and for maximum latitude ($T = 10$, $n = 7, 3$, $p = .04$). The 10 pairs of taxa, with the more vulnerable listed first, are: *Cercopithecus diana* and the mean of *C. ascanius* and *C. mitis; Procolobus badius* and *Colobus guereza; Pan* and *Gorilla; Presbytis comata* and *P. melalophos; Nasalis larvatus* and *P. obscura; Hylobates syndactylus* and *H. lar; Callicebus torquatus* and *C. moloch; Cebus albifrons* and *C. apella;* mean Ateline and mean remaining New World taxa; and *Pongo* and mean remaining Cercopithecines. For site density, the *Nasalis,* Hylobatid, and *Pongo* comparisons are missing.

ural forest at three of four sites, and yet at all sites it did extremely poorly in altered forest; by contrast, in two sites where *Colobus guereza* was at the lowest, or equal lowest density in the original forest, it survived the best in the altered forest. The contrasts of site density did not correlate significantly ($p < .05$) with any other measure's contrasts, although, as might be expected, the correlation with geographic density was high (table 3-3).

Table 3-3 Spearman rank correlation coefficients among contrasts.

	Geographic density	Body weight	% Frugivory	Home range	Altitudinal range	Latitudinal max.	Geographic range
Site density	0.71	−0.18	0.39	0.00	−0.14	0.32	0.64
Geographic density		−0.35	0.50	−0.09	−0.18	0.32	0.35
Body weight			−0.68*	−0.18	0.81*	−0.01	−0.20
% Frugivory				0.24	−0.49	0.15	0.66*
Home range					−0.54	0.64	0.72*
Altitudinal range						−0.2	−0.38
Latitudinal max.							0.68*

*$p < .05$, two tailed. (Note: a positive correlation can be obtained between two measures that correlate in opposite directions with vulnerability, because contrasts are being compared, and sign is retained.)

Geographic Density

Comparisons of geographic densities indicate a possible effect, if the exceptional *P. badius* and *C guereza* contrast is removed. Thus, although the 10 comparisons do not indicate an effect (T = 13, n = 7, 3, p = .07), removal of this extreme exception (along with, for balance, the opposite extreme (*Presbytis comata* vs. *P. melalophos*) produces a significant result (T = 2, n = 6, 2, p = .025). The lack of significant correlation of contrasts in geographic density with contrasts in any other measure (table 3-3) indicates an independent effect of geographic density, if there is an effect.

Body Weight

The analysis here indicates no significant effect of body size (T = 16, n = 7, 3, p = .12; table 3-2, fig. 3-1). However, removal of the two opposite extremes, as above, produces a significant result (T = 3, n = 6, 2, p = .02), with larger species being more vulnerable. Contrasts in body size correlated significantly (and negatively) with contrasts in frugivory (as expected; Clutton-Brock and Harvey, 1977a), and (positively) with contrasts in altitudinal range (table 3-3). The correlation with altitude remains when the extreme pair of *Pan/Gorilla* is removed. Neither frugivory or altitudinal range correlated independently with vulnerability, and their association with contrasts in body size is in the opposite direction to that expected were they influencing the correlation of body size with vulnerability. Thus in all three exceptions to the association between body size and vulnerability (*Procolobus* and *Colobus*, *Pan* and *Gorilla*, and *Macaca* and *Pongo*), the smaller but more vulnerable species was more frugivorous and had a smaller altitudinal range than did the species it was compared with. Body size might therefore be an independent correlate with vulnerability in this sample. *Alouatta*, a relatively large South American genus, is somewhat of an anomaly because, although it seems to do poorly in secondary forest (but see Pinto et al., 1993), it apparently survives well in extremely small forest fragments (Bernstein et al., 1976; Lovejoy et al., 1986; Ferrari and Diego, 1995).

Frugivory and Diversity of Diet

The proportion of fruit in the diet appeared to have no consistent relation to vulnerability (T = 24, n = 5, p = .4; table 3-2, fig. 3-1). Thus, while the frugivorous *Pan troglodytes* was more vulnerable than its relative *Gorilla,* the highly folivorous *Presbytis comata* was more vulnerable than its more frugivorous congener, *P. melalophos.* Nor did a low dietary diversity correlate with risk: although *C. diana* is probably more of a specialist than *C. mitis,* for none of the other pairs was the more vulnerable species more obviously specialist than the other. Some frugivores could, however, be seen as being more specialist than the folivores, if the difference between them is that in nonfruiting seasons, the frugivores can less easily switch to foliage than can the folivores; the African apes could fall into this category (Yamagiwa et al., 1992, 1994), as well as the two *Cebus* (C. Janson, personal communication). Contrasts in frugivory correlated significantly with contrasts in body weight and geographic range (table 3-3), which, if they independently affected vulnerability, would have canceled each other's effects on the relationship of frugivory with vulnerability.

Arboreality

Almost all the species in the analysis were classified as arboreal, and thus there was little chance to test this character (table 3-2). The two comparisons that allowed a test produced opposite results. Not only is this measure of dispersing ability extremely crude, but dispersing ability is probably not important in determining survival in secondary forest that is contiguous with primary forest. A better measure would probably have been a separation between obligate canopy dwellers and others. Also, primates can markedly change their habits in response to a change in their habitat (Johns, 1986). For instance, the otherwise highly arboreal *Procolobus badius* (Oates, 1994b) is unusually terrestrial in the tiny, scattered, disturbed forests of Zanzibar.

Home-Range Size

Home-range size was the strongest correlate of vulnerability in this analysis (table 3-2, fig. 3-1): in 8 of the 10 independent comparisons, the taxon with the larger home range was the more vulnerable [$T = 1, n = 8, 1$ (one contrast $= 0$); $p = .006$]. The exception was the contrast of the *Hylobates.* The significant effect remains even if the most extreme contrast in the expected direction is removed ($T = 1, n = 7, 1, p = .009$). Contrasts in home-range size correlated significantly with contrasts in geographic range (table 3-3). In the latter, the *Pan/Gorilla* contrast was an extreme exception. If it is removed, the correlation remains strong, but insignificant ($p = .08$). Because the correlation was not in the direction expected from the measures' independent associations with vulnerability, it seems as if home range might correlate independently with vulnerability.

Territoriality

As for arboreality, territoriality was classified categorically. If we take the contrasts identified as independent by CAIC and count the number of pairs that showed the predicted direction of difference or not (nonterritorial are vulnerable), no effect of territoriality is detected: five showed no difference, four were in the predicted direction; one (the African apes) was opposite (table 3-2).

Altitudinal and Latitudinal Limits

Altitudinal and latitudinal limits are lumped under one heading because both have the same explanation for their association with vulnerability. Taxa with a narrow range of altitudinal limits were not more at risk ($T = 15, n = 6, 4, p = .1$), nor were taxa with a low maximum altitudinal limit. Taxa with low latitudinal limits were significantly more vulnerable ($T = 11, n = 7, 3, p = .05$). The *Cebus* and *Callicebus* comparisons most clearly showed this result. Contrasts in both measures correlated with contrasts in other measures, but not with each other (table 3-3). The correlation of body weight with altitudinal range was in the opposite direction to that expected were contrasts on body weight influencing the association of altitudinal range with vulnerability. However, the correlation of contrasts in geographic range with contrasts in maximum latitude is in the direction expected (not surprisingly). By the argument here, it is more likely that contrasts in maximum latitude affect correlations of geographic range with vulnerability than vice versa, especially as geographic range did not correlate with vulnerability.

Geographic Range

Taxa with small geographic ranges were not, as predicted, more vulnerable than those with large geographic ranges ($T = 15, n = 6, 4, p = .1$; table 3-2, fig. 3-1). The extreme exception was the African ape pair: *Pan* has a far larger range than *Gorilla,* but is more vulnerable. If this pair is removed, along with the opposite extreme pair for balance, the association between vulnerability and geographic range remains nonsignificant ($T = 7, n = 5, 3, p = .07$). The contrasts in geographic range correlate significantly with contrasts in percent frugivory, home range, and maximum latitude (table 3-3). Only maximum latitude correlated in the direction expected if it influenced the correlation of geographic range with vulnerability. As I argued above, associations between vulnerability and geographic range size, if any, might work through the correlation of vulnerability with maximum latitude.

Summary

The strongest correlate of vulnerability to logging, in other words, of inability to survive in altered forest at as high a density as in primary forest, is a large home-range size. Species with low latitudinal limits are also significantly more vulnerable, but the two measures together do not provide better prediction than the home-range size comparison alone. Species with low average density over their geographic range and species with large body size might also be at risk. The other potential correlates of risk, density in primary forest at the site at which densities in secondary and primary forest was compared, diet, arboreality, territoriality, altitudinal limits, and geographic range size, were not significantly correlated with vulnerability. Correlations among contrasts in the measures indicate that home range size and maximum latitude might have been independent correlates of vulnerability and that maximum latitude might have influenced any correlation of vulnerability with geographic range.

The statistical method used for comparing taxa while accounting for phylogeny is not above criticism. The alternative is simply to compare closely related species that differ in their response to logging. Although the sample size is not as great, the comparisons might be more biologically valid, having less confounding influences. With the current data set, such a comparison produces answers similar to those of the CAIC, with geographic density, home range, altitudinal range, and maximum latitude showing the most obvious correlates with vulnerability (table 3-2).

Discussion and Caveats

This study searched for correlates, especially nondemographic correlates ("ecological correlates") of vulnerability of primate species to alteration of habitat, as measured by the ratio of population density in secondary forest compared to primary forest. The search has suggested that indicators of risk appear to exist. How do we explain them?

Given the limitations of the data, in both quantity and quality, an obvious explanation is that the results are mere artifact. Alternatively, it could be argued that any consistent result arising from such messy data indicates a strong effect. At the very least, the "messiness" should not introduce bias (Harvey and Pagel, 1991). Taking the results at face value (i.e., ignoring for the moment all the limitations of the data), the correlation of risk with narrow latitudinal limits was as predicted, which provides some hope for the correctness of its explanation, namely, that species with narrow latitudinal limits cannot adapt well to a changed habitat. By contrast, the stronger correlation with home-range size is puzzling, even though it was predicted on the assumption that home-range size is some indication of area needed for survival. Happel et al.'s (1987) finding that size of home range was the strongest (negative) correlate of a species' average population density implies that home-range size might be an index of population density. If so, why does home range size correlate with risk better than does geographic population density? The main puzzle, though, is that home range size is such a labile character, as are several of the strong correlates of home-range size, such as the nature of the habitat, density of the population, and group size (Clutton-Brock, 1977; Clutton-Brock and Harvey, 1977a; Isbell et al., 1990). Why cannot species that had large groups and hence large home ranges simply reduce the size of the group, therefore reducing the size of the home range, and so survive in small fragments or in disturbed forest? Of course, there probably are lower limits to group size, set by intergroup competition and antipredator tactics (Bertram, 1978; Wrangham, 1980; van Schaik, 1983; Dunbar, 1988; Mangel, 1990; Isbell, 1991), but they do not explain why a highly labile character should correlate best with vulnerability. It seems likely, therefore, that home-range size is an index of risk, not a cause. As an index, it would match Skorupa's (1986) identification of reliance on widely dispersed, rare food items as a risk factor.

An obvious further factor to consider is hunting. Hunting intensity increases enormously while logging is occurring and can continue for years afterwards if the logging attracts permanent settlement around the area (Oates, 1996). In such cases, hunting can have more of an adverse influence than logging (Tutin and Fernandez, 1984; Oates, 1996). However, most of the study sites that produce data on densities in secondary and primary forest are in conservation areas of some sort, often national parks, and thus hunting might be less of a confounding factor in the results than might otherwise be the case. The warnings about the quantity and quality of the data used in this chapter need repeating. For instance, although *Cercopithecus mitis* averaged only a quarter of the more vulnerable *C. diana*'s home-range size (table 3-2), the 3.5 km^2 home ranges of *C. mitis* groups at one site were three times the average for *C. diana*. We clearly need more than just one record for a species.

An important influence in future comparisons could be the changing taxonomy of primates. For instance, the highly vulnerable *Procolobus badius* might be four or more species (Oates et al., 1994), and Hylobatid taxonomy is still in flux (Groves, 1993). Primate taxonomy is going through a splitting phase at present, which seems to be driven in part by conservation politics. The effects remain to be seen, but splitting will influence latitudinal maxima and geographic ranges.

With respect to the parameters that might be most usefully correlated with risk, it might eventually be more meaningful to compare a measure such as variability in home-range size rather than a single value for home-range size. Finally, given that some of the contrasts correlated with one another in this analysis, future work will have to progress to multifactorial analyses that take account of potential or actual interdependence, once sufficient data are available.

Recommendations

The results suggest two recommendations. First, we should pay particular attention to species with narrow latitudinal ranges and whose social groups have large home ranges, perhaps especially if the large home ranges result from an obligatory diet of widely dispersed, rare food items. Second, we must collect more data. We still have far too few studies of the response of primates to logging.

It was hoped that the search for ecological indicators of risk would produce an indicator that was easier and more accurate to measure than population size, or any of the other demographic parameters we might need to assess risk from demographic indicators alone. Although maximum latitude is easily measured, home range size is probably just as difficult to assess accurately for forest-living taxa as is population size and so also obligate reliance on widely dispersed, rare food items. Both are, however, easier to measure than most of the other parameters needed in a demographic viability analysis model.

Caughley (1994; Caughley and Gunn, 1996) suggested that a main problem with application of ecological models (including those from behavioral ecology) to conservation is that ecological theory cannot match the substantiated generalizability of demographic theory. Conservation therefore still works on a case-by-case basis, although some important advances are being made (Sibly and Smith, 1985; Sutherland, 1996). A main advantage of a search for ecological indicators of threat should be a greater concentration on understanding the process of population change, as opposed to simply predicting the outcome (Clutton-Brock and Albon, 1985; Hassell and May, 1985; Lomnicki, 1988; Sinclair, 1989, 1992; Caughley, 1994; Sutherland, 1996). If home-range size is substantiated as an indicator of risk, and if others are as puzzled as I am about why it predicts risk, then clearly we have a long way to go in understanding the process.

Summary

Small population size is a strong predictor of imminent extinction, and therefore conservation biologists have rightly concentrated on demographic modeling as a powerful means of assessing risk. However, application of the models can be problematical, especially because they often require an amount and quality of data that will be unobtainable for most species. In addition, certain types of species seem more prone to extinction than others, independently of population size.

In this chapter, therefore, I searched for ecological (as well as demographic) measures that might correlate with risk of extinction. I did this by comparing the characteristics of primate species that do poorly in previously logged forest with those that do well (n = 28 species). The exact measure of risk was density in secondary forest over density in contiguous primary forest. Where possible, phylogeny was controlled for statistically. Density in the primary forest (a measure of population size) did not predict survival in the secondary forest at the same site in Africa, Asia, or South America.

Only two of nine ecological characteristics correlated significantly with risk in comparisons of ten phylogenetically independent pairs of taxa, one member of which did well in disturbed forest and the other did poorly. Taxa with naturally large home ranges and low maximum latitude were significantly at risk. If the outlier at either end of the distribution was removed, low average density of a species across its geographic range and large body size also predict risk. The measures that did not correlate with risk were diet type or breadth, arboreality, territoriality, altitudinal range, and size of geographic range. Correlation coefficients of the contrasts with one another indicated that home-range size and maximum latitude might have been independent correlates, and that maximum latitude might have influenced any correlation of geographic range with risk.

Although maximum latitude, a potential indicator of range of habitats usable by the species, is a logical correlate of vulnerability, and an easily measured one, it was not as powerful a correlate as home-range size, and it added no more predictive power than did home-range size. However, it is not clear why home-range size, an extremely labile character, should be a predictor. Nevertheless, it is easier and quicker to measure than some demographic variables, if not total population size.

At present, therefore, we still lack a general ecological model that is as powerful as demographic models in assessing risk, at least as measured here. Also, we are still sorely lacking data for these types of comparisons.

Acknowledgments I thank Jessica Gorin, Lynne Isbell, Charles Janson, John Oates, Andy Purvis, Joe Skorupa, Kelly Stewart, Truman Young, and two anonymous referees for extremely useful comments on previous versions and Gary Fanucchi and Suzanne Kokel for collating much of the information.

References

Abacus Concepts, 1990–91. Statview SE+. Berkeley, California: Abacus Concepts, Inc.

Akçakaya HR, Burgman M, 1995. PVA in theory and practice. Conserv Biol 9:705–707.

Angermeier PL, 1995. Ecological attributes of extinction-prone species: loss of freshwater fishes of Virginia. Conserv Biol 9:143–158.

Barnes RFW, 1990. Deforestation trends in tropical Africa. Afr J Ecol 28:161–173.

Bernstein IS, Balcaen P, Dresdale L, Gouzoules H, Kavanagh M, Patterson T, Neyman-Warner P, 1976. Differential effects of forest degradation on primate populations. Primates 17:401–411.

Bertram BCR, 1978. Living in groups: predators and prey. In: Behavioural ecology: an evolutionary approach, 1st ed (Krebs JR, Davies NB, eds). Oxford: Blackwell Scientific; 64–96.

Bibby CJ, 1995. Recent past and future extinctions in birds. In: Extinction rates (Lawton JH, May RM, eds). Oxford: Oxford University Press; 98–110.

Brown JH, 1971. Mammals on mountaintops: nonequilibrium insular biogeography. Am Nat 105:467–478.

Brown JH, Kodric-Brown A, 1977. Turnover rates in insular biogeography: effect of immigration on extinction. Ecology 58:445–449.

Brown LH, 1981. The conservation of forest islands in areas of high human density. Afr J Ecol 19:27–32.

Burbidge AA, McKenzie NL, 1989. Patterns of modern decline of western Australia's vertebrate fauna: causes and conservation implications. Biol Conserv 50:143–198.
Burgman MA, Ferson S, Akçakaya HR, 1993. Risk assessment in conservation biology. New York: Chapman and Hall.
Butynski TM, Koster SH, 1994. Distribution and conservation status of primates in Bioko Island, Equatorial Guinea. Biodivers Conserv 3:893–909.
Caldecott JO, 1986. An ecological and behavioral study of the pig-tailed macaque. Basel: Karger.
Caughley G, 1977. Analysis of vertebrate populations. London: John Wiley and Sons.
Caughley G, 1994. Directions in conservation biology. J Anim Ecol 63:215–244.
Caughley G, Gunn A, 1996. Conservation biology in theory and practice. Cambridge, Massachusetts: Blackwell Scientific.
Clutton-Brock TH (ed), 1977. Primate ecology. London: Academic Press.
Clutton-Brock TH, Albon SD, 1985. Competition and population regulation in social mammals. In: Behavioural ecology. Ecological consequences of adaptive behaviour (Sibly RM, Smith RH, eds). Oxford: Blackwell Scientific; 577–592.
Clutton-Brock TH, Harvey PH, 1977a. Primate ecology and social organization. J Zool 183:1–39.
Clutton-Brock TH, Harvey PH, 1977b. Species differences in feeding and ranging behaviour in primates. In: Primate ecology (Clutton-Brock TH, ed). London: Academic Press; 557–584.
Crouse DH, Crowder LB, Caswell H, 1987. A stage-based population model for loggerhead sea turtles and implications for conservation. Ecology 68:1412–1423.
Davies AG, Oates JF (eds), 1994. Colobine monkeys. Their ecology, behaviour and evolution. Cambridge: Cambridge University Press.
Davies NB, Houston AI, 1984. Territory economics. In: Behavioural ecology. An evolutionary approach, 2nd ed. (Krebs JR, Davies NB, eds). Oxford: Blackwell Scientific; 148–169.
Diamond JM, 1984. "Normal" extinctions of isolated populations. In: Extinctions (Nitecki MH, ed). Chicago: University of Chicago Press; 191–246.
Dobson AP, Lyles AM, 1989. The population dynamics and conservation of primate populations. Conserv Biol 3:362–380.
Dunbar RIM, 1988. Primate social systems. London: Croom Helm.
Felsenstein J, 1985. Phylogenies and the comparative method. Am Nat 125:1–15.
Ferrari SF, Diego VH, 1995. Habitat fragmentation and primate conservation in the Atlantic Forest of eastern Minas Gerais, Brazil. Oryx 29:192–196.
Fimbel C, 1994. Ecological correlates of species success in modified habitats may be disturbance- and site-specific: the primates of Tiwai Island. Conserv Biol 8:106–113.
Ford SM, Davis LC, 1992. Systematics and body size: implications for feeding adaptations in New World monkeys. Am J Phys Anthropol 88:415–468.
Frankham R, 1995. Inbreeding and extinction: a threshold effect. Conserv Biol 9:792–799.
Fuentes A, 1994. The socioecology of the Mentawi Island langur (PhD dissertation). Berkeley: University of California.
Ganzhorn JU, 1995. Low-level forest disturbance effects on primary production, leaf chemistry, and lemur populations. Ecology 76:2084–2096.
Gaston KJ, 1994. Rarity. London: Chapman and Hall.
Gevaerts H, 1992. Birth seasons of *Cercopithecus, Cercocebus* and *Colobus* in Zaire. Folia Primatol 59:105–113.
Goodman D, 1987. The demography of chance extinction. In: Viable populations for conservation (Soulé ME, ed). Cambridge: Cambridge University Press; 11–34.
Groves CP, 1993. Speciation in living hominoid primates. In: Species, species concepts, and

primate evolution (Kimbel WH, Martin LB, eds). New York: Plenum Press; 109–121.

Happel RE, Noss JF, Marsh CW, 1987. Distribution, abundance, and endangerment of primates. In: Primate conservation in the tropical rain forest (Marsh CW, Mittermeier RA, eds). New York: Alan R. Liss; 83–107.

Harcourt AH, 1995a. Population viability estimates: theory and practice for a wild gorilla population. Conserv Biol 9:134–142.

Harcourt AH, 1995b. PVA in theory and practice. Conserv Biol 9:707–708.

Harcourt AH, 1996. Is the gorilla a threatened species? How should we judge? Biol Conserv 75:165–176.

Harris LD, 1984. The fragmented forest. Chicago: University of Chicago Press.

Harrison P, 1987. The greening of Africa. London: Paladin.

Harvey PH, Pagel MD, 1991. The comparative method in evolutionary biology. Oxford: Oxford University Press.

Hashimoto C, 1995. Population census of the chimpanzees in the Kalinzu forest, Uganda: comparison between methods with nest counts. Primates 36:477–488.

Hassell MP, May RM, 1985. From individual behaviour to population dynamics. In: Behavioural ecology. Ecological consequences of adaptive behaviour (Sibly RM, Smith RH, eds). Oxford: Blackwell Scientific; 3–32.

Hill CM, 1994. The role of female diana monkeys, *Cercopithecus diana,* in territorial defense. Anim Behav 47:425–431.

Isbell LA, 1991. Contest and scramble competition: patterns of female aggression and ranging behavior among primates. Behav Ecol 2:143–155.

Isbell LA, Cheney DL, Seyfarth RM, 1990. Costs and benefits of home range shifts among vervet monkeys (*Cercopithecus aethiops*) in Amboseli National Park, Kenya. Behav Ecol Sociobiol 27:351–358.

IUCN, 1994. IUCN Red List Categories. Gland: International Union for the Conservation of Nature.

Johns AD, 1986. Effects of selective logging on the behavioral ecology of West Malaysian primates. Ecology 67:684–694.

Johns AD, 1988. Effects of "selective" timber extraction on rain forest structure and composition and some consequences for frugivores and folivores. Biotropica 20:31–37.

Johns AD, 1992. Species conservation in managed tropical forests. In: Tropical deforestation and species extinction (Whitmore TC, Sayer JA, eds). London: Chapman and Hall; 15–53.

Johns AD, Skorupa JP, 1987. Responses of rain-forest primates to habitat disturbance: a review. Int J Primatol 8:157–191.

Johns AG, Johns BG, 1995. Tropical forest primates and logging: long-term coexistence? Oryx 29:205–211.

Jolly A, 1986. Lemur survival. In: Primates. The road to self-sustaining populations (Benirschke K, ed). New York: Springer-Verlag; 71–98.

Karr JR, 1977. Ecological correlates of rarity in a tropical forest bird community. Auk 94:240–247.

Kattan GH, Alvarez-López H, Giraldo M, 1994. Forest fragmentation and bird extinctions: San Antonio eighty years later. Conserv Biol 8:138–146.

Lack D, 1968. Ecological adaptations for breeding in birds. London: Methuen.

Lande R, 1993. Risks of population extinction from demographic and environmental stochasticity and random catastrophes. Am Nat 141:911–927.

Laurance WF, 1990. Comparative responses of five arboreal marsupials to tropical forest fragmentation. J Mammal 71:641–653.

Laurance WF, 1991. Ecological correlates of extinction proneness in Australian tropical rain forest mammals. Conserv Biol 5:79–89.

Laurance WF, 1994. Rainforest fragmentation and the structure of small mammal communities in tropical Queensland. Biol Conserv 69:23–32.

Leach MK, Givnish TJ, 1996. Ecological determinants of species loss in remnant prairies. Science 273:1555–1558.

Leader-Williams N, Albon SD, 1988. Allocation of resources for conservation. Nature 336:533–535.

Lomnicki A, 1988. Population ecology of individuals. Princeton, New Jersey: Princeton University Press.

Lovejoy TE, Bierregaard JR, Rylands AB, Malcolm JR, Quintela CE, Harper LH, Brown JR, Powell AH, Powell GVN, Schubart HOR et al., 1986. Edge and other effects of isolation on Amazon forest fragments. In: Conservation biology. The science of scarcity and diversity (Soulé ME, ed). Sunderland, Massachusetts: Sinauer Associates; 257–285.

Lovejoy TE, Rankin JM, Bierregaard RO, Brown KS, Emmons LH, Voort Van der ME, 1984. Ecosystem decay of Amazon forest fragments. In: Extinctions (Nitecki MH, ed). Chicago: University of Chicago Press; 295–325.

MacArthur RH, 1972. Geographical ecology. Patterns in the distribution of species. Princeton, New Jersey: Princeton University Press.

MacArthur RH, Wilson EO, 1967. The theory of island biogeography. Princeton, New Jersey: Princeton University Press.

Mace G, Stuart S, 1994. Draft IUCN Red List categories, version 2.2. Species 21-22:13–24.

Mangel M, 1990. Resource divisibility, predation and group formation. Anim Behav 39:1163–1172.

McNab BW, 1963. Bioenergetics and the determination of home range size. Am Nat 97:133–140.

Milton K, May ML, 1976. Body weight, diet and home range area in primates. Nature 259:459–462.

Mitani M, 1991. Niche overlap and polyspecific associations among sympatric cercopithecids in the Campo Animal Reserve, southwestern Cameroon. Primates 32:137–151.

Mitchell CL, Boinski S, van Schaik CP, 1991. Competitive regimes and female bonding in two species of squirrel monkeys (*Saimiri oerstedi* and *Saimiri sciureus*). Behav Ecol Sociobiol 28:55–60.

Mittermeier RA, Cheney DL, 1987. Conservation of primates and their habitats. In: Primate societies (Smuts BB, Cheney DL, Seyfarth RM, Wrangham RW, Struhsaker TT, eds). Chicago: University of Chicago Press; 477–490.

Myers N, 1984. The primary source. Tropical forests and our future. New York: W.W. Norton.

Newmark WD, 1995. Extinction of mammal populations in western American national parks. Conserv Biol 9:512–526.

Oates JF, 1994a. Africa's primates in 1992: conservation issues and options. Am J Primatol 34:61–71.

Oates JF, 1994b. The natural history of African colobines. In: Colobine monkeys. Their ecology, behaviour and evolution (Davies AG, Oates JF, eds). Cambridge: Cambridge University Press; 75–128.

Oates JF, 1996. Habitat alteration, hunting and the conservation of folivorous primates in African forests. Aust J Ecol 21:1–9.

Oates JF, Davies AG, Delson E, 1994. The diversity of living colobines. In: Colobine monkeys. Their ecology, behaviour and evolution (Davies AG, Oates JF, eds). Cambridge: Cambridge University Press; 45–73.

Oates JF, Whitesides GH, 1990. Association between olive colobus (*Procolobus verus*), diana guenons (*Cercopithecus diana*), and other forest monkeys in Sierra Leone. Am J Primatol 21:129–146.

Oates JF, Whitesides GH, Davies AG, Waterman PG, Green SM, Dasilva GL, Mole S, 1990. Determinants of variation in tropical forest primate biomass: new evidence from West Africa. Ecology 71:328–343.

Olupot W, Chapman CA, Brown CH, Waser PM, 1994. Mangabey (*Cercocebus albigena*) population density, group size, and ranging: a twenty-year comparison. Am J Primatol 32:197–205.

Pearl M, 1992. Conservation of Asian primates: aspects of genetics and behavioral ecology that predict vulnerability. In: Conservation biology. The theory and practice of nature preservation and management (Fiedler PL, Jain SK, eds). New York: Chapman and Hall; 297–320.

Peres CA, 1994. Diet and feeding ecology of gray woolly monkeys (*Lagothrix lagotricha cana*) in Central Amazonia: comparisons with other Atelines. Int J Primatol 15:333–372.

Peres CA, Terborgh JW, 1995. Amazonian nature reserves: an analysis of the defensibility status of existing conservation units and design criteria for the future. Conserv Biol 9:34–46.

Pinto LPS, Costa CMR, Strier KB, Fonseca GAB, 1993. Habitat, density and group size of primates in a Brazilian tropical forest. Folia Primatol 61:135–143.

Plumptre AJ, Reynolds V, 1994. The effect of selective logging on the primate populations in the Budongo Forest Reserve, Uganda. J Appl Ecol 31:631–641.

Purvis A, 1995. A composite estimate of primate phylogeny. Phil Trans R Soc Lond B 348:405–421.

Purvis A, Rambaut A, 1995. Comparative analysis by independent contrasts (CAIC): an Apple Macintosh application for analysing comparative data. Computer Appl Biosci 11:247–251.

Rabinowitz D, Cairns S, Dillon T, 1986. Seven forms of rarity and their frequency in the flora of the British Isles. In: Conservation biology. The science of scarcity and diversity (Soulé ME, ed). Sunderland, Massachusetts: Sinauer Associates; 182–204.

Robinson JG, Redford KH, 1989. Body size, diet, and population variation in Neotropical forest mammal species: predictors of local extinction? Adv Neotrop Mammal 1989:567–594.

Rodman PS, 1978. Diets, densities, and distributions of Bornean primates. In: The ecology of arboreal folivores (Montgomery GG, ed). Washington, DC: Smithsonian Institution Press; 465–478.

Rylands AB (ed), 1993. Marmosets and tamarins. Systematics, behaviour, and ecology. Oxford: Oxford University Press.

Schoener TW, 1968. Sizes of feeding territories among birds. Ecology 49:123–141.

Sibly RM, Smith RH (ed), 1985. Behavioural ecology. Ecological consequences of adaptive behaviour. Oxford: Blackwell Scientific.

Siegel S, 1956. Nonparametric statistics for the behavioral sciences. Tokyo: McGraw-Hill Kogakusha Ltd.

Simberloff DS, Wilson EO, 1969. Experimental zoogeography of islands. The colonization of empty islands. Ecology 50:278–296.

Sinclair ARE, 1989. Population regulation in animals. In: Ecological concepts (Cherrett JM, ed). Oxford: Blackwell Scientific; 197–241.

Sinclair ARE, 1992. Do large mammals disperse like small mammals? In: Animal dispersal. Small mammals as a model (Stenseth NC, Lidicker WZ, eds). London: Chapman and Hall; 229–242.

Skole D, Tucker C, 1993. Tropical deforestation and habitat fragmentation in the Amazon: satellite data from 1978 to 1988. Science 260:1905–1910.

Skorupa JP, 1986. Responses of rainforest primates to selective logging in Kibale Forest, Uganda: a summary report. In: Primates. The road to self-sustaining populations (Benirschke K, ed). New York: Springer-Verlag; 57–70.

Smuts BB, Cheney DL, Seyfarth RM, Wrangham RW, Struhsaker TT (eds), 1987. Primate societies. Chicago: University of Chicago Press.

Stevens GC, 1989. The latitudinal gradient in geographical range: how so many species coexist in the tropics. Am Nat 133:240–256.

Struhsaker TT, 1997. Ecology of an African rain forest. Gainesville: University Press of Florida.

Sutherland WJ, 1996. From individual behaviour to population ecology. Oxford: Oxford University Press.

Taylor BL, 1995. The reliability of using population viability analysis for risk classification of species. Conserv Biol 9:551–558.

Thomas SC, 1991. Population densities and patterns of habitat use among anthropoid primates of the Ituri forest, Zaïre. Biotropica 23:68–83.

Tilman D, May RM, Lehman CL, Nowak MA, 1994. Habitat destruction and the extinction debt. Nature 371:65–66.

Tutin CEG, Fernandez M, 1984. Nationwide census of gorilla (*Gorilla g. gorilla*) and chimpanzee (*Pan t. troglodytes*) populations in Gabon. Am J Primatol 6:313–336.

van Schaik CP, 1983. Why are diurnal primates living in groups? Behaviour 87:120–144.

Wahlberg N, Moilanen A, Hanski I, 1996. Predicting the occurrence of endangered species in fragmented landscapes. Science 273:1536–1538.

Watanabe K, 1981. Variation in group composition and population density of the two sympatric Mentawaian leaf monkeys. Primates 22:145–160.

Watson A, Moss R, 1970. Dominance, spacing behaviour and aggression in relation to population limitation in vertebrates. In: Animal populations in relation to their food resources (Watson A, ed). Oxford: Blackwell Scientific; 167–220.

Weber AW, 1987. Socioecologic factors in the conservation of Afromontane forest reserves. In: Primate conservation in the tropical rain forest (Marsh CW, Mittermeier RA, eds). New York: Alan R. Liss; 205–229.

Weisenseel K, Chapman CA, Chapman LJ, 1993. Nocturnal primates of Kibale Forest: effects of selective logging on prosimian densities. Primates 34:445–450.

Wennergren U, Ruckelshaus M, Kareiva P, 1995. The promise and limitations of spatial models in conservation biology. Oikos 74:349–356.

Western D, 1982. Patterns of depletion on a Kenyan rhino population and the conservation implications. Phil Trans R Soc Lond B 24:147–156.

White FJ, Overdorff DJ, Belko EA, Wright PC, 1995. Distribution of ruffed lemurs (*Varecia variegate*) in Ronomafana National Park, Madagascar. Folia Primatol 64:124–131.

White L, Tutin CEG, in press. Why chimpanzees and gorillas respond differently to logging: a cautionary tale from Gabon. In: African rain forest ecology and conservation (Weber B, Vedder A, Simons Morland H, eds). New Haven, Connecticut: Yale University Press.

Williamson EA, Tutin CEG, Rogers ME, Fernandez M, 1990. Composition of the diet of lowland gorillas at Lopé in Gabon. Am J Primatol 21:265–277.

Wolfheim JH, 1983. Primates of the world: distribution, abundance and conservation. Seattle: University of Washington Press.

World Resources Institute, 1996. World resources, 1996–1997. Oxford: Oxford University Press.

Wrangham RW, 1980. An ecological model of female-bonded primate groups. Behaviour 75:262–300.

Yamagiwa J, Mwanza N, Yumoto T, Maruhashi T, 1992. Travel distances and food habits of eastern lowland gorillas: a comparative analysis. In: Topics in primatology, vol 2. Behavior, ecology, and conservation (Itoigawa N, Sugiyama Y, Sackett GP, Thompson RKR, eds). Tokyo: University of Tokyo Press; 267–281.

Yamagiwa J, Mwanza N, Yumoto T, Maruhashi T, 1994. Seasonal change in the composition of the diet of eastern lowland gorillas. Primates 35:1–14.

4

Future Prey: Some Consequences of the Loss and Restoration of Large Carnivores

Joel Berger

The Problem of Predation-Free Environments

Both Charles Darwin (1859) and Alfred Russell Wallace (1876) suspected that predation shaped individual behavior and ecological communities, ideas that have subsequently been examined in vertebrates and invertebrates (Edmunds, 1974; Reznick et al., 1990; Sih et al., 1992; Estes and Duggin, 1995). While in the past organisms of all sizes must regularly have had to deal with predation as a selective force, the loss of large terrestrial carnivores is particularly conspicuous in today's increasingly human-dominated world. Such change characterizes areas from arctic tundra and circumpolar boreal zones to tropical savannas, forests, and deserts. Natural systems have been altered to such an extent that Lineaeus apparently never saw moose *Alces alces* in Scandinavia, yet more than 300,000 now live in Sweden alone (Clutton-Brock and Albon, 1992). The role that the loss of predators and other factors have played in these rebounding populations is not clear. But it is evident that in the contiguous United States and Mexico, carnivores such as grizzly bears *Ursus arctos* and wolves *Canis lupus* are currently absent from more than 99% of their original range, losses which have freed many large herbivores from natural predation. These losses have also led, perhaps through interference competition, to range expansions in mesocarnivores including coyotes *Canis latrans* (Peterson, 1995; Johnson, et al., 1996) and changes in abundance of red and kit foxes *Vulpes vulpes* and *V. macrotis* (Sargeant and Allen, 1989; Ralls and White, 1995) and subsequently to locally high ungulate densities, which damage vegetation (Wagner and Kay, 1993). In tropical and subtropical systems, losses have also been substantial. Wild dogs *Lycaon pictus* once occurred in 34 African countries, but populations now exceed 100 or more individuals in only 6 (Creel and Creel, 1996); a similar pattern of constricted range characterizes some felids (cheetahs *Acinonyx jubatus;* jaguars *Panthera onca;* tigers *P. tigris*), a canid (the dhole *Duon alpinus*), and other taxa (Ginsberg and Macdonald, 1990; Seidensticker and Lumpkin, 1991; Caro, 1994a; Schaller, 1996; Weber and Rabinowitz, 1996).

Although natural recolonization by large carnivores has occurred in both the Voronezh (Russia) and Glacier (United States) parks, even after initial exterminations by humans, and wolves were reintroduced into Yellowstone (United States), most ecosystems are more likely to be depopulated of their large carnivores rather than recolonized by them (but see Corbett, 1995). In less developed countries as well as in northern boreal and arctic systems, antipathy for large carnivores has generally been high because jaguars, wolves, and even grizzly bears compete for the same prey desired by subsistence and trophy hunters (Gasaway et al., 1983, 1992; Jorgenson and Redford, 1993). As a result, the killing of large carnivores may be encouraged (see Cluff and Murray, 1995). Because hunting by humans with modern weapons is a comparatively recent phenomenon and, in many areas has replaced predation by native carnivores, prey may also be exposed to somewhat different selection forces from those in the past. For example, human and nonhuman hunters are known to select prey differently (Ginsberg and Milner-Gulland, 1994).

My goal in this chapter is to explore some of the more subtle nondemographic components of predator–prey relationships that involve large mammalian carnivores and their primary food. Specifically, I ask to what extent ungulate prey vary in response to the loss and restoration of predation as an ecological and evolutionary force.

Traditional Conservation Biology in Environments With and Without Large Carnivores

During the earlier part of this century when conservation efforts focused on the prey of large carnivores, the loss of "top" carnivores was a concern to only a small minority of biologists (Skinner, 1928; Murie, 1940). Although naturalists like John Muir and wildlife managers such as Aldo Leopold were disturbed about the rampant destruction of predators (Muir, 1894; Leopold, 1925, 1942), the prevailing view, particularly in North America, was that predators destroyed prey valued by humans. In 1914, for instance, some 40 years after the establishment of the world's first national park, Yellowstone, the acting superintendent granted permission "to shoot trap, or poison mountain lions, wolves, and coyotes in the Park . . . [and] great care will be used not to shoot so as to injure anything except the animals mentioned" (quoted in Phillips and Smith, 1996). Not only have campaigns to decimate carnivores been commonplace in the past, but many continue today. Thus, these conservation efforts can hardly be considered very active; it seems more reasonable to say that past conservation focused on the prey of carnivores, and the removal of carnivores was largely ignored. Fortunately, a different ethic is emerging now that more fundamental relationships between predators and prey are understood.

Environments with and without large predators have been used to assess direct impacts of predation and/or the loss of such processes. Where, for instance, herbivores have been released from predation, trophic cascades may occur, as seen in northern Pacific Ocean kelp forests which were severely impacted by the grazing of marine invertebrates after human-induced localized extinctions of sea otters *Enhydra lutris* (Estes and Duggin, 1995). In boreal and arctic systems, wolves and grizzly bears may depress prey populations, particularly moose, although environmental factors such as snow or cold interact with prey density to affect population size (Nelson and Mech, 1981; Ballard et al., 1987; Fuller, 1989; Messier, 1991, 1994). As a result of predator–prey studies, it has been possible to place limitations on human harvests so that local prey populations are not depleted by hunting (Gasaway et al., 1983, 1992). Recent research on prey abundance and population dynamics in the presence and absence of predators also suggests how wildlife management can contribute directly to local economies in developing countries through ecotourism and direct harvest (Sinclair and Arcese, 1995; Gasaway et al., 1996; Cunningham and Berger, 1997) and to the welfare of indigenous people (Metcalfe, 1994; Robinson and Redford, 1994).

Understandably, conservation involves more than science alone, and public perception plays a prominent role in how some components of biodiversity are managed (Brussard, 1991). This is particularly evident in North America and Europe, where species reintroductions have occurred and mammalian carnivores have a disproportionate potential to impact public opinion most strongly.

Effects of Large Carnivores on Prey Biology

Predation has ecological, evolutionary, demographic, and behavioral effects on prey populations (Endler, 1986; Strauss, 1991). Table 4-1 shows some examples of natural and human-assisted colonization in which it is likely that prey have been (or will be) affected at different time scales. Some of these potential effects are considered below.

Where large carnivores are present, prey often occur at low densities, an ecological situation referred to as the "predator pit" (Bergerud, 1980; Peterson, 1988); this situation is supported empirically by data on wolves and moose. In lightly exploited North American boreal and arctic systems, wolves depress prey densities, but where wolves have been exterminated or are heavily exploited, moose can be up to three or four times more abundant (Gasaway et al., 1992). Changes in predation pressure may also affect prey population density indirectly through compression of prey into safe zones due to human predation, or by giving them fewer competitors (Crowell, 1981; Owen-Smith, 1983; Douglas-Hamilton and Douglas-Hamilton, 1992).

In evolutionary terms, guppy life-history parameters have changed genetically in less than 15 generations with alterations in predation pressure (Reznick and Endler, 1982; Reznick et al., 1990). Among other taxa, predation regime has altered sexual advertisements (Ryan et al., 1987), mate guarding (Ward, 1986), synchronous swarming (Sweeney and Vannote, 1982), and habitat use (Sih et al., 1992). In environments where large predators have been eradicated, changes in prey demography can be pronounced. For instance, where predation has been relaxed, individuals which might otherwise die experience higher survival (Albon and Clutton-Brock, 1988; Caughley and Sinclair, 1994); and even physically deformed individuals may reproduce (Berger and Cunningham, 1994).

Finally, study of prey in small reserves lacking natural predation has facilitated the study of several paradigms in behavioral ecology. In both red deer *Cervus elaphus* and bison *Bison bison,* individuals have been followed throughout their lifetimes on the Scottish Island of Rhum and on American grassland parks in Montana and South Dakota, respectively, and from these studies, topics like dominance (Clutton-Brock et al., 1982a,b; Clutton-Brock, 1986; Rutberg, 1986; Green and Rothstein, 1993b), reproductive synchrony (Rutberg, 1987; Green and Rothstein, 1993a), and parental investment (Silk, 1983; Wolff, 1988; Green and Berger, 1990; Clutton-Brock, 1991; Green and Rothstein, 1991a,b) have led to the refinement of evolutionary theory. But the elimination of large carnivores has also removed the possibility of understanding ecological complexities that may affect social processes; for smaller species such as pronghorn *Antilocapra americana,* it is possible to consider predation as a selective force because medium-sized predators still occur (Byers and Byers, 1983; Byers and Kitchen, 1988; Byers and Hogg, 1995; Byers, 1998) but for larger prey species, this is not always the case.

Table 4-1 Examples of four categories of potential changes in mammalian predation as a selective force and their putative relationships (novel or not historically novel) to prey in relatively recent times.

Category and species[a]	Potential prey	Location
Distant past (novel species)		
Human[1]	Megafauna	Worldwide expansions
Dingo[2]	Varied marsupials	Australia
Range expansion (novel species)		
Coyote[3]	Caribou	Gaspesie Park, Quebec, Canada
Puma[4]	Porcupines	Great Basin Desert, Nevada, USA
Recolonization (not novel)		
Grizzly bear[5]	Elk, moose	Grand Teton Park, Wyoming, USA
Badger[6]	Ground squirrels	Southern Alberta, Canada
Wolf[7]	White-tailed deer, beaver	Northern Wisconsin, USA
	Varied ungulates	Glacier National Park, Montana, USA
	Elk (red deer)	Voronezh Reserve, Usman Forest, Russia
Restoration by humans (not novel)		
Eurasian lynx[8]	Various-sized mammals	Alps, Switzerland, and elsewhere in Europe
Black-footed ferret[9]	Prairie dogs	Northern central prairies, USA
Red wolf[10]	Various-sized mammals	Southeastern USA
Grey wolf[11]	Varied ungulates	Yellowstone Park, central Idaho, USA
Fisher[12]	Various-sized mammals	Alaska and other North American sites, USA
River otter[13]	Varied	Pine Creek, Pennsylvania, USA
Sea otter[14]	Marine invertebrates	Alaska and other northern sites, USA
Eurasian otter[15]	Varied	East Anglia, England
Serval[16]	Varied	Transvaal, South Africa
Cheetah[17]	Waterbuck, varied	Pilannesberg reserve, South Africa
	Varied	Hluhluwe/Umfolozi reserve, South Africa
Lion[18]	Ungulates, marine mammals	Skeleton Coast Park, Namibia
	Varied, ungulates	Hluhluwe/Umfolozi reserve, South Africa
Wild dog[19]	Varied ungulates	Hluhluwe/Umfolozi reserve, South Africa

[a]References: [1]Martin and Klein (1984); [2]Flannery (1994), Corbett (1995); [3]Crete and Desrosiers (1995); [4]Berger and Wehausen (1991), Sweitzer et al. (1997); [5]Recolonization of northern Tetons and adjacent areas during 1995, Cain (unpublished data), Berger (in preparation); [6]Michener (1996); [7]Wydeven et al. (1996), Boyd et al. (1994), Ryabov et al. (1994); [8]Breitenmoser and Breitenmoser-Wyrsten (1990); [9] Miller et al. (1996); [10]Moore and Smith (1991); [11]Phillips and Smith (1996); [12]Berg (1982); [13]Serfass and Rymon (1985); [14]Jameson et al. (1982); [15]Jeffries et al. (1986); [16]Van Aarde and Skinner (1986); [17]Anderson (1984, 1992); [18]Save the Rhino Trust Newsletter (Windhoek, Namibia), Anderson (1981); [19]Anderson (1992). Additional cases in Reading and Clark (1996).

Bison on the National Bison Range (NBR), a protected reserve in Montana, highlight the dilemma. There they have no natural predators and are guaranteed relatively abundant food because they are rotated among paddocks. Excess bison are rounded up and sold annually, which results in relatively low density, and a fecundity rate (calves/cow) of 80–90%, which exceeds that in less managed South Dakotan populations in Wind Cave (WCNP) and Badlands (BNP) national parks. The NBR population is also characterized by more synchronous births, with 50% occurring with 13 days (Rutberg, 1984); at WCNP, the same 50% point occurs in 21 days (Green and Berger, 1990), and at BNP it requires from 20 to 27 days (Berger and Cunningham, 1994). Depending on the density at a given site, births can be interpreted as being synchronous (Rutberg, 1984, 1987) or asynchronous (Green and Rothstein, 1993a). If loss of predation affects behavioral phenomena, then part(s) of the repertoires of wild animals may be changed or lost even before being described. For instance, animals are remarkably deft in discriminating among dangerous sounds (Owings and Leger, 1980) and smells (Sullivan et al., 1995a,b; Duvall et al., 1990; Engelhardt and Muller-Schwarze, 1995), distinguishing visual from olfactory cues (Phillips and Alberts, 1992), and differentiating aerial from terrestrial predators (Seyfarth et al., 1980; Cheney and Seyfarth, 1990). These behaviors might disappear in the absence of predation.

Effects of Large Carnivores on Prey Behavior

Prey might respond to different levels of predation pressure in one of several principal ways (fig. 4-1). Responsiveness in this case is any measurable behavioral change and can vary from that associated with strictly behavioral events such as recognition of predators, transmission of cultural patterns of predator avoidance, or active defense to those more tightly linked to ecological conditions such as group formation and group size, vigilance, or shifts in habitat use.

In scenario A (fig. 4-1), predation pressure is constant, though it may operate at levels varying from high to low. For simplicity, I assume that predation has been intense, as characterized by specialists like black-footed ferrets *Mustela nigripes* on prairie dogs *Cynomys ludovicianus* (Clark, 1989; Hoogland, 1995), cheetahs on Thomson's gazelles *Gazella thomsoni* (Fitzgibbon, 1990; Caro, 1994a), and grizzly bears on moose calves (Van Ballenberge, 1987, 1991; Ballard, 1991; Ballard et al., 1991). Scenario B depicts the truncation (extirpation) of effective predation, a situation found in many national parks or protected areas where predators such as lions *Panthera leo,* spotted hyenas *Crocuta crocuta,* or wild dogs are locally extinct and human hunting is prohibited. Examples include comparatively small reserves like Matopos in Zimbabwe or Waterberg Plateau in Namibia, or places where wolves and grizzly bears are extinct such as in Yosemite in California and Big Bend in Texas.

Scenario C of fig. 4-1 involves the recent restoration of large carnivores (the time frame remains unspecified). The restitution of predation as a selective force can be of two primary types: predators to which the prey have been exposed in the past, or those which are totally novel (i.e., no evolutionary history of sympatry). Wolves reinvading Russia's Voronezh Reserve after an absence of nearly two decades (Ryabov et al., 1994) represent the former (table 4-1). Whether prey have immediate (proximate) memory or some form of innate recognition of their historic predators is uncertain, but responses might be very different from those of prey encountering a new predator for the first time. Komodo dragons *Varanus komodoensis,* which may have once preyed on extinct dwarf elephants (Diamond, 1987), now kill rusa deer *Cervus timorensis,* which did not evolve with this reptilian carnivore and may show inappropriate antipredator responses (Auffenberg, 1981; Quammen, 1996). Both pumas *Felis concolor* and coyotes have expanded their ranges in areas of the Great Basin Desert and southeastern Quebec (Berger and Wehausen, 1991; Crete and Desrosiers, 1995) during the last 100 years, where local populations of porcupines *Erethizon dorsatum,* bighorn sheep *Ovis canadensis,* and caribou *Rangifer tarandus* may be ineffective at deal-

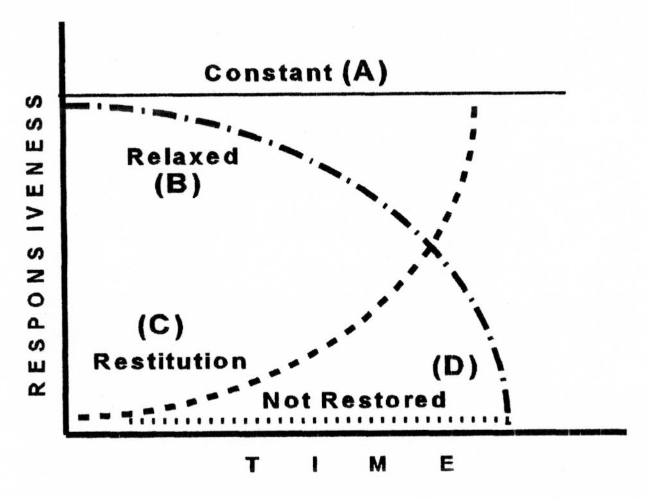

Figure 4-1 Hypothetical scenarios depicting expected prey responsiveness (antipredator behavior) as a function of predation pressure—constant, relaxed, restituted (restored), and not restored—over time. In scenario A predation pressure is assumed to be intense; in D it is trivial.

ing with them (table 4-1). While such range expansions by mammalian carnivores are due to recent anthropogenic events, the exposure of naive prey to novel suites of predators is common throughout history. Human colonization of the Americas, the spread of large felids (*Panthera* spp.) across Beringea into the Yukon and Alaska (Pielou, 1991), and/or the replacement of thylacines by dingoes after introduction into Australia by humans some 4000 years ago (Corbett, 1995) all represent cases in which prey were forced, often unsuccessfully, to respond to a predation pressure that presumably differed from that of their ancestors.

Among the methods for detecting behavioral changes associated with the loss of or change in predation, the comparative method is one of the best (Dobson and Murie, 1987; Hayes and Jenkins, 1997). For instance, in areas where predation has been relaxed, ground squirrels and salamanders respond differently than do their predator-savvy conspecifics (Coss and Owings, 1985; Ducey and Brodie, 1991; Coss et al., 1993). Contrasts between island populations of macaques with and without tigers and leopards *Panthera pardus* has led to the suggestion that macaque group sizes have changed where predators have been lost. Yet it is difficult to exclude the role that other factors, notably food quality or distribution (Mitani et al., 1991), may have played in influencing group size (table 4-2). Another difficulty arises when predators are extinct but exerted strong selection in the past (Byers, 1998). The loss of eagles for about 4000 years from Madagascar's contemporary fauna is a case in point. Two lemurs, ring-tailed *Lemur catta* and Verreaux's sifaka *Propithecus verreauxi,* show strong antipredator responses to birds, behavior that has been interpreted as the result of past selection (Goodman, 1994). But without excluding other possibilities such as contemporary predation by other birds (Csermely, 1996), it is not possible to say much about predation-related effects. Another possibility for examining how predation affects behavior is to contrast responses in closely related species, as has been done for endemic flightless birds in New Zealand in order to determine responsiveness to introduced mammalian predators (Bunin and Jamieson, 1996).

Table 4-2 Behavioral changes in mammalian prey as a result of relaxation of predation pressure.

Species[a]	Predators	System	Change	Comment on data
Macaques[1]	Tigers and leopards	Isles with and without predators	Group size	Possible habitat variation
Bison[2]	Wolves vs. none	Parks with and without predators	None in vigilance or foraging	Hunting not witnessed
Bighorn[3]	Wolves, coyotes	Contrast sexes in different habitats	Habitat shifts	Experimental
Pronghorn[4]	Humans vs. none	Contiguous study sites	Grouping and vigilance	Reasonably strong
Musk ox[5]	Wolves vs. none	Similar study areas	Group size	Strong
Lemurs[6]	Extinct eagles	Predator-free isles	Reactions to hawks	No controls
Squirrels[7]	Snakes vs. none	Historic; snake-free	Vocal and other behaviors	Experimental; excellent

[a]References: [1]van Schaik and van Noordwijk (1985); [2]Berger and Cunningham (1994); [3]Berger (1991); [4]Berger et al. (1983); [5]Heard (1992); [6]Goodman (1994), Csermely (1996); [7]Owings and Owings (1979), Rowe et al. (1986), Coss and Owings (1985), Coss (1991).

In short, although components of mammalian prey behavior have been linked to the relaxation of predation as a selective force (table 4-2), appropriate controls are usually absent. Empirically grounded field experiments are therefore the best means of understanding how prey respond to both the absence and presence of predators.

Experimental Data

As a result of the widespread extermination of large terrestrial carnivores on all continents, opportunities exist to examine issues in prey behavior where predation has been relaxed. Moose, for example, have a circumpolar distribution in environments that vary from predator-rich to predator-depauperate. In the Sikhote-Alin Mountains of the Russian Far East, moose live in a three-predator system that includes grizzly bears, wolves, and Siberian tigers (Jia and Faber, 1995; Miquelle et al., 1997), and in Alaska they coexist with two predators (grizzly bears and wolves). At sites in the Rocky Mountains, moose live where these same two native predators have been eliminated, exist, or have been replaced by subsistence or trophy hunters (Boyd et al., 1994). Predation on calves has been intense in Alaska's Talkeetna Mountains, with about 67% dying before 2 months of age (W. Testa, personal communication), but calf survival exceeds 95% at two sites in the southern greater Yellowstone ecosystem, one with human hunting of adults only and the other in the absence of predation because grizzly bears and wolves are lacking (fig. 4-2). Assuming predation intensity has been relaxed where large carnivores have been eliminated, to what extent might moose behavior have changed?

If predation has shaped the ability of prey to recognize predators (Coss, in press), then

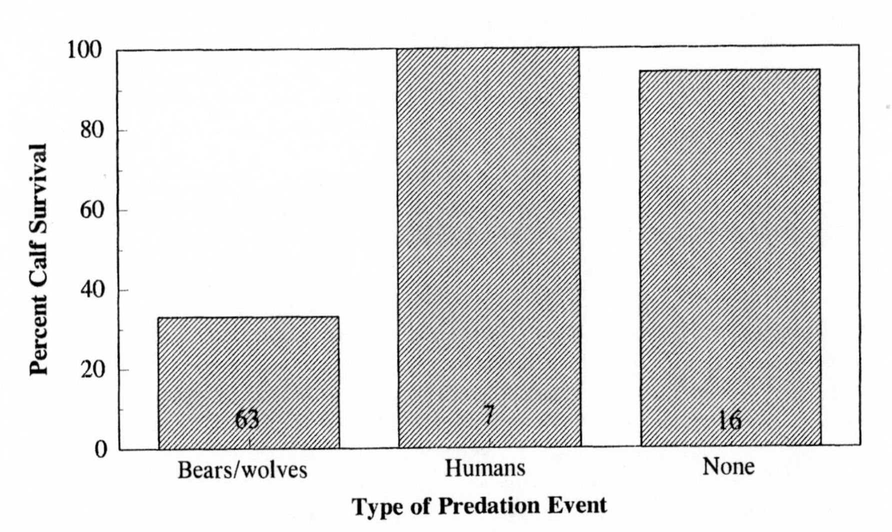

Figure 4-2 Calf survivorship to 2 months in Talkeetna Mountains, Alaska, where grizzly bears and wolves are present, in Grey's River region where grizzly bears and wolves have been replaced by human hunters (humans), and in the southern Greater Yellowstone Ecosystem where they have no impact (none; Grand Teton National Park). Sources of data as follows: bears/wolves (W. Testa, personal communication); humans and none (Berger, unpublished). Numbers on bars indicate sample sizes.

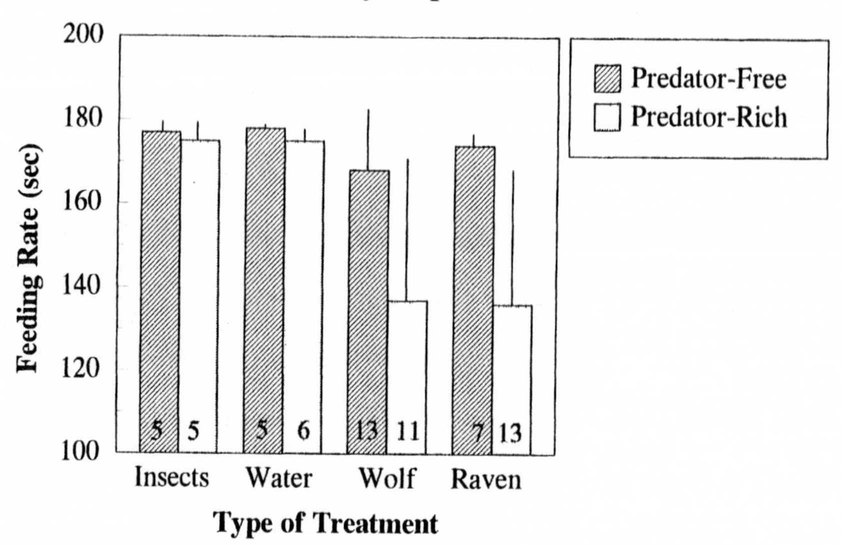

Figure 4-3 Effects of auditory playbacks on feeding rate (time per 180 s bout) in female moose from predator-rich (Talkeetna Mountains, Alaska) and predator-free (Teton Range, Wyoming) sites. As controls, foraging female moose were exposed to the taped sounds of cricket chirps and running water, which were subsequently played through a speaker situated from 50–200 m away. Numbers on histograms indicate sample sizes; bars are coefficients of variation. The use of parametric and nonparametric tests reflects unequal variance in samples (tests two tailed). Contrasts between moose in predator-free and predator-rich sites as follows: wolf howls, $p < .05$ ($t = 2.16$; df = 22); raven quorks, $p < .002$ ($U = 5$, Mann-Whitney test).

in its absence naive prey should fail to respond to cues suggestive of their major predators. If unfalsified, the hypothesis suggests that the immediate behavior of naive prey has changed in the absence of predation. Should the antipredator behaviors of savvy and naive prey (i.e., scenarios A and C in fig. 4-1) remain similar, the responses may either be obligatory or prey have had insufficient time to respond to selection.

Among the ways to detect prey responsiveness to potential predators are observations of direct encounters and through field playback experiments (McComb et al., 1994), since the chances of witnessing direct interactions between large predators and their prey are possible in only a few areas of the world (Packer et al., 1991; Scheel, 1993; Berger and Cunningham, 1995). More indirect procedures might include assessing prey behaviors associated with predator detection in naturally varying environments while controlling possible confounding variables (Caro, 1994b). Thus, to examine potential behavioral changes associated with differing predation regimes, I have relied on the behavior of nonhunted female moose after presenting auditory and olfactory signals of wolves, grizzly bears, and other species.

The experiments were conducted by playing different sounds (fig. 4-3) from a speaker placed 50–200 m from moose. The olfactory trials were conducted by soaking snowballs in urine or wrapping them with scat in biodegradable toilet paper or netting and then tossing

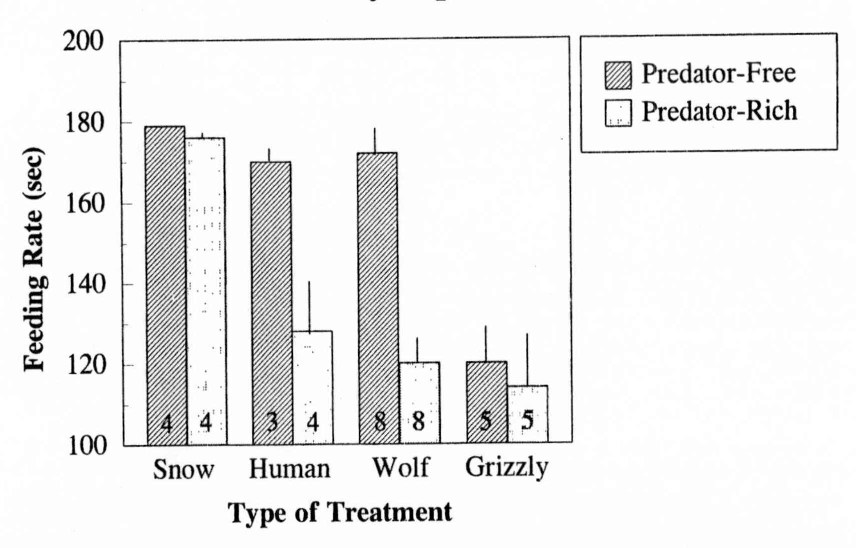

Figure 4-4 Effects of olfactory experiments on feeding rate (time per 180 s bout) in female moose from predator-rich (Talkeetna Mountains, Alaska) and predator-free (Teton Range, Wyoming) sites. The response of moose to wolf or human urine or to grizzly bear feces was tested (see text). As a control, snow was handled without gloves and tossed to within 8 m of foraging moose. Numbers on histograms indicate sample sizes; bars are coefficients of variation (tests two tailed). Contrasts between moose in predator-free and predator-rich sites as follows: wolf urine, $p < .02$ ($t = 2.75$; df $= 14$); grizzly feces, not significant.

them to within 5–10 m of prey subjects. Both types of field trials were conducted on windless days, when ambient temperatures varied from $-2°$ to 5°C and when animals were feeding. The response variable was the proportion of time that animals fed per 180 s.

Preliminary results are tantalizing, for they suggest that in the absence of active predation for about 50–75 years, female moose in Grand Teton National Park show only weak responses to the sounds and odors of wolves (figs. 4-3, 4-4). This is not the case in Alaska, where wolf predation is still present (see fig. 4-2 and previously cited references concerning the moose–wolf–grizzly bear system). Systematic playbacks of both wolf howls and exposure to wolf urine depressed feeding rates of the predator-savvy Alaskan females more than it did the predator-naive Teton females (fig. 4-3). Similarly, raven vocalizations depressed feeding rates more in predator-savvy than in predator-naive moose. This is intriguing because predator-savvy prey might be expected to associate ravens with the presence of carcasses, perhaps indicating that wolves or other large carnivores are nearby (see Craighead, 1979; Heinrich, 1989). But where raven densities are the highest recorded such as in the southern Greater Yellowstone region of my study area (Dunk et al., 1994) and large carnivores are absent, moose failed to show feeding reductions (fig. 4-3). Exposure to grizzly bear feces, irrespective of the past associations of female moose with carnivores, depressed feeding rates similarly, and more so than exposure to wolf urine (fig. 4-4). But there were no such reductions in Teton moose when exposed to wolf urine. If moose respond more intensively to one type of predator than another despite 50–75 years of no exposure to either, the possibility of innate recognition may persist. It may also be that one type of predator, in this case grizzly bears, has had a stronger evolutionary influence than others. In Alaska, grizzly bears continue to prey more heavily on moose calves than do wolves (Ballard et al., 1991; W. Testa, personal communication).

While preliminary, these results have implications for understanding how prey recognize potential threats from predators. White-tailed deer *Odocoileus virginianus* congregate along wolf territorial boundaries (Nelson and Mech, 1981) and chital *Axis axis*, moose, and vervet monkeys *Cercopithecus aethiops* all decrease predation risk by associating near human dwellings where tigers, leopards, and wolves are less likely to occur (Stephens and Peterson, 1984; Sunquist and Sunquist, 1989; Isbell and Young, 1993). If effective predators are lost from these and other systems, the clustering of individuals or sexes to decrease predation risk (Stacey, 1987), patterns of spatial association, and habitat use may all change.

Strengths and Weaknesses of a Behavioral Ecological Approach

As a construct for understanding immediate issues in conservation, large carnivores offer advantages and disadvantages. On the positive side, they elicit extraordinary attention from the lay public and politicians, are economically and ecologically valuable and, as an educational tool, they are usually unsurpassed. But, as subjects of scientific study, large carnivores are problematic because individuals range over large areas, populations are small, and experimental manipulations are difficult even where feasible. Pragmatically speaking, it is obvious that as human impacts increase globally, progressively fewer environments will maintain assemblages of large carnivores, and it would be surprising if the biology of their prey has not changed or is not changing. If we want to understand our future ecosystems without large carnivores or develop serious plans to restore carnivores, we must anticipate behavioral and ecological relationships in prey by seizing opportunities for comparative approaches while study is still possible (Berger, 1996).

Among the key issues that involve predator–prey dynamics once carnivores are either restored or lost are the functional responses of prey populations. With carnivores restored, their primary prey are expected to decline, whereas prey should increase if their major predators are eliminated. Opportunities also exist to understand how prey respond to predation; it is here that approaches steeped in behavioral ecology can play a role. For instance, rufous-hare wallabies *Lagorchestes hirsutus* became extinct on the Australian mainland not because they were unable to cope with predation by native species, but because of introduced carnivores—red foxes and feral cats—with which they were unfamiliar (McClean et al., 1996). Knowledge of antipredator behavior has been critical in designing plans to achieve viable wallaby populations; in this case, the effort has focused on teaching wallabies to avoid predators they have not evolved with, a situation also employed with other endangered mammals (Kleiman, 1989; Miller et al., 1996; Curio, chapter 7, this volume).

In today's anthropogenically disturbed world, predation comes in two primary forms, those in which predators are either familiar or nonfamiliar to possible prey. The former, involving species in which predator and prey evolved together, was the typical situation, whereas the latter is likely to become increasingly common. Behavioral ecologists have generally focused on natural systems with native species because the discipline has traditionally been concerned with evolutionary aspects of behavior. What eventually may prove more central to conservation will be an increased focus on exotic predators—the countless cases of foxes, dogs, cats, mongooses, birds, snakes, and fish that have been disseminated by our global carelessness and which have devastating effects on local prey populations. An understanding of behavioral variation and responsiveness to introduced and native species would do much toward enhancing conservation. For instance, populations of the California newt *Taricha torosa* apparently remain viable where exotics have not entered the system. Their larvae are able to avoid predation by native species but cannot avoid predation by introduced mosquitofish *Gambusia affinis* and crayfish *Procambarus clarkii* (Gamradt and Kats, 1996). Additionally, where the aim is to assure the persistence of local prey populations, it will be important to predict both behavioral and subsequent demographic responses to specific types of carnivores, irrespective of whether they are recolonizing or novel. And, in cases where prey populations are reintroduced into former habitats, it may be possible to enhance their viability by knowing about predator avoidance capabilities. Numerous anecdotes have consistently been raised in my discussions with management agency personnel in which reintroductions involving species as diverse as eland *Taurotragus oryx* and porcupines were unsuccessful because individuals were predator-naive. Although anecdotes may differ from reality and be susceptible to memory failure over the years, the message is clear: more can be done about enhancing population viability by designing research that takes into account the social history of founding populations and the extent to which individuals may be familiar with past and present predators.

I have argued that behavioral ecology has a role in the conservation biology of prey assemblages, but there are obvious constraints in developing appropriate study designs with large carnivores and large prey. Also, a lack of replication is likely to characterize most field studies, but this problem may be partially circumvented through the use of meta-analyses. And it is not totally clear how field experiments, such as the ones I described using the sounds and smells of novel and familiar carnivores, simulate real predator recognition, nor is it clear the extent to which behavioral responses are linked to demographic ones. If colonizing (native or exotic) or recolonizing carnivores encounter naive prey, the potential for local extinction exists. Indeed, evidence from porcupines and caribou suggests that native prey may be differentially susceptible to steep population reductions when encountering native but novel carnivores for the first time (Crete and Desrosiers, 1985; Sweitzer et al., 1997). Large macropods protected by fences from dingoes may also suffer disproportionate mortality relative to predator-savvy conspecifics when dingo reinvasion occurs, although little is known of how prior history affects individual kangaroo antipredator responses (Jarman and Wright, 1994; P. Jarman, personal communication).

Recommendations

There are at least three tactics that biologists and managers of reserves should consider with respect to prey populations when planning for reintroduction of prey or the restoration, recolonization, first-time arrival, or loss of large carnivores.

Plan for Restoration

Efforts should include anticipating natural recolonization and/or human-assisted efforts to restore previously extirpated species. If restoration efforts proceed, the effects of predation and when or if the system attains some equilibrium point need to be determined. Two endangered species offer real-world examples. Both golden lion tamarins *Leontopitheus rosalia* and black-footed ferrets *Mustelia nigripes* have been reintroduced at sites where populations were jeopardized or already extinct. Because individuals targeted for reintroduction lacked experience with predators in their new natural environments, a primary concern was whether these individuals would deal effectively with potential predators. Behavioral ecologists were instrumental in designing programs to reestablish antipredator behaviors (Kleiman, 1989; Miller et al., 1996).

Consider the Effects of Colonization by Native but Novel Carnivores

Where restoration of past communities is not feasible (due to agricultural or other economic interests) but colonization by different carnivores is a possibility, it will be important to anticipate responses of prey in order to maintain viable populations. Elaboration of two examples is illustrative.

One hundred years ago, mule deer *Odocoileus hemionus* were rare, and pumas were presumed virtually nonexistent in the Great Basic Desert (Berger and Wehausen, 1991). As domestic stock and feral species altered ecological communities through their effects on vegetation, deer populations erupted, followed by pumas. A variety of prey are now consumed, including some which may be highly susceptible to predation, such as bighorn sheep (Wehausen, 1996). Despite quills for defense, porcupines were driven to near local extinction in one population (Sweitzer et al., 1997). Thus, while still a native species, the range of pumas has expanded in part due to the indirect effects of pastoralists, which caused habitats to change.

Carnivore range expansion has occurred elsewhere as well. Expansion of coyote ranges coincides with the extinction or reduction of wolves (Johnson et al., 1996; Bekoff, 1977). In southeastern Quebec, the viability of remnant caribou is jeopardized, perhaps because coyotes are effective predators of juvenile caribou (Crete and Desrosiers, 1995). If further study substantiates this, knowledge about caribou antipredator responses could lead to important management considerations. For instance, where prey do not possess the ability to avoid novel predators and the viability of the former assumes precedence, control of predators should be an option.

Understand Consequences of Replacing Native Carnivores with Humans

Where restoration has not been possible and effective predation no longer occurs, pragmatic conservation biologists will have to deal with a politically sensitive issue: how to balance the abundance of single species with ecological diversity. If the latter is important, then population control will be critical whether by replacing mammalian carnivores with human predation or by contraception or translocation. With regulation by humans, antipredator behavior in the remaining population may change.

From a purely conservation perspective, a critical ecological issue concerns the extent to which the replacement of large carnivores by humans alters other components of the community (Terborgh, 1988, 1990; Wright et al., 1994). In the absence of some regulation by either natural predators or humans, populations even in large areas like the greater Yellowstone ecosystem may overshoot their food ceiling (Varley, 1988; Wagner and Kay, 1993). At some point, decision makers will be forced to make difficult choices between protecting the primary prey of large (and currently extinct) carnivores or protecting that ecosystem's biodiversity. For instance, abundant herbivores damage the vegetation of riparian zones (Chadde and Kay, 1991; Kay, 1994), which in arid environments may harbor most of the region's diversity, and as a consequence both avian and lepidopteran species may decline (Stacey, 1995).

If the goal of managers, then, is to understand the spatial dynamics of prey or causes of juvenile mortality, to predict susceptibility to new or past predators, or to restore prey populations, then approaches employing behavioral ecology will yield insights into how systems operate. By understanding decisions that individuals make about foraging, socializing, and reproducing, it will be possible to understand population-level responses to predation.

Summary

The earth is likely to lose large carnivores in the near future, a result that inevitably will alter pressures on prey. In the United States and Mexico alone, large and economically important species such as grizzly bears and wolves have experienced a 99% reduction in their historic range. Because predation is a major selective force and affects ecological dynamics, the altered environments in which we now work are either missing dominant selective forces or have been replaced by humans. The lack of natural carnivores at the top of food webs is likely to have repercussions on behavioral phenomena of prey. These may include reproductive synchrony and grouping as antipredator strategies, altered spatial distribution, and altered responsiveness to potential predators.

Comparative data on the behavior of female moose at a predator-rich site in Alaska and a predator-depauperate site in the Rocky Mountains imply that where predation has been relaxed, prey reduce their reliance on auditory and olfactory cues associated with predators that were extirpated only 50–75 years ago. Although preliminary, such data suggest that prey may have to learn how to avoid some predators.

If the loss of some large carnivores alters selection pressures, then so must the addition, range expansion, and restoration of others. Expanding populations of native species such as pumas and coyotes and the introduction of exotics such as dogs, cats, foxes, and mongooses have disproportionate effects on prey that have evolved in their absence.

To enhance options for implementing conservation, biologists can adopt some of the approaches used by behavioral ecologists. These include establishing field and experimental protocols to determine how prey might respond to past, present, and future predators, be they evolutionarily novel or familiar; exploring options for teaching prey to avoid predators; and where predation by nonhumans is unlikely, designing programs to evaluate how both human predation (hunting, road kills, etc.), and no predation affect life history, demographic, and behavioral variables.

Acknowledgments For discussion and suggestions, I thank Marc Bekoff, Carol Cunningham, Bill Gasaway, Paule Gros, and Peter Stacey. Karen McComb graciously provided advise on playback experiments. I am especially grateful to Tim Caro for advice, persistence, and patience.

References

Albon SD, Clutton-Brock TH, 1988. Climate and population dynamics of red deer in Scotland. In: Ecological changes in the uplands (Usher MB, Thompson BA, eds). Oxford: Blackwell Scientific; 93–107.

Anderson JL, 1981. The re-establishment and management of a lion *Panthera leo* population in Zululand, South Africa. Biol Conserv 19:107–117.

Anderson JL, 1984. A strategy for conservation in Africa. In: The extinction alternative (Mundy, PJ, ed). Johannesburg: Endangered Wildlife Trust; 127–135.

Anderson J, 1992. South African carnivores. Re-introduction News 4:8–9.

Auffenberg W, 1981. The behavioral ecology of the Komodo monitor. Gainesville: University of Florida Press.

Ballard WB, 1991. Management of predators and their prey: the Alaskan experience. Trans N Am Wildl Nat Res Conf 56:527–538.

Ballard WB, Whitman JS, Gardner CL, 1987. Ecology of an exploited wolf population in southcentral Alaska. Wildl Monogr 98:1–54.

Ballard WB, Whitman JS, Reed DJ, 1991. Population dynamics of moose in southcentral Alaska. Wildl Monogr 114:1–49.

Bekoff M, (ed), 1977. Coyotes, New York: Academic Press.

Berg WE, 1982. Reintroduction of fisher, pine marten, and river otter. In: Midwest furbearer management (Sanderson GC, ed), Wichita: Kansas Chapter of the Wildlife Society; 159–173.

Berger J, 1991. Pregnancy incentives, predation constraints, and habitat shifts: experimental and field evidence from wild bighorn sheep. Anim Beh 41:61–77.

Berger J, 1996. Animal behaviour and plundered mammals: is the study of mating systems a scientific luxury or a conservation necessity? Oikos 77:207–216.

Berger J, Cunningham C, 1994. Bison: mating and conservation in small populations. New York: Columbia University Press.

Berger J, Cunningham C, 1995. Predation, sensitivity, and sex: why female black rhinoceroses outlive males. Behav Ecol 6:57–64.

Berger J, Daneke D, Johnson J, Berwick SH, 1983. Pronghorn foraging economy and predator avoidance in a desert ecosystem: implications for the conservation of large mammalian herbivores. Biol Conserv 25:193–208.

Berger J, Wehausen JD, 1991. Consequences of a mammalian predator-prey disequilibrium in the Great Basin. Conserv Biol 5:244–248.

Bergerud AT, 1980. A review of the population dynamics of caribou and wild reindeer in North America. Reindeer-Caribou Symp 2:556–581.

Boyd DK, Ream RR, Pletscher DH, Fairchild MW, 1994. Prey taken by colonizing wolves and hunters in the Glacier National Park areas. J Wildl Manage 58:289–295.

Breitenmoser U, Breitenmoser-Wyrsten C, 1990. Status, conservation needs, and reintroduction of the Lynx (*Lynx lynx*) in Europe. Strasbourg: Council of Europe.

Brussard PF, 1991. The role of ecology in biological conservation. Ecol Appl 1:6–12.

Bunin JS, Jamieson IG, 1996. Responses to a model predator of New Zealand's endangered takahe and its closest relative, the pukeko. Conserv Biol 10:1463–1466.

Byers JA, 1998. American pronghorn: social adaptations and ghosts of predators past. Chicago: University of Chicago Press.

Byers JA, Byers KZ, 1983. Do pronghorn mothers reveal the locations of their hidden fawns? Behav Ecol Sociobiol 13:147–156.

Byers JA, Kitchen DW, 1988. Mating system shift in a pronghorn population. Behav Ecol Sociobiol 22:355–360.

Byers, JA, Hogg JT, 1995. Environmental effects on prenatal growth rate in pronghorn and bighorn: further evidence for energy constraint on sex-biased maternal expenditure. Behav Ecol 6:451–458.

Caro TM, 1994a. Cheetahs of the Serengeti plains. Chicago: University of Chicago Press.

Caro TM, 1994b. Ungulate anti-predator behaviour: preliminary and comparative data from African bovids. Behaviour 128:189–229.

Caughley G, Sinclair ARE, 1994. Wildlife ecology and management. Oxford: Blackwell Scientific.

Chadde SW, Kay CE, 1991. Tall-willow communities on Yellowstone's northern range: a test of the "natural regulation" paradigm. In: The greater Yellowstone ecosystem (Keiter RB, Boyce, MS). New Haven, Connecticut: Yale University Press; 231–262.

Cheney DL, Seyfarth RM, 1990. How monkeys see the world. Chicago: University of Chicago Press.

Clark T, 1989. Conservation biology of the black-footed ferret. Philadelphia: Wildlife Preservation Trust.

Cluff HD, Murray DL, 1995. Review of wolf control methods in North America. In: Ecology and conservation of wolves in a changing world (Carbyn LN, Fritts SH, Seip DR, eds). Edmonton: Canadian Circumpolar Institute, 491–504.

Clutton-Brock TH, 1986. Great expectations: dominance, breeding success and offspring sex ratios in red deer. Anim Behav 34:460–471.

Clutton-Brock TH, Albon SD, 1992. Trial and error in the Highlands. Nature 358:11–12.

Clutton-Brock TH, 1991. The evolution of parental care. Princeton, New Jersey: Princeton University Press.

Clutton-Brock TH, Albon SD, Guinness FE, 1982a. Competition between female relatives in a matrilocal mammal. Nature 300:178–180.

Clutton-Brock TH, Guinness FE, Albon SD. 1982b. Red deer; ecology and behavior of two sexes. Chicago: University Chicago Press.

Corbett L, 1995. The dingo in Australia and Asia. Ithaca, New York: Comstock/Cornell.

Coss RG, 1991. Evolutionary persistence of ground squirrel antisnake behavior: reflections on Burton's commentary. Ecol Psychol 5:171–194.

Coss RG, in press. Effects of relaxed natural selection on the evolution of behavior. In: Geographic variation of behavior: an evolutionary perspective (Foster SA, Endler JA, eds). New York: Oxford University Press.

Coss RG, Guse KL, Poran NS, Smith DG, 1993. Development of antisnake defenses in California ground squirrels (*Spermophilus beecheyi*): II. Microevolutionary effects of relaxed selection from rattlesnakes. Behaviour 124:137–164.

Coss RG, Owings DH, 1985. Restraints on ground squirrel anti-predator behavior: adjustments over multiple time scales. In: Issues in the ecological study of learning (Johnson TD, Pietrewicz AT, eds). Hillsdale, New Jersey: Lawrence Erlbaum Associates; 167–200.

Craighead F, 1979. Track of the grizzly. San Francisco: Sierra Club Books.

Creel S, Creel NM, 1996. Limitation of African wild dogs by competition with larger carnivores. Conserv Biol 10:526–538.

Crete M, Desrosiers A, 1995. Range expansion of coyotes, *Canis latrans,* threatens a remnant herd of caribou, *Rangifer tarandus,* in southeastern Quebec. Can Field Nat 109:227–235.

Crowell KL, 1981. Islands—insight or artifact? Population dynamics and habitat utilization in insular rodents. Oikos 41:442–454.

Csermely D, 1996. Antipredator behavior in lemurs: evidence of an extinct eagle on Madagascar or something else? Int J Primatol 17:349–354.

Cunningham C, Berger J, 1997. Horn of darkness; rhinos on the edge. New York: Oxford University Press.

Darwin C, 1859. On the origin of species by means of natural selection. London: John Murray.

Diamond JM, 1987. Did Komodo dragons evolve to eat pygmy elephants? Nature 326:832.

Dobson FS, Murie JO, 1987. Interpretation on intraspecific life history patterns: evidence from Columbian ground squirrels. Am Nat 129:382–397.

Douglas-Hamilton I, Douglas-Hamilton O. 1992. Battle for the elephants. London: Transworld Publishers.

Ducey PK, Brodie, ED Jr, 1991. Evolution of antipredator behavior: individual and population variation in a neotropical salamander. Herpetology 47:89–95.

Dunk JK, Cain SL, Reid ME, Smith RN, 1994. A high breeding density of common ravens in northwestern Wyoming. Northwest Nat 75:70–73.

Duvall DD, Chiszar D, Hayes WK, Leonhardt JK, Goode MJ, 1990. Chemical and behavioral ecology of foraging in prairie rattlesnakes (*Croatulus virdis*). J Chem Ecol 16:87–101.

Edmunds M, 1974. Defense in animals: a survey of anti-predator defenses. London: Longmans.

Endler JE, 1986. Natural selection in the wild. Princeton, New Jersey: Princeton University Press.

Engelhart A, Muller-Schwarze D, 1995. Responses of beaver (*Castor canadensis* Kuhl) to predator chemicals. J Chem Ecol 21:1349–1365.

Estes JA, Duggin DO, 1995. Sea otters and kelp forests in Alaska: generality and variation in a community ecological paradigm. Ecol Monogr 65:75–100.

Fitzgibbon CD, 1989. A cost to individuals with reduced vigilance in groups of Thomson's gazelles hunted by cheetahs. Anim Behav 37:508–510.

Fitzgibbon CD, 1990. Antipredator strategies of immature Thomson's gazelles: hiding and the prone response. Anim Behav 40:846–855.

Flannery T, 1994. The future eaters. Chatswood, New South Wales: Reed Books.

Fuller T, 1989. Population dynamics of wolves in north-central Minnesota. Wildl Monogr 105:1–41.

Gamradt SC, Kats LB, 1996. Effect of introduced crayfish and mosquitofish on California newts. Conserv Biol 10:1155–1162.

Gasaway WC, Boertje RD, Bravgaard DV, Kellyhouse DG, Stephenson RO, Larsen DG, 1992. The role of predation in limiting moose at low densities in Alaska and Yukon and implications for conservation. Wildl Monogr 120:1–59.

Gasaway WC, Gasaway KT, Berry HH, 1996. Persistent low densities of plains ungulates in Etosha National Park, Namibia: testing the food-regulating hypothesis. Can J Zool 74:1556–1572.

Gasaway WC, Stephenson RO, Davis JL, Shepherd PEK, Burris OE, 1983. Interrelationships of wolves, prey, and man in interior Alaska. Wildl Monogr 84:1–50.

Ginsberg JR, Macdonald DW, 1990. Foxes, wolves, jackals, and dogs: an action plan for the conservation of canids. Gland: World Conservation Union.

Ginsberg JR, Milner-Gulland EJ, 1994. Sex-biased harvesting and population dynamics in ungulates: implications for conservation and sustainable use. Conserv Biol 8:157–166.

Goodman SM, 1994. The enigma of antipredator behavior in lemurs: evidence of a large extinct eagle on Madagascar. Int J Primatol 15:129–134.

Green WCH, Berger J, 1990. Maternal investment in sons and daughters: problems of methodology. Behav Ecol Sociobiol 27:99–102.

Green WCH, Rothstein A, 1991a. Sex bias or equal opportunity: patterns of maternal investment in bison. Behav Ecol Sociobiol 29:373–384.

Green WCH, Rothstein A, 1991b. Trade-offs between growth and reproduction in female bison. Oecologia 86:521–527.

Green WCH, Rothstein A, 1993a. Asynchronous parturition in bison. J Mammal 74:920–925.

Green WCH, Rothstein A, 1993b. Persistent influences of birth date on dominance, growth, and reproductive success in bison. J Zool 230:177–186.

Hayes JP, Jenkins SH, 1997. Individual variation in mammals. J Mammal 78:274–293.

Heard DC, 1992. The effect of wolf predation and snow cover on musk-ox group size. Am Nat 139–204.

Heinrich B, 1989. Ravens in winter. New York: Random House.

Hoogland JL, 1995. The black-tailed prairie dog: social life of a burrowing mammal. Chicago: University of Chicago Press.

Isbell LA, Young TP, 1993. Human presence reduces predation in a free-ranging vervet monkey population in Kenya. Anim Behav 45:1233–1235.

Jameson RJ, Kenyon KW, Johnson AM, Wright HM, 1982. History and status of translocated sea otter populations in North America. Wildl Soc Bull 10:100–107.

Jarman PJ, Wright SM, 1994. Macropod studies at Wallaby Creek. IX*. Exposure and responses of eastern Grey kangaroos to dingoes. Wildl Res 20:833–843.

Jeffries DJ, Wayre, P, Jessop RM, Mitchell-Jones AJ, 1986. Reinforcing the native otter *Lutra lutra* population in East Anglia: an analysis of the behavior and range development of the first release group. Mammal Rev. 16:65–79.

Jia J, Faber WE, 1995. The habitat of *Alces alces cameloides*—a review. Alces 31:125–138.

Johnson WE, Fuller TK, Franklin WL, 1996. Sympatry in canids: a review and assessment. In: Carnivore behavior, ecology, and evolution, vol 2 (Gittleman JL, ed), Ithaca, New York: Cornell University Press; 188–218.

Jorgenson JP, Redford KH, 1993. Humans and big cats as predators in the Neotropics. Symp Zool Soc Lond 65:367–390.

Kay CE, 1994. The impact of native ungulates and beaver on riparian communities in the intermountain west. Nat Res Environ Issues 1:23–44.

Kleiman DG, 1989. Reintroduction of captive mammals for conservation: guidelines for reintroducing endangered species into the wild. Bioscience 39:152–161.

Leopold A, 1925. The last stand of the wilderness. Am For For Life 31:599–604.

Leopold A, 1942. The grizzly—a problem in land planning. Outdoor America 7:11–12.

McClean IG, Lundie-Jenkins G, Jarman PJ, 1996. Teaching an endangered mammal to recognize predators. Biol Conserv 75:51–62.

McComb K, Packer C, Pusey A, 1994. Roaring and numerical assessment in calls between groups of female lions, *Panthera leo*. Anim Behav 47:379–387.

Martin PS, Klein RG, 1984. Quaternary extinctions. Tucson: University of Arizona Press.

Messier F, 1991. The significance of limiting and regulating factors on the demography of moose and white-tailed deer. J Anim Ecol 60:377–393.

Messier F, 1994. Ungulate population models with predation: a case study with North American moose. Ecology 75:478–488.

Metcalf S, 1994. The Zimbabwe communal areas management programme for indigenous resources (CAMPFIRE). In: Natural connections (Western D, Wright RM, Strum SC, eds). Washington DC: Island Press, 161–192.

Michener GR, 1996. Badger predation limits population growth of Richardson's ground squirrels. Paper presented at The American Society of Mammalogists Meetings, Grand Forks, North Dakota.

Miller B, Reading RP, Forrest S, 1996. Prairie nights: black-footed ferrets and the recovery of endangered species. Washington, DC: Smithsonian Institution Press.

Miquelle DG, Smirnov EN, Quigley HG, Hornocker MG, Nikolaev I, Matyushkin EN, in press. Food habits of amur tigers in Sikotae-Alin Zapovednik and the Russian Far East, and implications for conservation. J Wildl Res.

Mitani JC, Grether GF, Rodman PS, Priatna D, 1991. Associations among wild organ-utans: sociality, passive aggregations or chance? Anim Behav 42:33–46.

Moore DE, Smith, 1991. The red wolf as a model for carnivore re-introduction. Symp Zool Soc Lond 62:263–278.

Muir J, 1894. The mountains of California. New York: Century.

Murie A, 1940. Ecology of the coyote in Yellowstone. Fauna Ser 4:1–206.

Nelson ME, Mech LD, 1981. Deer social organization and wolf predation in northeastern Minnesota. Wildl Monogr 77:1–53.

Owen-Smith RN, 1983. Dispersal and the dynamics of large herbivores in enclosed areas: implications for management. In: Management of large mammals in African conservation areas (Owen-Smith RN, ed). Pretoria: Haum; 127–143.

Owings DH, Owings S, 1979. Snake-directed behavior by black-tailed prairie dogs. Z Tierpsychol 49:35–54.

Owings DH, Leger DW, 1980. Chatter vocalizations of California ground squirrels: predator-and social role specificity. Z Tierpsychol 54:163–184.

Packer C, Pusey AF, Rowley H, Gilbert DA, Martenson J, O'Brien SJ, 1991. Case study of a population bottleneck: lions of the Ngorongoro Crater. Conserv Biol 5:219–230.

Peterson RO, 1988. The pit or the pendulum: issues in large carnivore management in natural ecosystems. Ecosystem management for parks and wilderness (Agee JK, Johnson DR, eds). Seattle: University of Washington Press; 105–117.

Peterson RO, 1995. The wolves of Isle Royale, a broken balance. Willow Creek Press, Minocqua.

Phillips JA, Alberts AC, 1992. Naive ophiphagus lizards recognize and avoid venomous snakes using chemical cues. J Chem Ecol 18:1775–1783.

Phillips MK, Smith DW, 1996. The wolves of Yellowstone. Stillwater: Voyageur Press.

Pielou EC, 1991. After the ice age: the return of life to glaciated North America. Chicago: University of Chicago Press.

Quammen D, 1996. The song of the dodo; island biogeography in an age of extinctions. New York: Scribner.

Ralls K, White PJ, 1995. Predation on San Joaquin kit foxes by larger canids. J Mammal 76:723–729.

Reading RP, Clark TW, 1996. Carnivore re-introductions: an interdisciplinary examination. In: Carnivore behavior, ecology, and evolution, vol 2 (Gittleman JL, ed). Ithaca, New York: Cornell University Press; 296–336.

Reznick DA, Endler JA, 1982. The impact of predation on life-history evolution in Trinidian guppies. Evolution 36:160–177.

Reznick DA, Bryga H, Endler JA, 1990. Experimentally induced life-history evolution in a natural population. Nature 346:357–359.

Robinson JG, Redford KH, 1994. Community-based approaches to wildlife conservation in neotropical forests. In: Natural connections (Western D, Wright RM, Strum SC, eds). Washington, DC: Island Press, 300–322.

Rowe MP, Coss RG, Owings DH, 1986. Rattlesnake rattles and burrowing owl hisses; a case of acoustic Batesian mimicry. Ethology 72:53–71.

Rutberg AT, 1984. Birth synchrony in American bison (*Bison bison*): response to predation or season? J Mammal 65:418–423.

Rutberg AT, 1986. Dominance and its fitness consequences in American bison cows. Behaviour 96:62–91.

Rutberg AT, 1987. Adaptive hypotheses of birth synchrony in ruminants: an interspecific test. Am Nat 130:692–710.

Ryabov LS, Likhatskii YP, Nikitin NM, 1994. The relationship of wolves (*Canis lupus* L.) in Voronezh Reserve to wild and domestic ungulates. Russ J Ecol 25:45–50.

Ryan MJ, Tuttle MD, Rand AS, 1987. Bat predation and sexual advertisement in a neotropical anuran. Am Nat 119:136–139.

Sargeant AB, Allen SH, 1989. Observed interactions between coyotes and red foxes. J Mammal 70:631–633.

Schaller GB, 1996. Introduction: carnivores and conservation biology. In: Carnivore behavior, ecology, and evolution, vol 2 (Gittleman JL, ed). Ithaca, New York: Cornell University Press; 1–10.

Scheel D, 1993. Profitability, encounter rates and prey choice of African lions Behav Ecol 4:90–97.

Seidensticker J, Lumpkin S, 1991. Great cats: majestic creatures of the wild. Emmaus, Pennsylvania: Rodale Press.

Serfass TL, Rymon LM, 1985. Success of river otter introduced in Pine Creek drainage in northcentral Pennsylvania. Trans Northeast Sect Wildl Soc 41:138–149.

Seyfarth RM, Cheney DL, Marler P, 1980. Monkey responses to three different alarm calls: evidence for predator classification and semantic communication. Science 210:801–803.

Sih A, Kats LB, Moore RD, 1992. Effects of predatory sunfish on the density, drift, and refuge use of stream salamander larvae. Ecology 73:1418–1430.

Silk JB, 1983. Local resource competition and facultative adjustment of sex ratios in relation to competitive abilities. Am Nat 121:56–66.

Sinclair ARE, Arcese P, 1995. Serengeti II: dynamics, management, and conservation of an ecosystem. Chicago: University of Chicago Press.

Skinner MP, 1928. The elk situation. J Mammal 9:309–317.

Stacey PB, 1987. Group size and foraging efficiency in yellow baboons. Behav Ecol Sociobiol 19:175–187.

Stacey PB, 1995. Diversity of rangeland bird populations. In: Biodiversity of rangelands (West N, ed). Logan: Utah State University Press; 33–41.

Stephens PW, Peterson RO, 1984. Wolf-avoidance strategies of moose. Hol Ecol 7:239–244.

Strauss SY, 1991. Indirect effects in community ecology: their definition, study, and importance. Trends Ecol Evol 6:206–210.

Sullivan TP, Nordstrom LO, Sullivan DS, 1985a. Use of predator odors as repellants to reduce feeding damage by herbivores. I. Snowshoe hares. J Chem Ecol 11:903–919.

Sullivan TP, Nordstrom LO, Sullivan DS, 1985b. Use of predator odors as repellants to reduce feeding damage by herbivores. II. Black-tailed deer. J Chem Ecol 11:921–935.

Sunquist FC, Sunquist ME, 1989. Tiger moon. Chicago: University of Chicago Press.

Sweeney BW, Vannote RL, 1982. Population synchrony in mayflies: a predator satiation hypothesis. Evolution 36:810–821.

Sweitzer RA, Jenkins SH, Berger J, 1997. Near-extinction of porcupines by mountain lions and consequences of ecosystem change in the Great Basin Desert. Conserv Biol 11:1407–1417.

Terborgh J, 1988. The big things that run the world—a sequel to EO Wilson. Conserv Biol 2:402–403.

Terborgh J, 1990. The role of felid predators in neotropical forests. Vida Silvest Neotrop 2:3–5.

Van Aarde RJ, Skinner JD, 1986. Pattern of space use by relocated *Felis serval*. S Afr J Ecol 24:97–101.

Van Ballenberghe V, 1987. Effects of predation on moose numbers: a review of recent North American studies. Viltrevy Suppl 1:431–460.

Van Ballenberghe V, 1991. Forty years of wolf management in the Nelchina Basin, south-central Alaska; a critical review. Trans N Am Wildl Nat Res Conf 56:561–566.

van Schaik CP, von Noordwijk MA. 1985. Evolutionary effect of the absence of felids on the social organization of the Macaques on the island of Simeulue. Folia Primatol 44:138–147.

Varley JD, 1988. Managing Yellowstone National Park into the twenty-first century: the park as an aquarium. In: Ecosystem management for parks and wilderness (Agee JK, Johnson DR, eds). Seattle: University of Washington Press; 216–225.

Wagner FH, Kay CE, 1993. "Natural" or "healthy" ecosystems: are U.S. National parks providing them? In: Humans as components of ecosystems: the ecology of subtle human effects and populated areas (McDonnell MJ, Pickett STA, eds). New York: Springer-Verlag, 257–270.

Wallace AR, 1876. The geographic distribution of animals. New York: Harper.

Ward PI, 1986. A comparative field study of the breeding behavior of a stream and pond population of *Gammarus pulex*. Oikos 46:29–36.

Weber W, Raminowitz A, 1996. A global perspective on large carnivore conservation. Conserv Biol 10:1046–1054.

Wehausen JD, 1996. Effects of mountain lion predation on bighorn sheep in the Sierra Nevada and Granite Mountains of California. Wildl Soc Bull 24:471–479.

Wolff JO, 1988. Maternal investment and sex ratio adjustment in American bison calves. Behav Ecol Sociobiol 23:127–133.

Wright SJ, Gompper ME, De Leon B, 1994. Are large predators keystone species in Neotropical forests? The evidence from Barro Colorado Island. Oikos 71:279–294.

Wydeven AP, Schultz RN, Thiel RP. 1996. Monitoring of a recovering gray wolf population in Wisconsin, 1979–1991. In: Ecology and conservation of wolves in a changing world (Carbyn LN, Fritts SH, Seip DR, eds). Edmonton: Circumpolar Press; 141–146.

Part II

Baseline Behavioral Ecological Data and Conservation Intervention

Conservation management practices can be divided into protecting species in their natural habitat, breeding them in captivity, and reintroducing them back into the wild. Each of these different spheres of intervention uses behavioral and ecological information, often implicitly rather than explicitly. The three chapters in this section attempt to spell out the significance of such information for these different types of management practice.

In the first chapter, Durant introduces some relatively novel forms of management intervention that could be used to bolster small population sizes in the wild. She examines three mammal species with different social structures to highlight these issues: Mediterranean monk seals *Monachus monachus*, mountain gorillas *Gorilla gorilla beringei*, and roan antelope *Hippotragus equinus*. The monk seal is endangered in the wild (Osterhaus et al., 1997), as it is restricted to remote beaches and caves in the eastern Mediterranean and Mauritanian coast (Harwood et al., 1996). The mountain gorilla population is threatened and has existed at low numbers in Rwanda, Uganda, and Eastern Zaire for the last 30 years (Harcourt, 1996). Roan antelope are found at low densities across a large swathe of sub-Saharan Africa (East, 1988–1990).

Durant uses behavioral and ecological data at three stages of her analysis. First, she models the growth rates of the three species using demographic data collected in the course of field studies and explores which reproductive parameters have the greatest effect on population growth rates and N_e. She finds that juvenile and adult survival have the largest impact on population growth rates in all three species (although gorillas have the lowest capacity to increase numerically) and that N_e is particularly influenced by juvenile survival in each species. Second, she divides the population into 10 subpopulations and investigates how management of different numbers of subpopulations alters population persistence, thereby drawing attention to the role that population subdivision may have on extinction probability. This analysis shows that intervention is more effective the more subpopulations are managed, at least in gorillas and roan antelope. Third, Durant investigates how dispersal rates between subpopulations affect the success of interventions. Models for seals and gorillas show that when dispersal rates are high, management of subpopulations becomes increasingly effective in enabling the whole population to persist. Thus, although the efficacy of managing populations in this way remains untested at present, the models demonstrate the importance of reproductive parameters, social structure, and dispersal in fine-tuning management strategies for populations *in situ*.

In chapter 6, Wielebnowski shows how knowledge of species-specific behavior and ecology in the wild (quantitative natural history) is used by managers charged with breeding mammals in captivity. Zoos play several roles in conservation (Kleiman et al., 1996): education, fund raising, resources for basic research, (Gibbons et al., 1994), and sources for reintroductions (Sarrazin and Barbault, 1996), but each of these goals relies on maintaining mammals in good condition and on regular breeding. Wielebnowski reviews seven areas in which captive programs have benefited from understanding social and ecological conditions that these species experience in the wild: (1) enriching the captive environment with platforms and hiding places, (2) creating awareness of potentially stressful interspecific interactions, (3) providing naturalistic foraging opportunities, (4) determining the role of grouping and intraspecific interactions in reproductive activity, (5) determining the effects of familiarity on patterns of mating, (6) determining the factors that drive appropriate maternal behavior, and (7) being cognizant of the way in which the social environment during development influences subsequent mating behavior. Unfortunately, there are still few common principles that apply even within a given taxonomic group that enable us to predict the type of environmental or social setting most conducive to keeping and breeding an endangered species in captivity.

Returning to *in situ* conservation, but this time from an empirical perspective, Curio, in the first part of chapter 7, sketches how free-living, endangered bird species can benefit from protective measures that thwart ecological and behavioral processes such as predation and siblicide. Then, moving to behavioral issues in captive settings, he reiterates some of Wielebnowski's themes in mammals, pointing out the importance of reducing stress by paying attention to interspecific interactions, allowing birds to choose mates, and, in particular, exploring the effects of hand-rearing on subsequent reproductive activity and predator avoidance. In the main part of the chapter, Curio demonstrates how knowledge of animal behavior can affect reintroduction attempts by referring to the large ethological literature on predator avoidance (Curio, 1993). The problem here is that many captive-born, released individuals suffer from predation, so how can we condition released animals to recognize and respond to predators? Curio reviews some of the ways in which birds learn about predators—for example through parent alarm calls, by cultural transmission in which pupils learn from the aversive behavior of teachers, and by way of traumatic experience. Managers intent on releasing birds can tap into these mechanisms, for instance, by releasing knowledgable (common) heterospecifics with endangered released animals, or by exposing animals that will be released to stuffed predators that apparently "attack" them (e.g., McLean et al., 1996). Behavioral modification of this sort is still in its infancy, however, and currently demands labor-intensive effort to be successful.

The studies presented in these chapters demonstrate the extent to which knowledge of animal behavior affects intensive management practice in the wild and in captivity. Given the numerous examples of species' breeding performance improving after bringing the captive physical or social environment more into line with the natural environment, it is obviously important for behavioral ecologists to present their findings in a form and in a place where they are picked up by people raising animals in captivity (see, for example, Laurenson, 1993), so that these findings can be shared more widely.

References

Curio E, 1993. Proximate and developmental aspects of antipredator behavior. Adv Study Behav 22:135–238.

East R, 1988–1990. Antelopes: global surveys and regional action plans. Gland, Switzerland: International Union for the Conservation of Nature and Natural Resources.

Gibbons EF, Wyers EJ, Waters E, Menzel EW Jr, 1994. Naturalistic environments in captivity for animal behavior research. Albany: State University of New York Press.

Harcourt AH, 1996. Is the gorilla a threatened species? How should we judge? Biol Conserv 75:165–176.

Harwood J, Stanley H, Beudels M-O, Vanderlinden C, 1996. Metapopulation dynamics of the Mediterranean monk seal. In: Metapopulations and wildlife conservation (McCullough DR, ed). Covelo, California: Island Press; 214–257.

Kleiman DG, Allen ME, Thompson KV, Lumpkin S (eds), 1996. Wild mammals in captivity: principles and techniques. Chicago: Chicago University Press.

Laurenson MK, 1993. Early maternal behavior of wild cheetahs: implications for captive husbandry. Zoo Biol 12:31–43.

McLean IG, Lundie-Jenkins G, Jarman PJ, 1996. Teaching an endangered mammal to recognise predators. Biol Conserv 75:51–62.

Osterhaus A, Groen J, Niesters H, van de Bildt M, Martina B, Vedder L, Vos J, van Egmond H, Abou Sidi B, Ely Ould Barham M, 1997. Morbillivirus in monk seal mass mortality. Nature 388:838–839.

Sarrazin F, Barbault R, 1996. Reintroduction: challenges and lessons for basic biology. Trends Ecol Evol 11:474–478.

5

A Minimum Intervention Approach to Conservation: the Influence of Social Structure

Sarah Durant

Minimum Viable Populations and Population Viability Analyses

Over the last decade, research within conservation biology has centered on the concept of a minimum viable population (MVP), originally defined as a population that has a 95% probability of persisting for 100 years without intervention (Shaffer, 1983). The number of individual animals constituting an MVP has become a fundamental question within conservation biology (Soulé, 1987) because, with this knowledge, it might be possible to establish reserves of a suitable size to sustain a population of a particular species. Minimum viable population research largely targets endangered species (e.g., grizzly bears *Ursus arctos*: Shaffer, 1983; black-footed-ferrets *Mustela nigripes*: Groves and Clark, 1986; red cockaded woodpeckers *Picoides borealis*: Reed et al., 1988; spotted owls *Strix occidentalis*: Barrowclough and Coats, 1985; Leadbeaters possums *Gymnobelideus leadbeateri*: Lindenmayer et al., 1993). More recently this approach to conservation has expanded into population viability analysis, or PVA, which is used to assess the viability of populations across a whole array of population variables rather than just population size (Gilpin and Soulé, 1986). Both PVA and MVP theory hinge on being able to establish whether a reserve can support viable populations. Yet they neglect the increasingly urgent issue of how to maintain existing small populations or metapopulations in protected and sometimes unprotected areas (Caughley, 1994).

The Inadequacy of Current Reserves

For many of these species, current reserves are not large enough to maintain viable populations in the long term (Newmark, 1985; Woodruff, 1989). For example, Shaffer (1983) calculated that grizzly bears need a minimum area of 2500 to 4000 km^2 to support a demographically viable population of 50 bears; this is a large size for a protected area. The Serengeti National Park in Tanzania, one of the largest protected areas of open savannah in

East Africa, contains only 250 cheetahs *Acinonyx jubatus* at most (Caro and Durant, 1995). Moreover, even populations in the largest protected areas are unlikely to be genetically viable: recent estimates suggest that effective population sizes need to be of the order of 5000 if they are to maintain adaptive potential in the face of mutation and random genetic drift (Lande, 1995).

In addition, low-density vertebrate species often persist in smaller, scattered protected areas as well as single large ones. This collection of populations forms a metapopulation of isolated populations of widely disparate sizes. As further acquisition of land to enlarge existing small parks is often impossible because surrounding areas are taken over by agriculture or urbanization, managers will have to work within an existing framework of protected areas. For some species, these reserves are too small to ensure their persistence.

An alternative approach for maintaining populations that are below their MVP size is through efficient monitoring and management. This should take into account stochastic factors that can influence extinction and should be valid for species with different social structures. For widest applicability, the approach should not depend on precise measurements of demographic parameters. Instead, it should depend on management intervention at critical moments identified during routine demographic monitoring. This approach would be relevant for species that occur at low density and hence seldom attain viable population sizes.

Conservation by Intervention

Efficient population monitoring is an essential first step toward developing an effective management plan. Unfortunately, monitoring programs are commonly developed for large, high-density populations and are difficult to implement for small, low-density populations. However, complete data may not be essential; for example, monitoring the number or characteristics of subpopulations may be just as effective at predicting extinction as absolute numbers. For example, the number of different stations in America that recorded sightings of migrant bird species has been shown to be strongly positively correlated with population size (Bart and Klosiewski, 1989). These and other characteristics can be used as *indicators of extinction* to target management intervention at particular times of vulnerability for the populations. Indicators of extinction should be capable of predicting the extinction of a small population sufficiently far in advance to allow effective management intervention, and they should not be observed when a population is healthy. Such indicators should thus enable populations below the MVP level to be maintained by implementing a monitoring scheme that can detect indicators and initiate management intervention. A useful indicator of extinction should (1) be easily measurable, (2) give reasonable warning to allow time for management intervention to take effect, (3) have high predictive power to indicate with reasonable confidence that the population is about to go extinct, and (4) occur during the history of most populations (there is little point in monitoring an indicator that rarely occurs before extinction, no matter how accurately it predicts extinction).

The latter three traits are referred to as the warning, power, and chance of occurrence of the indicator, respectively (Durant, 1991; Durant and Mace, 1994). Indicators of extinction have been previously shown to fall into two categories. *Low-risk indicators* were defined as having a good warning of extinction and a high chance of occurrence before extinction, whereas *high-risk indicators* had a high predictive power of extinction (Durant, 1991; Durant et al., 1992; Durant and Mace, 1994). However, low-risk indicators generally had a low predictive power, whereas high-risk indicators might not occur before extinction or might give little warning of extinction, providing little opportunity for management to act. Total population size, population sex ratio, number of subpopulations, and annual production of offspring were all found to be useful for predicting extinction (Durant, 1991; Beudels et al., 1992; Durant and Harwood, 1992; Durant and Mace, 1994). By their very nature, indicators of extinction are not sensitive to the population ceiling and initial conditions. Therefore they form a basis for a pragmatic approach to conservation which is particularly suited to situations where little information is available. Although the precise nature of management solutions may be species specific (Caughley, 1994), identifying and making use of indicators of extinction may be a useful first step toward generalizing the characteristics of declining populations and finding management solutions.

In this chapter I use population models of three mammal species to examine the effectiveness of such a strategy at maintaining persistence of populations that are below viable levels. I start by describing the demographies of the three species and outline the stochastic models used to simulate their dynamics. These models are tailored to the different social structures of the species investigated. I then explore the potential effectiveness of different management strategies at increasing growth rate and effective population size. Next, I examine whether intervention at these indicators is an effective management option for each species and whether management of a core number of subpopulations is an effective means of increasing persistence of the whole population. Finally, I discuss whether a minimum intervention approach to management is appropriate for each species and suggest means of implementing such a strategy.

Demography and Social Structure of Three Mammals

All species in this study were assumed to have stationary growth rates. This was achieved by using published data to estimate demographic parameters and setting either juvenile or adult survival to give a growth rate equal to 1.0 using eq. 2. Stochastic simulations were conducted using the model POPGEN, a population simulation model (Durant, 1991; Durant et al., 1992). In this model I assumed that the number of individuals giving birth or dying each year was distributed binomially. The fate of each individual in the population was determined annually by sampling from a uniform distribution on the interval (0,1) using a pseudo-random number generator. Each population was assumed to number 100 individuals initially, separated across 10 subpopulations, each of which could grow to four times this initial level but could not exceed this limit. All subpopulations initially had a stable age structure (Durant, 1991). Movement between subpopulations was modeled separately for each species according to their different social systems as described below.

Mediterranean Monk Seal

The Mediterranean monk seal *Monachus monachus* is sparsely distributed throughout the Mediterranean and North African coast. Unfortunately, little is known about its demography and behavior (Kenyon, 1981; Kouroutos et al., 1985; Marchessaux, 1989). Demographic parameters for this species were therefore assumed to be similar to its closest extant relative, the Hawaiian monk seal *Monachus schauinslandi*, about which there is more information. Different authors variously estimate the birth rate of both species as between 0.38 and 0.7 (Rice, 1960; Wirtz, 1968; Johnson et al., 1982). Mediterranean monk seals have not been observed to produce pups in successive years (Kouroutos et al., 1985), and so an annual birth rate of 0.5 was used. Female Hawaiian monk seals produce their first pup at 5 years of age (Johnson et al., 1982), and this was set at the age of first reproduction. Annual survival rates have been estimated at 0.87 for adults and 0.8 for juveniles (Johnson and Johnson, 1984) and longevity at around 30 years (Kenyon, 1981). These parameters result in a declining population and so, to aid comparison between species, adult survival was set at 0.9 and juvenile survival to 0.844 to produce a stationary population with deterministic growth rate of 1.0. This is optimistic, as evidence suggests that the Mediterranean monk seal population is declining (Marchessaux, 1989).

The monk seal population in the Mediterranean is scattered around numerous islands and rocky coastlines (Marchessaux, 1989). Individual seals are capable of moving over large distances (Sergeant et al., 1978; Reijnders and Ries, 1989). It was assumed that each individual had an equal probability of migrating in any particular year, and whether or not an individual migrated was modeled in the same way as survival (see above). The number of emigrations from each subpopulation therefore had a binomial distribution, which approximated to the Poisson for low migration probabilities. The destination of a migrating individual was chosen randomly with equal probability from all subpopulations, including those with no resident individuals. An individual was only allowed to migrate more than once in any particular year if the first destination subpopulation was at its population ceiling, when it was assumed to have insufficient space for an immigrant. In this case a new destination was chosen.

Mountain Gorilla

Age of first reproduction of mountain gorillas *Gorilla gorilla beringei* was estimated at 9 years (Harcourt et al., 1980) and longevity at 35 years (Harcourt et al., 1980). Mean annual survival rate was between 0.94 and 0.98, and the annual birth rate was 0.148 per female (Harcourt et al., 1981). A median adult survival rate of 0.96 was chosen for this study, and an annual juvenile survival rate of 0.977 was used to give a population growth rate of 1.0.

A mountain gorilla population is organized into social groups of between 3 and 27 adult males and females and their offspring (Schaller, 1963; Weber and Vedder, 1983). Some adult males are solitary (Schaller, 1963; Caro, 1976; Weber and Vedder, 1983). Movement is common between groups; however, only subadults and adults migrate (Harcourt et al., 1981). The youngest individual observed to migrate was 6.5 years (Harcourt et al., 1976). Females only leave a breeding group to join bachelor adult males or to join another breeding group and are never solitary (Harcourt et al., 1976). Males may also migrate into breeding groups, but they can become solitary (Caro, 1976; Harcourt et al., 1981). Migration is therefore not random but dependent on age and sex and was simulated using the monk seal model by placing the following additional conditions upon dispersal: Individuals had to be 7 years old before they could migrate; a female could only migrate to join another breeding group (with a resident adult male) or a solitary adult male; an adult male would not migrate if he was the only adult male in his group; a group which has lost all adult males would either merge with another group or would be joined by a solitary adult male; and if the population was near its ceiling level and a migrating male could not find space in any group, then he would die.

Roan Antelope

Roan antelope *Hippotragus equinus* produce their first calf at 3 years of age (Beudels et al., 1992). Females then have the potential to produce a calf once every 10 months (Wilson and

Hirst, 1977), allowing a female to produce a maximum of 1.20 offspring per year. Calves are hidden for 4–6 weeks after birth (Wilson and Hirst, 1977), during which period they suffer heavy mortality. The annual birth rate, combined with the calf survival over the period of concealment, averaged .04 per year (Beudels et al., 1992). This figure was used as the reproductive rate. Juvenile survival beyond the period of concealment had an annual geometric mean of 0.765 over the first 3 years until age of first reproduction (Durant, 1991; Beudels et al., 1992). The average life span has been reported to be between 13 and 15 years (Joubert, 1974), and 14 years by Spinage (1986); in neither case was it explained how these figures were derived. Because my models use maximum life span for longevity, I chose a higher figure of 20 years. From the third year onward, annual survival was set at 0.923 using eq. 2, to set the annual growth rate at 1.

Populations of roan antelope are separated into breeding herds of between 10 and 30 individuals (Joubert, 1974; Wilson and Hirst, 1977). All females tend to remain within their natal herd for life, forming a cohesive unit (Joubert, 1974). Each herd is held by a single adult male (Joubert, 1974). Male offspring are driven out by the resident male when they are around 2.5 years old, before they reach sexual maturity (Joubert, 1974). These males then remain in bachelor herds until they reach 5–6 years of age, when they separate and become solitary (Joubert, 1974). It has been hypothesized that if the number of females within a herd rises above a certain level, then the herd can fragment and join a solitary adult male (Starfield and Beloch, 1986).

The following assumptions were made in the model of roan social organization: only one male within any herd could reproduce; a herd of females could only be dominated by a new bull upon the death of the old one; if the number of breeding females was above a certain size, termed the "split level," then, provided there was at least one solitary adult male and the number of herds was below maximum capacity, the adult females split into two equal-sized groups. One of these groups, along with the young animals in the herd, remained with the resident bull, while the other formed a new herd with a new male chosen at random. On the death of the resident bull, the herd was immediately taken over by a solitary adult male selected at random with equal probability, and, finally, if the number of adult females in any herd exceeded a ceiling level, then reproduction ceased. Note that this model gives an inaccurate reflection of the total number of males in the population. In addition, the migration rate is fixed by the rate of turnover of males and females within herds.

Predicted Effectiveness of Different Management Options

Management must act by increasing one or more of the demographic parameters in a population. In any population there are five such parameters: age of first reproduction, longevity, juvenile survival, adult survival, and reproductive rate. If a population has a stable age-structure, the following equation (Charlesworth, 1980) can be used to calculate the deterministic growth rate:

$$\sum_{i=\alpha}^{d} \frac{1}{2} \lambda^{-i} m_i \prod_{j=1}^{i} p_j = 1 \tag{1}$$

where p_i and m_i are the annual survival rate for an individual from age $i-1$ to i and the annual reproductive rate for an individual of age i, respectively; d is the longevity in years, and α is the age of first reproduction in years. When annual adult survival and reproductive rates do not vary with age, this equation simplifies to:

$$l^{\alpha}m\lambda^{d-\alpha+1} - 2\lambda^{d+1} + 2p\lambda^{d} - 1^{\alpha}mp^{d-\alpha+1} = 0 \tag{2}$$

where l is the annual juvenile survival rate, p is the annual adult survival rate, and m is the annual reproductive rate. Solving this equation for the growth rate, λ, enables us to predict which of these demographic parameters might have the greatest impact on the population growth.

Similarly, examination of the series of equations relating the effective population size, N_e, to demographic parameters can show which parameters might have the greatest impact on maintenance of genetic diversity. For the purpose of this exercise, I assumed that each population had an equal sex ratio, and each sex had equal mean and variance in family size. This is unlikely to be the case for many species; however, in the absence of alternative data on lifetime reproductive success, it is likely to provide a useful first approximation. Effective population size (Kimura and Crow, 1963) is then given by:

$$N_e = \frac{Nk - 1}{k + (\sigma_k^2 / k) - 1} \tag{3}$$

where k is the mean and σ_k^2 is the variance of lifetime family size for an individual, and N is the population size. This formula assumes that generations are discrete, which is not true for most mammals. However, the formula generally fits species with overlapping generations provided populations have stable age-structures and fairly constant population size (Hill, 1972, 1979).

Because the species covered in this study do not produce more than one offspring per year, then $P(i,k)$, the probability of dying in the ith year and producing k offspring, is:

$$l^{i}(1 - l) \qquad i \leq \alpha,\ k = 0$$

$$_{i-k-\alpha}C_k l^{\alpha}(1 - p)p^{i-1-\alpha}m^{k}(1 - m)^{i-k-\alpha} \qquad \alpha \leq i \leq d,\ k < i - \alpha$$

$$_{i-k-}{}^{\alpha}C_k l^{\alpha} p^{i-1-\alpha}m\,(1 - m)^{i-k-\alpha} \qquad i = d + 1,\ k < d - \alpha, \tag{4}$$

where $_jC_i$ *is the combination* $j!/[i!(j-i)!]$. The expected lifetime contribution of offspring is given by:

$$E\{k\} = \sum_{i=\alpha}^{d-\alpha} \sum_{k=0}^{i-\alpha} kP(i,k), \tag{5}$$

and the variance in lifetime contribution of offspring, σ_k^2, is given by:

$$\sigma_k^2 = E\{k^2\} - E\{k\}^2. \tag{6}$$

$E\{k^2\}$ can be calculated from:

$$E\{k^2\} = \sum_{i=\alpha}^{d-\alpha} \sum_{k=0}^{i-\alpha} k^2 P(i,k). \tag{7}$$

Equations 3, 5, 6, and 7 can then be used to estimate effective population size.

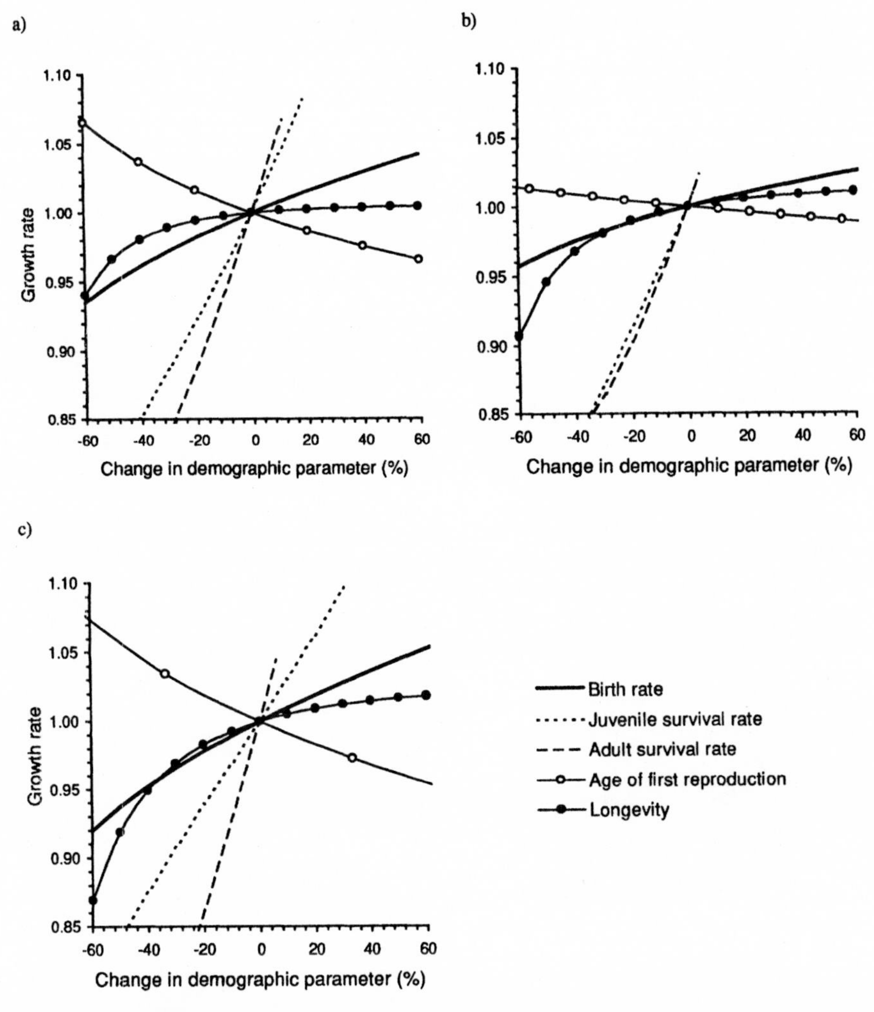

Figure 5-1 Sensitivity of the growth rate, calculated from equation 2, to changes in each of the demographic parameters for (a) Mediterranean monk seals, (b) mountain gorillas, and (c) roan antelopes.

Mediterranean Monk Seal

The deterministic growth rate, as calculated from eq. 2, was most sensitive to changes in the survival rates, particularly adult survival, and was much less sensitive to changes in longevity, birth rate, and age of first reproduction (fig. 5-1a). Nonetheless, because survival and birth rates cannot exceed 1, this places a maximum limit on the growth rate. The maximum growth rate achieved by increasing adult survival was only 1.07, slightly less than the value of 1.08 achieved by increasing juvenile survival, and not much greater than the value of 1.06 achieved by increasing the birth rate.

Results for effective population size are quoted in terms of the percentage of the census

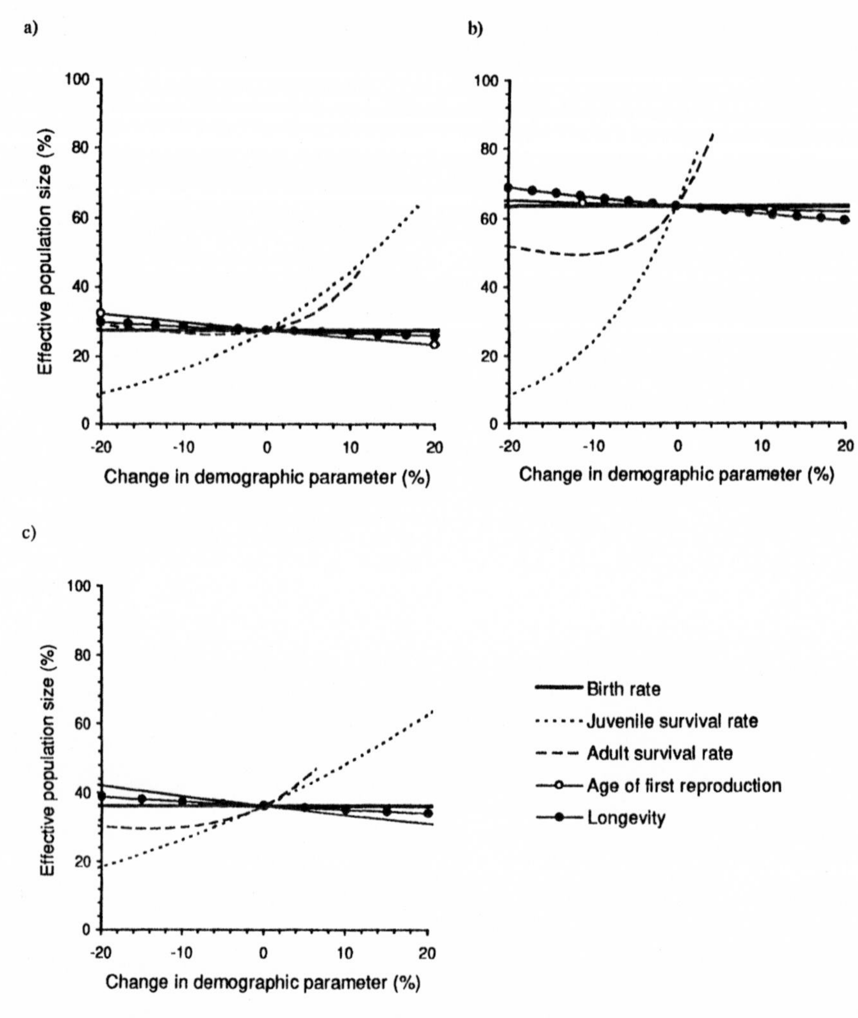

Figure 5-2 Sensitivity of the percent effective population size to changes in the demographic parameters for (a) Mediterranean monk seals, (b) mountain gorillas, and (c) roan antelopes. Effective population size was calculated using equations 3, 5, 6, and 7.

population; these values are equivalent to actual effective size of a population of 100 individuals. For Mediterranean monk seals, effective population size was only 28% of the census population size (fig. 5-2a). Percentage effective population size did not decrease substantially as adult survival decreased and was completely unaffected by changes in the birth rate. This was because, as birth rate increased, variance in lifetime family size increased at a slower rate than mean lifetime family size, resulting in an approximately constant effective population size. The capacity for increase in percentage effective size was largest when juvenile survival rates were increased. Increasing juvenile survival to 1.0 more than doubled the effective population size, whereas when adult survival was increased to 1.0, effective population size only attained a maximum of 44%.

These results illustrate the importance of the incorporation of demographic as well as genetic factors into conservation strategies. The measure of percentage effective size used here has widespread use in conservation biology; however, it should be remembered that total effective size could be declining in a population despite a concurrent increase in the percentage effective size. For example, a population of monk seals with an adult survival rate of 0.72 has a slightly higher percentage effective size as one with a rate of 0.9. However, figure 5-1a shows that such a population would decrease by 12% per year.

Mountain Gorilla

The deterministic growth rate of mountain gorilla populations was most sensitive to changes in the survival rates as in monk seals (fig. 4-1b). It was less sensitive to changes in the age of first reproduction and birth rate compared with monk seals, but was more sensitive to longevity. The overall capacity for population increase was low. The growth rate did not exceed 1.02 when survival rates were increased because they were already very close to their maximum. In addition, because female gorillas cannot produce more than one surviving offspring every 4 years (Harcourt et al., 1980), the maximum birth rate is 0.25, resulting in a maximum growth rate of 1.03.

Gorilla percentage effective population size was 64%, much higher than for monk seals. This was because survival rates of gorillas were high; therefore most individuals survived to reproduce and contribute to the gene pool. Effective population size responded to changes in the demographic parameters in a manner similar to the monk seal (fig. 5-2b). It responded most to changes in the survival rates, particularly juvenile survival, while it was little affected by changes in any of the other parameters. Effective population size showed a slight negative correlation with longevity since an increase in longevity increased the variance in family size and resulted in unequal contributions to the gene pool, decreasing effective population size. The maximum percentage effective size was 84%, and was attained by increasing adult survival to 1.0. The value was substantially higher than any obtained for the monk seal. Maximum juvenile survival gave a slightly lower percentage effective population size than that obtained through increasing adult survival but, at 79%, this value was still higher than for monk seals.

Roan Antelope

The shorter longevity and lower age of first reproduction in this species resulted in slightly higher sensitivity to changes in reproductive parameters compared with the longer-lived monk seals and gorilla (fig. 5-1c). The sensitivity to changes in the juvenile survival rate was lower, while the sensitivity to changes in adult survival was higher. Nevertheless, the maximum growth rates, attained when survival or birth rates increased to 1.0, were achieved by altering juvenile survival or the birth rate, not by altering adult survival. Increasing adult survival to 1.0 resulted in a growth rate of only 1.04, whereas increasing juvenile survival or birth rate to 1.0 resulted in growth rates of 1.10 and 1.11, respectively.

Percentage effective population size of roan antelope was 36%, a value intermediate between that of monk seals and gorillas (fig. 5-2c); sensitivity to changes in the demographic parameters was similar to that of the monk seal. As with both monk seals and gorillas, effective size showed no sensitivity to changes in the birth rate. The maximum achievable percentage effective population size was 81%, obtained by increasing juvenile survival to 1.0. Increasing adult survival to 1.0 only increased percentage effective size to 47%.

Conclusions

Changes in the survival rates were most effective at increasing the growth rate of populations of all three species. Similarly, percentage effective population size was more sensitive to changes in the survival rates than in any other demographic parameter. This suggests that increasing survival rates will increase both the growth rate and the effective population size. However, higher sensitivity of the growth rate to survival does not necessarily imply that this is the parameter best suited to manipulation through management. Management of these species will be constrained by the upper limit of 1.0 imposed on both survival and birth rates. If management can attain these upper limits, then this may make juvenile survival or birth rate the preferred parameter for manipulation, as these rates generally result in a higher maximum growth rate. Interestingly, the growth rate shows lower sensitivity to those parameters that are not amenable to management, longevity, and age of first reproduction.

Of all three species, gorillas appear have the lowest capacity for population increase. Management intervention therefore is likely to be less workable for this species. However, gorillas do have the highest effective population size, which implies that genetic management is less likely to be necessary than for the other two species.

Effects of Subpopulation Management on Population Persistence

In this and following sections, I investigate the consequences of management directed at increasing juvenile survival, since, in practice, this is often the demographic parameter most amenable to manipulation. Two indicators, which have previously been shown to be effective at predicting imminent extinction (Durant and Mace, 1994), were used to initiate management intervention. The low-risk indicator (juveniles <23% or >90% female, and population size <20) and the high-risk indicator (population <23% or >80% female, and population size <20) were chosen for this purpose because they were robust to changes in model parameters (Durant, 1991).

Populations were divided into 10 subpopulations. Each subpopulation initially contained 10 individuals and had an upper ceiling of 40 individuals, a size compatible with those observed for both free-living monk seal subpopulations and gorilla groups (Weber and Vedder, 1983; Marchessaux, 1989). Roan antelope, whose herds were modeled in terms of the number of females rather than total population size, initially numbered 5 females and each herd had an upper ceiling of 14 adult females; again, these values are compatible with those observed in the wild (Joubert, 1974; Wilson and Hirst, 1977). In addition, under a stable age distribution 14 adult female roan correspond to a herd size of 20 females (including female offspring), allowing this model to be directly compared with the monk seal and gorilla models.

Two, four, six, eight, and the entire ten subpopulations were selected for management in advance of simulation. Whenever an indicator was observed within one of the selected subpopulations, juvenile survival was raised to 1 within that subpopulation and was maintained at this level until 10 consecutive years had passed without exhibition of the indicator within the subpopulation. A management strategy such as this has advantages in that only a proportion of the subpopulations in a population need be monitored and managed, and management is directed at individual subpopulations when they need it. This strategy only makes sense if migration occurs between subpopulations since the managed subpopulations can then supplement unmanaged subpopulations. Results are reported from 1000 simulations over 750 years.

Mediterranean Monk Seals

If all 10 subpopulations of a population of monk seals were managed individually, levels of persistence were high (fig. 5-3). Under these circumstances persistence only dropped below 100% over the 750 years of simulation if the migration rate was low and management was high-risk. In addition more time was spent managing this population than under a low-risk management strategy (table 5-1). This was partially due to the fact that, under this model, time could be expended while trying to manage a subpopulation which was no longer extant. High migration rates made this less likely since recolonialism rates were higher.

When the migration rate was high, persistence of populations remained above 90% until less than half the subpopulations were managed (fig. 5-3b,d). Low-risk management was more effective than high risk when persistence was more than 60%, even if only two subpopulations were managed. High-risk management was less effective, but still markedly increased persistence above the value of 2% obtained without management. Persistences of only 5% were attained when the populations were managed under an identical strategy, but using indicators measured across the population rather than within subpopulations (Durant et al., 1992).

If the migration rate was low, management was less effective (fig. 5-3a,c). Here persistences of more than 90% were only achieved under low-risk management. Nonetheless, even here, management of two subpopulations resulted in a persistence of 42%, markedly higher than a persistence of zero without management. Under high-risk management, persistence was always below 25% unless all 10 subpopulations were managed. In this situation only if more than half the subpopulations were managed were persistences markedly higher than those obtained without management.

As the number of subpopulations under management decreased, the frequency of management interventions of each one tended to increase (table 5-1). This resulted in an overall increase in the time spent managing each subpopulation. For example, if 8 subpopulations were taken under low-risk management, then, for a low and a high migration rate, respectively, 19 and 9 years out of every 100 were spent in managing each one. This increased to 47 and 36 years when only two subpopulations were managed. The sole exception to this pattern was a population under high-risk management with a low migration rate (i.e., one where management was least effective). Here the time spent managing each subpopulation stayed roughly constant as fewer subpopulations were managed. Generally more years were expended in management if the migration rate was low than if it was high, and, for the most part, rather surprisingly, slightly more years were expended in high-risk management than in low-risk management.

Overall, therefore, the time expended in management of each subpopulation increased as the number of subpopulations under management declined. However, a more accurate reflection of the resources expanded in management is the number of years spent in managing each subpopulation multiplied by the number of subpopulations managed. The resulting figure, here termed the number of "management subpopulation years," is given in table 5-1. This figure did not necessarily decline as fewer subpopulations were managed but depended on the migration rate and the type of management. If the migration rate was low, the number of management subpopulation years declined as fewer subpopulations were

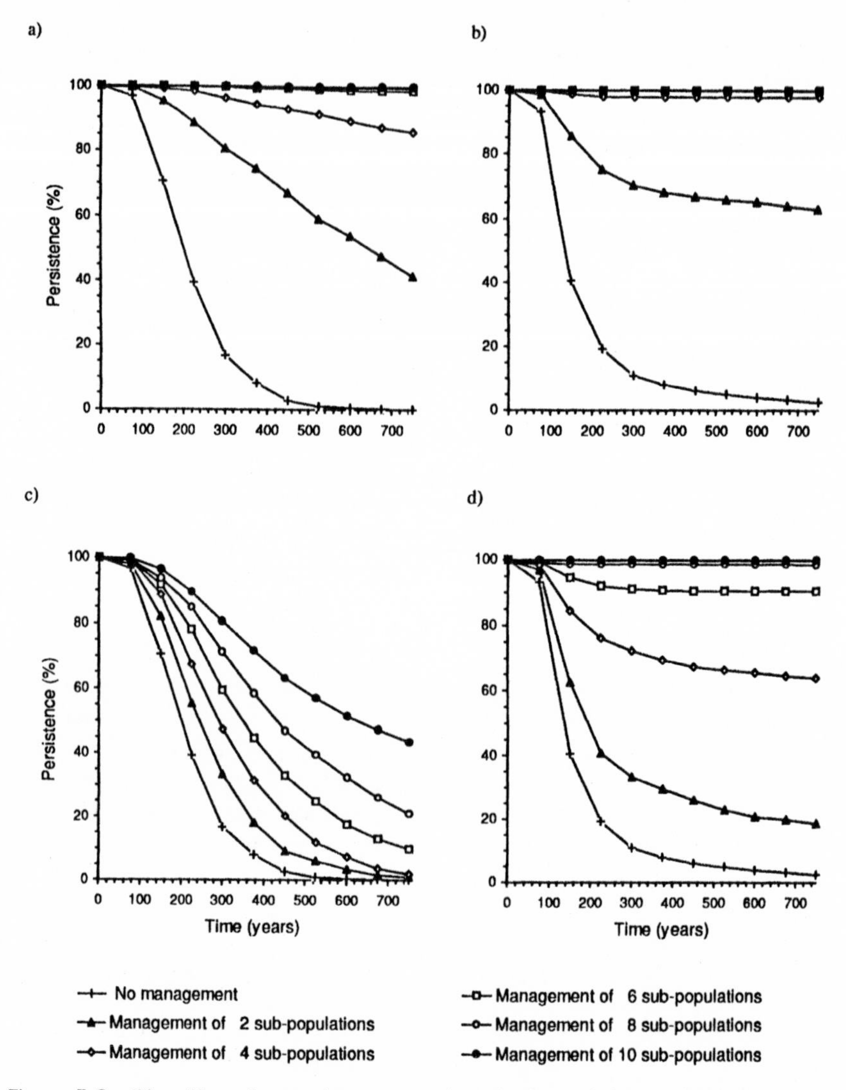

Figure 5-3 The effect of managing a proportion of subpopulations individually on percent persistence over time for populations of Mediterranean monk seals distributed across 10 subpopulations. Each subpopulation initially contained 10 individuals and had a ceiling of 40 individuals: the effect of low-risk management under migration probabilities of (a) 0.01 and (b) 0.1; the effect of high-risk management under migration probabilities of (c) 0.01 and (d) 0.1. Management, which increased juvenile survival to 1.0, was initiated on individual subpopulations whenever indicators were registered within one of the stated number of managed subpopulations and maintained until 10 years had passed without an indicator being exhibited.

Table 5-1 Frequency and length of management interventions at low- and high-risk indicators of extinction during 750 years of simulation for populations of monk seals subjected to demographic stochasticity under different migration rates.

	Low-risk management Migration probability										High-risk management Migration probability									
	0.01					0.1					0.01					0.1				
Number of managed subpopulations	2	4	6	8	10	2	4	6	8	10	2	4	6	8	10	2	4	6	8	10
Mean number of management interventions per 100 years per managed subpopulation	1.84	1.80	1.70	1.49	0.37	2.71	1.34	0.96	0.81	0.69	1.35	1.30	1.30	1.32	1.35	2.71	1.85	1.10	0.74	0.58
Mean number of years spent in managing each subpopulation out of every 100 years	46.7	38.4	24.7	19.2	17.0	36.3	15.4	10.6	9.0	7.6	53.3	54.7	56.5	54.3	47.9	39.1	24.8	13.8	8.5	6.5
Number of management subpopulation years (row 2 × number of subpopulations managed)	93	154	148	154	170	73	62	64	72	76	107	219	339	434	479	78	99	83	68	65
Mean duration of each management intervention	23.7	20.2	14.4	12.8	12.3	11.9	11.2	11.0	11.0	11.0	39.9	43.1	42.8	38.5	31.2	13.3	12.0	11.5	11.2	11.1

Management was initiated separately upon varying numbers of subpopulations whenever an indicator occurred within them. Populations initially numbered 100 individuals and were subdivided into 10 subpopulations, each with a ceiling of 40. The number of management interventions every 100 years was calculated as an average over the number of interventions during each simulation divided by the duration of a simulation, multiplied by 100. The mean number of years spent in managing a population out of every 100 years was calculated from an average of the total duration of management divided by the duration of simulation, multiplied by 100. The mean duration of each management intervention was calculated from the total duration of management divided by the total number of interventions across all simulations.

managed. However, if the migration rate was high, the number of management subpopulation years was lowest when four and six subpopulations were taken under low-risk management, whereas it peaked when four subpopulations were taken under high-risk management.

If a manager is to maintain persistence levels of 90% or more, then these results suggest that a monk seal population with a low migration rate had to be managed under a low-risk strategy. Such a population was most efficiently monitored if six subpopulations were managed. A population with a high migration rate could be maintained under either a low- or high-risk strategy; however, the most efficient strategy was low-risk management of four subpopulations.

Mountain Gorillas

Management of gorillas was not as effective at increasing persistence as was management of monk seals (fig. 5-4), as might be expected from the examination of the population growth rate above. As for monk seals, persistence was highest under low-risk management and responded most strongly to management when the migration rate was high. High-risk management was generally ineffective if the migration rate was low. Both types of management required longer management interventions than for monk seals (table 5-2).

When the migration rate was high, management of a proportion of groups in the population was effective (figs 5-4b,d). However, only low-risk management attained persistences of more than 90%, provided four or more groups were managed. This compares with a persistence of 65% without management. High-risk management was even less effective. When the migration rate was low, low-risk management had a limited effect but was unable to raise persistence above 90%, whereas high-risk management had virtually no effect.

The duration of management interventions was generally much longer than for monk seals and did not change as markedly as more groups were taken under management (table 5-2). Therefore, management subpopulation years increased with an increase in the number of groups under management. So, for both low and high migration rates, and both low- and high-risk management, more resources were expended in management as more groups were managed. Marginally less time was expended in high-risk than in low-risk management.

A manager could only achieve 90% persistence levels by managing all 10 gorilla groups when the migration rate was low, a huge management investment. Under this strategy a manager could expect to be actively engaged in management of each group 63% of the time. This compares to an expenditure of 25% of his or her time in management of six monk seal subpopulations to attain an identical persistence level. Circumstances are not quite so bad if the migration rate is high. Here a manager would be most efficient in achieving a 90% persistence if he or she managed four groups under a low-risk strategy, investing 40 years out of every 100 in managing each group.

Roan Antelope

If each of the 10 herds of roan antelope were managed under either low- or high-risk management, persistence was dramatically increased to 100% over 750 years (fig. 5-5). Even management of only two herds gave persistences of more than 90%. The duration of each management intervention did not change to any great extent as fewer herds were managed; however, the frequency of interventions increased (table 5-3). Overall, therefore, the num-

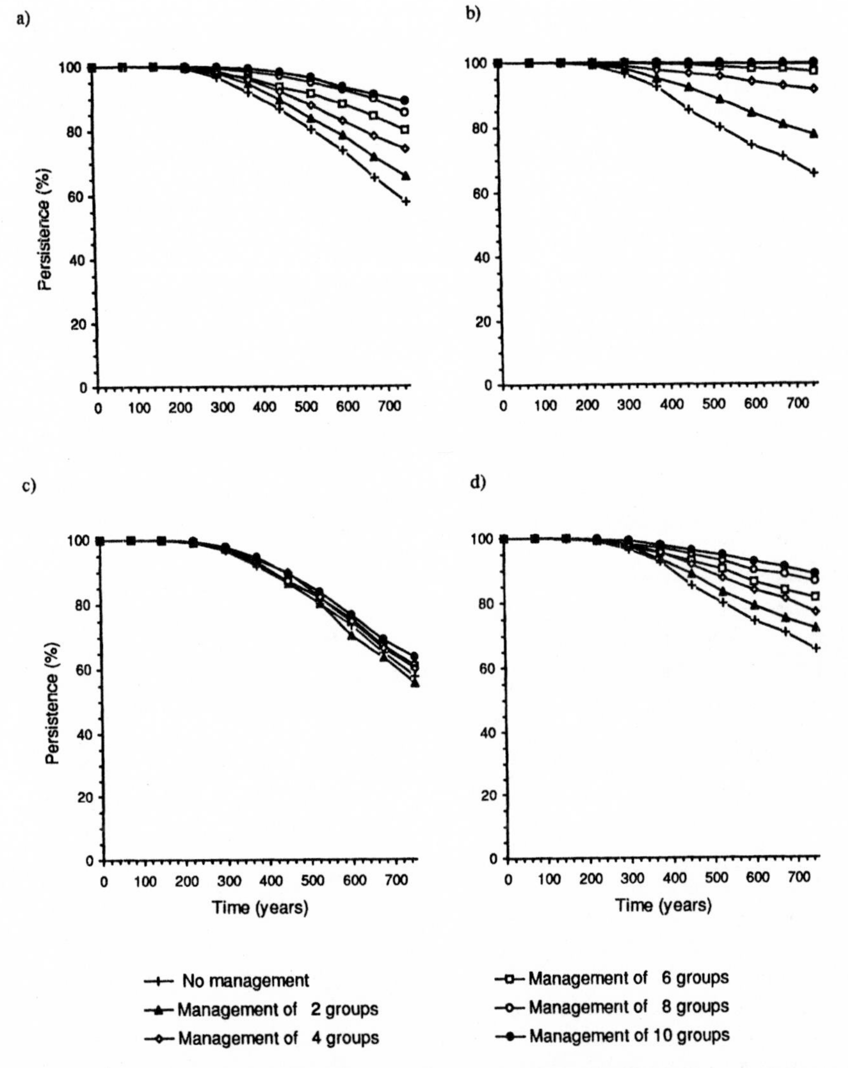

Figure 5-4 The effect of managing a proportion of groups individually on percent persistence over time for populations of mountain gorillas distributed across 10 groups. Each group initially contained 10 individuals and had a ceiling of 40 individuals: the effect of low-risk management under migration probabilities of (a) 0.01 and (b) 0.1; the effect of high-risk management under migration probabilities of (c) 0.01 and (d) 0.1. Management, as described in fig. 5-3, was initiated on individual groups whenever indicators were registered within one of the stated number of managed groups.

Table 5-2 Frequency and length of management interventions at low- and high-risk indicators of extinction during 750 years of simulation for populations of gorillas subjected to demographic stochasticity under different migration rates.

	Low-risk management Migration probability										High-risk management Migration probability									
	0.01					0.1					0.01					0.1				
Number of managed groups	2	4	6	8	10	2	4	6	8	10	2	4	6	8	10	2	4	6	8	10
Mean number of management interventions per 100 years per managed group	1.17	1.16	1.20	1.21	1.61	2.78	2.65	2.44	2.22	2.12	0.26	0.25	0.26	0.26	0.82	1.08	1.06	1.07	1.06	1.63
Mean number of years spent in managing each group out of every 100 years	63.7	65.3	64.0	64.1	62.7	46.9	40.3	34.3	29.1	25.9	60.7	60.6	61.1	60.9	61.2	60.8	29.1	27.1	25.1	22.8
Number of management subpopulation years (row 2 × number of groups managed)	127	261	384	513	627	94	161	206	233	259	121	242	367	487	612	62	116	163	201	228
Mean duration of each management intervention	53.4	55.2	52.7	52.2	38.6	16.3	14.8	14.0	13.0	11.9	234.3	238.9	239.4	233.5	74.5	26.6	25.7	23.9	22.7	13.7

Management was initiated separately on varying numbers of groups whenever an indicator occurred within them. Populations initially numbered 100 individuals and were subdivided into 10 groups, each with a ceiling of 40 individuals. See table 5-1 for calculation of reported statistics.

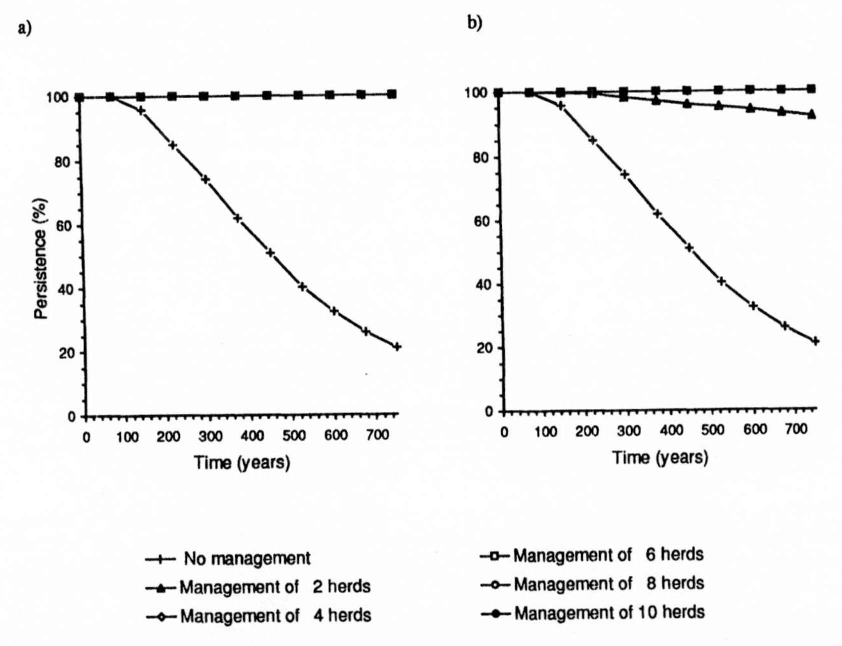

Figure 5-5 The effect of managing a proportion of herds individually on percent persistence over time for populations of roan antelope using (a) low-risk and (b) high-risk indicators. Populations were subjected to only demographic stochasticity and were made up of 10 herds, each initially containing 5 females and with a ceiling of 14 adult females. Herds could split when they contained at least 7 adult females. Management, as described in fig. 5-3, was initiated on individual herds whenever indicators were registered within one of the stated number of managed herds.

ber of years spent in managing each herd also increased. However, despite this increase, fewer management subpopulation years were expended as fewer herds were managed. Slightly less time was spent in high-risk management than in low-risk management.

Persistences of 90% or more were more easily achieved for roan than for any of the other species investigated here. The most efficient strategy was one using a high-risk strategy on two herds. Here a manager could expect to spend 27 years out of his or her time managing each herd. This is the lowest investment of time in management of all the species examined.

Conclusions

Interestingly, high-risk management sometimes required more management intervention than low-risk management, suggesting that a cautious approach to conservation can also be more economical in the long term. Monk seals and gorillas were both managed most easily if the migration rate was high. This result is perhaps surprising in the case of the monk seal where, without management, populations showed higher persistence under low migration rates (Durant and Harwood, 1992). The management strategy investigated here over-

Table 5-3 Frequency and length of management interventions at low- and high-risk indicators of extinction during 750 years of simulation for populations of roan antelope subjected to demographic stochasticity.

Number of managed herds	Low-risk management					High-risk management				
	2	4	6	8	10	2	4	6	8	10
Mean number of management interventions every 100 years per managed herd	2.68	2.31	2.12	1.99	1.91	1.51	1.37	1.26	1.19	1.14
Mean number of years spent in managing each herd out of every 100 years	30.0	25.8	23.7	22.2	21.4	27.1	21.6	19.1	18.0	17.2
Number of management subpopulation years (row 2 X number of herds managed)	60	103	142	178	214	54	86	115	144	172
Mean duration of each management intervention	11.2	11.2	11.2	11.2	11.2	17.5	15.6	15.2	15.1	15.1

Management was initiated separately on varying numbers of herds whenever an indicator occurred within them. Populations initially numbered 50 females, and were subdivided into 10 herds, each with a ceiling of 14 adult females. See table 5-1 for calculation of reported statistics.

turns this result because managing a proportion of subpopulations had more impact on the entire population if the migration rate was high.

Gorillas required the highest level of management intervention for a given increase in persistence, despite higher overall persistence without management, whereas roan needed the lowest level of intervention. However, whereas roan and gorillas showed an increasing relationship between the number of subpopulations managed and levels of intervention, monk seals did not. For monk seals the optimal strategy depended strongly on the migration rate, indicating that the random nature of migration in the monk seal model fundamentally changes the management conclusions.

Strengths and Weaknesses of the Models

Indicators of extinction are of little use within a management strategy if extinction cannot be prevented by management after their occurrence. Management intervention at a low-risk indicator gave populations a better chance of recovery than intervention at a high-risk indicator. Furthermore, although high-risk indicators occurred less frequently than low-risk indicators, the duration of each management intervention necessary for the recovery of the population after their occurrence was longer under a high-risk strategy. Therefore, the overall time expended in high-risk management was sometimes greater than the time expended in low-risk management. These results show that strategies where management was initiated at indicators which were most powerful at predicting extinction were not necessarily best at increasing persistence while minimizing the time spent on management.

The three different species showed different responses to management. Overall, gorillas were least responsive to all the management strategies investigated, and roan were the most responsive (Durant and Mace, 1994). In part this was a consequence of the low deterministic growth rate of gorillas and the high growth rate of roan during management. But the response to management also depended on the nature of migration and the migration rate. Both monk seals and gorillas were more responsive to individual subpopulation management if the migration rate was high because here a high migration rate increased the correlation of the growth rate between managed subpopulations and unmanaged subpopulations. This strategy was much more effective than one where the same indicators were used but were measured across the entire population rather than within individual subpopulations (Durant, 1991; Durant and Mace, 1994). However, it was similar in effectiveness to one using presence–absence indicators (registered when the population drops below a threshold number of subpopulations) to initiate management interventions (Durant and Mace, 1994).

The choice of a particular management strategy should be tailored to the practical limitations of its implementation. For a given level of persistence, intervention at presence–absence indicators required less management than when a few subpopulations were managed intensively (Durant and Mace, 1994). However, under presence–absence management, the entire population has to be scanned at frequent intervals in order to count extant subpopulations, and, if required, intervention has to be directed at all surviving breeding subpopulations. Some of these subpopulations may be inaccessible or unhabituated to humans, and it may therefore be more difficult to intervene in some subpopulations compared with others. In some circumstances it may be preferable to manage a few predetermined subpopulations, which are found in easily accessible locations and where animals can be habituated to humans during routine monitoring. An approach where resources are concentrated on only a small proportion of a population has been advocated for rhinoceroses *Diceros bicornis* and elephants *Loxondata africana* (Leader-Williams, 1985, 1990). For these species, management depends on effective antipoaching patrols, which are more effective if they can be concentrated on a small core area. This approach is particularly useful if animals can recognize protected areas, as is the case with elephants, when they may actively seek these areas (Leader-Williams and Albon, 1988).

This approach does have a number of drawbacks. First, it depends on some form of regular, routine monitoring. Such monitoring may take many forms but is, nonetheless, a prerequisite for the use of this strategy. Second, although the models do take account of group dynamics, they neglect many other features that are likely to be important for wild animals, such as interactions with other species and intraspecific aggression. It is inevitable that a modeling approach will neglect some aspects of a species' biology. Because of this, models should be used to guide between different options rather than to give definitive answers. In addition, for the analyses reported here, populations are assumed to be stable without management. If this is not the case and populations do in fact decline without management, then the interventions investigated here will be less effective. In these situations, reversal of long-term population decline should be the first management priority. Conversely, if populations tend to increase without management, persistence will be higher and management interventions less frequent than reported here. Finally, and perhaps most importantly, the approach is valid for populations where some degree of protection can be secured. It is obviously of little use for a population whose habitat is being destroyed or which is systematically persecuted. Here reducing human conflict is the obvious first priority.

Recommendations for Conservation

Because indicators of extinction are relatively insensitive to species-specific parameters, they can form the basis for a pragmatic approach to conservation which is particularly suited to situations where little information is available. Management at indicators of extinction can maintain populations below the MVP level, providing the appropriate strategy is chosen. A management strategy directed at increasing survival rates, rather than the birth rate, is best for increasing effective population size and hence for maintaining genetic diversity. It is not necessarily best to wait until a population is in immediate danger of extinction before intervention. In these circumstances, although management may be less frequently required, populations may not respond well to management, and more resources may be expended overall because the duration of management necessary to bring a population out of danger is much longer. In general, it is more efficient to manage a few individual subpopulations intensively using low-risk rather than high-risk indicators of extinction.

The indicators used here depend on low population size and skewed sex ratios. As populations become smaller, demographic stochasticity increases, and the sex ratio is more likely to become skewed. Years with no or low numbers of births can also be useful as an indicator of extinction (Durant, 1991) and can be easy to monitor. In addition, other factors not investigated here could be useful in this context. For example, frequencies of alleles at a locus may also become skewed as population size decreases and the effects of genetic drift predominate. Current advances in DNA technology enable single loci to be monitored using small samples of skin, hair, or feces. Further advances may make it possible to envisage a strategy in the future whereby a population, which is difficult to observe and monitor using conventional techniques, could be monitored demographically using genetic analyses.

Dispersal behavior and population structure are extremely important in determining persistence, and hence they determine optimal management strategies. For example, depending on the migration rate, it may be preferable to manage a population using presence–absence indicators (Durant and Mace, 1994) or to manage a few subpopulations intensively. Furthermore, random migration, such as observed in monk seals, resulted in counter-intuitive conclusions about the efficiency of management interventions because resources expended in management did not necessarily increase as more subpopulations were managed. In general, management strategies tailored to the social structure and demography of each species can assure the long-term survival of small populations and provide hope for populations of other species that are below the MVP level. Provided interventions are not delayed, management can also help slow the loss of genetic diversity (Durant, 1991).

Monk seals occur across a wide distribution in the Mediterranean Sea; therefore, it may not be feasible to manage the population as a whole by using presence–absence indicators, both because of logistical considerations and because subpopulations are located in different countries. In this situation, it may be preferable to monitor and manage a few subpopulations, which might be located within protected areas.

Mountain gorillas also span different countries: Rwanda, Uganda, and Zaire (Harcourt, 1986). The region they inhabit is mountainous, and the terrain is extremely difficult. Although censuses are conducted regularly at 4- or 5-year intervals, they are extremely arduous. The 1978 census took 68 days and 501 h in the field (Weber and Vedder, 1983). However, some groups are visited regularly by tourists and researchers, and no extra resources need be expended on monitoring these groups. My analyses showed that if all 10 groups within a metapopulation were monitored and managed individually at low-risk indicators, persistence was very high. If migration rates are high, then only four groups need to be managed. Within the larger Virunga population of about 300 gorillas (Aveling and Aveling, 1989), persistence would be even higher.

Monitoring using presence–absence indicators would be most feasible for roan antelope because this species occurs in open terrain and at a higher density than monk seals (Joubert, 1974). Animals are not easily recognized individually (Beudels et al., 1992), so it is more difficult to measure demographic indicators accurately. However, much of the surviving population occurs in national parks, where animals are often habituated to humans (Wilson and Hirst, 1977). For these protected populations, management of the entire population might be no more difficult than management of individual herds.

The means of implementing management will vary according to species. Captive rearing is one option, but it is only feasible when high survival rates can be obtained in captivity, reintroductions are possible, and capture involves minimal trauma to both the population and the individual. It is therefore not a feasible option for gorillas. While Fossey reintroduced a single juvenile gorilla into a wild population, the success of the reintroduction depended on the particular circumstances of the 'foster' gorilla group (Fossey, 1983). Furthermore, even if a breeding colony were to be established in captivity, the survival rates are not likely to be any better than in wild gorillas because the existing captive population of western lowland gorillas is just self-sustaining (Mace, 1988; Balmford et al., 1996).

Captive rearing could be successful for monk seals. A large captive-rearing plan recently initiated for female pups of the related species, the Hawaiian monk seal, has been effective (Gerrodette and Gilmartin, 1990). Survival rates of nearly 100% were achieved for pups captured off beaches and reared in captivity, a management intervention mimicked in my simulations. Furthermore, two Mediterranean monk seal pups were successfully reared and reintroduced into Greece (Reijnders and Ries, 1989). Captive rearing could also be effective for roan antelope, as this species does well in captivity (Joubert, 1974; Wilson and Hirst, 1977).

The management strategy of increasing juvenile survival to 1.0 which was investigated here is difficult to achieve in practice. However, if management can also increase adult survival, it may be as effective. Such management strategies might include the provision of extra food (implemented for rhinoceroses; Martin, 1984), inoculating and giving veterinary care to injured or diseased animals (implemented for rhinoceroses; Martin, 1984), preventing predation, and minimizing conflict with humans. All of these options will vary in their effectiveness across species. However, they are all likely to increase adult, as well as juvenile, survival, and, given the sensitivity of the growth rate to this parameter, may result in higher growth rates during management.

The mountain gorilla poses a special problem. In this species reproduction is slow—no more than one surviving offspring can be produced every 4 years (Harcourt et al., 1980)—and survival rates are high. It is therefore impossible to achieve high growth rates for this species. Furthermore, management has to be intensive to be effective, yet is likely to be extremely difficult to implement, and so it is hard to envisage a practical strategy for this species. Perhaps the best approach is that currently undertaken, where the population is constantly monitored for disease and injury, and veterinary interventions are made whenever necessary (Byers and Hastings 1991). Fortunately, of all the three species examined here, gorillas have the highest persistence rates without management.

Although this chapter hinges on theoretical results, it does show that information on social structure can be a priority for conservation research. If an effective, resource-efficient conservation strategy is to be devised, it is important to recognize individual subpopulations and breeding groups. Populations that no longer interbreed should be identified and managed separately. The nature of migration between groups needs to be determined. In particular, it is important to have qualitative information on whether individuals migrate to areas inhabited by others or if migration is random (see Dobson and Poole, chapter 8, this volume), since this affects the choice of particular management strategies. With such information it is possible to build minimum intervention management strategies that can maintain populations that are below their MVP level.

Summary

The demographies of Mediterranean monk seals, mountain gorillas, and roan antelope are described. The growth rate and percentage effective population size of these three species were most sensitive to changes in survival rates. However, because these rates could not exceed 1, maximal increase of the growth rate was generally achieved by altering either juvenile survival or the birth rate.

Stochastic simulation models were constructed to examine the effectiveness of different management strategies. A management strategy based on short-term management intervention at indicators of extinction was successful at increasing persistence of both monk seals and roan antelope populations, even when a small proportion of subpopulations within the populations were managed. This strategy was less effective at increasing persistence of gorilla populations. In general, management was most effective if the risks of imminent extinction at the initiation of management were lowest. This did not necessarily mean that more time was spent in management.

The social structure of the three species strongly influenced the effectiveness of management. The more stable dynamics and strong social cohesion resulted in higher overall persistence of mountain gorilla populations, but this species was least sensitive to management. Management was most effective if the migration rate was high.

By their very nature, indicators of extinction are not sensitive to the population ceiling and initial conditions. Therefore, they form a basis for a pragmatic approach to conservation which is particularly suited to situations where little information is available.

Acknowledgments This research was funded by the National Environmental Research Council, UK. I thank the Department of Zoology, University of Cambridge, for providing me with facilities for the duration of this research. I thank John Harwood and Rosaline Beudels for their support, encouragement, and discussions. The chapter was much improved by comments from Steve Albon, Georgina Mace, James Umbanhowar, and two anonymous referees.

References

Aveling C, Aveling R, 1989. Gorilla conservation in Zaire. Oryx 23:64–70.

Balmford A, Mace GM, Leader-Williams N, 1996. Designing the ark: setting priorities for captive breeding. Conserv Biol 10:719–727.

Barrowclough GF, Coats SL, 1985. The demography and population genetics of owls, with special reference to the conservation of the spotted owl (*Strix occidentalis*). In: Ecology and management of the spotted owl in the Pacific Northwest (Gutierrez RJ, Carey AB, eds). U.S. Forest Service General Technical Report PNW-185. City: U.S. Forest Service; Portland, OR 74–85.

Bart J, Klosiewski SP, 1989. Use of presence-absence to measure changes in avian density. J Wildl Manage 53:847–852.

Beudels R, Durant SM, Harwood J, 1992. Assessing the risks of extinction for local populations of roan antelope (*Hippotragus equinus*). Biol Conserv 61:107–116.

Byers AC, Hastings B, 1991. Mountain gorilla mortality and climatic factors in the Parc National des Volcans, Ruhengeri prefecture, Rwanda. Mount Res Dev 11:145–151.

Caro TM, 1976. Observations on the ranging behaviour and daily activity of lone silver back mountain gorillas (*Gorilla gorilla beringei*). Anim Behav 24:889–897.

Caro TM, Durant SM, 1995. The importance of behavioural ecology for conservation biology: examples from studies of Serengeti carnivores. In: Serengeti II: Research, management and conservation of an ecosystem (Sinclair ARE, Arcese P, eds). Chicago: University of Chicago Press; 451–472.

Caughley G, 1994. Directions in conservation biology. J Anim Ecol 63:215–244.

Charlesworth B, 1980. Evolution in age-structured populations. Cambridge: Cambridge University Press.

Durant SM, 1991. Individual variation and dynamics of small populations: implications for conservation and management (PhD dissertation). Cambridge: University of Cambridge.

Durant SM, Harwood J, 1992. Assessment of monitoring and management strategies for local populations of the Mediterranean monk seal *Monachus monachus*. Biol Conserv 61:81–92.

Durant SM, Harwood J, Beudels R, 1992. Monitoring and management strategies for endangered populations of marine mammals and ungulates. In: Wildlife 2001 (McCullough DR, Barrett RH, eds). New York: Elsevier Applied Science; 252–261.

Durant SM, Mace GM, 1994. Species differences and population structure in population viability analysis. In: Creative conservation: interactive management of wild and captive animals (Olney PJS, Mace GM, Feistner ATC, eds). London: Chapman and Hall; 67–91.

Fossey D, 1983. Gorillas in the mist. Boston: Houghton Mifflin.

Gerrodette T, Gilmartin WG, 1990. Demographic consequences of changed pupping and hauling sites of the Hawaiian monk seal. Conserv Biol 4:423–430.

Gilpin ME, Soulé ME, 1986. Minimum viable populations: processes of species extinction. In: Conservation biology: the science of scarcity and diversity (Soulé ME, ed). Sunderland, Massachusetts: Sinauer Associates; 19–34.

Groves DR, Clark TW, 1986. Determining minimum viable population size for the recovery of the black-footed ferret. Gt Basin Nat Mem 8:150–159.

Harcourt AH, 1986. Gorilla conservation: anatomy of a campaign. In: Primates, the road to self-sustaining populations (Benirschke K, ed). New York: Springer-Verlag; 31–46.

Harcourt AH, Fossey D, Sabater-Pi J, 1981. Demography of *Gorilla gorilla*. J Zool 195:215–233.

Harcourt AH, Fossey D, Stewart KJ, Watts DP, 1980. Reproduction of wild gorillas and some comparisons with chimpanzees. J Reprod Fertil 28 (suppl):59–70.

Harcourt AH, Stewart KJ, Fossey D, 1976. Male emigration and female transfer in wild mountain gorilla. Nature 263:226–227.

Hill WG, 1972. Effective size of populations with overlapping generations. Theor Popul Biol 3:278–289.

Hill WG, 1979. A note on effective population with overlapping generations. Genetics 92:317–322.

Johnson AM, Delong RL, Fiscus CH, Kenyon KW, 1982. Population status of the Hawaiian monk seal (*Monachus schauinslandi*), 1978. J Mammal 63:413–421.

Johnson BW, Johnson PA, 1984. Observations on the Hawaiian monk seal on Layson Island from 1977 through 1980. National Marine Fisheries Service-South West Fisheries Center, Honolulu. report 49.

Joubert SCT, 1974. The social organisation of the roan antelope (*Hippotragus equinus*) and its influence on the spatial distribution of herds in the Kruger National Park. In: The behaviour of ungulates and its relation to management (Geist V, Walther F, eds). Gland: International Union for the Conservation of Nature 661–675.

Kenyon KW, 1981. Monk seals *Monachus* (Fleming 1822). In: Handbook of marine mammals, vol 2. Seals (Ridgeway SH, Harrison RJ, eds). London: Academic Press; 195–220.

Kimura M, Crow JF, 1963. The measurement of effective population number. Evolution 17:279–288.

Kouroutos V, Papapanayotou D, Matsakis J, 1985. L'etat de la population du phoque moine. In: Programme 'Sporades du Nord'. Project report to Directorate General for the Environment, Consumer Protection and Nuclear Safety. Brussels: Commission of the European Communities; 45–73.

Lande R, 1995. Mutation and conservation. Conserv Biol 9:782–791.

Leader-Williams N, 1985. Black rhino in South Luangwa National Park: their distribution and future protection. Oryx 19:27–33.

Leader-Williams N, 1990. Black rhinos and African elephants: lessons for conservation funding. Oryx 24:23–29.

Leader-Williams N, Albon SD, 1988. Allocation of resources for conservation. Nature 336:533–535.

Lindenmayer DB, Lacy RC, Thomas UC, Clark TW, 1993. Predictions of the impacts of changes in population size and environmental variability on Leadbeaters possum, *Gymobelideus leadbeateri* McCoy (*Marsupialia Petauridae*) using population viability analysis—an application of the computer program Vortex. Wildl Res 20:67–86.

Mace G, 1988. The genetic and demographic status of the western lowland gorilla (*Gorilla g. gorilla*) in captivity. J Zool 216:629–654.

Marchessaux D, 1989. The biology, status and conservation of the monk seal *Monachus monachus*. Nature and Environment Series report 41.

Martin EB, 1984. They're killing off the rhino. Nat Geogr 165:298–299.

Newmark WD, 1985. Legal and biotic boundaries of western north American national parks: a problem of congruence. Biol Conserv 33:197–208.

Reed JM, Doerr PD, Walters JR, 1988. Minimum viable population size of the red-cockaded woodpecker. J Wildl Manage 52:385–391.

Reijnders PJH, Ries EH, 1989. Release and radio-tracking of two rehabilitated monk seals in the marine park 'Northern Sporades', Greece. Tescel, The Netherlands: Research Institute for Nature Management.

Rice DW, 1960. Population dynamics of the Hawaiian monk seal. J Mammal 41:376–385.

Schaller GB, 1963. The mountain gorilla, ecology and behaviour. Chicago: University of Chicago Press.

Sergeant D, Ronald K, Boulva J, Berkes F, 1978. The recent status of *Monachus monachus,* the Mediterranean monk seal. Biol Conserv 14:259–287.

Shaffer ML, 1983. Determining minimum viable population sizes for the grizzly bear. Intl Con Bear Res Manage 5:133–139.

Soulé ME, 1987. Where do we go from here? In: Viable populations for conservation (Soulé ME, ed). Cambridge: Cambridge University Press; 175–183.

Spinage CA, 1986. The natural history of antelopes. London: Croom Helm.

Starfield AM, Beloch AL, 1986. Building models for conservation and wildlife management. New York: Macmillan.

Weber AW, Vedder A, 1983. Population dynamics of the Virunga gorillas: 1959–1978. Biol Conserv 26:341–366.

Wilson DE, Hirst SM, 1977. Ecology and factors limiting roan and sable antelope populations in South Africa. Wildl Monogr 54:1–111.

Wirtz WDI, 1968. Reproduction, growth and development and juvenile mortality in the Hawaiian monk seal. J Mammal 49:229–238.

Woodruff DS, 1989. The problems of conserving genes and species. In: Conservation for the twenty-first century (Western D, Pearl MC, eds). Oxford: Oxford University Press; 76–88.

6

Contributions of Behavioral Studies to Captive Management and Breeding of Rare and Endangered Mammals

Nadja Wielebnowski

Captive Breeding of Rare and Endangered Mammals: Does It Work?

The importance of captive breeding as a viable tool for conservation of rare and endangered species has been emphasized by zoos, aquariums, and other captive facilities since the early 1970s. Zoos have repeatedly been portrayed as future "arks" for endangered species (Durrell, 1975; Martin, 1975; Brambell, 1977; Soulé et al., 1986; Durrell and Mallinson, 1987; Luoma, 1987; Tudge, 1991; DeBlieu, 1994; Gibbons et al., 1994; Rabb, 1994). In concordance with this perspective, the International Union for the Conservation of Nature (IUCN) has adopted a policy endorsing captive breeding as an important supportive intervention to avoid the loss of many species (IUCN, 1987). The policy calls for the establishment of captive-breeding programs and self-sustaining captive populations (i.e., populations that are viable in captivity without further genetic influx from the wild) before a species is reduced to critically low numbers in the wild. Captive-breeding programs are to be coordinated internationally according to genetic and demographic principles. The ultimate goal of captive breeding programs is the maintenance and/or reestablishment of viable populations in the wild (IUCN, 1987).

Although there is general agreement that as more species become extinct in the wild, more will persist only in captivity (Carpenter, 1983; Conway, 1986; MacKinnon and MacKinnon, 1991; May, 1991; Stuart, 1991; Hedrick, 1992; Ebenhard, 1995), the effectiveness and relative importance of captive-breeding programs for overall conservation efforts remains under debate (Warland, 1975; McKenna et al., 1987; Leader-Williams, 1990, 1993; MacKinnon, 1991; Balmford et al., 1995, 1996; Norton et al., 1995; Sunquist, 1995; Snyder et al., 1996). Today, there have been few success stories of species rescue and reintroduction. So far, 25 species have been preserved in captivity following extinction in the wild (Magin et al., 1994). Among these are eight mammal species, six of which have been reintroduced into areas of their former range: European bison *Bison bonasus;* Przewalski's horse *Equus przewalskii;* Pere David's deer *Elaphurus davidianus;* Arabian oryx *Oryx leucoryx;* red wolf *Canis lupus;* and black-footed ferret *Mustela nigripes* (Magin et al., 1994;

Table 6-1 Examples of threatened mammalian species in which wild populations have been restocked through release of captive-bred individuals (based on Wilson and Stanley Price, 1994; Beck et al., 1994).

Species	Red Data Book status[a]	Country
Marsupialia		
Brush-tailed bettong *Bettongia penicillata*	E	Australia
Rufous hare-wallaby *Lagorchestes hirsutus*	R	Australia
Numbat *Myrmecobius fasciatus*	E	Australia
Greater bilby *Macrotis lagotis*	E	Australia
Primates		
Golden lion tamarin *Leontopithecus rosalia*	E	Brazil
Rodentia		
Greater sticknest rat *Leporillus conditor*	R	Australia
Jamaican hutia *Geocapromys brownii*	I	Jamaica
Carnivora		
Eurasian otter *Lutra lutra*	V	United Kingdom
Red wolf *Canis rufus*	E	United States
Perissodactyla		
Indian rhinoceros *Rhinoceros unicornis*	E	India/Nepal
Black rhinoceros *Diceros bicornis*	E	African sanctuaries
White rhinoceros *Ceratotherium simum*	E	South Africa
Artiodactyla		
Barbary deer *Cervus elaphus barbarus*	V	Algeria
Formosan sika deer *Cervus nippon taiouanus*	E	Taiwan
Huemul *Hippocamelus bisculeus*	E	Chile
Addax *Addax nasomaculatus*	E	Niger, Tunisia
Cuvier's gazelle *Gazella cuvieri*	E	Algeria
Mountain gazelle *Gazella gazella*	E	Saudi Arabia
Dama gazelle *Gazella doma*	E	Senegal
Scimitar-horned oryx *Oryx dammah*	E	Tunisia
Abbruzzo chamois Rupicapra pyrenaica ornata	V	Italy

[a]IUCN (1993) categories for threatened species: E, endangered; V, vulnerable; R, rare; I, indeterminate.

Wilson and Stanley Price, 1994). In addition, wild populations of a number of threatened mammal species have been reinforced by captive breeding (Kleiman, 1989; Wilson and Stanley Price, 1994) (table 6-1). Despite considerable financial costs and animal losses, some of these programs are conservatively regarded as successful (Kleiman, 1989; Stanley Price, 1989; Beck et al., 1994; Miller et al., 1994).

Magin et al. (1994) recently estimated that specimens of approximately 34% (174/507) of all threatened mammal species are held in zoos worldwide, and about 15% of the zoos' 200,000 "mammal spaces" are devoted to threatened taxa. However, only 90 of the 174 zoo-held threatened species are kept at levels that could allow establishment of self-sustaining captive populations that would be viable in the absence of future genetic interchange (Magin et al., 1994). Similarly, a perusal of recent reports on North American Species Survival Plans (SSP) of mammals (AZA, 1994) showed that only 13 of the 56 species and subspecies managed under North American SSP programs had self-sustaining or stable captive populations (table 6-2). Two commonly encountered difficulties in attaining viable captive populations are unreliable or no reproductive success and high rates of infant mortality (table 6-2; see also Snyder et al., 1996). The question then is, how can we solve existing breeding problems and achieve sustainability of captive populations to turn captive breeding into a useful and reliable conservation tool where needed?

In this chapter, I first discuss the role of captive breeding for conservation efforts in general, then outline the importance of behavioral research for solving captive breeding problems. I present a list of areas of behavioral research in which examples can be found for a crucial effect of behavioral management on captive breeding success. Finally, strengths and weaknesses of the behavioral approach are identified and specific recommendations provided.

Historical Synopsis of the Role of Captive Breeding in Conservation

Over the past 10 years, the captive-breeding community has evolved into a powerful and influential component of the world's conservation scene (Olney et al., 1994). Many of today's zoos and captive breeding facilities changed from simple menageries, through a period of living museums with more naturalistic exhibits, to today's modern conservation centers with a vested interest in preserving *in situ* (i.e., in the wild) biodiversity (Luoma, 1987; Rabb, 1994; Hutchins and Conway, 1995; Sunquist, 1995; Hoage and Deiss, 1996). The recent publication of a World Zoo Conservation Strategy by the World Zoo Organization (IUZDG) and the Conservation Breeding Specialist Group (CBSG) of the Species Survival Commission of the IUCN emphasizes broad goals for zoo conservation and the integration of *in situ* and *ex situ* (i.e., in captivity) conservation efforts (IUDZG/CBSG, 1993).

Zoo contributions to conservation efforts may be grouped into four major areas. First, zoos can educate and influence a largely urban public to appreciate the value and the importance of species and ecosystem conservation (Rabb, 1994; Sunquist, 1995; Whitehead, 1995). This may represent their most powerful contribution to conservation efforts. An estimated 100 million people visit North American zoos every year, and approximately 600 million people, about 10% of the world's population, are thought to visit zoos worldwide per year (Sunquist, 1995; Wheater, 1995).

Second, zoos can contribute substantially to *in situ* conservation efforts through fundraising and sponsoring field projects (Hutchins and Conway, 1995). The use of endangered mammals as flagship species for *in situ* and *ex situ* protection efforts may be most effective to secure monetary, political, and public support for conservation efforts (Dietz et al., 1994; Rabb, 1994, Hutchins et al., 1995). Some organizations such as the Frankfurt Zoological Society, the New York Wildlife Conservation Society, the National Zoological Park of the Smithsonian Institution, and the San Diego Zoological Society, for example, have been involved in supporting field research and *in situ* conservation efforts for many years (Rabb, 1994). Now smaller zoos are also discovering the benefits of such projects (Sunquist, 1995).

Third, basic research carried out in captive facilities can contribute substantially to our

Table 6-2 Juvenile mortality, sustainability, and behavioral problems reported for mammalian species managed under the North American Species Survival Plan Program, compiled from the AZA annual report on conservation and science (Bowdoin et al., 1994).

Species	North American captive population 1993–94	Juvenile mortality, 1992–94 (%)	Population self-sustaining in 1994?	Behavioral problems past and present[a]	Additional references[b]
Marsupials					
Tree kangaroos *Dendrolagus dirianus*	1	na	No	—	
Dendrolagus inustus	7	na	No	—	
Dendrolagus goodfellowi	15	na	No	—	
Matschie's tree kangaroo *Dendrolagus matschiei*	77	na	No	S, P, BP	1
Bats					
Rodrigues fruit bat *Pteropus rodricensis*	128	48	No	S, O	
Primates					
Black lemur *Lemur macaco macaco*	131	37	No	S	
Lemur macaco flavifrons	23	42	No	—	
Ruffed lemur *Varecia variegata variegata*	256	38	Undetermined	S	
Varecia variegata rubra	214	3	Undetermined	—	
Cotton-top tamarin *Saguinus oedipus*	236	na	Undetermined	S, P	2, 3
Golden lion tamarin *Leontopithecus rosalia*	488	na	Yes	S, P, TP, BP	4, 5
Goeldi's monkey *Callimico goeldii*	202	na	Undetermined	—	
Lion-tailed macaque *Macaca silenus*	258	35	Yes	P, GA	6
Drill *Mandrillus leucophaeus*	23	25	No	S, U, BP	7
Gibbons *Hylobates gabriellae*	10	na	No	—	
Hylobates leucogenys	39	na	No	—	
Hylobates lar	175	na	No	—	
Hylobates pileatus	16	na	No	—	
Hylobates syndactylus	136	na	No	—	

(*continued*)

Table 6-2 Continued

Species	North American captive population 1993–94	Juvenile mortality, 1992–94 (%)	Population self-sustaining in 1994?	Behavioral problems past and present[a]	Additional references[b]
Orangutan *Pongo pygmaeus*	299	na	No	—	
Bonobo *Pan paniscus*	103	0	No	S	
Chimpanzee *Pan troglodytes*	239	45	Yes	S, P, U	
Lowland gorilla *Gorilla gorilla gorilla*	710 (world pop.)	19	No	S, P, MC, U	8
Carnivores					
Mexican gray wolf *Canis lupus baileyi*	92	29	No	—	
Red wolf *Canis rufus*	216	24	No	TP	
Maned wolf *Chrysocyon brachyurus*	81	51	No	P	
African wild dog *Lycaon pictus*	50	59	No	S, P	
Spectacled bear *Tremarctos ornatus*	80	30	No	—	
Giant panda *Ailuropoda melanoleuca*	113	na	No	MC, BP	9
Red panda *Ailurus fulgens*	149	39	Yes	S, P	4, 10
Asian small-clawed otter *Aonyx cinera*	73	6	No	P, BP	
Black-footed ferret *Mustela nigripes*	362	na	Yes	TP	11
Lion *Panthera leo nubica*	11	na	No	—	
Panthera leo krugeri	35	11	No	—	
Jaguar *Panthera onca*	87	na	No	—	
Tiger *Panthera tigris altaica*	163	26	Yes	P, MC	12
Panthera tigris sumatrae	63	20	No	P, MC	12
Snow leopard *Panthera uncia*	580 (world pop.)	26	Yes	P, MC	13

Clouded leopard *Neofelis nebulosa*	300	29	No	MC, BP	14
Cheetah *Acinonyx jubatus*	302	21	Yes	MC, BP	9, 15, 16
Ungulates					
African elephant *Loxodonia africana*	196	na	No	GA	17
Asian elephant *Elephas maximus*	376	9	No	—	
Przewalski's horse *Equus prszewalskii*	188	21	Yes	S, MC, GA	18
Grevy's zebra *Equus grevvi*	at least 231	14	Yes	—	
Hartmann's mountain zebra E*quus zebra hartmannae*	139	na	No	S, GA	
Greater one-horned Asian rhinoceros *Rhinoceros unicornis*	124	39	No	BP	
Sumantran rhinoceros *Dicerorhinus sumatrensis*	23	na	No	MC, BP	
White rhinoceros *Ceratotherium simum*	124	na	No	S, GA	18, 19
Black rhinoceros *Diceros bicornis michaeli*	69	13	No	MC, BP, GA	20
Diceros bicornis minor	32	67	No	—	
Chacoan peccary *Catagonus wagneri*	na	na		—	
Barasingha *Cervus duvauceli duvauceli*	1	na	No	—	
Okapi *Tragelaphus euryceros*	43	36	No	P	21
Gaur *Bos gaurus*	121	na	No	—	
Arabian oryx *Oryx leucoryx*	na	na		—	
Scimitar-horned oryx *Oryx dammah*	500	5	Yes	—	
Addax *Addax nasomaculatus*	739	16	Yes	—	

[a]Categories of reported behavioral problems: S, social; P, parental; MC, mate compatibility; O, overcrowding; TP, prerelease training problems; U, unspecified behavioral deficiencies; BP, poor breeding performance; GA, general aggression.

[b]References in AZA (1994) unless indicated otherwise. Additional references: 1, Hutchins et al. (1991); 2, Evans (1983); 3, Snowdon (1989); 4, Kleiman (1980); 5, Kleiman (1989); 6, Lindburg and Gledhill (1992); 7, Gadsby et al. (1994); 8, Beck and Power (1988); 9, Lindburg and Fitch-Snyder (1994); 10, Kleiman (1992); 11, Miller et al. (1994); 12, Van Bemmel (1968); 13, Marma and Yunchis (1968); 14, Yamada and Durrant (1989); 15, Caro (1993); 16, Wielebnowski (1996); 17, Schmidt (1982); 18, Dolan (1977); 19, T. Wagener (personal communication); 20, Smith and Read (1992); 21, Lindsey et al. (1994).

Table 6-3 Examples of associations, groups, committees, and workshops that are part of the worldwide zoo conservation effort and their acronyms (Hutchins and Conway, 1995; Ellis and Seal, 1995; Mallinson, 1993).

Worldwide and regional zoo and aquaria organizations	
IUDZG	International Union of Directors of Zoological Gardens
AZA	American Zoo and Aquarium Association
EAZA	European Association of Zoos and Aquaria
JMSG	Joint Management of Species Group, British Isles and Ireland
AZDANZ	Association of Zoo Directors of Australia and New Zealand
JAZGA	Japanese Association of Zoological Gardens and Aquaria
Examples of important regional committees	
SSPs	Species Survival Plan committees
EEPs	European Endangered Species Programmes
JMSPs	Joint Species Management Programmes
ASMPs	Australian Species Management Programmes
SSCJs	Species Survival Committees' in Japan
TAGs	Taxon Advisory Groups
SAGs	Scientific Advisory Groups
Worldwide coordination of captive conservation efforts	
CBSG	Conservation Breeding Specialist Group
Programmes central to the CBSG's global management function	
CAMPs	Conservation Assessment and Management Plan workshops
PHVAs	Population and Habitat Viability Assessment workshops
GCAPs	Global Captive Action Plans
GASPs	Global Animal Survival Plans

understanding and knowledge of biological processes (Rabb, 1994; Hardy and Krackow, 1995; Hutchins and Conway, 1995; Ryder and Feistner, 1995). Rigorous research on captive populations of endangered species may also provide us with improved knowledge for management of wild populations and help us to develop new technologies useful for conservation efforts (Hutchins and Conway, 1995).

The fourth, and over the past decade, most advertised, contribution is the envisioned role of zoos as "arks" for species survival by providing self-sustaining captive populations of endangered species and subspecies for future reintroduction into the wild. The frequently stated goal for captive populations is to preserve 90–95% of heterozygosity for 100–200 years (Rabb, 1994). The rationale for long-term captive propagation is based on the assumption that human population growth may level off in about 200 years, and restoration of habitat and species may then become possible (Tudge, 1991; Rabb, 1994). Critics, however, have pointed out that zoos do not have sufficient space to accommodate viable captive populations for all threatened species (Conway, 1986; Soulé et al., 1986), and the costs for captive-breeding programs can be staggering, up to a half-million dollars per year per species (Derrickson and Snyder, 1992; Leader-Williams, 1993; Balmford et al., 1995; Snyder et al., 1996). More recently, therefore, the so-called Noah's ark paradigm of captive breeding has been amended to provide a more realistic conservation goal which attempts to consider spatial and financial limitations in captive facilities (Hutchins, 1995; Hutchins et al., 1995). In this new endeavor, captive-breeding programs are envisioned as participating in an interactive management strategy largely focused on *in situ* conservation efforts with coordination on a global scale (IUDZG/CBSG, 1993; Olney et al., 1994; Conway, 1995; Hutchins and Conway, 1995; Hutchins et al., 1995; Sunquist, 1995).

Today's zoo conservation efforts have become a systematic and cooperative venture based on scientific principles largely derived from the areas of demography and population genetics (Mallinson, 1993, 1995). A complex bureaucracy with a dizzying array of acronyms, representing various associations, committees, groups, and conservation tools, has developed as part of this worldwide effort (Tudge, 1991; Wemmer and Derrikson, 1995) (table 6-3). Systematic census and inventory programs, such as the Animal Records Keeping System and the International Species Information System, are used to compile information on captive populations of all species from the majority of the world's large zoos. To coordinate these efforts and to integrate collected information the Conservation Breeding Specialist Group (CBSG, formerly Captive Breeding Specialist Group), a multidisciplinary specialist group, was organized by the IUCN Species Survival Commission. The CBSG advises, monitors, and catalyzes captive propagation on a global basis (Seal, 1991) and provides a link between *in situ* and *ex situ* conservation efforts worldwide (Ellis-Joseph and Seal, 1992).

However, genetic and demographic issues have so far predominated captive breeding management (Mellen, 1994; Ralls, 1995) and have provided the foundation for captive-breeding plans. Now the need for a broader approach to captive management has been recognized (e.g., Hutchins et al., 1995; Mallinson, 1995) and research efforts incorporating behavior, nutrition, disease, physiology, genetics, population biology, and various interdisciplinary studies are seen as necessary to facilitate responsible and successful captive propagation and conservation.

An examination of North American SSP reports and other publications relevant to the zoo community (e.g., *Zoo Biology, International Zoo Yearbook*) revealed that behavioral problems are frequently cited as major impediments to successful captive propagation for many species (table 6-2). Also, much evidence suggests that behavioral research has been instrumental in past breeding and husbandry improvements (Eisenberg and Kleiman, 1977; Kleiman, 1994; Lindburg and Fitch-Snyder, 1994). However, little information is available on the social and behavioral requirements and general biology of many endangered species when breeding programs are initiated (Eisenberg and Kleiman, 1977; Conway, 1985; Ralls, 1995). Solutions to captive breeding problems are often found by trial and error and by exchange of anecdotal and descriptive information among facilities rather than by systematic investigation. Behavioral studies that address the conservation needs of captive-breeding programs using systematic hypothesis testing are rare. As a consequence, animal behavioral research is seldom incorporated in *ex situ* conservation planning (Eisenberg and Kleiman, 1977; Kleiman, 1994; Ralls, 1995), and no general theory exists on which aspects of behavior are likely to be the most critical for conservation efforts (Ralls, 1995; but see Curio, 1996). Integrating systematic behavioral and husbandry research into conservation planning has been highlighted as crucial for *ex situ* conservation (Kleiman, 1994; Lindburg and Fitch-Snyder, 1994; Mellen, 1994).

The Role of Behavioral Studies in Zoo Conservation

Much early and influential research in animal behavior and comparative ethology has been carried out on captive-held animals and in zoo settings (e.g., see Eisenberg and Kleiman, 1977, for review; Koehler, 1931; Heinroth, 1938; Leyhausen, 1956; Meyer-Holzapfel, 1956; Tembrock, 1963; Fagen, 1981). Indeed, the development of ethology as a scientific discipline and the beginnings of zoo behavioral research complemented each other and occurred concurrently in the mid-1940s. The work of Hediger and Meyer-Holzapfel in the early 1950s pioneered behavioral zoo research in Europe (Kleiman, 1992). In the early 1960s, Desmond Morris started the first formal zoo behavioral research unit at the Zoological Society of London, Regent's Park (Kleiman, 1992). Soon thereafter, the New York Zoological Society, in association with the Rockefeller University, formed the Institute for Research in Animal Behavior, and in 1965 the National Zoological Park founded its research unit (Kleiman, 1992).

Behavioral studies have predominated in zoos for decades (Hediger, 1950, 1955; Eisenberg and Kleiman, 1977; Finlay and Maple, 1986; Kleiman, 1992; Schreiber et al., 1993). Animal behavior studies have contributed significantly to the improvement of management and reproduction of zoo animals (Hediger, 1950; Kleiman, 1975, 1980, 1992; Eisenberg and Kleiman, 1977) and have been regarded as a powerful tool for captive management (Hediger, 1950; Morris, 1966; Burghardt, 1975; Eisenberg and Kleiman, 1977; Kleiman, 1994; Lindburg and Fitch-Snyder, 1994; Stobbkopf, 1983). However, many behavioral research papers, although important in their findings, are the result of casual observation or purely descriptive case studies (Schreiber et al., 1993; Kleiman, 1994). They frequently arise from an effort to solve a particular crisis encountered with captive-held individuals or species. Solutions are applied over time until breeding success is obtained, and biological understanding of the obtained solution may or may not be achieved during this process. The scientific method using systematic hypotheses testing has rarely been employed in such studies (Berger, 1990; Kleiman, 1994).

Recent technological advances in wildlife immobilization and anesthesiology, biotechnology, genetics, and reproductive physiology have received increasing attention in the search for solutions to breeding problems. The development of techniques for assisted reproduction (e.g., artificial insemination, *in vitro* fertilization, and embryo transfer) has contributed much valuable information to our overall understanding of the reproductive biology and physiology of many species. Furthermore, in some cases the use of assisted reproduction may provide the only feasible way to save a species or subspecies from certain extinction. Theoretically, techniques for assisted reproduction can offer a way to increase species capacity and to facilitate the movement of animals without the hazards of disease, behavioral stress, and social problems (Soulé, 1992; Conway, 1995). Similarly, importing genetic material from wild into captive populations can be accomplished without further depletion of wild specimens. However, so far these technologies have only been worked out for less than three dozen wild species of mammals and birds, and broader application may not be feasible due to limited resources (Conway, 1995). Furthermore, for species in which such technologies are available, concurrent behavioral studies are urgently needed to examine possible drawbacks of the use of assisted reproduction.

Shepardson (1994) appropriately cautioned that there is more to preserving species than looking after their genes. The intended goal of captive-breeding programs is to preserve animals as representatives of their wild ancestors for posterity, research, education, and for potential release back to the wild. Thus, preservation of natural behaviors should be a key factor for all these endeavors (Hediger, 1982; Snowdon, 1989). Failure to reproduce naturally and raise young in the captive environment will most likely result in a loss of many forms and patterns of natural behavior (Shepardson, 1994). Many mammalian species show a large repertoire of complex learned and culturally transmitted behaviors (e.g., sexual and parental competence, social interactions, some aspects of foraging behaviors). Such behaviors may be lost permanently if no provisions are made to support their development and display in captivity (Lyles and May, 1987; Shepardson, 1994). Indeed, the diversity of culturally transmitted behaviors has the potential to be lost much faster than genetic diversity (May, 1991, in Shepardson, 1994; Snyder et al., 1995). It is therefore of utmost importance that captive animals develop behavioral competence in all areas important for living in the wild if the ultimate goal is reintroduction (Snowdon, 1989). Provision of more naturalistic environments (i.e., simulating a species' social and physical environment found in the wild) may help to minimize the effects of domestication or selection for a specific captive environment (Frankham and Loebel, 1992), and successful unassisted captive propagation of a species may serve as an indicator of our intuitive or scientific understanding of a particular species' requirements.

In short, systematic research efforts are urgently needed to examine various aspects of behavior pertinent to conservation efforts for captive and wild populations of rare and endangered species. Below, several areas of behavioral research that are particularly relevant for captive-breeding success in mammals are identified. Examples of successes and failures of captive propagation that can be attributed to behavioral factors and that illustrate the importance of behavioral research are presented in each section. The chosen examples are based on an extensive review of the zoo literature with particular emphasis on the *Journal of Zoo Biology* and the *International Zoo Yearbook.*

Major Aspects of Animal Behavior Contributing to Captive Breeding Success

Territoriality and Spatio-Structural Requirements

Hediger (1950) first noted the importance of territoriality and functional space for captive management of wild animals. Knowledge of species-specific movement patterns and territory requirements is essential for maintaining close-to-natural social and reproductive interactions in mammals (Eisenberg and Kleiman, 1977; Kleiman, 1994). In addition, perceived space, including olfactory and auditory components, should be considered together with specific niche requirements for successful maintenance and breeding of captive mammals (e.g., Meyer-Holzapfel, 1968; Stobbkopf, 1983).

Providing appropriate space objects (e.g., climbing structures, hiding places) has been shown to have substantial behavioral effects by reducing aggression and stress in several species. For example, in a pair of giant pandas *Ailuropoda melanoleuca,* a significant decrease in frequency of agonistic vocalizations occurred after addition of a complex series of platforms, swinging tires, and other mobile features to the enclosure (Kleiman and Peters, 1990). In leopard cats *Felis bengalensis,* reduced exploratory behavior was an indicator of chronically elevated adrenocortical activity (indicating stress), and hiding behavior increased with elevated adrenocortical output (Carlstead et al., 1993). Consequently, provision of appropriate hiding places stimulated exploratory behavior and reduced pacing and urinary cortisol levels. These results indicate that structural enrichment of enclosures may increase behavioral opportunities for coping with aversive stimuli (Carlstead et al., 1993).

Cage size has been shown to affect behavior and reproduction when crowding occurs (Nieuwenhuijsen and de Waal, 1982; Stobbkopf, 1983). Groups of rodents (Archer, 1970) and nonhuman primates (Rowell, 1967; Southwick, 1969; Elton and Anderson, 1977; Erwin, 1979) in captivity exhibited higher frequencies of aggressive behavior than less crowded or free-living conspecifics. Blake and Gillett (1984) found that an increase in cage size resulted in larger litter sizes and increased overall breeding success in Asian chipmunks *Tamias sibiricus*. In other rodents, cannibalism (Calhoun, 1962) and delayed implantation (Dickson, 1964) have been observed as a result of crowding.

Even when cage size is appropriate, sounds and odors from adjacent enclosures may lead to a perceived crowding effect or territory invasion (Stobbkopf, 1983). Kleiman (1994) reported that in an off-exhibit breeding facility for golden lion tamarins *Leontopithecus rosalia* where an air-handling system inadvertently also moved odors and sounds from enclosure to enclosure, neighboring family groups frequently seemed agitated and vocalized in a distressed manner.

Among primates an inappropriate captive environment is frequently associated with stereotyped movement, self-aggression, higher levels of aggression between cage mates, inactivity, and infertility (Erwin et al., 1976; Erwin and Dani, 1979; Maple, 1979; Maple and Hoff, 1982). Indeed, numerous studies on a variety of mammal species have documented an increase in natural behavior patterns and a decline is stereotypic movements after providing various enrichment devices and increasing the complexity of the captive environment (White, 1978; Yanofsky and Markowitz, 1978; Clarke et al., 1982; Markowitz, 1982; Wilson, 1982; McKenzie et al., 1986; Macedonia, 1987; Wiepkema, 1989; Shepardson et al., 1990; Carlstead et al., 1991). In some cases reproduction may be directly and positively affected by such changes. For example, several previously nonreproductive gorillas (*Gorilla gorilla*) became reproductively active after transfer to an enriched enclosure (Maple and Stine, 1982).

For some species certain enclosure features may be necessary simply to increase survivorship in captivity. For example, in several ungulate species fatalities have been reported due to inadequate enclosures or enclosure size. Death resulted from collision with fences or other enclosure structures (e.g., Moore, 1982; B. Williams personal communication). In pronghorn *Antilocapra americana* the importance of distant-visibility has been emphasized, and recommendations for fencing and shelter constructions have been made to accommodate natural behaviors and reduce fatalities (Müller-Schwarze and Müller-Schwarze, 1973; Moore, 1982). Furthermore, species-specific enclosure design needs to consider appropriate hiding places, denning sites, substrate, and other structural features (Holzapfel, 1939; Jolly, 1966; Eisenberg and Kleiman, 1977; Snowdon, 1989). For example, the giant panda spends little time auto-grooming but requires soil or grass and water to allow for regular rubbing and bathing to maintain a healthy fur condition (Eisenberg and Kleiman, 1977).

Interspecific Interactions

Some species are maintained in mixed-species exhibits. This practice is particularly common in ungulates (Popp, 1984). However, excessive aggression and even fatalities have sometimes occurred as a result (Popp, 1984). Zebras (*Equus* spp.), for example, have been observed to kill wildebeest (*Connochaetes* spp.) and giraffe calves (*Giraffa* spp.) at some facilities (K. Snodgrass and B. Williams, personal communication). In a multi-institutional study on interspecific aggression in 10 ungulate species, a correlation between aggressive behavior and taxonomic relatedness was found. Interestingly, aggression was found to be highest between distantly related species (Popp, 1984). A possible explanation offered for this phenomenon was the inability of distantly related species to recognize each others' threat displays and thus avoid further aggressive interaction (Popp, 1984). However, there are encouraging accounts of various attempts to establish mixed-species exhibits (e.g., Xanten, 1990), and some have reported an increase in the behavioral repertoire shown by species under mixed-housing conditions (e.g., Drüwa, 1986).

For small to medium-sized predators, such as cheetahs *Acinonyx jubatus* and a number of small felid species, it has been reported that housing next to large-sized predators and natural competitors, such as lions *Panthera leo* and leopards *Panthera pardus,* may decrease reproductive success (Rawlins, 1972; McKeown, 1991; Mellen, 1991).

Although recently there has been increased awareness that some species may not thrive well in captivity if in constant, close proximity to their natural predators (Stobbkopf, 1983) there have been few systematic studies to investigate this contention. One study indicates that ungulates may change their behavior as a result of constant visual presence of predators. However, the authors concluded that because the changes only occurred in rare behaviors, juxtaposing predators and their prey in exhibits did not result in serious behavioral changes (Stanley and Aspey, 1984).

Foraging Behavior

Wild mammals usually spend a significant amount of time foraging. For instance, wild orangutans *Pongo pygmaeus* spend about 46% of their time feeding (Rodman, 1979), wild gorillas about 45% (Harcourt and Stewart, 1984), and chimpanzees *Pan troglodytes* about 74% (Riss and Busse, 1977). During foraging wild animals face decisions about where, when, and how long to forage, and which items to choose. However, most captive mammals are deprived of such experiences and are frequently provided with uniform diets presented at the same time and place every day. At one facility foraging time in captive gorillas was reported to have increased from 11% to 27% after the provision of browse (Gould and Bres, 1986); another study on captive gorillas reported about 8% of time spent on feeding and foraging. It is likely that severely reduced levels of foraging behavior are prevalent for most captive mammals when compared with their wild counterparts, and variation of food presentation over space and time may allow increased foraging activity in captive specimens.

Foraging may be associated with developmental, parental, and sexual behavior for many species (Hoff, 1982). Provision of more varied diets and the possibility to exhibit natural foraging behavior may therefore be important for breeding success. Breeding success may also be influenced indirectly by enhancing the general well-being of the animal or by directly stimulating sexual activity. For example, exhibition of more complex feeding behaviors decreases stereotypic behaviors and increases general activity and well-being in different species (Shepardson, 1994) such as macaques (*Macaca* spp.; Chamove and Anderson, 1979), orangutan (Barbiers, 1985; Tripp, 1985), several other primate species (Chamove et al., 1982), and leopards (Markowitz et al., 1995). In the cheetah the stimulation provided by catching and killing live prey appears to enhance breeding success in some instances (Eaton and Craig, 1973).

Hediger (1950) noted that feeding wild animals in captivity should meet nutritional requirements but also needs to fulfill psychological and behavioral requirements. Detailed knowledge about natural foraging behavior and foraging patterns is necessary to provide appropriate variety and appropriate temporal and spatial presentation of food items. Aggression in a group of stump-tailed macaques *Macaca arctoides* was reduced by presenting food scattered on to a wood-chip litter substrate (Anderson and Chamove, 1984). Similar results were obtained for pig-tailed macaques *Macaca nemestris* (Boccia, 1989) and for chimpanzees (Bloomsmith et al., 1988). In the light of preserving natural behaviors for possible reintroduction, the importance of learning species-specific foraging patterns may be crucial for success. For example, adult dietary preferences of black-footed ferrets were found to be established at an early developmental stage during the third month of life (Vargas and Anderson, 1996).

Social Organization: Patterns of Group Formation and Dispersal

The importance of providing an appropriate social environment for successful captive propagation has frequently been emphasized (Hediger, 1950, 1955; Sadleir, 1975; Eisenberg and Kleiman, 1977; Kleiman, 1980, 1994; Stobbkopf, 1983). Nevertheless, many species are still held in arrangements that consist of forced social living of at least one adult of each gender being permanently housed together (Lindburg and Fitch-Snyder, 1994). Species that require different social conditions may show little to no reproductive success or will be unable to rear viable offspring as a consequence (Martin, 1975; Kleiman, 1994; Lindburg and Fitch-Snyder, 1994).

Our understanding of the social interactions of endangered species is frequently based on anecdotal information or few rigorous studies. For example, most felids are simply regarded as asocial, yet data from field studies only exist for a small number of felid species (Bennett and Mellen, 1983). In one case, extensive field studies revealed an unexpected pattern of sociality given the species' solitary nature: in the cheetah, males formed long-term stable bonds termed "male coalitions" (Caro and Collins, 1987; Caro, 1993). Thus, patterns of sociality even in largely solitary species may be more intricate than assumed from minimal observations or extrapolation of findings in related species. Furthermore, extrapolation from closely related species may frequently prove inadequate (Eisenberg and Kleiman, 1977). For example, three zebra species, Burchell's zebra *Equus burchelli,* Grevy's zebra *Equus grevyi,* and mountain zebra *Equus zebra,* were found to behave quite differently from each other with regard to herd formation (Klingel, 1967, 1969; Dolan, 1977; Eisenberg and Kleiman, 1977).

False assumptions about the social nature of an endangered species, based on extrapolations and scant field data, are not uncommon, and detection of such errors is critical for long-term captive-breeding success. For example, the aye-aye *Daubentonia madagascariensis* was originally described as solitary (Petter and Petter, 1967), but later findings indicated a more social nature (DuBois and Izard, 1990; Haring et al., 1994). Matschie's tree kangaroos *Dendrolagus matschiei* were traditionally housed in pairs or small groups, but behavioral studies on captive and wild tree kangaroos suggest a largely solitary existence (Proctor-Gray and Ganslosser, 1986; Proctor-Gray, 1990; Hutchins et al., 1991). The brown hyena *Hyaena brunnea,* although belonging to its own family (Hyaenidae), had mostly been kept in pairs, a social structure successful for many canid species; however, field studies showed that this species is far more social than had been assumed (Shoemaker, 1983). Elephant shrews *Elephantulus rufescens* were often kept in groups but never reproduced in this situation (Kleiman, 1994). Field research indicated that most elephant shrew species might be monogamous (Rathbun, 1979). Once housing arrangements in captivity had been changed and elephant shrews were kept in pairs rather than in groups, successful reproduction occurred (Kleiman, 1994). Similarly, red pandas *Ailurus fulgens* were found to be unable to rear offspring when two or more females were housed together (Kleiman, 1980). Among marmoset and tamarin species, founding pairs were often allowed to develop into large family groups. The offspring, however, would never reproduce in these family settings. Periodic removal of adolescent young and their establishment as new pairs resulted in release of reproductive suppression and was therefore crucial for a successful breeding program (Eisenberg and Kleiman, 1977; Kleiman, 1980; Wasser and Barash, 1983; Abbott et al., 1993).

Various types of reproductive suppression have been documented for a number of species (Kleiman, 1980), where either offspring remained within family groupings past the time of sexual maturity—e.g., bush dog *Speothos venaticus* (Porton et al., 1987), cotton-top tamarin *Saguinus oedipus* (Ziegler et al., 1987), naked mole rat *Heterocephalus glaber* (Faulkes et al., 1991)—or where individuals with low dominance status were forced to remain in inappropriate social settings—e.g., black-howler monkey *Aloutta caraya* (Shoemaker, 1982a), cotton-top tamarin (Ziegler et al., 1987), Japanese macaques *Macaca fuscata* (Rendall and Taylor, 1991), marmosets and tamarins (Abbott et al., 1993). This may be part of a reproductive strategy for some species, where alloparental "helpers" are necessary to rear offspring successfully in the wild. However, for other species it may be a reflection of inappropriate social conditions in captivity. Therefore, detailed knowledge of natural patterns of dispersal and grouping is necessary to determine species-specific management decisions.

Inappropriate social composition has also been observed to lead to abortion or unsuccessful reproduction (e.g., mountain zebra, Przewalski's horse: Dolan, 1977). Furthermore, introductions of unfamiliar individuals into established groups without knowledge of dispersal patterns in the wild may lead to serious injuries (e.g., squirrel monkeys *Saimiri sciureus*: Williams and Abee, 1988; sable antelope *Hippotragus niger*: Thompson, 1993), unexpected fatalities (e.g., infanticide in Przewalski's horse: Ryder and Massena, 1988; and chimpanzees: Angst and Thommen, 1977), or death of the newcomer. For some ungulates, the establishment of stable female hierarchies has proven crucial for reduced aggression and successful reproduction: (e.g., addax *Addax nasomacolatus*: Reason and Laird, 1988; roan antelope *Hippotragus equinus*: Joubert, 1974; sable antelope *Hippotragus niger*: Thompson, 1993).

Furthermore, there is evidence that a minimum group size may be necessary for successful reproduction in some species (see also Kleiman, 1980). Such social facilitation of reproduction, also termed the Allee effect (Allee, 1958) has been documented for several bird species (e.g., Stevens, 1991; Pickering et al., 1992). However, it may also occur in mammals. For example, in black lemurs *Eulemur macaco macaco* and mongoose lemurs *Eulemur mongoz*, reproductive success may be enhanced by the presence of additional conspecific pairs (Hearn et al., 1996). In Przewalski's horse, white rhinoceroses *Ceratotherium simum*, and vicunas *Vicugna vicugna* larger groups have been found to dissipate aggression of males and avert fatalities in females, and they may also act as a stimulus necessary for breeding (Bareham, 1973; Dolan, 1977).

Mating Systems and Reproductive Behavior

Consideration of species' mating systems is one of the most basic requirements of captive propagation. Historically, however, it has been common practice in zoos to house only a single pair, male and female, of each species. Mating systems in mammals are highly diverse; 11 different mating systems have been described in one classification (Eisenberg, 1981), and only about 3% of mammals have been classified as monogamous. Thus, monogamous pairings were only successful for a small number of species and individuals. In many species the natural mating system may not be obvious from the social composition observed in the wild (Kleiman, 1980). For example, in packs of wolves (*Canis* spp.) and African wild dogs, only one pair will form a monogamous unit and reproduce; however, to a casual observer, polygyny, polyandry, or promiscuity may be inferred based on social composition alone.

Even when the general mating system is known (e.g., giant panda, cheetah), our understanding of the underlying behavioral–physiological mechanisms of mate location, mate choice, timing and induction of sexual arousal, and courtship behaviors may be insufficient to create the appropriate environment for successful reproduction. This is particularly apparent in asocial species, where mating represents an unusual event of "sociality" during which the tendencies to avoid conspecifics are overcome for a short time (Lindburg and Fitch-Snyder, 1994). Mating in these species may involve a substantial risk, and aggression is usually part of the courtship ritual. There are numerous accounts of captive animals where mates have killed or severely injured each other (e.g., clouded leopard *Neofelis nebulosa*: Yamada and Durrant, 1989; Law, 1991; cheetah: S. Wells, personal communication; kangaroo rats *Dipodomys hermanni*: Daly et al., 1984; Roest, 1991; black rhinoceros *Diceros bicornis*: Smith and Read, 1992; vicunas: Schmidt, 1975). Behavioral studies have offered successful solutions in some cases. For example, Thompson's study (1995) of captive kangaroo rats suggested that individuals destined for breeding should be kept within sensory contact of the opposite sex to allow for socialization and familiarization. Long-term familiarity appears to mitigate aggression among males and females. Similarly, in clouded leopards and leopards, one strategy of achieving compatible breeding pairs is to place sexually immature individuals together for bonding and familiarization (Yamada and Durrant, 1989).

However, long-term familiarity of breeding partners can also be detrimental to breeding success, particularly in species where avoidance of mating with closely familiar individuals may be part of a strategy to avoid inbreeding or to facilitate dispersal (Nadler and Collins, 1984; Watts, 1990; Kleiman, 1994; Lindburg and Fitch-Snyder, 1994). Introductions of an unfamiliar estrous female can lead to sudden sexual interest in previously sexually unresponsive males (see Bermant, 1976, for review). For the cheetah, an important provision for repeated successful breeding is to isolate females and only occasionally bring males and females together for breeding introductions (Manton, 1975). More generally, the presentation of novel stimuli such as live prey or an altered environment may also stimulate reproductive activity in some species (e.g., cheetah, red panda, white rhinoceros; Lindburg and Fitch-Snyder, 1994). Courtship chases or courtship-associated fighting may also be a necessary part of reproduction (e.g., cheetah: Benzon and Smith, 1975; Caro, 1993; Cupps, 1985; black rhinoceros: Jones, 1979). For several gregarious ungulate species (e.g., white rhinoceros, Asiatic wild ass *Equus hemionus*, Przewalski's horse), it has been reported that keeping males housed singly or only with one female may result in severe aggression and even death upon reintroduction to a herd (Dolan, 1977).

Understanding mechanisms of mate choice and patterns of individual variation is also crucial for successful breeding of many species (Lindburg and Fitch-Snyder, 1994). For example, a survey on random pairings of tree shrews revealed that only 20% resulted in successful pair formation (Schreiber et al., 1993). Mate incompatibility and rejection of apparently reproductively healthy individuals (e.g., gorilla: Nadler and Collins, 1984; Watts, 1990; cheetah: Brand, 1980; Schuman, 1991; Stearns, 1991; clouded leopard: Yamada and Durrant, 1989; leopard: Shoemaker, 1982a; snow leopard *Panthera unica*: Marma and Yunchis, 1968; giant panda: Kleiman, 1994; kangaroo rats: Thompson, 1995) presents a major problem for breeding facilities housing small numbers of individuals of a particular species. Mating recommendations in captive-breeding programs based solely on genetic and demographic considerations may therefore conflict with natural mating strategies (Schreiber et al., 1993; Lindburg and Fitch-Snyder, 1994) and may lead to unproductive management decisions (Curio, 1996; Grahn et al., chapter 13, this volume).

Parental Care

Management practices that provide the appropriate environmental and social requirements for natural parental behavior are critical for successful propagation of captive mammals. In particular, the survivorship of young may be affected when parental care strategies are not carefully considered for each species (Kleiman, 1980). Although maternal care is universal among mammals, the occurrence and extent of paternal and alloparental care varies substantially among species (Baker, 1994).

Many mammalian females require complete isolation from conspecifics and seclusion from potentially stressful events around the time of birth and early rearing (e.g., tiger *Panthera tigris*: Husain, 1966; Van Bemmel, 1968; cheetah: Manton, 1975; Laurenson, 1993; snow leopard: Koivisto et al., 1977; several small felid species: Eaton, 1984). This appears to be particularly important for species in which only maternal care is provided.

Maternal neglect, abandonment, or cannibalism of young has historically been one of the most common causes of death among captive-born mammals (Faust and Scherpner, 1967; Brady and Ditton, 1979; Peel et al., 1979; Meier, 1986; Courtenay, 1988). Indeed, it seems that a large portion of unusually high infant mortality rates reported for several captive-bred mammal species may be attributed to parental care problems (see table 6-2). In tree shrews, for example, it has been documented that about 50% of young are lost due to maternal neglect or cannibalism (Schwaier, 1973).

Changes in environmental factors such as the provision for more than one den site can provide solutions to maternal care problems in some species. For example, in the wild, cheetahs and several other carnivore species move their young when disturbed (Laurenson, 1993). Disturbance in captivity, such as a thunderstorm or other unpredictable events, may cause a similar reaction in the female. If only one den site is provided, cubs may be carried around excessively and eventually die of hypothermia or bite wounds (Roberts, 1975; Eisenberg and Kleiman, 1977; K. Snodgrass, personal communication). Behavioral studies on tree shrews showed that females will not share a nest with their young, but rather visit them and then return to their own den (Martin, 1968; Roberts, 1993). Provision of additional nest-boxes in captivity therefore led to an increase in breeding success (Martin, 1975). In several ungulates where mothers spend a large amount of time away from their young in the wild ("hider" species; Lent, 1974), appropriate hiding places are necessary to facilitate this separation of mother and infant. Reportedly, injuries and maternal care problems may result if constant visual contact of mother and infant is enforced due to inappropriate enclosure design (e.g., okapi *Okapia johnstoni*: Lindsey et al., 1994; agouti *Dasyprocta punctata*: Eisenberg and Kleiman, 1977).

Maternal incompetence has been reported for a number of species and may result from lack of experience (e.g., Chamove and Anderson, 1982; Couver and Schroeder; 1984; Snowdon, 1989; Schroeder, 1990; Waters, 1995). Recommendations for captive management have been provided based on these findings (Baker, 1994; Waters, 1995). For example, in bottlenose dolphins *Tursiops truncatus*, it has been suggested that first-time mothers should be kept with experienced females for enhanced rearing success (Cornell et al., 1987), and in gorillas and chimpanzees considerable effort has been undertaken to teach mothering skills to females (Joines, 1977; Keiter and Pichette, 1977; Hannah and Brotman, 1990). Overall, parental care behavior appears to be a particularly labile function and can be disrupted easily even at low levels of stress when other aspects of reproductive function are normal (Martin, 1968).

Hand rearing of captive-born offspring has often been employed to bypass some parental care problems encountered in many species (Chamove and Anderson, 1982; Tsang and Collins, 1985). For example, Nadler (1975) reported that more than 80% of captive-born gorillas were hand reared. Similarly, in maned wolves *Chrysocyon brachyurus*, the majority of offspring have been hand reared because captive females rarely care for their young (Rodden, 1994). However, this form of intervention may further complicate the longer-term prospects for obtaining a self-sustaining captive population (see next section).

In some species, where paternal care or alloparental care are part of the natural care pattern, the presence of conspecifics may be necessary for successful rearing. For example, the presence of the male seems to be essential for successful captive rearing in bush dogs (Jantschke, 1973). Survival of young was also highest for green acouchi *Myoprocta pratti* when male and female were present (Kleiman, 1970). Dudley (1984) showed that pups of mice *Peromyscus californicus parasiticus* were heavier during early development when the male was present.

In a number of mammal species females rear their young cooperatively or communally (Baker, 1994; Kleiman, 1980). Colobine monkeys, for example, often transfer their young soon after birth to other conspecific females for prolonged periods of time (Horwich and Manski, 1975; Blaffer-Hrdy, 1976), and in several ungulate species females alternate in guarding each other's offspring (Langman, 1977). Alloparental care is also common in canids (Malcolm and Marten, 1982), rodents (Hoogland, 1981), bats (McCracken, 1984), dolphins (Cornell et al., 1987), and many other species. Such helping experiences may be important for the adequate development of mothering skills (Hoage, 1977; Snowdon et al., 1985; Schapiro et al., 1995; Waters, 1995) and may be crucial for reproductive success (Kleiman and Malcolm, 1981).

Early Development and Sociosexual Competence

Transmission of intraspecific, learned behaviors from one individual to another is an essential part of early development in mammals. Experience with captive breeding of many mammals suggests that behavioral anomalies can result from rearing individuals in socially deficient environments (e.g., Mason et al., 1968; Nadler, 1981; Maple and Hoff, 1982; Shepardson, 1994). Social deprivation during early development is an underlying cause for breeding and parental care problems in captive-bred species (e.g., Nadler, 1981; Beck and Power, 1988; Mellen, 1992; King and Mellen, 1994). For example, social deprivation of infant gorillas has led to deficits in social, sexual, and parental behavior, and these effects appear to be difficult, if not impossible, to reverse (Maple et al., 1977; Nadler, 1981; Beck and Power, 1988). Social access during the first year of life is associated with higher reproductive success, and mother-reared females may be reproductively more successful than hand-reared ones (Beck and Power, 1988). Hand rearing has been documented to result in behavioral deficiencies in adult individuals of several mammals (Harlow and Harlow, 1962; Goldfoot, 1977; Maple, 1980; Roberts, 1982; King and Mitchell, 1987; Beck and Power, 1988; Mellen, 1992). This practice may lead to a vicious cycle by producing more behaviorally deficient adults, which may either not breed or not care for young and therefore necessitate further hand rearing of offspring (Snowdon, 1989). Furthermore, even in cases where maternal care is provided, premature removal of young from their mother, peers, or family group may have a negative impact on sexual and social competence because development may not be completed at the time of removal (Shepardson, 1994). Deficiencies in motor patterns may also result from inadequate social exposure during developmental stages. Improper orientation and mounting by males and failure to assume the lordosis posture in females are some of the problems observed in a variety of mammals (see Larsson, 1978, for review; Mason, 1968).

In addition to the social environment, the complexity of the physical environment can have far-reaching effects on the developing animal (Carlstead and Shepardson, 1994). Both neural and muscle tissue have certain "windows of vulnerability" during which they are particularly sensitive to environmental disruption (Grand, 1992). Appropriate stimuli are necessary to facilitate optimal development. Laboratory research, mainly on rodents, found a number of morphological, physiological, and behavioral consequences as a result of rearing in more complex environments. Some of these consequences include changes in brain morphology and differences in size, number, and complexity of neural synapses (see Widman et al., 1992, for review). Important behavioral changes include greater behavioral diversity and complexity when exploring objects (Renner and Rosenzweig, 1986). Such differences in exploratory behavior appear to reflect functional differences in learning and may be important for an animal's ability to behave adaptively when faced with a novel situation (Carlstead and Shepardson, 1994; Renner, 1988). Lack of environmental complexity may lead to severe behavioral deficiencies and development of rigid and abnormal behavior patterns (stereotypic behaviors) (see Mason, 1991, for review; Shepardson, 1994). For example, when reared in enriched environments, black-footed ferrets were able to kill hamsters more efficiently than individuals reared in less complex environments (Miller et al., 1994). As a consequence, it has been suggested that animals reared in deficient environments may also be less able to cope with novel and stressful situations such as mate introduction and caring for young (Carlstead and Shepardson, 1994).

Strengths and Weaknesses of Behavioral Research for Solving Captive Breeding Problems

Strengths

Behavioral research can be noninvasive, of relatively low cost, and can be carried out by trained zoo personnel, making it an ideal first approach to management problems. The provision of an environment in which natural (or close to natural) behaviors can be exhibited enhances the educational and research potential of a captive population. In addition, this practice is likely to decelerate adaptation to the captive environment and to increase breeding success, which in turn will permit the establishment of self-sustaining captive populations.

Insufficient knowledge about husbandry and behavioral requirements are frequently the main impediments to successful captive propagation for many mammals. Systematic behavioral studies on wild and captive specimens can therefore allow us to provide the appropriate environment for a species to breed naturally (unassisted) in captivity. This will also facilitate the development of behavioral competence and preservation of the behavioral repertoire associated with breeding and courtship.

Learned and culturally transmitted behaviors are frequently an important part of early development in most mammal species. Although some behaviors may require little experience to be exhibited, others need to be acquired during various stages of early development in more or less complex social groupings in interactions with parents, siblings, and other conspecifics. Such behaviors may be lost permanently if no provisions are made for their preservation. Behavioral research can help identify problems and allow us to provide the appropriate social and physical environment for successful development.

Finally, behavioral studies are necessary for reintroduction efforts. Appropriate social structure for release, foraging skills, predator avoidance, orientation, and dispersal are some of the important behavioral strategies that need to be examined carefully before any reintroduction attempts to enable the provision of adequate prerelease training.

Weaknesses

Behaviors may vary widely over time and between individuals. A trade-off between time spent investigating individuals in one area and overall number of individuals is necessary. Therefore, results are frequently based on relatively small sample sizes and on data obtained from one facility. However, innovative research approaches employing a variety of techniques, such as keeper questionnaires, remote sensing, and video recording, in combination with direct behavioral observations, can provide data from multiple institutions over the same time period (e.g., Mellen, 1991; Pickering et al., 1992; Gold and Maple, 1994; King and Mellen, 1994; Carlstead and Kleiman, unpublished data). Nevertheless, great care is required to standardize behavioral studies across individual researchers and facilities. The high variability in behavior makes it difficult for behavioral studies to generate simple guidelines comparable to demographic and genetic guidelines for captive breeding. Instead, a species-specific approach will have to be applied, and even within a species, individual variation and individuals' needs may have to be considered to achieve reliable reproduction in captivity. For example, individual cheetahs in captivity showed quantifiable behavioral variation with regard to fearfulness and other characteristics. A correlation was found between the degree of fearfulness and breeding status. Nonbreeders scored significantly higher than breeders on the fear component (N. Wielebnowski, unpublished data). Studies on domestic dogs and laboratory animals have shown a high heritability of fearfulness (e.g., Goddard and Beilharz, 1983, 1984), and selection may therefore occur for such traits in captivity. Temperamental differences and patterns in individual behavioral variation may therefore require consideration to counteract the adaptation to a specific captive environment. The captive environment may need to be adjusted accordingly to allow different types of individuals to reproduce successfully in captivity.

Behavioral studies are time intensive and often do not produce conclusive short-term results. Alone they may not provide all the information necessary to understand behaviorally based captive management problems such as mate choice, reproductive suppression, stress-related problems; an interdisciplinary approach may be advisable in most cases.

Recommendations

This chapter generated six specific recommendations. First, appropriate enclosure design and furnishing are often established based on anecdotal and descriptive information and frequently on scant field data. More systematic investigation of the effects of various enclosure features on reproductive behavior and parental care are urgently needed for numerous captive-held species.

Second, the dynamics of interspecific competition and the trade-off of close coexistence among different species in the wild and their possible ramifications for captive management have rarely been systematically examined and are in need of further investigation (see also Stobbkopf, 1983). In addition, the effect of housing prey and predators in close proximity needs further attention. Considering the preservation of naturally occurring behavior patterns, it may be beneficial for prey species to have some degree of exposure to predators as long as ample hiding spaces and room for retreat are provided.

Third, studies on wild and captive populations of many species are necessary to identify species-specific foraging patterns and to investigate their relationship to other important behavioral variables, such as reproductive behavior and parental care.

Fourth, there are many captive species in which problems associated with mating, incompatibility, and unknown reproductive behavior are regarded as a major impediment to captive propagation (Schreiber et al., 1993) (see also table 6-2). Although behavioral studies have provided solutions for some of these breeding problems, an understanding of underlying mechanisms and causes has not been achieved for many captive species, and behavioral–physiological research is urgently needed in this area.

Fifth, establishing time budgets and behavioral repertoires for wild populations is needed to provide a baseline for comparison to behaviors and time budgets found in captivity and to allow identification of abnormal behavior patterns.

Sixth, combined genetic and behavioral studies identifying parentage in wild populations can help to establish the true mating systems of a species, which may in some cases differ from previous findings based on behavioral observations alone.

The above recommendations are based on the literature and are intended to provide a few salient examples of necessary future research. Many more could certainly be added to this list. In addition, four general recommendations can be made. First, information from descriptive and anecdotal accounts on "do's and don't's" in captive breeding needs to be synthesized for each species and used to generate testable hypotheses. Second, the development of standardized methods and training procedures for behavioral research in captive studies is necessary to allow for broader application of results. These methods should then be applied across facilities and countries. In particular, multi-institutional studies and innovative methodological approaches to the study of behavior should be encouraged. Third, integration of behavioral research with other research disciplines is becoming increasingly necessary. Interdisciplinary studies, for example, combining behavior and physiology (increasingly feasible through the recently developed noninvasive techniques of fecal, urinary, and saliva steroid monitoring), will allow us to test hypotheses about complex aspects of behavior and the interplay of behavior, physiology, and various environmental factors. Finally, guidelines and provisions concerning behavioral management of endangered species need to be an official part of captive-breeding management plans. Consultation with behavioral ecologists should occur at the beginning of any planned breeding program.

Summary

Captive breeding is gaining increased attention as a valuable conservation tool and has been endorsed by the IUCN as part of a World Zoo Conservation Strategy. However, most captive populations of threatened mammalian species are not self-sustaining or viable (i.e., they cannot persist in captivity without further genetic influx from the wild). Behavioral and husbandry problems are frequently reported to be major impediments to successful captive propagation and management, but systematic behavioral studies addressing these management issues are still rare, and more detailed information on behavior of wild and captive populations is needed for most species. Numerous examples from the zoo literature illustrate the importance of various aspects of behavior for captive breeding success.

Several areas of behavioral research that are crucial for captive propagation and management have been highlighted: territoriality and spatio-structural requirements, interspecific interactions, foraging, social organization, mating systems and reproductive behavior, parental care, early development, and sociosexual competence.

A major strength of the behavioral approach is that it allows us to identify the necessary provisions for a close-to-natural environment in which as many of the naturally exhibited behaviors of a species as possible can be preserved and natural breeding can be facilitated. In particular, provisions can be made to help maintain learned and culturally transmitted behaviors crucial for possible future reintroduction efforts. Furthermore, behavior research can be noninvasive, of low cost, and may be carried out by trained zoo personnel. A main weakness of the behavioral approach may be that behavioral studies are often time intensive and require relatively large sample sizes for reliable results. Furthermore, the variability of behaviors across individuals, populations, and species makes it difficult to generate broad conservation principles similar to the ones provided by other research areas.

More emphasis on systematic behavioral studies on captive mammals is urgently needed to improve the effectiveness of captive conservation efforts. Behavioral studies need to become an integral part of captive management plans, and guidelines and provisions concerning behavioral management of each species should be included at every stage of captive management and conservation planning.

Acknowledgments I thank Tim Caro, Eberhard Curio, Matt Gompper, Peter Marler, Joel Pagel, and Dirk Van Vuren for reviewing the manuscript and for their many helpful comments. In particular, I thank Joel Pagel for his help with the compilation of table 6-2 and the references.

References

Abbott DM, Barrett J, George LM, 1993. Comparative aspects of the social suppression of reproduction in female marmosets and tamarins. In: Marmosets and tamarins, systematics, behavior, and ecology (Rylands AB, ed). New York: Oxford University Press; 152–163.

Allee WC, 1958. The social life of animals, 2nd ed. Boston: Beacon Press.

Anderson JR, Chamove AS, 1984. Allowing captive primates to forage. In: Standards in laboratory animal management, part 2. Potters Park, England: Universities Federation for Animal Welfare; 252–256.

Angst W, Thommen D, 1977. New data and a discussion of infant killing in Old World monkeys and apes. Folia Primatol 27:198–229.

Archer J, 1970. Effects of population density on behavior in rodents. In: Social behavior in birds and mammals (Crook JH, ed). New York: Academic Press; 169–210.

AZA, 1994. AZA annual report on conservation and science (Bowdoin J, Wiese RJ, Willis K, Hutchins M, eds). Bethesda, Maryland: American Zoo and Aquarium Association.

Baker A, 1994. Variation in the parental care systems of mammals and the impact on zoo breeding programs. Zoo Biol 13:413–421.

Balmford A, Leader-Williams N, Green MJB, 1995. Parks or arks: where to conserve large threatened mammals? Biodivers Conserv 4:595–607.

Balmford A, Mace GM, Leader-Williams N, 1996. Designing the ark: setting priorities for captive breeding. Conserv Biol 10:719–727.

Barbiers RB, 1985. Orang utans' color preference for food items. Zoo Biol 4:287–290.

Bareham JR, 1973. General behavior patterns of wild animals in captivity. In: The welfare and management of wild animals in captivity. Potters Park, England: Universities Federation for Animal Welfare; 90–97.

Beck BB, Power ML, 1988. Correlates of sexual and maternal competence in captive gorillas. Zoo Biol 7:313–327.

Beck BB, Rapaport LG, Stanley Price MR, Wilson AC, 1994. Reintroduction of captive-born animals. In: Creative conservation: interactive management of wild and captive animals (Olney PJS, Mace GM, Feistner ATC, eds). London: Chapman and Hall; 265–286.

Bennett SW, Mellen J, 1983. Social interactions and solitary behaviors in a pair of captive sand cats (*Felis margarita*). Zoo Biol 2:39–46.

Benzon TA, Smith RF, 1975. A technique for propagating cheetah breeding. Int Zoo Yrbk 15:154–157.

Berger J, 1990. Behavioral insights and endangered species conservation. Zoo Biol 9:331–333.

Bermant G, 1976. Sexual behavior: hard times with the Coolidge effect. In: Psychological research: the inside story (Siegel MH, Zeigler HP, eds). New York: Harper and Row; 76–103.

Blaffer-Hrdy S, 1976. The care and exploitation of nonhuman primate infants by conspecifics other than the mother. Adv Study Behav 6:101–158.

Blake GH, Gillett KE, 1984. Reproduction of Asian chipmunks (*Tamias sibiricus*) in captivity. Zoo Biol 3:47–63.

Bloomsmith MA, Alford PL, Maple TL, 1988. Successful feeding enrichment for captive chimpanzees. Am J Primatol 16:155–164.

Boccia ML, 1989. Preliminary report on the use of a natural foraging task to reduce aggression and stereotypies in socially housed pig-tailed macaques. Lab Primate Newslett 28:3–4.

Brady CA, Ditton MK, 1979. Management and breeding of maned wolves at the National Zoological Park, Washington. Int Zoo Yrbk 19:171–176.

Brambell MR, 1977. Reintroduction. Int Zoo Yrbk 17:112–116.

Brand DJ, 1980. Captive propagation at the National Zoological Gardens of South Africa. Int Zoo Yrbk 20:107–112.

Burghardt GM, 1975. Behavioral research on common animals in small zoos. In: Research in zoos and aquariums. Symposium at the 49th Conference of the American Association of Zoological Parks and Aquariums, Houston, TX, October 1973. Bethesda, Maryland: AAZPA; 103–133.

Calhoun JB, 1962. Population density and social pathology. Sci Am 206:139–148.

Carlstead K, Brown JL, Seidensticker J, 1993. Behavioral and adrenocortical responses to environmental changes in leopard cats (*Felis bengalensis*). Zoo Biol 12:321–331.

Carlstead K, Shepardson D, 1994. Effects of environmental enrichment on reproduction. Zoo Biol 13:447–458.

Carlstead K, Seidensticker J, Baldwin R, 1991. Environmental enrichment for zoo bears. Zoo Biol 10:3–16.

Caro TM, 1993. Behavioral solutions to breeding cheetahs in captivity: insights from the wild. Zoo Biol 12:19–30.

Caro TM, Collins DA, 1987. Ecological characteristics of territories of male cheetahs (*Acinonyx jubatus*). J Zool 211:89–105.

Carpenter JW, 1983. Species decline: a perspective on extinction, recovery, and propagation. Zoo Biol 2:165–178.

Chamove AS, Anderson JR, 1979. Wood chip litter in macaque groups. J Inst Anim Techn 30:69–74.

Chamove AS, Anderson JR, 1982. Hand-rearing infant stump-tailed macaques. Zoo Biol 1:323–331.

Chamove AS, Anderson JR, Jones SC, Jones SP, 1982. Deep woodchip litter: hygiene, feeding and behavioral enrichment in eight primate species. Int J Study Anim Prob 3:308–313.

Clarke SA, Juno CJ, Maple TL, 1982. Behavioral effects of a change in the physical environment: a pilot study of captive chimpanzees. Zoo Biol 1:371–380.

Conway W, 1985. The species survival plan and the conference on reproductive strategies for endangered wildlife. Zoo Biol 4:219–223.

Conway WG, 1986. The practical difficulties and financial implications of endangered species breeding programmes. Int Zoo Yrbk 24/25:210–219.

Conway W, 1995. Wild and zoo animal interactive management and habitat conservation. Biodivers Conserv 4:573–594.

Cornell LH, Asper ED, Antrim JE, Searles SS, Young WG, Goff T, 1987. Progress report: results of a long-range captive breeding program for the bottlenose dolphin, *Tursiops truncatus* and *Tursiops truncatis gilli*. Zoo Biol 6:41–53.

Courtenay J, 1988. Infant mortality in mother-reared captive chimpanzees at Toronga Zoo, Sydney. Zoo Biol 7:61–68.

Couver RL, Schroeder JP, 1984. A status report on a survey of *Tursiops* breeding programs. In: Progress summary, Annual Symposium of the European Association. Aquat Mamm 34.

Cupps W, 1985. The cheetah program at the Columbus Zoo: what's worked and what hasn't. In: Annual Proceedings of The National Conference of The American Association for Zoological Parks and Aquariums, 1985. Wheeling, West Virginia: AAZPA; 552–557.

Curio E, 1996. Conservation needs ethology. Trends Ecol Evol 11:260–263.

Daly M, Wilson MI, Behrends P, 1984. Breeding of captive kangaroo rats, *Dipodomys merriami* and *Dipodomys microps*. J Mammal 65:338–341.

DeBlieu J, 1994. Meant to be wild: The struggle to save endangered species through captive breeding. Golden CO: Fulcrum Publishing.

Derrickson SR, Snyder NFR, 1992. Potentials and limits of captive parrot conservation. In: New world parrots in crisis: solutions from conservation biology (Beissinger SR, Snyder NFR, eds). Washington, DC: Smithsonian Institution Press; 133–163.

Dickson AD, 1964. Delay of implantation in super-ovulated mice subjected to crowded conditions. Nature 201:839–840.

Dietz JM, Dietz LA, Nagagata EY, 1994. The effective use of flagship species for conservation of biodiversity: the example of lion tamarins in Brazil. In: Creative conservation: interactive management of wild and captive animals (Olney PJS, Mace GM, Feistner ATC, eds). London: Chapman and Hall; 32–49.

Dolan JM, 1977. The saiga, *Saiga tatarica*: a review as a model for the management of endangered species. Int Zoo Yrbk 17:25–30.

Drüwa P, 1986. Maintaining maned wolves and giant anteaters, *Chrysocyon brachyurus* and *Myrmecophaga tridactyla,* together in one enclosure. Int Zoo Yrbk 24/25:271–274.

DuBois C, Izard MK, 1990. Social and sexual behaviours in captive aye-ayes (*Daubentonia madagascariensis*). J Psychol Behav Sci 5:1–10.

Dudley D, 1984. Contributions of paternal care to the growth and development of the young in *Peromyscus californicus.* Behav Biol 11:156–166.

Durrell G, 1975. Foreword. In: Breeding endangered species in captivity (Martin RD, ed). London: Academic Press.

Durrell L, Mallinson J, 1987. Reintroduction as a political and educational tool for conservation. Dodo 24:6–19.

Eaton RL, 1984. Survey of smaller felid breeding. Zool Garten 54:101–120.

Eaton RL, Craig SJ, 1973. Captive management and mating behavior of the cheetah. In: The world's cats, vol 1. Ecology and conservation (Eaton RL, ed). Winston, Oregon: World Wildlife Safari; 217–254.

Ebenhard T, 1995. Conservation breeding as a tool for saving animal species from extinction. Trends Ecol Evol 10:438–443.

Eisenberg JF, 1981. The mammalian radiations. Chicago: University of Chicago Press.

Eisenberg JF, Kleiman DG, 1977. The usefulness of behavior studies in developing captive breeding programmes for mammals. Int Zoo Yrbk 17:81–89.

Ellis S, Seal US, 1995. Tools of the trade to aid decision-making for species survival. Biodivers Conserv 4:553–572.

Ellis-Joseph S, Seal US, 1992. Conservation assessment and management plan (CAMP) progress report. CBSG Strategic Planning Reference Materials, section 4. Gland: International Union for the Conservation of Nature; 1–10.

Elton RH, Anderson BV, 1977. The social behavior of a group of baboons (*Papio anubis*) under artificial crowding. Primates 18:225–234.

Erwin J, 1979. Aggression in captive macaques: interaction of social and spatial factors. In: Captivity and behavior. (Erwin J, Maple TL, Mitchell G, eds). New York: Van Nostrand Reinhold; 139–171.

Erwin J, Anderson B, Erwin N, Lewis L, Flynn D, 1976. Aggression in captive groups of pigtail monkeys: effects of provision of cover. Percep Motor Skills 42:219–224.

Erwin J, Deni R, 1979. Strangers in a strange land: abnormal behaviors or abnormal environments? In: Captivity and behavior (Erwin J, Maple TL, Mitchell G, eds). New York: Van Nostrand Reinhold; 1–28.

Evans S, 1983. Breeding of the cotton-top tamarin *Saguinus oedipus oedipus*: comparison with the common marmoset. Zoo Biol 2:47–54.

Fagen R, 1981. Animal play behavior. Oxford: Oxford University Press.

Faulkes CG, Abbott DH, Liddell CE, George LM, Jarvis JUM, 1991. Hormonal and behavioral aspects of reproductive suppression in female naked mole-rats. In: The biology of the naked mole rat (Sherman PW, Jarvis JUM, Alexander RD, eds). Princeton, New Jersey: Princeton University Press; 426–445.

Faust R, Scherpner C, 1967. A note on breeding of the maned wolf. (*Chrysocyon brachyurus*) at Frankfurt Zoo. Int Zoo Yrbk 7:119.

Finlay TW, Maple TL, 1986. A survey of research in American zoos and aquariums. Zoo Biol 5:261–268.

Frankham R, Loebel DA, 1992. Modeling problems in conservation using captive *drosophila* populations: rapid genetic adaptation to captivity. Zoo Biol 11:333–342.

Gadsby EL, Feistner ATC, Jerkins PD Jr, 1994. Coordinating conservation for the drill

(*Mandrillus leucophaeus*): Endangered in forest and zoo. In: Creative conservation: interactive management of wild and captive animals (Olney PJS, Mace GM, Feistner ATC, eds). London: Chapman and Hall; 439–454.

Gibbons EF Jr, Durrant BS, Demarest J, eds, 1994. Conservation of endangered species in captivity: an interdisciplinary approach. Ithaca, New York: University of New York Press.

Goddard ME, Beilharz RG, 1983. Genetics of traits which determine the suitability of dogs as guide-dogs for the blind. Appl Anim Behav Sci 9:299–315.

Goddard ME, Beilharz RG, 1984. A factor analysis of fearfulness in potential guide dogs. Appl Anim Behav Sci 12:253–265.

Gold KC, Maple TL, 1994. Personality assessment in the gorilla and its utility as a management tool. Zoo Biol 13:509–522.

Goldfoot DA, 1977. Rearing conditions which support of inhibit later sexual potential of laboratory-born rhesus monkeys: hypothesis and diagnostic behaviors. Lab Anim Sci 27:548–556.

Gould E, Bres M, 1986. Regurgitation and re-introduction in captive gorillas: description and intervention. Zoo Biol 5:241–250.

Grand TI, 1992. Altricial and precocial mammals: a model of neural and muscular development. Zoo Biol 11:3–15.

Hannah AC, Brotman B, 1990. Procedures for improving maternal behavior in captive chimpanzees. Zoo Biol 9:233–240.

Harcourt AH, Stewart KJ, 1984. Gorilla's time feeding: aspects of methodology, body size, competition and diet. Afr J Ecol 22:207–215.

Hardy ICW, Krackow S, 1995. Does sex appeal to zoos? Trends Ecol Evol 10:478–479.

Haring DM, Hess WR, Coffman BS, Simons EL, Owens TM, 1994. Natural history and captive management of the aye-aye, *Daubentonia madagascariensis,* at the Duke University Primate Center, Durham. Int Zoo Yrbk 33:201–219.

Harlow HF, Harlow HK, 1962. Social deprivation in monkeys. Sci Am 207:136–146.

Hearn GW, Berghaier RW, George DD, 1996. Evidence for social enhancement of reproduction in two *Eulemur* species. Zoo Biol 15:1–12.

Hediger H, 1950. Wild animals in captivity. London: Butterworths Scientific Publications.

Hediger H, 1955. Psychology and behavior of captive animals in zoos and circuses. London: Butterworths Scientific Publications.

Hediger H, 1982. Zoo biology: retrospect and prospect. Zoo Biol 1:85–88.

Hedrick PW, 1992. Genetic conservation in captive populations and endangered species. In: Applied population biology (Jain SK, Botsford LW, eds). Dordrecht: Kluwer; 426–445.

Heinroth O, 1938. Aus dem Leben der Vögel. Berlin. Fischer verlag.

Hoage RJ, 1977. Parental care in *Leontopithecus rosalia rosalia,* sex and age differences in carrying behavior and the role of prior experience. In: The biology and conservation of the Callitrichidae (Kleiman D, ed). Washington, DC: Smithsonian Institution Press; 293–305.

Hoage RJ, Deiss WA (eds), 1996. New worlds, new animals. From menagerie to zoological park in the nineteenth century. Baltimore, Maryland: The John Hopkins University Press.

Hoff MP, 1982. On eating and being eaten: foraging behavior. Zoo Biol 1:273–279.

Holzapfel M, 1939. Über Bewegungstereotypien bei gehaltenen Säugern III. Zool Garten 10:5–6.

Hoogland JL, 1981. Nepotism and cooperative breeding in the black-tailed prairie dog

(Scuridae: *Cynomys ludovicianus*). In: Natural selection and social behavior (Alexander RD, Tinkle DW, eds). New York: Chiron Press; 283–310.

Horwich RH, Manski D, 1975. Maternal care and infant transfer in two species of Colobus monkeys. Primates 16:49–73.

Husain D, 1966. Breeding and hand-rearing of white tiger cubs, *Panthera tigris,* at Delhi Zoo. Int Zoo Yrbk 6:187–188.

Hutchins M, 1995. An expanded role for zoo- and aquarium-based conservation. Am Zoo Aquar Assoc Comm 12 (Dec):15.

Hutchins M, Conway WG, 1995. Beyond Noah's ark: the evolving role of modern zoological parks and aquariums in field conservation. Int Zoo Yrbk 34:117–130.

Hutchins M, Smith GM, Mead DC, Elbin S, Steenberg J, 1991. Social behavior of Matschie's tree kangaroos (*Dendrolagus matschiei*) and its implications for captive management. Zoo Biol 10:147–164.

Hutchins M, Willis K, Wiese RJ, 1995. Strategic collection planning: theory and practice. Zoo Biol 14:5–25.

IUCN. Translocation of living organisms: Introductions, re-introductions, and re-stocking. IUCN position statement. Gland: International Union for the Conservation of Nature.

IUCN, IUCN Red List of Threatened Animals. Gland: International Union for the Conservation of Nature.

IUDZG/CBSG, International Union for the Conservational of Nature, 1993. The world zoo conservation strategy: the role of zoos and aquaria in global conservation. Chicago: Chicago Zoological Society.

Jantschke F, 1973. On the breeding and rearing of bush dogs, *Speothos venaticus,* at the Frankfurt Zoo. Int Zoo Yrbk 13:141–143.

Joines SA, 1977. A training programme designed to induce maternal behavior in a multiparous female lowland gorilla. Int Zoo Yrbk 17:185–188.

Jolly A, 1966. Lemur behavior. Chicago: University of Chicago Press.

Jones DM, 1979. The husbandry and veterinary care of captive rhinoceros. Int Zoo Yrbk 19:239–352.

Joubert SCJ, 1974. The social organisation of the roan antelope, *Hippotragus equinus,* and its influence on the spatial distribution of herds in the Kruger National Park. In: The behavior of ungulates and its relation to management (Geist V, Walther F, eds). Gland: International Union for the Conservation of Nature; 661–675.

Kastelein RA, Wiepkema PR, 1989. A digging trough as occupational therapy for Pacific walruses (*Odobonus rosmarus divergens*) in human care. Aquat Mamm 15:9–17.

Keiter M, Pichette P, 1977. Surrogate infant prepares a lowland gorilla, *Gorilla g. gorilla,* for motherhood. Int Zoo Yrbk 17:188–189.

King N, Mellen J, 1994. The effects of early experience on adult copulatory behavior in zoo-born chimpanzees (*Pan troglodytes*). Zoo Biol 13:51–60.

King NE, Mitchell G, 1987. Breeding primates in zoos. In: Comparative primate biology, vol. 2, part B. Behavior, cognition, and motivation (Mitchell G, Erwin J, eds). New York: Alan R. Liss; 219–261.

Kleiman DG, 1970. Reproduction in the green acouchi, *Myoprocta pratti.* Reprod Fertil 23:55–60.

Kleiman DG, 1975. The management of breeding programs in zoos. In: Research in zoos and aquariums. Washington, DC: National Academy Press; 157–177.

Kleiman DG, 1980. The sociobiology of captive propagation. In: Conservation biology: an evolutionary-ecological perspective (Soulé ME, Wilcox BA, eds). Sunderland, Massachusetts: Sinauer Associates 243–261.

Kleiman DG, 1989. Reintroduction of captive mammals for conservation: guidelines for reintroducing endangered species into the wild. Biol Sci 39:152–161.

Kleiman DG, 1992. Behavior research in zoos: past, present, and future. Zoo Biol 11:301–312.

Kleiman DG, 1994. Animal behavior studies and zoo propagation programs. Zoo Biol 13:411–412.

Kleiman DG, Malcolm JR, 1981. The evolution of male parental investment in mammals. In: Parental care in mammals (Gubernick DJ, Klopfer PH, eds). New York: Plenum Press; 347–387.

Kleiman DG, Peters G, 1990. Auditory communication in the giant panda: Motivation and function. In: Proceedings of the 2nd International Symposium on Giant Panda (Asakura S, Nakagawa S, eds). Tokyo: Tokyo Zoological Park Society; 107–122.

Klingel H, 1967. Soziale Organisation und Verhalten frielebender Steppenzebras. Z Tierpsychol 24:580–624.

Klingel H, 1969. Reproduction in the plains zebra, *Equus burchelli boehmi*: behaviour and ecological factors. J Reprod Fertil (suppl) 6:339–345.

Koehler W, 1931. The mentality of apes. New York: Harcourt, Brace and Company.

Koivisto I, Wahlberg C, Muroonen P, 1977. Breeding the snow leopard *Panthera uncia* at the Helsinki Zoo 1967–1976. Int Zoo Yrbk 17:39–44.

Langman VA, 1977. Cow-calf relationships in giraffe (*Giraffa camelopardalis giraffa*). Tierpsychol 43:264–286.

Larsson K, 1978. Experiental factors in the development of sexual behaviour. In: Biological determinants of sexual behaviour (Hutchinson JB, ed). New York: John Wiley and Sons; 55–86.

Laurenson KM, 1993. Early maternal behavior of wild cheetahs: implications for captive husbandry. Zoo Biol 12:31–43.

Law G, 1991. Clouded leopard management. In: Management guidelines for exotic cats (Partridge J, ed). Bristol: Association of British Wild Animal Keepers; 77–81.

Leader-Williams N, 1990. Black rhinos and African elephants: lessons for conservation funding. Oryx 24:23–29.

Leader-Williams N, 1993. Theory and pragmatism in the conservation of rhinos. In: Rhinoceros biology and conservation (Ryder OA, ed). San Diego, California: Zoological Society of San Diego; 69–81.

Lent PC, 1974. Mother-infant relationships in ungulates. In: Behavior of ungulates and its relation to management, vol 1 (Geist V, Walther F, eds). IUCN new series no. 24. Gland: International Union for the Conservation of Nature; 14–55.

Leyhausen P, 1956. Verhaltensstudien an Katzen. Berlin: Paul Parey.

Lindburg DG, Gledhill L, 1992. Captive breeding and conservation of lion-tailed macaques. Endangered Species Update 10:1–4.

Lindburg DG, Fitch-Snyder H, 1994. Use of behavior to evaluate reproductive problems in captive mammals. Zoo Biol 13:433–455.

Lindsey SL, Bennet C, Pyle E, Willow M, Yang A, 1994. Calf and management and the collection of physiological data for Okapi, *Okapia johnstoni,* at Dallas Zoo. Int Zoo Yrbk 33:263–268.

Luoma JR, 1987. A crowded ark. Boston: Houghton Mifflin.

Lyles AM, May RB, 1987. Problems in leaving the ark. Nature 326:245–246.

Macedonia JM, 1987. Effects of housing differences upon activity budgets in captive sifakas (*Propithecus verreauxi*). Zoo Biol 6:55–67.

MacKinnon K, 1991. How about letting them breed by themselves? BBC Wildlife 9:454–455.

MacKinnon K, MacKinnon J, 1991. Habitat protection and re-introduction programmes. Symp Zool Soc Lond 62:173–198.

Magin CD, Johnson TH, Groombridge B, Jenkins M, Smith H, 1994. Species extinctions, endangerment and captive breeding. In: Creative conservation: interactive management of wild and captive animals (Olney PJS, Mace GM, Feistner ATC, eds). London: Chapman and Hall; 3–31.

Malcolm JR, Marten K, 1982. Natural selection and the communal rearing of pups in African wild dogs (*Lycaon pictus*). Behav Ecol Sociobiol 10:1–13.

Mallinson JJ, 1993. Coordinated breeding programmes for endangered species with special reference to the continental Europe and the British Isles. Dodo 29:11–22.

Mallinson JJC, 1995. Conservation breeding programmes: an important ingredient for species survival. Biodivers Conserv 4:617–635.

Manton VF, 1975. Captive breeding of cheetahs. In: Breeding endangered species in captivity (Martin RD, ed). London: Academic Press; 337–344.

Maple T, 1979. Great apes in captivity: the good, the bad, and the ugly. In: Captivity and behavior (Erwin J, Maple T, Mitchell G, eds). New York; Van Nostrand Reinhold; 239–272.

Maple T, 1980. Breaking the hand-rearing syndrome in captive apes. In: AAZPA 1980 regional workshop proceedings. Wheeling, West Virginia: American Association for Zoological Parks and Aquariums; 200–201.

Maple T, Hoff MP, 1982. Gorilla behavior. New York: Van Nostrand Reinhold.

Maple T, Stine W, 1982. Environmental variables and great ape husbandry. Am J Primatol (suppl) 1:67–76.

Maple T, Zucker EL, Hoff MP, Wilson ME, 1977. Behavioral aspects of reproduction in the great apes. In: Proceedings of the Annual National Conference of the American Association of Zoos and Aquariums. Bethesda, Maryland: AAZPA; 194–200.

Markowitz H, 1982. Behavioral enrichment in the zoo. New York: Van Nostrand Reinhold.

Markowitz H, Aday C, Gavazzi A, 1995. Effectiveness of acoustic "prey": environmental enrichment for a captive African leopard (*Panthera pardus*). Zoo Biol 14:371–379.

Marma BB, Yunchis VV, 1968. Observations on the breeding and physiology of snow leopards, *Panthera uncia,* at the Kaunas Zoo from 1962 to 1967. Int Zoo Yrbk 8: 66–74.

Martin RD, 1968. Reproduction and ontogeny in the tree-shrews (*Tupaia belangeri*) with reference to their general behaviour and taxonomic relationships. Z Tierpsychol 25:409–532.

Martin RD, 1975. General principles for breeding small mammals in captivity. Breeding endangered species in captivity (Martin RD, ed). London: Academic Press; 143–166.

Mason GJ, 1991. Stereotypies: a critical review. Anim Behav 41:1015–1037.

Mason WA, Davenport RK, Menzel EW, 1968. Early experience and the social development of rhesus monkeys and chimpanzees. In: Early experience and behavior (Newton G, Levine S, eds). Springfield, Illinois: Charles C. Thomas; 440–480.

May RM, 1991. The role of ecological theory in planning-reintroduction of endangered species. Symp Zool Soc Lond 62:145–163.

McCracken GF, 1984. Communal nursing in Mexican free-tailed bat maternity communities. Science 223:1090–1091.

McKenna V, Travers W, Wray J. 1987. Beyond the bars: the zoo dilemma. Wellingborough, North Hamptonshire: Thorsons Publishing Group.

McKenzie SM, Chamove AS, Feistner ATC, 1986. Floor-coverings and hanging screens alter arboreal monkey behavior. Zoo Biol 5:339–348.

McKeown S, 1991. The cheetah. In: Management guidelines for exotic cats (Partridge J, ed). Bristol: The Association of British Wild Animal Keepers; 82–91.

Meier J, 1986. Neonatology and hand-rearing of carnivores. In: Zoo and wild animal medicine, 2nd ed (Fowler ME, ed). Philadelphia: Saunders.

Mellen JD, 1991. Factors influencing reproductive success in small captive exotic felids (*Felis* spp): a multiple regression analysis. Zoo Biol 10:95–110.

Mellen JD, 1992. Effects of early rearing experience on subsequent adult sexual behavior using domestic cats (*Felis catus*) as a model for exotic small fields. Zoo Biol 11: 17–32.

Mellen JD, 1994. Survey and interzoo studies used to address husbandry problems in some zoo vertebrates. Zoo Biol 13:459–470.

Meyer-Holzapfel M, 1956. Das Spiel beii Säugetieren. Handbk Zool 8(5):1–36.

Meyer-Holzapfel M, 1968. Abnormal behavior in zoo animals. In: Abnormal behavior in animals (Fox MW, ed). Philadelphia: Saunders; 476–503.

Miller B, Biggens D, Hanebury L, Vargas A, 1994. Reintroduction of the black-footed ferret (*Mustela nigripes*). In: Creative conservation: interactive management of wild and captive animals (Olney PJS, Mace GM, Feistner ATC, eds). London, Chapman and Hall; 455–464.

Moore D, 1982. History and assessment of future management of pronghorn (*Antilocapra americana*) in captivity "East of the Mississippi." In: AAZPA Regional Conference Proceedings. Wheeling, West Virginia: American Association of Zoological Parks and Aquariums.

Morris D, 1966. Animal behaviour studies at London Zoo. Int Zoo Yrbk 6:288–291.

Müller-Schwarze D, Müller-Schwarze C, 1973. Behavioral development of hand-reared pronghorn, *Antilocapra americana*. Int Zoo Yrbk 13:217–220.

Muñiz RS, 1984. Nipple contact in captive black-faced chimpanzees (*Pan troglodytes* [Blumenbach, 1779]). Zoo Biol 3:267–271.

Nadler RD, 1975. Determinants of variability in maternal behavior of captive female gorillas. In: Proceedings of the Symposia of the Fifth Congress of the International Primatological Society (Kondo S, Kawai M, Ehara E, Kawamara, eds). Tokyo: Japan Science Press; 207–216.

Nadler RD, 1981. Laboratory research on sexual behavior of the great apes. In: Reproductive biology of the great apes: comparative and biomedical perspectives (Graham CE, ed). New York: Academic Press; 191–238.

Nadler RD, Collins DC, 1984. Research on reproductive biology of gorillas. Zoo Biol 3:13–25.

Nieuwenhuijsen K, de Waal FBM, 1982. Effects of spatial crowding on social behavior in a chimpanzee colony. Zoo Biol 1:5–28.

Norton BG, Hutchins M, Stevens E, Maple TL (eds, 1995. Ethics on the ark: zoos, animal welfare and wildlife conservation. Washington, DC: Smithsonian Institution Press.

Olney PJS, Mace GM, Feistner ATC, eds, 1994. Creative conservation: interactive management of wild and captive animals. London: Chapman and Hall.

Peel RR, Price J, Karsten P, 1979. Mother-rearing of a spectacled bear cub (*Tremarctos ornatus*) at Calgary Zoo. Int Zoo Yrbk 19:177–182.

Petter JJ, Petter A, 1967. The aye-aye of Madagascar. In: Social communication among primates (Altman SA, ed). Chicago: University of Chicago Press; 195–205.

Pickering S, Creighton E, Stevens-Wood B, 1992. Flock size and breeding success in flamingos. Zoo Biol 11:229–234.

Popp JW, 1984. Interspecific aggression in mixed ungulate species exhibits. Zoo Biol 3:211–219.

Porton I, Kleiman DG, Rodden M, 1987. Aseasonality of bush dog reproduction and the influence of social factors on the estrous cycle. J Mammal 68:867–871.

Proctor-Gray E, 1990. Kangaroos up a tree. Nat Hist 1:61–67.

Proctor-Gray E, Ganslosser U, 1986. The individual behaviors of Lumholtz's tree kangaroo: Repertoire and taxonomic implications. J Mammal 67:343–352.

Rabb GB, 1994. The changing roles of zoological parks in conserving biological diversity. Am Zool 34:159–164.

Ralls K, 1995. Why is behavior so rarely incorporated into conservation planning? In: Twenty-fourth International Ethological Conference Abstracts, August 1995. Honolulu, HI: University of Hawaii.

Rathbun GB, 1979. The social structure and ecology of elephant-shrews. Z Tierpsychol (suppl) 20:1–76.

Rawlins CGC, 1972. Cheetahs (*Acinonyx jubatus*) in captivity. Int Zoo Yrbk 12:119–120.

Reason RC, Laird EW, 1988. Determinants of dominance in captive female addax (*Addax nasomaculatus*). J Mammal 69:375–377.

Rendall D, Taylor LL, 1991. Female sexual behavior in the absence of male-male competition in captive Japanese macaques (*Macaca fuscata*). Zoo Biol 10:319–328.

Renner MJ, 1988. Learning during exploration: the role of behavioral topography during exploration in determining subsequent adaptive behavior. Int J Comp Psychol 2:43–45.

Renner MJ, Rosenzweig MR, 1986. Object interactions in juvenile rats (*Rattus norvegicus*): effects of different experiential histories. J Comp Psychol 100:229–236.

Riss DC, Busse CD, 1977. Fifty-day observation of a free-ranging male chimpanzee. Folia Primatol 28:283–297.

Roberts M, 1975. Growth and development of mother-reared red pandas, *Ailurus fulgens*. Int Zoo Yrbk 15:57–63.

Roberts M, 1982. Demographic trends in a captive population of red pandas (*Aiiurus fulgens*). Zoo Biol 1:119–126.

Roberts MS, 1993. The minimalist motherhood of tree shrews. Zoogoer 22:6–11.

Rodden M, 1994. Maned wolf (*Chrysocyon brachyurus*). In: AZA annual report on conservation and science (Bowdain J, Wiese RJ, Willis K, Hutchins M, eds). Bethesda, Maryland: American Zoo and Aquarium Association: 48–50.

Rodman PS, 1979. Individual activity patterns and the solitary nature of orang utans. In: The great apes (Hamburg DA, McCowan ER, eds). Menlo Park, California: Benjamin Cummings; 235–255.

Roest AI, 1991. Captive reproduction in Heerman's kangaroo rat, *Dipodomys heermanni*. Zoo Biol 10:127–137.

Rowell TE, 1967. A quantitative comparison of the behavior of a wild and a caged baboon group. Anim Behav 15:499–505.

Ryder OA, Feistner ATC, 1995. Research in zoos: a growth area in conservation. Biodivers Conserv 4:671–677.

Ryder OA, Massena R, 1988. A case of male infanticide in *Equus przewalskii*. Appl Anim Behav Sci 21:187–190.

Sadleir RMFS, 1976. Role of the environment in the reproduction of mammals in zoos. In: Research on zoos and aquariums. Symposium at the 49th Annual National Conference of the American Association of Zoological Parks and Aquariums, AAZPA Houston, Texas, October 1973. Wheeling, West Virginia: AAZPA; 151–156.

Schapiro SJ, Bloomsmith MA, Suarez SA, Porter LM, 1995. Maternal behavior of primiparous rhesus monkeys: effects of limited social restriction and inanimate environmental enrichment. Appl Anim Behav Sci 45:139–149.

Schmidt CR, 1975. Captive breeding of the vicuna. In: Breeding endangered species in captivity (Martin RD, ed). London: Academic Press; 271–283.

Schmidt MJ, 1982. Studies on Asian elephant reproduction at the Washington Park Zoo. Zoo Biol 1:141–147.

Schreiber A, Kolter L, Kaumanns W, 1993. Conserving patterns of genetic diversity in endangered mammals by captive breeding. Acta Theriol 38 (suppl 2):71–88.

Schroeder JP, 1990. Breeding bottlenose dolphins in captivity. In: The bottlenose dolphin (Leatherwood S, Reeves RR, eds). New York: Academic Press; 435–446.

Schuman ME, 1991. Breeding strategy and field observations of captive cheetahs. Int Zoo News 38(8):9–19.

Schwaier A, 1973. Breeding tupaias (*Tupaia belangeri*) in captivity. Z Versuch 15:255–271.

Seal US, 1991. Life after extinction. Symp Zool Soc Lond 62:39–55.

Shepardson D, 1994. The role of environmental enrichment in the captive breeding and reintroduction of endangered species. In: Creative conservation: interactive management of wild and captive animals (Olney PJS, Mace GM, Feistner ATC, eds). London: Chapman and Hall; 167–177.

Shepardson DJ, Brownback T, Tinkler D, 1990. Putting the wild back into zoos: enriching the zoo environment. Appl Anim Behav Sci 28:300.

Shoemaker AH, 1982a. The effect of inbreeding and management on propagation of pedigree leopards *Panthera pardus* ssp. Int Zoo Yrbk 22:198–206.

Shoemaker AH, 1982b. Fecundity in the captive howler monkey, *Alouatta caraya*. Zoo Biol 1:149–156.

Shoemaker AH, 1983. 1982 Studbook report on the brown hyena, *Hyaena brunnea*: Decline of a pedigree species. Zoo Biol 2:133–136.

Smith RL, Read B, 1992. Management parameters affecting the reproductive potential of captive, female black rhinoceros, *Diceros bicornis*. Zoo Biol 11:375–383.

Snowdon CT, 1989. The criteria for successful captive propagation of endangered primates. Zoo Biol (suppl) 1:149–161.

Snowdon CT, Savage A, McConnel PB, 1985. A breeding colony of cotton-top tamarins, *Saguinus oedipus*. Lab Anim Sci 35(5):477–481.

Snyder NRF, Derrickson SR, Beissinger SR, Wiley JW, Smith TB, Toone WD, Miller B, 1996. Limitations of captive breeding in endangered species recovery. Conserv Biol 10:338–348.

Soulé ME, 1992. Ten years ago, ten years from now: summary remarks for the symposium. Symp Zool Soc Lond 64:225–234.

Soulé M, Gilpin M, Conway W, Foose T, 1986. The millennium ark: how long a voyage, how many staterooms, now many passengers? Zoo Biol 5:101–113.

Southwick CH, 1969. Aggressive behavior of rhesus monkeys in natural and captive groups. Excerpta Medica 5:32–43.

Stanley ME, Aspey WP, 1984. An ethometric analysis in a zoological garden: modification of ungulate behavior by the visual presence of a predator. Zoo Biol 3:89–109.

Stanley Price MR, 1989. Animal re-introductions: the Arabian oryx in Oman. Cambridge Studies in Applied Ecology and Resource Management. Cambridge: Cambridge University Press.

Stearns MJ, 1991. Mate selection by a female cheetah at Fossil Rim Wildlife Center. Cheetah News 3:3.

Stevens EF, 1991. Flamingo breeding: the role of group display. Zoo Biol 10:53–63.

Stobbkopf MK, 1983. The physiological effects of psychological stress. Zoo Biol 2:179–190.

Stuart SN, 1991. Re-introductions: to what extent are they needed? Symp Zool Soc Lond 62:27–37.

Sunquist F, 1995. End of the ark? Int Wildl 6:23–29.

Tembrock G, 1963. Acoustic behaviour of mammals. In: Acoustic behavior of animals (Busnel RG, ed). Amsterdam: Elsevier; 751–788.

Thompson KV, 1993. Aggressive behaviour and dominance hierarchies in female sable antelope, *Hippotragus niger*: implications for captive management. Zoo Biol 12:189–202.

Thompson KV, 1995. Factors affecting pair compatibility in captive kangaroo rats, *Dipodomys heermanni*. Zoo Biol 14:317–330.

Tripp JK, 1995. Increasing activity in captive orangutans: Provision of manipulable and edible materials. Zoo Biol 4:225–234.

Tsang WSN, Collins PM, 1985. Techniques for hand-rearing tree-shrews (*Tupaia belangeri*) from birth. Zoo Biol 4:23–31.

Tudge C, 1991. Last animals at the zoo. Washington, DC: Island Press.

Van Bemmel ACV, 1968. Breeding tigers, *Panthera tigris*, at Rotterdam Zoo. Int Zoo Yrbk 8:60--63.

Vargas A, Anderson SH, 1996. Effects of diet on captive black-footed ferret (*Mustela nigripes*) food preference. Zoo Biol 15:105–113.

Warland MAG, 1975. A cautionary note on breeding endangered species in captivity. In: Breeding endangered species in captivity (Martin RD, ed). London: Academic Press; 373–377.

Wasser SK, Barash DP, 1983. Reproductive suppression among female mammals: implications for the biomedicine and sexual selection theory. Q Rev Biol 58:513–538.

Waters SS, 1995. A review of social parameters which influence breeding in white-faced saki, *Pithecia pithecia*, in captivity. Int Zoo Yrbk 34:147–153.

Watts DP, 1990. Mountain gorilla life histories, reproductive competition, and sociosexual behavior and some implications for captive husbandry. Zoo Biol 9:185–200.

Wemmer C, Derrickson S, 1995. Collection planning and the modern zoo. Zoo Biol 14:38–40.

Wheater R, 1995. World zoo conservation strategy: a blueprint for zoo development. Biodivers Conserv 4:544–552.

White L, 1978. Behavioral response of orangutans to remodeling of their exhibit. Anim Keepers For 5:179–180.

Whitehead M, 1995. Saying it with genes, species and habitats: biodiversity education and the role of zoos. Biodivers Conserv 4:664–670.

Widman DR, Abrahamsen GC, Rosellini RA, 1992. Environmental enrichment: the influences of restricted daily exposure and subsequent exposure to uncontrollable stress. Physiol Behav 51:309–318.

Wielebnowski N, 1996. Reassessing the relationship between juvenile mortality and genetic monomorphism in captive cheetahs. Zoo Biol 15:353–369.

Williams LE, Abee CR, 1988. Aggression with mixed age-sex groups of Bolivian squirrel monkeys following single animal introductions and new group formations. Zoo Biol 7:139–145.

Wilson AC, Stanley Price MR, 1994. Reintroduction as a reason for captive breeding. In: Creative conservation: interactive management of wild and captive animals (Olney PJS, Mace GM, Feistner ATC, eds). London: Chapman and Hall; 243–269.

Wilson SF, 1982. Environmental influences on the activity of captive apes. Zoo Biol 1:201–209.

Xanten WA, 1990. Marmoset behaviour in mixed-species exhibits at the National Zoological Park, Washington. Int Zoo Yrbk 29:143–148.

Yanofsky R, Markowitz H, 1978. Changes in general behavior of two mandrills (*Papio sphinx*) concomitant with behavioral testing in the zoo. Psychol Rec 28:369–373.

Yamada JK, Durrant BS, 1989. Reproductive parameters of clouded leopards (*Neofelis nebulosa*). Zoo Biol 8:223–231.

Ziegler TE, Savage A, Scheffler G, Snowdon CT, 1987. The endocrinology of puberty and reproductive functioning in female cotton-top tamarins (*Saguinus oedipus*) under varying social conditions. Biol Reprod 37:618–627.

7

Behavior as a Tool for Management Intervention in Birds

Eberhard Curio

Conservation in the Future

A number of species have become extinct in the wild and survive only in captivity (Magin et al., 1994). Over the next century, therefore, it will be necessary to finesse our abilities to breed highly endangered species in captivity because there will be no room for error. In addition, as conservation areas in the wild become more isolated and smaller in size, we may have to manage *in situ* populations along similar lines to captive populations.

Eventually captive individuals will have to be reintroduced into the wild, yet there are difficulties in reintroducing captive-bred animals into nature. For example, captive-born individuals survive significantly less well than wild-caught ones when released into the wild (Griffith et al., 1989; Beck et al., 1994; Ginsberg, 1994; Miller et al., 1994). Often releases have to be repeated, with the success rate increasing with the number of releases (Beck et al., 1994; see also Veltman et al., 1996) and the number of places where release occurs (Veltman et al., 1996). Based on this poor record, some postulate spreading releases over many years, but it may be more cost effective to understand the reasons that releases fare so poorly so that captive breeding, now called conservation breeding, and prerelease and postrelease training can be improved.

It is here that behavioral ecology, or more particularly, behavioral study, can play a pivotal role, as many of the deficiencies currently impeding captive breeding are behavioral in nature (Curio, 1996; Beissinger, 1997). The purpose of this chapter is to demonstrate the pivotal role of behavioral ecology and ethology for species conservation in the wild, in captive breeding, and prerelease training, and release (see also Clemmons and Buccholz 1997).

Captive Breeding: A Brief Synopsis

Until recently, zoological institutions primarily obtained animals for public exhibition and education rather than for breeding purposes. In the last 15 years, the focus of zoos has changed to being safe havens for endangered species (Wielebnowski, Chapter 6, this volume), and, in order to establish self-sustaining populations, consistency in breeding has taken center stage in management. The principal focus of research and management in zoos nowadays is establishing studbooks so that individuals' pedigrees are documented (de Boer, 1994) and developing high-tech physiological methods for breeding species. Studbooks are viewed as essential for preventing mating between close relatives and for equalizing genetic representation in future generations so as to maximize effective population size. High-tech breeding is seen as an important tool for promoting breeding of individuals housed apart or showing reproductive difficulties, and it may provide an avenue for saving space through cryopreservation of gametes and embryos (Wildt et al., 1997).

In contrast, little systematic attention has been paid to behavioral problems associated with captive breeding (e.g., Edwards, 1995; but see Kleiman, 1994). Behavioral difficulties are usually tackled in an ad hoc fashion on a case-by-case basis, relying on word of mouth and trial and error to induce individuals to mate and rear offspring successfully.

Although zoological institutions often claim interest in *in situ* conservation, only a few make contributions in this area, and almost entirely on a financial basis. Although reintroduction programs are often touted as the ultimate goal of captive breeding programs, few reintroductions have been attempted compared to the number of endangered species housed in captivity.

Behavioral Issues Involved in *In Situ* Conservation

The success of conservation measures taken in the wild has been arbitrarily defined by Beck et al. (1994: 273) as the establishment of a population in the wild of "at least 500 individuals which are free of provisioning or other human support," or where a "formal demographic analysis," e.g., a population viability analysis, "predicts that the population will be self-sustaining." Here the goal is to increase population *numbers* rather than to improve the quality (see below) of individuals. A number of behavioral measures have been used to bolster populations in the wild.

Protection from Predation

When the population size of a prey species has declined to dangerously low levels, predation may wipe it out. This has occurred frequently when native species are faced with exotic predators with which they have no previous experience or with which they have no history of potential coevolution (see Berger, Chapter 4, this volume). Many of the most famous examples are flightless birds on oceanic islands (Savidge, 1987). Hence, shielding vulnerable individuals of an endangered species from predation has often proved a straightforward means of protection. Such protective methods have taken the form of guarding birds' nests, providing predator-safe nest-boxes, or even placing nest-boxes in ways that exploit the territoriality of other species (table 7-1). The pearly-eyed thrasher *Margarops fuscatus* attempts to use the nest holes of the endangered Puerto Rican parrot *Amazona vittata*. In so doing, the eggs or chicks of the parrot are killed and eaten. By providing an extra nest-box near a parrot's box, the thrasher can be induced to occupy it without harming the parrot's brood. Moreover, through its territorial behavior the thrasher keeps other thrashers away from both nests. Thus a detailed knowledge of the thrasher's behavior helps protect the target species' brood. Knowledge of the behavior of other nest competitors has aided in designing still other countermeasures, thus boosting successful reproduction of the parrot (table 7-1).

Table 7-1 Examples of *in situ* measures used in preserving endangered bird species in the wild

Species of concern	Threat	*In situ* measures	Success	Reference
Puerto Rican parrot (*Amazona vittata*)	Diverse predators and parasites	Nest guarding, artificial nestboxes	66% with guarding, 38% without guarding	Lindsay (1992)
	Nest predation/competition	Nest-boxes to shield from predators/competitors	+	Wiley (1985)
	Honey bees (nest hole competitors)	Sealing nest holes outside breeding season	+	Wiley (1985)
Rarotonga monarch (*Pomarea dimidiata*)	Ship rat (*Rattus rattus*) predation	Metal cuffs around nest tree	Population increasing	McCormack and Künzle (1990)
Many New Zealand bird species	Predation by introduced predators	Translocation to predator-free islands, eradication of predators	+ in many cases	Baker (1991)
Polynesian megapode (*Megapodius pritchardii*)	Overharvesting of eggs; predation by cats; "co-extinction" of ectoparasites	Translocation of eggs (and adults) to predator-free islands (= benign introduction, see text)	+ (hatching demonstrated)	Göth and Vogel (1995)
Wood duck (*Aix sponsa*)	Intraspecific brood parasitism	Nest boxing less visible and less dense	Increased hatchability	Semel et al. (1988)
Lesser spotted eagle (*Aquila pomarina*)	Siblicide	Temporary cross-fostering to buzzard (*Buteo buteo*) and black kite (*Milvus migrans*) and return to original nest-sibling	+	Meyburg (1978)
Spanish imperial eagle (*A. heliaca adalberti*)	Siblicide (?), parental imcompetence (?)	Fostering to other pairs until fledging	+	Meyburg (1978)
Chatham Islands black robin (*Petroica traversi*)	Extinction	Cross-fostering to Chatham Islands tit (*P. macrocephala chathamensis*) until fledging; nest-boxes, exclusion of nest competitors, supplementary feeding	+	Butler and Merton (1992)
	Hybridization due to misimprinting on foster species	Removal of foster species from island with *P. traversi*; translocation of cross-fostered *P. traversi* to island devoid of foster species	+	Butler and Merton (1992)

Island faunas are among the most vulnerable, as evidenced by extinction rates (e.g., Johnson and Stattersfield, 1990), and the main reason for extinctions has been the presence of introduced animals (Atkinson, 1989). Various transfer methods have proved useful in avoiding introduced predators. According to a recent catalogue (IUCN, 1995) these methods are: reintroductions or attempts to establish a species in an area which was once part of its historical range from which it has been extirpated or become extinct; translocations or deliberate movements of wild individuals or populations from one part of their range to another; reinforcements or supplementations (the addition of individuals to an existing population of conspecifics); and "benign" introductions (attempts to establish a species outside its recorded distribution but within an appropriate habitat and ecogeographical area).

Protection from Intraspecific Brood Parasitism

In many insect and bird species, females lay their eggs in other females' nests. Extra parental effort jeopardizes the reproductive success of the parasitized broods. As Semel et al. (1988) and Eadie et al. (Chapter 12, this volume) have shown for wood ducks *Aix sponsa*, judiciously placing nest-boxes in more secluded places and spacing boxes farther apart can markedly reduce intraspecific brood parasitism.

Protection from Siblicide and Rescue Efforts through Cross-Fostering

Right after hatching, many species of birds of prey and herons kill their sibling(s), and their parents do not intervene (Mock and Forbes, 1994). In pioneering experiments, Meyburg (1978) cross-fostered endangered birds of prey and thus increased their fledging success (table 7-1). Due to the presence of conspecific nestlings before and after cross-fostering, misimprinting may be of minor concern (see below).

Since its discovery by Baron Adam von Pernau in 1702, imprinting has been widely studied, but its relevance for captive breeding has been appreciated only recently (ten Cate, 1995). For instance, it becomes a problem when threatened species are cross-fostered. Whooping cranes *Grus americana* that had been raised by sandhill cranes *Grus canadensis* migrated and even flocked with conspecifics, but they failed to breed properly (Hutchins et al., 1995). When Chatham Islands robin *Petroica traversi* numbers had been reduced to one female and four males, cross-fostering the offspring to the Chatham Islands tit *P. macrocephala* was attempted. Misimprinting on the wrong species, a real threat to pure breeding of the robin, was countered by taking the focal species and its foster species out of reach of each other (Butler and Merton, 1992).

Forestalling Infanticide

Both in the wild and in captivity, infanticide can decrease numbers of offspring reaching independence. Infanticide has been recorded in a wide variety of species (Hausfater and Hrdy, 1984) and occurs as part of a species' normal behavior that is easily absorbed in a large population. Because conservation efforts in the wild aim at increasing numbers, it may be unwise to remove potentially infanticidal individuals. As yet we have not developed methods of helping individuals defend themselves against infanticidal attacks.

Conservation of Top Predators

Top predators tend to depress the densities of their prey and of predator species at lower trophic levels ("mesopredators"); mesopredators in turn depress prey at lower trophic levels. The top predator therefore assumes the role of a "keystone" species in the community (Paine, 1966; but see Schoener and Spiller, 1996), and its removal leads to losses of still further species. For example, the removal of top carnivores from a community benefits the mesopredators to the point that their prey come under strong predation pressure. Eventually the latter may be driven to extinction. Removal of coyotes *Canis latrans* boosted the populations of both gray foxes *Urocyon cinereoargenteus* and feral domestic cats so that the songbird prey of these latter mesopredators suffered accordingly (Soulé et al., 1988; see also Diamond, 1989). Consequently, it has been argued that keystone predators should receive more conservation attention than other members of the community. The same reasoning applies to "keystone mutualists" such as pollinators or seed dispersers (Meffe and Carroll, 1994).

Behavioral Issues in Captivity

For introductions of any sort to be successful, there must be large numbers of animals to be introduced (see Beck et al., 1994; Veltman et al., 1996), preferably of high *quality*. Assessing an individual's phenotypic and genotypic attributes is difficult, however, and in reality managers will have to be satisfied with an assessment of the health and nutritional status of captive animals that will be released.

Environmental Stress in Captivity

Generally, removal of stress can promote breeding in captivity. For example, Shepherdson (1994) has emphasized the pivotal role of environmental enrichment in captive breeding. It may well be that this factor exerts its beneficial effect via a general reduction of fear of predation because enrichment entails providing objects that subjects can use for hiding and feeling safe. Zoo animals that are continually disturbed by visitors or that have a constant view of predators may not breed. Thus caretakers should behave unobtrusively or even engage in friendly relations with their subjects. Similarly, the removal of threat from natural predators may have the same effect. For example, a pair of pigmy owls *Glaucidium passerinum* that failed to breed for several years bred immediately once it was shielded from view of a pair of eagle owls *Bubo bubo* in an adjacent aviary (W. Scherzinger, personal communication). However, the generality of this observation is open to question. A pair of great tits *Parus major* bred with impunity in an unoccupied nest-box of a captive pair of Tengmalm's owls *Aegolius funereus,* flying freely in and out the aviary (W. Scherzinger, personal communication). Habituation to a predator may at times overcome fear that would typically suppress reproduction.

The evidence for or against habituation to captive predators in captive prey subjects is ambiguous. Under natural conditions, fear of predators does not abate as a consequence of frequent stimulation, whereas fear of novel objects or places readily habituates (Curio, 1993). By contrast, under captive conditions some animals seem to lose their fear of predators, though this has not been explored systematically. Some animals may assess that predators constrained in enclosures or aviaries pose no real threat. Klinghammer (1992: 40) observed similarly that hand-raised birds and mammals "are less fearful of many other stimuli and situations than normally raised animals." He cautions that this deficiency renders hand-raised animals more vulnerable to accidents in new surroundings. In an insightful but widely ignored paper, Barlow (1968) drew attention to the fact that to behave normally (i.e., similar to that in nature), animals need a moderate level of stimulation with both complete isolation and full exposure leading to fearful behavior (see also Shepherdson, 1994). He eliminated stress in captive pairs of fish by presenting to them "dither" fish (or birds) behind a transparent partition. Treated this way, reluctant pairs started to breed successfully. Finding a balance between moderate levels of stimulation and reducing fear is a challenge for zoo breeding programs. Other beneficial effects of environmental enrichment, apart from breeding, include successful killing of prey after release (see below).

Free Choice of Mates

In captivity, potential mates are chosen on the basis of genetic and demographic considerations for both maintaining maximum genetic diversity and avoiding inbreeding (e.g., Kleiman, 1980; de Boer, 1994). These needs may directly conflict with free choice of mates (*Amazona vittata:* Anonymus, 1994; Wilson et al., 1994; see also Grahn et al., Chapter 13, this volume). Free choice is often the key to success in propagating a target species (e.g., *Gymnogyps californianus:* Curio 1992; 1996; Cox et al., 1993), or enhancing its breeding success (e.g., *Falco sparverius:* Bird, 1982). For example, in the monogamous cockatiel *Nymphicus hollandicus,* freely mated pairs had higher breeding success than force-paired birds. Nevertheless, prior association of force-paired mates before the onset of breeding conditions (long days) improved breeding success; pairs were better synchronized and performed breeding activities in less time. Species may be differently susceptible to forced pairing, ranging from the detrimental effects mentioned to immunity in other species (Yamamoto et al., 1989); mammals seem less affected than birds (M. Stanley Price, personal communication).

Long-lived birds that form life-long pairs are expected to be more choosy than short-lived species in selecting a mate. This is because errors committed in selecting the "wrong" mate multiply with each bout of breeding. "Divorce" may be preceded by lower breeding success (Coulson and Thomas, 1983), suggesting that errors committed in pair formation can be corrected. This possibility is largely denied to the majority of birds that are short lived.

Ideally, potential pair mates should gauge their choice of partner on both compatibility and mate quality. There is scant evidence for either. Failed breeders tended to fail because they did not coordinate their activities well in time (Coulson and Thomas, 1983); and female blue tits *Parus caeruleus* seem to gear their rate of extrapair copulations to the quality of their mates as measured by their mates' survival (Kempenaers et al., 1992). But how do potential mates assess compatibility and/or quality in advance? In a valiant effort to discover which of 13 morphological features females use in mate choice, Ostermeier (1994) found in the sexually monomorphic Javanese mannikin *Lonchura leucogastroides* that any one or a combination of three morphological characters and older age rendered a male successful against a rival in choice tests. The results might be interpreted in many different ways. Some of the morphological traits, including body weight, correlated significantly with male age. It is plausible to assume that what females were after in a mate was older age (see also Burley, 1981), and perhaps heavier weight, and that the characters chosen were merely indicators of age. Age is assumed either to be an indicator of "good genes" (but see Hansen and Price, 1995) or of being a good father because of greater experience. This may be true up to a point (review in Nol and Smith, 1987), but the choice of an older male may be compromised by its senescence (review in Curio, 1989), impairing its parental competence. Domestic pigeons *Columba livia* do in fact exhibit morphological signs of senescence at old ages, and old males seem to be less often chosen by females than males of prime age, although they should excel in breeding experience (Burley and Moran, 1979).

In the mannikin study (Ostermeier, 1994), female choice was based on multiple criteria as judged from the female population as a whole. Similar results have been obtained in the sexually dimorphic guppy *Poecilia reticulata* (Kodric-Brown, 1993). In both cases some condition-dependent criteria (weight and age in the mannikin; courtship and dominance rank in the guppy) may allow females to gauge male quality. The "redundant signal" hypothesis suggests that several traits in combination provide a better estimate of general condition than does any single ornament (Møller and Pomiankowski, 1993). Although conceived for explaining multiple ornaments, the hypothesis can also explain rules of pair formation in a monomorphic (i.e., ornamentless) species, as shown by the mannikin study.

In laboratory mice inbreeding is avoided by olfactory cues connected to the major histocompatibility locus (Grafen, 1992). The odors linked with this highly variable locus offer to a choosy mate a convenient handle by which it can compare its own odor ("self") and, hence, genes, with the odor of a potential mate. As a result, phenotype matching translates into genotype matching with the aim of choosing, in this case, a genotype least susceptible to inbreeding.

So far, not a single study permits one to predict which individuals of a species would or would not make a good pair. This is because multiple criteria may underlie female choice, and still others have surely been overlooked by researchers. In species with pair bonding, the prediction is even more difficult: even if the criteria of choice were known, the necessary harmony among pair mates is an emergent property of the phenotypic profiles of both.

Removing Reproductive Suppression

At times, the reproductive potential of a group of potential breeders cannot be realized because of reproductive suppression by high-ranking group members. This occurs, for example, in golden lion tamarins *Leontopithecus rosalia* (Kleiman, 1980). Artificial fissioning of groups may help to alleviate this impasse, thus permitting suppressed yet physically mature animals to reproduce.

Preventing Siblicide

Captive breeding of birds in zoos can be severely marred by siblicide. In the endangered bearded vulture *Gypäetus barbatus,* the older of the two chicks habitually kills the second hatching sibling. Because siblicidal aggression wanes within a few weeks, fledging success could be markedly enhanced by placing a baffle between the two chicks (Thaler and Pechlaner, 1980).

Hand-Rearing and Solutions to Its Drawbacks

Captivity often impairs the quality of animals (e.g., Beck et al., 1994; Snyder et al., 1994) and adversely affects their potential for release (Griffith et al., 1989). In particular, artificial rearing by hand or by a foster-parent or surrogate, a popular means of increasing population numbers, exacerbates this problem. It produces animals that are often incompetent breeders and hence perpetuates the very problem that gave rise to it (e.g., Shepherdson, 1994; Hutchins et al., 1995). On the other hand, hand-rearing may be essential for keeping a species in captivity at all. For example, certain birds can only be familiarized with their artificial diet by being raised on it (Hutchins et al., 1995), and green iguanas *Iguana iguana* acclimate to captive conditions well only when growing up under them (Burghardt and Milostan, 1995).

Reproductive Abnormalities Through Hand-Rearing

In a systematic study, Myers et al. (1988) found that hand-reared cockatiels had a strikingly lower breeding success than did parent-raised birds. Although hand-reared females laid more eggs, they failed to fledge more young because they misplaced eggs on the cage floor. For breeding to be successful, a parent-raised male had to be present, but this condition was not always sufficient.

Absence of the right nest or appropriate building materials during hand-rearing may later inhibit constructing nests as adults (Hutchins et al., 1995). In some pairs the behavior deficit can be remedied by one failed breeding experience. For reasons unknown, some females that had laid first on cage floors used nest-boxes in their second cycles, and hand-reared males that had been infertile produced fertile eggs. In addition, the type of nest-box used affected reproductive output of cockatiels in several ways. Breeding in cavity-type boxes (with a normal entrance hole) as compared to shelf-type boxes (with an open front) improved fertility, hatchability, and fledging success. Despite this inferiority of shelf-type boxes, they may facilitate subsequent breeding (Martin and Millam, 1995). These results warrant further experiments and inclusion of other species, for they have implications for the propagation of cavity-nesting species and other birds.

In mammals, the ill-effects of hand-rearing pertain to impaired behavior during birth and mothering (Shepherdson, 1994; see also Gittleman and McMillan, 1995; Wielebnowski, Chapter 6, this volume). Hand-reared males of certain primates orient themselves in the wrong way and show other deficits during copulatory behavior (Harlow and Harlow, 1962).

Impairment of Predator Avoidance and Its Relation to Misimprinting

Predator avoidance mechanisms in animals are either innate (i.e., require no individual experience with the predator in question) or require learning of various sorts, or a combination of the two (review in Curio, 1993). Artificial rearing of birds and mammals, in particular, tends to disrupt the normal development of innate enemy recognition. There is no problem as long as captive stock is held in captivity, but predator recognition and avoidance becomes an issue when captive stock is released into the wild.

Klinghammer (1967) showed that ring (barbary) doves *Streptopelia roseogrisea* develop fear of humans when raised by their parents, but are tame ("blind" to a human predator) when raised by hand before they are 8 days old. A second effect of this early hand-rearing is misimprinting on humans as the objects of later sexual responses. The two effects may be the outcome of the same underlying mechanism, as the fear of humans is shown only after the young are mature at 8–9 months of age. In mourning doves *Zenaida macroura*, misimprinting is similarly affected by postfledging experience (Klinghammer, 1967). When removed from their parents by their 8th day of life, kept in isolation until 52 days of age, and then given access to conspecifics in a community cage, mourning doves invariably mated with their species' members and rejected humans in a free-choice test. Furthermore, mourning doves that are misimprinted on humans can be made to shift their sexual responses to conspecifics if given continual access to doves across four breeding seasons. Hence, experience with species' members after 52 days of age can still reverse the adverse imprinting on humans and forestall any fear of species' members. Yet the abolishment of fear of humans caused by hand-rearing is permanent in both species of doves.

The "evil duo" of predator blindness and misimprinting is, at first sight, bad news for planners of release programs. However, it may be tempered first by the observation that ring doves lose their tameness toward humans immediately in strange surroundings, as do ravens *Corvus corax* (E. Curio, personal observation). Second, when hand-reared with a sibling, ring doves do not become misimprinted on humans (Klinghammer, 1967; see also Hutchins et al., 1995). Although this protocol does not work with ravens and jackdaws *Corvus monedula* (Lorenz, 1935), it does work with parrots, where sibs can even be replaced by other species' nestlings (*Amazona* spp. with *Psittacus erithacus* and vice versa) (W. Wüst, personal communication). Third, in other species of birds, imprintability may be more malleable than in the two dove species. In ravens, imprintability depends on the length of exposure to the imprinting object (Gwinner, 1964). Also, a classic criterion of imprinting, irreversibility, has recently been found to be less stringent than previously thought (ten Cate, 1995; McLean 1997). Fourth, it has been suggested (E. Thaler, personal communication) that in hand-reared grouse (*Alectoris graeca, Lagopus mutus, Bonasia bonasus*), the innate fear of humans will be maintained into adulthood when only a single human caretaker has raised them and toward whom the birds have no fear. This can be altered with more caretakers. A similar situation has been observed at crocodile farms (M. Stanley Price, personal communication), although these animals perceive humans in the wild rather as prey than as a predator. A further caveat concerns the number of caretakers for foster-parenting animals. In species with biparental care, imprinting may become established more firmly if two caretakers instead of only one are involved (ten Cate, 1995).

The lesson to be learned from these studies is twofold. First, there may be a critical point in time when exposure to a species renders the animal perceptually blind toward potential predators; that juncture may or may not be related to sexual imprinting depending on the species involved (*Canis lupus:* Woolpy and Ginsburg, 1967; *Dama dama:* Gilbert, 1968; *Alectoris rufa:* Csermely et al., 1983). Second, during development of their animals, captive-breeding specialists should minimize contact with species that may act both as imprinting objects and potential predators. Remedies include hand-rearing individuals in groups or with other species or with a mechanical model parent (see, e.g., Toone and Wallace, 1994; Hölzer et al., 1995) and in the group-raised category with conspecifics or with other species (Klinghammer, 1992). These refinements need to be tested with species from various taxonomic groups.

Behavioral Abnormalities Due to Food Deficiencies

Although nutrition has been at the center stage of animal husbandry for a long time, its role is just now becoming more appreciated by conservationists. There is anecdotal evidence that a deficient diet for young birds may result in death when going through the first molt of the flight feathers or result in the loss of the whole brood due to impaired parenting. This is despite the fact that full-grown, ill-nourished birds may actually look healthy to an expert's eyes (Curio, 1992). The first systematic study exploring the ill effects of a deficient diet on behavior was conducted with European blackbirds *Turdus merula* by Pollack (1994). She raised two groups of nestlings from age 4 days to independence, with one group receiving an optimal diet and the other a diet deficient in terms of six species of insects and rennet. The two diets were otherwise identical with regard to animal protein, carbohydrates, and snail shells from day 7 on, and body weights did not differ significantly. After fledging the restricted-diet birds displayed reduced neophobia, greater tameness toward their human foster-parent, increased intraspecific aggression, and lower activity in terms of location calls, vigilance, and flight behavior. Similar though less varied behavioral changes have been found in a similar food deprivation experiment with great tits (H. Putz, personal communication). Two interrelated questions remain: do the behavior changes persist into adulthood, and can they be remedied later on by supplementary feeding?

The lesson from these still isolated findings for the captive breeder is clear: aside from the many direct effects of hand-rearing on behavior, indirect effects operating through nutrition must now be heeded (Hutchins et al., 1995). Apart from being detrimental to any release program, a deficient diet may also hold the key for improving captive breeding.

Reintroduction of Captive-Bred Animals

Releasing animals from captivity into the wild can be classified as "hard" when introduced animals have no prior familiarization with the release area, and "soft" when the animals have some familiarity with the area before final release (Beck et al., 1994). There are many problems faced by managers in both sorts of reintroductions (table 7-2).

Exploring the Environs Needs Tutelage

One reason for the poor success of releases is that introducing large numbers of animals precludes allowing them to explore the release area under the tutelage of their parents. In a pioneering study, Thaler et al. (1992) developed a successful protocol with hand-reared bald ibis *Geronticus eremita* that took the long-lasting family ties of this species into account. Continuous guidance of only six hand-reared fledglings by two alternating human caretakers familiarized the birds with an ever larger area around their acclimation cage in the countryside. This tutelage continued until the birds were entirely independent. In this way the panic flights seen in hundreds of captive-bred bald ibis precipitated by hard releases in the Near East were avoided. Behavioral deficiencies that hindered the release of golden lion tamarins were alleviated by providing them with wild-born individuals as tutors (Beck et al., 1991).

Experience with Live Prey and Other Factors Facilitate Hunting

Raptors and carnivores often need to be trained to hunt. In captivity, opportunities for such training are restricted so that captive-born predators often lack the necessary skills to recognize, locate, subdue, and handle their prey. Wild avian and mammalian predators increase their hunting success dramatically with age (Marchetti and Price, 1989; Caro, 1994). In pre-release training, individuals need to be primed (pre-exposed to prey) to perfect hunting skills (Polsky, 1977a), and reinforcement must occur through killing (Polsky, 1975, 1977b). Dietary cues can also be transmitted via olfaction (Galef, 1988), or even through conspe-

Table 7-2 Behavioral problems associated with releasing birds and mammals.

Behavior deficit	Species	Reference
Unfamiliarity with home range	Bald ibis (*Geronticus eremita*)	Thaler et al. (1992)
Hunting incompetence	Eagle owl (*Bubo bubo*)	Herrlinger (1973)
Predator blindness	Various species	Beck et al. (1991)
Locomotory deficits	Golden lion tamarin (*Leontopithecus rosalia*)	Beck et al. (1991, 1992)
Nonavoidance of poisonous food	Golden lion tamarin (*Leontopithecus rosalia*)	Beck et al. (1991, 1992)
Loss of knowledge (e.g., migratory routes) acquired by tradition	Trumpeter swan (*Cygnus buccinator*); Arabian oryx (*Oryx leucoryx*)	Baskin (1993), Stanley-Price (1989)
Loss of migratory restlessness in captivity	Trumpeter swan; white stork (*Ciconia ciconia*)	Baskin (1993); refs. in Stanley-Price (1989)
Loss of correct social organization (and social competence?)	Various mammals	Stanley-Price (1989); Parker and Waite (1995)
Adverse dispersal upon release	Thick-billed parrot (*Rhynchopsitta pachyrhyncha*)	Snyder et al. (1989)

cific alarm calls (Rothschild and Lane, 1959). When captive-born eagle owls are given live avian and mammalian prey before 220 days of age, they hunt their full array of prey when later released. If denied live prey until that age, they refuse to kill these prey later on. Releases can therefore take place only when live prey has been introduced early enough (i.e., between day 110 and 220) (Herrlinger, 1973). Also, release of eagle owls needs to occur before a critical period of live prey-hunting ends.

In mammals, many factors affect the development of killing behavior (Martin and Caro, 1985). For example, in polecats *Mustela putorius,* social play with litter mates, maturational processes, and a critical period to exercise the neck bite may all interact in the perfection of the killing response (Wüstehube, 1960; Eibl-Eibesfeldt, 1963; Gossow, 1970). Imprinting on olfactory and other stimuli associated with the prey may further complicate the matter (Apfelbach, 1973). In the endangered black-footed ferret *Mustela nigripes* with no killing experience before release, environmental enrichment of maintenance pens contributed to the proper killing of hamsters after release (Shepherdson, 1994; see also Miller et al., 1994). At present there are no general rules for turning captive-born predators into competent hunters.

Predator Avoidance May Require Learning

Many released animals suffer unnaturally high mortality from predation (e.g., Hill and Robertson, 1988; Dowell, 1990; Marcstrøm, 1990). It is therefore imperative that one understands why captivity renders introduced animals so vulnerable to predators even after being raised by their parents (Dowell, 1992; *Rhynchopsitta pachyrhyncha:* Snyder et al., 1994; *Dipodomys heermanni:* Chiszar et al., 1995; Yoerg, 1995). There are at least three ways in which the social environment affects proper antipredator behavior, while the predator itself may instill avoidance as well.

Pseudoconditioning

Juvenile three-spined sticklebacks *Gasterosteus aculeatus* fail to avoid pike *Esox lucius*, an important predator, when raised in captivity without their fathers. When sticklebacks from pike-threatened populations are reared by their fathers, they develop pike avoidance even in the absence of experience of pike (Tulley and Huntingford, 1987). The mechanism underlying pike recognition is unknown, but it may be that when fry are retrieved from the nest by their father, they experience an aversive stimulus that they later generalize to the pike. Because being reared without a father does not impair the escape response per se but rather the perceptual system underlying it, the normal development seems to imply pseudoconditioning rather than sensitization (McFarland, 1985).

Improvement and Maintenance of Alarm Responses Through Parents

Many birds and mammals deploy innate antipredator behavior (Curio, 1993). Yet the social environment modulates such responsiveness in ways crucial for release projects. The effects of hand-rearing versus parent raising range from a complete failure to develop appropriate alarm responses, to maturing later, to needing longer stimulation or disappearing rapidly (table 7-3). Decay of responsiveness to a predator demonstrates that the presence of or guidance by the mother is indispensable for the maintenance of the response, as, for example, in hazel grouse *Bonasia bonasus* (Thaler, 1987; see also Russock and Hale, 1978, for a similar effect in domestic fowl, *Gallus gallus*).

Table 7-3 shows that hand-rearing has a detrimental effect on the antipredator behavior of Tetraoninae and Perdicinae. In no instance has hand-rearing led to a response that is superior to that of hen-raising (table 7-3). Yet the ways in which the presence of the parent hen brings about improvements have gone unanalyzed so far (see also Adler, 1975, for *Clethrionomys glareolus*). The effects of the parent hen must be rather specific. Bantam hens, a domestic breed popular as foster parents among pheasant breeders, fail to instill in gray partridge the appropriate antipredator and avoidance behavior (Dowell, 1989). Equally alarming is the fact that there are differences among the maintenance procedures of different captive breeders (table 7-3). Extraordinarily, in black grouse, the behavior of hand-reared chicks toward a fox was more similar to chicks raised by a hen-reared hen than to chicks raised by a hand-reared hen (fig. 7-1 Költringer et al., 1995). The hand-reared hen was more docile and more chick-directed, whereas the hen-reared hen was more fearful and rarely gave contact or warning calls. Therefore, the chicks of this latter hen grew up more similar to the hand-reared chicks. This indicates that hand-rearing may produce a good mother but behaviorally handicapped chicks. Hence, the history of the parent may be of paramount importance to any prerelease training. Ignorance of the precise mechanisms underlying the normal development of antipredator behavior (Curio, 1975) has precluded any successful mass-breeding technique and has thus prevented mass release programs from being successful. The outcome of normal development should be calibrated against the behavior of wild-caught, well-acclimated adults.

Cultural Transmission via Conditioning to Predator Stimuli

When innate enemy recognition does not occur, animals can learn in other ways (reviewed in Curio, 1993) which lend themselves well to prerelease training. Although repeatedly sug-

Table 7-3 Comparison of innate antipredator responses to various predators of hand-reared versus hen-reared chicks of gallinaceous birds.

	Response characteristic of chick raised by		
Species	Hand	Hen	Reference
Hazel grouse (Bonasia bonasus)	Decay of crouching to perched hawk before independence[a]	No decay in same period[a]	Thaler (1987)
	Rapid habituation to hen alarm call to red fox	Slow habituation	Thaler (1987)
	No alarm response to terrestrial or aerial predators	"Wild" toward humans	Scherzinger (1982)
	Response to hen alarm call?	Crouching to hen alarm call	Scherzinger (1982)
Black grouse (Lyrurus tetrix)	Later maturation of crouching response	Early maturation to same stimulus ($\leq$ 4 weeks of age)	Költringer et al. (1995)
	Long latency of crouching response	Short latency to same stimulus	Költringer et al. (1995)
	No alarm response to any predator in males	?	Krätzig (1940)
Gray partridge (Perdix perdix)	Short duration of crouching to flying hawk	Long duration	Dowell (1989)
	Low stimulus specificity to hen alarm call (respond also to any conspecific adult call)	High stimulus specificity	Dowell (1989)

[a]Similar results for both *Lagopus rufus helveticus* and *Celectoris graeca* (Thaler, 1987).

gested by observations (Lorenz, 1931; von de Wall and Rajala, 1969; Rasa, 1989), this type of social learning has been experimentally demonstrated in only three taxa: fish (Suboski et al., 1990; Chivers et al., 1995), birds (reviewed in Curio, 1988), and primates (*Macaca mulatta:* Mineka and Cook, 1988). In these experiments, a knowledgeable animal, the "teacher," is made to respond fearfully to a harmful object that to the novice, the "learner," is at first a neutral stimulus. As a consequence of being paired with the fear response emitted by the teacher, the neutral stimulus acquires predator qualities for the learner. The underlying mechanism, vicarious conditioning (see Mineka and Cook, 1988), is a variant of Pavlovian conditioning in that the fear response emitted by the teacher is also induced during learning in the learner.

In European blackbirds, a teacher was made to mob a neutral stimulus, an Australian honeyeater *Philemon corniculatus,* by juxtaposing it beside an owl that was not visible to the learner, while the neutral stimulus was shielded off from the teacher (fig. 7-2). As a consequence of perceiving the teacher mob the place of the honeyeater, the learner came to mob the honeyeater. Certain stimuli, such as a stuffed honeyeater, are more effective in producing this response than other stimuli, such as a plastic bottle, or are learned to the exclusion

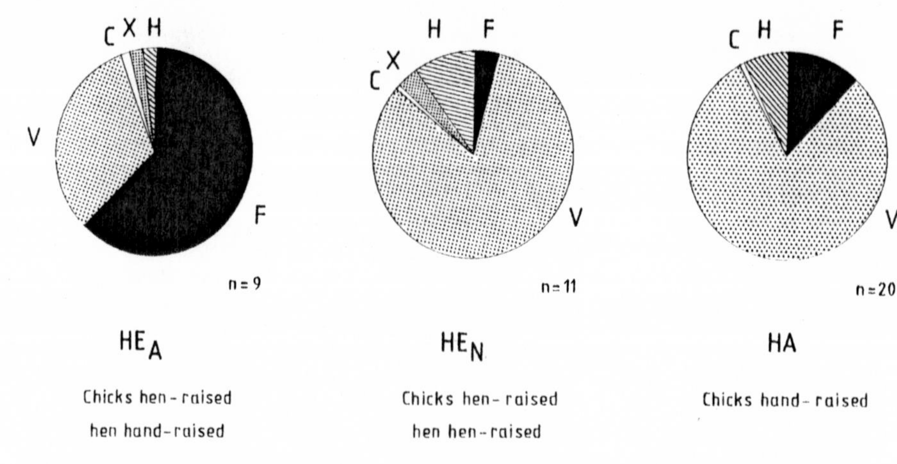

Figure 7-1 Antipredator behavior to live fox of black grouse chicks (pooled responses of ages 4–10 weeks) from three different rearing regimes (n = number of chicks). Behavior elements: F, freezing; V, vigilance; C, crouching; H, hiding in cover; X other kinds of antipredator behavior (modified from Költringer et al., 1995, courtesy World Pheasant Association and Istituto Nazionale per la Fauna Selvatica).

of others (snake versus flowers; Mineka and Cook, 1988). Hence, there is "preparedness for learning" (Seligman and Hager, 1972); biologically relevant stimuli are more effective than others.

In blackbirds, the live conspecific can be replaced with taped alarm calls (see Herzog and Hopf, 1984; Mineka and Cook, 1988), including those of other syntopic species (see Nuechterlein, 1981; Chivers et al., 1995). Training with the help of a surrogate therefore renders the experimental protocol of cultural transmission especially amenable to release training (Curio et al., 1978). An innovative approach that employed a related species as tutor was adopted by Carpenter et al. (1991; see also Beck et al., 1994). They released the endangered masked bobwhite *Colinus virginianus texanus* together with knowledgeable members of a nonthreatened subspecies, *C. v. ridgwayi*. As a result, masked bobwhites acquired the necessary skills in avoiding predators and finding food. The tutor species' members had been previously sterilized to forestall hybridization.

Acquisition of Predator Avoidance via Traumatic Experience

A simpler, more straightforward way of acquiring knowledge of predators is to learn to avoid them as a result of being stared at, silently approached, or pursued at full speed (see Curio, 1993). This relies on some perceptual mechanism decoding generalized threat stimuli that triggers avoidance behavior. As a consequence of being targeted in this way, details of the entire stimulus become associated with its unconditioned aspects. Reinforcement obviously arises from the performance of the avoidance response and as a consequence, predators may come to be avoided that originally were not recognized. In such an experiment, Curio (1969) produced fear of cats by having a live feral cat harass a caged Galapagos ground finch *Geospiza fortis*. Without this treatment this species and other geospizines are "blind" to cats; carnivores have never been part of the Galapagos fauna.

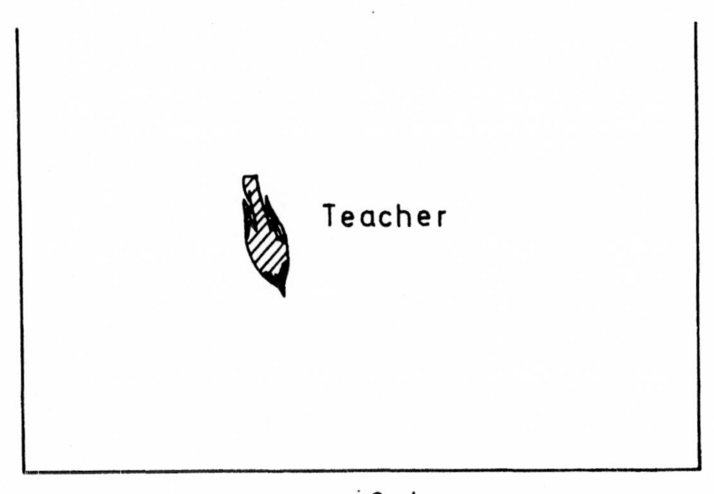

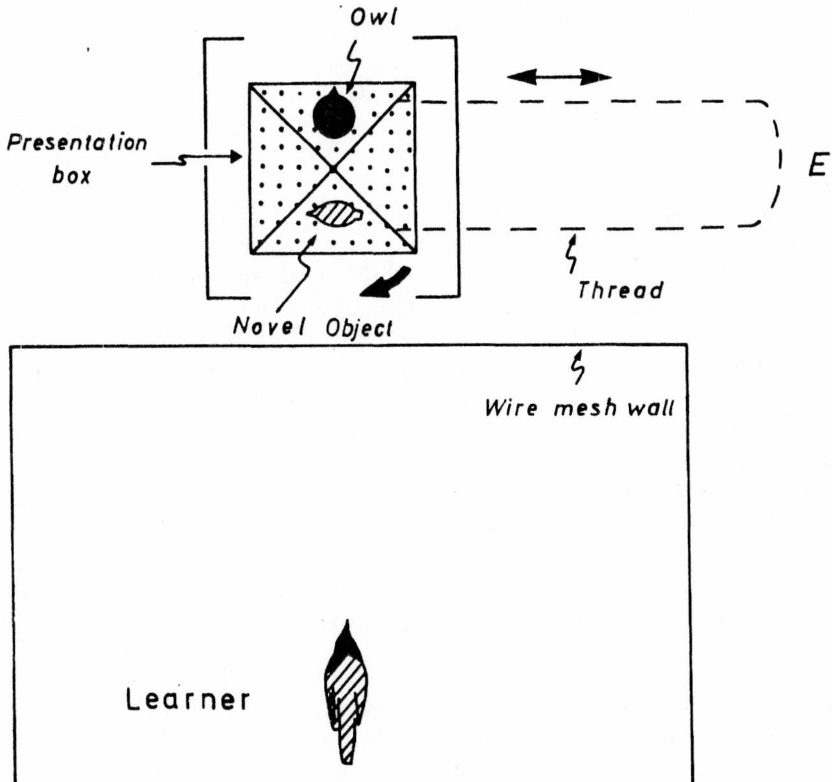

Figure 7-2 Experimental apparatus with presentation box between the aviaries of teacher and learner blackbirds. Experimenter, E, is hidden from both behind a white canvas. Owl (*Athene noctua*) and novel object are shown in position for training the learner (modified from Curio, 1988, courtesy of Lawrence Erlbaum Associates, Inc.).

In regions where prey have not coevolved with predators, such as New Zealand, predator blindness is common. Consequently, the native fauna suffers from introduced predators, and many species have become extinct (Johnson and Stattersfield, 1990). The takahe *Phorphyrio mantelli,* a flightless rail in New Zealand, has dramatically declined due to climate change and the introduction of stoats *Mustela erminea.* The rail's evolution in a mammal-free environment has led to a loss of carnivore recognition. To prepare captive-bred takahe for release, Hölzer et al. (1995) trained 10 chicks to avoid stoats in their rearing pens. Birds were first tested with a stuffed stoat and then frightened using a "stick stoat," manipulated by a hidden observer, and a "string-stoat," suspended puppet-style from an overhead line, which was made to attack the chicks in various ways. This treatment was interspersed by two others in which the stoat attacked a replica of an adult takahe emitting distress calls until it was "killed." In another treatment designed to engender self-defense, the takahe replica strongly pecked at the stoat until it was ("killed.") As a result of these treatments, takahe chicks became more vigilant, approached the stoat significantly less, stayed farther away, hid more often, and showed more signs of discomfort, such as standing alert and making escape panic runs. Coping with a stoat by retaliation was not learned, however, although all birds had intensely watched the takahe replica hitting the stoat. The length of time that these responses to the stoat would persist, to what extent they would persist after release into the wild, and how they would translate into enhanced survival have recently received a tentative answer. Captive-reared and released takahe that had been followed by radiotelemetry survived about as well as wild-reared ones. However, the number of birds and years of study are small (Maxwell and Jamieson, 1997). Finally, the behavioral modification would have to be culturally transmitted across generations to have a lasting effect on the well-being of the released population. The same questions apply to the successfully conditioned avoidance of introduced predators (fox, cat) of the Australian rufous hare-wallaby *Lagorchestes hirsutus;* individuals were subjected to a similar treatment using dummies (McLean et al., 1996).

Ellis and Serafin (1977) trained masked bobwhite quail for release by harassing captive-bred individuals with dogs, humans, and a trained Harris's hawk *Parabuteo unicinctus.* Before release, the birds showed greater general mobility, covey coordination, and predator avoidance skills (see also Carpenter et al., 1991; Beck et al., 1994). As a result of this prerelease training, released bobwhites survived better than nontrained birds (see also Ellis et al., 1977). Only in this study has it been documented that survival is enhanced in the wild as a consequence of prerelease training of a bird (see also Dowell, 1988; Beck and Kleiman, 1992; Hutchins et al., 1995).

There may be a fundamental difference between training an animal that has lost (or never had) the ability to cope with a certain predator over its evolutionary history and an animal that has become predator-blind due to captive breeding (table 7-3; see also Ellis et al., 1977; Hill and Robertson, 1988; Yoerg, 1995). In captive animals, depending on the degree of their deprivation (hand-reared, reared by foster parent, parent-reared with or without siblings; see Klinghammer, 1992), proper coping behavior may be instilled through minute improvements of the rearing regime. Yet this may not apply to an animal that evolved in a predator-free environment. For example, Darwin's finches on islands devoid of avian raptors show less fear of them, but they have a pronounced fear of snakes, and finch-eating snakes do occur on some islands (Curio, 1969; 1993). Perhaps faunas that have gone through extinction filters in the past have developed more resilience to novel threat (Balmford, 1996), for example, from introduced predators. As a case in point, more than 90% of bird species that have gone extinct in historic times lived on oceanic islands

(Johnson and Stattersfield, 1990). These islands are characterized by a paucity of vertebrate predators. Because predation has played a prime role in the extinction process (Savidge, 1987; Johnson and Stattersfield, 1990), it appears that both innate coping mechanisms and avoidance learning via traumatic experience had been impaired or lost during evolution on the islands. In these species, learning by traumatic experience may be the only way to train individuals for release.

Unidentified Learning Mechanisms

Some very simple forms of treatment may prove useful for releases. For example, laboratory mice of various strains that prove notoriously insensitive to predatory threat (e.g., movement, noise etc.) turn almost wild by being allowed to dig burrows in natural soil. After a few hours of digging and moving in and out of their burrows they respond with instantaneous flight below ground when a human approaches (E. Curio, personal observation). Similar observations have been made with domestic guinea pigs *Cavia aperea* running free along tunnels in lawns (W. Scherzinger, personal communication). Further, lizards (*Agama* spp.) adapted to indoor maintenance conditions where they have been tame to their caretakers turn wild under outdoor conditions, receiving their caretaker with their mouth agape in defense (E. Curio, personal observation). Behavioral changes accompanying both planned and accidental releases of animals are still unstudied.

Strength and Weaknesses of the Behavioral Approach

Knowledge of a species' behavior may be indispensable for management intervention to succeed. Such knowledge may save valuable resources and reduce mistakes in captive breeding and reintroductions. Understanding the function of behavior is of great help in gauging the consequences of any prerelease training on released animals' survival and reproduction. It may be profitable to consult behavioral ecologists in this endeavor, as they are used to viewing behavior as adapted to the species' environment and are therefore especially qualified to spot a mismatch between the ill effects of management and fitness-maximizing behavior.

Nevertheless, behavior studies are time intensive, and rescuing a species is often a race against time. In both captive breeding and release programs, the cost of time may be minimized when intervention measures are tried out with nonthreatened species as proxies of the endangered species (Toone and Wallace, 1994; Yoerg, 1995). Another difficulty is that behavior studies conducted at different breeding facilities can be compared only with great caution because of lack of standardization across both observers and institutions (e.g., table 7-3), although these problems can be ironed out in the long term.

Recommendations

While *in situ* activities are largely concerned with maximizing the numbers of individuals of a species, both *ex situ* and release programs need to attend to both numbers *and* the quality of breeding individuals or releases. Findings in this chapter suggest that programs should seek to encourage behavioral research both in the wild and in breeding facilities. Zoo staff need to be convinced that allocating part of its maintenance facilities to behavioral research will benefit breeding or release programs. A number of specific recommendations follow from this. First, in regard to captive breeding of birds, stress from handling and exposure to predators should be removed and enclosures enriched. Noninvasive methods should be used in capture and handling. When breeding species in captivity, females should have access to a male of her choice. This pertains only to situations where genetic screening has shown that a pair is suitable for breeding. Such a proviso needs to be relaxed when the captive population size is so low that propagation of whatever individuals must have precedence over not breeding the species at all. In addition, managers should remove sources of reproductive suppression and prevent siblicide. Reproductive suppression may be alleviated or avoided by finding the appropriate social structure conducive to breeding. Finally, when subjects are hand-reared, steps must be taken to avoid behavioral abnormalities that may affect reproduction.

In regard to captive breeding and release, hand-rearing should be avoided. Instead, subjects should be raised in the company of siblings or even other taxa if research suggests that misimprinting on those siblings does not occur.

Before release, carnivorous species must be competent at killing live prey and releases should be afraid of potentially dangerous novelty and humans. In particular, releasees must develop their innate fear of predators or be trained by some appropriate procedure with conspecifics or aversive stimuli.

Summary

A sound knowledge of a species' behavioral ecology is important in three areas of conservation management: in the wild, in captive breeding (conservation breeding) and in prerelease training and release.

Management of bird populations *in situ* may require protection from predation, prevention of intraspecific brood parasitism, siblicide and infanticide, cross-fostering, and sustaining predation at normal levels. The ill-effects of cross-fostering may be overcome by withholding the species of the foster-parent to facilitate mating with conspecifics.

Although it is a major objective of captive breeding, the establishment of endangered species in their native habitat has met with varying success. Major obstacles have been breeding sufficient numbers of releasees and release of individuals that are poorly adapted to the natural habitat, such as losing traditions of migratory routes or the readiness to migrate at all.

In captivity, removal of stress facilitates breeding; this includes shielding prey from seeing predators. Free choice of mates has been essential for successful breeding in some species.

Hand-rearing or cross-fostering individuals often renders them unsuitable for release because of deficiencies in breeding, not recognizing or coping with food, or becoming misimprinted on the foster-parent and predator blind to humans. Behavioral deficits may be forestalled by raising the animal with siblings or chicks of related species, minimizing contacts with humans, or restricting contact to only one care-taker; or remedied by breeding with a normally raised mate, or by tiding over a period failed breeding. The sexes may be affected by hand-rearing differently. Deficient nutrition may generate far-reaching but subtle ill-effects on behavior. These deficits are among the least known and the most difficult to control for.

In prerelease training and after release, experience with live prey during a sensitive period and environmental enrichment have been found to bring about appropriate hunting and killing behavior in raptors and carnivorous species. Similarly, lack of predator avoidance has perhaps been the most important behavioral deficit in released animals. An innate avoidance response (i.e., one functioning without prior experience with the predator concerned) may need parental (maternal) guidance to be maintained or, perhaps, to come about at all. Both cultural transmission of enemy recognition and an individual's traumatic experience with predator models, and other learning paradigms, can produce predator avoidance. To design an optimal pre-release training program, the adverse effects of captivity per se need to be separated from those of rearing animals in isolation.

The strength of the behavioral approach lies in that it is cost effective, but its major weakness is its lack of generalizations across species.

Acknowledgments I thank Rüdiger Cordts for critically reading a first draft of this chapter. The chapter benefited from discussions with many ethologists, conservationists, and conservation-breeding specialists and from the constructive criticism of Mark Stanley Price, Nadja Wielebnowski, an anonymous referee, and the editor.

References

Adler EM, 1975. Genetic and maternal influences on docility in the skomer vole, *Clethrionomys glareolus skomerensis.* Behav Biol 13:241–255.

Anonymous, 1994. Puerto Rican Parrots increase. CBSG News 5:11.

Apfelbach R, 1973. Olfactory sign stimulus for prey selection in polecats (*Putorius putorius* L.). Z Tierpsychol 33:270–273.

Atkinson I, 1989. Introduced animals and extinctions. In: Conservation for the twenty-first century (Western D, Pearl M, eds.). New York: Oxford University Press; 54–75.

Baker AJ, 1991. A review of New Zealand ornithology. Curr Ornithol 8:1–67.

Balmford A, 1996. Extinction filters and current resilience: the significance of past selection pressures for conservation biology. Trends Ecol Evol 11:193–196.

Barlow GW, 1968. Dither–a way to reduce undesirable fright behavior in ethological studies. Z Tierpsychol 25:315–318.

Baskin Y, 1993. Trumpeter swans relearn migration. Bioscience 43:76–79.

Beck BB, Kleiman DG, 1992. Preparation of captive-born golden lion tamarins for release into the wild. In: A case study in conservation biology: the golden lion tamarin (Beck BB, Kleiman DG, Castro I, Rettberg-Beck B, Carvalho C, eds.). Washington, DC: Smithsonian Institution Press.

Beck BB, Kleiman DG, Dietz JM, Castro I, Carvalho C, Martins A, Rettberg-Beck B, 1991. Losses and reproduction in reintroduced golden lion tamarins *Leontophithecus rosalia.* Dodo J Jersey Wildl Preserv Trust 27:50–61.

Beck BB, Rapaport LG, Stanley Price MR, Wilson AC, 1994. Reintroduction of captive-born animals. In: Creative conservation (Olney PJS, Mace GM, Feistner ATC eds.). London: Chapman and Hall; 265–303.

Beissinger SR. 1997. Integrating behavior into conservation biology: Potentials and limitations. In: Behavioral approaches to conservation in the wild. (Clemmons JR, Buchhold R eds.). Cambridge: Cambridge University Press, 23–47.

Bird DM. 1982. The American kestrel as a laboratory research animal. Nature 299:300–301.

Burley N, 1981. Mate choice by multiple criteria in a monogamous species. Am Nat 117:515–528.

Burley N, Moran N, 1979. The significance of age and reproductive experience in the mate preferences of feral pigeons, *Columbia livia*. Anim Behav 27:686–698.

Burghardt G, Milostan MA, 1995. Ethological studies on reptiles and amphibians: lessons for species survival plans. In: Conservation of endangered species in captivity (Gibbons EF, Durrant BS, Demarest J, eds.). Albany: State University of New York Press; 187–204.

Butler D, Merton D, 1992. The black robin. Oxford: Oxford University Press.

Caro TM, Durant SM, 1995. The importance of behavioral ecology for conservation biology: examples from Serengeti carnivores. In: Serengeti II: Dynamics, management, and conservation of an ecosystem (Sinclair ARE, Arcese P, eds.). Chicago: University of Chicago Press; 451–472.

Caro TM, 1994. Cheetahs of the Serengeti plains. Chicago: University of Chicago Press.

Carpenter JW, Gabel RR, Goodwin JG, 1991. Captive breeding and reintroduction of the endangered masked bobwhite. Zoo Biol 10:439–449.

Chiszar D, Rodda G, Odum RA, 1995. Two problems in ophidian conservation biology: an explosive exotic on Guam and endangered endemic on Aruba. In: Abstracts of the Animal Behavior Society Meeting, Lincoln, Nebraska, 1995; A 159.

Chivers DP, Brown GE, Smith RJF, 1995. Acquired recognition of chemical stimuli from pike, *Esox lucius*, by brook sticklebacks, *Culaea inconstans* (Osteichthyes, Gasterosteidae). Ethology 99:234–242.

Clemmons JR, Buchholz R, 1997. Behavioural approaches to conservation in the wild. Cambridge: Cambridge University Press.

Coulson JC, Thomas CS, 1983. Mate choice in the kittiwake gull. In: Mate choice (Bateson P, ed.). Cambridge: Cambridge University Press; 361–376.

Cox CR, Goldsmith VI, Engelhardt HR, 1993. Pair formation in California condors. Am Zool 33:126–138.

Csermely D, Mainardi D, Span S, 1983. Escape-reaction of captive young red-legged partridges (*Alectoris rufa*) reared with or without visual contact with man. Appl Anim Ethol 11:177–182.

Curio E, 1969. Funktionsweise und Stammes-geschichte des Flugfeinderkenuens einiger Darwinfinken (Geospizinae). Z. Tierpsychol. 26:394–487.

Curio E, 1975. The functional organization of antipredator behavior in the pied flycatcher: a study of avian visual perception. Anim Behav 23:1–115.

Curio E, 1988. Cultural transmission of enemy recognition by birds. In: Social learning: psychological and biological perspectives (Zentall T, Galef BG, eds.). Hillsdale, New Jersey: Lawrence Erlbaum Associates; 75–97.

Curio E, 1989. Some aspects of avian mortality patterns. Mitt Zool Mus Berlin 65(suppl) Ann Ornithol 13:47–70.

Curio E, 1992. Ethologie und Artenschutz. In: Erhaltungszucht programme für bedrohte Papageien. Mimeographed Sympos. Report, Detmold-Heiligenkirchen. 2–12.

Curio E, 1993. Proximate and developmental aspects of antipredator behavior. Adv Study Behav 22:135–238.

Curio E, 1996. Conservation needs ethology. Trends Ecol Evol 11:260–263.

Curio E, Ernst U, Vieth W, 1978. The adaptive significance of avian mobbing. II. Cultural transmission of enemy recognition in blackbirds: effectiveness and some constraints. Z Tierpsychol 48:184–202.

de Boer LEM, 1994. Development of coordinated genetic and demographic breeding programmes. In: Creative conservation (Olney PJS, Mace GM, Feistner ATC, eds.). London: Chapman and Hall; 304–311.

Diamond J, 1989. Overview of recent extinctions. In: Conservation for the twenty-first century (Western D, Pearl M, eds.). New York: Oxford University Press; 37–41.

Dowell SD, 1988. Rearing and predation. Game Conservancy Rev. 20: 85–88.

Dowell SD, 1989. Differential behaviour and survival of hand-reared and wild gray partridge in the United Kingdom. In: Pheasants in Asia. Proceedings Fourth Intern. Pheasant Sympos. (Hill D, Garson P, Jenkins D, eds.). Beijing, China: World Pheasant Association; 230–239. Emporia: Kansas Department of Wildlife and Parks; 230–244.

Dowell SD, 1992. Problems and pitfalls of gamebird reintroduction and restocking: an overview. Gibier Faune Sauvage (December): 773–780.

Edwards P, 1995. Ecological progress to meet the challenge of environmental change. Trends Ecol Evol 10:261.

Eibl-Eibesfeldt I, 1963. Angeborenes und Erworbenes im Verhalten der Säuger. Z Tierpsychol 20:705–754.

Ellis DH, Dobrott SJ, Goodwing Jr JG, 1977. Reintroduction techniques for masked bobwhites. In: Endangered birds (Temple SA, ed.). Madison: University of Wisconsin Press; 345–354.

Ellis DH, Serafin JA, 1977. A research program for the endangered masked bobwhite. J World Pheasant Assoc 2:16–33.

Galef BG Jr, 1988. Communication of information concerning distant diets in a social central-place foraging species: *Rattus norvegicus* In: Social learning (Zentall TR, Galef BG, eds.). Hillsdale, New Jersey: Lawrence Erlbaum Associates; 119–139.

Gilbert BK, 1968. Development of social behavior in the fallow deer (*Dama dama*). Z Tierpsychol 25:867–876.

Ginsberg JR, 1994. Captive breeding, reintroduction and the conservation of canids. In: Creative conservation (Olney PJS, Mace GM, Feistner ATC, eds.). London: Chapman and Hall; 364–383.

Gittleman JL, McMillan GC, 1995. Mammalian behavior: lessons from captive studies. In: Conservation of endangered species in captivity (Gibbons EF, Durrant BS, Demarest J, eds.). Albany: State University of New York Press.

Gossow H, 1970. Vergleichende Verhaltensstudien an Marderartigen. I. Über Lautäussrungen und zum Beuteverhalten. Z Tierpsychol 27:405–480.

Göth A, Vogel U, 1995. Status of the Polynesian megapode (*Megapodius pritchardii*) on Niuafo'ou (Tonga). Bird Conserv Int 5:117–128.

Grafen A, 1992. Of mice and the MHC. Nature 360:530.

Griffith B, Scott JM, Carpenter JW, Reed C, 1989. Translocation as a species conservation tool: status and strategy. Science 245:477–480.

Gwinner E, 1964. Untersuchungen über das Ausdrucks- und Sozialverhalten des Kolkraben (*Corvus corax corax* L.). Z Tierpsychol 21:657–748.

Hansen TF, Price DK, 1995. Good genes and old age: do old mates provide superior genes? Journ Evol. Biol. 8:759–778.

Harlow HF, Harlow MK, 1962. Social deprivation in monkeys. Sci Am 207:136–146.

Hausfater G, Hrdy SB, 1984. Infanticide. New York: Aldine.

Herrlinger E, 1973. Die Wiedereinbürgerung des Uhus *Bubo bubo* in der Bundesrepublik Deutschland. Bonn Zool Monogr 4:1–151.

Herzog M, Hopf S, 1984. Behavioral responses to species specific warning calls in infant squirrel monkeys reared in social isolation. Am J Primatol 7:99–106.

Hill D, Robertson P, 1988. Breeding success of wild and hand-reared ring-necked pheasants. J Wildl Manage 52:446–450.

Hölzer C, Bergmann HH, McLean I, 1995. Training captive-raised, naive birds to recognise their predator. In: Research and captive propagation (Ganslosser U, Hodges JK, Kaumanns W, ed.). Fürth: Filander; 198–206.

Hutchins M, Sheppard C, Lyles AM, Casadei G, 1995. Behavioral considerations in the captive management, propagation, and reintroduction of endangered birds. In: Conservation of endangered species in captivity (Gibbons EF, Durrant BS, Demarest J, eds.). Albany: State University of New York Press; 263–289.

IUCN, 1995. IUCN/SSC guidelines for re-introductions. Gland: International Union for the Conservation of Nature/SCC Re-Introduction Specialist Group; 1–4.

Johnson TH, Stattersfield AJ, 1990. A global review of island endemic birds. Ibis 132:167–180.

Kempenaers B, Verheyen GR, van den Broeck M, Burke T, van Broeckhoven C, Dhondt AA, 1992. Extra-pair paternity results from female preference for high-quality males in the blue tit. Nature 357:494–496.

Kleiman DG, 1980. The sociobiology of captive propagation. In: Conservation biology. An evolutionary-ecological perspective (Soulé ME, Wilcox BA, eds.); Sunderland, Mass.: Sinauer, Associates; 243–261.

Kleiman DG, 1994. Animal behavior studies and zoo propagation programs. Zoo Biol 13:411–412.

Klinghammer E, 1967. Factors influencing choice of mate in altricial birds. In: Early behavior: comparative and developmental approaches (Stevenson HW, ed.). New York: Wiley and Sons; 1–42.

Klinghammer E, 1992. Imprinting and early experience: how to avoid problems with tame animals. In: Wildlife Rehabilitation, vol. 9 (Ludwig DR ed.). Selected papers Ninth Sympos. National Wildlife Rehabilitators Ass. Schaumburg, Illinois; 135–143.

Kodric-Brown A, 1993. Female choice of multiple male criteria in guppies: interacting effects of dominance, coloration and courtship. Behav Ecol Sociobiol 32:415–420.

Költringer C, Sodeikat G, Curio E, 1995. Antipredator behavior of black grouse *Tetrao tetrix* chicks as influenced by hen-rearing versus hand-rearing. In: Proc Intern Symp Grouse 6: World Pheasant Association, Reading, UK; and Istituto Nazionale 1993 per la Fauna Selvatica, Ozzano dell'Emilia, Italy (Jenkins D, ed.); 81–83.

Krätzig H, 1940. Untersuchungen zur Lebensweise des Moorschneehuhns (*Lagopus l. lagopus*) während der Jugendentwicklung. J Ornithol 88:139–165.

Lindsay GD, 1992. Nest guarding from observation blinds: strategy for improving Puerto Rican parrot nest success. J Field Ornithol 63:466–472.

Lorenz K, 1931. Beiträge zur Ethologie sozialer Corviden. J Ornithol 79:67–127.

Lorenz K, 1935. Der Kumpan in der Umwelt des Vogels. J Ornithol 83:137–213; 289–413.

Magin CD, Johnson TH, Groombridge B, Jenkins M, Smith H, 1994. Species extinctions, endangerment and captive breeding. In: Creative conservation (Olney PJS, Mace GM, Feistner ATC, eds.). London: Chapman and Hall; 3–31.

Marchetti K, Price T, 1989. Differences in the foraging of juvenile and adult birds: the importance of developmental constraints. Biol Rev 64:51–70.

Marcström V, 1990. Wild and released pheasants: a comparison of their survival and breeding success. In: Vortragssammlung Conseil Intern de la Chasse et de la Conservation du Gibier (CIC ed.), Symposium 1990, Wien: (CIC); 52–62.

Martin SG, Millam JR, 1995. Nest box selection by floor laying and reproductively naive captive cockatiels (*Nymphicus hollandicus*). Appl Anim Behav Sci 43:95–109.

Maxwell JM, Jamieson IG, 1997. Survival and recruitment of captive-reared and wild-reared takahe in Fjordland, New Zealand. Conserv. Biol. 11:683–691.

McCormack G, Künzle J, 1990. Käkeröri—Rarotonga's endangered flycatcher. Rarotonga: Cook Islands Conservation Service.

McFarland D, 1985. Animal behaviour. Essex: Addison Wesley Longman Ltd.

McLean IG, 1997. Conservation and the ontogeny of behavior. In: Behavioral approaches to conservation in the wild (Clemmons JR, Buchholz R, eds.). Cambridge: Cambridge University Press; 132–156.

McLean IG, Lundie-Jenkins G, Jarman PJ, 1996. Teaching an endangered mammal to recognise predators. Biol Conserv 75:51–62.

Meffe GK, Carroll CR, 1994. Principles of conservation biology. Sunderland, Massachusetts: Sinauer Associates.

Meyburg BU, 1978. Sibling aggression and cross-fostering of eagles. In: Endangered birds: management techniques for preserving threatened species (Temple SA, ed.). Madison: University of Wisconsin Press; 195–200.

Miller B, Biggins D, Hanebury L, Vargas A, 1994. Reintroduction of the black-footed ferret (*Mustela nigripes*). In: Creative conservation (Olney PJS, Mace GM, Feistner ATC, eds.). London: Chapman and Hall; 364–383.

Mineka S, Cook M, 1988. Social learning and the acquisition of snake fear in monkeys. In: Social learning (Zentall TR, Galef BG, eds.). Hillsdale, New Jersey: Lawrence Erlbaum Associates; 51–53.

Mock DW, Forbes LS, 1994. Life-history consequences of avian brood reduction. Auk 111:115–123.

Møller AP, Pomiankowski, 1993. Why have birds got multiple sexual ornaments? Behav Ecol Sociobiol 32:167–176.

Moors PJ, 1985. Conservation of island birds. ICBP Technical Publication no 3. Cambridge: International Council for Bird Preservation.

Myers SA, Millam JR, Roudybush TE, Grau CR, 1988. Reproductive success of hand-reared vs parent-reared cockatiels (*Nymphicus hollandicus*). Auk 105:536–542.

Nol E, Smith JNM, 1987. Effects of age and breeding experience on seasonal reproductive success in the song sparrow. J Anim Ecol 56:301–313.

Nuechterlein GL, 1981. 'Information parasitism' in mixed colonies of western grebes and Forster's terns. Anim Behav 29:985–989.

Ostermeier G, 1994. Was macht männliche Java-Bronzemännchen (*Lonchura leucogastroides*) für Weibchen sexuell attraktiv? (Diploma thesis). Bochum: Ruhr-Universität Bochum.

Paine RT, 1966. Food web complexity and species diversity. Am Nat 100:65–76.

Parker PG, Waite TA, 1995. Mating systems and effective population size. In: Abstracts of the Animal Behavior Society Lincoln, Nebraska, 1995 A 254.

von Pernau A. 1702. Unterricht, was mit dem lieblichen Geschöpf, denen Vögeln, auch ausser den Fang, nur durch die Ergründung deren Eigenschaften, und Zahmmachung, oder anderer Abrichtung/man sich vor Lust und Zeit=Vertreib machen könne. Reprint 1982, Monographs; Coburg: Natur-Museum, suppl 3.

Pollack V, 1994. Verändern verschiedene Futterqualitäten während der Nestlingszeit das Verhalten von Amseln? (Diploma thesis). Innsbruck: University of Innsbruck.

Polsky RH, 1975. Hunger, prey feeding, and predatory aggression. Behav Biol 13:81–93.

Polsky RH, 1977a. The ontogeny of predatory behaviour in the golden hamster (*M. a. auratus*). III. Sensory preexposure. Behaviour 63:175–191.

Polsky RH, 1977b. The ontogeny of predatory behaviour of the golden hamster (*Mesocricetus a. auratus*). II. The nature of the experience. Behaviour 61:58–81.

Rasa OAE, 1989. Behavioural parameters of vigilance in the dwarf mongoose: social acquisition of a sex-biased role. Behaviour 110:125–145.

Rothschild M, Lane C, 1959. Warning and alarm signals by birds seizing aposematic insects. Ibis 102:328–330.

Russock HI, Hale EB, 1978. Functional validation of the *Gallus* chick's response to the maternal food call. Z Tierpsychol 49:250–259.

Savidge JA, 1987. Extinctions of an island forest avifauna by an introduced snake. Ecology 68:660–668.

Scherzinger W, 1982. Trials with natural broods of grouse. Proc. 2nd Internat. Sympos. on Grouse (Lovel TWI ed.). Dalhousie Castle: Bonnyrigg, Edinburgh; 199–201.

Schoener TW, Spiller DA, 1996. Devastation of prey diversity by experimentally introduced predators in the field. Nature 381:691–694.

Seligman M, Hager J, 1972. Biological boundaries of learning. New York; Meredith.

Semel B, Sherman PW, Byers SM, 1988. Effects of brood parasitism and nest-box placement on wood duck breeding ecology. Condor 90:920–930.

Shepherdson D, 1994. The role of environmental enrichment in the captive breeding and reintroduction of endangered species. In: Creative conservation (Olney PJS, Mace GM, Feistner ATC, eds.). London: Chapman and Hall; 167–177.

Snyder NF, Snyder HA, Johnson TB, 1989. Thick-billed parrot reintroduction. Birds Int 1:3–15.

Snyder NFR, Koenig SE, Koschman J, Snyder HA, Johnson TB, 1994. Thick-billed parrot releases in Arizona. Condor 96:845–862.

Soulé ME, Bolger DT, Alberts AC, Wright J, Sorice M, Hill S, 1988. Reconstructed dynamics of rapid extinctions of chaparral-requiring birds in urban habitat islands. Conserv Biol 2:75–92.

Stanley Price MR, 1989. Reconstructing ecosystems. In: Conservation for the twenty-first century (Western D, Pearl M, eds.). New York: Oxford University Press; 210–218.

Suboski MD, Bain S, Carty AE, McQuoid LM, Seelen MI, Seifert M, 1990. Alarm reaction in acquisition and social transmission of simulated-predator recognition by zebra danio fish (*Brachydanio rerio*). J Comp Psychol 104:101–112.

ten Cate C, 1995. Behavioural development in birds and the implications of imprinting and song learning for captive propagation. In: Research and captive propagation (Ganslosser U, Hodges JK, Kaumanns W, eds.). Fürth: Filander-Verlag; 187–197.

Thaler E, Pechlaner H, 1980. Cainism in the lammergeier or bearded vulture. Int Zoo Yrbk 20:278–280.

Thaler E, 1987. Studies on the behavior of some Phasianidae-chicks at the Alpenzoo-Innsbruck. J Sci Fac CMU 14:135–149.

Thaler E, Pegoraro K, Stabinger S, 1992. Familienbindung und Auswilderung des Waldrapps *Geronticus eremita*—ein Pilotversuch. J Ornithol 133:173–180.

Toone WD, Wallace MP, 1994. The extinction in the wild and reintroduction of the California condor (*Gymnogyps californianus*). In: Creative conservation (Olney PJS, Mace GM, Feistner ATC, eds.). London: Chapman and Hall; 411–419.

Tulley JJ, Huntingford FA, 1987. Paternal care and the development of adaptive variation in antipredator responses in sticklebacks. Anim Behav 35:1570–1572.

Veltman CJ, Nee S, Crawley MJ, 1996. Correlates of introduction success in exotic New Zealand birds. Am Nat 147:542–557.

von de Wall W, Rajala P, 1969. Zum Feindverhalten finnischer Rentiere. Ann Acad Sci Fenn Ser A Biol 147:1–18.

Wildt DE, Rall WF, Critser JK, Monfort SL, Seal US, 1997. Genome resource banks. BioScience 47:689–98.

Wiley JW, 1985. The Puerto Rican Parrot and competition for its nest sites. In: Conservation of island birds (Moors PJ, ed.). Cambridge: International Council for Bird Preservation; 213–223.

Wilson MH, Kepler C, Snyder NFR, Derrickson S, Dein FJ, Wiley JW, Wunderle JM, Lugo

AE, Graham DL, Toone WD, 1994. Puerto Rican Parrots and potential limitations of the metapopulation approach to species conservation. Conserv Biol 8:114–123.

Woolpy JH, Ginsburg BE, 1967. Wolf socialization: a study of temperament in a wild social species. Am Zool 7:357–363.

Wüstehube C, 1960. Beiträge besonders des Spiel- und Beuteverhaltens einheimischer Musteliden. Z Tierpsychol 17:579–613.

Yamamoto JT, Shields KM, Millam JR, Roudybush TE, Grau CR, 1989. Reproductive activity of force-paired cockatiels (*Nymphicus hollandicus*). Auk 106:86–93.

Yoerg SI, 1995. Born to be wild: captive-born and wild-caught kangaroo rats respond differently to a live snake. In: Abstracts of the Animal Behavior Society Meeting, Lincoln, Nebraska, 1995; A 156.

Part III

Mating Systems and Conservation Problems

The ways in which individuals acquire mates, the number of partners they acquire, their type of pair bonds, and the form and division of parental care are all components of a mating system (Emlen and Oring, 1977). Differences in mating systems are therefore reflected in differences in many facets of animal behavior, including male and female reproductive tactics (Andersson, 1994), aspects of parental care (Clutton-Brock, 1991), variance in reproductive success (Clutton-Brock, 1988), and adult sex ratios (Hamilton, 1967). Each of these factors has the potential to influence N_e, population growth rates, and hence the resilience of a population to exploitation and speed of population recovery. The first two chapters in this section investigate some of these connections, while the third tests the importance of different aspects of mating systems in influencing N_e. The strength of these chapters, taken together, is in highlighting the diversity of ways in which animal mating systems are pertinent to problems in conservation biology and in suggesting numerous avenues of further research.

At the most basic level, most animals have to come together to mate, and the mating system dictates the different types of association between individuals in terms of cooperation, permanency, and even location. In chapter 8, Dobson and Poole address the ramifications of conspecific aggregations for several aspects of conservation biology first by drawing attention to the importance of conspecific aggregations for population demography. For years, ecologists have realized that population growth rates can be nonlinear with respect to population densities; Allee (1931) pointed out that there may be a critical density threshold below which a population may experience a precipitous decline. There are still few empirical examples of the Allee effect, however, and the mechanisms by which this could occur remain obscure. Next, Dobson and Poole show how the spread of disease is influenced by patterns of aggregation in a host population, with infection more likely to be transmitted when hosts live close together. Since diseases are an important but relatively unexplored topic in conservation (Dobson and May, 1986; Dobson, 1995); and because populations will become increasingly tightly packed in isolated habitat patches, this topic needs further modeling and experimental work.

Dobson and Poole then go on to examine how selective harvesting affects a population's response to exploitation and its potential to recover. In their example from elephants *Loxondata africana*, ivory poaching removes first large-tusked males, then females, and finally juveniles from a population, resulting in highly female-biased adult sex ratios. Because males gain access to mates by searching over large distances for clumped groups of related females (Barnes, 1982), and because females preferentially mate with older males, selective removal of large-tusked males may prevent females from becoming pregnant. In addition, the distribution of female group sizes also affects the rate of offspring production, with fewer calves born as females aggregate in smaller and smaller herds. This story is now one of the classic examples linking animal behavior and conservation. Finally, Dobson and Poole point to a growing literature on individuals using conspecifics as cues as to where to settle and breed (Stamps, 1988). This has repercussions for both metapopulation dynamics, reintroductions, and the efficacy of reserves because all assume that a habitat is equally attractive whether or not conspecifics are present.

Vincent and Sadovy, in chapter 9, examine the ways in which seven different aspects of mating systems in marine fishes could affect responses to fisheries exploitation. They suggest that fecundity could be limited if larger fish were taken in nets, as size and fertility are often correlated; that fishing could disrupt spawning activity; that removal of fish with long-term pair bonds could disrupt offspring production; that the operational sex ratio might be biased if one sex is larger and preferentially taken; and that removal of males could impact paternal care of offspring.

In their last two examples of the ways in which individual adaptive strategies could affect populations in surprising ways, Vincent and Sadovy focus on species in which individuals show more than one reproductive tactic. In coho salmon *Oncorhynchus kisutch*, males either mature when small and young (jacks) or delay maturation (hooknoses) in a frequency-dependent manner. Because hooknoses spend more time at sea and are therefore at greater risk from fishing, jacks may be at a selective advantage under intense fishing practice. Eventually this will alter the phenotypic distribution of the male population (Gross, 1991) and will additionally lower fishing returns. In a second example of hermaphroditic fish species in which adult fish change sex either at a given size and age (uncompensated sex change) or in response to social conditions (compensated), Vincent and Sadovy compare how these different mechanisms will lead to different adult sex ratios in exploited populations. For example, in protogynous fish species, removal of large individuals (here males) may lead to female-biased sex ratios in uncompensated systems but to equal sex ratios in compensated situations, which in turn could affect population growth rates in the remaining population. In essence, Vincent and Sadovy suggest that models of population response to harvesting (or population recovery) will be too simplistic without considering the subtleties of mating systems.

Creel examines the factors affecting N_e in the last chapter in this section using long-term demographic data collected on seven carnivore species, cheetahs *Acinonyx jubatus*, dwarf mongooses *Helogale parvula*, Ethiopian wolves *Canis simiensis*, gray wolves *Canis lupus*, lions *Panthera leo*, spotted hyenas *Crocuta crocuta*, and wild dogs *Lycaon pictus*, which have different mating systems. Theory predicts that variability in population size, sex ratios, and variation in reproductive success all influence N_e (Franklin, 1980), but the relative importance of these factors is rarely tested using real data. Comparative records show that a large skew in reproductive success has the greatest impact on N_e and that this is pronounced in reproductively suppressed species. Interestingly, however, N_e is lower in lions than in dwarf mongooses because, whereas in lions all females but only some males breed, in dwarf mongooses suppression occurs in both sexes. Creel also examines demographic N_e as defined as the size of an ideal population with an even sex ratio and stable age distribution that has the same net change in numbers over a year as the population of interest (Caughley, 1994). Creel finds that the demographic N_e is only 60% of the actual population size in dwarf mongooses.

The conservation implications of Creel's analyses are depressing: his genetic N_e values for carnivores are lower than for other mammals, around 20% of the census population size. Thus for a target N_e of 500, the population size would have to reach 2500. For wild dogs living at average population density, one would need a reserve of 142,500 km^2 to hold this number of wild dogs, yet the largest area protecting a wild dog population, the Selous Game Reserve in Tanzania, is only 43,000 km^2.

Given the multiplicity of ways in which animal mating systems affect a population's response to economic exploitation in numerical and genetic terms, the studies presented in this section should be viewed as the tip of an emerging iceberg of links between patterns of mating and conservation biology.

References

Allee WC, 1931. Animal aggregations. Chicago: University of Chicago Press.

Andersson M, 1994. Sexual selection. Princeton, New Jersey: Princeton University Press.

Barnes RFW, 1982. Mate searching behaviour of elephant bulls in a semi-arid environment. Anim Behav 30:1217–1223.

Caughley G, 1994. Directions in conservation biology. J Anim Ecol 63:215–244.

Clutton-Brock TH, 1988. Reproductive success: studies of individual variation in contrasting breeding systems. Chicago: University of Chicago Press.

Clutton-Brock TH, 1991. The evolution of parental care. Princeton, New Jersey: Princeton University Press.

Dobson AP, 1995. The ecology and epidemiology of rinderpest virus in Serengeti and Ngorongoro conservation area. In: Serengeti II: dynamics, management, and conservation of an ecosystem (Sinclair ARE, Arcese P, eds.). Chicago: University of Chicago Press; 485–505.

Dobson AP, May RM, 1986. Disease and conservation. In: Conservation biology: the science of scarcity and diversity (Soulé ME, ed.). Sunderland, Massachusetts: Sinauer Associates; 345–365.

Emlen ST, Oring LW, 1977. Ecology, sexual selection, and the evolution of mating systems. Science 197:215–223.

Franklin IR, 1980. Evolutionary change in small populations. In: Conservation biology: an evolutionary-ecological perspective (Soulé ME, Wilcox BA, eds.). Sunderland, Massachusetts: Sinauer Associates; 135–149.

Gross MR, 1991. Salmon breeding behavior and life history evolution in changing environments. Ecology 72:1180–1186.

Hamilton WD, 1967. Extraordinary sex ratios. Science 156:477–487.

Stamps JA, 1988. Conspecific attraction and aggregation in territorial species. Am Nat 131:329–347.

8

Conspecific Aggregation and Conservation Biology

Andy Dobson
Joyce Poole

The Ubiquity of Conspecific Aggregation

Nearly all animals aggregate for some period of their lives, for mating, for raising offspring, at feeding sites, or because they live in temporary or permanent groups (Wilson, 1975). Many years of behavioral ecological research have shown that there are costs and benefits for each individual in associating with conspecifics, and that these may be important for predator defense (e.g., Thomson's gazelles *Gazella thomsoni:* FitzGibbon, 1990), increasing access to food (weavers *Quelea quelea:* De Groot, 1980; wild dogs *Lycaon pictus:* Creel and Creel, 1995), or for reproduction (many cooperatively breeding birds; Stacey and Koenig, 1990). Because many benefits of grouping affect survival and reproduction, and many of these benefits fall disproportionately on certain segments of the population a priori, it seems likely that variations in patterns of conspecific aggregation will affect population growth rates and thus the ability of populations to withstand exploitation, environmental insult, and to make a recovery. The difficulty is in determining the precise ways in which conspecific aggregation affects growth rates, as well as understanding ramifications for other aspects of conservation biology including habitat selection, metapopulation dynamics, and monitoring strategies. Despite these acknowledged difficulties, it is increasingly recognized that models and management plans will continue to remain crude if such knowledge is not incorporated (see Durant, Chapter 5, this volume).

Conspecific Aggregation Disregarded

Although the causes and consequences of group living in animals have been central questions in behavioral ecology, conservation biologists have given patterns of grouping scant attention. Admittedly, those long interested in conservation, such as wildlife biologists, have known that birds, for example, are attracted to conspecifics (Darling, 1952), and naturalists have noted the importance of conspecifics in contexts such as offspring defense. Nevertheless, conservation biologists, concerned primarily with the behavior of populations rather than the behavior of individuals, have almost disregarded the phenomenon of group living (but see Reed and Dobson, 1993). Only in one context have they embraced the importance of conspecific aggregation or grouping: in terms of the Allee effect. Allee (1931, 1938) described situations in which per capita growth rate at low population densities can be an increasing function of population density and in which conspecific attraction can increase local density without affecting population size. Population growth rates may thereby increase disproportionately in locally aggregated species or may decline rapidly where a minimum aggregation size is necessary for reproduction. Unfortunately, observations often rest on unsubstantiated case studies. For instance, some authors argue that the passenger pigeon *Ectopistes migratorius* finally became extinct in the wild when pairs stopped reproducing in colonies that had been much reduced in size through shooting and trapping (Blockstein and Tordoff, 1985). Nevertheless, it is now clear that patterns of conspecific aggregation have ramifications that extend to many other areas of conservation biology. The purpose of this chapter is to highlight, with selected examples, five areas of conservation biology in which grouping or conspecific aggregation affects the outcome of conservation predictions or management plans. These areas are (1) the effect of grouping on population growth rates, (2) the effect of grouping on the susceptibility of populations to disease, (3) the effect of grouping on populations' responses to exploitation, (4) the effect of grouping on population recovery, and (5) the importance of conspecific attraction for patch recolonization.

Conspecific Aggregation and the Allee Effect

Individuals of colonial species are better able to find a mate, avoid predation, and forage efficiently if they are members of a colony (Greene, 1987). If a colony provides a net advantage, we might expect colony size to be associated with breeding success. For example, Birkhead (1977) found that reproductive success was associated with colony size in common guillemots *Uria aalge*. This suggests that conspecific attraction is important in determining whether individuals in a population establish successful breeding colonies or locate habitats in which to establish breeding territories. Conspecific attraction also can affect population dynamics and persistence, particularly in small or establishing populations. For example, declining populations in species where conspecific attraction is important will show a greater decline in reproductive success than will populations of other species. The effect of this decline in reproductive success on recruitment to the breeding population and on the longer term dynamics of the population can be examined using a simple model. We assume some animals are required to be present in a colony or habitat for individuals in the population to achieve their maximum potential reproductive success. The reproductive rate of individuals in the population at any time is thus modified by the function

$$f(N_t, G) = 1 - e^{-(N_t/G)}. \tag{1}$$

Here, N_t is the total population size at time t, and G is the approximate number of breeding adults required for successful reproduction. The shape of this function is sketched for three different minimum population sizes in fig. 8-1. The function was then used in a simple stochastic simulation of population growth for a population that would otherwise exhibit a small positive rate of population growth (annual survival = 0.34, annual fecundity = 3). The population was initialized with 50 individuals and run for a maximum of 100 years in each of 30 simulations. Two figures were calculated for each value of critical group size:

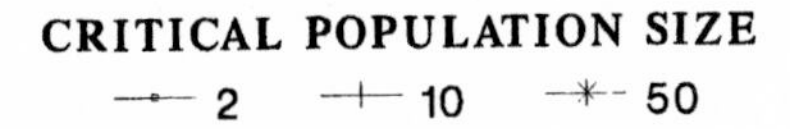

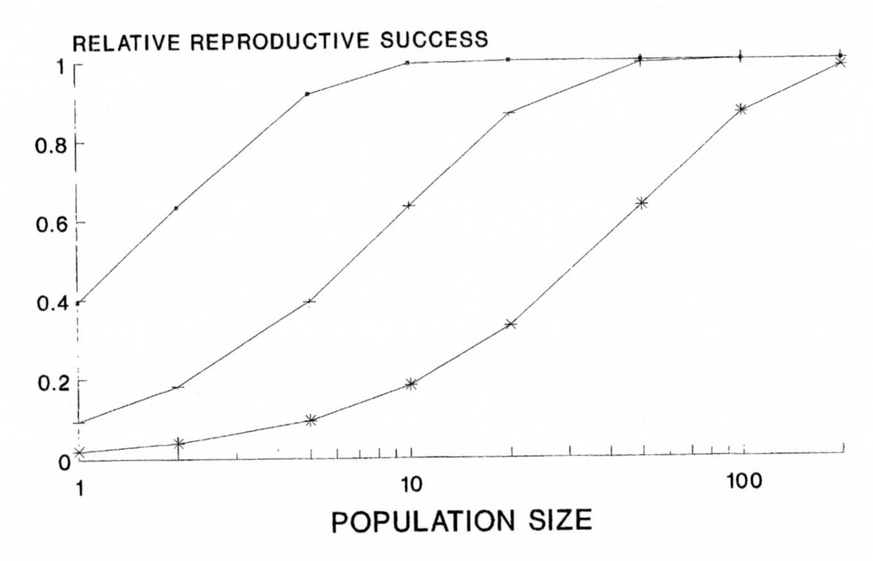

Figure 8-1 The relationship between population size and relative population growth rate for three different values of G. The figure is drawn for $G = 2$ (top line), $G = 10$ (center line), and $G = 50$ (bottom line); these figures correspond to populations which only attain their maximum attainable growth rates when in excess of 10, 50, and 250 individuals, respectively.

the proportion of populations that declined to extinction and the time it took them to do so. Figure 8-2 illustrates that, as critical group size increases, a higher proportion of populations decline to extinction, and the time it takes them to do so is an inverse function of critical group size. Plainly, behavioral mechanisms can cause effects that considerably increase the vulnerability of small populations to extinction.

Dobson and Lyles (1989) have examined the effect of primate social system on primate population viability (see also Harcourt, chapter 3, this volume). Their work showed that different types of social systems tend to produce threshold population densities—analogous in some ways to the establishment of infectious diseases (see below). When population size drops below these densities, it is likely that the population will decline fairly rapidly to local extinction. This work suggested that viability thresholds would be higher for monogamous species than for polygamous species that lived in larger social groups.

A further aspect of this work indicated that the life-history of most primate species showed relatively invariant properties when rescaled by interbirth interval. Thus, the age of first-reproduction of mountain gorillas *Gorilla gorilla beringei* initially appears very different from that of pygmy marmosets *Cebuella pygmaea,* until it is realized that in both cases, age at first reproduction is approximately three interbirth intervals. This allows demographic parameter estimation, a problem that has beset many population viability analyses, to be simplified and undertaken with minimal data. This approach allowed Dobson and Lyles (1989) to predict that most primates require survival rates *per interbirth interval* to

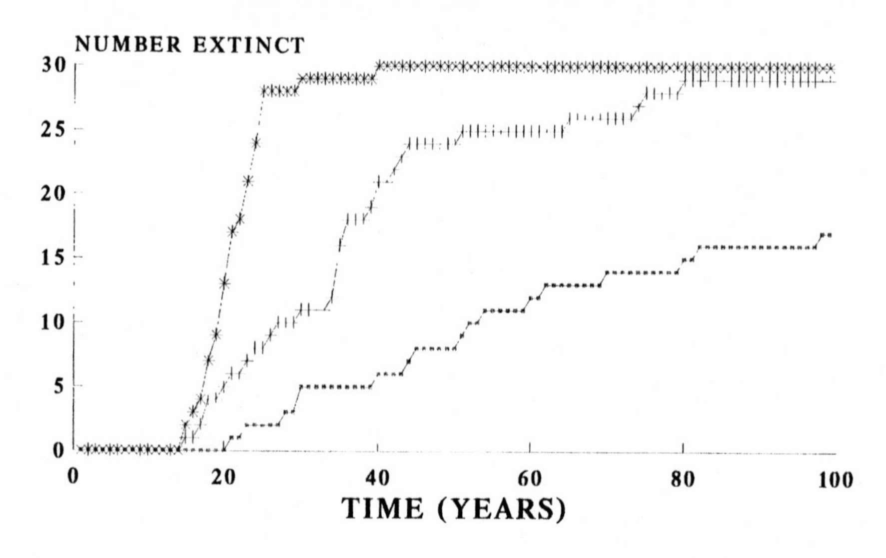

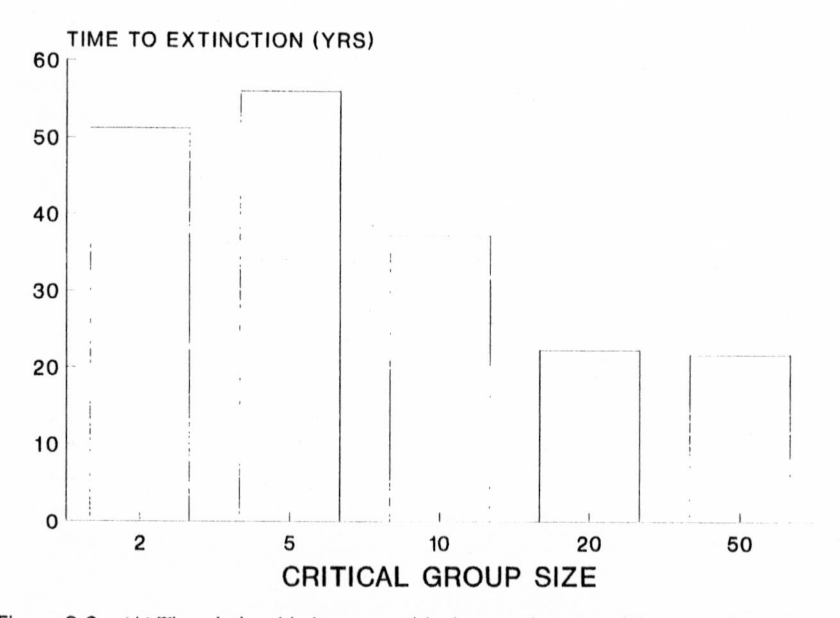

Figure 8-2 (A) The relationship between critical group size, G, and the proportion of populations that decline to extinction in one year. (B) The average time taken for populations of initial size 50 to decline to extinction for five different values of G.

exceed 70%; otherwise they were likely to decline in numbers. Harcourt (1995) showed that this was a useful approximation for the mountain gorillas in Rwanda. Such allometric approaches could be more widely used in conservation biology (Peters, 1983; Charnov, 1993).

Grouping and Pathogen Transmission

The ability of parasites and disease to become established in host populations and the impact they have when accidentally introduced is crucially determined by the spatial distribution of the host population. For most species, this spatial distribution is determined by the social organization of the species, which is in turn a consequence of the underlying distribution of food and other vital resources (Rubenstein and Wrangham, 1986). There have recently been a number of dramatic examples of disease outbreaks in wild populations of large vertebrates (rabies or distemper in wild dogs: Gascoyne et al., 1993; morbilliviruses in lions: Roelke-Parker et al., 1996, in Atlantic white-sided dolphins *Lagenorhynchus acutus* and in harbor seals *Phoca vitulina:* Kennedy, 1990). In each case the social system of the hosts determined their spatial distribution and the impact of the disease. In the Serengeti lions and wild dogs, individual prides and packs were infected, but the rate of spread between packs was slow. In contrast, in the North Sea seal colonies, large numbers of individuals died as a result of living in large groups.

The consequence of social aggregation can be fairly easily added to the classic SIR model for an infectious disease where the host population is divided into susceptible, infectious, and recovered individuals:

$$dS/dt = (a - b)\,S - f(S, I)\,S. \tag{2}$$

$$dI/dt = f(S, I)\,S - bI - \alpha I - \gamma I \tag{3}$$

$$dR/dt = \gamma I - bR \tag{4}$$

Here S, I, and R are the densities of susceptible, infected, and resistant individuals, respectively, a and b are the birth and death rates of the host in the absence of the pathogen, α is the parasite-induced host mortality rate, and γ is the rate at which infected individuals recover. The transmission term, $f(S,I)$, can take a number of alternate forms. Unfortunately, the behavior of many epidemiological models is sensitive to the assumptions made about whether transmission proceeds as a simple pseudomass action function, IS, or a true mass action function, IS/H, where $H = S + R$. (In some publications these are designated density-dependent and frequency-dependent transmission). Here we will consider a simple modification to the pseudomass action formulation which gives rise to an expression for the basic reproductive ratio of the pathogen, R_0, that contains an expression for the numbers of susceptible individuals in the population. When the expression for R_0 is rearranged at unity to obtain an estimate of the threshold number of hosts which are required to just sustain an infection of the pathogen, H_T, this can only be obtained for the pseudomass action transmission model.

$$H_T = (\alpha + b + v)/\beta. \tag{5}$$

We can now add an additional degree of complexity into the transmission term that considers the spatial distribution of the host population. Here we generalize this approach by using a mixture of terms based on Lloyd's mean crowding (Lloyd, 1967) and Taylor's power law (Taylor, 1961; Taylor and Taylor, 1977). Lloyd's index of mean crowding was derived to provide an estimate of the number of individuals likely to co-occur in a quadrat centered on a randomly chosen individual. In its simplest form, mean-crowding is measured as

$$\hat{x} = \bar{x} + \frac{s^2}{\bar{x}} - 1. \tag{6}$$

Here x is the mean number of individuals found in a quadrat, s^2 is the variance of the numbers of individuals found in the quadrat, and $\hat{x}$ is Lloyd's mean crowding. As the population becomes more aggregated in its spatial distribution, the variance increases faster than the mean, and each individual is likely to be surrounded by a greater number of other individuals. The index has interesting epidemiological applications; if we consider the randomly chosen individual to be an infected individual, then $\hat{x}$ will be the number of susceptible individuals in its vicinity. As the host population becomes more aggregated, any infected individual is likely to transmit the infection to more susceptible individuals.

If we are to include information about the spatial distribution of the host population into simple epidemiological models, it is also important to consider how spatial aggregation will change with changes in population density. An important empirical property of most biological populations is the variance in their spatial density is usually a simple power function of their mean density (Taylor, 1961; Taylor and Taylor, 1977). It allows us to replace the expression for the variance in eq. 6 with one that describes the variance as a function of the mean density:

$$\hat{x} = \bar{x} + \frac{c\bar{x}^d}{\bar{x}} - 1. \tag{7}$$

This allows us to obtain an expression for the transmission of a pathogen that captures phenomonologically some of the details of the spatial aggregation of the host population. Hence the transmission term in eqs. 2–4 becomes

$$f(S,I) = \beta I(\bar{s} + c\bar{s}^{d-1} - 1). \tag{8}$$

Taylor and colleagues (Taylor and Taylor, 1977; Taylor et al., 1978) provide evidence that the slope of the power law relationship for most species is between 1 and 3, with unity corresponding to a randomly distributed population and 3 to a species that is highly aggregated in its spatial distribution.

As the host population becomes more aggregated in its distribution, a significant increase occurs in the number of susceptible hosts that come into contact with any infected individual. This will lead to a significant decline in the threshold for establishment of the pathogen into the host population (fig. 8-3). Obviously, this approach to understanding the role of host spatial distribution on epidemic outbreaks and pathogen establishment ignores many details. In particular, it disregards the explicit spatial distribution of the individuals in the host population, but to examine this would require detailed case-specific computer models. However, the approach provides an intermediate step that uses information that can be quite readily quantified empirically on the spatial distribution of hosts. Most importantly, the results suggest that the spatial distribution of the host population is important in determining the critical community density for pathogen establishment. More detailed spatial models can then be developed to examine the importance of where the parasite is introduced.

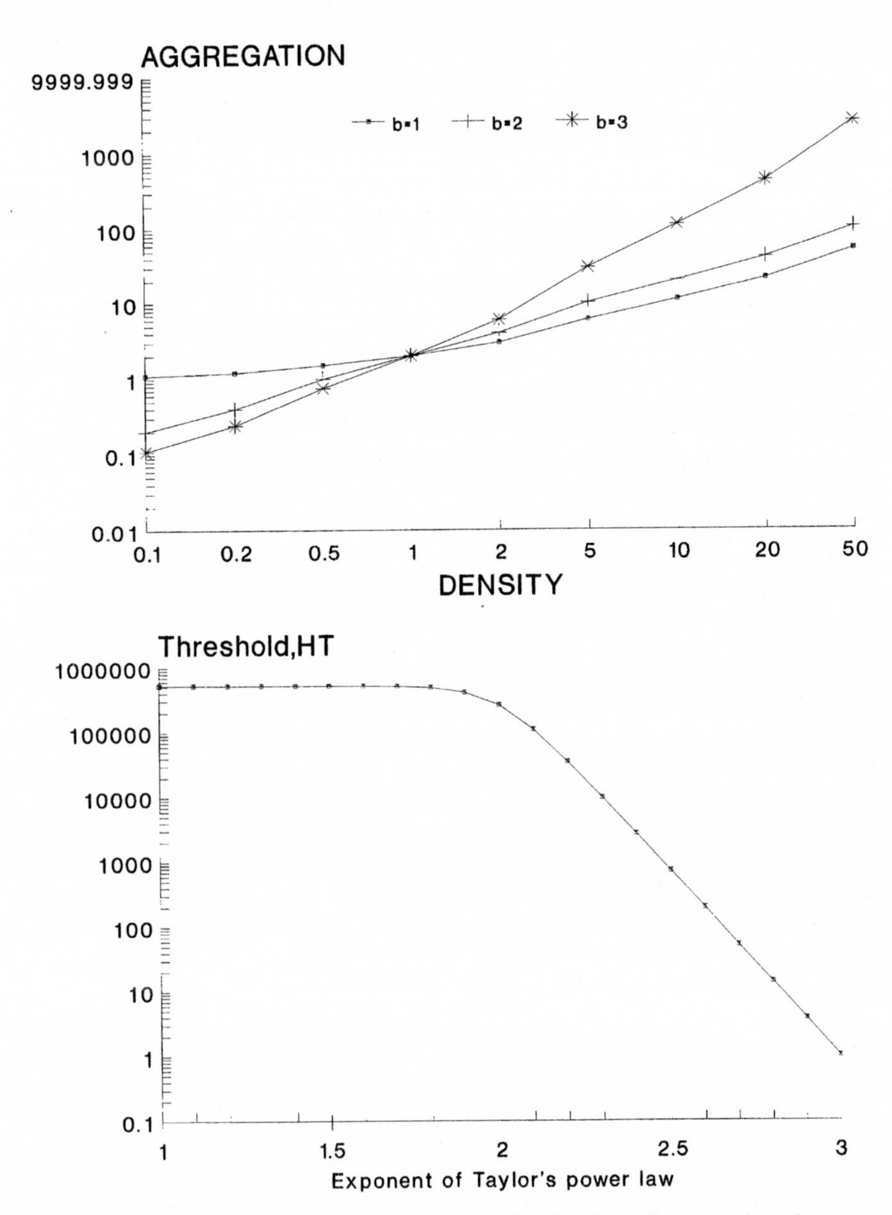

Figure 8-3 (A) The relationship between population density and aggregation, the number of individuals surrounding a single infective individual. Changes in aggregation with population density are determined by the exponent of the power function in Taylor's power law; this is then substituted into the modified expression for Lloyd's mean crowding in the main text (eq. 6). In this figure the exponent b has been set to 1 (dots), 2 (crosses), or 3 (stars). (B) The relationship between threshold for establishment of a pathogen and degree of aggregation of the host population. The degree of host aggregation is quantified using the slope of Taylor's power law for the relationship between the variance of conspecifics in a transmission circle surrounding an infected host and the mean number of susceptibles within the transmission circle. The area of the circle is assumed to increase with the magnitude of β_A.

Exploitation of Species

The harvesting of natural resources in particular populations of fish, whales, elephants, and other vertebrates has as an underlying assumption that there is a surplus of individuals in the population. In many long-lived vertebrates, certain individuals often play a crucial role in maintaining recruitment levels, particularly by feeding and protecting young and immature relatives. There are numerous anecdotal examples: old female elephants lead groups to food sources (Moss and Poole, 1983); wildebeest *Connochaetes taurinus* may require a minimum group size to defend their offspring against wild dog attack (Creel and Creel, 1995); and nonbreeding black-backed jackals *Canis mesomelas* increase the reproductive success of their parents (Moehlman, 1979). In short, the excess of individuals is often more apparent than real. While these observations are in urgent need of more systematic study, the idea that wildlife can only be protected by harvesting it remains prevalent in the game departments of many countries and international wildlife organizations.

Our work on African elephants provides a clear example of this problem. Rapid exploitation of elephant populations has led to a continental decline of more than 95% in the last 200 years (Douglas-Hamilton, 1987; Milner-Gulland and Beddington, 1993). Available data show that poachers selectively remove elephants with tusks larger than a minimum size, and this results in a more complex relationship between ivory yield and population size than if poachers removed elephants at random. The reason is that tusk size is dependent on both the age and sex of an elephant (fig. 8-4a). Initially, the larger tusked adult males will be selected in preference to females and younger males. Later, selective poaching at lower tusk weights rapidly produces populations with distorted sex ratios and age structures (fig. 8-4b). Surveys conducted on East African elephant populations confirm these highly skewed sex ratios: in some areas mature females outnumber mature males by a factor of $> 50:1$ (Poole, 1989a). Furthermore, the proportion of females showing signs of pregnancy or accompanied by recent offspring was diminished in these populations. Because females preferentially mate with older males (Poole, 1989b), there is a danger of reducing the population's growth rate. To examine how these changes in social structure affect the ability of populations to recover, we have extended the model to ascertain the conditions when a paucity of mature males limits conception rates.

Recovery from Exploitation

We can model the social system of elephants by assuming that males searching for females have the dynamics of a simple predator–prey system. We assume that M males search at a

Figure 8-4 The effect of selective poaching for large tusks on ivory yield and elephant population size. Here we assume that poachers select elephants with tusks larger than a certain size. As males have substantially larger tusks than females, they will be more heavily exploited in the early years. As mature males become rare, poachers include females and immature males in the harvest, producing a more complex relationship between yield and population size. It is particularly important to note that once females have been included in the harvest, small decreases in average tusk weight produce large decreases in population size. The irregularity that occurs in the recruitment function close to carrying capacity reflects the switching point when poachers begin to include females in their harvest. (B) The effect of selective poaching on the sex ratio of elephants in the population (mature females per mature male).

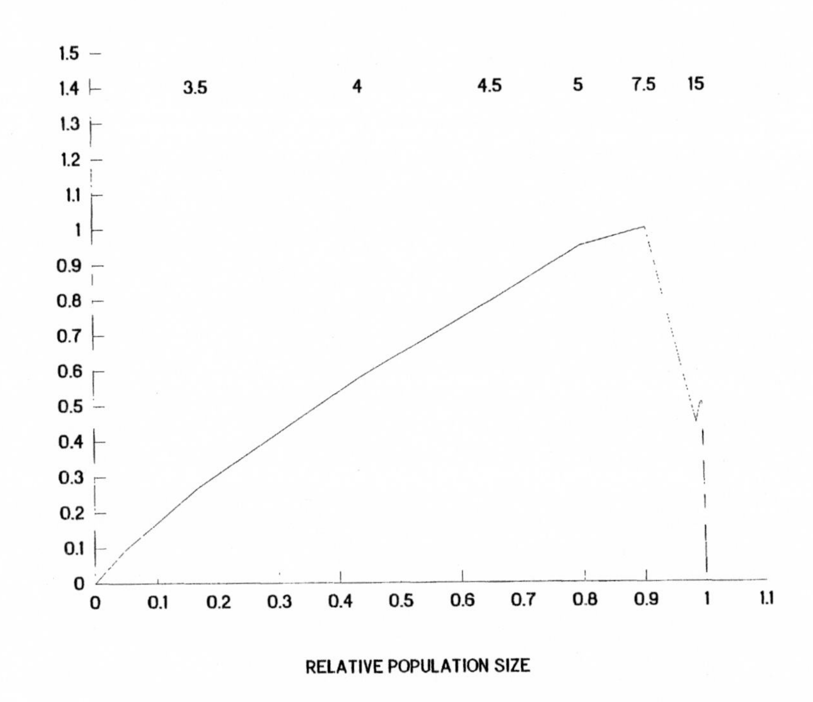
MINIMUM TUSK SIZE POACHED
RELATIVE IVORY YIELD
RELATIVE POPULATION SIZE
3.5
4
4.5
5
7.5
15
1.5
1.4
1.3
1.2
1.1
1
0.9
0.8
0.7
0.6
0.5
0.4
0.3
0.2
0.1
0
0.1
0.2
0.3
0.4
0.5
0.6
0.7
0.8
0.9
1
1.1

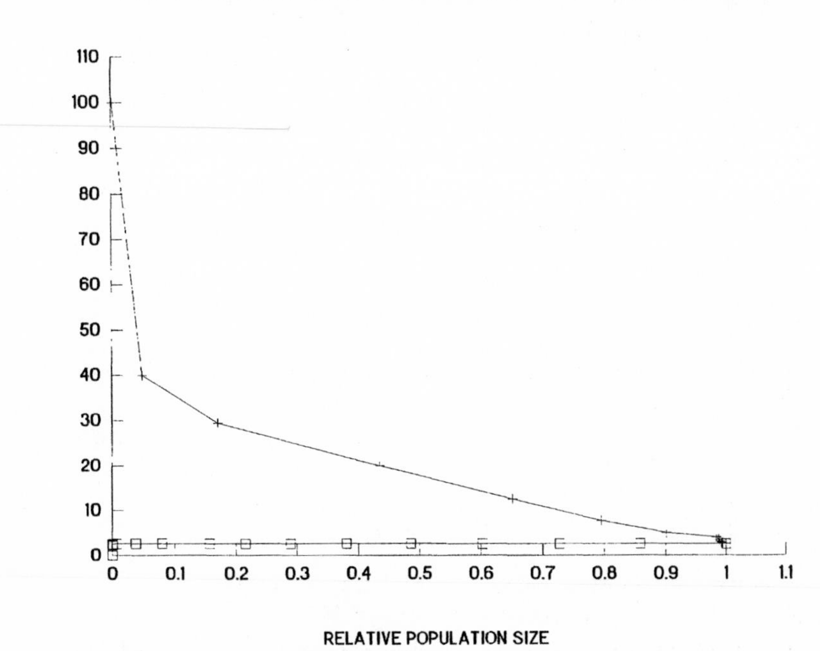
SEX-RATIO IN EXPLOITED POPULATIONS
FEMALE / MALE
RELATIVE POPULATION SIZE
"CONTROL"
SELECTIVE
110
100
90
80
70
60
50
40
30
20
10
0
0.1
0.2
0.3
0.4
0.5
0.6
0.7
0.8
0.9
1
1.1

rate, α, in a population of F females, divided into G groups (the rate at which males find females in estrus is thus $\alpha \epsilon F/G$, where ϵ is the proportion of time for which a female is receptive). Each male takes a constant period of time, h, to consort with each estrus female. The number of females that any male successfully consorts with in a period of time T, equal to one estrus cycle, is given by the following equation:

$$\frac{N_e}{M_t} = \frac{\alpha T \epsilon F/G}{1 + \alpha h \epsilon F/G}. \tag{9}$$

This expression for the instantaneous rate at which males encounter females may be used to estimate the probability that any individual female is not located by the available males during any one estrous period. This probability is given by substituting eq. 1 into the zero-th term of a Poisson distribution:

$$p_0 = e^{-(N_e/F_t)}. \tag{10}$$

The proportion of females who eventually conceive during the time equal to one interbirth period is a function of the probability that a female is not mated, p_0, the probability, c, that she conceives if mated, and the number of times a female would come into estrus every 4 years if not successfully mated. Thus the proportion of females who are pregnant or produce offspring in any 4-year interval is given by

$$f(F,M) = 1 - [(1 - c)(1 - p_0) + p_0]^{\tau}. \tag{11}$$

Substitution of eq. 2 into eq. 3 allows us to obtain an expression for the proportion of females producing young in any 4-year interbirth interval:

$$f(F,M) = 1 - [1 - c(1 - e^{-N_e/F})]^{\tau}. \tag{12}$$

This expression is a modification of one described by Dobson and Lyles (1989) and adds important biological detail (if we assume that females who do not conceive will continue cycling). Here we assume that males attain sexual maturity at age 20 and are sexually active for a proportion of time, ϵ, each year; they are promiscuous and will mate with several females during this period. The mating function suggests that the probability of a female conceiving during any one reproductive cycle is strongly dependent on the numbers of males available and the number of groups into which the female population is divided (fig. 8-5). As the sex ratio becomes more distorted toward females (as is the case in heavily poached elephant populations), or as the females become divided into many small groups, the proportion of females producing offspring declines (fig. 8-6).

The mating function (eq. 11) is most sensitive to estimates of male search rate and the number of times a female enters estrus. We assume that a female enters estrus three times a year; the actual incidence may be lower than this, as females will fail to cycle if resources are limiting. Data collected from studies of bull elephants at Amboseli (Poole, 1989b) were used to quantify the rates at which males locate females. Because females find mature males more attractive than younger, inexperienced males (Poole, 1989b), these data may overestimate the success rate of males in populations where poaching will have reduced the numbers of older males. Both of these effects may cause the model to overestimate expected conception rates in heavily exploited populations.

Inclusion of the mating function (eq. 11) into the fecundity terms of a demographic viability model considerably increases the susceptibility of exploited elephant populations to extinction. Essentially, the elephant's social behavior determines a threshold population size where successful mating becomes limiting and below which the population always col-

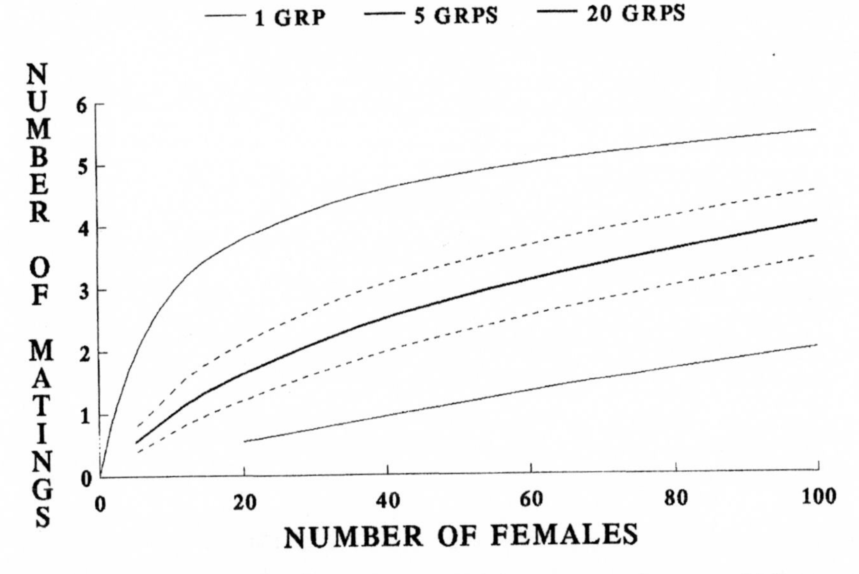

Figure 8-5 The effect of population size and grouping pattern on the rate at which estrous female elephants are located by mature males. The probability that an estrous female will be mated is assumed to have the dynamics of a simple predator–prey system; the numbers of females located depends on encounter rate, α, female group size, F/G, and the proportion of females in estrus, ϵ. Each estrous female is fertile for 2 days, and if she does not conceive, she will cycle three times a year (thus $\epsilon = 6/360$). The duration of estrus also determines the period of time, h, that a male spends consorting with any individual female and is unavailable to others. The observed pregnancy rates at Amboseli may be used to obtain a conservative estimate for encounter rate α ($\alpha = 3$). The dotted lines on either side of the line for five groups illustrate the sensitivity of the mating function to this parameter (upper line $\alpha = 2$, lower line $\alpha = 4.5$). When either sex is limiting, encounter rate is likely to be a function of habitat size. We have therefore estimated α for other parks by modifying the Amboseli estimate by the relative area of each park (we assume α is a function of distance traveled and that this will scale with the square root of habitat area, thus $\alpha \approx 1.2$ for Queen Elizabeth and $\alpha \approx 1.8$ for Mikumi).

lapses to extinction. The presence of such a threshold means that the viability of elephant populations is dependent not only on past and present levels of poaching, but also on the sex ratio and group structure of the exploited populations.

Empirical data from surveys of East African elephant populations provide support for this model of mate limitation (Poole, 1989a). In Mikumi National park, Tanzania, where the surveyed elephants were widely dispersed in many small groups, only 39% of the adult females had developed breasts (and were, therefore, either pregnant or lactating). In contrast, in Queen Elizabeth National Park, Uganda, most of the elephants had grouped together to form a single permanent large aggregation, and 87% of the adult females had developed breasts. The number of mature females to mature males in the two parks was 74:1 and 26:1, respectively. Using the observed numbers of males of breeding age and female group-size structures in these populations (the mean number of females per group for Mikumi is 3, while mean number of females per group in Queen Elizabeth is 67), our model predicts 40%

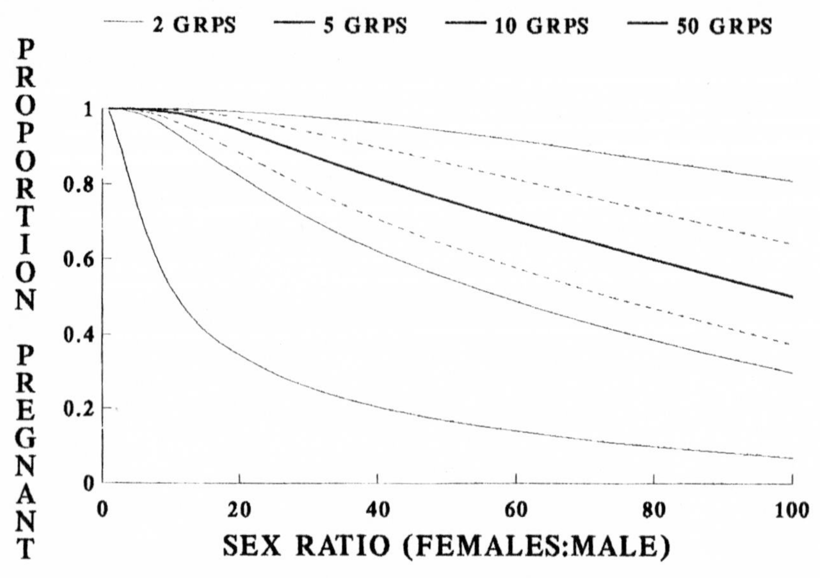

Figure 8-6 The effect of variation in sex ratio on expected proportion of females pregnant or lactating in a population containing 200 sexually mature female elephants. The figure assumes that males search for females with a success rate determined by the functional response depicted in fig. 8-4. Successfully mated females have only a 50% chance of conceiving. The expected proportions of pregnant or lactating females are drawn for populations split into 2, 5, 10, and 50 groups. The lines are discontinued below densities where it would be impossible to subdivide the population to the given extent. The dashed lines on either side of the five-group contour illustrate the sensitivity of the function to α; it is again allowed to vary by a factor of 2. Included in the figure are the data from surveys of pregnant or lactating females in Queen Elizabeth National Park, Uganda; Mikumi National Park, Tanzania; and the relatively unpoached population at Amboseli National Park, Kenya.

and 95% of females pregnant or lactating for Mikumi and Queen Elizabeth Parks, respectively.

Patch Recolonization and Conspecific Attraction

Although it is possible that some species use food abundance, or perhaps the density of tree snags greater than 3.5 cm in diameter, or the young/mature ratio of secondary successional deciduous tree species as cues to determine potential breeding habitat, most species probably follow simpler rules when attempting to locate suitable nest sites. In particular, "attempt to breed where you were born or where you bred successfully last year" and "attempt to breed as close as possible to someone else who is already defending a potential site" appear to be rules that apply to a range of species (Reed and Dobson, 1993). In both cases, species use conspecific cues to recognize suitable habitat; the presence of appropriate resources (nest sites, food, and protection from natural enemies) will eventually ensure that individuals in the best sites will survive, reproduce, and be present as cues for subsequent colonists. However, should the residents in a patch of suitable habitat go extinct, then potential immigrants may not recognize the habitat as suitable. Thus patches of habitat classified as ideal by habitat quality indices may remain completely unoccupied when animals require songs or scent marks from conspecifics to initiate territorial defense and other behavioral patterns that attract mates.

The importance of recognizing that animals may use conspecific cues to locate suitable habitat can be quickly perceived from work on *Anolis* lizards. Stamps (1987) observed lizards breeding at natural densities at a number of sites. The boundaries of each male's territories were delineated by fights with the male from the adjoining territory. However, when Stamps removed all the *Anolis* from an area and reintroduced a solitary male who had previously held a territory, he wandered through the previously occupied habitat and failed to establish a territory. The absence of other males with whom to fight meant that the lizard had no means of determining the boundary of its territory and was therefore unable to establish a territory with which to attract a mate. These observations have ramifications for patch recolonization in metapopulations and in initial reintroductions because conspecifics will be absent in both situations.

Strengths and Weaknesses of Conspecific Aggregation

Information about the size and type of aggregations and the circumstances under which they occur give models of population growth added realism. In addition, parameters of models can be changed quite simply to incorporate different patterns of grouping in different populations or species (see Durant, chapter 5, this volume). Information about aggregations can also be useful for management programs that seek to bolster populations in the wild or to reintroduce individuals. Fortunately, information on conspecific aggregations is relatively easy to tabulate because it simply involves collecting an adequate sample of group sizes and their composition.

Unfortunately, such data are difficult to gather for aquatic, fossorial, or secretive species; for example, we know little about grouping patterns of most fish species that are heavily exploited (Vincent and Sadovy, chapter 9, this volume). Further, little information is available on changes in aggregation patterns in response to exploitation or to environmental disturbance such as disease. Sustained changes in group size or composition in the face of perturbation will alter models' predictions and management programs in ways that we do not yet understand.

Recommendations

There are a number of recommendations that emerge from considering the phenomenon of conspecific aggregation in conservation biology. We will mention just three.

Game managers have known for centuries that conspecific attraction to decoys is a useful tool for trapping some bird and mammal species when hunting. Conspecific cues are also known to attract birds to breeding sites, either for colonization or to bolster local populations. This suggests that a better understanding of conspecific attraction could provide an important tool for conservation biologists and wildlife managers. In areas where populations have gone locally extinct, conspecific attraction has been used experimentally to start new colonies of seabirds. For these species, the presence of conspecifics, even painted decoys in the guise of seabirds, was an indicator of acceptable settling sites, particularly for juveniles (Fisher, 1954). In other studies, wildlife managers have often commented on the absence of a species from patches of habitat they regard to be adequate to support small populations (Wilcove, 1994). If the absence of the species is due to failure of the species to recognize the habitat, then artificial cues could be provided that might attract potential colonists. For example, taped recordings of territorial defense calls might attract birds to colonize patches of deserted woodland.

Conspecific cues can also be an important consideration when designing reserves. Verner (1992) points out that large blocks of habitat can attract disproportionate number of birds because of synergistic action in singing rates. Song attracts conspecifics, and song rates are higher in the presence of singing neighbors, so sites rich in birds might become richer. This suggests that larger reserves might contain proportionally more individuals than smaller reserves because of conspecific attraction and interaction. Given the current population fragmentation of many species, decoys and playback might be useful tools for attracting animals to newly recovered or recently abandoned sites. One concern for repopulating sites is that artificial means might attract potential breeders, but adequate social interactions might be required to retain them. Only testing in the field can determine the potential utility of these methods for different species. Social facilitation is an important aspect of biology, and conservation biologists have a good opportunity to make it work for species recovery and maintenance.

In regard to using big game hunting as a management tool (see Greene et al., chapter 11, this volume), our models indicate the importance of maintaining large, old males in the elephant populations. Until recently these individuals were thought to be reproductively senile: young males were originally thought to do all the mating, old males were regarded as dead wood and could be culled (Parker, 1979). This opinion prevails today in some isolated bureaucracies; for example, as recently as 1995, the U.S. Fish and Wildlife Service passed a nondetriment finding to support the culling of older male elephants by sports hunters in Northern Tanzania. Both modeling and empirical data now suggest that old males are important players in rapid recovery of exploited elephant populations. The degree to which this phenomenon may operate in other exploited game populations needs study.

Summary

Although individuals of some species live together permanently, others form temporary aggregations for at least portions of their lives. If population growth rate depends on a critical group size (the Allee effect), then conspecific aggregations will have conservation consequences. Modeling indicates, for example, that viability thresholds are higher for monogamous primate species than polygamous ones. Modeling also shows that as a host population becomes more aggregated, infection is more likely to be transmitted to susceptible individuals. Aggregation also changes the effect of harvesting for population growth rates. Among elephant populations, modeling shows that the proportion of females producing offspring declines as females become divided into small groups.

Behavioral observations suggest that individuals use conspecifics as cues for where to settle and may avoid uninhabited areas. This will affect habitat patch recolonization and reintroductions. In practical terms, conspecifics can be used to attract individuals to particular areas, and in designing reserves.

Information about aggregations can make models more realistic, but such data can be difficult to collect; in addition, grouping patterns may change in response to exploitation.

Acknowledgments We thank Scott Carroll, Dan Rubenstein, Cathy Toft, and an anonymous reviewer for constructive comments. We are particularly grateful to Tim Caro for help in redrafting and reorganizing this chapter.

References

Allee WC, 1931. Animal aggregations. Chicago: University of Chicago Press.

Allee WC, 1938. The social life of animals. Norton.

Birkhead TR, 1977. The effect of habitat and density on breeding success in the common guillemot (*Uria aalge*). J Anim Ecol 46:751–764.

Blockstein DE, Tordoff HB, 1985. Gone forever—a contemporary look at the extinction of the Passenger pigeon. Am Birds 39:845–851.

Charnov EL, 1993. Life history invariants. Some explorations of symmetry in evolutionary ecology. Oxford: Oxford University Press.

Creel SR, Creel NM, 1995. Communal hunting and pack size in African wild dogs *Lycaon pictus*. Anim Behav 50:1325–1339.

Darling FF, 1952. Social behavior and survival. Auk 69:183–191.

De Groot P, 1980. Information transfer in a socially roosting weaver bird (*Quelea quelea:* Ploceinae): an experimental study. Anim Behav 28:1249–1254.

Dobson AP, AM Lyles, 1989. The population dynamics and conservation of primate populations. Conserv Biol 3:362–380.

Douglas-Hamilton I, 1987. African elephant population trends and their causes. Oryx 21:11–34.

Fisher J, 1954. Evolution and bird sociality. In: Evolution as a process (Huxley J, Hardy AC, Ford EB, eds). London: Allen and Unwin; 71–83.

FitzGibbon CD, 1990. Why do hunting cheetahs prefer male gazelles? Anim Behav 40:837–845.

Gascoyne SC, Laurenson MK, Lelo S, Bomer M, 1993. Rabies in African wild dogs (*Lycaon pictus*) in the Serengeti region, Tanzania. J Wild Dis 29:396–402.

Greene E, 1987. Individuals in an osprey colony discriminate between high and low quality information. Nature 329:239–241.

Harcourt AH, 1995. Population viability estimates: theory and practice for a wild gorilla population. Conserv Biol 9:134–142.

Kennedy S, 1990. A review of the 1988 European seal morbillivirus epizootic. Vet Rec 127:563–567.

Lloyd M, 1967. Mean crowding. J Anim Ecol 36:1–30.

Milner-Gulland EJ, Beddington JR, 1993. The exploitation of elephants for the ivory trade: an historical perspective. Proc R Soc Lond B 252:29–37.

Moehlman PD, 1979. Jackal helpers and pup survival. Nature 277:382–383.

Moss CJ, Poole JH, 1983. Relationships and social structure in African elephants. Primate social relationships: an integrated approach (Hinde RA, ed). Oxford: Blackwell; 315–325.

Parker ISC, 1979. The ivory trade. Washington, DC: U.S. Fish and Wildlife Service.

Peters RH, 1983. The ecological implications of body size. Cambridge: Cambridge University Press.

Poole JH, 1989a. The effects of poaching on the age structure and social and reproductive patterns of selected East African elephant populations. In: The Ivory Trade and the Future of the African Elephant, report no. 2 (Ivory Trade Review Group, ed).

Poole JH, 1989b. Mate guarding, reproductive success and female choice in African elephants. Anim Behav 37:842–849.

Reed JM, Dobson AP, 1993. Behavioural constraints and conservation biology: conspecific attraction and recruitment. Trends Ecol Evol 8:253–255.

Roelke-Parker ME, Munson L, Packer C, Kock R, Cleaveland S, Carpenter M, O'Brien SJ, Pospischil A, Hofmann-Lehmann RH, Lutz H, Mwamengele GLM, Mgasa MN, Machange GA, Summers BA, Appel MJG, 1996. A canine distemper virus epidemic in Serengeti lions (*Panthera leo*). Nature 379:441–444.

Rubenstein DI, Wrangham RW (eds), 1986. Ecological aspects of social evolution: birds and mammals. Princeton, New Jersey: Princeton University Press.

Stacey PB, Koenig WD, 1990. Cooperative breeding in birds: long-term studies of ecology and behavior. Cambridge: Cambridge University Press.

Stamps JA, 1987. Conspecifics as cues to territory quality. 1. A preference for previously used territories by juvenile lizards (*Anolis aeneus*). Am Nat 129:629–642.

Taylor LR, 1961. Aggregation, variance and the mean. Nature 189:732–735.

Taylor LR, Taylor RAJ, 1977. Aggregation, migration and population dynamics. Nature 265:415–421.

Taylor LR, Woiwood IP, Perry JN, 1978. The density-dependence of spatial behaviour and the rarity of randomness. J Anim Ecol 47:383–406.

Verner J, 1992. Data needs for avian conservation biology: have we avoided critical research? Condor 94:301–303.

Wilcove DS, 1994. Turning conservation goals into tangible results: the case of the spotted owl and old-growth forests. In: Large-scale ecology and conservation biology (Edwards PJ, May RM and Webb NR, eds). Oxford: Blackwell Scientific; 313–329.

Wilson EO, 1975. Sociobiology: the new synthesis. Harvard, Massachusetts: Belknap Press.

9

Reproductive Ecology in the Conservation and Management of Fishes

Amanda Vincent
Yvonne Sadovy

Conservation Status of Fishes

Overexploitation, habitat degradation, physical change, and introductions of exotic species menace fishes around the world. The 1996 International Union for the Conservation of Nature (IUCN) Red List of Threatened Animals included 734 species of fishes (IUCN, 1996), up from 576 eight years previously (IUCN, 1988), with most species still not assessed (IUCN, 1996). In North America, fishes and other aquatic taxa are up to eight times more likely to be in peril than birds and mammals (Master, 1990). Despite the level of threat, it will often be particularly difficult to implement conservation measures for fishes because so many people depend on them for livelihoods and food. The prevailing attitude that fishes are food meant, for example, that Australia exempted all marine fishes from monitoring and controls under the Wildlife Protection Act until 1998, relying instead on fisheries legislation.

In this chapter, we concentrate on the direct effects of exploitation on fishes in the marine environment, though reference is made to other damaging activities and to freshwater species. Our justification for such a focus is that the real threat of human predation to the biological diversity of marine fish populations and ecosystems is only now being recognized and still has not been adequately addressed in fisheries management. The 1994 IUCN Red List included fewer than 20 of the 15,000 marine fishes (Groombridge, 1993), whereas the 1996 IUCN Red List had at least 6 times as many, after only limited further assessment (IUCN, 1996). Some of those species meeting criteria for listing were important commercially, such as haddock *Melanogrammus aeglefinus,* rockfish *Sebastes paucispinus,* and several tunas (*Thunnus* spp.).

Overfishing is a major threat to biodiversity because it depletes numbers, disturbs communities, damages habitats, and can lead to loss of genetic integrity, local populations, and even species (Pitcher and Hart, 1982; Norse, 1993; Huntsman, 1994; Safina, 1994; Boehlert, 1996, cited in Bohnsack and Ault, 1996). Most fishes can be eaten, so those that are not exploited are either difficult to catch (e.g., small, rare, or awkward to access) or culturally distasteful (Moyle and Moyle, 1995). Fishing yields peaked during the 1970s but are now in decline, with approximately 60% of the 200 major fishery resources monitored by the Food and Agriculture Organization (FAO) either fully exploited or overexploited (FAO, 1996). Long-established fisheries have collapsed with little or no sign of recovery for decades thereafter (e.g., Peruvian anchovy *Engraulis ringens*). The major fishing ground for North Atlantic cod was closed indefinitely in 1994. The small yellow croaker *Pseudosciaena polyactis* of the China Seas is now caught at only a tiny fraction of its former abundance (Deng and Yang, 1993). Shark fisheries require urgent management the world over, to the extent that many species may be headed toward extinction (Manire and Guber, 1990). Commercial fisheries can terminate in conservation crises: catches of one large North American marine sciaenid, the totoaba *Totoaba macdonaldi,* declined sufficiently between 1942 and 1975 for listing on CITES (Convention on International Trade in Endangered Species) Appendix I, in danger of extinction (Barrera Guevara, 1990). Clearly, the widely held notions that entirely marine fishes are relatively secure (e.g., Moyle and Leidy, 1992; Huntsman, 1994) and that fisheries will inevitably go extinct before the fish (Graham, 1935), need serious reevaluation. Indeed, direct exploitation is credited with pushing fishes past five recognized milestones to extinction: continuous declines in numbers, profound reduction in reproductive success, overexploitation and crash, curtailment of seasonal cycles through loss of spawning aggregations, and commercial extinction (King, 1987).

The depletion of marine resources and new market demands are compelling fishers to move onto nontraditional species. Particularly large growth areas are high-value fisheries for medicines, luxury food items, the aquarium trade, curios, mariculture, and recreation. For example, the rapidly growing Chinese economy is prompting ever larger global consumption of most products used in traditional Chinese medicines, including at least 59 marine fish species ranging from herring (Clupeidae) to seahorses (*Hippocampus* spp; Tang, 1987). A greatly increased demand for luxury food items such as live reef fishes or shark fins comes from affluent Chinese markets and is threatening species such as the Napoleon (humphead) wrasse *Cheilinus undulatus* (Johannes and Riepen, 1995) and certain sharks (Dayton, 1991). Aquarium fish enthusiasts in the West have contributed to the commercial extinction of some freshwater fishes (Bayley and Petrere, 1990) and the depletion of several tropical marine species (Wood, 1985); the problem will worsen for the latter because virtually all marine aquarium fishes are wild-caught, as compared to only 10% of freshwater aquarium fishes (Andrews, 1990). The curio trade for some species is large enough to cause conservation concern (Hawkins and Roberts, 1994; Vincent, 1996). Fry fish for mariculture have also become an internationally traded commodity as wild supplies are depleted in certain areas (e.g., groupers and milkfish: Sadovy, in press; M. Pajaro, personal communication) and demand grows. Though their impact is seldom assessed, sports anglers take significant numbers of marine fishes, operating very selectively (e.g., Slaney et al. 1996).

Fishing pressure is set to worsen. Catches are declining while human populations grow, and it is anticipated that the world may need twice as much food fish per annum in 2010 as can be supplied by capture fisheries (those other than aquaculture; FAO, 1996). World fisheries already operate at a global loss of about $54 billion per annum (FAO, 1993); the difference is made up by government subsidies that exacerbate the problem and mask the imbalance between effort and yield. Small-scale artisanal and subsistence fishers are seldom included in FAO statistics, yet they account for approximately 90% of the world's fishers (Bailey, 1988). They extract a wide range of fishes from inshore habitats in increasingly unsustainable manners, their overfishing compounded by destructive fishing techniques including driftnets, dynamite, poisons, and drive-net fishing with weighted scare lines (e.g., Johannes and Riepen, 1995). Fishing also damages nontarget species through incidental bycatch, ecosystem disruption, and habitat degradation: for example, shrimp trawlers scour the bottom for hauls that can consist of at least 95% nontarget species, much of which is discarded (Andrew & Pepperell, 1992; Norse, 1993; Safina, 1994).

We focus on exploitation, but habitat degradation and introductions of exotic species are also serious threats. Pollution and silt runoff damage freshwater and marine habitats alike. The world's major fish nurseries are threatened when seagrasses are subject to dredging or sea filling (land reclamation), mangroves are removed or inundated with freshwater, and coral reefs are blasted or poisoned. Spawning ground destruction and water diversion threaten species from salmonids to desert pupfish (Nehlson et al., 1991; Contreras-Balderas and Lozano-Vilano, 1994). The range of introduced fishes in marine and estuarine systems is increasing because of intentional and inadvertent releases, the latter sometimes through ballast water from ocean-going vessels (Baltz, 1991; Carlton and Geller, 1993). Several such introductions have contributed to the depletion of native fishes, particularly in the former Soviet Union; sometimes the destructive agents were fish parasites or other invertebrates that altered community structure (Baltz, 1991).

The premise of this chapter is that identifying conservation threats and finding solutions will benefit from a better understanding of fish behavioral ecology. Our concern is not merely with proximate extinctions per se, but also with processes that make fishes rare and that can be the ultimate causes of extinction, perhaps through loss of genetic diversity (Simberloff, 1986; Lawton, 1994). The new IUCN criteria consider severe declines in fish abundance or declines in age at maturity and reproductive potential as indicative of threats to populations and species (IUCN, 1994). Thus, the concerns of conservationists interested in maintaining biological diversity overlap considerably with those of fishery managers who need to sustain fisheries.

Seeking Persistence of Fish Populations

Both conservation biology and fisheries management ideally seek to ensure long-term viability of fish populations. The two disciplines can differ greatly in focus—the former seeks to maintain biological diversity, while the latter emphasizes production—but their interests overlap substantially. Conservation biologists do some work that is solely conservation-focused as with attempts to save the Devil's Hole pupfish *Cyprinodon diabolis* (Pister, 1985), but most fish conservation research is also broadly relevant to management, as in the following examples; attempts to predict extinction-prone species (Frissell, 1993; Vermeij, 1993; Angermeier, 1995); genetic analyses to estimate effective population sizes to manage stocks or to assess the relation of changes in heterozygosity with fishing pressure (Allendorf and Leary, 1988; Smith et al., 1991; Bartley et al., 1992); considerations of biological and commercial extinctions and ecological collapses of exploited species (Lowe-McConnell, 1993; Sadovy, 1993; Camhi, 1995); investigations of the risk posed by fish extraction for Chinese medicine (Vincent, 1996, 1997); reports on reductions in species diversity through selective trawling (Turner, 1977); or evaluations of changing fish numbers in and around marine reserves (Russ and Alcala, 1989; Roberts, 1995).

Historically, conservation of marine fishes has been the preserve of fishery managers seeking to ensure healthy persistence of commercial species, races, and stocks for exploitation. Such managers employ a range of assessment and management approaches to estimate population size and structure of target species in the past, present, and future (see Pitcher and Hart, 1982). They then establish a policy that determines the number and type of fish to be extracted (i.e., controls mortality). Management options include limiting the number of fishers or boats, restricting gear, establishing seasonal closures for fishing, or proscribing the fishes retained. The problem is that implementation of such measures currently fails to ensure long-term persistence of fisheries.

In theory, a fishery is sustainable on a long-term basis if exploitation is limited to the maximum sustainable yield (MSY) of the stock (but see below). The logic is that catching older fish allows more rapid reproduction and growth of younger fish (density-dependent responses), with a consequent yield of biomass which is "surplus" to the virgin, or unfished, state. Theoretically, the population will persist as long as only this surplus is removed by fishing (e.g., Ricker, 1975).

Several generations of fisheries models have been devised during this century to estimate MSY and to plan how to attain it. The early class of fisheries models, surplus yield models, estimated current and potential yield based on simple relationships between fishing effort (number of boats, number of fishers, etc.) and fish yield or catch. The models were concerned only with changes in biomass in terms of number of fish ("numeric effects;" fig. 9-1). They were not concerned with details of how the biomass was generated or whether mortality was selective with respect to size, sex, or genotype.

It has long been established that fishing can be selective, that yields are influenced by factors such as gear, time, and place of fishing (fig. 9-1), and that such factors can have profound impacts on population responses to exploitation (Beverton and Holt, 1957; Ricker, 1975; Leaman, 1991). Yet early fisheries models paid no attention to the specific impacts of fishing on four key population processes as identified by Russell (1931): recruitment, growth, natural mortality, or fishing mortality. Thus neither "demographic effects," which occur through selective removals, nor their "causal links" with reproductive potential, growth, or survival were considered (fig. 9-1).

The later models, dynamic pool models, incorporate Russell's four processes and examine the effects of selective removal upon them (Beverton and Holt, 1957). Parameters are evaluated independently for each age group (cohort) to produce detailed age- or size-structured assessments of stocks. These models operate on the assumption that, under equilibrium conditions, some optimal combination of fishing effort and minimal capture size can be found to extract maximum yield from any particular cohort without compromising stocks. Newer dynamic pool models incorporate even more parameters by submodeling the four basic processes (e.g., Goodyear, 1993; Mace and Sissenwine, 1993), but they still fail to integrate the full range of behavior exhibited by exploited species.

Among the ongoing problems with dynamic pool models is that they assume an equal sex ratio and generally ignore sex-specific growth and mortality changes (Bannerot et al., 1987; Goodyear, 1993). This may result in incorrect estimates of spawning potential or of optimal exploitation levels. For example, species which do not conform to the above assumptions, such as the many hermaphroditic (sex-changing) species are completely overlooked by current models, even though theoretical treatments suggest that they lose reproductive capacity more rapidly than gonochores as fishing effort increases and also fail reproductively at lower fishing effort (Bannerot et al., 1987; Shepherd and Idoine, 1993; Huntsman and Schaaf, 1994).

The inadequacies of fisheries models partly explain why most fisheries today are managed unsustainably, with exploited fishes suffering from declines in number, size, or density. In the United States, for example, management ambitions have retreated for about 60% of stocks; where formerly managers sought to maximize yield, they now seek to ensure sufficient spawning biomass for population maintenance (e.g., Rosenberg, 1993). The prob-

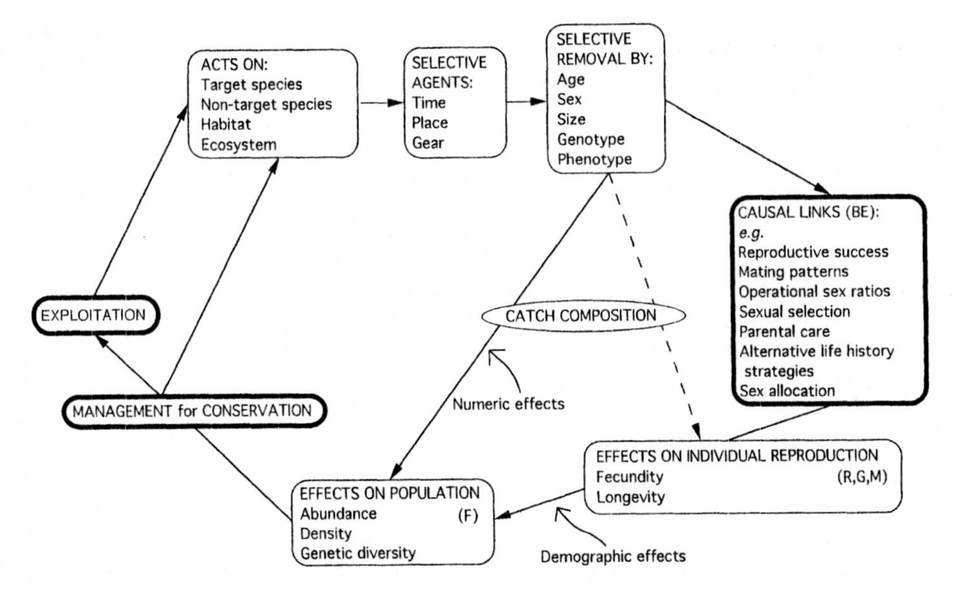

Figure 9-1 Flow chart to illustrate how exploitation acts on individuals and populations through selective removals by the agents of time, place, and gear. The numeric response to fishing can be evaluated readily, but demographic effects cannot be understood fully without the causal insights provided by behavioral ecology and integration of four key parameters which influence longevity, reproductive output, and reproductive success, as used in fisheries models: R, recruitment; G, growth; M, natural mortality; and F, fishing mortality.

lems are not surprising given the substantial doubts over the concept of MSY (e.g., Larkin, 1977; Ludwig et al., 1993; but see Rosenberg et al., 1993) and the relatively recent incorporation of more realistic parameters, especially those relevant to reproduction, into many fishery assessments.

We still have little understanding of the more subtle effects of selective removals on population persistence and genetic diversity or of causal behavioral links between the two (fig. 8-1). For example, genotypic and phenotypic effects of persistent and selective removal of larger individuals may result in life-history shifts such as reduction in size at sexual maturation with a concomitant reduction in mean female fecundity because of size-dependent fecundity (Gulland, 1974; Ricker, 1981; Kasperski and Kozlowski, 1993). Indeed, the long-term fecundity consequences of selectively removing large fish can be grave; differential phenotypic and genetic shifts drove Atlantic salmon *Salmo salmar* into early spawning within two decades (e.g., Montgomery, 1983; Rago and Goodyear, 1987). Behavioral ecology is the tool through which impacts of such selectivity can be examined in detail and their conservation consequences predicted.

The Role of Behavioral Ecology

The real effects of exploitation will depend vitally on which and how many individuals are extracted and on how the remaining individuals within a population respond. Behavioral ecology examines how behavior, ecology, and genetics interact to determine survival and reproduction (e.g., Gross, 1994). Such adaptive studies should play an integral role in understanding the long-term consequences of exploitation at the population level.

We concentrate on reproductive parameters because they are vital determinants of recovery from exploitation (see fig. 9-1). Here we discuss behavioral ecological research related to fecundity limitation, disruption of spawning activity, mating patterns, mate competition, parental care, alternative life-history strategies, and sex allocation. Table 9-1 provides summaries of behavioral ecological research that might bear on aspects of fish conservation, including further details and references for some of the examples cited later in the text. The goal is to consider how such work might help plan sustainable fishing and aquaculture, and mitigate their impacts.

Fecundity Limitation

Fecundity and spawning biomass are key parameters in fisheries management, and are often measured, yet they are vulnerable to misinterpretation without an understanding of the behavioral ecology of the species, population, or individual. Direct exploitation usually places greatest pressures on larger individuals, whether target species or bycatch, if only because of gear selectivity such as the mesh size of nets (e.g., Ricker, 1975; Munro, 1983; Pankhurst, 1988; Bohnsack et al., 1989). Extracting larger females can have a disproportionately large impact on reproductive output relative to change in biomass because egg number tends to increase allometrically with female body size (Bagenal, 1978). Moreover, remaining females may mature when younger and smaller, with consequent decrease in fecundity (Gulland, 1974; Pitt, 1975; Ricker 1981; Trippel, 1993). We therefore need to understand behavior patterns that might increase the probability of extracting larger fishes or influence our estimates of egg production and fertilization rates.

Understanding the relationship between large female removal and reproduction often relies on behavioral work. If females of different sizes breed at different times or places, then fishing schedules could greatly affect the impacts of exploitation on populations (e.g., embiotocids; see table 9-1). An extreme example of the penalties of removing large females comes from anemonefishes (Pomacentridae), which are protandrous (sex changing; male first): repeatedly removing larger anemonefish (here females) will force males to change sex at ever smaller sizes, becoming smaller females who produce fewer eggs (Fricke, 1979). Conversely, removing large (female) wrasse or female pipefish will reduce egg output less than might be expected because larger females reproductively suppress smaller females.

Accurate estimates of egg output also require behavioral knowledge. Only thus can we know, for example, that not all adult females spawn every year in the yellowfin bream, even though they develop ripe eggs, and that female seahorses discard ripe eggs when a partner is lacking. On the other hand, multiple spawners may release considerably more eggs each year than was originally estimated from limited sampling techniques (Davis and West, 1993; see also Sadovy, 1996).

Differential male harvest and associated sperm limitation are known to lead to greatly reduced female fecundity and population collapse in some ungulates (Ginsberg and Milner-Gulland, 1994). In fishes, too, possible sperm limitation—seldom considered in fisheries models—may mean that not all released eggs are fertilized (see table 9-1). In porgy and gag grouper, extreme differential male mortality through fishing is also inferred to lead to potential sperm limitation. Moreover, behavioral ecological studies have revealed that male fertilization rates can be limiting, as in the bluehead wrasse, even where no sperm limitation per se is evident (Warner et al., 1995) (fig. 9-2).

Table 9-1 Summaries of selected studies in reproductive ecology, listing specific behaviors likely to magnify or diminish immediate impacts of exploitation.

Species	Behavioral ecology study	Reference
Fecundity limitation		
Lemon tetra (*Hyphessobrycon pulchripinnis*)	Under experimental conditions, multiple spawning reduces the fertilization rates of successive females. Sperm limitation indicated.	Nakatsuru and Kramer (1982)
Coho salmon (*Oncorhynchus kisutch*)	Ranched females produce larger eggs than wild females and ranched males have larger testes. Egg size and number heritable, favoring ranched salmon.	Fleming and Gross (1992), Gall and Gross (1978)
Seahorses (*Hippocampus fuscus*)	Females discard ripe eggs if they fail to meet a male after hydrating them. Dissections may overestimate reproductive output.	Vincent (1994)
Pipefish (*Syngnathus typhle*)	Large females suppress reproduction in smaller females, which produce fewer and smaller eggs. Loss of large females could reduce egg output, although smaller females would provide some compensation.	Berglund (1991)
Bluehead wrasse (*Thalassoma bifasciatum*)	Males have two color phases (initial and terminal). The one that is relatively more abundant at lower densities (terminal phase) has lower fertilization rates. Declines in density will reduce number of eggs fertilized, making fecundity estimates misleading.	Warner (1984), Shapiro et al. (1994)
Wrasse (*Pseudolabrus celidotus*)	Removing large females will reduce egg output less than expected because larger females reproductively suppress smaller females.	Jones and Thompson (1980)
Razorfish (*Xyrichthys novacula*)	Males show declining fertilization rates with increasing numbers of matings. Sperm limitation indicated.	Marconato et al. (1995)
Yellowfin bream (*Acanthopagrus australis*)	Not all adult females spawn every year, so simple biomass or fecundity calculations are misleading	Pollock (1984)
Gag grouper (*Mycteroperca microlepis*), porgy (*Chrysoblephus puniceus*)	Selective fishing for males in this protogynous species produces a heavily female-biased sex ratio. Sperm may be limited as a result of reduced male numbers.	Koenig et al. (1996), Garratt (1986)
Embiotocid (*Micrometrus minimus*)	Smaller females delay breeding until later in the season, so fishing early may mean they never breed.	Schultz et al. (1991)
Disruption of spawning activity		
Atlantic salmon (*Salmo salar*)	Sea-ranched fish are less successful at spawning than wild fish partly because ranched females move much more and stay in spawning areas for shorter periods.	Jonsson et al. (1990), Økland et al. (1995)
Pacific pollock (*Theragra chalcogramma*)	Depth segregation by sexes around time of spawning, so fishing could exert selective removal, disrupting spawning activity.	Maeda (1986)

(*continued*)

Table 9-1 (*Continued*)

Species	Behavioral ecology study	Reference
Orange roughy (*Hoplostethus atlanticus*)	Sex-segregated aggregations come together only to spawn. Selective fishing could remove excessive numbers of one sex at critical time in life history.	Pankhurst (1988)
Sea moth (*Eurypegasus draconis*)	Form pairs that rise off the bottom to spawn every few days. Trawls catch these fishes, perhaps differentially during spawning, and disrupt pairs(?)	Herold and Clark (1993)
Seahorse (*Hippocampus* spp.)	Male–female pairs court actively, releasing holdfasts and rising off the bottom to mate. They could thus be more easily caught. Trawls in Florida catch more seahorses around the time they are mating.	Vincent and Sadler (1995, personal observation)
Nassau grouper (*Epinephelus striatus*)	Spawns in aggregations. Courtship activity appears less intense in smaller (fished) aggregations. Some aggregations have disappeared entirely because of fishing pressure.	Colin (1992), Sadovy (1993)
Groupers (Serranidae)	Larger species of groupers spawn in aggregations; smaller ones do not. Aggregating species may be more vulnerable to aggregation-targeted fishing.	Samoilys and Squire (1994), Sadovy et al. (1994)
Tobaccofish (*Serranus tabacarius*)	Simultaneous hermaphroditism partly maintained by pairs engaging in egg trading. Regular disruption by fishing could decrease egg trading and theoretically promote sequential hermaphroditism.	Ghiselin (1969), Petersen (1995)
Surgeonfishes (Acanthuridae)	Aggregation spawn as mass groups, probably to reduce egg predation. Reduction in density of aggregations could increase egg predation.	Robertson (1983)
Mating patterns		
Yellowfin shiners (*Notropis lutipinnis*	Obligate symbionts with bluehead chubs, *Nocomis leptocephalus,* and must spawn in their nests. Decline in chub population would jeopardize shiners.	Wallin (1992)
Seahorses (*Hippocampus whitei*)	Rigidly monogamous, take a long time to re-pair if widowed. Fishers recognize that they pair and on finding one seahorse, they seek its partner.	Vincent and Sadler (1995), Vincent (1994)
Damselfish (*Acanthochromis polyacanthus*)	If one partner is removed, rapid re-pairing occurs. More flexible than some other monogamous species.	Nakazono (1993)
Anemonefishes (*Amphiprion* spp.)	Anemone size and distribution determines breeding unit. Fishing often removes anemones.	Hattori (1991, 1994)
Filefish (*Paramonacanthus japonicus*)	Facultative monogamy or polygamy, thus allowing a flexible response to depredations.	Nakazone and Kawase (1993)

(*continued*)

Table 9-1 (*Continued*)

Species	Behavioral ecology study	Reference
Sexual selection		
Atlantic salmon (*Salmo salar*); coho salmon (*Oncorhynchus kisutch*)	Both sexes of hatchery fish less able to compete for mates than wild fish because of reduced aggression. Significant heritability of aggressive behavior. May limit hatchery introgression on wild populations.	Jonsson et al. (1990), Fleming and Gross (1992, 1993), Rosenau and McPhail (1987)
Seahorses (*Hippocampus whitei*)	Competition for mates usually rare because most animals are paired. Would increase when more animals are seeking mates (e.g., after fishing losses). This would be socially disruptive and make seahorses more visible to predators.	Vincent and Sadler (1995); Vincent (1994)
Convict cichlid (*Cichlasoma nigrofasciatum*)	Spawning sites limit number of fish breeding. There could initially be compensatory breeding if fish were removed.	Wisenden (1995)
Coral trout (*Plectropomus areolatus*)	Spatial segregation of sexes before spawning. Females are selectively removed. Consequent male biases in the sex ratio of spawning aggregations. Territorial males harass the few females to the extent that they may not spawn.	Johannes et al. (1994)
Sand goby (*Pomatoschistus minutus*)	Female expression of preference for large, colorful mates was lessened in the presence of predators. Fishing could thus influence sexual selection pressures.	Forsgren (1992)
Parental care		
Three-spined stickleback (*Gasterosteus aculeatus*)	Males vary in the risks they take to raise broods of different sizes. Take fewer risks and produce smaller broods in unpredictable environments. Fishing is unpredictable.	Pressley (1981), Lachance and Fitzgerald (1991)
Seahorses and pipefishes (Syngnathidae)	Male broods the young, so killing the male condemns the young.	Vincent et al. (1992)
Pipefishes (*Nerophis ophidion*)	Brooding males more likely to be depredated than nonbrooding males, making parental care costly.	Svensson (1988)
Cardinalfish (*Apogon doederleini*)	Males mouthbrood their young. Egg cannibalism of young increases as male condition deteriorates, usually at the end of the season. Harassment (even indirectly) through fishing could promote cannibalism.	Okuda and Yanagisawa (1996)
Haplochromine cichlids (Cichlidae)	Males mouthbrood their young and lose them to waiting predatory fishes when caught.	Witte and Goudswaard (1985), in Ribbink (1987) Fryer (1984), in Ribbink (1987)
Haplochromine cichlids (Cichlidae)	Males build prominent sand castle nests which trawls flatten, catching adults.	
Convict cichlid (*Cichlasoma nigrofasciatum*)	Biparental care of eggs and young lasts 4–6 weeks. Survival of young decreases if male is removed < 7 days into their free-swimming	Keenleyside and Mackereth (1992)

(*continued*)

Table 9-1 *(Continued)*

Species	Behavioral ecology study	Reference
	period. Premature removal of males would compromise survival of offspring.	
Princess of Burundi cichlid (*Lamprologus brichardi*)	Offspring stay at nest to help parents with next broods. Fishing could reduce reproductive output by removing helpers.	Taborsky and Limberger (1981), Taborsky (1984)
St. Peter's fish (*Sarotherodon galilaeus*)	Either sex can care for the young. Thus may be buffered against some impacts of fishing.	Balshine Earn (1995)
Damselfish (Stegastes spp.)	Some species provide obligate egg-guarding. Removal of guarding males results in loss of nest of eggs. Thus more vulnerable than others.	Petersen (1990), Itzkowitz and Mackie (1986)
Damselfish (*Acanthochromis polyacanthus*)	Re-pairing after loss of a partner often results in loss of young, probably through cannibalism.	Nakazono (1993)
Anemonefish (*Amphiprion clarkii*)	Paternal care but young may survive loss of their father because incoming males will step-father a brood to obtain access to the female.	Yanagisawa and Ochi (1986)
Anemonefish (*Amphiprion akallopisos*)	Smaller fishes assist with parental care. Removal of these helpers could reduce reproductive success.	Fricke (1979)
Peacock wrasse (*Symphodus tinca*)	Parental care is facultative. Females either give eggs to male or disperse them. More flexible response to male removals.	Warner et al. (1995)
Alternative strategies		
Salmon (Salmonidae)	Males mature when young or small (jacks) or delay maturation until older and larger (hooknoses). Fishing can lead to greater frequency of jacks relative to hooknoses.	Gross (1984), Gross (1991)
Sunfishes (Centrarchidae)	Different male tactics; smaller cuckolders and larger parental males. Fishery takes primarily parental males because they guard nests. Consequently both population size and male size are reduced more quickly than anticipated by simple numeric estimates.	Gross (1982, personal communication)
Tilefish (*Lopholatilus chamaeleonticeps*)	Removal of large males permits smaller sneaker males to spawn. Could lead to diminished size.	Grimes et al. (1988)
Anemonefish (*Amphiprion clarkii*)	Widowed males change sex at low anemone density but remain male at higher densities, moving to neighboring female. Thus anemone density determines consequence of fishing for social system.	Hattori (1994)

Resilience will depend on the interaction between behavior and fishing pressure, with some species more vulnerable than others. Most studies focused on individuals, so the population level responses presented here are usually our extrapolations. Within each category, species are listed in taxonomic order.

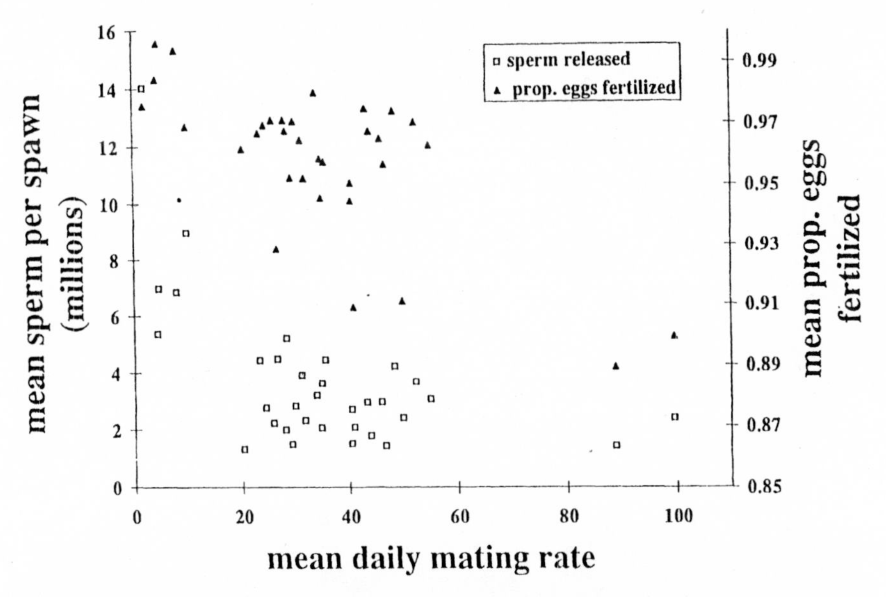

Figure 9-2 Reduction in the number of males in a population because of differential fishing mortality could lead to sperm limitation. Male bluehead wrasse *Thalassoma bifasciatum* produce pelagic eggs and spawn in pairs up to 100 times per day, but the mean sperm per spawn and the mean fertilization rates both decline highly significantly with successive matings (from Warner et al., 1996).

Behavioral ecological research has provided information about the reproductive possibilities and problems with culturing, often suggested as a solution to conservation concerns. Ranched salmon, for example, may have higher fecundity than wild salmon, although the long-term impact of this apparent advantage is mitigated by the frequently lower mating success of ranched salmon.

Disruption of Spawning Activity

The way in which fishes spawn may well affect their vulnerability to fishing. We might expect exploitation to penalize more heavily fishes whose spawning makes them more accessible to fishers (e.g., sea moths and seahorses; fig. 9-3). In some species, spawning disruption may have long-term effects on reproductive strategies; for example, it could alter selection pressures on sex allocation in simultaneous hermaphrodites that trade eggs (e.g., tobaccofish).

Species that aggregate briefly to spawn in large numbers and at specific locations may be directly or indirectly highly susceptible to disruption by heavy fishing pressure (e.g., groupers). For some species, a threshold size of spawning group may be necessary, below which courtship may decline until the aggregation disappears entirely (e.g., groupers). Moreover, spawning aggregations are often highly structured by sex, time, or depth, such that selective removal could disrupt spawning activity (e.g., orange roughy, Pacific pollock, grouper). Reduction of group spawning size with population depletion could also acceler-

Figure 9-3 Seahorses mating. Fishes can become more vulnerable to capture during mating. Seahorses usually hide cryptically on the bottom, virtually immobile, with their tails grasping a holdfast firmly. However, they become active for many hours during courtship and rise off the bottom to mate, when they could be caught in trawls. This species is *Hippocampus breviceps,* and the male is on the left. (Photograph: R. Kuiter.)

ate numeric decline if group size itself confers any protection from natural predators (e.g., acanthurids).

Culturing fishes may also affect their spawning activity, with severely reduced mating in ranched salmon, for example. The long-term implications of such differences in reproductive output are unknown, whether for escaped sea-ranched fish or for hybrids from ranched and wild parents.

Mating Patterns

The social structure of a fish population will probably bear directly on its resilience to exploitation. The real survivors may be species with flexible mating patterns (e.g., filefish). Individuals with fewer potential partners (e.g., faithful to one mate or living at low densities) are probably more likely to cease breeding if a partner is lost through fishing than are those with many potential partners (Greene et al., Chapter 11, this volume). This will be

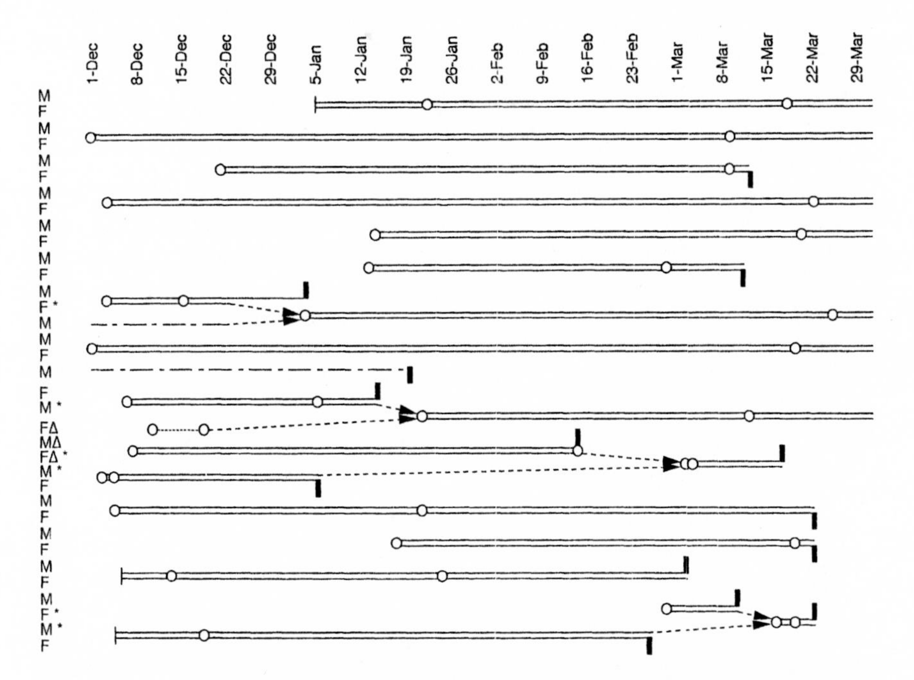

Figure 9-4 An overview of *Hippocampus whitei* seahorse pairings from December through March. The seahorses are organized such that, roughly speaking, adjacent individuals in the figure are neighbors in the seagrass meadow. Each seahorse's pair status is indicated by a single line. A solid line represents a paired seahorse, a dashed line indicates a seahorse changing partners (after its former partner disappears), and a broken line represents a male whose partner was unknown (we believe that both these males had lost partners by the time we started watching them). A triangle (△) denotes a male who greeted (but did not mate) bigamously, his mate, and his second female. Seahorses that re-paired during the season are marked with an *asterisk. Thick vertical lines indicate the date when a seahorse disappeared. Animals that lost their partners took several weeks to find new mates (adapted from Vincent and Sadler, 1995).

particularly true if the partners form long-term bonds (e.g., seahorses; fig. 9-4), rather than associating ephemerally. Indeed, the speed with which fishes resume mating after social disruption could be a vital determinant of fishing impact; some damselfishes re-pair very quickly, whereas others do not (see table 9-1), leading us to predict that the former will be more resilient. Among anemonefishes, ease of re-pairing depends on the importance and degree of male/female size differences in mated pairs (Hirose, 1995).

The degree to which fishes associate physically as breeding units could also influence the probability of both sexes being removed simultaneously. For example, fishers that hand-collect seahorses realize that they are paired and, upon finding one seahorse (Syngnathidae), seek its partner nearby (A. Vincent, personal observation). Aquarium collectors, too, tend to obtain both members of a butterflyfish pair (Chaetodontidae), since they are usually in close proximity (A. Vincent, personal observation). The sexes are particularly likely to be in close proximity, albeit briefly, when in spawning aggregations, although these too can be sex-segregated.

Fishes that rely on particular biological or physical features for mating and reproduction may be disturbed if these features are damaged incidentally by fisheries or by direct exploitation. Anemonefishes depend on anemones (Hattori, 1991, 1994) and are thus vulnerable to their damage or removal, for the aquarium trade, for example. Excessive loss of anemones may reduce breeding because suitable anemones can become saturated with adult fish (Hattori, 1994). The corollary, that removing some fishes will free others to breed (density-dependent response), may initially be correct but will not necessarily hold true with repeated disturbance of a complex mating system. Reproduction in other fishes might be disrupted if the species with which they have an obligate symbiosis for mating is removed through exploitation (e.g., yellowfin shiners). Damselfish rely on particular coral substrates to breed, so they may cease reproducing when fishing damages corals (e.g., by dynamiting the coral reef, applying cyanide randomly, pounding the sea bottom with heavy lead lines (*muro-ami*), or removing mollusc shells used as fish nests).

Sexual Selection

Differential fishing pressure on the two sexes (if, for example, one sex is larger than the other or is geographically more accessible to fishing) may result in biased operational sex ratios (OSR; total number of males ready to mate relative to females ready to mate). The consequently more abundant sex will probably become less particular in mate choice and more aggressive in competition for mates (Emlen and Oring, 1977). Even where exploitation does not distort sex ratios, fishing pressure could reduce mate choice if it somehow shortened time for courtship or otherwise affected the ability to evaluate potential mates freely (as perhaps in sand gobies; fig. 9-5). Presumably, mate choice might also begin to favor new attributes (e.g., lower conspicuousness or more sheltered nest location) if fishing pressure were extreme. Either diminished selectivity or a shift in the preferred features for a mate could alter sexual selection pressures.

The OSR can become biased in either direction through fishing pressure, promoting increases in mate competition. For example, where differential exploitation means that aggregations have become male-biased, males may harass the few remaining females to the extent that mating is impeded (e.g., groupers). In other species, sex differences in distribution or behavior mean that more males than females are caught (e.g., scalloped hammerhead shark *Sphyrna lewini:* Branstetter, 1987), with uncertain reproductive consequences. Even species in which mate competition is usually rare could experience increased competition if many individuals were seeking mates simultaneously, as after fishing depredations (e.g., seahorses). It is important to emphasize that the net impact of exploitation should depend on the change in OSR, not in the simple numeric change in the two sexes. Simply removing adults need not cause quite the anticipated decrease in reproduction if reduced population pressure makes spawning sites available to individuals that could not previously breed (e.g., convict cichlids).

Culturing can also reduce effectiveness in competition for mates. Behavioral ecological studies show that hatchery coho salmon suffered a breeding disadvantage under experimental conditions (see table 9-1), although interpreting these results is difficult without knowing the magnitude of direct competition in depleted populations of conservation concern; it could be rather low.

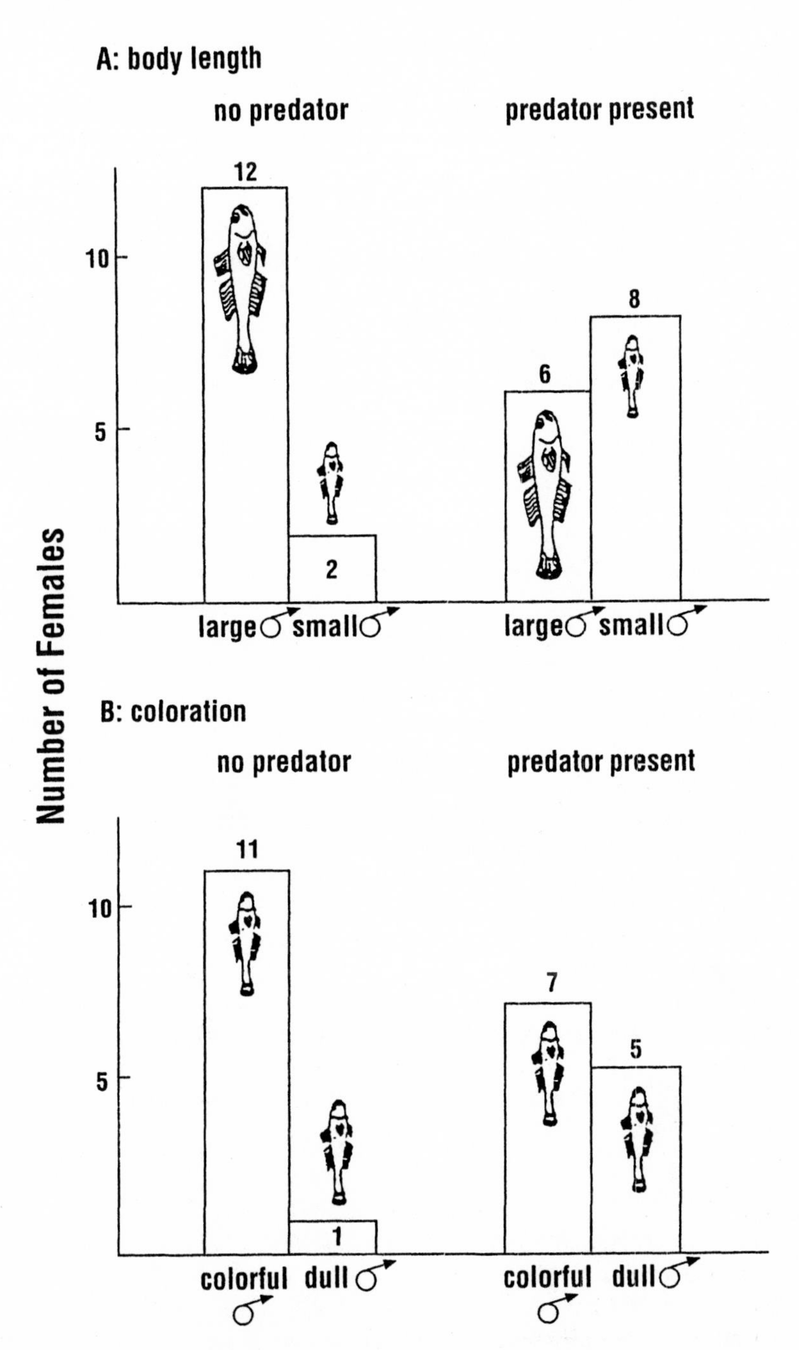

Figure 9-5 The number of sand goby *Pomatoschistus minutus* females in mate-choice experiments that were found closest to males of (A) different sizes or (B) colorations in the absence and presence of a fish predator (Forsgren, 1992). Mate choice may be affected by the risk of predation (and perhaps the risk of being fished). Female sand gobies were less likely to exert a preference for large or colorful males when predators were present.

Parental Care

Parental fishes may be more at risk from fishing, perhaps because they are rather sedentary and restricted in their ranges. Twelve of 36 fish species that are listed as recently extinct in the 1994 IUCN Red Book (Groombridge, 1993) belonged to guarding or bearing reproductive guilds (Bruton, 1995), even though only 21% of fish species provide parental care. Caring species may just be better-studied, but the suggestion is that parental care may put fish at risk because they are exposed and/or unable to swim away from threats (Bruton, 1995). Such risks will place particular pressures on males, as they are the primary caregivers in fishes, caring alone in about half the species with care and with females in another quarter of the species (Gross and Sargent, 1985). Most parental care takes the form of nest guarding and fanning, but fishes also bear eggs on or in their bodies (Blumer, 1979).

The form of care will be important in determining risk both to the parent and to the offspring. For example, species making prominent nests are more likely to be caught in nets than those nesting less conspicuously (e.g., haplochromine cichlids). Young of species that carry eggs in or on their bodies will be killed when the egg-bearing adults are fished (e.g., cardinalfishes, haplochromine cichlids, seahorses, pipefishes). In contrast, young from nest-making species can sometimes survive if far enough advanced when the male is removed (e.g., convict cichlid). Among nesting species, those where new partners cannibalize eggs (e.g., some damselfish) will probably be more affected by fishing than those where step-parenting occurs (e.g., anemonefish).

Removing necessary parental care would be expected to reduce rates of population recovery, but the impact may well depend on the lability of parental care. Resilience may be greater if either sex can care facultatively (e.g., some cichlids) and greatest if parental care is optional (e.g., peacock wrasse) rather than obligate (both are found among damselfish in the genus *Stegastes*). It should be noted that the quality of parental care may be affected by fishing pressures (e.g., sticklebacks) and may deteriorate or yield fewer eggs if smaller, non-breeding fishes are removed in species where these provide alloparental care (e.g., cichlids and anemonefish; fig. 9-6).

The loss of parental males could have a particularly large effect on population reproductive output in species where male care is sufficiently costly to limit female reproductive rate. These are most commonly the egg-bearing species (Berglund et al., 1989; Vincent, 1992, 1994). Extracting parental males will probably be less damaging to the population where potential male reproductive rate exceeds that of females through the male's capacity to care for multiple broods more quickly than a female can produce them (Clutton-Brock and Vincent, 1991).

Alternative Strategies

Fishes that exhibit alternative reproductive strategies at the level of the individual and/or the population offer some of the best examples of behavioral ecological studies informing management decisions. Male fishes commonly diverge in their reproductive strategies: fertilization stealing (sneaking), female mimicry, cuckoldry, and precocious maturity provide alternatives to territory holding and aggressive competition for mates. Such territoriality and competition are often associated with parental care (e.g., sticklebacks, gobies, sunfish, salmon; see review by Gross, 1984; see also Parker, 1992). Fishery removal of large males can alter reproductive success of males with other tactics, thus eventually affecting mean male size (e.g., tilefish), and even leading to dwarf males (Parker, 1992). Exploitation can

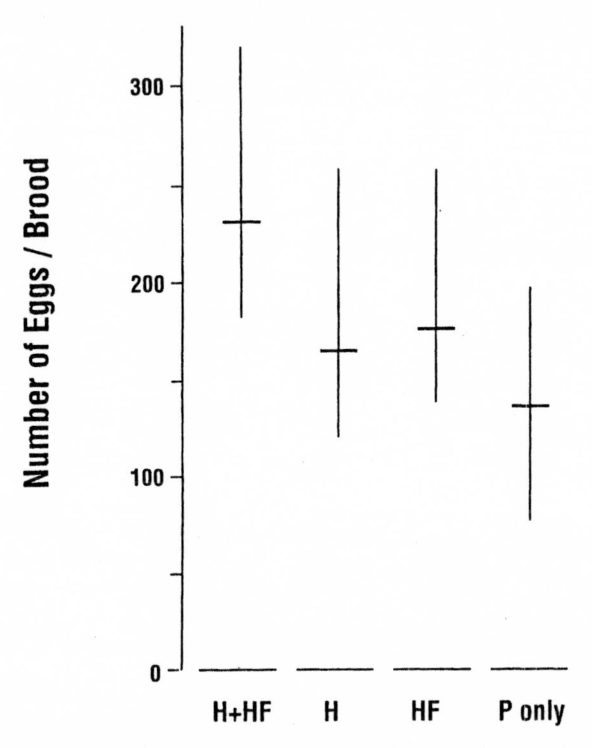

Figure 9-6 The median number of eggs per brood of females of male–female pairs (P) with or without helpers (H) and helper fry (HF). Selective removal of broodcare helpers in the cichlid fish, *Lamprolagus brichardi,* from their parents' territories is expected to reduce reproductive success of parents because females with helpers or helper fry produce more eggs per brood and more free-swimming fry than those without helpers and helper fry. Vertical lines represent interquartile ranges (Taborsky, 1984).

also influence life-history pathways in sex-changing species, sometimes by altering habitat (e.g., anemonefish).

Behavioral ecological studies on sunfishes (Centrarchidae) have already influenced management techniques in the United States, where these fishes provide important sport fishing. Sunfish males can (1) take on a parental, nest-guarding role and court females to spawn; (2) act as satellite males that mimic females and release sperm when intruding on spawnings; or (3) sneak copulations by darting into nests of parental males during spawning (Gross, 1982). The fishery was taking primarily larger (nesting) males, thus reducing survival of young because care is removed and sunfish will cannibalize unguarded eggs and larvae (see table 9-1). Management policies were revised to control numbers of cuckolders (satellite and sneaker males; M. Gross, personal communication).

Understanding alternative life-history strategies may be helpful in salmon management. Direct exploitation and habitat destruction have led to the decline of many salmon species and stocks and to decreased fish size through selective harvesting of larger individuals (e.g., Williams et al., 1989; Nehlson et al., 1991). Salmon are anadromous, migrating from fresh-

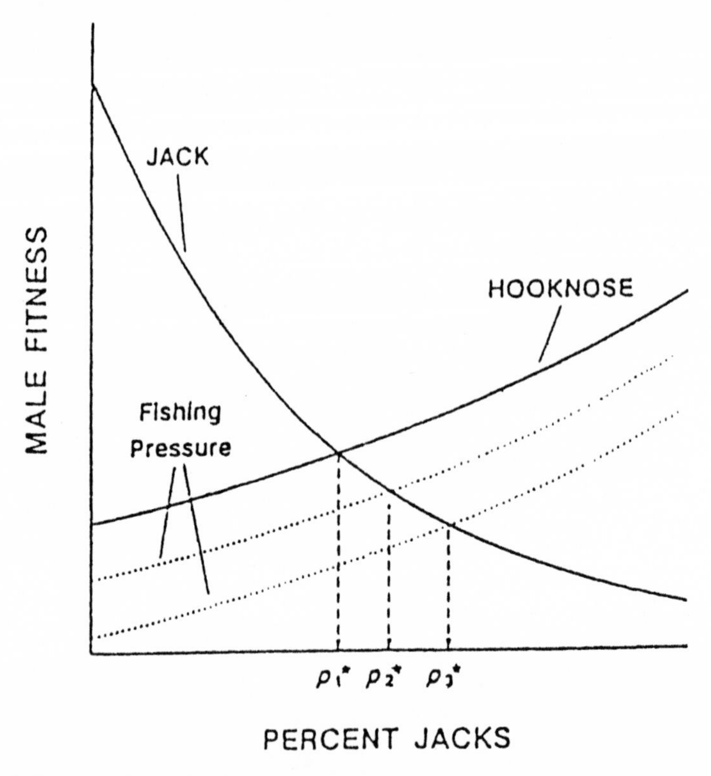

Figure 9-7 Selective harvesting of hooknose male salmon, *Oncorhynchus kisutch*, decreases their relative fitness and favors an evolutionary increase in the frequency of jack salmon as a proportion of the entire breeding male population. (p1* = no harvesting; p2* and p3* are increasing levels of selective harvest) (from Gross, 1991).

water to the sea and then returning to natal streams to breed. Male coho salmon *Oncorhynchus kisutch* either mature when young and small or delay maturation until they are older and larger. The small males (parr or jacks) usually sneak to obtain unobtrusive fertilizations while the larger males (adults or hooknoses) usually fight for matings (Gross, 1984). The reproductive advantage of choosing either strategy depends on the probability of surviving to maturity and the probability of obtaining matings. Both average body size and average frequency of jack males may be altered when salmon are fished. Hooknoses spend more time at sea, so fishing at sea may increase jack success and lead to a higher proportion of jacks in the long term (Gross, 1991) (fig. 9-7). Selectively catching larger fish, whether at sea or in rivers, will also kill more hooknoses and increase jack success (Gross, 1991). The decision to breed as a jack is influenced by body size, which is at least partly genetically determined (Iwamoto et al., 1983), so greater success of the jack tactic should favor a change in the switch point of the conditional strategy regulating the decision to become a jack (Gross, 1996). The increased jack frequency and presumed consequent loss of genetic diversity may have conservation and management consequences, although these are still unclear.

Sex Allocation

Approximately half of exploited reef fish families include sequentially hermaphroditic (sex-changing) species, justifying considerable attention to the issue of sex allocation. Individuals of some species can change sex during their lives to take advantage of increased reproductive opportunities with growth. Sex change from female to male occurs in at least 15 fish families [e.g., groupers (Serranidae), wrasses (Labridae), sweetlips (Lethrinidae), gobies (Gobiidae)], and from male to female in at least 6 families [e.g., anemonefishes (Pomacentridae), barramundi (Centropomidae), and porgy (Sparidae)] (see review by Warner, 1984; Sadovy, in preparation). In general, an organism should change sex at a size or age that would maximize its reproductive success relative to spending a lifetime as one sex (Warner, 1975), and in the direction that offers maximum benefits (Ghiselin, 1969). The precise mechanism determining sex allocation is unknown, but the consequence is that larger animals tend to be all of one sex (see below). We now explore the consequences of rapid, repeated large fish removals on sex-changing species using a theoretical approach with supporting empirical data.

Small sex-changing fishes are sought for aquarium pets and caught by artisanal fishers for food, sometimes using dynamite or poisons (Rubec, 1986; Derr, 1992). Larger sex-changing fishes are caught by both artisanal and commercial operators throughout the tropics and subtropics. Overexploitation by both scales of fishery is depleting individuals and damaging habitats (Munro, 1983; Sadovy, 1994; Johannes and Riepen, 1995). Some of these fishes are already under local protection; others are being considered for protective status (Sadovy, 1993). Many groupers are included in the 1996 IUCN Red List (IUCN, 1996).

Most fishing operations target larger individuals. In sex-changing fishes, consequent effects on operational sex ratio will depend on how sex change is determined (see Sadovy, 1996). We conceive of sex change regulation as operating along a gradient from (1) controlled either purely by social factors and hence responsive to local changes in population or social unit composition to (2) controlled purely by genetic factors, such that sex change can only take place at a given size or age. For illustration, we develop a simplified scheme to explore possible outcomes at the two extremes of such a hypothetical regulatory cline in terms of sex ratios and sizes of males and females under fishing pressure.

Social control is characterized as compensated sex change because it can respond to local changes, whereas genetic control is labeled as uncompensated because it cannot. Behavioral ecological studies can discover where on this continuum a species functions, thereby contributing to the evaluation of the effects of fishing on sequential hermaphrodites. Although we concentrate on protogynous species (sex changing, female first) the scheme is also applicable to protandrous species. For our purposes, we assume that only larger individuals (here males) are removed, recruitment into the fishery is negligible, and removal has no impact on the size of female sexual maturation. These assumptions are meant for illustration but are not unreasonable for the relatively short time scales over which the processes in our scheme would operate.

In compensated sex change, local changes in social structure determine sex change. Removing a male, for example by fishing, from the social group leads a female in the group—usually the largest—to change sex (within weeks or months). We assume a one-to-one replacement for the purposes of argument here, but this will in reality depend on the OSR of the social group, which can fluctuate slightly (Shapiro and Lubbock, 1980). Therefore, although the absolute number of females will decline on male removal through

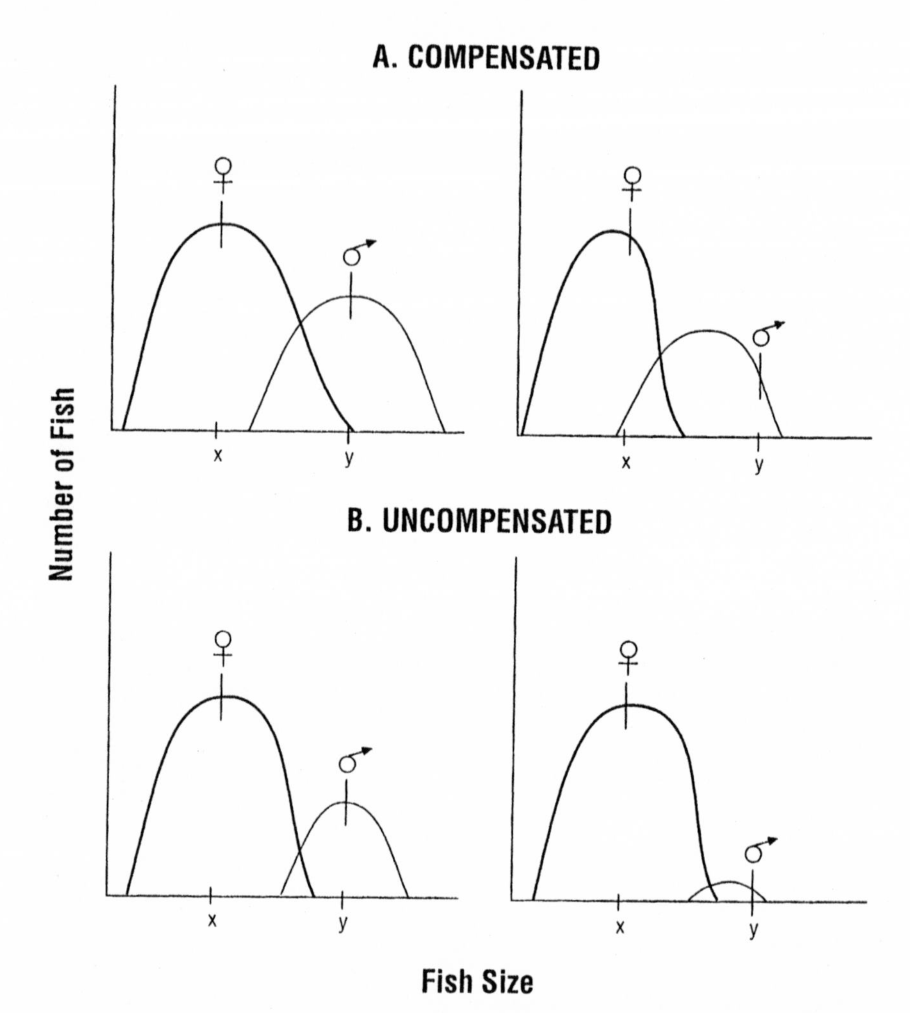

Figure 9-8 Hypothetical size-frequency distributions of male and female protogynous fishes, before and after removal of large males through selective catch of large size fishes, in compensated and uncompensated species. For illustration, it is assumed that there is no recruitment or change in size of female maturation; x and y are mean female and male lengths, respectively. (A) If one or more males is removed from a species where sex change is socially controlled, then the loss is compensated by an equivalent number of females changing sex. Sex ratios remain essentially stable, and both abundance and mean male and female sizes decline with continued removals, as they would in any fishery selectively extracting large individuals. Mean fecundity also declines as female size declines. (B) If sex change is uncompensated, lost males are not replaced by females because these must attain a specific size or age before they can change sex, sex ratios become female biased, and female abundance, size and fecundity remain little altered. Empirical examples given in figs. 9-9 and 9-10 appear to conform closely to the uncompensated end of our hypothetical uncompensated–compensated continuum.

female to male sex change, selectively removing large fish (males) should not greatly alter the OSR. Mean size and abundance of both sexes, and mean female fecundity, will decline as larger females change sex (fig. 9-8a).

In uncompensated sex change, change occurs at a fixed size or at a fixed age and is genetically determined. Timing of sex change is not influenced by local population changes, so selective fishing of larger individuals (here males) will not automatically mean that other females can change sex to take their place. Thus, removing males means that the OSR will become more female biased, creating potential for sperm limitation, while the absolute number of females will change little. Mean male size will decline as larger males are taken selectively, but mean female size remains unaltered (fig. 9-8b).

All behavioral studies thus far have examined social units in which removal rates were low or male removals were few, as might happen naturally. What would happen if we fished a population intensively, removing multiple males at a much higher rate than is likely to occur in nature? Our scheme predicts that even species with compensated sex change would be unable to respond rapidly enough (remember sex change takes several weeks to months) to provide a one-to-one replacement, particularly since sex change by females takes progressively longer with successive removals of males (Shapiro, 1980). Sex ratios would be expected to become increasingly female biased, mean male size and number would decline, and the number and mean size of females would remain constant for some time before ultimately declining (fig. 9-8b). Thus a heavily fished population that usually relied on compensated sex change would soon be indistinguishable from a population with uncompensated sex change.

Most detailed behavioral ecological research on sex-changing species has focused on small fishes living in small social groups with sustained social interactions and somewhat limited movements. These studies have found sex change that is compensated, or nearly so. Prime determinants of sex change were found to be social organization and mating system, mediated by some combination of sex ratios, size ratios, behavioral interactions (types and rates), growth rates, and population density (Shapiro, 1979, 1984; Aldenhoven, 1986; Ross et al., 1983; Cowen, 1990; Ross, 1990; Warner and Swearer, 1991; Lutnesky, 1994). Sex ratios never became very biased because fishes changed sex in response to male removal or female recruitment (i.e., addition of females). Experimental sequential removals of male seabass *Pseudoanthias-Anthias squamipinnis* and wrasse *Thalassoma bifasciatum*, for example, created temporary increases in female sex ratios that quickly disappeared as successive females changed sex in compensation (Shapiro, 1980; Warner and Swearer, 1991). Thus removing large males resulted in absolutely fewer and smaller females and males, but maintenance of the OSR (fig. 9-8a).

The small protogynous species have close relatives that are heavily exploited (e.g., groupers, porgies, and emperors: Serranidae, Sparidae, and Lethrinidae, respectively). Theoretical treatments would predict similar mechanisms of sex change (Ghiselin, 1969; Shapiro, 1989), even though the exploited species tend to be larger, more wide-ranging, and (apparently) lack such prolonged social groupings. Yet, in contrast to the small protogynous species, sex change in more exploited species *appears* to lie toward the uncompensated end of the compensated–uncompensated continuum. For example, fished populations of protogynous grouper or porgy demonstrated extreme sex ratio biases and declines in mean sizes compared to lightly fished or unfished populations (see figs. 9-9, 9-10).

The implication is that either sex change in such exploited species really is uncompensated or, more probably, that it appears so because the rapid and stochastic nature of fisheries depredations has disturbed sex change induction that is dependent on social cues

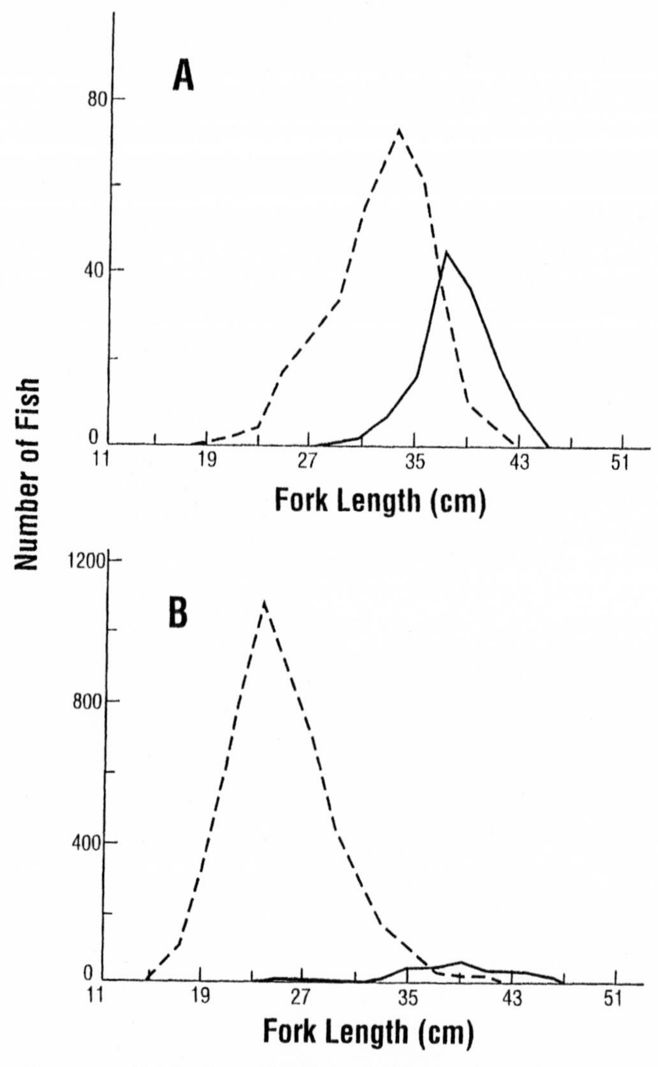

Figure 9-9 Frequency distributions of male (solid line) and female (dotted line) porgy *Chrysoblephus puniceus* of different lengths from (A) Mozambique, where fishing pressure is light and (B) Natal, a heavily fished locale. The figure illustrates a case where removals are effected too rapidly for sex change of females to replace removed males, thus producing extreme female bias and appearing uncompensated. In this example, male size does not drop below a certain threshold, implying a possible minimum size for sex change, and females have apparently responded to heavy fishing by initiating maturation at a smaller size (from Garratt, 1986).

(Sadovy, 1996). Such persistent and repeated male removals, as under continuous fishing, appear to deny the population the opportunity to adjust, thus making compensated sex change appear uncompensated. Compensation and associated population recovery might be expected to succeed only if fishing pressure were reduced or removed or fishing selectivity minimized.

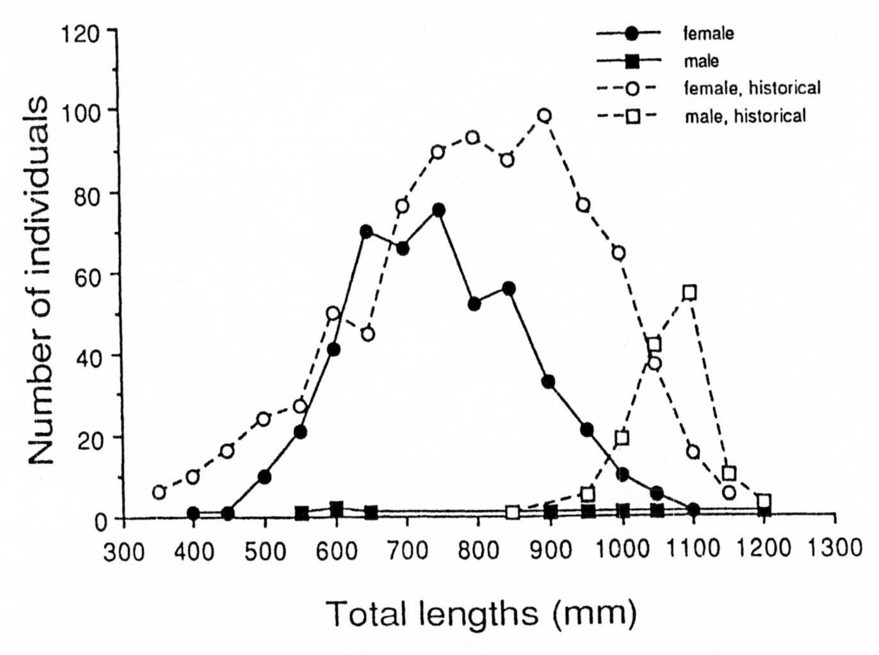

Figure 9-10 Size-frequency distributions of male (squares) and female (circles) gag grouper *Mycteroperca microlepis* from the Gulf of Mexico, USA. Historical data come from Hood and Schlieder in 1977–1980 (open symbols; Hood and Schlieder 1992) and more recent data come from Florida State University in 1991–1992 (closed symbols; Koenig et al., 1996). Gag grouper, particularly larger ones, were heavily fished in the intervening 11–15 years. It appears that sex change is uncompensated or that it is compensated but removals are effected too rapidly for sex change of females to replace removed males, producing extreme female bias (figure redrawn from Koenig et al., 1996). Males lost to fishing are not replaced proportionally (dropping from 17% to 2% of adult population), and their mean size has declined.

Our scheme for sex-changing species shows that disruption of sex change processes may occur under severe size-selective fishing pressure, even in species that are compensated. Other parameters such as recruitment and size at sexual maturation need to be incorporated to increase realism. However, it is already clear that a detailed understanding of factors governing sex change could be vital, just as an appreciation of alternative life-history strategies in gonochores is critical, for fully appreciating the consequences of fishing and adjusting management accordingly.

The Utility of Behavioral Ecology

Strengths of the Approach

The particular strength of reproductive ecology lies in explaining how selective removals of individuals influence the reproduction of remaining individuals and thus population persistence. Without this information, one is in danger of interpreting the impact of extraction in numerical terms alone, rather than through a recognition of demographic effects (fig. 9-1). Even the newest generation of fisheries models fails to consider behavioral parameters or prerecruit survival. Yet predicting the nonrandom effects of differential mortality—exercised through the time, place, and agent of death—should improve fisheries management and conservation measures alike. Farmers and herders have long managed their domestic mammals with an awareness that not all individuals are alike and that population responses vary when different individuals are removed. Such a recognition is similarly needed in managing fish populations.

Behavioral ecology can contribute to the risk assessment that should be an intrinsic part of fisheries management (Rosenberg et al., 1993) and can help to identify suites of characteristics that render species more or less vulnerable to overexploitation (e.g., Adams, 1979; Parrish, 1995). Current fisheries and conservation management are outcome oriented and often work on the basis of analyzing what went wrong and recovering from imbalance and loss. By understanding processes (fig. 9-1), behavioral ecology can help with this retrospective analysis and also make it possible to anticipate conservation problems under different management regimes. Specifically, behavioral ecology could be useful in helping to understand why some species are more resilient under fishing pressure (e.g., Atlantic cod *Gadus morhua;* flatfish, Pleuronectiformes) than others (e.g., small gadoids and pelagic species); in these cases, the former can apparently maintain reproductive output at higher levels of fishing (Mace and Sissenwine, 1993). (It should be noted, however, that not even Atlantic cod has proven resilient to overfishing.)

Behavioral ecology already has a history of focusing on reproductive parameters of immediate interest to management. Fisheries biologists looking at a spawning aggregation, for example, may worry primarily about when and where it occurs and whether certain types of aggregation are more vulnerable than others (Sadovy et al., 1994; Samoilys and Squire, 1994). To understand the impact of fishing and devise the most effective management approach, however, fisheries biologists need also to know the sex ratio, time of spawning, mating pattern, and proportion of the population's spawning effort represented by that aggregation. In carrying out such studies, behavioral ecology can also contribute to new paradigms in fisheries.

Oversimplification can be avoided through behavioral ecology studies. Though fecundity is a key parameter in dynamic pool fishery models, it is often simplistically estimated as adult biomass or, occasionally, as age- or size-specific batch fecundity (the number of hydrated eggs present in the ovary; Goodyear, 1993; Sadovy, 1996). Such crude estimates have potential for error, given all the behavioral and ecological factors that could influence egg release, spawning frequency, and fertilization. Sometimes, protecting smaller fish from fishing may be less effective than protecting larger fish, given that fecundity is an exponential function of size in many species (Plan Development Team, 1990).

Behavioral ecological studies can provide a comparative framework for anticipating suites of characters and behaviors across species that would lend themselves to similar management approaches (see Harvey and Pagel, 1991). They can also identify appropriate management units by revealing the diversity of life-history patterns in apparently similar species (e.g., pipefishes, Syngnathidae; Vincent et al., 1992) and thus encouraging appropriate planning for each. In so doing, behavioral ecological research can contribute to the determination of the "evolutionarily significant unit," proposed as the basis for distinguishing stocks, to be managed individually (Dixon et al., 1992).

Thus behavioral ecology has a definite role in the assessments of Minimum Viable

Population size (MVP) that are often employed in trying to prevent extinctions. These assessments seek to quantify the minimum number of individuals required to ensure population persistence for a particular time period—for example, 1000 years (Schaffer, 1981; Gilpin and Soulé, 1986). However, theoretical advances in the concept have outstripped empirical data, particularly for fishes. Moreover, estimates of MVP are purely numeric values, ignoring the fact that understanding demographic data on behavioral, phenotypic, and fitness variability can help project population persistence, by identifying which and how many individuals will breed and survive at a particular time and place (fig. 9-1). Minimum viable population theory based strictly on numeric abundance was generally developed for organisms with internal fertilization, such as mammals, and may not be useful where threshold densities or numbers are required (see above; Hockey and Branch, 1994). Indeed, the argument that low densities could result in disproportionately low fertilization rates and lead to population demise, regardless of the absolute number of animals (e.g., marine invertebrates; Levitan and Petersen, 1995), is recognized in the newly formulated guidelines for interpreting IUCN Red Listing criteria for marine fishes (Hudson et al., 1997).

Quite apart from reproductive studies, other research in behavioral ecology on, for example, foraging, can also contribute to planning fish conservation and management. For instance, puffer fish *Takifugu niphobles* are size-segregated with respect to habitat utilization; young puffers live inshore, whereas adults are mostly found offshore, though young puffers moved with tides to feed in both regions (Yamahira et al., 1996). Thus, demographic outcomes of fishing could depend on where and when it takes place. Site fidelity may also be a factor influencing response to exploitation and design of marine reserves. For example, species with small home ranges and low mobility, such as seahorses (Vincent and Sadler, 1995) and pegasids (Herold and Clark, 1993) will be slow to recolonize depleted areas.

Weaknesses of the Approach

For all their merits, current behavioral ecological studies face some serious shortcomings as management and conservation tools. First, and most seriously, relatively few such studies focus on fishes, particularly on exploited or marine species. Those that do commonly provide little assistance in ensuring long-term sustainability of fish populations, because they are too local and short-term to investigate spatial and temporal variation in selective pressures, they do not extrapolate from studies of few individuals to population scale responses, and/or they do not distinguish between phenotypic and genotypic responses to exploitation.

Second, existing behavioral work on fishes has focused on small species with small populations and low mobility largely because these are the most tractable for research (e.g., sticklebacks, guppies, damselfishes, seahorses: Gasterosteidae, Poeciliidae, Pomacentridae, Syngnathidae, respectively). Because many of these species are not in danger from exploitation, such research has hitherto played little role in conservation and management. Fisheries biologists have traditionally directed most effort toward larger fishes with higher mobility (e.g., tuna, haddock, sharks, snappers: Scombridae, Gadidae, elasmobranchs, Lutjanidae, respectively), which are often more difficult to study in detail at the individual level. However, the growing awareness that small-scale fishers exert considerable pressures on smaller inshore species, with their structured populations, will promote the role of behavioral ecology in management, while technological innovation may open the door to studies on the larger, wide-ranging species.

Third, behavioral ecological research gains the clearest understanding of how natural selection has shaped behaviors through studying undisturbed populations, seldom available in species of conservation concern. In disturbed populations, individuals may no longer operate within their original ecological context, and their behaviors will respond to new selective pressures, for which they are often not adapted. Even populations in protected areas will have been affected by exploitation or habitat destruction to some extent, even if only because of range limitation or intermigration with other, affected, populations. Where extreme exploitation has produced permanent loss of genetic diversity, then it may be difficult to apply a behavioral ecological approach.

Fourth, behavioral ecology often fails to document the fundamental life histories parameters most important in conservation. For example, studies on reproductive ecology often do not report the fecundity or reproductive rate of the two sexes, or else they treat all females as equivalent in reproductive output (see Sadovy, 1996). Moreover, behavioral ecological research has had a tendency to report statistical generalities, minimizing or dismissing exceptions or rarities in behavior. For example, species may be categorized as exhibiting a certain mating pattern (e.g., monogamy), even though some individuals or populations may behave differently (e.g., polygamy). This is particularly problematic in fishes, which show much more intraspecific phenotypic variation than most other taxa (Allendorf et al., 1987).

Fifth, studies are commonly not explicit about the physical agents or processes that may have structured behavior. For most studies we cite in this chapter, it was necessary to guess the natural or man-made process that the research might have imitated (e.g., selective removals of larger or smaller fishes, reductions in density, death of spawning adults) and to speculate about potential population level outcomes (see table 9-1). Such extrapolation currently limits confidence in this behavioral research for conservation and management, although findings can provide warnings about declines in long-term viability and integrity of populations.

Finally, behavioral ecological studies tend to be very slow indeed. For African lakes, the development of scientific understanding of species and communities is so slow relative to rates of exploitation and manipulation that accurate predictions regarding the future are impossible (Ribbink, 1987). The same may well be true of many marine ecosystems, particularly in enclosed bodies of water. Given the pace of fish declines, we may still have to rely on conventional management without nuance or detail or draw inferences from related species.

A Future for Behavioral Ecology in Fish Conservation?

Fishes will always be highly vulnerable to exploitation because of their essential role as food. A vast array of fishes is now of commercial value, with ever more being exploited as traditional fishing stocks decline and demand increases. Every possible tool will be needed for fish conservation. Fisheries biologists have already begun to document fish reproductive ecology and to explore the genetics of exploited species in order to understand better the effects of fishing (e.g., Davis, 1984; Thresher et al., 1986; Grimes et al., 1988; Leaman, 1991; Smith et al., 1991; Darwall et al., 1992). There is growing recognition that individuals vary and that this variability should play an important role in population dynamics (e.g., Cury, 1994). The knowledge gained in behavioral ecology can also be incorporated into models of exploitation and population viability analyses; these commonly deal primarily in numeric, often highly theoretical, values and ignore the phenotypic and fitness variability of biological reality that behavioral ecologists have identified (Rose, 1997).

Behavioral ecological studies can contribute to a more holistic approach to fisheries management. For example, a network of marine reserves would be important both as a conservation measure and as a source for fishery recruits (Bohnsack and Ault, 1996); designing such reserves and predicting their recovery will partly depend on an understanding of the reproductive activities, mobility, and interconnectedness of species concerned (e.g., Roberts and Polunin, 1991; Russ, 1991). An increased emphasis on sustainable mariculture, that is, not dependent on wild-caught juveniles, could also reduce pressure on wild populations (Johannes and Riepen, 1995; Sadovy, in press). Such improvements in captive breeding will come from a knowledge of the behavior of the species involved and how it relates to their ecology. Culture of depleted species might also help reseed former ranges but the effectiveness of this reseeding will also partly depend on how well we understand possible impacts of releases on the target individuals and the wild populations that receive them.

Research in behavioral and ecology is helping to formulate criteria appropriate for assessing the conservation status of exploited marine species. The newly composed guidelines for applying IUCN Red List criteria to marine fishes recognize the need to assess the potential and mechanism for sex change, the relative abundance of the rate-limiting sex, and density effects on reproductive output (Hudson et al., 1997). The guidelines also note that the behavior and ecology of given populations and species should be considered when evaluating how changes in abundance might reflect a population's risk of extinction (Hudson et al., 1997).

To play a full role in fish conservation, behavioral ecologists should consider the following recommendations:

1. Carry out research on fishes. Take on the difficult challenge of studying key, highly mobile, commercial species as well as marketable inshore species with more structured populations.
2. Explicitly consider conservation needs when structuring research. Where a choice can be made between two theoretical precepts of equal interest or between two species, then studies should focus on the one with most conservation urgency.
3. Extend studies over long enough periods to incorporate change or revisit populations after change. Study more than one population per species, preferably comparing disturbed with undisturbed or less disturbed. Consider more studies of relationships among species.
4. Incorporate genetic studies. Distinguishing between genotypic and phenotypic effects of extraction is extremely difficult but necessary to understand the long-term consequences of exploitation at the population level (Smith, 1994).
5. Devise and expand methodologies to distinguish between recent anthropogenic selective pressures and natural selection pressures in order to facilitate understanding of disturbed populations. Ensure that these are empirically supported.
6. Report data necessary for management as they are obtained during fundamental research. Management needs to understand parameters of recruitment, growth, natural mortality, and anthropogenic mortality as a bare minimum (fig. 9-1).
7. Explain and explore exceptions to generalities, as these provide the template for adaptation to new pressures.
8. Make use of the comparative method to predict how unstudied species will re-

spond based on available research on other species, thus increasing the impact of behavioral research; identify suites of characteristics that might make species more or less resilient to fishing pressure.

We said earlier in this chapter that the ultimate goals of conservation and fisheries management can be similar. The difference is that conservation should seek ideally to ensure that human disturbance is so slight as not to alter selection pressures on a population, while fisheries inevitably inflict new pressures on populations (Ribbink, 1987). Given the ever-increasing demands to extract fishes, we must hope that behavioral ecologists can at least help minimize the impact of new anthropogenic selection pressures through understanding original natural selection pressures.

Summary

Fishes are overexploited, subject to habitat damage, and disrupted by introductions. They have historically attracted little conservation attention, largely because of widespread, but unfounded, assumptions regarding high resilience to fishing and other perturbations. However, recent overexploitation and stock collapses have produced a catalog of fishes of conservation concern and a recognition that they are inadequately managed by current paradigms. This chapter has focused on the effects of selective exploitation of marine fishes.

Both conservation biology and fisheries science aim to ensure long-term persistence of fish populations and to preserve the genetic diversity of exploited populations. Until recently, conservation was largely the preserve of fishery managers. However, traditional fishery assessments and management approaches, concentrating as they do on the number and size of extracted fishes, do not incorporate many of the causal links between fish removal and population outcomes.

Behavioral ecology examines how behavior, ecology, and genetics interact to determine survival and reproduction. This approach can be used to understand the effects of human predation on fish populations and to project the long-term consequences of such perturbations at the individual and population levels. Reproductive ecology, a branch of behavioral ecology, focuses particularly on the ways in which individuals replace themselves. It can elucidate how fishing pressure—especially that which is selective for size, sex, time, or place—compromises fecundity, mating, or parental care or alters selective pressures on alternative life histories and sex allocation. It confers an appreciation that individuals and species are differently vulnerable to exploitation.

Behavioral ecology can open the door to more biologically realistic approaches to management and fishery enhancement. Its strengths include its broad applicability in explaining the effects of selective removals on population persistence, its contributions to risk assessment, its importance in identifying appropriate units for management, and its potentially important role in MVP analyses. The weaknesses of behavioral ecology lie in its current focus on a subset of exploited species, those that are small and relatively immobile, and the practical difficulties of applying behavioral ecological approaches to the larger, highly mobile fishes that are more heavily exploited. Moreover, such studies are most applicable to undisturbed populations, tend not to be explicit about physical processes that may structure behavior, are time consuming, have seldom focused on fishes of conservation concern, and are frequently too local or narrow in their scope.

Behavioral ecology could (1) play a key role in modifying fishery selectivity; (2) better tailor fishery management to different suites of reproductive characteristics; (3) enhance fish production by identifying key physical and biological features for breeding; (4) contribute to marine reserve design by providing much needed information about reproduction, recruitment, movement and survivorship; (5) identify the role of genetics in reproductive strategies and reproductive output; and (6) help improve fish culturing in attempts to take pressure off wild populations. Achieving these objectives will require a specific awareness of conservation needs when planning and conducting behavioral ecological research.

Acknowledgments We are grateful to Don Kramer and Karl Evans for important early help with this chapter and to Mart Gross, Bob Johannes, Peter Moyle, and an anonymous referee for valuable comments on the draft manuscript. A.V. appreciates the support of the Darwin Initiative for the Survival of Species (United Kingdom Department of the Environment) and British Airways Assisting Conservation and is grateful for the hospitality of the Swire Institute of Marine Sciences at The University of Hong Kong. We both thank George Mitcheson for his many contributions.

References

Adams PB, 1979. Life history patterns in marine fishes and their consequences for fisheries management. Fish Bull 78:1–12.

Aldenhoven JM, 1986. Different reproductive strategies in a sex-changing coral reef fish *Centropyge bicolor* (Pomacanthidae). Austr J Mar Freshwat Res 37:353–360.

Allendorf FW, Leary RF, 1988. Conservation and distribution of genetic variation in a polytypic species, the cutthroat trout. Conserv Biol 2:170–184.

Allendorf FW, Ryman N, Utter FM, 1987. Genetics and fishery management. Past, present and future. In: Population genetics and fishery management (Ryman N, Utter F, eds). Seattle: University of Washington Press; 1–19.

Andrew NL, Pepperell JG, 1992. The by-catch of shrimp trawl fisheries. Oceanogr Mar Biol Annu Rev 30:527–565.

Andrews C, 1990. The ornamental fish trade and fish conservation. J Fish Biol 37: 53–59.

Angermeier PL, 1995. Ecological attributes of extinction-prone species: loss of freshwater fishes of Virginia. Conserv Biol 9:143–158.

Bagenal TB, 1978. Aspects of fish fecundity. In: Methods of assessment of ecology of freshwater fish production (Gerking SD, ed). Oxford: Blackwell; 75–101.

Bailey C, 1988. Optimal development of Third World fisheries. In: North-South Perspectives on Marine Policy (Morris MA, ed). Boulder, Colorado: Westview Press; 105–128.

Balshine Earn, S. 1995. The costs of parental care in Galille St. Peter's fish, *Sarotherodon galilaeus*. Anim Behav 50:1–7.

Baltz DM, 1991. Introduced fishes in marine systems and inland seas. Biol Conserv 56:151–177.

Bannerot SP, Fox WW Jr, Powers JE, 1987. Reproductive strategies and the management of snappers and groupers in the Gulf of Mexico and Caribbean. In: Tropical snappers and groupers: biology and fisheries management (Polovina JJ, Ralston S, eds). Boulder, Colorado: Westview Press; 561–603.

Barrera Guevara JC, 1990. The conservation of *Totoaba macdonaldi* (Gilbert), (Pisces: Sciaenidae), in the Gulf of California, Mexico. J Fish Biol 37(suppl A):201–202.

Bartley D, Bagley M, Gall G, Bentley B, 1992. Use of linkage disequilibrium data to estimate effective size of hatchery and natural fish populations. Conserv Biol 6:365–375.

Bayley PB, Petrere M Jr., 1990. Amazon fisheries: assessment methods, current status and management options. In: Proceedings of the International Large River Symposium: Honey Harbour, Ontario, Canada, September 14–21, 1986 (Dodge DP, ed). 385–398.

Berglund A, 1991. Egg competition in a sex-role reversed pipefish: subdominant females trade reproduction for growth. Evolution 45:770–774.

Berglund A, Rosenqvist G, Svensson I, 1989. Reproductive success of females limited by males in two pipefish species. Am Nat 133:506–516.

Beverton RJH, Holt SJ, 1957. On the dynamics of exploited fish populations. In: Fishery Investigation Series 2 (Great Britain, Ministry of Agriculture, Fisheries and Food) Series II, v. 19.

Blumer LS, 1979. Male parental care in the bony fishes. Q Rev Biol 54:149–161.

Boehlert GW, 1996. Biodiversity and the sustainability of marine fisheries. Oceanography 9:28–35.

Bohnsack JA, Ault JS, 1996. Management strategies to conserve marine biodiversity. Oceanography 9:73–82.

Bohnsack JA, Sutherland DL, Harper DE, McClellan DB, Hulsbeck MW, Holt CM, 1989. The effects of fish trap mesh size on reef fish catch off southeastern Florida. Mar Fish Rev 51:36–46.

Branstetter S, 1987. Age, growth and reproductive biology of the silky shark, *Carcharhinus falciformis,* and the scalloped hammerhead, *Sphyrna lewini,* from the northwestern Gulf of Mexico. Environ Biol Fishes 19:161–173.

Bruton MN, 1995. Have fishes had their chips? The dilemma of threatened fishes. Environ Biol Fishes 43:1–27.

Camhi M, 1995. Industrial fisheries threaten ecological integrity of the Galapagos Islands. Conserv Biol 9:715–724.

Carlton JT, Geller JB. 1993. Ecological roulette: the global transport of nonindigenous marine organisms. Science 261:78–82.

Clutton-Brock TH, Vincent ACJ, 1991. Sexual selection and the potential reproductive rates of males and females. Nature 351:58–60.

Colin PL, 1992. Reproduction of the Nassau grouper, *Epinephelus striatus* (Pisces: Serranidae) and its relationship to environmental conditions. Environ Biol Fishes 34:357–377.

Contreras-Balderas S, Lozano-Vilano ML, 1994. Water, endangered fishes, and development perspectives in arid lands of Mexico. Conserv Biol 8:379–387.

Cowen RK, 1990. Sex change and life history patterns of the labrid, *Semicossyphus pulcher,* across an environmental gradient. Copeia 1990: 787–795.

Cury P, 1994. Obstinate nature: an ecology of individuals. Thoughts on reproductive behaviour and biodiversity. Can J Fish Aquat Sci 51:1664–1673.

Darwall WRT, Costello MJ, Donnelly R, Lysaght S, 1992. Implications of life-history strategies for a new wrasse fishery. J Fish Biol 41(suppl B):111–123.

Davis, TLO, 1984. A population of sexually precocious barramundi, *Lates calcarifer,* in the Gulf of Carpentaria, Australia. Copeia 1984:144–149.

Davis TLO, West GJ, 1993. Maturation, reproductive seasonality, fecundity, and spawning frequency in *Lutjanus vittus* (Quoy and Gaimard) from the north west shelf of Australia. Fish Bull 91:224–236.

Dayton L, 1991. Save the sharks. New Sci 1773(15 June):28–32.

Deng JY, Yang CH, 1993. Marine fishery. In: Marine science study and its prospect in China

(Tseng CK, Zhou HO, Li BC, eds). Qingdao, China: Qingdao Publishing House; 681–688.

Derr M, 1992. Raiders of the reef. Audubon 94:48–54.

Dixon AE, Lockyer C, Perrin WF, Demastser DP, Sisson J, 1992. Rethinking the stock concept: a phylogeographic approach. Conserv Biol 6:24–36.

Emlen ST, Oring LW, 1977. Ecology, sexual selection and the evolution of mating systems. Science 197:215–223.

FAO, 1993. Review of the state of world marine fishery resources. FAO Fisheries Technical Paper no. 335. Rome: Food and Agriculture Organization.

FAO, 1996. The state of the world's fisheries and aquaculture. Rome: Food and Agriculture Organization.

Fleming IA, Gross MR, 1992. Reproductive behavior of hatchery and wild coho salmon (*Oncorhynchus kisutch*): does it differ? Aquaculture 103:101–121.

Fleming IA, Gross MR, 1993. Breeding success of hatchery and wild coho salmon (*Oncorhynchus kisutch*) in competition. Ecol Appl 3:230–245.

Forsgren E, 1992. Predation risk affects mate choice in a gobiid fish. Am Nat 140: 1041–1049.

Fricke HW, 1979. Mating system, resource defense, and sex change in the anemonefish, *Amphiprion akallopisos*. Z Tierpsychol 50:313–326.

Frissell CA, 1993. Topology of extinction and endangerment of native fishes in the Pacific Northwest and California (U.S.A.). Conserv Biol 7:342–354.

Fryer G, 1984. The conservation and rational exploitation of the biota of Africa's great lakes. In: Conservation of threatened natural habitats (Hall AV, ed). South Africa National Scientific Programmes Report 92. Pretoria: Council for Scientific and Industrial Research; 135–154.

Gall GAE, Gross SJ, 1978. A genetic analysis of the performance of three rainbow trout broodstocks. Aquaculture 15:113–127.

Garratt PA, 1986. Protogynous hermaphroditism in the slinger, *Chrysoblephus puniceus* (Gilchrist and Thompson, 1908) (Teleostei: Sparidae). J Fish Biol 28:297–306.

Ghiselin MT, 1969. The evolution of hermaphroditism among animals. Q Rev Biol 4:189–208.

Gilpin ME, Soulé ME, 1986. Minimum viable populations: processes of species extinction. In: Conservation biology; the science of scarcity and diversity (Soulé, ME). Sunderland, Massachusetts: Sinauer Associates; 19–34.

Ginsberg JR, Milner-Gulland EJ, 1994. Sex-biased harvesting and population dynamics in ungulates: implications for conservation and sustainable use. Conserv Biol 8:157–166.

Goodyear CP, 1993. Spawning stock biomass per recruit in fisheries management: foundation and current use. In: Risk evaluation and biological reference points for fisheries management (Smith SJ, Hunt JJ, Rivard D, eds). Canadian Special Publications in Fisheries and Aquatic Science no. 120; 67–81. Ottawa: Fisheries and Oceans.

Graham M, 1935. Modern theory of exploiting a fishery and application to North Sea trawling. J Conserv Int Explor Mer 10:264–274.

Grimes, CB, Idelberger CF, Able KW, Turner SC, 1988. The reproductive biology of tilefish, *Lopholatilus chamaeleonticeps* Goode and Bean, from the United States Mid-Atlantic Bight, and the effects of fishing on the breeding system. Fish Bull 86:745–761.

Groombridge B (ed), 1993. IUCN Red List of threatened animals. Cambridge: World Conservation Monitoring Centre.

Gross MR, 1982. Sneakers, satellites and parentals: polymorphic mating strategies in North American sunfishes. Z Tierpsychol 60:1–26.

Gross MR, 1984. Sunfish, salmon and the evolution of alternative reproductive strategies and tactics in fishes. In: Fish reproduction: strategies and tactics (Potts G, Wootton RJ, eds). London: Academic Press; 55–75.

Gross MR, 1991. Salmon breeding behavior and life history evolution in changing environments. Ecology 72:1180–1186.

Gross MR, 1994. The evolution of behavioural ecology. Trends Ecol Evol 9:358–360.

Gross MR, 1996. Alternative reproductive strategies and tactics: diversity within sexes. Trends Ecol Evol 11:92–98.

Gross MR, Sargent RC, 1985. The evolution of male and female parental care in fishes. Am Zool 25:807–822.

Gulland JA, 1974. The management of marine fisheries. Bristol: Scientechnia.

Harvey PH, Pagel MD, 1991. The comparative method in evolutionary biology. New York: Oxford University Press.

Hattori A, 1991. Socially controlled growth and size-dependent sex change in the anemonefish *Amphiprion frenatus* in Okinawa, Japan. Jpn J Ichthyol 38:165–178.

Hattori A, 1994. Inter-group movement and mate acquisition tactics of the protandrous anemonefish, *Amphiprion clarkii,* on a coral reef, Okinawa. Jpn J Ichthyol 41:159–165.

Hawkins JP, Roberts CM, 1994. The growth of coastal tourism in the Red Sea: present and future effects on coral reefs. Ambio 23:503–508.

Herold D, Clark E, 1993. Monogamy, spawning and skin-shedding of the sea moth, *Eurypegasus draconis* (Pisces: Pegasidae). Environ Biol Fishes 37:219–236.

Hirose Y, 1995. Patterns of pair formation in protandrous anemonefishes, *Amphiprion clarkii, A. frenatus* and *A. perideraion,* on coral reefs of Okinawa, Japan. Environ Biol Fishes 43:153–161.

Hockey PAR, Branch GM, 1994. Conserving marine biodiversity on the African coast: implications of a terrestrial perspective. Aquat Conserv 4:345–362.

Hood PB, Schlieder RA, 1992. Age, growth, and reproduction of gag *Mycteroperca microlepis* (Pisces: Serranidae) in the eastern Gulf of Mexico. Bull Mar Sci 51:337–352.

Hudson E, et al. 1997. Threatened fish? Initial guidelines for applying the IUCN Red List criteria to marine fishes. Gland: IUCN.

Huntsman GR, 1994. Endangered marine finfish: neglected resources or beasts of fiction? Fisheries 19:8–15.

Huntsman GR, Schaaf WE, 1994. Simulation of the impact on fishing of reproduction of a protogynous grouper, the graysby. N Am J Fish Manage 14:41–52.

Itzkowitz M, Mackie D, 1986. Habitat structure and reproductive success in the beaugregory damselfish. J Exper Mar Biol Ecol 97:305–312.

IUCN, 1988. Red List of Threatened Animals. Gland: International Union for the Conservation of Nature.

IUCN, 1994. IUCN Red List Categories. Gland: International Union for the Conservation of Nature.

IUCN, 1996. IUCN Red List of Threatened Animals. Gland: International Union for the Conservation of Nature.

Iwamoto RN, Alexander BA, Hershberger WK, 1985. Genotypic and environmental effects on the incidence of sexual precocity in Coho Salmon. In: Salmonid reproduction: review papers: an international symposium, Bellvue, Washington, October 31–November 2, 1983 (Iwamoto RN, Sower S, eds). Seattle, Washington: Washington Sea Grant Program, University of Washington.

Johannes RE, Riepen M, 1995. Environmental, economic, and social implications of the live reef fish trade in Asia and the Western Pacific. Report to the Nature Conservancy and South Pacific Forum Fisheries Agency, October 1995; 1–82. Arlington, Virginia: The Nature Conservancy.

Johannes RE, Squire L, Graham T, 1994. Developing a protocol for monitoring spawning aggregations of Palauan serranids to facilitate the formulation and evaluation of strategies for their management. South Pacific Forum Fisheries Agency Report 94/28. Honiara, Solomon Islands.

Jones GP, Thompson SM, 1980. Social inhibition of maturation in females of a temperate wrasse, *Pseudolabrus celidotus* and a comparison with the blennioid *Tripterygion varium*. Mar Biol 59:247–256.

Jonsson B, Jonsson N, Hansen LP, 1990. Does juvenile experience affect migration and spawning of adult Atlantic salmon? Behav Ecol Sociobiol 26:225–230.

Kasperski W, Kozlowski J, 1993. The effect of exploitation on size at maturity in laboratory populations of guppies *Poecilia reticulata* (Peters). Acta Hydrobiol 35:65–72.

Keenleyside MHA, Mackereth RW, 1992. Effects of loss of male parent on brood survival in a biparental cichlid fish. Environ Biol Fishes 34:207–212.

King FW, 1987. Thirteen milestones on the road to extinction. In: The road to extinction (Fitter R, Fitter M, eds). Cambridge; UK: International Union for the Conservation of Nature/United Nations Environmental Protection; 7–18.

Koenig CC, Coleman FC, Collins LA, Sadovy Y, Colin PL, 1996. Reproduction in gag, *Mycteroperca microlepis* (Pisces: Serranidae) in the eastern Gulf of Mexico and the consequences of fishing spawning aggregations. In: Biology, fisheries, and culture of tropical groupers and snappers (Arregui'n-Sanchez F, Munro JL, Balgos MC, Pauly D, eds). ICLARM Conf. Proc. 48.

Lachance S, Fitzgerald GJ, 1992. Parental care tactics of three-spined sticklebacks living in a harsh environment. Behav Ecol 3:360–366.

Larkin PA, 1977. An epitaph for the concept of maximum sustainable yield. Trans Am Fish Soc 106:1–11.

Lawton JH, 1994. Population dynamic principles. Phil Trans R Soc Lond B 344:61–68.

Leaman BM, 1991. Reproductive styles and life history variables relative to exploitation for fisheries management of *Sebastes* stocks. Environ Biol Fishes 30:253–271.

Levitan DR, Petersen C, 1995. Sperm limitation in the sea. Trends Ecol Evol 10:228–231.

Lowe-McConnell R, 1993. Fish faunas of the African great lakes: origins, diversity and vulnerability. Conserv Biol 7:634–643.

Ludwig D, Hilborn R, Walters C, 1993. Uncertainty, resource exploitation and conservation: lessons from history. Science 260:17–36.

Lutnesky MD, 1994. Density-dependent protogynous sex change in territorial-haremic fishes: models and evidence. Behav Ecol 5:375–383.

Mace PM, Sissenwine MP, 1993. How much spawning is enough? In: Risk evaluation and biological reference points for fisheries management. Can Spec Publ Fish Aquat Sci 120:101–118.

Maeda T, 1986. Life cycle and behaviour of adult pollock (*Theragra chalcogramma* (Pallas)) in water adjacent to Funka Bay, Hokkaido Island, Internat. N Pacific Fish Comm Bull 45:39–65.

Manire CA, Gruber S, 1990. Many sharks may be headed toward extinction. Conserv Biol 4:10–11.

Marconato A, Tessari V, Marin G, 1995. The mating system of *Xyrichthys novacula:* sperm economy and fertilization success. J Fish Biol 47:292–301.

Master L, 1990. The imperiled status of North American aquatic animals. Biodivers Network News 3:1–2, 7–8.

Montgomery WL, 1983. Parr excellence. Nat Hist 1983:59–67.

Moyle PB, Leidy, RA, 1992. Loss of biodiversity in aquatic ecosystems: evidence from fish faunas. In: Conservation biology (Fiedler PL, Jain SK, eds). New York: Chapman and Hall; 127–169.

Moyle PB, Moyle PR, 1995. Endangered fishes and economics: intergenerational obligations. Environ Biol Fishes 43:29–37.

Munro JL (ed), 1983. Caribbean coral reef fishery resources. Manila: International Center for Living Aquatic Resources Management.

Nakatsuru K, Kramer DL, 1982. Is sperm cheap? Limited male fertility and female choice in the lemon tetra (Pisces, Characidae). Science 216:753–755.

Nakazono A, 1993. One-parent removal experiment in the brood-caring damselfish, *Acanthochromis polyacanthus,* with preliminary data on reproductive biology. Aust J Mar Freshwat Res 44:699–707.

Nakazono A, Kawase H, 1993. Spawning and biparental egg-care in a temperate filefish, *Paramonacanthus japonicus* (Ponacanthidae). Environ Biol Fishes 37:245–256.

Nehlson W, Williams JE, Lichatowich JA, 1991. Pacific salmon at the crossroads: stocks at risk from California, Oregon, Idaho and Washington. Fisheries 16:4–21.

Norse EA (ed), 1993. Global marine biological diversity. Washington, DC: Island Press.

Økland F, Heggberget TG, Jonsson B, 1995. Migratory behaviour of wild and farmed Atlantic salmon (*Salmo salar*) during spawning. J Fish Biol 46:1–7.

Okuda N, Yanagisawa Y, 1996. Filial cannibalism by mouthbrooding males of the cardinalfish, *Apogon doederleini,* in relation to their physical condition. Environ Biol Fishes 45:397–404.

Pankhurst NW, 1988. Spawning dynamics of the orange roughy, *Hoplostethus atlanticus,* in mid-slope waters of New Zealand. Environ Biol Fishes 21:101–116.

Parker GA, 1992. The evolution of sexual size dimorphism in fish. J Fish Biol (suppl B) 41:1–20.

Parrish RH, 1995. Lanternfish heaven: the future of world fisheries? NAGA, The ICLARM Quarterly (July): 7–9.

Petersen CW, 1990. The occurrence and dynamics of clutch loss and filial cannibalism in two Caribbean damselfishes. J Exper Mar Biol Ecol 135:117–133.

Petersen CW, 1995. Reproductive behavior, egg trading, and correlates of male mating success in the simultaneous hermaphrodite, *Serranus tabacarius.* Environ Biol Fishes 43:351–361.

Pister EP, 1985. Desert pupfishes: reflections on reality, desirability, and conscience. Environ Biol Fishes 12:3–12.

Pitcher TJ, Hart PJB, 1982. Fisheries ecology. London: Croom Helm.

Pitt TK, 1975. Changes in abundance and certain biological characters of Grand Bank American plaice, *Hippoglossoides platessoides.* J Fish Res Board Can 32:1383–1398.

Plan Development Team, 1990. The potential of marine fishery reserves for reef fish management in the U.S. Southern Atlantic. Contribution no. CRD/89-90/04. Coastal Resources Division, National Oceanographic and Atmospheric Administration, National Marine Fisheries Service, Washington, DC.

Pollock BR, 1984. Relations between migration, reproduction and nutrition in yellowfin bream *Acanthopagrus australis.* Mar Ecol Prog Ser 19:17–23.

Pressley PH, 1981. Parental effort and the evolution of nest guarding tactics in the threespine stickleback *Gasterosteus aculeatus.* Evolution 35:282–295.

Rago PJ, Goodyear CP, 1987. Recruitment mechanisms of striped bass and Atlantic salmon: Comparative liabilities of alternative life histories. Am Fish Soc symp 1:402–416.

Ribbink AJ, 1987. African lakes and their fishes: conservation scenarios and suggestions. Environ Biol Fishes 19:3–26.

Ricker WE, 1975. Computation and interpretation of biological statistics of fish populations. Bulletin of the Fisheries Research Board of Canada no. 191. Ottawa, Canada: Department of Fisheries and Oceans.

Ricker WE, 1981. Changes in the average size and average age of Pacific salmon. Can J Fish Aquat Sci 38:1636–1656.

Roberts CM, 1995. Effects of fishing on the ecosystem structure of coral reefs. Conserv Biol 9:988–995.

Roberts CM, Polunin NVC, 1991. Are marine reserves effective in management of reef fisheries? Rev Fish Biol Fish 1:67–91.

Robertson DR, 1983. On the spawning behavior and spawning cycles of eight surgeonfishes. Environ Biol Fishes 9:193–223.

Rose GA, 1997. The trouble with fisheries science! Rev Fish Biol Fisheries 7:365–370.

Rosenau ML, McPhail JD, 1987. Inherited differences in agonistic behaviour between two populations of coho salmon. Trans Am Fish Soc 116:646–654.

Rosenberg AA, 1993. Defining overfishing—defining stock rebuilding. In: NOAA Technical Memorandom NMFS-F/SPO-8 (Rosenberg AA, ed). Washington, DC: U.S. Department of Commerce; 1–68.

Rosenberg AA, Fogarty MJ, Sissenwine MP, Beddington JR, Shepherd JG, 1993. Achieving sustainable use of renewable resources. Science 262:828–829.

Ross RM, 1990. The evolution of sex change mechanisms in fishes. Environ Biol Fishes 29:81–93.

Ross RM, Losey GS, Diamond M, 1983. Sex change in a coral-reef fish: dependence of stimulation and inhibition on relative size. Science 221:574–575.

Rubec PJ, 1986. The effects of sodium cyanide on coral reefs and marine fish in the Philippines. In: Proceedings of the First Asian Fisheries Forum (Maclean JL, Dizon LB and Hosillos L, eds). Manila: Asian Fisheries Society; 297–302.

Russ GR, 1991. Coral reef fisheries: effects and yields. In: The ecology of fishes on coral reefs. (Sale PF, ed). San Diego, Califfornia: Academic Press; 601–635.

Russ GR, Alcala AC, 1989. Effects of intense fishing pressure on an assemblage of coral reef fishes. Mar Ecol Prog Ser 56:13–27.

Russell ES, 1931. Some theoretical considerations on the overfishing problem. J Conserv Int Explor Mer 6:3–20.

Sadovy Y, 1993. The Nassau grouper, endangered or just unlucky? Reef Encounter 13:10–12.

Sadovy Y, 1994. Grouper stocks of the western central Atlantic: the need for management and management needs. Proc Gulf Carib Fish Inst 43:43–64.

Sadovy, Y, 1996. Reproduction of reef fishery species. In: Reef fisheries (Polunin NVC, Roberts CM, eds). London, Chapman and Hall; 15–59.

Sadovy Y, in press. Problems of sustainability in grouper fisheries. In: Proceedings of the Fourth Asian Fisheries Forum, October 1995. Beijing, China: Asian Fisheries Society.

Sadovy Y, Rosario A, Roman A, 1994. Reproduction in an aggregating grouper, the red hind, *Epinephelus guttatus*. Environ Biol Fishes 41:269–286.

Safina C, 1994. Where have all the fishes gone? Issues Sci Technol 10:37–43.

Samoilys MA, Squire LC, 1994. Preliminary observations on the spawning behavior of coral trout, *Plectropomus leopardus* (Pisces: Serranidae), on the Great Barrier Reef. Bull Mar Sci 54:332–342.

Schultz ET, Cliffton LM, Warner RR, 1991. Energetic constraints and size-based tactics:

the adaptive significance of breeding schedule variation in a marine fish (Embiotocidae: *Micrometrus minimus*). Am Nat 138:1408–1430.

Shaffer ML, 1981. Minimum population sizes for species conservation. Bioscience 31:131–134.

Shapiro DY, 1979. Social behavior, group structure, and the control of sex reversal in hermaphroditic fish. Adv Study Behav 10:43–102.

Shapiro DY, 1980. Serial female sex changes after simultaneous removal of males from social groups of a coral reef fish. Science 209:1136–1137.

Shapiro DY, 1984. Sex reversal and sociodemographic processes in coral reef fishes. In: Fish reproduction: strategies and tactics (Potts GW, Wootton RJ, eds). Orlando, Florida: Academic Press; 103–118.

Shapiro DY, 1989. Sex change as an alternative life-history style. In: Alternative life-history styles of animals (Bruton MN, ed). Dordrecht: Kluwer Academic; 177–195.

Shapiro DY, Lubbock R, 1980. Group sex ratio and sex reversal. J Theor Biol 82:411–426.

Shapiro DY, Marconato A, Yoshikawa T, 1994. Sperm economy in a coral reef fish, *Thalassoma bifasciatum*. Ecology 75:1334–1344.

Shepherd BR, Idoine JS, 1993. Length-based analyses of yield and spawning biomass per recruit for black sea bass *Centropristis striata*, a protogynous hermaphrodite. Fish Bull 91:328–337.

Simberloff D, 1986. The proximate causes of extinction. In: Patterns and processes in the history of life (Raup D, Jablonski D, eds). Berlin: Springer-Verlag; 259–276.

Slaney TL, Hyatt KD, Northcote TG, Fielden RJ, 1996. Status of anadromous salmon and trout in British Columbia and Yukon. Fisheries 21(10):20–35.

Smith PJ, 1994. Genetic diversity of marine fisheries resources: possible impacts of fishing. FAO Fisheries Technical Paper no. 344. Rome: Food and Agriculture Organization.

Smith PJ, Francis RICC, McVeagh M, 1991. Loss of genetic diversity due to fishing pressure. Fisheries Res 10:309–316.

Svensson I, 1988. Reproductive costs in two sex-role reversed pipefish species (Syngnathidae). J Anim Ecol 57:929–942.

Taborsky M, 1984. Brood care helpers in the cichlid fish *Lamprologus brichardi:* their costs and benefits. Anim Behav 32:1236–1252.

Taborsky M, Limberger D, 1981. Helpers in fish. Behav Ecol Sociobiol 8:143–145.

Tang W-C, 1987. Chinese medicinal materials from the sea. Abstr Chin Med 1:571–600.

Thresher RE, Sainsbury KJ, Gunn JS, Whitelaw AW, 1986. Life history strategies and recent changes in population structure in the lizardfish genus, *Saurida,* on the Australian Northwest Shelf. Copeia 1986:876–885.

Trippel EA, 1993. Relations of fecundity, maturation, and body size of lake trout, and implications for management in northwestern Ontario lakes. N Am J Fish Manage 13:64–72.

Turner JL, 1977. Changes in size structure of cichlid populations of Lake Malawi resulting from bottom trawling. J Fish Res Board Can 34:232–238.

Vermeij GJ, 1993. Biogeography of recently extinct marine species: implications for conservation. Conserv Biol 7:391–397.

Vincent ACJ, 1992. Prospects for sex role reversal in teleost fishes. Neth J Zool 42:392–399.

Vincent ACJ, 1994. Operational sex ratios in seahorses. Behaviour 128:153–167.

Vincent ACJ, 1996. The international trade in seahorses. Cambridge: TRAFFIC International.

Vincent ACJ, 1997. Trade in pegasid fishes (sea moths), primarily for traditional Chinese medicine. Oryx. 31:199–208.

Vincent A, Ahnesjo I, Berglund A, Rosenqvist G, 1992. Pipefishes and seahorses: are they all sex role reversed? Trends Ecol Evol 7:237–241.

Vincent ACJ, Sadler LM, 1995. Faithful pair bonds in wild seahorses, *Hippocampus whitei.* Anim Behav 50:1557–1569.

Wallin JE, 1992. The symbiotic nest association of yellowfin shiners, *Notropis lutipinnis,* and bluehead chubs, *Nocomis leptocephalus.* Environ Biol Fishes 33:287–292.

Warner RR, 1975. The adaptive significance of sequential hermaphroditism in animals. Am Nat 109:61–82.

Warner RR, 1984. Mating behavior and hermaphroditism in coral reef fishes. Am Sci 72:128–136.

Warner RR, Shapiro DY, Marconato A, Petersen CW, 1995. Sexual conflict: males with highest mating success convey the lowest fertilisation benefits to females. Proc R Soc Lond B. 262:135–139.

Warner RR, Swearer SE, 1991. Social control of sex change in the bluehead wrasse, *Thalassoma bifasciatum* (Pisces: Labridae). Biol Bull 181:199–204.

Warner RR, Wernerus F, Lejeune P, van den Berghe E, 1995. Dynamics of female choice for parental care in a fish species where care is facultative. Behav Ecol 6:73–81.

Williams JE, Johnson JE, Hendrickson DA, et al. 1989. Fishes of North America endangered, threatened, or of special concern. Fisheries 14:2–20.

Wisenden BD, 1995. Reproductive behavior of free-ranging convict cichlids, *Cichlasoma nigrofasciatum.* Environ Biol Fishes 43:121–134.

Witte F, Goudswaard P, 1985. Aspects of the haplochromine fishery in southern Lake Victoria. In: CIFA report of the third session of the sub-committee for the development and management of the fisheries of Lake Victoria, Jinja, Uganda, 4–5 October 1984. FAO Fisheries Report no. 335. Rome: Food and Agriculture Organization.

Wood E, 1985. Exploitation of coral reef fishes for the aquarium trade. Ross-on-Wye, UK: Marine Conservation Society.

Yamahira K, Kikuchi T, Nojima S, 1996. Age specific bond utilization and spatial distribution of the puffer, *Takifugu niphobles,* over an intertidal sand flat. Environ Biol Fishes 45:311–318.

Yanagisawa Y, Ochi H, 1986. Step-fathering in the anemonefish, *Amphiprion clarkii:* a removal study. Anim Behav 34:1769–1780.

10

Social Organization and Effective Population Size in Carnivores

Scott Creel

What Is a Typical Ratio of Effective Population Size to Real Population Size?

Real populations rarely match the assumptions of population genetic theory. The concept of effective population size (N_e) was developed to describe how a population can lose genetic variability (through genetic drift) more quickly than would be predicted by population size alone (Wright, 1931, 1938, 1943). Effective population size is defined as the size of an ideal population that would lose genetic variation at the same rate as a given real population (Wright, 1969; Lande and Barrowclough, 1987). In this definition, an ideal population meets five assumptions: (1) population size is constant through time, (2) the number of breeding males and females are equal, (3) individuals' reproductive success is Poisson distributed, (4) individuals mate at random, and (5) generations do not overlap. Because few real populations meet these assumptions, N_e is usually lower than the censused population size (Nunney and Elam, 1994). Quantitative methods exist to measure the reduction in N_e when each assumption is not met (Wright, 1943, 1969; Crow and Kimura, 1970; Lande and Barrowclough, 1987) and these methods are well developed in theory (Wright, 1943, 1969; Crow and Kimura, 1970; Lande and Barrowclough, 1987), but they are not often applied to data from the wild. A recent review by Nunney and Elam (1994) includes estimates of N_e for 14 species, with only one carnivore (the grizzly bear *Ursus horribilis*: Harris and Allendorf, 1989) and one endangered species (the spotted owl *Strix occidentalis*).

Behavioral ecologists, with their focus on studying known individuals over long periods, often have the data needed to determine effective population sizes. In this chapter I use data from long-term studies of social carnivores to determine the effects of social organization on N_e. I do not discuss the underlying theory, which has been well reviewed by others (Lande and Barrowclough, 1987; Nunney and Campbell, 1993). I consider four factors: (1) variability in population size through time, (2) skew in the sex-ratio, (3) variability among individuals in reproductive success and reproductive suppression of social subordinates, and (4) dispersal and its effect on geographic patterns of mating. First, I examine each factor independently to determine which factors cause the greatest reduction in N_e. Second,

I combine the impacts of several factors to determine the total reduction in N_e. Third, I determine the effects of social organization on demographic effective population size, N_{ed} (Caughley, 1994). The term N_{ed} is analogous to N_e, but it measures the effect of social structure on a population's growth rate, rather than its loss of genetic variability. The concept of N_{ed} is an important advance because the probability of extinction depends mainly on the mean and variance of a population's growth rate (Franklin, 1980), which in turn depend on its demography. Moreover, demographic problems can eliminate a small population before loss of genetic variability becomes a problem (Lande, 1988).

Finally, I discuss conclusions for the conservation of social carnivores. For example, the largest protected population of African wild dogs *Lycaon pictus* is only one-third of the widely accepted "target" N_e of 500 individuals, despite living in a well-protected area of 43,000 km^2. Lande (1995) has recently suggested that a target N_e of 5000 is more realistic. If this target is correct, most populations of large carnivores fall short by a wide margin.

Variation in Population Size through Time

Among social carnivores, population density can fluctuate dramatically (fig. 10-1). In some cases, the fluctuations appear to be random [e.g., Serengeti dwarf mongooses *Helogale parvula* between 1977 and 1983 (Fig. 10-1a) and Superior National Forest wolves *Canis lupus* between 1975 and 1985 (Fig. 10-1c). In other cases population size fluctuates but also shows an increase [Ngorongoro lions *Panthera leo* (Fig. 10-1e], a decrease [Serengeti wild dogs (fig. 10-1b)], or both in sequence [Isle Royale wolves (Fig. 10-1d)]. In a population with discrete generations, N_e is equal to the harmonic mean of population size for each generation (table 10-1):

$$N_e = \frac{n}{\Sigma 1 / N_i} \qquad (1)$$

where N = population estimate, i = subscripts censuses (years); and r = number of censuses (years). Because the harmonic mean is strongly influenced by small values, N_e falls below the (arithmetic) mean population size. Equation 1 does not apply directly to species with overlapping generations, so the estimates in Table 10-1 give effective population size per unit time following Lande and Barrowclough (1987).

Table 10-1 summarizes the effect of fluctuating population size on N_e for six carnivore populations. Of these data sets, only lions in Ngorongoro and wolves on Isle Royale are closed populations (Packer et al., 1991a,b; Peterson and Page, 1988). For other species, the population under study was an artificial subset of a larger true population (defined by study-site borders). For example, the wolves in Superior National Forest were estimated to be 3% of Minnesota's total wolf population (Mech, 1986), and the dwarf mongooses studied in Serengeti National Park formed <1% of the park's population (Waser et al., 1995), but the study population was two to four times larger than the genetic neighborhood size (Wright, 1943; Barrowclough, 1980; see below). I assume that trends within the study population reflect changes in the population at large, but this assumption may be false. Ethiopian wolves *Canis simensis* in the Bale Mountains were included despite a short time-series because they are the most endangered of the social carnivores (Sillero-Zubiri, 1994). To allow comparisons among species, I express N_e as a proportion (Δ) of the mean adult population size (Nunney and Elam, 1994).

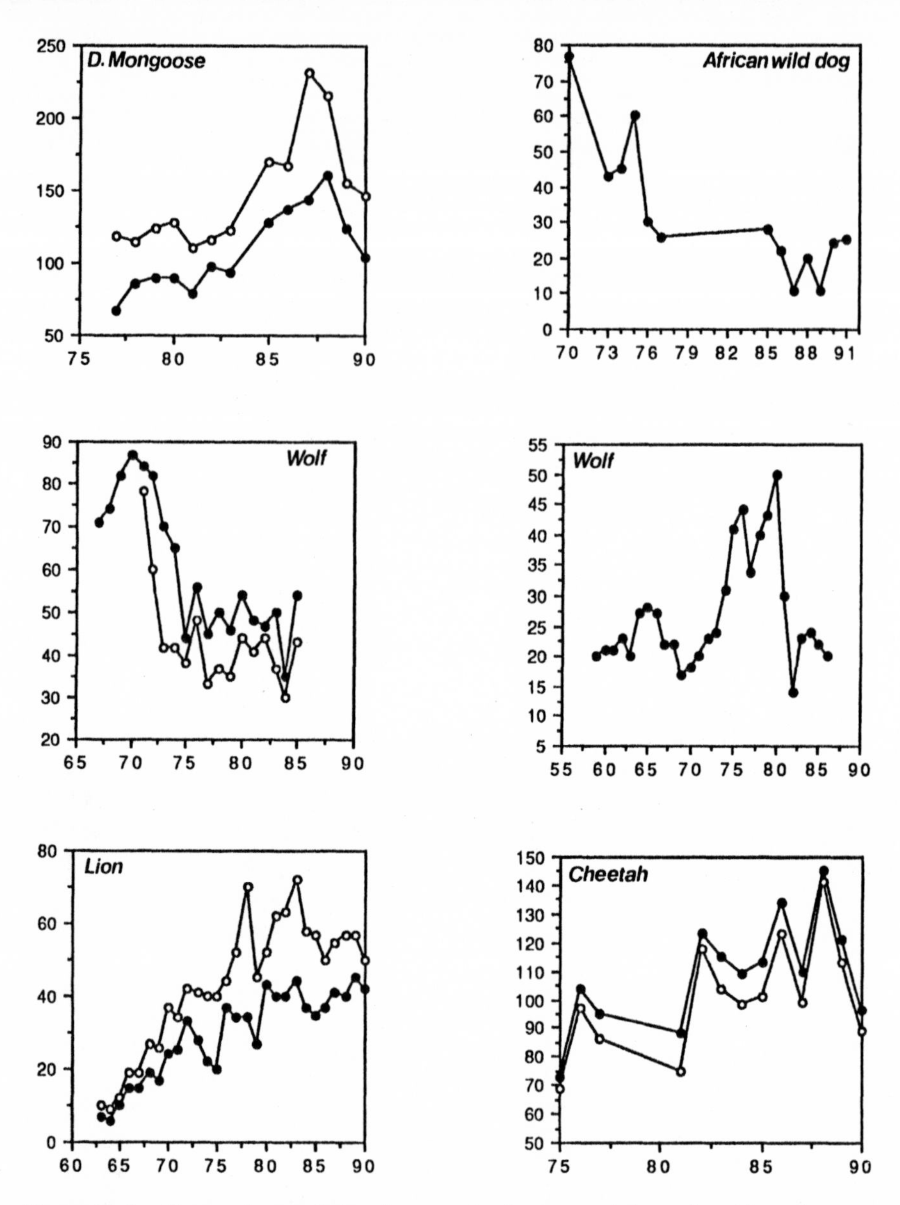

Figure 10-1 Changes in population size over time in six populations of social carnivores. (A) Dwarf mongooses in Serengeti National Park; (closed circles) adults; (open circles) adults and subadults. (B) African wild dogs in Serengeti National Park, adults and subadults. (C) Grey wolves in Superior National Forest; (closed circles) winter adults and subadults; (open circles) spring adults and subadults. (D) Grey wolves in Isle Royale National Park, adults and subadults. (E) Lions in Ngorongoro Crater; (closed circles) adults; (open circles) adults and subadults. (F) Cheetahs on Serengeti Plains; (closed circles) adults seen and adults inferred to be alive; (open circles) adults seen. See table 10-1 for the effects of variation in population size on N_e and for data sources.

Table 10-1 Carnivore effective population size [$N_e = n/\Sigma(1/N_i)$]: consequences of variation in population size through time.

Species	Location	Measure[a]	n	Mean N	N_e[b]	Δ[c]	Reference
Ethiopian wolf	Bale Mtns, Ethiopia	A in packs	4	44.3	43.5	0.98	Sillero-Zubiri (1994)
(*Canis simensis*)		All A and SA	4	61.3	60.6	0.99	
Gray wolf	Superior NF, Minnesota	Winter packs	19	60.2	56.3	0.94	Mech (1986)
(*Canis lupus*)		Spring packs	19	43.5	41.3	0.95	
Gray wolf	Isle Royale, Michigan	A	28	26.8	24.3	0.91	Peterson and Page (1988)
African wild dog	Serengeti NP, Tanzania	A and SA	13	32.5	24.0	0.74	Burrows et al. (1994)
(*Lycaon pictus*)							
Cheetah	Serengeti NP, Tanzania	A observed	13	101.0	97.5	0.97	M. Kelly and T. M. Caro
(*Acinonyx jubatus*)		A observed/ inferred	13	109.7	106.4	0.97	(personal communication)
Lion	Ngorongoro, Tanzania	A	28	29.2	21.6	0.74	Packer et al. (1991a, b)
(*Panthera leo*)		A and SA	28	42.9	31.7	0.74	
Dwarf mongoose	Serengeti NP, Tanzania	A	13	107.5	101.0	0.94	Waser et al. (1995)
(*Helogale parvula*)		A and SA	13	147.5	139.6	0.95	

[a]A, adult; SA, subadult.

[b]Correction for variation across years gives N_e per unit time following Lande and Barrowclough (1987).

[c]$\Delta = N_e/N$, the ratio of effective population size to census population size.

The value of Δ ranges from 0.74 to 0.99 for these data sets. Not surprisingly, the strongest effect on Δ comes with a systematic change in population size: it is smallest in Serengeti wild dogs (0.74), which have steadily decreased (fig. 10-1b), and in Ngorongoro lions (0.74), which have steadily increased (fig. 10-1e). Ngorongoro lions show that even an increasing population may suffer from its history (Packer et al., 1991a,b). Interestingly, the value of Δ is relatively insensitive to the way that population size was measured. Where two measures of population size were available (five studies), both measures gave similar estimates of Δ.

Uneven Sex Ratios

An imbalanced sex ratio reduces N_e according to eq. 2 (table 10-2) (Lande and Barrowclough, 1987; Caughley, 1994):

$$N_e = \frac{4N_m N_f}{N_m + N_f} \tag{2}$$

where N_m and N_f = the numbers of breeding males and females, respectively. In this equation, only males and females that breed should be considered, so the equation cannot be directly applied to the adult sex ratio for species with reproductive suppression of subordinates (including many social carnivores). Table 10-2 summarizes data for three species without overt reproductive suppression: cheetahs *Acinonyx jubatus*, spotted hyenas *Crocuta crocuta*, and lions. Because social carnivores often have adult sex ratios near equality (table 10-2; also many canids: Moehlman, 1986; and social mongooses: Rood, 1986), sex ratios by themselves have little impact on N_e. Values of Δ between 0.95 and 1.0 are apparently typical (table 10-2).

N_e is reduced to a greater extent when reproductive suppression skews the effective sex ratio. If subordinates are suppressed in one sex only, the effective sex ratio can be heavily skewed even if the adult sex ratio is not. For example, the low Δ of 0.75 for lions (table 10-2) arises partly from a skewed adult sex ratio (0.33 male:0.67 female; Packer et al., 1988) and partly because some males in large coalitions do not breed, which skews the effective sex ratio to 0.25 male:0.75 female (Packer et al., 1991b, C. Packer, personal communication). This effect could be dramatic in species such as dwarf mongooses or African wild dogs, where reproduction is physiologically suppressed in subordinates (Creel et al., 1992, 1997). In practice, the effective sex ratio has little impact on N_e in these two species (Δ = 0.88 for wild dogs, Δ = 0.99 for dwarf mongooses). Subordinate males and females are suppressed to different degrees (Creel and Waser, 1996), but the difference is not sufficient to reduce N_e dramatically.

For spotted hyenas, four different sex ratios have been reported from the Serengeti ecosystem, from a span of more than 50 years (table 10-2). Comparison of the Δ values for these hyena studies suggests random fluctuations in the sex ratio do not have much impact on N_e, in practice. The data from Kruuk (1972), Hofer and East (1993), and Frank et al. (1995) probably differ primarily by chance and give Δ values of 0.97–1.00. Matthews's (1939) data show the only male-biased sex ratio for hyenas, probably because his sample was composed of hyenas shot while scavenging from humans. Males have poor access to kills because they are subordinate to females (Frank, 1986), so they are more likely to scavenge. Comparing Matthews's study (Δ = 0.95) to the others suggests that, for hyenas, N_e is not highly sensitive to sex ratio sampling biases.

Table 10-2 Carnivore effective population size [$N_e = (4N_m N_f)/(N_m + N_f)$]: consequences of bias in the effective sex ratio, restricted to species without overt reproductive suppression.

Species	Location	Sex ratio (M:F)	N	N_e	Δ[a]	Reference
Spotted hyena[b] (*Crocuta crocuta*)	Serengeti NP, Tanzania	0.48:0.52	2000	1999	1.00	Kruuk (1972)
Spotted hyena	Serengeti NP, Tanzania	0.61:0.39	2000	1900	0.95	Matthews (1939)
Spotted hyena	Serengeti ecosystem	0.41:0.59	5200	5020	0.97	Frank et al. (1995)
Spotted hyena	Serengeti ecosystem	0.46:0.54	5200	5167	0.99	Hofer and East (1993)
Cheetah[b] (*Acinonyx jubatus*)	Serengeti NP, Tanzania	0.44:0.56	210	206.8	0.98	M. Kelly and T. M. Caro (personal communication)
Lion[c] (*Panthera leo*)	Ngorongoro, Tanzania	0.25:0.75	29	21.7	0.75	Packer et al. (1991a, b)

[a] $\Delta = N_e/N$, the ratio of effective population size to census population size.
[b] Hyena and cheetah data equate population sex ratio with effective sex ratio.
[c] Lion data use an effective sex ratio of 3F:1M as estimated by C. Packer (personal communication).

Variation in Reproductive Success

For N_e to equal N, reproductive success must be Poisson distributed. Skewed reproductive success reduces N_e according to eq. 3 (Wright, 1940; Lande and Barrowclough, 1987):

$$N_e = \frac{NF - 1}{F + (s^2/F) - 1} \tag{3}$$

where N = population size; F = mean lifetime reproductive success; and s^2 = variance in lifetime reproductive success. Among social carnivores, variance in reproductive success has two main sources. First, when social subordinates do not breed, variance in reproductive success will increase dramatically, and N_e will decrease accordingly. Second, even among those that breed, reproductive success will rarely be equal. For instance, reproductive success increases with age and experience in female dwarf mongooses (Creel et al., 1996).

Table 10-3 summarizes the effect of non-Poisson lifetime reproductive success on N_e for dwarf mongooses and lions. N_e is calculated separately for males and females because the variances (s^2) are not equal. For dwarf mongooses, N_e is less than one-fifth of N (Δ = 0.16–0.18). Because subordinates of both sexes are reproductively suppressed, the Δ values for males and females are similar. This calculation assigns a reproductive success of zero to life-long subordinates (64% of males and 62% of females), but calculations incorporating more detailed information on reproduction by subordinates are presented below.

For lions, the Δ value for males (0.61) is lower than that of females (0.78), but both values are higher than those of dwarf mongooses. Female lions have no dominance hierarchy (Schaller, 1972; Bertram, 1975) and thus lack reproductive suppression, yielding a high Δ (0.78). For males, the mean and variance in reproductive success were calculated assuming that all males of a coalition share paternity equally (Packer et al., 1988). Recent genetic data have shown this assumption to be wrong (Packer et al., 1991a), so the Δ of 0.61 is inflated. More detailed data information on patterns of paternity are incorporated below.

The small Δ values in table 10-3 suggest that skewed reproductive success is a critical determinant of N_e in social carnivores. Unfortunately, data on lifetime reproductive success

Table 10-3 Carnivore effective population size $\{N_e = (NF - 1)/[F + (s^2/F) - 1]\}$: consequences of variation in reproductive success among individuals, applications to data on lifetime reproductive success for both sexes.

Species	Location	F	s^2	N	N_{eg}	Δ[a]	Reference
Dwarf mongoose (*Helogale parvula*)	Serengeti NP, Tanzania						
Males[b]		3.29	59.81	75	12.0	0.16	Creel, Rood, and Waser (unpublished data)
Females[b]		3.76	67.90	82	14.8	0.18	
Both sexes				157	26.8	0.17	
Lion (*Panthera leo*)	Serengeti NP, Tanzania						
Males[c]		4.48	17.11	990	607	0.61	Packer et al. (1988, 1991a, b), Packer (1990)
Females		3.81	7.80	2010	1576	0.78	
Both sexes				3000	2183	0.73	

[a] $\Delta = N_e/N$, the ratio of effective population size to census population size.
[b] Dwarf mongoose data assign zero reproductive success to lifelong subordinates.
[c] Lion data assumes all males of a coalition have equal reproductive success.

are needed to estimate this impact correctly, and these data are difficult, expensive, and time consuming to collect.

Incorporating Behavioral and Genetic Data

The results in table 10-3 suggest that skewed reproductive success is a critical factor in determining N_e for social carnivores, but the calculations relied on simplifying assumptions that are not completely correct. For example, the lion data assume that males within a coalition share paternity equally, based on observations of mating behavior (Packer et al., 1988). DNA fingerprints reveal that reproduction within lion coalitions is skewed, and some males do not reproduce (Packer et al., 1991a). For dwarf mongooses, 64% of males ($n = 436$) and 62% of females ($n = 378$) died without attaining dominance, and the calculations of table 10-3 assume that these mongooses left no offspring. However, behavioral, endocrine, and genetic data all suggest that some subordinates of both sexes do reproduce (Keane et al., 1994; Creel and Waser, 1996). How does this information affect estimates of Δ?

Direct data on reproductive skew can be used to determine effective numbers of breeding males (N_{em}) and females (N_{ef}), allowing N_e to be calculated with eq. 2. This approach lends itself to incorporating behavioral, physiological, and genetic data, as shown by three examples.

Dwarf Mongooses

For dwarf mongooses in Serengeti, the adult sex ratio was 48% male ($n = 1676$ mongoose-years). The mean population size was 108 adults ($n = 13$ years; Waser et al., 1995), so the population in an average year would be 52 males and 56 females. The mean pack size was 9.0 (Creel and Waser, 1994), so the population comprised 12 packs in an average year. Assuming that subordinates do not breed, this yields 12 breeders (N_{ef}) out of 56 females (N_f) and 12 breeders (N_{em}) out of 52 males (N_m). Using these data in eq. 2 yields an N_e of 24 and a Δ of 0.22. How does reproduction by subordinates alter this result?

DNA fingerprints show that between 12% and 40% (95% CI) of dwarf mongooses have subordinate fathers, with a best estimate of 24% subordinate paternity (Keane et al., 1994). Using the best estimate of 24%, there are

$$\frac{0.24}{0.76} = \frac{\text{subordinate fathers}}{\text{dominant fathers}}$$

or 0.32 subordinate fathers per dominant father. Thus there are 1.32 breeding males per pack, on average. To drop the assumption that subordinate males do not breed, the 12 alpha males are multiplied by 1.32, giving $N_{em} = 15.8$. Applying the same logic with the upper and lower 95% confidence limits for subordinate paternity gives boundary estimates for N_{em} of 20.0 and 13.6.

DNA fingerprints show that between 6% and 30% of dwarf mongooses have subordinate mothers, with a best estimate of 15% subordinate maternity (Keane et al., 1994). Following the logic outlined for males, these percentages modify the N_{ef} of 12 (assuming no reproduction by subordinate females) to a best estimate of 14.2, with upper and lower boundaries of 17.2 and 12.7. Combining N_{em} and N_{ef} values in eq. 2, the pair of lower limits gives an N_e of 26.3 and $\Delta = 0.24$ (table 10-4). The pair of upper limits gives $N_e = 37.0$ and $\Delta = 0.34$, and the pair of best estimates gives $N_e = 29.9$ and $\Delta = 0.28$. Comparing these results with table 10-3 shows that adjusting for subordinate reproduction increases the estimate of N_e by a factor of 1.3 (range of 1.1–1.5).

The logic just described is based on a static snap-shot view of the individuals that breed at a specific time. However, data on annual patterns of reproduction will generally underestimate N_e for species such as the dwarf mongoose. Some suppressed subordinates will later become dominant (Creel and Waser, 1994; Creel et al., 1997), so lifetime reproductive success will be less skewed than annual reproductive success. In dwarf mongooses, 64% of males and 62% of females were lifelong subordinates. To create a reference population with this lifetime pattern of social dominance, I took the N_m of 52 males for an average year (as before), with 33 subordinates (64%) and 19 dominants. Likewise, I took the average N_f of 56 females, with 35 subordinates (62%) and 21 dominants. Genetic information on paternity and maternity by subordinates was incorporated using the same multipliers as above. For example, the best estimate for N_{em} is 19 dominants × 1.32 breeders/dominant, or 25.1. The best estimate for N_{ef} is 21 dominants × 1.18 breeders/dominant, or 24.8.

The N_{em} and N_{ef} values for this reference population are given in table 10-4. Accounting for lifelong patterns of reproduction, the best estimate of Δ is 0.46, (minimum = 0.41, maximum = 0.57), which is considerably larger than the previous estimates of 0.22 (annual patterns assuming no subordinate reproduction), and 0.28 (annual patterns with subordinate reproduction). Reproduction by subordinates is often ignored for species in which suppression of subordinates is typical, and the mating system is considered effectively monogamous (Kleiman, 1977; Greene et al., Chapter 11, this volume). Relating reproductive suppression to effective population size, Caro and Durant (1995) suggest that the number of breeding individuals is equivalent to the number of groups. Table 10-4 shows that this is too dire an assessment, particularly when lifetime data are considered. For dwarf mon-

Table 10-4 Carnivore effective population size: consequences of social organization, estimated with data combined from demography, behavior, endocrinology, and molecular genetics (see text for assumptions and methods of calculation).

Species[a]	Location	Measure of population	N_m	N_f	N	N_{em}	N_{ef}	N_e[b]	Δ[c]
Dwarf mongoose	Serengeti NP								
Annual data		Adults							
Lower limit			52	56	108	13.6	12.7	26.3	0.24
Best estimate			52	56	108	15.8	14.2	29.9	0.28
Upper limit			52	56	108	20.0	17.2	37.0	0.34
Lifetime data		Adults							
Lower limit			52	56	108	21.7	22.3	44.0	0.41
Best estimate			52	56	108	25.1	24.8	49.9	0.46
Upper limit			52	56	108	31.7	30.0	61.7	0.57
Lion	Serengeti NP	Adult	990	2010	3000	322	1576	1069	0.36
	Ngorongoro	Adult	9.6	19.4	29	3.1	15.0	10.3	0.36
African wild dog	Selous GR	Adult and subadult	170	127	297	46	23	61.3	0.21

[a]References: Dwarf mongoose: Creel et al. (1991, 1992), Creel and Waser (1994), Keane et al. (1994), Waser et al. (1995); lion: Packer et al. (1988, 1991a, 1991b), C. Packer (personal communication); wild dog: S. Creel and N. Marusha Creel (unpublished data).

[b]$N_e = 4N_mN_f/(N_m + N_f)$.

[c]$\Delta = N_e/N$, the ratio of effective population size to census population size.

gooses, N_e is almost half of N, when reproduction by subordinates and lifetime turnovers in dominance are considered. An encouraging conclusion is that more detailed descriptions of social organization produce larger estimates of N_e than do simple assessments. A discouraging conclusion is that N_e is nonetheless roughly one-half of N in this case.

Lions

For Serengeti lions, the adult sex ratio is 0.67 females to 0.33 males among adults aged 4–10 years (n = 1384 lion-years; data from Packer et al., 1988: fig. 23.3). Packer (1990) estimates that roughly 3000 lions live in Serengeti, or 990 males and 2010 females (table 10-4). The lion population in Ngorongoro averaged 29 adults over 28 years (Packer et al., 1988). If the sex ratio from Serengeti is representative, this gives 9.6 males and 19.4 females in an average year (table 10-4).

Both endocrine and demographic data suggest that all lionesses breed (Packer et al., 1988; Brown et al., 1993). Thus, N_{ef} is calculated using eq. 3 to account for variation in breeding females' reproduction without adjustment for reproductive suppression. For male lions, the situation is more complex. Packer et al. (1988) calculated the mean and variance of males' lifetime reproductive success under the assumption that all males in a pride-holding coalition share equally in paternity of the pride's cubs. DNA fingerprinting later showed that two males in each coalition typically fathered cubs, regardless of coalition size (Packer et al., 1991a; C. Packer, personal communication). This information can be combined with the frequency distribution of male coalition sizes and the coalition-size–specific probability of gaining access to females (Packer et al., 1988) to give the proportion of males that breed. Of 49 coalitions (holding 114 males), 19 coalitions (61 males) gained membership in a pride of females. Of these 61 pride-holders, 24 are likely to have been nonbreeders, based on the degree of reproductive skew expected for their coalition sizes. Thus, 37 breeders are expected from the total of 114 males, or 0.33 breeding males/adult male. Multiplying N_m by 0.33 gives N_{em} (table 10-4).

Using eq. 2 to determine N_e from N_{em} and N_{ef} yields an effective population size of 1069 lions in Serengeti and 10.3 in Ngorongoro. The Δ value is 0.36 for both populations. This result is based on a mixture of annual and lifetime data. Because lions do not have an overt dominance hierarchy (unlike most social carnivores), annual and lifetime data are likely to produce similar results because turnover of dominance is not a factor. A surprising conclusion emerges from table 10-4: Δ is smaller for lions than for dwarf mongooses. This arises because reproductive suppression in dwarf mongooses affects both sexes and creates little skew in the effective sex ratio. Reproductive skew in lions is pronounced for males but not for females, which imbalances the effective sex ratio and brings N_e down substantially. The genetic data on reproductive skew in male lions from Packer et al. (1991) are based on partial lifetimes. If the degree that a given male lion is excluded from reproduction changes over a lifetime, the effect on N_e will be less severe than estimated here. This caveat itself is worrisome. After decades of research, the data on demography, behavior, and genetics of dwarf mongooses and lions are not sufficient to confidently state which species has a smaller Δ.

Wild Dogs

Data from wild dogs in the Selous Game Reserve (n = 22 pack-years, 10 packs) show an adult sex ratio of 170 males to 127 females (0.57:0.43). What is the effective sex ratio? Like

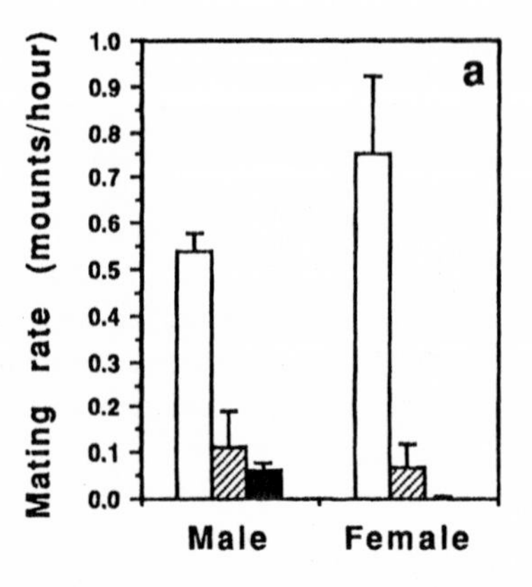

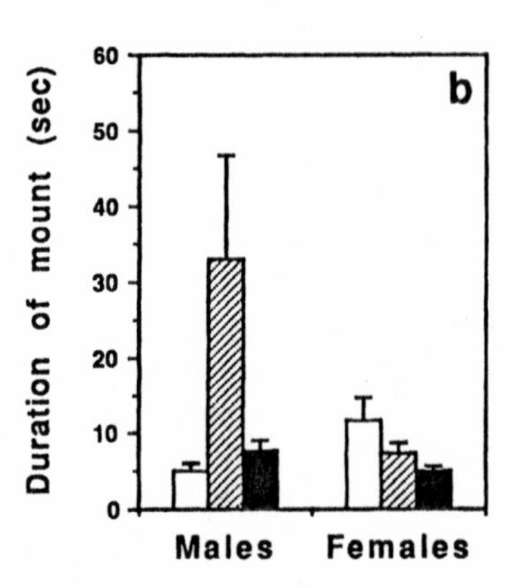

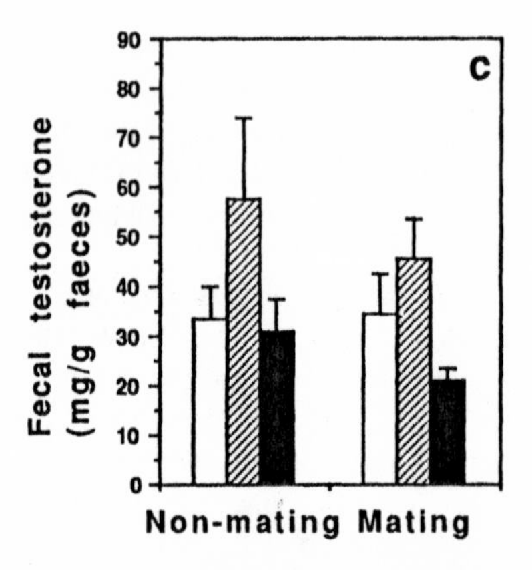

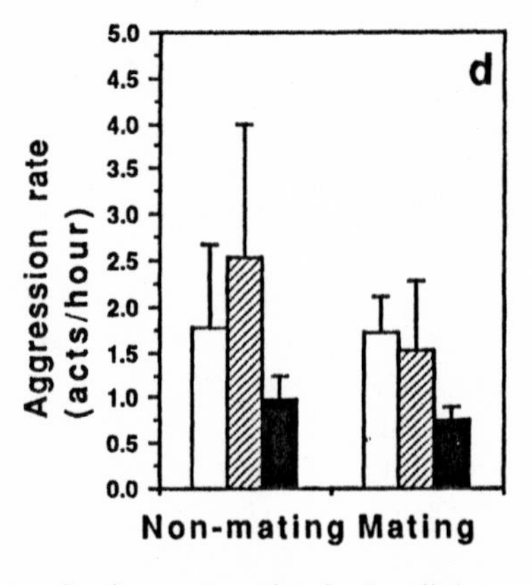

Figure 10-2 Endocrine and behavioral data on reproductive suppression in free-living wild dogs in the Selous Game Reserve suggest that two males may breed within a pack. Data shown are mean and SE in all cases. (A) Mating rates for alpha (open), beta (hatched), and lower-ranking (filled histograms) wild dogs. Alpha males mate at the highest rates, but beta males also mate at nontrivial rates. (B) The duration of mounts. The mounts of beta males are significantly more likely to end in a copulatory tie indicative of ejaculation. (C) Testosterone levels. During mating periods, both alpha and beta males have higher testosterone levels than do lower ranking males. Beta males tend to have higher testosterone levels than alphas, during and away from mating periods. (D) Aggression rates. Particularly during mating periods, alpha and beta males are more aggressive than lower ranking males, improving access to females.

dwarf mongooses, subordinate wild dogs typically do not breed. In 22 pack-years, 18 dominant females and 5 subordinate females produced litters that were raised. Treating the 22 pack-years as a reference population of 22 packs, $N_{ef} = 23$ and $N_f = 127$.

Behavioral and endocrine data from Selous wild dogs suggest that two males may breed per pack. During mating periods, alpha males mount females more often than others (fig. 10-2a), but the mounts of beta (second-ranking) males are longer and more likely to end in a copulatory tie (fig. 10-2b). During mating periods, alphas and betas have higher testosterone levels than lower-ranking males (fig. 10-2c). For wild dog males, aggression controls access to breeding females, and serious fights are common during mating periods. Beta males are as aggressive as alpha males, while both are more aggressive than males of lower rank (fig. 10-2d). Together, these data suggest that two breeding males per breeding female is a reasonable estimate (to be tested with genetic data). This estimate yields $N_{em} = 46$ males (out of $N_m = 170$ males).

Using eq. 2 with $N_{ef} = 23$ and $N_{em} = 46$ yields $N_e = 61.3$ and $\Delta = 0.21$ for wild dogs. This Δ value is an underestimate because it does not take lifelong turnovers in dominance into account. For male wild dogs in particular, fights in which the dominant male is deposed by a pack mate are common. Female wild dogs are rarely deposed by pack mates but are occasionally evicted by an immigrating coalition of females. Both of these processes will raise the true value of Δ above 0.21 by an unknown amount. If turnovers have an effect on Δ similar to their effect in dwarf mongooses, Δ for wild dogs would fall near 0.35.

Dispersal and Genetic Isolation by Distance

N_e will be less than N when individuals do not mate randomly within the population as a whole. Social carnivores normally mate with members of their own group, a pattern far from random-mating. Exceptions exist; for example, in aardwolves *Proteles cristatus* (Richardson, 1987), brown hyenas *Hyena brunnea* (Mills, 1989), and Ethiopian wolves (Sillero-Zubiri, 1994), females commonly mate with males from another social group. However, dispersal is the primary mechanism promoting random mating for most social carnivores (Waser, 1996). Large social carnivores can disperse great distances, with records of 866 km in wolves (Fritts, 1983), 544 km in coyotes *Canis latrans* (Carbyn and Paquet, 1986), and 250 km in African wild dogs (Fuller et al., 1992). Small social carnivores do not move as far. Median dispersal distance is 0.5 km for males and 1.0 km for females in dwarf mongooses with maximums of <10 km (Waser et al., 1994). Emigration is male biased in most species (Waser, 1996) but female biased in some (Frame and Frame, 1976). For dwarf mongooses, females are less likely to disperse, but move farther (Waser et al., 1994).

Wright (1943, 1969) showed that the effect of dispersal on N_e is given by:

$$N_e = 4\pi\sigma^2 d F_k \tag{4}$$

where σ^2 = one-way variance in dispersal distance; d = population density (sex-specific); and F_k = correction factor for skewed (leptokurtic) dispersal distribution (table 10-5). In this context, N_e gives the "genetic neighborhood size," (i.e., the size of an ideal population within which mating is expected to be random; Wright, 1969; Lande and Barrowclough, 1987). To apply eq. 4, one must know the mean, variance, and kurtosis of the dispersal distance distribution, and few studies report these data (Barrowclough, 1980). Dispersal is a difficult topic to study (Van Vuren, Chapter 14, this volume), and few studies of social carnivores report even the mean dispersal distance. Thus table 10-5 calculates the effect of dispersal distance on N_e only for dwarf mongooses. For dwarf mongooses, the genetic neigh-

Table 10-5 Genetic neighborhood size in the Serengeti dwarf mongoose ($N_e = 4\pi\sigma^2 dF_k$): consequences of dispersal distance using Wright's isolation-by-distance model.

Variable	Females	Males	Both sexes
σ^{2a} (km^2)	1.13	0.54	
d (individuals/km^2)	3.1	2.8	
$F_k^{\,b}$	0.91	0.88	
N_e	40.2	16.6	56.8

[a]σ^2 differs from value expected from Waser et al. (1994) because dispersed distances of zero were included here.

[b]F_k values interpolated from Wright (1969: 304).

borhood size is 57 individuals. Maruyama (1977) shows that a population can be considered panmictic (for the purpose of estimating loss of genetic variability) if $d\sigma^2 > 1$. For dwarf mongooses, $d\sigma^2 > 1$, so limited dispersal should not accelerate genetic drift.

Combined Impacts on Effective Population Size

Tables 10-1–10-5 suggest that most populations of social carnivores violate many of the assumptions of the ideal population in which N_e equals N. Some of these effects are substantial even when viewed in isolation, so their collective effect could be severe. Table 10-6 shows the combined effect of skew in the sex ratio, skewed reproductive success, and variation in population size through time for Serengeti dwarf mongooses, following Lande and Barrowclough (1987). The data for this analysis (n = 11 years) include only years in which census effort was high, so parameter values differ slightly from those above. Combining effects, N_e is only 16% of the census population size.

Several endangered social carnivores (e.g., African wild dogs and Ethiopian wolves) have group sizes and within-group social organization similar to those of dwarf mongooses. Other aspects of their demography (e.g., annual survivorship) and social organization (e.g., dispersal patterns) also resemble the dwarf mongoose pattern. The Δ values for these species probably fall near the value for dwarf mongooses, and this is cause for concern (see below).

Demographic Effective Population Size

Tables 10-1–10-6 deal exclusively with the effects of social organization on the loss of genetic variability. But populations will often go extinct due to demographic problems (stochastic or deterministic) before loss of genetic variability can become a problem (Lande, 1988). This certainly does not mean that we need not concern ourselves with genetic diversity, if our aim is to maintain the potential for adaptive evolution. It does mean that demography will often be a more pressing conservation problem than genetics.

In response to this issue, Caughley (1994) defined the demographic equivalent of genetic effective population size. He defined the demographic effective population size, N_{ed},

Table 10-6 Serengeti dwarf mongoose effective population size, accounting for variation in reproductive success among individuals, skew in the sex ratio, and variation in population size through time.[a]

	Censused population size (adults)		Skew in reproductive success[b]		Skew in sex ratio[c]
Year	Males	Females	Males	Females	N_{ei}
1977	52	53	8.3	9.5	17.7
1978	44	43	7.0	7.7	14.7
1979	46	48	7.3	8.6	15.8
1980	35	41	5.6	7.4	12.8
1981	29	39	4.6	7.0	11.1
1982	39	37	6.2	6.6	12.8
1985	60	48	9.6	8.6	18.1
1987	60	61	9.6	11.0	20.5
1988	73	59	11.7	10.6	22.2
1989	60	46	9.6	8.3	17.8
1990	63	53	10.1	9.5	19.6

Variation through time $N_e = n/\Sigma(1/N_{ei}) = 11/0.69 = 15.9$

$\Delta = N_e/N_i = 15.9/99 = 0.16$

[a]Excludes years with low census effort and does not adjust for census effort in remaining years. Mean and variance in reproductive success used were from data pooled across years, assigning 0 to subordinates. Correction for variation among years gives N_e per unit time following Lande and Barrowclough (1987).

[b]$N_e = (NF - 1)/[F + (s^2/F) - 1]$

[c]$N_e = 4N_m N_f/(N_m + N_f)$

as "the size of an "ideal" population with an even sex-ratio and a stable age-distribution that has the same net change in numbers over a year as the population of interest" (Caughley, 1994:223). The effect of a skewed sex-ratio on N_{ed} is given by Caughley as:

$$N_{ed} = \frac{N(P_f sb + p - 1)}{[(0.5)sb] + s - 1} \qquad (5)$$

Effects on N_{ed} due to deviation from the stable age-distribution, coupled with age-specific reproduction, are given by:

$$N_{ed} = \frac{N[\Sigma f_x(s_x m_x + s_x - 1)]}{\Sigma l_x e^{-rx}(s_x m_x + s_x - 1)}$$

where N = population size; x = age class (years); f_x = true proportion of population in age class x, scaled so that $f_0 = 1$; s_x = annual survival at age x; m_x = annual reproduction at age x; r = annual population growth rate. Following eq. 5, even the mildly skewed sex ratio of dwarf mongooses causes a surprising reduction in N_{ed} to 86% of N. Deviation from the stable age-distribution also has a pronounced effect on N_{ed} in dwarf mongooses (table 10-7). Compared to the stable age-distribution, the actual population has an excess of females in young age-classes with low fecundity and a deficiency of middle-aged females with high fecundity (table 10-7). As a result, N_{ed} is only 70% of actual population size. Combining the effects of the age distribution and sex ratio, N_{ed} is only 60% of the actual population size.

Table 10-7 Demographic effective population size $\{N_{ed} = N[\Sigma f_x(s_x m_x + s_x - 1)]/\Sigma l_x e^{-rx}(s_x m_x + s_x - 1)\}$ in female dwarf mongooses: consequences of age-specific fecundity and deviation from the stable age distribution.[a]

x	N_x	l_x	s_x	m_x	$l_x m_x$	$l_x e^{-rt}$	f_x	$s_x m_x + s_x - 1$	True output[b]	Ideal output[c]
0	402	1.00	0.41	0	0	1	1	—	—	—
1	165	0.41	0.80	0	0	0.406	0.410	—	—	—
2	134	0.33	0.77	0.21	0.069	0.323	0.333	—	—	—
3	101	0.25	0.72	0.39	0.098	0.243	0.250	0.001	0.000	0.000
4	63	0.18	0.78	0.95	0.171	0.173	0.157	0.521	0.082	0.090
5	40	0.14	0.60	1.32	0.185	0.133	0.100	0.392	0.039	0.052
6	18	0.09	0.67	1.48	0.133	0.085	0.045	0.662	0.030	0.056
7	11	0.06	0.55	2.45	0.147	0.056	0.026	0.898	0.023	0.050
8	9	0.03	0.67	3.78	0.083	0.028	0.021	2.203	0.046	0.062
9	6	0.02	0.67	2.56	0.051	0.018	0.014	1.385	0.019	0.025
10	3	0.01	0.33	4.07	0.041	0.009	0.007	0.673	0.005	0.006
11	2	0.01	1.00	3.76	0.038	0.009	0.005	3.760	0.019	0.034
12	2	0.01	0.50	3.00	0.030	0.009	0.005	1.000	0.005	0.009
13	1	0.00	1.00	2.00	0.020	0	0	2.000	0.000	0.000
14	1	0.00	0.00	0	0	0	0	0	0.000	0.000
15	0									
$\Sigma =$					1.066				0.268	0.384

$N_{ed} = N(0.268/0.384)$
$\Delta^d = 0.70$

[a] $R_0 = 1.066$, $G = 6.04$, $r = .010$.
[b] True output $= f_x[s_x m_x + (s_x - 1)]$.
[c] Ideal output $= l_x e^{-rt}[s_x m_x + (s_x - 1)]$.
[d] $\Delta = N_{ed}/N$, the ratio of effective demographic population size to census population size.

Relevance to Conservation

Genetics

Franklin (1980; see also Lande and Barrowclough, 1987) suggests that an N_e of 500 is needed to maintain typical levels of heritable genetic variation (in the absence of selection). Lande (1995) argues that 5000 is a more logical target. The Δ values for carnivores (tables 10-1–10-6) are generally lower than Δ values for other taxa—Nunney and Elam (1994) report a mean Δ of 0.73 for 14 species. In general, social carnivore populations would have to number 600–3000 individuals to meet the lower target of $N_e = 500$ (6,000–30,000 for the upper target). Many populations of large carnivores fall far short of both targets.

For example, African wild dogs in optimal habitat live at an average density of one individual per 57 km^2 (Creel and Creel, 1996). With a target N_e of 500 and a Δ of 0.2, the target population size is 2500. At average population density, a protected area of 142,500 km^2 is needed. The largest known wild dog population (~880 adults) is in the 43,000-km^2 Selous Game Reserve. The next largest protected population (~350 adults) lives in the 21,000-km^2 Kruger National Park (Maddock and Mills, 1994). Neither population approaches the size needed to conserve the wild dog's current genetic diversity. Areas bordering the Selous are relatively wild and harbor some additional wild dogs. These unprotected but undeveloped areas also allow occasional immigrants from disjunct populations. For example, a wild dog was killed in 1995 on a highway 80 km from the nearest edge of Selous, probably dispersing in or out of the population. Kruger is partially fenced and distant from other sizable wild dog populations (perhaps excepting parts of Mozambique), so it is less likely to receive gene flow (Maddock and Mills, 1994).

These calculations give a concrete example of two problems. First, large, protected areas are essential for the conservation of large carnivores (Clark et al., 1996). Second, even the largest protected areas are inadequate for some species, so freedom of movement between protected areas is needed. For large carnivores, movement is problematic due to conflict with animal agriculture and (in some cases) danger for people. Consider the uproar over the return of grey wolves to Wyoming, or mountain lion *Felis concolor* attacks in California. If restricted movements prevent a metapopulation from forming, managed movements will be necessary in some cases.

Demography

The likelihood of extinction depends on the size and growth rate of a population (Goodman, 1987). Almost universally, estimates of population size and growth rate come from standard census and life-table methods. Combining the methods suggested by Caughley (1994) to determine N_{ed}, the demographic effective population size of Serengeti dwarf mongooses is only 60% of the censused population size. It would be foolish to generalize from a single species, but this result suggests that social carnivores may be subject to considerably greater demographic stochasticity than one would expect from their population sizes.

Recommendation

At least in Africa, it is not unusual for wildlife managers and conservation personnel to see little value in the detailed data gathered by behavioral ecologists. Simple data on population size, for example, are considered more concrete and practical, and in many ways, they are. Yet the examples presented here show that data on population size can be misleading if considered in a vacuum. The effect of social organization on N_e will substantially affect risk assessments for most social carnivores. For a given species, quantitative data on N_e and N_{ed} could be used to assess short-term demographic risk, long-term genetic risk, and the adequacy of a given reserve considering its size and isolation from other populations.

Summary

Despite many long-term studies in behavioral ecology, we still have few estimates of N_e/N (Δ). This ratio is essential to assess the conservation of genetic variability. Although we cannot say what a typical Δ is for most taxonomic groups, Δ values are low in social carnivores.

Empirically, unequal reproductive success caused the greatest reduction in N_e for social carnivores. Skewed sex ratios had little impact on N_e, unless reproductive suppression altered the effective sex ratio. Variation in population size appeared to be less important. Isolation by distance did not accelerate genetic drift in the one case examined. Complex data are needed to assess Δ accurately because the most easily assessed factors had relatively small effects on N_e. For management, this implies that detailed monitoring may be needed for species of special concern. The alternative is to generalize results from better-studied species with similar biology. Long-term demographic data are difficult and costly to collect, particularly for carnivores. This is an impediment to gathering estimates of N_e/N. However, published estimates of N_e are fewer than expected (given the number of long-term studies).

For dwarf mongooses, the demographic effective population size (N_{ed}) was 60% of the adult population size. This effect was primarily due to a surplus of young individuals with low fecundity relative to the stable age distribution. Estimates of N_{ed} require a full life table, so they will be rare for carnivores.

A target of $N_e = 500$ (Franklin, 1980), combined with the Δ values calculated here, suggests that carnivore populations would have to number 600–3000 adults to retain typical genetic variation. Lande's (1995) target of $N_e = 5000$ would require populations of 6,000–30,000 adults. Many populations of large social carnivores do not meet the lower target range. For managers, this implies that preserving freedom of movement among populations is necessary. Where practical limitations prevent this (e.g., carnivore–livestock or carnivore–human conflicts), managed movement should be considered.

Acknowledgments My thanks to Nancy Marusha Creel for data collection and analysis in the studies of Selous wild dogs and Serengeti dwarf mongooses. I thank the Frankfurt Zoological Society (Help for Threatened Animals Project 1112/90), the National Science Foundation (IBN-9419452), and the Rockefeller University Meyer Fund for support of research on Selous wild dogs and support during preparation of this chapter. I am grateful to those who kindly contributed unpublished data for other species: M Kelly, B Bertram, TM Caro, A Collins, SM Durant, CD FitzGibbon, G Frame, and MK Laurenson (cheetah); LG Frank, KE Holekamp, and L Smale (spotted hyena); C Sillero-Zubiri and D Gottelli (Ethiopian wolf); and N Marusha Creel, J Rood, and P Waser (dwarf mongoose).

References

Barrowclough GF, 1980. Gene flow, effective population sizes, and genetic variance components in birds. Evolution 34:789–798.

Bertram BCR, 1975. Social factors influencing reproduction in wild lions. J Zool 177:463–482.

Brown JL, Bush M, Packer C, Pusey AE, Monfort SL, O'Brien SJ, Janssen DL, Wildt DE, 1991. Developmental changes in pituitary-gonadal function in free-ranging lions *Panthera leo leo* of the Serengeti Plains and Ngorongoro Crater. J Reprod Fertil 91:29–40.

Brown JL, Bush M, Packer C, Pusey AE, Monfort SL, O'Brien SJ, Janssen DL, Wildt DE, 1993. Hormonal characteristics of free-ranging female lions *Panthera leo* of the Serengeti Plains and Ngorongoro Crater. J Reprod Fertil 97:107–114.

Carbyn LN, Paquet PC, 1986. Long distance movements of a coyote from Riding Mountain National Park. J Wildl Manage 50:89.

Caro TM, Durant SM, 1995. The importance of behavioural ecology for conservation biology: examples from Serengeti carnivores. In: Serengeti II: Research, management and conservation of an ecosystem (Sinclair ARE, Arcese P, eds). Chicago: University of Chicago Press; 451–472.

Caughley G, 1994. Directions in conservation ecology. J Anim Ecol 63:215–244.

Clark TW, Paquet PC, Curlee AP, 1996. General lessons and positive trends in large carnivore conservation. Conserv Biol 10:1055–1058.

Creel S, Creel NM, 1996. Limitation of African wild dogs by competition with larger carnivores. Conserv Biol 10:526–538.

Creel SR, Creel NM, Monfort SL, 1997. Rank and reproduction in cooperatively breeding African wild dogs: behavioral and endocrine correlates. Behav Ecol. 8:298–306.

Creel SR, Creel NM, Wildt DE, and Monfort SL, 1992. Behavioral and endocrine mechanisms of reproductive suppression in Serengeti dwarf mongooses. Anim Behav 43:231–245.

Creel S, Monfort SL, Creel NM, Wildt DE, Waser PM, 1996. Pregnancy, oestrogens and future reproductive success in Serengeti dwarf mongooses. Anim Behav 50: 1132–1135.

Creel SR, Waser PM, 1994. Inclusive fitness and reproductive strategies in dwarf mongooses. Behav Ecol 5:333–348.

Creel SR, Waser PM, 1996. Variation in reproductive suppression in dwarf mongooses: interplay between evolution and mechanisms. In: Cooperative breeding in mammals (Solomon N, French J, eds). Cambridge, Cambridge University Press, 150–170.

Crow JF, Kimura M, 1970. An introduction to population genetics theory. Minneapolis, Minnesota: Burgess Publishing.

Frame LH, Frame G, 1976. Female African wild dogs emigrate. Nature 263:227–229.

Frank LG, 1986. Social organization of the spotted hyaena *Crocuta crocuta*. II. Dominance and reproduction. Anim Behav 34:1510–1527.

Frank LG, Holekamp KE, Smale L, 1995. Dominance, demography and reproductive success of female spotted hyenas. in: Serengeti II: Research, management and conservation of an ecosystem (Sinclair ARE, Arcese P, eds). Chicago: University of Chicago Press; 364–384.

Franklin IR, 1980. Evolutionary changes in small populations. In: Conservation biology: an evolutionary-ecological perspective (Soulé M, Wilcox BA, eds). Sunderland, Massachusetts: Sinauer Associates; 135–149.

Fritts SH, 1983. Record dispersal by a wolf from Minnesota. J Mammal 64:166–167.

Fuller TK, Mills MGL, Borner M, Laurenson MK, Kat PW, 1992. Long distance dispersal by African wild dogs in East and South Africa. J Afr Zool 106:535–537.

Goodman D, 1987. The demography of chance extinction. In: Viable populations for conservation (Soulé M, ed). Cambridge: Cambridge University Press; 11–34.

Harris RB, Allendorf FW, 1989. Genetically effective population size of large mammals: an assessment of estimators. Conserv Biol 3:181–191.

Hofer H, East ML, 1993. The commuting system of Serengeti spotted hyaenas: how a predator copes with migratory prey. I. Social organization. Anim Behav 46:547–557.

Keane B, Waser PM, Creel SR, Creel NM, Elliott LF, Minchella DJ, 1994. Subordinate reproduction in dwarf mongooses. Anim Behav 47:65–75.

Kleiman DG, 1977. Monogamy in mammals. Q Rev Biol 52:39–69.

Kruuk H, 1972. The spotted hyaena: a study of predation and social behavior. Chicago: University of Chicago Press.

Lande R, 1988. Genetics and demography in biological conservation. Science 241: 1455–1460.

Lande R, 1995. Mutation and conservation. Conserv Biol 9:782–791.

Lande R, Barrowclough GF, 1987. Effective population size, genetic variation, and their use in population management. In: Viable populations for conservation (Soulé M, ed). Cambridge: Cambridge University Press; 87–123.

Maddock AH, Mills MGL, 1994. Population characteristics of African wild dogs Lycaon pictus in the eastern Transvaal lowveld, South Africa, as revealed through photographic records. Biol Conserv 67:57–62.

Maruyama T, 1977. Stochastic problems in population genetics. Lecture notes in biomathemataics. Berlin: Springer-Verlag.

Matthews LH, 1939. Reproduction in the spotted hyaena, *Crocuta crocuta* (Erxleben). Phil Trans R Soc Lond B 230:1–78.

Mech LD, 1986. Wolf population in the central Superior National Forest, 1967–1985. USDA Research Paper NC-270. St. Paul, Minnesota: U.S. Department of Agriculture.

Mills MGL, 1989. The comparative behavioral ecology of hyaenas: the importance of diet and food dispersion. In: Carnivore behavior, ecology and evolution (Gittleman JL, eds). Ithaca, New York: Cornell University Press; 125–152.

Moehlman PD, 1986. Ecology of cooperation in canids. In: Ecological aspects of social evolution (Rubenstein DI, Wrangham RW, eds). Princeton, New Jersey: Princeton University Press; 64–86.

Nunney L, Campbell KA, 1993. Assessing minimum viable population size: demography meets population genetics. Trends Ecol Evol 8:234–239.

Nunney L, Elam DR, 1994. Estimating the effective population size of conserved populations. Conserv Biol 8:175–184.

Packer C, 1990. Serengeti lion survey: report to TANAPA, SWRI, Mweka and the Game Department. Serengeti Wildlife Research Center, Sesonea, Tanzania.

Packer C, Gilbert DA, Pusey AE, O'Brien SJ, 1991a. A molecular genetic analysis of kinship and cooperation in African lions. Nature 351:562–565.

Packer C, Herbst L, Pusey AE, Bygott JD, Hanby JP, Cairns S, Borgerhoff Mulder M, 1988. Reproductive success of lions. In: Reproductive Success, (Clutton-Brock TH, eds). Chicago: University of Chicago Press; 363–383.

Packer C, Pusey AE, Rowley H, Gilbert DA, Martenson J, O'Brien SJ, 1991b. Case study of a population bottleneck: lions of the Ngorongoro Crater. Conserv Biol 5:219–230.

Peterson RO, Page RE, 1988. The rise and fall of Isle Royale wolves, 1975–1986. J Mammal 69:89–99.

Richardson PRK, 1987. Overt cuckoldry in an apparently monogamous mammal. S Afr J Sci 83:405–410.
Rood JP, 1986. Ecology and social evolution in the mongooses. In: Ecological aspects of social evolution (Rubenstein DI, Wrangham RW, eds). Princeton, New Jersey: Princeton University Press; 131–152.
Schaller GB, 1972. The Serengeti lion. Chicago: University of Chicago Press.
Sillero-Zubiri C, 1994. Behavioural ecology of the Ethiopian wolf (PhD dissertation). Oxford: University of Oxford.
Waser PM, 1996. Patterns and consequences of dispersal in gregarious carnivores. In: Carnivore behavior, ecology and evolution (Gittleman JL, eds). Ithaca, New York: Cornell University Press; 267–295.
Waser PM, Creel SR, Lucas JR, 1994. Death and disappearance: estimating mortality risks associated with philopatry and dispersal. Behav Ecol 5:135–141.
Waser PM, Elliott LF, Creel NM, Creel SR, 1995. Habitat variation and mongoose demography. In: Serengeti II: Research, management and conservation of an ecosystem (Sinclair ARE, Arcese P, eds). Chicago: University of Chicago Press; 421–447.
Wright S, 1931. Evolution in Mendelian populations. Genetics 16:97–159.
Wright S, 1938. Size of population and breeding structure in relation to evolution. Science 87:430–431.
Wright S, 1940. Breeding structure of populations in relation to speciation. Am Nat 74:232–248.
Wright S, 1943. Isolation by distance. Genetics 28:114–138.
Wright S, 1969. Evolution and the genetics of populations: II. The theory of gene frequencies. Chicago: University of Chicago Press.

Part IV

Mating Systems and Conservation Intervention

The chapters in this section demonstrate some of the ways in which individual reproductive strategies that underlie mating systems can affect the outcome of management intervention schemes. Three issues emerge from these contributions: that different mating systems affect a population's response to anthropogenic perturbation in surprising ways; that extrapolations about such effects between closely related species are risky if species' reproductive strategies differ; and that the success of captive-breeding programs might hinge on allowing females to choose mates. These studies show that the effects of completely different forms of intervention can be influenced by nuances in animal mating systems.

In many parts of the world, hunting is used as a conservation tool in the sense that areas are protected from agricultural development and deforestation because of the revenue generated from people hunting in that area (Taylor and Dunstone, 1996). In some situations, local people are given incentives to protect wilderness areas by allowing them to hunt; in other situations, tourist hunting is encouraged (Metcalfe, 1994; Robinson and Redford, 1994; Lewis and Alpert, 1997). The form of hunting includes trophy hunting of adult males; hunting adults of both sexes, for example, for food; or indiscriminate hunting of all age and sex classes. In the first chapter in this section, Greene, Umbanhowar, Mangel, and Caro model the effects of different forms of offtake on population growth rates. They find that hunting adults reduced population growth rates at a lower level of hunting than does indiscriminate hunting.

The mating system of exploited species has an additional and important effect. Greene et al.'s models show that monogamous and weakly polygynous species are much more susceptible to hunting of males than are highly polygynous species; in the latter case harem holders are rapidly replaced by floaters, but not in the monogamous situation. Population growth rates of infanticidal species such as lions *Panthera leo* are particularly affected by male offtake at small harem sizes because removal of harem-holding males allows floating males to take over the harem and kill the offspring. In reproductively suppressed species, hunting of both sexes causes rapid population declines because removal of the breeding female may prevent an entire group from reproducing.

Results of these models raise issues for sustainable hunting and for population monitoring. For example, hunting quotas for monogamous species such as duikers (Cephalophinae) should be set at lower levels than for polygynous species; and, in infanticidal species, quotas allowing hunting of males and females should be lower than those sanctioning male hunting only. Due to their susceptibility to offtake, monogamous species require closer monitoring than polygynous species; for mammals, this invites development of new monitoring techniques that can census small edible species on the ground.

A more hands-on, *in situ* management technique concerns habitat restoration, which includes eradicating exotic species, encouraging native species to return, actively reintroducing selected species, and bolstering existing populations (Meffe and Carroll, 1997). One aspect of bolstering existing populations is to enhance breeding opportunities through the provision of nest-boxes for birds. Until recently, nest-box supplementation had been viewed uncritically; it was assumed that additional nest sites increased population size in cavity-nesting birds. In chapter 12, Eadie, Sherman, and Semel use empirical data and modeling to show how a reproductive tactic used by female wood ducks *Aix sponsa*, namely, laying eggs in other females' nests, can affect a population's response to nest-box provisioning. As more nest-boxes are provided in visible locations, breeding density increases and, with it, levels of conspecific brood parasitism. As a result, hatching success declines and population crashes may occur. In the closely related Barrow's goldeneye *Bucephala islandica*, however, where males and females defend feeding territories, opportunities for conspecific brood parasitism are far more limited and addition of nest-boxes does not result in population declines. In sum, differences in female reproductive tactics and territoriality between cavity-nesting waterfowl can significantly influence the population's response to the same management technique, suggesting that understanding intraspecific social interactions is critical to the success of such projects.

A third management endeavor is captive breeding. The principal strategy in captive-breeding programs has been to maintain genetic heterozygosity by minimizing inbreeding between relatives (Ralls et al., 1988; Lacy et al., 1993). This involves constructing pedigrees, moving potential breeding partners between institutions, and increasingly, artificial insemination. In many cases, these methods disregard or circumvent mate choice. In the third chapter in this section, Grahn, Langefors, and von Schantz review burgeoning evidence to support the claim that females choose mates on the basis of males' condition and perhaps on the genetic variability of the major histocompatability complex, which affects resistance to pathogens or even antioxidant defense mechanisms. They then examine a case study of the Atlantic salmon *Salmo salar,* which are bred in hatcheries along the Baltic coast. In these captive facilities, juvenile mortality is extremely high as a result of a syndrome that is cured by adding thiamine supplement. Given that salmon alevins hatched from clutches of more pigmented roe are less likely to suffer from juvenile mortality syndrome and that females may choose males on the basis of carotenoid pigmentation, as occurs in other fish (Milinski and Bakker, 1990), Grahn et al. argue that prevention of mate choice may make offspring susceptible to juvenile mortality syndrome. While these suggestions are speculative, they suggest that the adaptive consequences of mate choice, now the focus of enormous attention in behavioral ecology, can affect a population's ability to combat pathogens. Because disease poses a particular threat to small populations (Thorne and Williams, 1988), this suggestion holds considerable conservation significance and needs to be investigated experimentally.

These three studies examine only a subsample of animal mating systems. For example, the outcome of hunting pressure on polyandrous species or species with high levels of extrapair copulations are not addressed. In addition, the effects of hunting pressure contingent on which individuals are removed from a population are barely examined. For example, the removal of harem-holding or floating males radically affects the outcome of hunting pressure on lion population growth rates (Starfield and Bleloch, 1991). Thus there is much room for additional modeling, as well as incorporating density dependence and physiological state, as Greene at al. suggest. Moreover, we urgently need empirical work to examine the impact of hunting on species with different breeding systems.

It is clear from these studies that generalizations between closely related species cannot be made in the absence of information about social interactions. One of the key findings to emerge from research on breeding systems is that they are acutely sensitive to ecological and social conditions and consequently show great variability both between and within species (Davies, 1992). This outlook is depressing because it means that the effects of management intervention on population growth rates cannot be predicted easily using comparative data from different species or even populations. Again, we need a wider comparative database of the response of species with different mating systems to anthropogenic disturbance in order to pick out common themes.

Finally, these three chapters, closely juxtaposed, are instructive in that they show how different forms of conservation practice—*in situ* conservation, restoration, and captive breeding—are influenced by animal mating systems and mating tactics. Human disturbance and animal reproductive strategies interacting to affect population growth rates can no longer be considered an isolated phenomenon that we can more or less ignore in species management practice.

References

Davies NB, 1992. Dunnock behaviour and social evolution. Oxford: Oxford University Press.

Lacy RC, Petric AM, Warneke M, 1993. Inbreeding and outbreeding depression in captive populations of wild species. In: The natural history of inbreeding and outbreeding (Thornhill NW, ed). Chicago: University of Chicago Press; 352–374.

Lewis SM, Alpert P, 1997. Trophy hunting and wildlife conservation in Zambia. Conser Biol 11:59–68.

Metcalfe S, 1994. The Zimbabwe communal areas management programme for indigenous resources (CAMPFIRE). In: Natural connections: perspectives in community-based conservation (Western D, Wright RM, Strum SC, eds). Washington, DC: Island Press; 161–192.

Meffe GK, Carroll RC, 1997. Principles of conservation biology, 2nd ed. Sunderland, Massachusetts: Sinauer Associates.

Milinski M, Bakker TCM, 1990. Female sticklebacks use male colouration in mate choice and hence avoid parasitized males. Nature 344:330–333.

Ralls K, Ballou JD, Templeton AR, 1988. Estimates of lethal equivalents and the cost of inbreeding in mammals. Conserv Biol 2:185–193.

Robinson JG, Redford KH, 1994. Community-based approaches to wildlife conservation in neotropical forest. In: Natural connections: perspectives in community-based conservation (Western D, Wright RM, Strum SC, eds). Washington, DC: Island Press; 300–319.

Starfield AM, Bleloch AL, 1991. Building models for conservation and wildlife management, 2nd ed. Edina, Minnesota: Burgess International Group.

Taylor VJ, Dunstone N, 1996. The exploitation of mammal populations. London: Chapman and Hall.

Thorne ET, Williams ES, 1988. Disease and endangered species: the black-footed ferret as a recent example. Conserv Biol 2:66–74.

11

Animal Breeding Systems, Hunter Selectivity, and Consumptive Use in Wildlife Conservation

Correigh Greene
James Umbanhowar
Marc Mangel
Tim Caro

Multiple-Use Areas

Wild animal and plant populations are protected from development and agriculture in a number of ways, ranging from strict nature reserves and national parks that are managed for wilderness protection and nonconsumptive recreation, to multiple-use areas from which species or products are extracted (IUCN/UNEP/WWF, 1991). In the face of the growing human population, much of the world's animal and plant diversity will in the future be found within protected areas, so the efficacy with which different forms of protection conserve biodiversity is now of central concern (Meffe and Carroll, 1997). In the long term, we will need to determine whether wildlife populations protected in national parks suffer less attrition than those protected in multiple-use areas and which types of multiple-use areas are best.

To date, most studies of consumptive use of plants and animals have been from an economic perspective. These studies have attempted to quantify resource availability (e.g., Lawrence et al., 1995) and assign economic weighting to products (e.g., Peters et al., 1989; Balick and Mendelsohn, 1992), and hence examine whether exploitation is sustainable in economic terms (Clark, 1985, 1990; Redford and Padoch, 1992). Viewing natural resources solely from an economic standpoint does not necessarily ensure biological sustainability, and therefore, if multiple-use areas are to be an effective conservation tool, they must support viable populations of species in the long term.

Consumptive Use of Animals: Classical Conservation Approaches

Consumptive use of animals includes subsistence hunting by indigenous people (Marks, 1973; Kaplan and Kopischke, 1992), local market hunting (Glanz, 1991), ranching (Ojasti, 1991), sport hunting (Anderson, 1983; Cumming, 1989; Metcalfe, 1994), and commercial harvest (Lewis et al., 1990; Thomsen and Brautigam, 1991). To assess the effects of consumptive use on terrestrial animal populations, some researchers have compared densities in areas of differing hunting intensity (Freese et al., 1982; Peres, 1990; Fragoso, 1991). Although prey populations are often smaller in hunted areas (but see Harcourt, chapter 3, this volume), reduction in density does not mean that hunting is unsustainable. To address this possibility, studies have examined how kills per unit effort (assumed to be a measure of abundance) change over time. Vickers (1991), for example, concluded that the Siona-Secoya people in northeastern Ecuador were not depleting their mammalian or avian prey because, for most species, returns per unit effort remained fairly constant over a decade.

Other studies have employed indirect methods and comparative data to determine reproductive parameters of prey populations and hence sustainable offtake levels. Robinson and Redford (1986, 1991a) calculated species densities in neotropical forests using body weights and intrinsic rates of natural increase and hence estimated production rates per unit area. Several studies have used Robinson and Redford's model to estimate potential harvests. For example, Fa et al. (1995) surmised that five primates and one ungulate species were being overhunted on the Island of Bioko, equatorial Guinea, but that few species were being overhunted on mainland Rio Muni. FitzGibbon et al. (1995) determined that yellow baboons *Papio cynocephalus* and Sykes' monkey *Cercopithecus mitis* were being overhunted in Kenya's Arabuko-Sokole Forest, but offtake rates for elephant shrews, squirrels, and duikers (*Cephalophus* spp.) were sustainable.

Using more direct methods, Bodmer (1994; see also Bodmer et al., 1994) examined whether Amazonian wildlife could be harvested commercially without rendering species vulnerable to local extinction. Bodmer gathered data on mammal densities in the Reserva Comunal Tamshiyacu-Tahuayo in northeast Peru by walking line-transects and calculated reproductive rates by autopsying shot females. This allowed him to determine the magnitude of harvest that the population might sustain. In this study he also measured the actual number of animals hunted in an area. By comparing this with animal densities and production rates, it was possible to determine which species were being harvested on a sustainable basis and which were overexploited. He found that tapirs *Tapirus terrestris* and primates were overhunted, whereas peccaries *Tayassu tajacu* and *T. pecari* and brocket deer *Mazama americana* and *M. gouazoubira* were probably being hunted sustainably. Bodmer's study is unusual because it measured mammal densities, calculated production rates, and measured offtake simultaneously.

Data on animal densities and hunting rates are difficult to collect, but they represent a major step in both assessing the biological sustainability of consumptive animal use and in recommending appropriate offtake levels in multiple-use areas. A tentative conclusion emerging from studies in the Neotropics is that seasonal habitats are more likely to contain large-bodied species with high rates of population increase that can tolerate some sort of sustainable harvest. Species-diverse habitats, such as tropical rainforests, on the other hand, do not contain single species at high enough densities and with the population growth rates to be commercially exploitable. Neotropical forests are therefore more likely to be important for subsistence hunters who harvest a diversity of species (Robinson and Redford, 1991b). These generalizations should be treated cautiously, however, because aside from Bodmer's study, most attempts to derive estimates of harvest potential are crude and fail to take account of numerous human factors or the biology of prey animals (FitzGibbon, chapter 16, this volume).

Two important factors are hunter selectivity and animal breeding system. In general, animal harvest in buffer zones or multiple-use conservation areas takes three forms: (1) non-selective hunting, where subsistence hunters (or poachers) kill the first individual they encounter or that fall into their traps, regardless of age or sex (see Arcese et al., 1995; Campbell and Hofer, 1995); (2) hunting of adult males, as exemplified by tourist hunters shooting ungulates (Ginsberg and Milner-Gulland, 1994); and (3) selectively hunting adults of either sex (Marks, 1973; Alvard, chapter 17, this volume).

The range of breeding systems of commonly exploited species include harem and resource-defense polygyny, as exhibited by most mammals, and monogamy as found in many birds (Greenwood, 1980). Within these systems, species may exhibit particular behavioral and life-history strategies that have population consequences. In some species, resource competition or kin selection may cause females in larger harems to have lower fecundity than those in smaller harems (Clutton-Brock et al., 1982). Certain exploited carnivores and primates exhibit high levels of infanticide, usually by males (Hausfater and Hrdy, 1984). Other species show reproductive suppression in which only the dominant male and female normally breed (Creel and Creel, 1989). The ability of a population to sustain a given level of harvest depends, in part, on the interaction of both hunter selectivity and breeding system. This level directly affects the economic returns of harvest and hence its efficacy as a conservation strategy.

In this chapter, we explore the interactions of breeding systems and hunting selectivity on the ability of mammals to withstand different forms of hunting and hence highlight differences in sustainable offtake levels for commonly hunted species. We first develop a series of models showing how population growth rate responds to different hunting regimes, and then we examine how polygyny, infanticide, and reproductive suppression affect growth rates of hunted populations. Next, we use these models to examine the effect of legal hunting on three carnivore and three ungulate species hunted in Africa. These species were chosen because they have different breeding systems, their reproductive parameters are reasonably well documented, and two of them bring in substantial revenue from hunting. In the final section, we apply this information to the Selous Game Reserve in Tanzania, the world's largest hunting reserve. In Selous, population sizes of ungulates have been surveyed from the air (Caro et al., in press a,b) and those of carnivores estimated from the ground (Creel and Creel, 1996). We use our models to evaluate whether current tourist hunting levels will allow these six species to replace themselves sustainably in the Selous (i.e., whether they have a population growth rate > 1) and to make general recommendations for management of large mammal populations.

Examining Effects of Mating Systems Using Demographic Models

Classical Approaches Based on Life Tables

The classical approach based on life tables would typically begin with schedules of female fecundity (usually assuming a 50:50 natal sex ratio) and survivorship. These give the reproduction at each age and survival to that age. From fecundity and survivorship schedules, one can compute expected lifetime reproduction, R_0, of an individual and the instantaneous rate of increase, which is the solution of the Euler-Lotka equation (Gotelli, 1995).

The classical approach lacks any explicit recognition of mating system, except possibly in the inclusion of sex ratio. Depending on the specific mating system, R_0 computed independently for females and males might differ, especially if mean or variance in reproductive success differs between sexes (cf. Trivers, 1972). For example, Waser et al. (1995) calculated R_0 for dwarf mongooses *Helogale parvula* to be 1.06 based on females but 0.94 based on males. Under such a circumstance, one can only say that the population growth rate is close to 1, and more precise statements, especially regarding whether the population is slightly growing or declining, are not possible if the variance in these rates is unknown.

Sex differences in population growth rate would not be a problem if, as is generally assumed, females solely limit population growth due to control of offspring production and recruitment. However, current theory on the evolution of mating systems predicts several circumstances in which males may limit population growth: when males provide significant parental care, as is often the case in monogamous, polyandrous, and some polygynous systems (Emlen and Oring, 1977), or when male dispersion is large relative to female dispersion (Clutton-Brock, 1989). In such cases, differential offtake of males could reduce population growth but would not be detected by classical life-table analysis.

The second problem with classical approaches is that, except for monogamous species, the empirical fecundity and survivorship schedules incorporate the result of the mating system. Because of that, they are not useful for predicting how the population characteristics will respond to changes. For example, a 10% hunting offtake of male lions *Panthera leo* will reduce survivorship accordingly, but the response of the mating system to this reduction may also be increased infanticide due to greater turnover in harem-holding males, which is difficult to predict a priori. Thus, one cannot simply reduce the survivorship schedule by 10% and assert that the new growth rates have not been otherwise affected.

Incorporating Breeding Systems into Age- and Stage-Structured Models

The age-based or stage-based approach to life history and conservation described in the previous section treats all individuals at the same age or stage as if they were identical. However, individuals of the same age may vary in physiological condition (reviewed in Mangel and Clark, 1988; Mangel and Ludwig, 1992; McNamara and Houston, 1996), and subpopulations may vary in the type of breeding system. While other population models have explicitly examined how both sexes influence population dynamics (Beddington, 1974; Beddington and May, 1980; Charlesworth, 1980; Burgoyne, 1981; Caswell and Weeks, 1986; Starfield et al., 1981; Starfield and Bleloch, 1991), few of these models incorporate the specific breeding system (however, see Caswell and Weeks, 1986). In this section, we show how breeding system can be incorporated into classical age- and stage-based approaches to life history and conservation. Like classical approaches, we use a population-level model without spatial dynamics, stochasticity, or density dependence to study the interaction of breeding system and hunting mortality. Our intention is produce a relatively simple, general model; possible modifications such as density dependence and spatial structure are discussed later.

The Fundamental Variables

The fundamental time unit in the model is the length of the birth interval (interbirth interval; IBI). In some species, the IBI is 1 year. In other species that have multiple litters per year, the IBI can be less than 1 year. We measure time, t, in multiples of the IBI. The fundamental population variables are the sizes of birth, juvenile, and adult populations at any time t (table 11-1). We denote the birth populations by

Table 11-1 Variables used in the text and their definitions.

Variable	Definition
a	Adult age class
$A_M(t)$, $A_F(t)$	Total male and female adults at interval t
$A_m(t,a)$, $A_f(t,a)$	Number of male or female adults in class a at interval t
$B_m(t)$, $B_f(t)$	Number of male or female birth class individuals at interval t
F, f	Females
H	Total hunting intensity
h	Harem size
H_{jm}, H_{jf}, H_{am}, H_{af}	Hunting intensity of male and female juveniles and adults
i	Probability of takeover following death of a male in a coalition
j	Juvenile age class
$J_M(t,j)$, $J_F(t,j)$	Total male and female juveniles at interval t
$J_m(t,j)$, $J_f(t,j)$	Number of male and female juveniles in class j at interval t
λ	Population growth rate
$m(a)$	Fecundity of class a
M, m	Males
$P(t)$	Total population at interval t
P_H	Total hunted population
θ_{jm}, θ_{jf}, θ_{am}, θ_{af}	Hunting selectivity of male and female juveniles and adults
r	Proportion of birth class born that are male
R	Total reproduction
$\rho_m(a)$, $\rho_f(a)$	Survivorship of male and female adults of class a
s_f, s_m	Survivorship of birth class individuals
$\sigma_m(j)$, $\sigma_f(j)$	Survivorship of male and female juveniles of class j
t	Interbirth interval
τ	Average number of takeovers per male per interbirth interval

$$B_f(t) = \text{number of birth-class females at the start of interval } t$$
$$B_m(t) = \text{number of birth-class males at the start of interval } t. \quad (1)$$

After one birth interval, surviving offspring move into the juvenile class, where they remain for j_{max} birth intervals. Hence, if juvenile populations are described by

$$J_f(t,j) = \text{number of female juveniles in class } j \text{ at the start of interval } t$$
$$J_m(t,j) = \text{number of male juveniles in class } j \text{ at the start of interval } t \quad (2)$$

the total juvenile populations are

$$J_F(t) = \sum_{j=1}^{j_{max}} J_f(t,j) \text{ and } J_M(t) = \sum_{j=1}^{j_{max}} J_m(t,j) \quad (3)$$

for females and males, respectively. Because we track sexes separately, we need not assume that j_{max}, the number of IBIs an individual spends as a juvenile, is the same for both sexes, but we do not add that complication here.

After j_{max} IBIs, individuals become adults. To track the different ages and sexes, we set

$$A_f(t,a) = \text{number of female adults in class } a \text{ at the start of interval } t$$
$$A_m(t,a) = \text{number of male adults in class } a \text{ at the start of interval } t. \quad (4)$$

Here $a = 1, \ldots a_{max}$, where a_{max} can be interpreted as the maximum adult life span or the age of reproductive senescence. As with j_{max}, it is possible that a_{max} differs between sexes. As with the juveniles, we denote the total adult male and female populations by $A_M(t)$ and $A_F(t)$, respectively.

In addition to j_{max} and a_{max}, we must specify fecundity and survivorship. Female fecundity depends, in principle, upon age, so we let

$$m(a) = \text{fecundity (number of offspring weaned) by a female of age } a. \quad (5)$$

To characterize survival in the absence of hunting, we assume that survivorship may vary among different stages, among age classes within each stage, and between sexes. For birth class individuals

$$s_f = \text{fraction of birth class females that survive to juvenile class 1}$$

$$s_m = \text{fraction of birth class males that survive to juvenile class 1.} \quad (6)$$

The survival of juveniles and adults in the absence of hunting depends in principle not only upon sex, but also upon age class:

$$\sigma_f(j) = \text{fraction of juvenile females in IBI class } j \text{ that survive to class } j + 1$$
$$\sigma_m(j) = \text{fraction of juvenile males in IBI class } j \text{ that survive to class } j + 1$$

$$\rho_f(a) = \text{fraction of adult females in IBI class a that survive to class } a + 1$$
$$\rho_m(a) = \text{fraction of adult males in IBI class a that survive to class } a + 1. \quad (7)$$

In these equations, IBI class $j_{max} = 1$ corresponds to adult class 1, and survival of adults past a_{max} is minuscule. In the simplest case, survivorship is constant within age class for juveniles and adults.

Hunting intensity may depend upon stage (birth class individuals are not hunted) and sex, but not upon IBI class within a stage, and is characterized by

$$H_{jf} = \text{hunting intensity on juvenile females}$$
$$H_{jm} = \text{hunting intensity on juvenile males}$$
$$H_{af} = \text{hunting intensity on adult females}$$
$$H_{am} = \text{hunting intensity on adult males.} \quad (8)$$

Hunting intensity is the proportion of individuals in a particular stage-sex class hunted, and the survivorship associated with hunting intensity, H_{ik} is $\exp(-H_{ik})$. We adopt this form because it is common in wildlife management and fisheries (e.g., Clark, 1990; Beverton and Holt, 1993) and because it reflects diminishing returns as hunting efforts increase.

The Population Dynamics and the Growth Rate in the Basic Polygynous Model

We now describe the population dynamics, beginning with adults. Adults of age class a at time t must survive natural mortality and hunting intensity. In addition, adults of age class 1 are represented by the surviving juveniles of class j_{max}. Thus, for $a = 1$

$$A_f(t + 1,1) = \sigma_f(j_{max}) \exp(-H_{jf}) J_f(t,j_{max})$$

$$A_m(t + 1,1) = \sigma_m(j_{max}) \exp(-H_{jm}) J_m(t,j_{max}). \quad (9)$$

Subsequent ($a > 1$) adult age classes are determined by the number of adults surviving from the previous age class, so that for $a > 1$

$$A_f(t+1,a) = \rho_f(a-1)\exp(-H_{af})\,A_f(t,a-1)$$

$$A_m(t+1,a) = \rho_m(a-1)\exp(-H_{am})\,A_m(t,a-1). \qquad (10)$$

Similarly, juveniles of the first IBI class ($j = 1$) are those surviving from the birth class

$$J_f(t+1,1) = s_f B_f(t)$$

$$J_m(t+1,1) = s_m\,B_m(t) \qquad (11)$$

and for $j > 1$

$$J_f(t+1,j) = \sigma_f(j-1)\exp(-H_{jf})\,J_f(t,j-1)$$

$$J_m(t+1,j) = \sigma_m(j-1)\exp(-H_{jm})\,J_m(t,j-1). \qquad (12)$$

If there are no differences in survival among age classes, then eqs. 9–12 simplify to dynamics in terms of the total population.

The computation of the birth age class proceeds in three steps. First, we compute the total number of males and females surviving through the time period:

$$A_M(t) = \sum_{a=1}^{a_{max}} \rho_m(a-1)\exp(-H_{am})A_m(t,a)$$

$$A_F(t) = \sum_{a=1}^{a_{max}} \rho_f(a-1)\exp(-H_{af})A_f(t,a). \qquad (13)$$

Second, we compute the total reproduction, R, according to

$$R = \sum_{a=1}^{a_{max}} m(a)\rho_f(a)\exp(-H_{af})A_f(t,a). \qquad (14)$$

Hence, reproduction is assumed to follow episodes of natural and harvest mortality.

Third, we correct for the effects of the mating system. The fundamental parameters describing the mating system in our model is harem size (see Caswell and Weeks, 1986), loosely defined as the number of females a male can fertilize. The basic model assumes that all females that are in harems reproduce and all adult males can hold harems. Thus, when harem size is h, reproduction will be limited by the smaller of two values: number of females, $A_F(t)$, or the number of females in harems, $hA_M(t)$. That is, all females reproduce if $hA_M(t) > A_F(t)$, and only a fraction, $[hA_m(t)]/[A_F(t)]$, reproduce otherwise. Assuming that neonatal sex ratio is a proportion, r, that is male,

$$B_f(t+1) = (1-r)R\min\left\{1, \frac{hA_M(t)}{A_F(t)}\right\}$$

$$B_m(t+1) = rR\min\left\{1, \frac{hA_M(t)}{A_F(t)}\right\} \qquad (15)$$

Note that

$$R\min\left\{1,\frac{hA_{\mathrm{M}}(t)}{A_{\mathrm{F}}(t)}\right\}=\sum_{a=1}^{a_{\max}}m(a)\rho_{\mathrm{f}}(a)\exp(-H_{\mathrm{af}})A_{\mathrm{f}}(t,a)\min\left\{1,\frac{hA_{\mathrm{M}}(t)}{A_{\mathrm{F}}(t)}\right\}. \tag{16}$$

Thus, if the minimum is 1, eq. 16 is

$$R=\sum_{a=1}^{a_{\max}}m(a)\ \rho_{\mathrm{f}}(a)\exp(-H_{\mathrm{af}})A_{\mathrm{f}}(t,a) \tag{17}$$

which is the standard linear model for cohort analysis. However, the mating system introduces a nonlinearity via eq. 15. Caswell and Weeks (1986) provide a similar method for incorporating both males and females into cohort analysis using a harmonic mean function instead of the minimum function that we use.

At any time, t, the total population size is

$$P(t)=B_{\mathrm{f}}(t)+B_{\mathrm{m}}(t)+\sum_{j=1}^{j_{\max}}J_{\mathrm{f}}(t,j)+J_{\mathrm{m}}(t,j)+\sum_{a=1}^{a_{\max}}A_{\mathrm{f}}(t,a)+A_{\mathrm{m}}(t,a). \tag{18}$$

The growth rate of the population is

$$\lambda(t)=\frac{P(t)}{P(t-1)} \tag{19}$$

which may approach a constant λ as time increases if the population is increasing. Because of the nonlinearity due to the mating system, the growth rate changes over time when the population is decreasing. To account for this, we report the average growth rate of the population after 20 IBIs for up to 300 IBIs.

Equations 9–19 constitute the "basic model." We first present results from the basic model, then results of various modifications: fecundity depends on harem size, infanticide, and reproductive suppression. For each modification, we consider a completely protected population and then three different types of hunting: trophy hunting of adult males (hereafter termed "male hunting"), trophy hunting of adults of both sex ("adult hunting"), and hunting of juveniles and adults of both sex ("subsistence hunting"). In order not to confound adding sex and stage classes with increasing harvest intensity, we kept harvest intensity on the population as a whole constant by proportioning harvest to specific stages based on their size. We introduce hunting selectivities $\theta_{ik}=1$ if a particular sex and age class is hunted and $\theta_{ik}=0$ otherwise, where i = a or j (adult or juvenile) and k = m or f (male or female). Suppose that the total hunting intensity is H. The total hunted population is

$$P_H=J_{\mathrm{F}}(t)\theta_{\mathrm{jf}}+J_{\mathrm{M}}(t)\theta_{\mathrm{jm}}+A_{\mathrm{M}}(t)\theta_{\mathrm{am}}+A_{\mathrm{F}}(t)\theta_{\mathrm{af}} \tag{20}$$

and hunting intensity for each stage-sex class is

$$\begin{aligned}H_{\mathrm{jf}}&=H\,J_{\mathrm{F}}(t)\,\theta_{\mathrm{jf}}/P_H\\H_{\mathrm{jm}}&=H\,J_{\mathrm{M}}(t)\,\theta_{\mathrm{jm}}/P_H\\H_{\mathrm{af}}&=H\,A_{\mathrm{F}}(t)\,\theta_{\mathrm{af}}/P_H\\H_{\mathrm{am}}&=H\,A_{\mathrm{M}}(t)\,\theta_{\mathrm{am}}/P_H\end{aligned} \tag{21}$$

Several assumptions of this model could affect our predictions of the population dynamics. First, the deterministic population approach assumes that values of survivorship and fecundity are constant within a particular age–sex class and that all adult males have an equal probability of obtaining harems independent of age (but see appendix 1). In addition, harem size is invariable, although we show how this assumption can be partially relaxed below. Because birth class individuals are not tied to specific adults, this model assumes no parental care. In addition, this model assumes no selectivity of hunting mortality within particular age–sex classes. For example, harem-holding males are no more likely to be hunted than males lacking harems.

The discrete time approach assumes that reproduction occurs after natural and hunting mortality and that the rate of harvest mortality is in terms of the interbirth interval. In general, harvest mortality is on a per annum basis, while a particular species may reproduce more or less than once per year. This difference can be corrected by modifying harvest mortality accordingly; hence, if a given species reproduces four times per year, its harvest rate should divided by four to match its reproductive rate.

In the basic model, we assume that sex ratio at birth is 50:50 and $j_{max} = 2$, $a_{max} = 10$. For simplicity of presentation, we also assume that survivorship does not vary for different juvenile or adult IBI classes and that adult fecundity remains constant across different adult age classes. Thus, in the absence of hunting mortality, we still must specify the birth, juvenile, and adult sex-dependent survival and fecundity. For most mammals, males have lower survivorship than females (Clutton-Brock, 1988). We take this into account for juvenile and adult age classes, such that $s_f = s_m = 0.7$, $\sigma_f(j) = 0.8$, $\sigma_m(j) = 0.5$ (for all $j = 1$ to j_{max}), $\rho_f(a) = 0.8$, and $\rho_m(a) = 0.7$ (for all $a = 1$ to a_{max}). We assigned $m(a) = 3$ for all models except reproductive suppression, in which case $m(a) = 10$. All models detailed below were simulated with all cohorts of each sex having 100 individuals at $t = 1$.

Model Results

The Basic Model

In the absence of hunting, the population growth rate for the basic model increases for small harem sizes and levels off at $h = 5$ (fig. 11-1). The lower population growth rates for small harem sizes are the result of the lower survivorship of males. This is due to an Allee effect: there are not enough males to fertilize females when harem size is less than five and male survivorship is low. If survivorship were the same for both sexes, population growth would be equal for all harem sizes (resulting in a straight line). Although the Allee effect controls reduced growth rates for populations having small harems in our model, other mechanisms including absence of necessary paternal care could account for this pattern in reality.

For all methods of hunting ($H = 0.15$), small harem sizes (less than six females) have lower growth rates than large harem sizes, again a reflection of the fact that lower male survivorship results in some females not reproducing when harem size is small. The impact of hunting is particularly evident for monogamous species ($h = 1$), which, as shown in fig. 11-1, decline even in the absence of hunting. Male hunting is especially detrimental to monogamous species because fertilizations are equally limited by males and females. However, male hunting exerts less of an impact on large harem sizes; indeed, it approaches levels of no hunting. In contrast, adult hunting and subsistence hunting slightly reduce growth rates

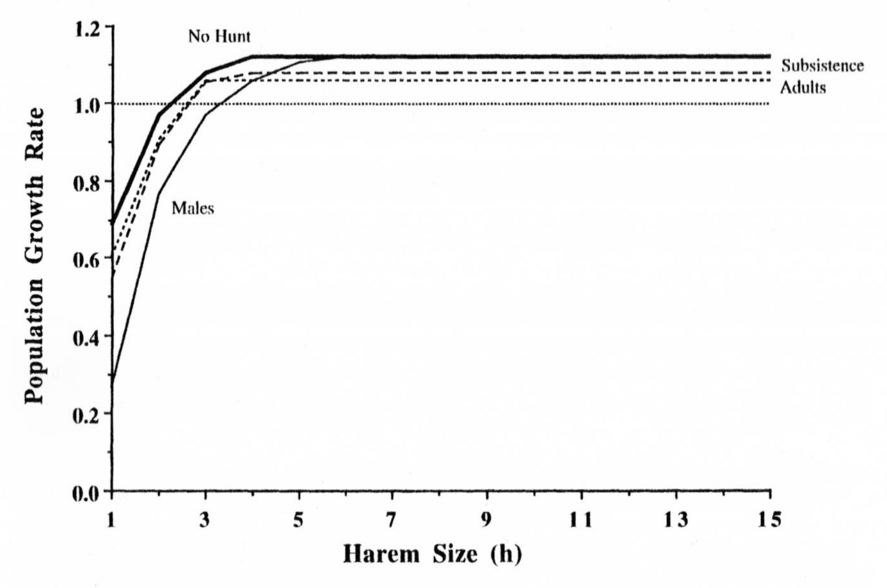

Figure 11-1 Population growth rate as a function of harem size using the basic model when hunting intensity (H) is either zero (no hunt) or 0.15 (all other lines). We consider three different types of hunting: adult males (male), adults of either sex (adult) and juveniles and adults of either sex (subsistence). The dashed horizontal line indicates the boundary between increasing and decreasing population change.

for large harem sizes. Differences between hunting methods are magnified for smaller harem sizes and reduced at larger harem sizes (table 11-2).

When harem size is constant at a moderate size ($h = 5$), male hunting is more sustainable: it has replacement population growth rates over a wider range of hunting intensities than adult hunting and is only slightly less sustainable than subsistence hunting (fig. 11-2). This results primarily because females are less affected when only males are hunted. Note that at this harem size, growth rates are near the maximum for any harem size (compare with fig. 11-1); at smaller harem sizes, male hunting is less sustainable because hunting effort is spread over fewer classes of individuals and because not enough males survive to fertilize all available females for a larger number of harem sizes.

These results have several implications. First, population sustainability depends on the particular breeding strategy and the hunting effort on the population. In particular, while highly polygynous species are relatively unaffected by male hunting, monogamous species are much more susceptible to such hunting. Furthermore, if natural survivorship is lower for males than for females, male hunting will reduce growth rates of monogamous species more than polygynous species. These generalizations depend to some extent upon sex- and age-specific survivorship patterns, sex-specific j_{max} and a_{max}, and age-specific fecundity. For example, results in fig. 11-1 show that monogamous species cannot have positive population growth rates, but if the fecundity and survivorship schedules had higher values, λ could exceed 1.0.

The basic model shows that while different types and intensities of hunting result in different reductions in population growth, harem size affects population growth only for low harem sizes. Figure 11-1 suggests that harem size may change in response to increased mor-

Table 11-2 Effects of different aspects of breeding system and harem size on populations' responses to three types of hunting, shown in order of the greatest population impact to the least impact.

Breeding system	Harem size[a]	Negative effects
Basic polygyny	Small	**Male**>Adult>Subsistence
	Large	Adult>Subsistence>Male
Fecundity depends on harem size	Small	**Male**>Adult>Subsistence
	Large	**Adult**>Subsistence>Male
Infanticide	Small	**Male**>Adult>Subsistence
	Large	**Adult**>Subsistence>Male
Reproductive suppression	Small	Male>Adult>Subsistence
	Large	**Adult**>Subsistence>Male

A hunting method is in boldface when the breeding system is particularly sensitive to it.

[a]Small = < 4 or 5 females, large = > 4 or 5 females.

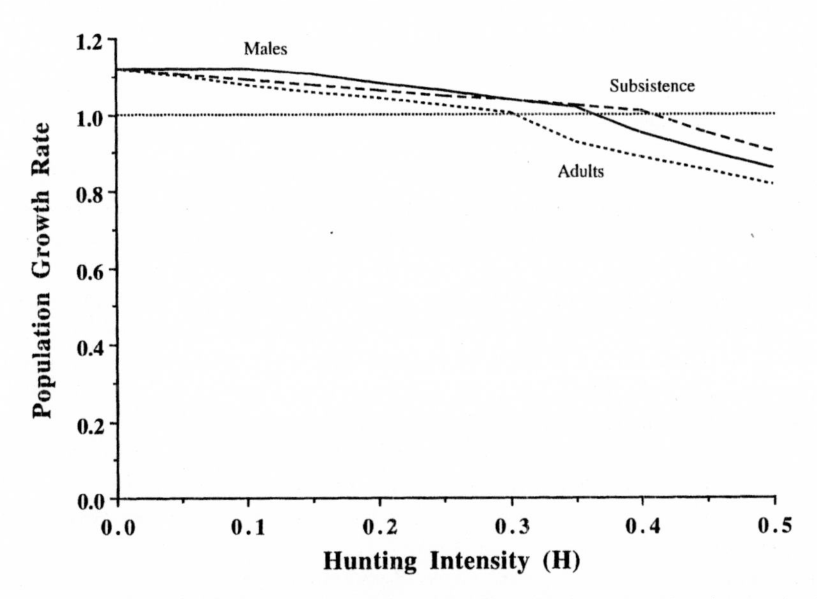

Figure 11-2 Population growth rate as a function of hunting intensity when harem size is held constant at five females and adult males (males), adults of either sex (adults), and juveniles and adults of either sex (subsistence) are hunted. The dashed horizontal line indicates the boundary between increasing and decreasing population change. Demographic parameters are identical to those used for fig. 11-1.

tality due to the fact that larger harems have higher growth rates. This change depends largely on the variability of the mating system. For the sake of generality, we assume that harem size remains stable, but we recognize that intraspecific variation may result in changes of breeding system (and hence harem size) in response to demographic changes. We now consider variations of the basic model: fecundity varies as a function of harem size, infanticide by males, or reproductive suppression where only one male and one female in a group breed.

Fecundity Depends on Harem Size

Fecundity may decrease in larger harems as females compete for resources or partition effort among other related individuals in the group (Downhower and Armitage, 1971; Armitage, 1986). Declining relationships between fecundity and group size have been noted in a number of species, including marmots *Marmota flaviventris* (Downhower and Armitage, 1971), wolves *Canis lupus* during prey shortages (Harrington et al., 1983), red deer *Cervus elaphus* (Clutton-Brock et al., 1982), and many primates (Van Schaik, 1987). In these and other species for which fecundity varies inversely with harem size, this relationship is likely a case of density dependence. For the purposes of modeling, however, we take a frequency-dependent approach. For species in which fecundity decreases with harem size,

$$\text{fecundity at age } a = m(a)\left(1 - \frac{h}{h_{\max}}\right) \tag{22}$$

where $h_{\max}$ is the harem size at which no females are reproductive ($h \leq h_{\max}$). We use this modification in eq. 15. Demographic parameters are identical to those used in the basic model.

When fecundity declines with harem size (fig. 11-3), the population growth rate is lower than for the basic model, especially at large harem sizes, because of the density dependence in fecundity associated with large harem sizes. In this case, a window of population growth exists, bounded by the lowest harem size at which males can fertilize all females and the largest harem size in which females cease to be productive. As in the frequency-independent case, monogamous species or populations are more sensitive to male mortality than polygynous ones, and if natural mortality is higher for males than for females, a proportionally lower growth rate will exist relative to polygyny. In this case, low degrees of polygyny are most favorable for population growth.

Although the same dome-shaped pattern of population growth rate across harem sizes exists for all hunting methods, the harem size at which hunting is most sustainable varies. For a particular hunting intensity, male hunting is again less sustainable at low harem sizes because many females remain unfertilized, but population growth rates approach natural mortality at high harem sizes. Adult hunting and subsistence hunting become sustainable at lower harem sizes because more females are fertilized, but the existence of fewer females reproducing in the population results in both hunting methods never approaching natural levels, in contrast to male hunting. At a moderate harem size ($h = 5$), subsistence hunting is more sustainable across different hunting intensities than male hunting and adult hunting (fig. 11-4) because the hunting mortality is spread over individuals.

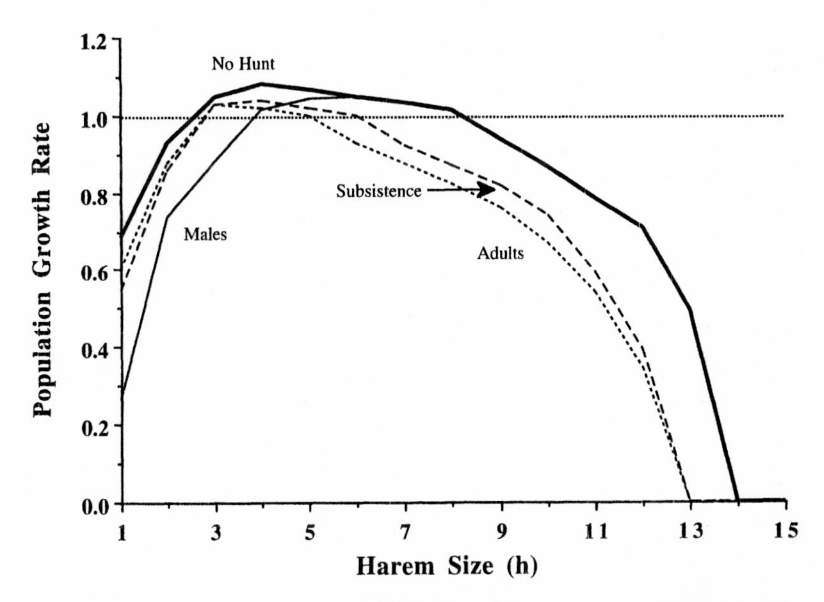

Figure 11-3 Population growth rate as a function of harem size when fecundity varies inversely with harem size and males, male and female adults, or all adults and juveniles (subsistence) are hunted ($H = 0.15$) or when there is no hunting ($H = 0$). The dashed horizontal line indicates the boundary between increasing and decreasing population change.

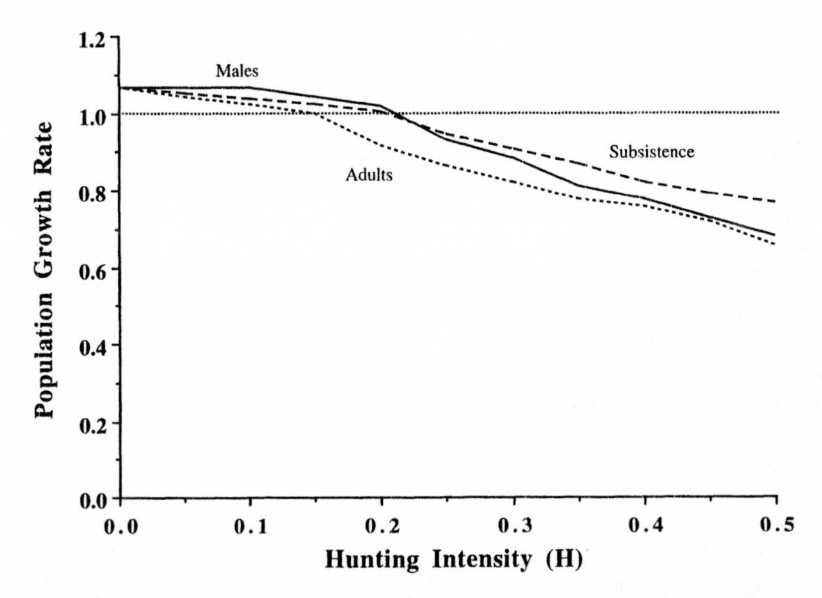

Figure 11-4 Population growth rate as a function of hunting intensity when fecundity varies inversely with harem size and harem size is held constant at five females and there are three different types of hunting. The dashed horizontal line indicates the boundary between increasing and decreasing population change.

Infanticide

In some infanticidal species, the death of a male controlling a harem leads to reproductive failure for that harem because the incoming male kills some or all the infants of the harem (Hausfater and Hrdy, 1984; Parmigiani et al., 1992). Because of this, female reproduction is tied to male survivorship, which is

$$S_M(t) = \frac{\sum_{a=1}^{a_{max}} p_m(a)\exp(-H_{am})A_m(t,a)}{A_M(t)}. \tag{23}$$

In this case, we find it easier to separate the analysis into two cases. First, if $A_F(t) > hA_M(t)$, then there is an "excess" of females, in the sense that all males hold harems. In such a case (e.g., when male mortality is extremely high), only a fraction $[hA_M(t)/A_F(t)]$ of the females are reproductive and there is no infanticide, so number of birth class individuals is

$$B_F(t+1) = (1-r)R\frac{hA_M(t)}{A_F(t)}$$

$$B_M(t+1) = rR\frac{hA_M(t)}{A_F(t)}. \tag{24a}$$

However, if $A_F(t) < hA_M(t)$, all females are reproductively active, but the death of a male leads to the loss of the reproduction of all females in that harem, so that

$$B_f(t+1) = (1-r)R\,S_M(t)$$

$$B_m(t+1) = r\,R\,S_M(t). \tag{24b}$$

We assume that survival does not differ within juvenile or adult age classes and demographic parameters are identical to those used for the basic model. In contrast to other models of infanticide (e.g., Starfield et al., 1981; Starfield and Bleloch, 1991), our model assumes that harem holders and nomads have no differences in either natural or hunting mortality.

The effects of harem size on population growth in an infanticidal species are qualitatively similar to those for the basic model. Without a harvest, larger harem sizes generally have higher growth rates, leveling off at four females (fig. 11-5). When harem size is low, population growth is less than two, due mostly to males limiting reproduction. When harem size is greater than two, the population is increasing, and increases in harem size have little effect on population growth. Note that unlike the basic model, the maximum population growth rate occurs at $h = 3$ and levels off at a lower population growth rate thereafter, a result of the change in the birth equation from eq. 24a to 24b.

The effects of hunting on the infanticidal species are qualitatively different from the basic model. Harvests of females and juveniles still reduce population growth as in the basic model, but hunting only males reduces growth rate over all large harem sizes. These patterns contrast with the basic model (fig. 11-1), which shows that at large harem sizes, populations are very resilient to hunting of males. At $h = 5$, hunting males can cause the population growth rate to be greater than in an unhunted population, due to the fact that with moderate hunting of males, there are no extra males and consequently, no infanticide. When harem size is fixed ($h = 5$) and hunting intensity is increased (fig. 11-6), harvest of all adults

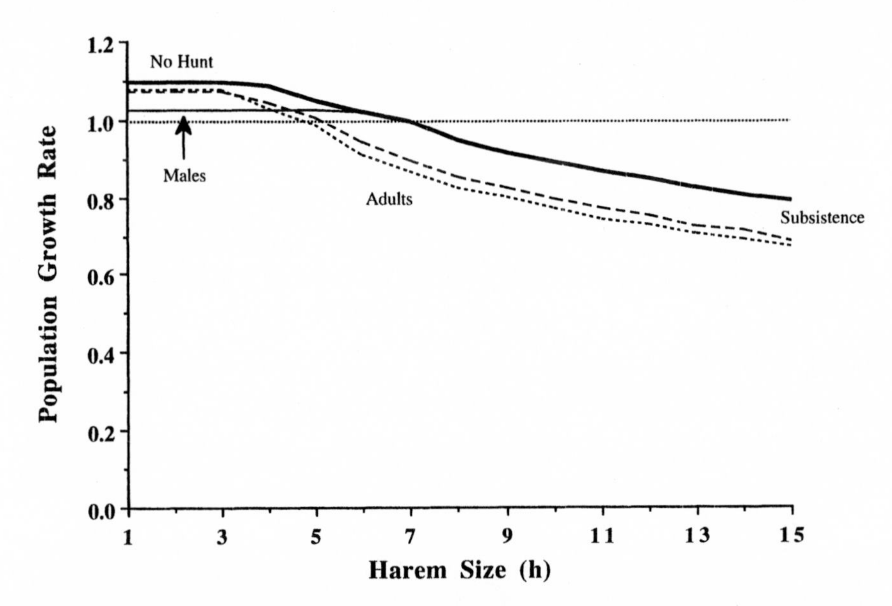

Figure 11-5 Population growth rate of an infanticidal species as a function of harem size when hunting intensity (H) is either zero (no hunt) or 0.15 (all other lines). We consider three different types of hunting: adult males (male), adults of either sex (adult) and juveniles and adults of either sex (subsistence). The dashed horizontal line indicates the boundary between increasing and decreasing population change.

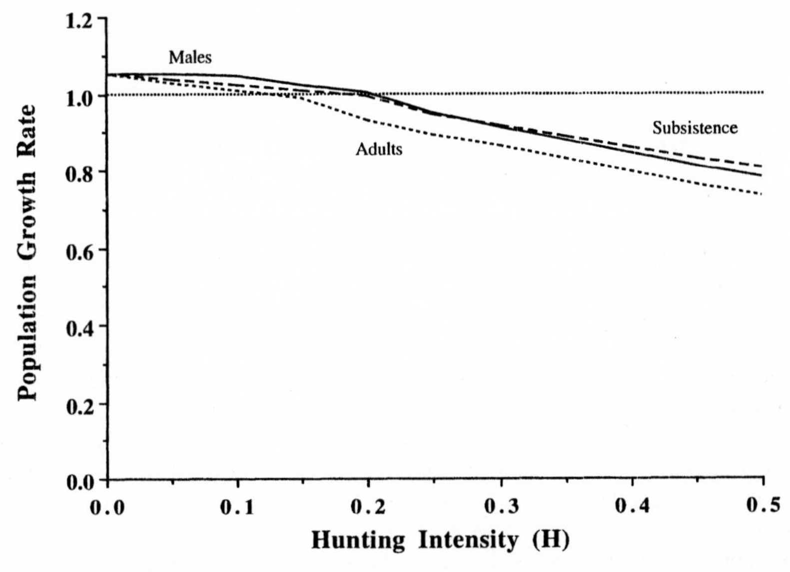

Figure 11-6 Population growth rate in an infanticidal species as a function of hunting intensity (H). We consider three different types of hunting: adult males (male), adults of either sex (adult) and juveniles and adults of either sex (subsistence). Harem size is constant at five females. The dashed horizontal line indicates the boundary between increasing and decreasing population change.

is the less sustainable than other hunting methods because adult hunting results in populations declining over a wider range of hunting intensity values. This is true because (1) females control fecundity by bearing young and (2) the maximum population growth rate occurs when $A_F(t) > hA_M(t)$ (i.e., there is no infanticide), which is less likely to occur when females are hunted in addition to males. Subsistence hunting is less sustainable than male hunting for similar reasons, although the proportionally lower hunting mortality of females causes subsistence hunting to be more sustainable than hunting of both sexes.

The population dynamics of infanticidal species are changed if there are male coalitions (as in lions) or if takeovers occur without the death of a harem-holding males. These additional factors are discussed in appendix 2.

Reproductive Suppression

In the case of reproductive suppression, only one female in the harem is reproductively active. In such a case, each harem receives only $1/h$ of the potential reproduction. Consequently, eq. 15 becomes

$$B_f(t) = (1-r)R\frac{1}{h}\min\left\{1, \frac{hA_M(t)}{A_F(t)}\right\}$$

$$B_m(t) = rR\frac{1}{h}\min\left\{1, \frac{hA_M(t)}{A_F(t)}\right\}. \tag{25}$$

Note that unlike the basic model, we divide the number of females by harem size to determine the level of reproductive output. We can combine the mating system components into

$$\frac{1}{h}\min\left\{1, \frac{hA_M(t)}{A_F(t)}\right\} = \min\left\{\frac{1}{h}, \frac{A_M(t)}{A_F(t)}\right\} \tag{26}$$

Thus, for moderately sized harems, the minimum in eq. 26 is almost certainly going to be $1/h$ unless the survivorship of males is very low. This has important consequences for the effects of harem size on population growth rate and for the effect of different types of harvesting on population growth rate.

In the absence of hunting, the population growth rate decreases monotonically as a function of harem size, falling below one for harems of seven or larger (fig. 11-7). In essence, the effect of reproductive suppression in such large harems is to reduce the number of potentially reproductive females to $1/h$ of the actual number (Caughley, 1994). Moderate hunting ($H = 0.15$) of adult males reduces the population growth rate when harems are small but has little effect for larger harems. On the other hand, including females in the hunt can cause the population growth rate to drop below one at a smaller harem size than if only males were hunted. The pattern for subsistence hunting is virtually indistinguishable from the pattern when all adults are taken.

When harem size is fixed (fig. 8-11), the population growth rate is essentially constant for moderate levels of hunting intensity on only males because it is the female population size that determines the growth rate. However, if females are included in the hunt, then the growth rate drops below one for modest levels of hunting. The implication is that if the harvest is guaranteed to be male only, then the population can sustain a relatively heavy har-

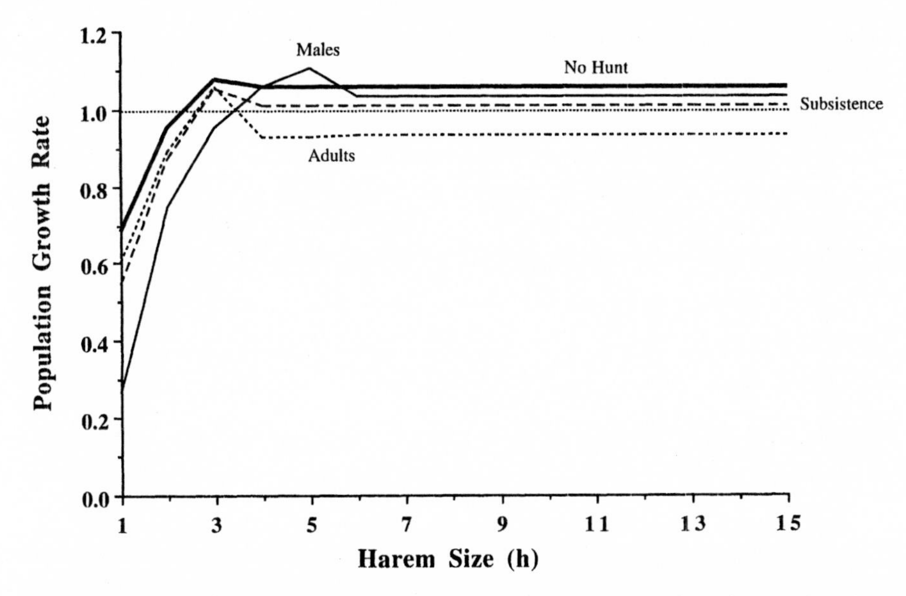

Figure 11-7 Population growth rate of a reproductively suppressed species as a function of harem size when hunting intensity is zero (no hunt) or 0.15. Demographic parameters are identical to those used for fig. 11-1, except that $m(a) = 10$. We consider three different types of hunting: adult males (male), adults of either sex (adult) and juveniles and adults of either sex (subsistence). The dashed horizontal line indicates the boundary between increasing and decreasing population change.

vest. However, if it is impossible to separate males and females in the hunt, then the harvest level must be much lower. For the parameters used to obtain the results shown in fig. 11-8, the harvest on adult males and females must approximate 0.12 to sustain a growth rate > 1. If only adult males were taken, the harvest could be as high as 0.2.

Summary

These models indicate that several breeding system attributes influence the magnitude of a population's response to hunting pressure (table 11-2). The response of the population to hunting pressure may vary greatly depending on the degree of polygyny and the extent of particular life-history strategies (e.g., reproductive suppression). Monogamous and weakly polygynous species are much more susceptible to male hunting than species characterized by large harems. Reduction of fecundity in larger harems results in less rapid population increases than weakly polygynous populations, although patterns for different hunting methods follow the basic model. Infanticide and reproductive suppression reduce the population's ability to withstand hunting, in the first case because loss of harem-holding males results in their replacement by infanticidal males, and in the second because groups containing many nonbreeders are effectively similar to monogamous situations, which are themselves sensitive to offtake. Depending on harem size, hunting only males or adults of both sexes pose particular problems for infanticidal species with small and large harems, respectively. Even at large harem sizes, hunting adult males still reduces population growth

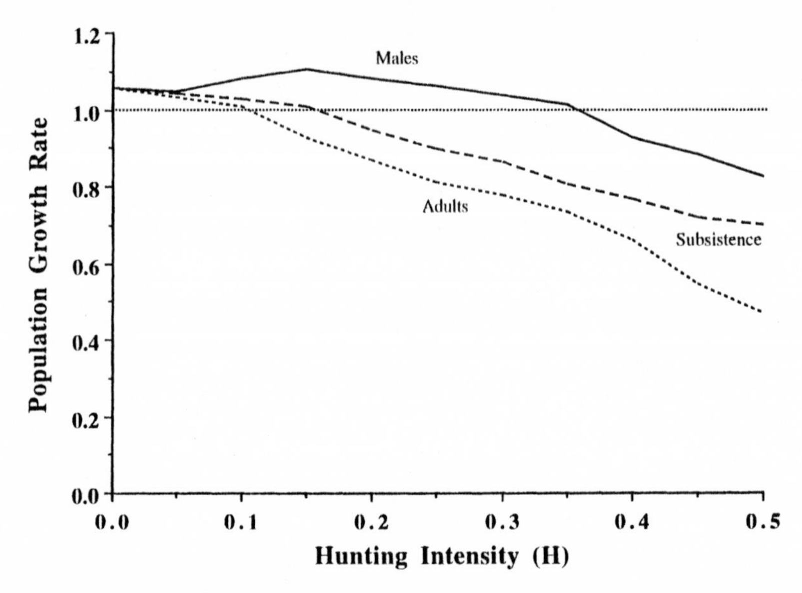

Figure 11-8 Population growth rate of a reproductively suppressed species as a function of hunting intensity (H) when harem size is five females. We consider three different types of hunting: adult males (male), adults of either sex (adult) and juveniles and adults of either sex (subsistence). The dashed horizontal line indicates the boundary between increasing and decreasing population change. Demographic parameters are identical to those used for fig. 11-1, except that $m(a) = 10$.

compared to species lacking infanticide. Hunting adults of both sexes also causes rapid population declines for reproductively suppressed species, but for different reasons: increasing hunting intensity on females may prevent an entire group from breeding if the breeding female is killed.

Applications

We now apply the models to six species of hunted African mammals that have reasonably well-documented demographic parameters and whose breeding systems vary. Although not all these species are regularly hunted (e.g., mongooses and hyenas), use of these species shows how modifications of the basic model can be used for particular animal populations.

Ungulates

Ungulates have an enormous diversity of breeding systems ranging from monogamy in the smaller species (Kleiman, 1977) to polygyny and lekking (Clutton-Brock, 1989; Clutton-Brock et al., 1993). We focus on three species, all of which are hunted only for males, but each having a different breeding system: impala *Aepyceros melampus,* buffalo *Syncerus caffer,* and dikdik *Madoqua kirki.*

Impala are characterized by resource-defense polygyny (Jarman, 1979) in which females form groups that wander between territories defended by dominant males. These males mate with an average of 15 females per year (Murray, 1982), whereas many bachelors obtain no mates. We calculated a positive growth rate ($\lambda > 1.00$) based on this mating system, litter size of one young per female, and sex- and age-specific survivorship from Murray (1982) with longevity of females lengthened to 15 years (Skinner, 1989). We note that Murray's life-history data, in which female longevity is 10 years, do not produce a positive growth rate and that growth rates in general are extremely sensitive to this parameter.

Given life-history traits that result in a positive growth rate, impala populations could still grow even when adult males were completely hunted because of replacement by juvenile males capable of holding large harems. However, any inclusion of density dependence could make these very resilient populations sensitive to parameter changes. For example, if impala fecundity declines as a function of harem size, populations with large harem sizes generally decline. In addition, our model assumes that newly recruited adult males are equally capable of holding large harems as older males, an assumption that may well be false for polygynous species (Clutton-Brock et al., 1982). This assumption can be relaxed by introducing a parameter into eq. 15 capturing the lack in experience, $E(a)$, of newly recruited males relative to older males (see appendix 1). When this modification is used and hunting mortality, H, is set at 1.0 for all older adult males (all adult males hunted), growth rates become negative at $E(1) = 0.72$. In other words, if newly recruited males could achieve $> 72\%$ the number of matings older males achieve, the population would still increase. Given that new recruits are often much less capable of fertilizing females than are older males (e.g., Poole, 1989), the apparent resilience of the impala population to hunting in our model is misleading.

African buffalo form large mixed-sex herds (Sinclair, 1977; Mloszewski, 1983; Prins, 1996). Dominant males constitute about 10% of the herd and obtain most of the matings (Sinclair, 1977), with an average harem size of 4.4 per male (calculated from Mloszewski, 1983). Based on age- and sex-specific survivorship and average fecundity of 1.0 offspring per adult female per IBI estimated by Sinclair (1977), our population model predicted slightly growing populations ($\lambda = 1.07$) that were resilient to high levels of hunting effort because of high survivorship, the large number of adult age classes, and polygyny. Again, inclusion of density dependence in fecundity or lack of experience of newly recruited males can cause population declines.

Dikdik are small, territorial antelopes that form monogamous pair bonds (Komers, 1996). As shown earlier, monogamous species appear to be at a distinct population disadvantage when males are hunted; therefore, dikdik and other antelope such as duikers (*Cephalophus* spp., *Sylvicapra* spp.), suni *Neotragus moschatus,* and klipspringer *Oreotragus oreotragus* represent an especially interesting set of species to examine. Unfortunately, population data such as age- and sex-specific survivorship are extremely scanty for all these species. Hence, we used an amalgamation of dikdik and klipspringer demographic parameters to estimate population growth rates of dikdik. Dikdik females give birth twice per year (Komers, 1996), and we estimated that offspring have a survivorship of 0.55 using data within the range reported for several klipspringer populations (Dunbar, 1990). Tilson and Tilson (1986) observed that 85% and 69% adult male and female dikdiks, respectively, disappeared from territories over 25 months. As these data do not indicate which age classes experienced this mortality, we assumed this was a population average for adults. Calculations of population growth rates are further complicated because the number of juvenile and adult age classes and juvenile survivorship are unknown. We constructed age-specific survivorship by assuming no juvenile age classes, that adults lived up to 6 years, and that adult age-specific survivorship was high for both sexes over the first 2 years of life ($\rho_f = \rho_m = 0.9$) and correspondingly lower at the remaining four age classes ($\rho_f = 0.36$ and $\rho_m = 0.16$). When the population age structure is taken into account, the average adult survivorship for the entire population closely matches the values reported by Tilson and Tilson (1986). These designations resulted in a slightly larger than replacement population growth rate ($\lambda = 1.01$).

Spotted Hyenas

Spotted hyenas *Crocuta crocuta* group in large clans of multiple males and females, with a few dominant males mating with the females (Frank et al., 1995). Even though the number of males per group is high, the harem size per male is still rather large (8–12) because most males in the group do not mate (Frank, 1986). Female spotted hyenas have high levels of androgens compared to other mammals, and as a result are dominant over most males and have masculinized genitalia (Frank, 1986). Because of these characteristics, it is difficult to distinguish male and female hyenas in the field, and as a result, individuals of both sexes are often shot.

As in most carnivore studies, collecting demographic data for both sexes has been difficult because male dispersal cannot usually be distinguished from death. Kruuk (1972) and Hofer and East (1995) nevertheless suggest a relatively constant mortality rate of 13% and 15% for males and female adults, respectively. We combined these estimates with known infant siblicide of 30% for the common litter size of two (Frank et al., 1995) to construct a life table for both sexes. Using these data, we estimate population growth rate of 1.03.

Lions

We adjusted the model with infanticide to fit the form of the data for lions in Packer et al. (1988). In particular, published data on fecundity and mortality of cubs generally include the effects of infanticide that occur in the population. Therefore, we could only model the additional infanticide that would occur with hunting of a population. We modified eq. 23 to reflect additional infanticide caused by hunting by multiplying total reproduction by $\exp(-H_{am})$, which reflects the change in survivorship as a result of hunting. We found that the population has a positive growth rate with no hunting ($\lambda = 1.10$) and that growth rate declines but remains 1.0 up to a harvest mortality of about 19% of the male population. This is equivalent to approximately 6% of the entire population due to the female-biased adult sex ratio. These models assume there is no population response to harvesting other than total changes in numbers of individuals. It is nevertheless known that lions produce male-biased sex ratios in larger litters (Packer and Pusey, 1987), and recent data indicate that populations of lions hunted for males produce an excess of male cubs (Creel and Creel, 1997). Incorporating this response into the model would substantially reduce the number of male lions that could be sustainably harvested (from 19% down to 12% with a natal sex ratio of 0.55) because the population would quickly become limited by females. Furthermore, male lions usually control harems in coalitions of individuals (Packer et al., 1988), and sustainable offtake is much reduced when this factor is incorporated in the model (appendix 2).

Dwarf Mongooses

Dwarf mongooses show strong reproductive suppression (Rood, 1980, 1990). We used data in Waser et al. (1995) in the model of reproductive suppression but made two changes. First, we assumed that total reproduction is determined by the number of females at the start of interval t, rather than the number of surviving females. Hence, we modified eq. 14 to

$$R = \sum_{a=1}^{a_{max}} m(a) \exp(-H_{af}) A_f(t,a). \tag{27}$$

Second, in order to use the empirical values for $m(a)$ (Waser et al., 1995), we recognize that the reported fecundities include the effect of reproductive suppression in harems. Consequently, we used the birth class equations of the basic model (eq. 15) instead of eq. 25. We assume that the harvest includes adults and juveniles of both sex. For our analyses, we assumed that only dominant individuals breed. Although both subordinate males and females occasionally breed (Keane et al., 1994), relaxing the above assumption (see Creel, chapter 10, this volume) effectively changes harem size from nine to eight females and hence has little effect on population growth rates. Our result is that $\lambda > 1.00$ as long as $H < 0.1$.

Hunted Populations in the Selous Game Reserve

We used the data on population growth rates to calculate rates of sustainable offtake for the six species described above in the Selous Game Reserve in Tanzania. Table 11-3 shows estimated population sizes for each species in the 41,245-km² area of the Selous Game Reserve, over which species were counted or estimated. Table 11-4 gives annual hunting levels and the percentage of the population that is legally hunted by tourists.

Ungulates were resilient to tourist hunting levels in the Selous Game Reserve, with the particular exception of monogamous antelope. Buffalo and impala have large harems and a relatively low hunting intensity in the Selous which prevent tourist hunting from having a significant impact. Current hunting effort (i.e., number of animals taken) is well below that which their population could sustain assuming no additional offtake through illegal

Table 11-3 Estimated population size of species in the Selous Game Reserve.

Species	Mating system	Estimated total adult population size	Estimated hunted population size[a]	Reference for population size
Impala	Highly polygynous	32,287	10,762[b]	Caro et al (in press a)
Buffalo	Mildy polygynous	69,219	23,073[b]	Caro et al (in press a)
Small antelope	Monogamous	436[c]	218[b,c]	Caro et al (in press a)
Spotted hyena	Highly polygynous	13,198	13,198	Creel and Creel (1996)
Lion	Polygynandrous and infanticide	4,537	1,747[b]	Creel and Creel (1996)
Dwarf mongoose	Monogamous and reproductive suppression	138,171	138,171	S. Creel (personal communication)

[a]Based on published estimates of adult sex ratios.

[b]As only adult males are hunted, these numbers represent the estimated number of adult males in the reserve.

[c]Estimated population size for all small antelopes (dikdik, duikers, oribi, klipspringer, and steenbok). Because these cryptic species are difficult to count in aerial censuses, the population size is certainly a great underestimate.

Table 11-4 Hunting parameters of species hunted in the Selous Game Reserve.

Species	Stage and sex hunted	Average offtake/year[a]	Hunting Mortality (%)[b]	Will legal offtake cause decline?
Impala	Adult males	118	1.10	No
Buffalo	Adult males	163	0.71	No
Small antelope	Adult males	46	21.10	Yes
Spotted hyena	Adult males and females	17	0.13	No
Lion	Adult males	42	2.40	No
Mongoose	Adult males and females	10[c]	0.01	No

[a]Derived from individuals legally shot between 1988 and 1992, from Caro et al. (in press b).
[b]Mortality is expressed as percent of stage and sex hunted per year.
[c]Estimated, based on information from S. Creel.

hunting. Small antelope populations were greatly affected by even slight changes in hunting effort; hunting effort of approximately 21% (that occurring in the Selous; see table 11-4) reduced population growth rates to < 1.0 ($\lambda = 0.52$). In fact, mortality rates > 1% cause populations to drop below replacement growth. Although these results may stem from lack of accurate demographic data and from seriously underestimating population sizes of small antelopes during aerial censuses, it is clear from the models that monogamous species are particularly sensitive to male offtake.

All three carnivore species modeled maintained positive growth rates under current Selous hunting levels. For spotted hyenas, our calculations indicate that hunting both sexes at hunting effort > 1% of the adult population will cause the population to decline. Thus, our model suggests that spotted hyenas are currently hunted at very conservative levels in the Selous. Dwarf mongooses live at high densities and are rarely hunted in the Selous. These two factors mean that offtake is far below what the population could sustain.

Lions live at relatively low densities compared to ungulates and are subject to reasonably strong hunting pressure. Nevertheless, our model suggests they can withstand current levels of hunting. These results run contrary to earlier findings which suggested that lions were being hunted at high levels throughout Tanzania (Caro et al., in press b) and that quotas in the northern part of the Selous were set too high (Creel and Creel, 1997). Modeling efforts by Starfield and Bleloch (1991), Starfield et al. (1981), and Venter and Hopkins (1988) predict that hunting harem-holding males increased infanticide and led to much lower population growth rates. Our model does not differentiate between harem-holding and bachelor males, and therefore harvest is proportioned to them according to their relative population sizes. If harem-holding males are actually more easily detected by hunters than nomads due to their association with prides, they may suffer a disproportionate amount of hunting, thereby reducing population growth. However, common hunting practices in which males are shot at bait stations may actually take harem holders and bachelors in proportion to their availability as assumed by our model.

In summary, our models suggest that these animals are being removed conservatively and therefore, from a conservation standpoint, are being hunted appropriately. The important exception is the hunting of small antelope, which may be sustainably hunted only if population estimates (table 11-3) are too small by a factor of at least 20. Our analyses are at a large scale because data on ungulate populations sizes were available only for the whole reserve, which forced us to combine hunting levels in different areas into one figure. At a small scale, it is known that hunting pressures differ according to region in the game reserve, with the eastern and northern section having a long history of sustained hunting pressure. Although our analyses examined only six species of mammals, it appears that tourist hunting in the Selous is an effective conservation tool because it generates substantial revenue for the Tanzanian government and occurs at levels that appear sustainable despite the different forms of hunting and variability of breeding systems in hunted species. We hesitate to recommend large increases in hunting intensity that our models suggest are possible for reasons that we discuss below.

Discussion: Strengths and Weaknesses of the Models

In this chapter we used a modeling approach to examine the impact of different types of hunting on animal populations and to investigate how breeding system modifies a population's ability to withstand hunting pressure. Models force one to formalize the logic of the relationships between the given parameters. Doing so may lead to unanticipated conclusions that are relatively straightforward after the fact. For example, the susceptibility of monogamy to male trophy hunting is a result that has long been overlooked by managers. The models we present are extremely flexible in that they are amenable to the numerous behavioral and life-history variations that complicate the population dynamics of real organisms. Other complications, such as environmental stochasticity and density dependence, could be incorporated to address the particular system that a researcher or manager is interested in modeling. Because the aim was to highlight the generality and flexibility of our approach, we did not explore the depth to which our model can be applied to population dynamics, but we encourage others to test its limits.

The problem of predicting natural population changes has been a continual challenge for biologists. These problems are further complicated by human-induced effects such as hunting. We argue that single-sex models do not characterize population dynamics very well because they do not take into account population-level effects of breeding systems and because current hunting practices may be targeted at different age–sex classes, and often those that are not incorporated into single-sex models. For example, a classical life-table model of lions not incorporating male hunting would be particularly egregious because it ignores consequences of the removal of harem-holding males (leading to infanticide) and because it ignores the number of available breeding males. While our model predicts that in such circumstances approximately 19% of adult males can be sustainably hunted, a naive life-table model would predict a much larger offtake without serious consequences. The need to examine population changes in terms of the dynamics of both sexes has been suggested by others (Beddington, 1974; Beddington and May, 1980; May and Beddington, 1980; Starfield et al., 1981; Starfield and Bleloch, 1991). These researchers have shown that population growth rates can be affected by changes in male and female availability and hence the operant sex ratio. In addition to effects caused by changes in sex ratio, our results reveal that the specific breeding system can play a critical role in population growth rates.

Our first result is that hunter selectivity at a given offtake level can be important. In particular, except for the case of reproductive suppression, we found that as hunting intensity increases, offtake of males or all adults causes a population to decline earlier than when offtake is spread over all juvenile and adult age and sex classes. Second, the models show that different types of hunting may have contrasting effects, depending on aspects of the breeding system. Monogamous and weakly polygynous species were more susceptible to hunting of males than strongly polygynous species, and species exhibiting reduced fecundity at large harem sizes were more affected by hunting adults of both sexes than species for which this characteristic does not apply. Infanticidal species and reproductively suppressed species with moderate to large harem sizes were radically impacted by adult hunting, but could sustain higher intensities of hunting if hunters were able to select only males. These results suggests that other breeding systems not modeled here such as polyandry, polygynandry, protandry, and protogyny (see Vincent and Sadovy, chapter 9, this volume) need to be explored in a similar fashion. In particular, the effects of mate choice and alternative mating strategies on population growth rates in the face of hunting pressure need investigation. For example, if females are particularly choosy, populations may decline faster because the effective population size of males is much smaller than the actual size (see appendix 1).

Because our models lack certain complications, their application may not be appropriate for all conservation situations, and the models may require modification. First, density dependence can be important in exploited populations (Milner-Gulland, 1994; Milner-Gulland et al., 1995). In Tanzania, for example, population sizes of some mammals hunted in game reserves are high, matching those found in national parks (Caro et al., in press a). Density dependence could be added to the model if needed. For example, one version of our own models incorporating harem size dependence of fecundity involved a simple change of fecundity from an absolute value to a function of harem size. Although this change is technically a frequency-dependent change (because number of females in a harem, not female density, varies), true density-dependent modifications could be similarly added. In addition to fecundity, density dependence has been hypothesized to increase juvenile mortality, adult mortality, and even interbirth interval, and hence could have important population-level effects, especially when human harvest is considered (Beddington and Basson, 1994). However, density dependence should be tailored based on the specifics of the particular species; while inclusion of density dependence may make a model more realistic, it also makes it highly system specific.

Second, individual condition, which is averaged out in population-level models like ours, may have a large impact on population dynamics. For example, wounded individuals in hunted populations may survive but subsequently fail to breed. Other individual-based characteristics, such as alternative mating strategies, may be difficult to incorporate in our models (but see appendix 1 for an example of how age can be correlated with male mating success).

Third, the spatial structure of populations may be important. Many of the species favored by tourist and subsistence hunters live in groups and more than one individual is killed when a group is encountered, especially in the case of subsistence hunting. Selective removal of certain types of groups (bachelor males or harems) will affect a population's response to offtake. For example, female elephants in herds find it difficult to locate solitary roving males in heavily poached areas (Dobson and Poole, chapter 8, this volume). In addition, populations may become fragmented in multiple-use areas as regions of heavy exploitation expand around settlements (Alvard, chapter 17, this volume).

Fourth, random events, especially in small populations, could have large impacts on the sustainability of the population. This could be included in a straightforward manner by using frequency distributions for survivorship and then using Monte Carlo simulation to compute statistics for the population growth rate. In addition, several parameters in our model, especially harem size, fecundity, and survivorship, might be better characterized by distributions rather than mean values due to skewed reproductive success and survivorship in many polygynous species.

Our models are relatively data rich, and in applying them we found it difficult to obtain relevant demographic parameters. While age at first breeding and interbirth intervals were available from zoo and field records, survivorship curves for males and females, age-specific fecundity, and harem sizes had to be gleaned from field studies and were often unavailable. Indeed, our choice of species was in part restricted by the lack of high-quality, long-term field studies. We expect that similar difficulties would be faced in applying models to exploited mammals outside the tropics. Clearly, detailed life-history information is needed to make valid predictions of population dynamics that can be applied to real populations, although simplifications of our model are possible, as we have shown.

We were forced to make additional assumptions to apply the models to the Selous Game Reserve. First, we used reproductive data from populations studied in protected areas where exploitation is reduced (but see Hofer et al., 1993). The extent to which population parameters vary in different parts of a species' range and the influence of hunting on these parameters is recognized but poorly understood. The fact that hunted populations of lions produce male-biased sex ratios among cubs, (Creel and Creel, 1997) shows that this variation is likely to be important. Second, we assumed that sources of mortality in study populations from which reproductive data were obtained were the same as those in areas of exploitation, but this is unlikely. Poaching pressure is low in most protected areas but in multiple-use areas that are not protected by guards, poaching is higher (Caro et al., in press a), and this additional offtake will affect recruitment. While this may have relatively little impact on abundant populations, it could seriously impact species living at low density. In light of these assumptions, it seems prudent to conduct sensitivity analyses in future models that incorporate additional offtake by legal hunters and poachers and that allow for biologically sensible population responses to hunting pressure (Hilborn and Mangel, 1997). Other hunting practices, such as the culling of entire herds, should also be modeled.

Finally, these models consider the effects of hunting on animal populations alone. Tourist hunting is a large revenue source for many countries (Cumming, 1989), and the type of recommendations made and the degree to which they are implemented depend on the worth of individual animals removed from the population. For example, in Tanzania, lions and buffalo are critical species to tourist hunting; lions generate more than 12% of hunting revenue per annum (Creel and Creel, 1997). Incorporation of economic returns into models would have important effects on decision making. In particular, managers need to face the decision about how close to the boundary $\lambda = 1.00$ they are willing to operate for larger economic gain. When parameters vary, operating close to the boundary may bring in more revenue over the short term but could lead to long-term catastrophe.

Recommendations

A number of practical recommendations emerge from these findings. For moderately polygynous species, hunting that focuses on either males or on adults of both sexes has a greater effect on populations than hunting of all age–sex classes if hunting is carried out at high intensity (figs. 11-2, 11-4). Because tourist hunting is less sustainable than subsistence hunting from a population perspective, tourist hunting quotas should be set at a lower level than subsistence hunting quotas. However, as shown by our models, polygynous species with large harem sizes are more resilient to male trophy hunting than species with small harem sizes (figs. 11-1, 11-3). If managers lack specific demographic information but can influence quota limits for particular species, managers should curtail male trophy hunting of mongamous or weakly polygynous species and favor hunting of strongly polygynous species.

Furthermore, the decision to hunt either adult males or adult males and females is important in infanticidal and reproductively suppressed species. For both types of breeding systems, our conclusion that male hunting has less impact on population growth rates than hunting both sexes has important management consequences. In Tanzania, for example, male and female leopards *Panthera pardus* are sometimes shot because hunters find it difficult to distinguish the sexes. In contrast, only male lions are shot. Since both species are infanticidal (Caro and Durant, 1995), our model suggests that leopard populations are likely to suffer a greater reduction in growth rates than lion populations for a given number of hunting quotas (figs. 11-5, 11-6). As detailed studies of leopard demography are lacking, we recommend that tourist hunting of this species should be reduced to low levels until such data have been collected and evaluated.

Our models also suggest where future monitoring efforts should be directed. Monogamous and weakly polygynous species are particularly sensitive to hunting. Hence, we recommend that species in Africa with these breeding systems such as dikdik, klipspringer, oribi *Ourebia ourebi,* duiker, and reedbuck (*Redunca* spp.) should be monitored carefully in multiple-use areas. This will necessitate increasing use of ground counts rather than aerial surveys, which often fail to discern these species from the air (Caro et al., in press c). As noted above, infanticidal species that are hunted for both sexes should also be closely monitored.

Summary

Effectiveness of multiple-use areas as a conservation tool depends on harvesting populations sustainably. We incorporated mating systems into age- and stage-structured models to assess populations' responses to different types of hunting under different types of breeding systems. Hunting that removed either adult males or adults of both sexes reduced population growth rates at lower hunting intensities than did hunting of adults and juveniles of both sexes. Monogamous and weakly polygynous species were more sensitive to hunting offtake than strongly polygynous systems, although if fecundity declines with harem size, strongly polygynous species will have lower growth rates than weakly polygynous ones. Infanticidal species and reproductively suppressed species were particularly sensitive to offtake of adult males and females.

Models were applied to populations of polygynous spotted hyenas, impala, and buffalo, to monogamous small antelopes, infanticidal lions, and reproductively suppressed dwarf mongooses. Positive growth rates were predicted when empirical reproductive parameters were used in the model. Data on population growth rates were used to calculate sustainable offtake for these species in the Selous Game Reserve in Tanzania. Results showed that current tourist hunting levels there are conservative for most species. The modeling approach allows for the incorporation of mating system and life-history attributes into demographic analysis and can track interactions of such attributes and effects of different hunting methods. Our models do not include density dependence, physiological condition, or economic parameters, although in certain cases these variables could be added. The models may be difficult to apply to real-life situations because long-term demographic variables are available for only a few species. Nevertheless, our models do allow us to make recommendations about changing the form of hunting of certain infanticidal species and increasing the monitoring of monogamous species.

Acknowledgments We thank Scott Creel and Mike Fogarty for their helpful comments.

Appendix 1: Mate Choice and Intrasexual Competition

In many species, females choose mates among available males. Hence, although a particular male may be capable of breeding, it may actually achieve no matings because it lacked experience or capabilities that made it desirable to females. Such experience is age dependent in many species; older males achieve most of the matings, whereas younger males often have low reproductive success (Poole, 1989). This mating bias also is caused by intrasexual competition; very young and very old males may be excluded from mating by dominant, intermediately aged males.

These constraints could have population consequences. For example, our model of impala populations predicts that even when all adult males in the population are hunted in an interbirth interval, the population would be able to sustain itself because newly recruited juveniles could mate with all females in harems. In reality, female choice and lack of experience by these males would likely curtail the number of matings.

One way to model experience is to introduce a new parameter into the equations calculating number of birth-class individuals. Let experience, $E(a)$, represent the proportion of females in a harem that an adult male in age class a can actually fertilize, where $a = 1 \ldots a_{max}$. The appropriate modifications to the basic model are as follows. Because $E(a)$ affects the breeding system in an age-specific manner, experience is incorporated in the minimum rule of eq. 15:

$$B_{\rm f}(t+1) = (1-r)R\min\left[1, h\frac{\sum_{a=1}^{a_{\max}} E(a)\,A_{\rm m}(t,a)}{A_{\rm F}(t)}\right]$$

$$B_{\rm m}(t+1) = r\,R\min\left[1, h\frac{\sum_{a=1}^{a_{\max}} E(a)\,A_{\rm m}(t,a)}{A_{\rm F}(t)}\right]. \tag{28}$$

In the basic model, males are excluded when all harems are filled, but this exclusion is age independent. $E(a)$ weights the degree to which males of particular age classes are able to obtain harems and hence are included in the effective population.

Inclusion of experience has the effect of reducing population growth rate, depending on hunting mortality. Table 11-5 shows a manipulation of the impala population model that incorporates experience: $E(1)$ varies from 0.1 to 0.9 and is 1.0 for all other adult age classes. Thus, newly recruited adult males need one IBI to achieve full reproductive competency and are only a fraction as capable during their first breeding season. At high levels of hunting mortality (40% or higher), the impala population slowly declines when $E(1) < 0.5$. Experience should be regarded as a continuous trait that generally increases across age

Table 11-5 Population growth rate as a function of hunting mortality and $E(1)$, the experience of newly recruited adult males.

Hunting mortality	$E(1)$			
	0.2	0.4	0.6	0.8
0.3	1.000	1.000	1.000	1.000
0.4	0.969	1.000	1.000	1.000
0.5	0.955	1.000	1.000	1.000
0.6	0.946	1.000	1.000	1.000
0.7	0.937	1.000	1.000	1.000
0.8	0.932	0.964	1.000	1.000
0.9	0.928	0.954	1.000	1.000
1.0	0.924	0.947	0.970	1.000

groups. Consequently, the population growth rates reported in table 11-5 are probably too high. In general, a negative correlation between average harem size in a species and variation in $E(a)$ is expected because of increased mate choice and intrasexual competition that accompanies polygyny (Trivers, 1972). Measuring such parameters could be accomplished by comparing harem sizes, and perhaps more importantly, number of fertilizations, of adult males of different age.

Appendix 2: Infanticide Revisited

In the interest of brevity and generality, the initial model of infanticide only included a minimum of details concerning infanticide's population effects. Applying the model to lions revealed several ways that the model could accommodate added complexity. Lion males occur in coalitions of up to seven males (Packer et al., 1988). Adding coalitions alters the dynamics of infanticide in two ways. First, because more than one male controls a harem, a death does not lead automatically to a takeover and subsequent infanticide. Second, more males are tied to harems, so that each mortality has a higher chance of affecting a harem. We added two parameters to the model to accommodate male coalitions: c is the coalition size, and p is the probability that a death leads to a new coalition taking over a pride. In this case, assuming a linear relationship between mortality and probability of takeover, we replace eq. 24b with

$$B_f(t+1) = (1-r)\,R\,\{1 - pc[1 - S_M(t)]\}$$

$$B_m(t+1) = r\,R\,\{1 - pc[1 - S_M(t)]\}. \quad (29)$$

We used data from Bygott et al. (1979) on how length of tenure changes between lion coalitions of different sizes to make a rough estimate of p. Using the data they presented, we estimated the probability that a coalition of a given size would survive one IBI after a hunting mortality. We then used data from Bygott et al. (1979) on the relative number of various sized coalitions to create a weighted mean of these probability over all coalition sizes. Using this estimate ($p = 0.75$), we obtained a value of 10% male hunting mortality (approximately 3% of the adult population) as the maximal offtake in the Selous before the lion population will decline. This lower value compared to the simpler infanticide model suggests that the existence of male coalitions can make a population more susceptible to decline in the face of hunting mortality. Nonetheless, this value exceeds the 3–4% that Creel and Creel (1997) suggest is sustainable offtake in the Selous.

In many infanticidal species, infanticide also occurs when a bachelor male (or coalition, in the case of lions) displaces the harem-holding male (Hausfater and Hrdy, 1984). Undoubtedly, higher numbers of bachelor males (non–harem-holding males) will cause higher the rates of takeover attempts and subsequent infanticide. To predict how the relative number of males to females affects the rates of infanticide, consider a population that has A_F/h harems and A_M males, where there are $(A_M - A_F/h) = B$ bachelor males that do not hold a pride. If each male attempts on average τ takeovers per IBI and the attempts are evenly spread across all harems, then an individual harem faces $\{B/[A_F(t)/h]\}\tau$ turnover attempts, which simplifies to $[Bh\tau/A_F(t)]$. Furthermore, if each attempt succeeds with probability i, then a harem is held with probability $(1 - i)^{Bh\tau/A_F(t)}$. The resulting replacement to Eq. 24b is

$$B_f(t + 1) = (1 - r)\, R\, (1 - i)^{Bh\tau/A_F(t)}$$

$$B_m(t + 1) = r\, R\, (1 - i)^{Bh\tau/A_F(t)}. \qquad (30)$$

Using the same parameter values as before and setting τ, the number of takeover attempts per male per IBI = 1 and the takeover probability $i = 0.15$, we simulated population dynamics using this model (fig. 11-9). The most obvious qualitative effect is a decline

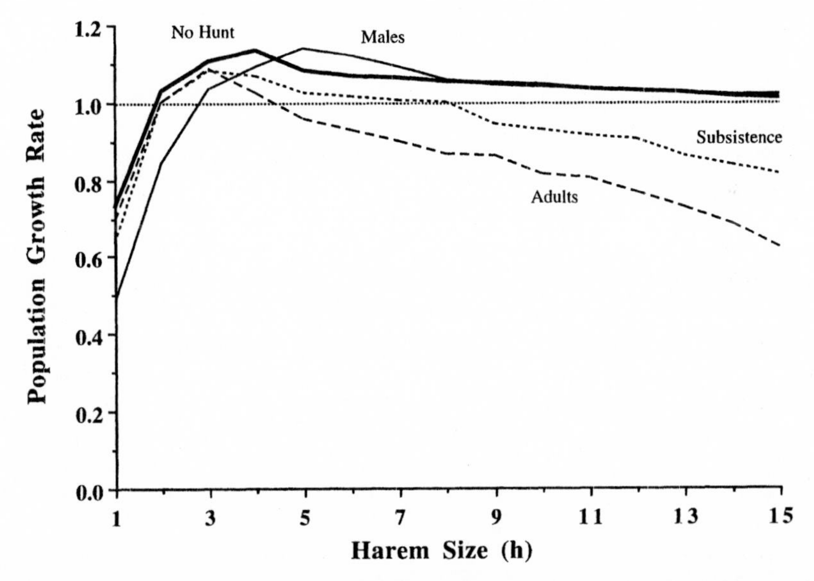

Figure 11-9 Population growth rate of an infanticidal species as a function of harem size when hunting intensity (H) is either zero (no hunt) or 0.15 (all other lines), and when τ = 1 and i = 0.15 (see text for details). We consider three different types of hunting: adult males (male), adults of either sex (adult) and juveniles and adults of either sex (subsistence). The dashed horizontal line indicates the boundary between increasing and decreasing population change.

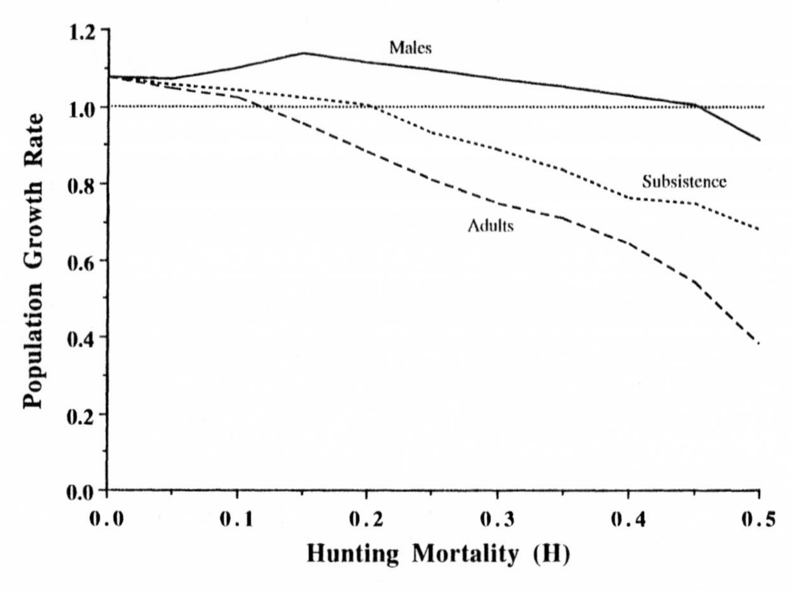

Figure 11-10 Population growth rate in an infanticidal species as a function of hunting intensity. Harem size is constant at five females, and $\tau = 1$ and $i = 0.15$ (see text for details). We consider three different types of hunting: adult males (male), adults of either sex (adult) and juveniles and adults of either sex (subsistence). The dashed horizontal line indicates the boundary between increasing and decreasing population change.

in population growth rate after harem size is greater than three or four females. At larger harem sizes, many males do not hold harems and increase the rate of takeover and subsequent infanticide. The effect of hunting (fig. 11-10) is qualitatively different from when infanticide only occurred after the death of the harem-holding male. Hunting of males is actually more sustainable than any other type of hunting, and in some cases may increase population growth rate. In this case, additional hunting reduces the large number of extra males, thereby reducing takeovers and infanticide. This result agrees with those of Starfield and Bleloch (1991), who showed that culling nomadic males led to increases in population growth (see also Venter and Hopkins, 1988). These results should be interpreted cautiously because the choice of values for τ and i were arbitrary. The actual dynamics of infanticide in the field have been difficult to analyze: it is rare to witness a takeover, and the results of the takeover are somewhat ambiguous (Packer and Pusey, 1984). However, this model can be used as a guide for further research into the dynamics of infanticide, especially for dealing with issues such as the frequency of takeover attempts and their success rate and how attempts and success change at different densities.

References

Anderson JL, 1983. Sport hunting in national parks: sacrilege or salvation? In: Management of large mammals in African conservation areas (Owen-Smith RN, ed). Pretoria: Haum; 271–280.

Arcese P, Hando J, Campbell K, 1995. Historical and present-day anti-poaching efforts in Serengeti. In: Serengeti II: Dynamics, management and conservation of an ecosystem (Sinclair ARE, Arcese P, eds). Chicago: University of Chicago Press; 506–533.

Armitage KB, 1986. Marmot polygyny revisited: determinants of male and female reproductive strategies. In: Ecological aspects of social evolution (Rubenstein DI, Wrangham RW, eds). Princeton, New Jersey: Princeton University Press; 303–331.

Balick MJ, Mendelsohn R, 1992. Assessing the economic value of traditional medicines from tropical rain forests. Conserv Biol 6:128–130.

Beddington JR, 1974. Age structure, sex ratio and population density in the harvesting of natural animal populations. J Appl Ecol 11:915–924.

Beddington JR, Basson M, 1994. The limits to exploitation on land and sea. Phil Trans R Soc B 343:93–98.

Beddington JR, May RM, 1908. A possible model for the effect of adult sex ratio and density on the fecundity of sperm whales. Rep Int Whal Comm 2:75–76.

Beverton RJH, and Holt SJ, 1993. On the dynamics of exploited fish populations. New York: Chapman and Hall.

Bodmer RE, 1994. Managing wildlife with local communities in the Peruvian Amazon: the case of the Reserva Comunal Tamashiyacu-Tahuayo. In: Natural connections: perspectives in community-based conservation (Western D, Wright RM, Strum SC, eds). Washington, DC: Island Press; 113–134.

Bodmer RE, Fang TG, Moya I I, Gill R, 1994. Managing wildlife to conserve Amazonian forest: population biology and economic considerations of game hunting. Biol Conserv 67:29–35.

Burgoyne GE Jr, 1981. Observations on a heavily exploited deer population. In: Dynamics of large mammal populations (Fowler CW, Smith TD, eds). New York: Wiley and Sons; 403–413.

Bygott JD, Bertram BCR, Hanby JP, 1979. Male lions in large coalitions gain reproductive advantages. Nature 282:839–841.

Campbell K, Hofer H, 1995. People and wildlife: spatial dynamics and zones of interaction. In: Serengeti II: Dynamics, management and conservation of an ecosystem (Sinclair ARE, Arcese P, eds). Chicago: University of Chicago Press; 534–570.

Caro TM, Durant SM, 1995. The importance of behavioral ecology for conservation biology: examples from Serengeti carnivores. In: Serengeti II: Dynamics, management, and conservation of an ecosystem (Sinclair ARE, Arcese P, eds.). Chicago: University of Chicago Press; 451–472.

Caro TM, Pelkey N, Borner M, Campbell KLI, Woodworth BL, Farm BP, ole Kuwai J, Huish SA, Severre ELM, in press a. Consequences of different forms of conservation for large mammals in Tanzania: Preliminary analyses. Afr J Ecol.

Caro TM, Pelkey N, Borner M, Severre ELM, Campbell KLI, Huish SA, ole Kuwai J, Farm BP, Woodworth BL, in press b. The impact of tourist hunting on large mammals in Tanzania: An initial assessment. Afr J Ecol.

Caro TM, Rejmanek M, Pelkey N, in press c. Which mammals benefit from protection in East Africa? In: Has the panda had its day? Future priorities for the conservation of mammal biodiversity (Entwhistle A, Dunstone N, eds). London: Chapman and Hall.

Caswell H, Weeks DE, 1986. Two sex models: chaos, extinction, and other dynamic consequences of sex. Am Nat 128: 707–735.

Caughley G, 1994. Directions in conservation biology. J Anim Ecol 63:215–244.

Charlesworth B, 1980. Evolution in age-structured populations. London: Cambridge University Press.

Clark CW, 1985. Bioeconomic modeling and fisheries management. New York: Wiley Interscience.

Clark CW, 1990. Mathematical bioeconomics, 2nd ed. New York: Wiley Interscience.

Clutton-Brock TH (ed), 1988. Reproductive success: studies of individual variation in contrasting breeding systems. Chicago: University of Chicago Press.

Clutton-Brock TH, 1989. Mammalian mating systems. Proc R Soc B 235:339–372.

Clutton-Brock TH, Albon SD, Guiness FE, 1982. Competition between female relatives in a matrilocal mammal. Nature 300:178–180.

Clutton-Brock TH, Deutsch JC, Nefdt RJC, 1993. The evolution of ungulate leks. Anim Behav 46:1121–1138.

Creel SR, Creel NM, 1989. Energetics, reproductive suppression, and obligate communal breeding in carnivores. Behav Ecol Sociobiol 28:263–270.

Creel SR, Creel NM, 1996. Limitation of African wild dogs by competition with larger carnivores. Conserv Biol 10:526–538.

Creel S, Creel NM, 1997. Lion density and population structure in the Selous Game Reserve: evaluation of hunting quotas and offtake. Afr J Ecol. 35:83–93.

Cumming DHM, 1989. Commercial and safari hunting in Zimbabwe. In: Wildlife production systems: economic utilization of wild ungulates (Hudson RJ, Drew KR, Baskin LM, eds). Cambridge: Cambridge University Press; 147–169.

Downhower JF, Armitage KB, 1971. The yellow-bellied marmot and the evolution of polygamy. Am Nat 105:355–370.

Dunbar RIM, 1990. Environmental determinants of fecundity in klipspringer (*Oreotragus oreotragus*). Afr J Ecol 28:307–313.

Emlen ST, Oring LW, 1977. Ecology, sexual selection, and the evolution of mating systems. Science 197:215–223.

Fa JE, Juste J, Perez de Val J, Castroviejo J, 1995. Impact of market hunting on mammal species in Equatorial Guinea. Conserv Biol 9:1107–1115.

FitzGibbon CD, Mogaka H, Fanshawe JH, 1995. Subsistence hunting in Arubuko-Sokole Forest, Kenya, and its effects on mammal populations. Conserv Biol 9:1116–1126.

Fragoso JMV, 1991. The effects of hunting tapirs in Belize. In: Neotropical wildlife use and conservation (Robinson JG, Redford KH, eds). Chicago: University of Chicago Press; 154–162.

Frank LG, 1986. Social organization of the spotted hyaena *Crocuta crocuta*. II. Dominance and reproduction. Anim Behav 34:1510–1527.

Frank LG, Holekamp KE, Smale L, 1995. Dominance, demography, and reproductive success of female spotted hyenas. In: Serengeti II: Dynamics, management and conservation of an ecosystem (Sinclair ARE, Arcese P, eds). Chicago: University of Chicago Press; 364–384.

Freese CG, Heltne PG, Castro N, Whitesides C, 1982. Patterns and determinants of monkey densities in Peru and Bolivia, with notes on distribution. Int J Primatol 3:53–90.

Ginsberg JR, Milner-Gulland EJ, 1994. Sex-biased harvesting and population dynamics in ungulates: implications for conservation and sustainable use. Conserv Biol 8:157–166.

Glanz WE, 1991. Mammalian densities at protected versus hunted sites in Central Panama. In: Neotropical wildlife use and conservation (Robinson JG, Redford KH, eds). Chicago: University of Chicago Press; 236–252.

Gotelli N, 1995. A primer of ecology. Sunderland Massachusetts: Sinauer Associates.

Greenwood PJ, 1980. Mating systems, philopatry and dispersal in birds and mammals. Anim Behav 28:1140–1162.

Harrington FH, Mech LD, Fritts SH, 1983. Pack size and wolf pup survival: their relationship under varying ecological conditions. Behav Ecol Sociobiol 13:19–26.

Hausfater G, Hrdy SB (eds), 1984. Infanticide: comparative and evolutionary perspectives. New York: Aldine.

Hilborn R, Mangel M, 1997. The ecological detective. Confronting models with data. Princeton, New Jersey: Princeton University Press.

Hofer H, East M, 1995. Population dynamics, population size, and the commuting system of Serengeti spotted hyenas. In: Serengeti II: Dynamics, management and conservation of an ecosystem (Sinclair ARE, Arcese P, eds). Chicago: University of Chicago Press; 332–363.

Hofer H, East M, Campbell KLI, 1993. Snares, commuting hyaenas, and migrating herbivores: humans as predators in the Serengeti. Symp Zoo Soc Lond 65:347–366.

IUCN/UNEP/WWF, 1991. Caring for earth: a strategy for sustainable living. Gland: International Union for the Conservation of Nature and Natural Resources.

Jarman MV, 1979. Impala social behaviour: territory, hierarchy, mating, and the use of space. Berlin: Verlag Paul Parey.

Kaplan H, Kopischke K, 1992. Resource use, traditional technology, and change among native peoples of lowland South America. In: Conservation of neotropical forests: working from traditional resource use (Redford KH, Padoch C, eds). New York: Columbia University Press; 83–107.

Keane B, Waser PM, Creel SR, Creel NM, Elliot LF, Minchella DJ, 1994. Subordinate reproduction in dwarf mongooses. Anim Behav 47:65–75.

Kleiman DG, 1977. Monogamy in mammals. Q Rev Biol 52:39–69.

Komers PE, 1996. Obligate monogamy without parental care in Kirk's dikdik. Anim Behav 51:131–140.

Kruuk H, 1972. The spotted hyena: a study of predation and social behavior. Chicago: University of Chicago Press.

Lawrence DC, Leighton M, Peart DR, 1995. Availability and extraction of forest products in managed and primary forests around a Dayak village in west Kalimantan, Indonesia. Conserv Biol 9:76–88.

Lewis D, Kaweche GB, Mwenya A, 1990. Wildlife conservation outside protected areas—lessons from an experiment in Zambia. Conserv Biol 4:171–180.

Mangel M, Clark CW, 1988. Dynamic modeling in behavioral ecology. Princeton, New Jersey: Princeton University Press.

Mangel M, Ludwig D, 1992. Definition and evaluation of behavioral and developmental programs. Annu Rev Ecol Syst 23:507–536.

Marks SA, 1973. Prey selection and annual harvest of game in a rural Zambian community. E Afr Wildl J 11:113–128.

May RM, Beddington JR. 1980. The effect of adult sex ratio and density on the fecundity of sperm whales. Rep Int Whal Comm 2:213–217.

McNamara JM, Houston AI, 1996. State dependent life histories. Nature 380:215–221.

Meffe GK, Carroll RC, 1994. Principles of conservation biology. Sunderland, Massachusetts: Sinauer Associates.

Metcalfe S, 1994. The Zimbabwe communal areas management programme for indigenous resources (CAMPFIRE). In: Natural connections: perspectives in community-based conservation (Western D, Wright RM, Strum SC, eds). Washington, DC: Island Press; 161–192.

Milner-Gulland EJ, 1994. A population model for the management of the Saiga antelope. J App Ecol 31:25–39.

Milner-Gulland EJ, Bekenov AB, Grachov YA, 1995. The real threat to Saiga antelopes. Nature 377:488–489.

Mloszewski MJ, 1983. The behaviour and ecology of the African buffalo. Cambridge: Cambridge University Press.

Murray MG, 1982. The rut of impala: aspects of seasonal mating under tropical conditions. Z Tierpsychol 59:319–337.

Ojasti J, 1991. Human harvest of capybara. In: Neotropical wildlife use and conservation (Robinson JG, Robinson KH, eds). Chicago: University of Chicago Press; 234–252.

Packer C, Pusey AE, 1984. Infanticide in carnivores. In: Infanticide: comparative and evolutionary perspective (Hausfater G, Hrdy SN, eds). New York: Aldine; 43–64.

Packer C, Pusey AE, 1987. Intrasexual cooperation and the sex ratio in African lions. Am Nat 130:636–642.

Packer C, Herbst L, Pusey AE, Bygott JD, Hanby JP, Cairns EJ, Borgerhoff Mulder M, 1988. Reproductive success of lions. In: Reproductive success: studies of individual variation in contrasting breeding systems (Clutton-Brock TH, ed). Chicago: University of Chicago Press; 363–383.

Parmigiani S, vom Saal F, Svare B (eds). 1992. Infanticide and parental care. London: Harwood Academic Press.

Peres CA, 1990. Effects of hunting on western Amazonian primate communities. Biol Conserv 54:47–59.

Peters CM, Gentry AH, Mendelsohn RO, 1989. Valuation of an Amazonian rainforest. Nature 339:655–656.

Poole JH 1989. Mate guarding, reproductive success and female choice in African elephants. Anim Behav 37:842–849.

Prins HHT, 1996. Ecology and behaviour of the African buffalo. London: Chapman and Hall.

Redford KH, Padoch C (eds), 1992. Conservation of neotropical forest: working from traditional resource use. New York: Columbia University Press.

Robinson JG, Redford KH, 1986. Intrinsic rate of natural increase in neotropical forest mammals: relationship to phylogeny and diet. Oecologia 68:516–520.

Robinson JG, Redford KH, 1991a. Sustainable harvest of neotropical forest mammals. In: Neotropical wildlife use and conservation (Robinson JG, Redford KH, eds). Chicago: University of Chicago Press; 415–429.

Robinson JG, Redford KH, 1991b. The use and conservation of wildlife. In: Neotropical wildlife use and conservation (Robinson JG, Redford KH, eds). Chicago: University of Chicago Press; 3–5.

Rood JP, 1980. Mating relationships and breeding suppression in the dwarf mongoose. Anim Behav 28:143–150.

Rood JP, 1990. Group size, survival, reproduction, and routes to breeding in dwarf mongooses. Anim Behav 39:566–572.

Sinclair ARE, 1977. The African buffalo: a study of resource limitation of populations. Chicago: University of Chicago Press.

Skinner JD, 1989. Game ranching in southern Africa. In: Wildlife production systems: economic utilization of wild ungulates (Hudson RJ, Drew KR, Baskin LM, eds). Cambridge: Cambridge University Press; 286–306.

Starfield AM, Bleloch AL, 1991. Building models for conservation and wildlife management, 2nd ed. Edina, Minnesota: Burgess International Group.

Starfield AM, Furniss PR, Smuts GL, 1981. A model of lion population dynamics as a function of social behavior. In: Dynamics of large mammal populations (Fowler CW, Smith TD, eds). New York: Wiley and Sons; 121–134.

Thomsen JB, Brautigam A, 1991. Sustainable use of neotropical parrots. In: Neotropical wildlife use and conservation (Robinson JG, Redford KH, eds). Chicago: University of Chicago Press; 359–379.

Tilson RL, Tilson JW, 1986. Population turnover in a monogamous antelope (*Madoqua kirki*) in Namibia. J Mammal 67:610–613.

Trivers R, 1972. Parental investment and sexual selection. In: Sexual selection and the descent of man 1871–1971 (Campbell B, ed). Chicago: Aldine; 136–179.

Van Schaik CP, 1987. Why are diurnal primates living in groups? Behaviour 87: 120–144.

Venter J, Hopkins ME, 1988. Use of a simulation model in the management of a lion population. S Afr J Widl Res 18:126–130.

Vickers WT, 1991. Hunting yields and game composition over ten years in an Amazon indian territory. In: Neotropical wildlife use and conservation (Robinson JG, Redford KH, eds). Chicago: University of Chicago Press; 53–81.

Waser PM, Elliott LF, Creel NM, Creel SR, 1995. Habitat variation and mongoose demography, In: Serengeti II: Dynamics, management and conservation of an ecosystem (Sinclair ARE, Arcese P, eds). Chicago: University of Chicago Press; 421–447.

12

Conspecific Brood Parasitism, Population Dynamics, and the Conservation of Cavity-Nesting Birds

John Eadie
Paul Sherman
Brad Semel

Conservation of Cavity-Nesting Birds

A substantial number of bird species nest or roost in cavities (9–18% of all species in Europe, North America, southern Africa, and Australia). More than half of these are obligate cavity-nesters (Newton, 1994a,b). A central conservation problem facing many species is loss of breeding habitat. Considerable evidence, both observational and experimental, indicates that populations of hole-nesting birds frequently are limited by shortages of suitable nesting cavities (von Haartmann, 1957; Brush, 1983; Brawn and Balda, 1988; Bock et al., 1992; Newton, 1994a,b). Development, and the removal of large, mature timber and snags (dead trees) for commercial purposes, exacerbate these shortages. The problem is most acute for secondary cavity-nesting birds, which rely on primary cavity excavators (mainly woodpeckers), physical damage (from storms), or natural tree rot to produce cavities of sufficient size and quality.

A review of the North American Blue List, the European EEC Directive Annex I list, and the Red Data Book list indicates that at least 25 species of primary cavity-nesting birds and 38 species of secondary cavity-nesters currently are or have recently been of conservation concern (table 12-1). An additional 57 species listed in the Red Data Book probably are secondary cavity-nesters, but we were unable to confirm this. Thus 63–120 species of known or assumed cavity-nesting birds have declined sufficiently to warrant national and international concern. At least two of these species, and possibly four, are already extinct.

Loss of nesting habitat is not the only threat to cavity-nesting species. Introduced predators and competitors, pesticide contamination, and overharvest of game species all have contributed to population declines (King, 1981). However, of the species listed in the Red Data Book for which a probable cause of decline was identified, more than half listed loss of breeding habitat. In North America, for example, this situation is typified by dwindling of the mature forested swamps required by the ivory-billed woodpecker *Campephilus principalis principalis,* reduction in the mature pine forests of the southeast essential to the red-cockaded woodpecker *Picoides borealis,* and decline in old-growth conifer forest of the northwest upon which the spotted owl *Strix occidentalis* depends.

Table 12-1 Cavity-nesting species of conservation concern (data from Ehrlich et al., 1988, 1994; King, 1981).

Common name	Species Name	Status
North American Bluelist		
Merlin	*Falco columbarius*	Bluelist 1972–81, special concern 1982–86
Carolina parakeet	*Conuropsis carolinensis*	Extinct
Barn owl	*Tyto alba*	Blue list 1972–81, special concern 1982–86
Spotted owl	*Strix occidentalis*	Red Data Book: endangered, Blue list 1980–86
Eastern screech owl	*Otus asio*	Blue list 1981, special concern 1982, 1986
Red-headed woodpecker	*Melanerpes erythrocephalus*	Blue list 1972, 1976–81, special concern 1982–86
Lewis' woodpecker	*Melanerpes lewis*	Blue list 1975–81, special concern 1982, local concern 1986, now stable
Hairy woodpecker	*Picoides villosus*	Blue list 1975–82, special concern 1986
Red-cockaded woodpecker	*Picoides borealis*	Red Data Book: endangered, 1979
American ivory-billed woodpecker	*Campephilus principalis principalis*	Extinct in N. America, Red Data Book 1979
Purple martin	*Progne subis*	Blue list 1975–81, special concern 1982–86
Carolina wren	*Troglodytes ludovicianus*	Blue list 1980–81, special concern 1982–86
Bewick's wren	*Troglodytes bewickii*	Blue list 1972–86
Eastern bluebird	*Sialia sialis*	Blue list 1972, 1978–82, special concern 1986
Western bluebird	*Sialia mexicana*	Blue list 1972, 1978–81, special concern 1982, local concern 1986
Prothonotary warbler	*Protonotaria citrea*	Not listed but noted as declining
European EEC Council Directive		
Ruddy shelduck	*Tadorna ferruginea*	Annex I list 1991
Pygmy owl	*Glaucidium passerinum*	Annex I list 1991
Tengmalm's owl	*Aegolius funereus*	Annex I list 1991
European roller	*Coracias garrulus*	Annex I list 1991
Grey-headed woodpecker	*Picus canus*	Annex I list 1991
Black woodpecker	*Dryocopus martius*	Annex I list 1991
Great-spotted woodpecker	*Dendocopus major*	Annex I list 1991
Syrain woodpecker	*Dendocopus syriacus*	Annex I list 1991
Middle spotted woodpecker	*Dendocopus medius*	Annex I list 1991
White-backed woodpecker	*Dendocopus leucotos*	Annex I list 1991
Three-toed woodpecker	*Picoides tridactylus*	Annex I list 1991
Red-breasted flycatcher	*Ficedula parva*	Annex I list 1991
Collared flycatcher	*Ficedula albicollis*	Annex I list 1991
Kruper's nuthatch	*Sitta krupperi*	Annex I list 1991
Corsican nuthatch	*Sitta whiteheadi*	Annex I list 1991

(*continued*)

Table 12-1 (*Continued*)

Common name	Species Name	Status
Red Data Book List 1979		
White-winged wood duck	*Cairina scutulata*	Vulnerable
Brazilian merganser	*Mergus octosetaceus*	Indeterminate
Chinese merganser	*Mergus squamatus*	Indeterminate
Cuban whistling duck	*Dendrocygna arborea*	Rare/vulnerable
Seychelles kestrel	*Falco araea*	Rare
Aldabra kestrel	*Falco newtoni aldabranus*	Rare
Mauritius kestrel	*Falco punctatus*	Critically endangered
St. Vincent parrot	*Amazona arausiaca*	Endangered
St. Lucia parrot	*Amazona versicolor*	Endangered
Puerto Rican parrot	*Amazona vittata*	Critically endangered
Seychelles lesser vasa parrot	*Coracopsis nigra barklyi*	Endangered
Maroon-fronted parrot	*Rhynchopsitta pachyryhncha terrisi*	Endangered
Norfolk boobook owl	*Ninox novaeseelandiae undulata*	Indeterminate
Seychelles owl	*Otus insularis*	Rare
Cuban ivory-billed woodpecker	*Campephilus principalis bairdii*	Critically endangered
Imperial woodpecker	*Camephilus imperialis*	Endangered
Owston white-backed woodpecker	*Dendrocopus leucotos owstoni*	Rare
Helmuted woodpecker	*Dryocopus galeatus*	Critially endangered/extinct
Tristram's woodpecker	*Dryocopus javensis richardsi*	Critically endangered
Grand Bahama red-bellied woodpecker	*Melanerpes superciliaris bahamensis*	Critically endangered (Extinct?)
San Salvador red-bellied woodpecker	*Melanerpes superciliaris nyeanus*	Rare
Inouye's woodpecker	*Picoides tridactylus inouyei*	Rare
Takalsukasa's woodpecker	*Picus awokera takatsukasae*	Rare
Okinawa woodpecker	*Sapheopipo noguchii*	Endangered
Trinidad straight-billed woodpecker	*Xiphorhynchos picus altirostris*	Rare
Jamaican golden swallow	*Kalochelidon euchrysea euchrysea*	Critically endangered
Algerian (Kabylian) nuthatch	*Sitta lendanti*	Rare
Ponopa mountain starling	*Aplonis pelzelni*	Vulnerable

Thirty-five species of Psittiformes (parrots, makaws, parokeets), 2 species of Piciformes (barbets), 11 species of Stigiformes (owls), 7 species of Coraciformes (6 rollers, 1 hornbill), and 2 starlings (Passeriformes) are on the Red list and are suspected of being cavity-nesters; however, nest-type was not confirmed.

Traditional Solutions

A variety of ways have been proposed to mitigate the effects of habitat loss for cavity-nesting species. The most obvious is to prevent destruction of potential nesting sites, but in many cases this option has already been foreclosed. Where suitable habitats remain, relatively minor modifications to resource-use practices can provide substantial benefits. Much recent effort has focused on snag management. Rather than removing old snags and dead timber (the traditional silviculture practice), foresters are now encouraged to leave snags to enhance habitat for wildlife (e.g., Davis, 1983; Raphael and White, 1984; Land et al., 1989; Caine and Marion, 1991). Lengthening the period between harvests to increase the abundance of mature trees and altering harvest rotation to increase heterogeneity of stand age structures have also been advocated (see Sandström, 1992).

A second widely used approach has been to supplement natural cavities with artificial nest sites (e.g., nest-boxes or artificial snags; Caine and Marion, 1991). Such an approach is appealing because it focuses on a critical habitat variable and it is relatively inexpensive. At least 65 species of cavity-nesting birds (primarily in North America and Europe) will nest in boxes (table 12-2). Population increases in response to the addition of artificial cavities were seen in 22 of the 23 species for which such information was recorded. In eastern North America, nest-boxes are at least partly responsible for the recovery of the eastern bluebird *Sialia sialis* and wood duck *Aix sponsa* from historical lows in the early 1900s (Bellrose, 1990). In light of these successes, nest-box programs have been recommended or implemented (King, 1981) for several rare and endangered species, including the Puerto Rican parrot *Amazona vittata,* the Seychelles lesser vasa parrot *Coracopsis nigra barkly,* the Aldabra kestrel *Falco newtoni aldabranus,* and the Jamaican golden swallow *Kalochelidon euchrysea.*

Despite their potential usefulness, however, nest-boxes may sometimes have effects opposite to those intended. Frequently, well-intentioned managers install as many nest-boxes as possible to accommodate the maximum number of individuals of the target species. Intuitively this approach makes sense, unless the costs of constructing and installing more boxes are prohibitive. However, there are hidden biological pitfalls. Consider, for example, the effects of adding large numbers of boxes to a limited area: the density of nesting sites soars, and their dispersion becomes clumped. How might such habitat alterations influence a species' social structure?

In most cases we do not know because so little research has focused on this question. Social behavior, nest-site selection, and reproductive competition are rarely considered when nest-box programs are implemented, other than to ensure that boxes are attractive to the target species. Yet, it is becoming increasingly apparent that boxes can alter population dynamics and intraspecific social interactions, particularly conspecific brood parasitism (CBP), in counterintuitive ways. In some species, these changes reduce individual reproductive success and lead to population instability or decline.

This chapter illustrates the advantages and disadvantages of nest-box programs highlighting two species of cavity-nesting waterfowl. For more than a decade we have studied wood ducks in Missouri, Illinois, and New York, and Barrow's goldeneyes *Bucephala islandica* in British Columbia. Our parallel research was conceived independently, in both cases to investigate the species' behavioral ecology, especially brood parasitism. We were drawn together, and into conservation issues, when it became evident that some current management practices, although well intentioned, might actually be counterproductive. In

Table 12-2 Species of birds that have been recorded to breed in nest-boxes.

Common name	Species name	Response to boxes	Comments
Black-bellied whistling duck	*Dendrocygna autumnalis*		
Wood duck	*Aix sponsa*	+	90% decline in 1900s
Mandarin duck	*Aix galericulata*		Declining in Asia
Barrow's goldeneye	*Bucephala islandica*	+	
Common goldeneye	*Bucephala clangula*	+	
Bufflehead	*Bucephala albeola*	–	
Smew	*Mergus albellus*		
Common merganser	*Mergus merganser*		
Hooded merganser	*Lophodytes cucullatus*		
American kestrel	*Falco sparverius*	+	
European kestrel	*Falco tinnunculus*	+	Major decline in Finland
Aldabra kestrel	*Falco newtoni aldabranus*		Rare (Red list 1979); nest-boxes recommended
Barn owl	*Tyto alba*	+	Blue list 1982–86; vulnerable in UK
Ural owl	*Strix uralensis*	+	
Western screech owl	*Otus kennicotti*		
Eastern screech owl	*Otus asio*		Blue list 1981, declining in E and SE N. America
Flammulated owl	*Otus flammeolus*		
Northern saw-whet owl	*Aegolius acadicus*		
Boreal owl	*Aegolius funereus*		
Northern hawk owl	*Surnia ulula*		
Puerto Rican parrot	*Amazona vittata*		Critically endangered (Red list 1979); nest-boxes provided
Seychelles lesser vasa parrot	*Coracopsis nigra barkly*		Endangered (Red List 1979); nest-boxes recommended
Northern flicker	*Colaptes auratus*		
Red-bellied woodpecker	*Melanerpes carolinus*		
Golden-fronted woodpecker	*Melanerpes aurifrons*		
Downy woodpecker	*Picoides pubescens*		
Great-crested flycatcher	*Myiarchus crinitus*		
Ash-throated flycatcher	*Myiarchus cinerascens*	+	
Dusky-capped flycatcher	*Myiarchus tuberculifer*		
Collared flycatcher	*Ficedula albicollis*	+	Annex I listing 1991; locally common
Pied flycatcher	*Ficedula hypoleuca*	+	Annex I listing 1991; locally common
Redstart	*Phoenicurus phoenicurus*		80% decline in UK in 1970s
Tree swallow	*Tachycineta bicolor*	+	
Violet-green swallow	*Tachycineta thalassina*	+	
Purple martin	*Progne subis*	+	Blue list 1975–81
Jamaican golden swallow	*Kalochelidon euchrysea*		Critically endangered (Red list 1979); nest-boxes recommended
Tufted titmouse	*Parus bicolor*		

(*continued*)

Table 12-2 *(Continued)*

Common name	Species name	Response to boxes	Comments
Plain titmouse	*Parus inornatus*		
Bridled titmouse	*Parus wollweberi*		
Black-capped chickadee	*Parus atricapillus*		
Carolina chickadee	*Parus carolinesis*		
Mexican chickadee	*Parus sclateri*		
Mountain chickadee	*Parus gambeli*		
Marsh tit	*Parus palustris*		
Siberian tit	*Parus circtus*		Declined in Finland
Coal tit	*Parus ater*		Local declines noted
Blue tit	*Parus caeruleus*	+	
Great tit	*Parus major*	+	Increased in UK due to boxes
White-breasted nuthatch	*Sitta carolinensis*	+	
Pygmy nuthatch	*Sitta pygmaea*	+	
Brown-headed nuthatch	*Sitta pusilla*		
European nuthatch	*Sitta europaea*		
Common tree creeper	*Certhia familiaris*		
Short-toed tree creeper	*Certhia brachydactyla*		
House wren	*Troglodytes aedon*	+	
Brown-throated wren	*Troglodytes brunneicollis*		
Carolina wren	*Troglodytes ludovicianus*		Blue list 1980–81
Bewick's wren	*Troglodytes bewicki*		Blue list 1972–86
Eastern bluebird	*Sialia sialis*	+	Blue list 1972, 1978–82
Western bluebird	*Sialia mexicana*	+	Blue list 1972, 1978–81
Mountain bluebird	*Sialia currucoides*	+	
European starling	*Sturnus vulgaris*	+	
Prothonotary warbler	*Protonotaria citrea*		May be declining
Eurasian tree sparrow	*Passer montanus*		Major declines in UK
House sparrow	*Passer domesticus*		

Data on use and response to nest-boxes were compiled from Newton (1994a, b), King (1981), and Ehrlich et al. (1988, 1994). A plus (+) indicates that an increase in the size of the population was observed when nest-boxes were provided, a minus (−) indicates no response.

our two species, at least, an understanding of the behavioral ecology underlying female reproductive behaviors is essential to developing more effective management protocols.

A Behavioral Ecological Perspective: Wood Ducks and Barrow's Goldeneyes

Wood ducks inhabit wooded swamps. They breed throughout eastern North America and in Pacific Coast states and overwinter along the Gulf Coast and in southern California. Wood duck numbers declined precipitously at the turn of the century, apparently due to the combined effects of deforestation, loss of wetland habitat, and overharvest (Bellrose, 1990). Ornithologists feared they would go extinct, although numbers apparently were never as low as many thought (Hepp and Bellrose, 1995). Protection under the Migratory Bird Treaty Act (1918), preservation of wetlands, reforestation, and, in some areas, provision of nest-boxes resulted in an impressive comeback, and wood duck numbers have increased steadily in the past 50 years. Bellrose and Holm (1994) estimated that there are presently 1.4 million breeding pairs of wood ducks in North America.

Barrow's goldeneyes, by contrast, are quite restricted in distribution and abundance. More than 90% of the world's population of 100,000–150,000 goldeneyes is found in the Pacific northwest (Oregon to Alaska); a small, secondary population (fewer than 2,000 birds) occurs in eastern North American (Labrador) and in Iceland. Barrow's goldeneyes breed in alkaline and freshwater lakes and overwinter along coastlines in estuaries and rocky intertidal seashores.

Wood duck and goldeneyes nest in tree cavities created when limbs break and there is subsequent heart rot of the trunk, as well as in cavities originally created by large woodpeckers (e.g., pileated woodpeckers *Dryocopus pileatus;* Soulliere, 1988, 1990b). Both ducks readily use nest-boxes. For about 40 years erection of nest-boxes has been a standard management practice for these and other cavity-nesting species. Often large numbers of boxes are concentrated in highly visible locations, such as on poles over open water (see photographs in Bellrose and Holm, 1994).

Under natural conditions, cavities that are large enough to accommodate goldeneye and wood duck nests are widely dispersed, well concealed in the forest canopy, and limited in number (Weier, 1966; Soulliere, 1990b; Bellrose and Holm, 1994; Eadie et al., 1995; Hepp and Bellrose, 1995). Females search extensively for these cavities before the breeding season, sometimes as much as a year in advance (Bellrose, 1976; Eadie and Gauthier, 1985). Nest searching is usually undertaken by females that are paired, although groups of young females sometimes prospect for nest sites together (Eadie and Gauthier, 1985; Bellrose and Holm, 1994).

Scarcity of suitable, high-quality nest sites has promoted the evolution of alternative nesting behaviors among females. Many females lay eggs in just one cavity, incubate them, and remain with the precocial young until they are fully independent. However, a substantial number of females lay eggs in more than one cavity and then either abandon all of them (some of which are incubated and cared for by conspecific females) or, sometimes, incubate the eggs in one cavity and abandon the others.

Of 28 wood duck nests recorded in natural cavities, 8 (29%) certainly contained eggs of more than one female (Semel and Sherman, 1986) and clutch sizes ranged from 7 to 31 eggs; individual female wood ducks lay fewer than 13 eggs (Semel and Sherman, 1992). However, in highly visible nest-boxes CBP is more common. In some populations 100% of nests are parasitized, and clutches occasionally contain more than 50 eggs (Clawson et al., 1979; Haramis and Thompson, 1985; Bellrose and Holm, 1994). In our own studies in Missouri, 95% of boxes were parasitized, and the largest clutch contained 37 eggs (Semel and Sherman, 1986), and in Illinois 46% of nests were parasitized, and the largest clutch contained 44 eggs (Semel et al., 1988, 1990).

Parasitism also occurs frequently in populations of Barrow's and common goldeneye *B. clangula* (Andersson and Eriksson, 1982; Eadie, 1989, 1991). Twelve of 44 goldeneye nests in natural cavities (27%) were suspected of being parasitized, compared with an average of 35% of nests in nest-boxes (Eadie, 1989). By following marked females and using DNA fingerprinting and automatic nest cameras, Eadie (1989) estimated that 17–20% of all eggs were laid parasitically. On some lakes, frequencies of parasitism in nest-boxes reached 100%. The largest clutch in a nest-box contained 28 eggs; individual females typically lay 8–10 eggs.

Linking Conspecific Brood Parasitism with Population Dynamics

Our understanding of CBP has developed in two important areas in the past decade. First, its ecological and evolutionary bases have been clarified (reviewed by Yom-Tov, 1980; Andersson, 1984; Eadie et al., 1988; Rohwer and Freeman, 1989; Petrie and Møller, 1991; Sayler, 1992). Second, we have begun to appreciate the links between CBP and population dynamics. Not long ago, CBP was dismissed as an aberrant behavior that was of little interest or reproductive consequence to hosts or parasites. Symptomatic of this attitude, brood parasitism was commonly referred to as "egg-dumping" or "dump-nesting" (particularly in ducks; e.g., Haramis, 1990; Bellrose and Holm, 1994), because females were thought to dump off and abandon their eggs.

This view is changing rapidly. It has now become clear that there are several reasons females might lay parasitically: (1) females that lay eggs in the nests of others potentially gain reproductive benefits without incurring the physiological costs or risks associated with incubation and parental care; (2) females that are unable to locate a suitable nest site of their own or are unable to lay an entire clutch could salvage some reproduction by laying a few eggs parasitically; (3) parasitism among conspecifics might be facilitated if parasites and hosts are closely related (Andersson, 1984; Eadie et al., 1988); and (4) "parasitism" might represent nest-site competition between females as each attempts to lay her eggs in the same high-quality cavity. Thus, parasitic egg-laying represents a viable and adaptive female reproductive strategy, selected because of the potential benefits of emancipation from parental care and because it allows females to breed when critical resources (high-quality nest sites) are limited.

The belief that CBP was aberrant arose because the frequency of parasitic egg-laying was often high in box-nesting wood duck populations, and it frequently resulted in low reproductive success (e.g., Jones and Leopold, 1967). However, in goldeneyes and wood ducks CBP occurs whether or not nest sites are limited (Morse and Wight, 1969; Andersson and Eriksson, 1982; Eadie, 1989, 1991; Semel et al., 1990). Moreover, in years or populations where CBP is infrequent, productivity per box increases with increasing clutch size, suggesting that CBP is sometimes beneficial to population productivity (Clawson et al., 1979). These opposite results imply a connection between the frequency of CBP and population density.

Haramis and Thompson (1985) demonstrated this link using data from a 7-year study of box-nesting wood ducks in an experimental greentree impoundment at Montezuma National Wildlife refuge in upstate New York. They found that when boxes were first erected and duck densities were low, the frequency of parasitism also was low and nesting success was high (fig. 12-1a). As the population increased and nest sites became saturated, increased "intraspecific density strife" occurred, resulting in increased "dump nesting" and reduced hatching success (proportion of eggs that hatched). By year 5 of their study, reproductive success had "crashed," with only 22% of all eggs hatching (compared to 79% at the start of the study; fig. 12-1a). In the final 2 years, flooding of the impoundment was discontinued. Interestingly, this resulted in reduction in population density and a corresponding increase in egg hatching success to 60%.

Haramis and Thompson's (1985) data indicated an inverse relationship between total number of eggs laid (an index of breeding population density) and hatching success (fig. 12-1b). Semel et al. (1988, 1990) documented similar effects in a box-nesting wood duck

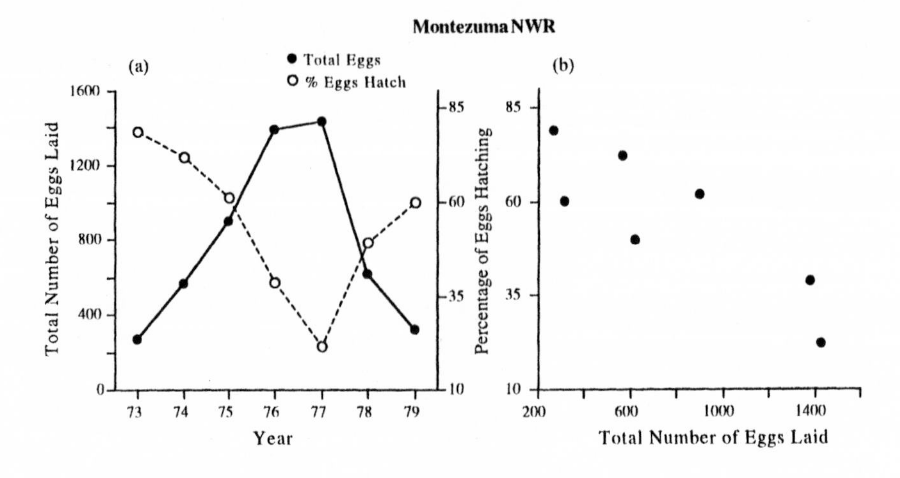

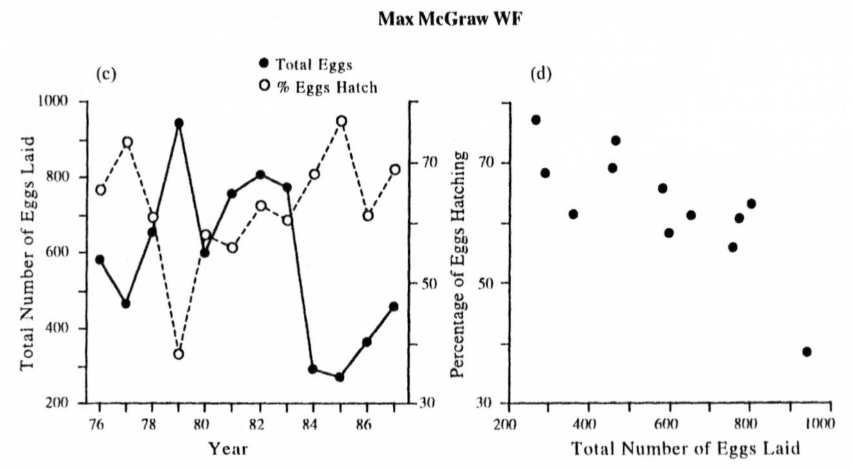

Figure 12-1 Relationship between total number of eggs laid (an index of the size of the breeding population) and hatching success (percentage of eggs hatching) in two wood duck populations. Panels a and c demonstrate the decline in hatching success (open circles) over time as the populations build up and the total numbers of eggs laid (solid circles) increases. Note that both populations "crashed" after reaching high densities. Panels b and d demonstrate the negative relationship between total eggs laid and hatching success (b) $r_s = -.82$, $p < .05$ (data for a and b from Haramis and Thompson, 1985); (d) $r_s = -.76$, $p < .02$ (data for c and d from Semel and Sherman, 1988).

population in Illinois (fig. 12-1c,d). Semel et al. also reported a direct relationship between population density and frequency of CBP and an inverse relationship between CBP and egg hatchability (see also Bellrose and Holm, 1994).

From these studies, a common pattern emerges. When a population is first established (i.e., nest-boxes are first erected), CBP is infrequent and hatching success is high. However, as the population increases, so does the frequency of parasitism, which raises the average clutch size. As a result, egg hatching success declines, often dramatically, due to inefficient incubation of supernormal clutches, broken eggs and subsequent fungal infections, disturbance of laying females by parasitic females, nest abandonment, and eggs laid after the onset of incubation (Semel et al., 1988, 1990; Semel and Sherman, 1995). These effects are apparently linked to population declines (fig. 12-1).

To explore the generality of these patterns, Semel et al. (1990) analyzed relationships between hatching success and average clutch size for six separate wood duck populations (each in a different state). They found that in all cases the relationships were negatively exponential. Following this lead, we compiled information on population densities and egg hatching success for wood ducks breeding in 10 national and state wildlife areas (table 12-3). This analysis revealed that, as the total number of eggs laid increased, hatching success declined in seven populations, significantly so in five (table 12-3); in the remaining populations, the relationship was positive, but not significant. Considering all 10 populations collectively, there was a (marginally significant) negative relationship between total eggs and proportions that hatched (fig. 12-2). We also reanalyzed data on egg hatching success and frequencies of parasitism that were tabulated by Bellrose and Holm (1994) for 17 studies in 10 states. As estimated frequencies of CBP increased, average clutch sizes increased significantly (fig. 12-3a), and proportions of eggs that hatched declined significantly (fig. 12-3b).

These relationships are not specific only to wood ducks. There were similar associations between frequencies of parasitism, population densities, and reproductive success among Barrow's goldeneyes in central British Columbia over a 9-year period. In these populations, egg hatching success was again a mirror image of population density (fig. 12-4); the inverse correlation between density and hatching success was significant (r_s = .78, $N = 9, p < .03$). The latter relationship was apparently driven by conspecific parasitism (fig. 12-5). As the proportion of parasitic eggs increased, total clutch size increased significantly ($r_s = .98, p < .01$), and hatch success declined significantly ($r_s = -.67, p = .05$). The parallels between goldeneyes and wood ducks are obvious (compare figs. 12-1 and 12-3 with 12-4 and 12-5).

Modeling the Effect of Conspecific Brood Parasitism on Population Dynamics

The relationships among population density, CBP, and reproductive success suggest that social behavior can play an important role in demography. At the most basic level, we envision that these factors interact as depicted in fig. 12-6. When population densities are high, CBP increases, leading to reduced egg hatching success and, hence, to population declines. At low population densities, parasitism is reduced, leading to enhanced reproductive success and increases in density. A variety of effects on population density is possible, depending on the lag between increases in CBP and population declines. Only short lags yield stable equilibria.

There have been three attempts to model these interactions (May et al., 1991; Eadie and Fryxell, 1992; Nee and May, 1993). All three are based on a behavioral ecological framework. May et al. (1991) described the evolutionary dynamics of populations composed of different strategists: parasitic females, vigilant nonparasitic females (i.e., vigilant against being parasitized), and naive (nonvigilant) nonparasitic females. Eadie and Fryxell (1992) examined the dynamics of a goldeneye population composed of parasitic and nonparasitic females in relation to density and frequency of parasitism. Nee and May (1993) studied the

Table 12-3 The relationship between the total number of eggs laid in a study area and nesting success (percentage of eggs hatching) for wood ducks (means ± 1 SE).

Refuge or wildlife area	Total eggs laid	% Eggs hatching	Regression of hatching success (% Hatch: Y) on the number total eggs laid (X)	R^2	Number of years (dates)	p value
Great Swamp NWR (New Jersey)	9405 ±1296	24.0 ±3.6	Y= 37.3 − 0.001 X	.253	17 (1970–86)	.04
Nauvoo Slough WA (Illinois)	1345 ±140	32.9 ±3.0	Y= 49.4 − 0.012 X	.335	8 (1983–90)	.14
Yazoo NWR (Mississippi)	6514 ±955	35.0 ±6.3	Y= 70.0 − 0.005 X	.659	11 (1966–78)	.002
Duck Creek WMA (Illinois)	1776 ±220	47.2 ±5.5	Y= 79.9 − .02 X	.629	9 (1966–74)	.011
Piedmont NWR (Georgia)	897 ±76	70.8 ±2.3	Y = 66.7 + 0.004 X	.022	14 (1971–84)	.62
Fort Lewis NWR (Washington)	661 ±73	63.4 ±2.8	Y = 61.6 + 0.002 X	.005	16 (1974–91)	.79
Montezuma NWR (New York)	1121 ±103	53.4 ±4.0	Y = 89.7 − .032 X	.709	19 (1973–91)	.001
Bombay Hook NWR (Delaware)	1728 ±369	49.5 ±6.1	Y = 42.0 + 0.004 X	.069	8 (1984–91)	.53
Union Slough NWR (Iowa)	2495 ±417	63.2 ±2.8	Y = 69.2 − 0.002 X	.134	17 (1975–91)	.14
Max McGraw WF (Illinois)	578 ±62	62.5 ±2.8	Y = 83.3 − .036 X	.617	12 (1976–87)	.003

NWR, National Wildlife Refuge; WA, Wildlife Area; WMA, Wildlife Management Area; WF, Wildlife Foundation.

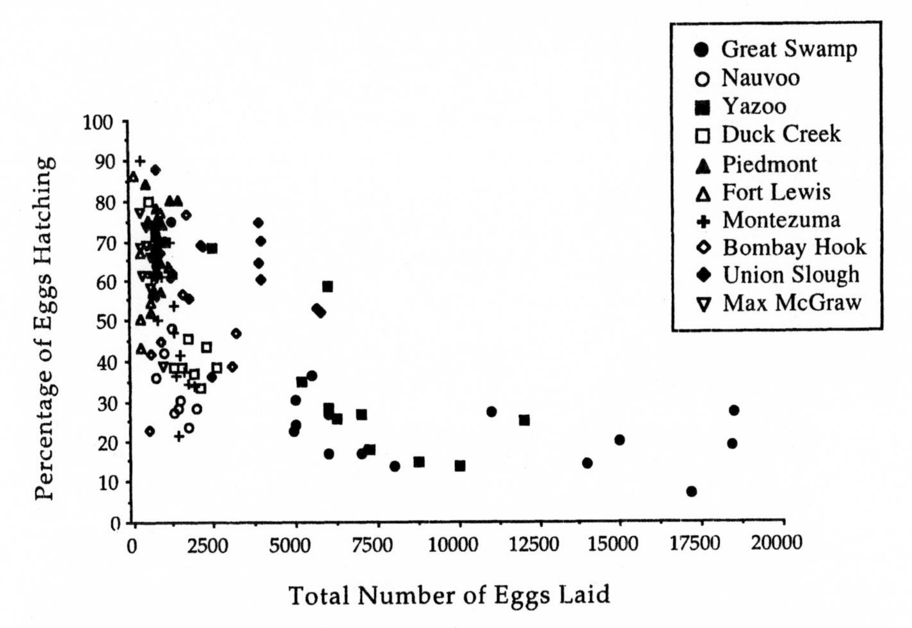

Figure 12-2 Relationship between the total number of eggs laid and egg hatching success of wood ducks at 10 national and state wildlife areas. For these data, Spearman $r_s = -.65$ and $p < .06$, using each site as an independent sample and averaging over all years for each location. Data from Semel et al. (1990) and Semel and Sherman (unpublished); see table 12-3.

conditions under which two different types of brood parasites—"professional" parasites (females that reproduce solely by parasitism) and "best-of-a-bad-job" parasites (females that parasitize only if they are unable to obtain a nest site)—might invade a population. Although these models made different assumptions and focused on different aspects of parasitic behavior, remarkably similar conclusions were obtained:

1. Conspecific brood parasitism can significantly impact population dynamics, leading to stable populations, populations that oscillate cyclically, or populations that fluctuate chaotically. Both Eadie and Fryxell (1992) and Nee and May (1993) found that CBP could lead to extinction of hosts and parasites (i.e., total population collapse). This occurred without the inclusion of any extraneous, stochastic factors (e.g., predators, bad weather, high water); it was driven entirely by interaction between hosts and parasites.
2. Understanding how CBP affects population dynamics depends critically on detailed knowledge of the behavioral ecology of brood parasitism. Important parameters of the models include (a) which females act as parasites, (b) whether females can act both as parasites and hosts, (c) numbers of eggs laid by hosts and parasites, (d) numbers of nests that parasites lay eggs in, (e) patterns of vigilance by hosts and their reproductive costs, and (f) how parasitism affects the reproductive success of hosts and parasites.
3. All three models explicitly link the evolutionary dynamics of the behavior (i.e., fitness consequences of females pursuing different behavioral tactics) with the

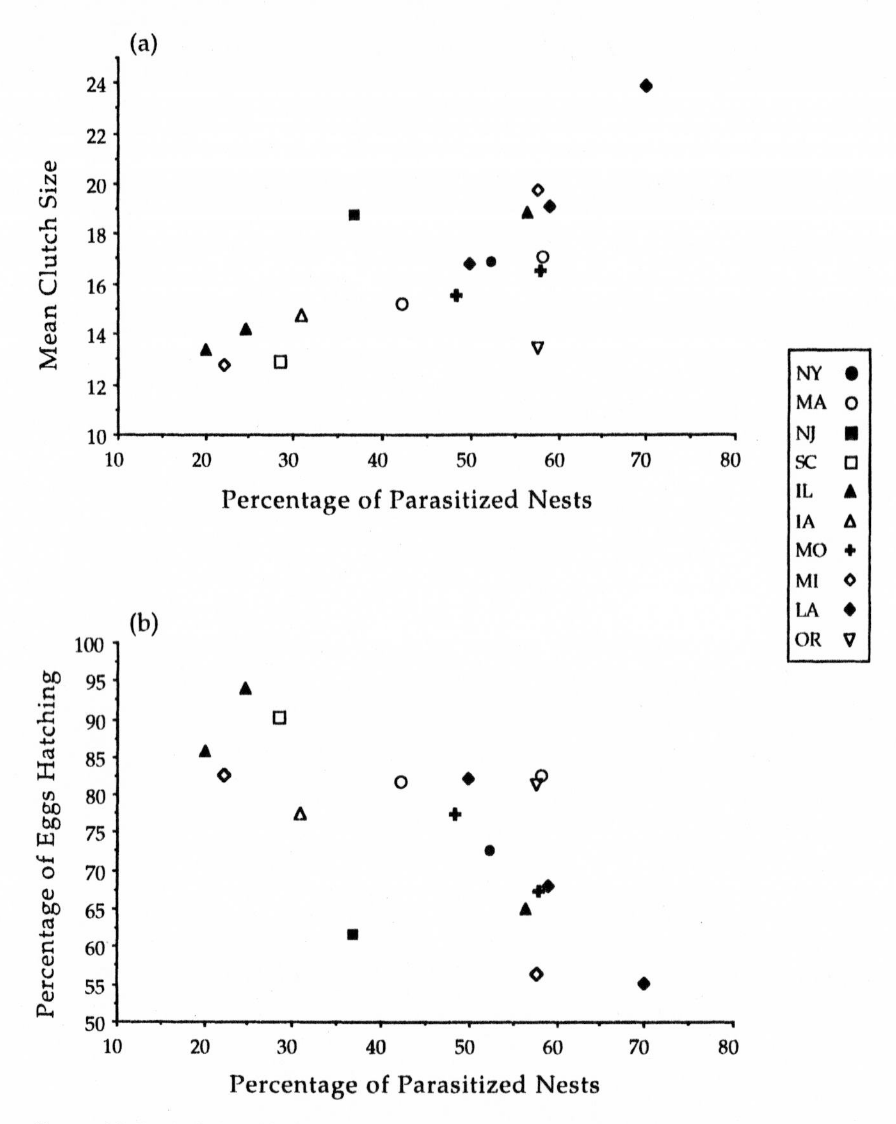

Figure 12-3 Relationship between the frequency of conspecific brood parasitism and (a) mean clutch size ($r_s = .76, p < .01$) and (b) average hatching success of eggs ($r_s = -.65, p < .01$) of wood ducks in 17 studies in 10 states. Data from Bellrose and Holm (1994).

population dynamics. Parasitism has a direct effect on reproductive success of females and therefore can regulate density of local populations; in turn, population density will influence the relative reproductive success of a parasite or host. Attempts to evaluate the fitnesses of parasites or hosts must therefore consider both the frequency-dependent and the density-dependent components of fitness (Eadie and Fryxell, 1992; Nee and May, 1993).

So far our thrust has been to emphasize the value of understanding the behavioral ecological dynamic in order to better predict and manage the population dynamic. But the op-

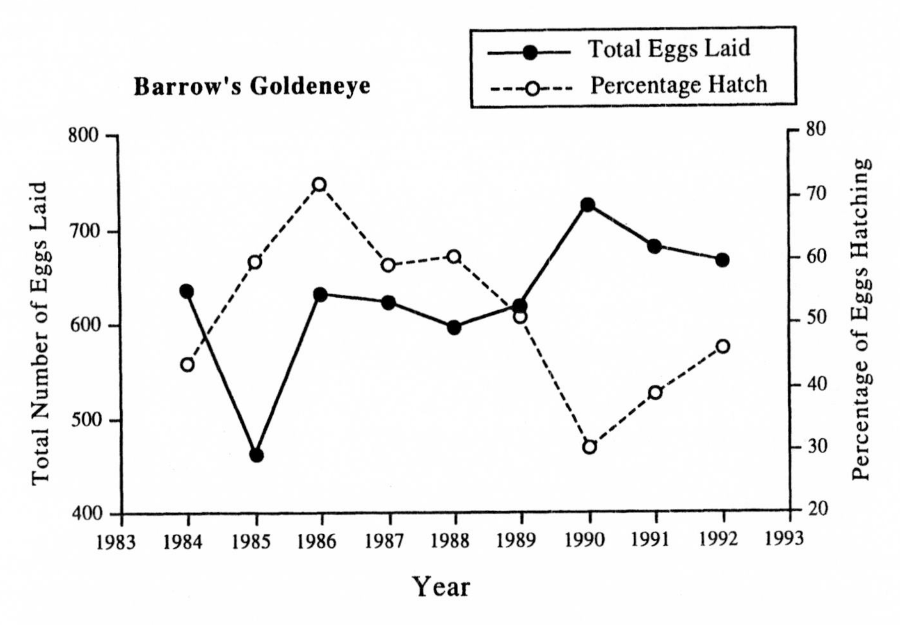

Figure 12-4 Relationship between the total number of eggs laid and hatching success (proportion of eggs hatching) of Barrow's goldeneyes in British Columbia. Data from Eadie (unpublished).

posite applies as well. Understanding the evolutionary dynamic of a behavior such as CBP requires that we also understand the endogenous and exogenous factors that influence population density and dispersion. These latter factors are perhaps most susceptible to anthropogenic perturbation and therefore should be of greatest management concern.

To visualize these points, consider Eadie and Fryxell's (1992) model of conspecific brood parasitism in goldeneyes. This model determines the reproductive success of parasitic and nonparasitic females over a range of population densities and frequencies of parasitism. For simplicity, the model assumes that a female acts either as a parasite or as a nonparasite but not both in any given breeding attempt (i.e., mixed strategies are not allowed). The reproductive success of a nonparasitic female is determined by (1) the number of eggs she can lay, (2) the probability of obtaining a nest site (a function of population density and the number of nest sites), (3) the probability of being parasitized by i parasitic females, and (4) the survivorship of offspring (hatching success), which, in turn, is a function of the number of eggs in the nest (fig. 12-7; a similar function was reported for wood ducks by Semel et al., 1988). The reproductive success of a parasite is likewise determined by (1) the number of eggs she can lay, (2) the probability of finding a host nest (a function of the number of hosts available, (3) the probability of $i - 1$ other parasites finding and parasitizing the same nest, and (4) the survivorship of parasitic offspring (again a function of number of eggs in the nest). To simulate the long-term population dynamic, Eadie and Fryxell (1992) used estimates of annual survival from field data to determine population growth rates (λ) as functions of the frequency of parasitism and the density of breeding females.

Essential findings of the model are shown in fig. 12-8. The top panel illustrates how reproductive success (RS) of parasitic and nonparasitic behavior varies with population den-

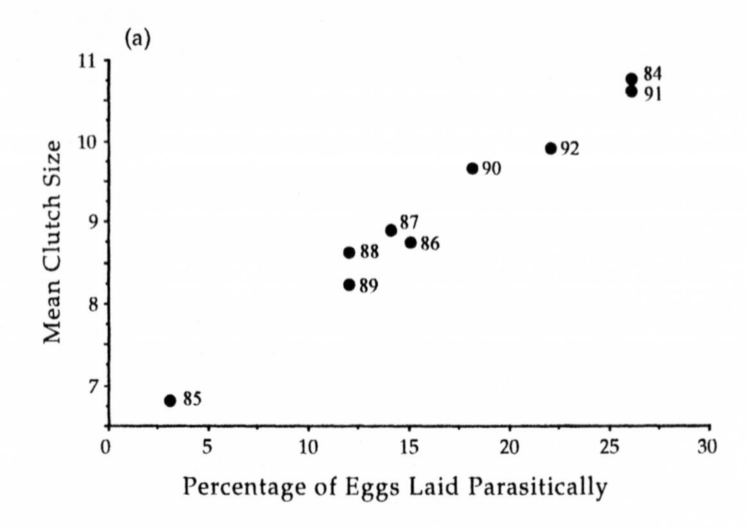

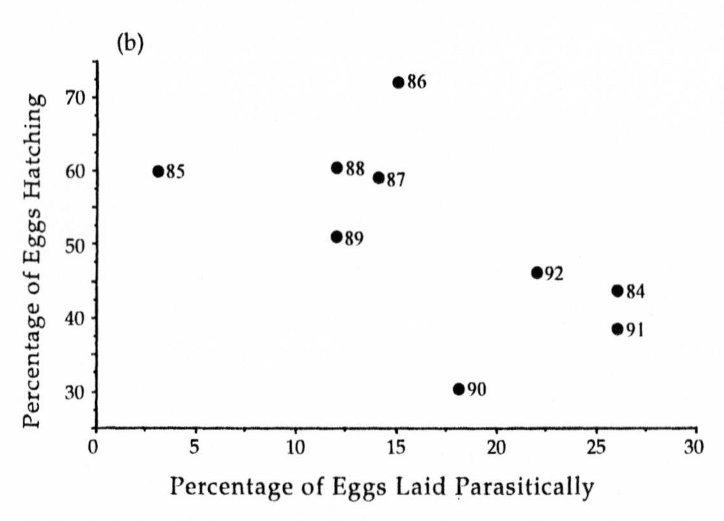

Figure 12-5 Relationship between the frequency of conspecific brood parasitism and (a) mean clutch size ($r_s = .98, p < .01$) and (b) average hatching success of eggs $r_s = -.67, p < .05$) of Barrow's goldeneyes in British Columbia. Data from Eadie (unpublished).

sities and frequencies of parasitism. The middle panel illustrates how the population growth rate varies (measured by λ, the ratio of the population size at time $t + 1$ to time t) with female density and frequency of parasitism. The lower panel combines these dynamics. We consider each of these briefly.

Reproductive Success of Parasites and Nonparasites

The relative RS of a parasitic female exceeds that of a nonparasitic female as the density of females becomes greater than the number of nest sites (i.e., moving up the ordinate in fig. 12-8a). The average RS of a nonparasitic female is devalued by limited nest-site availabil-

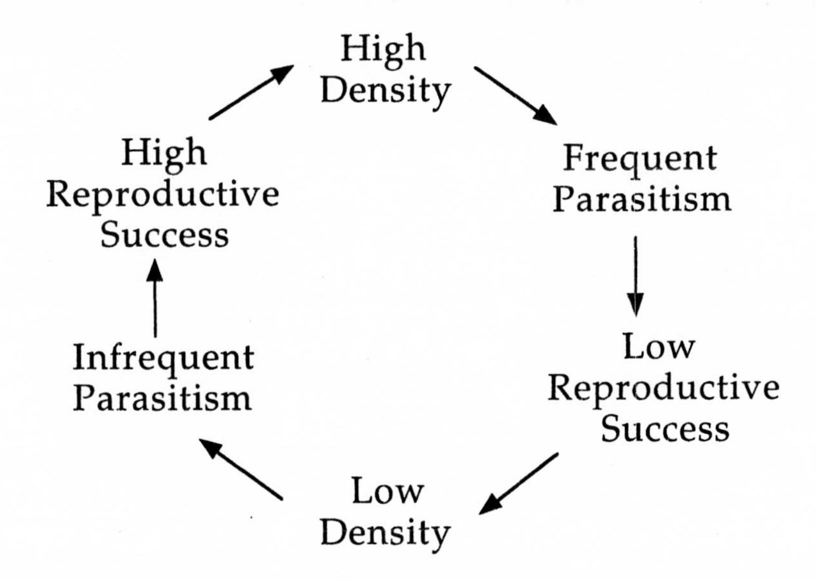

Figure 12-6 Hypothetical relationships between population density, frequency of conspecific brood parasitism, and reproductive success in a cavity-nesting species.

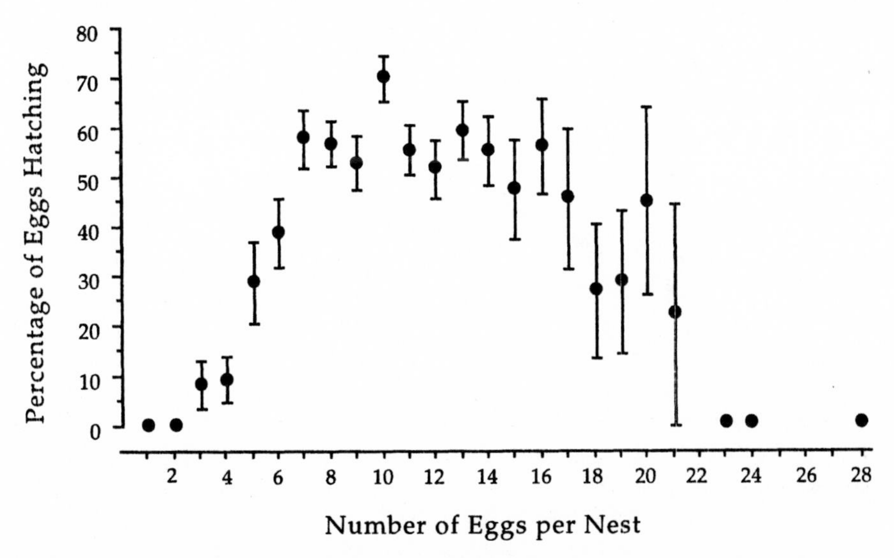

Figure 12-7 Hatching success of Barrow's goldeneye eggs as a function of the clutch size, based on a 9-year study in British Columbia. As clutch sizes become large, hatching success declines. Very small clutch sizes are often not incubated. Error bars are ± 1 SE. Data from Eadie (unpublished).

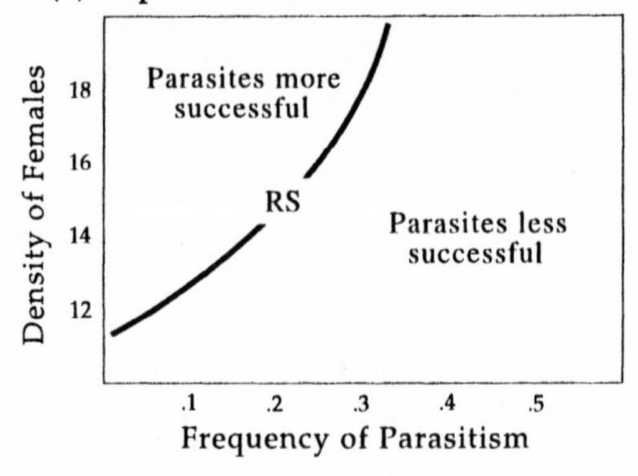

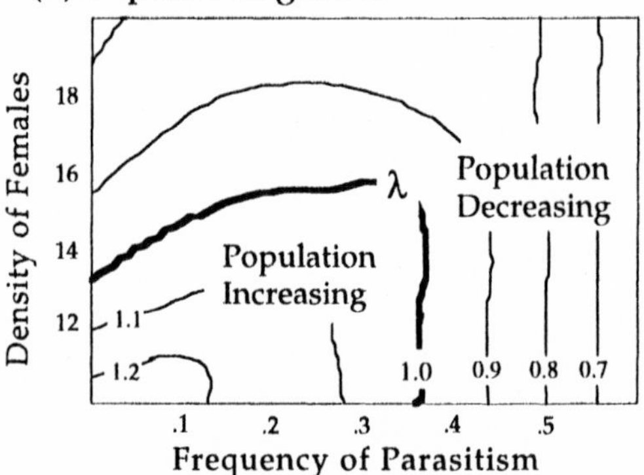

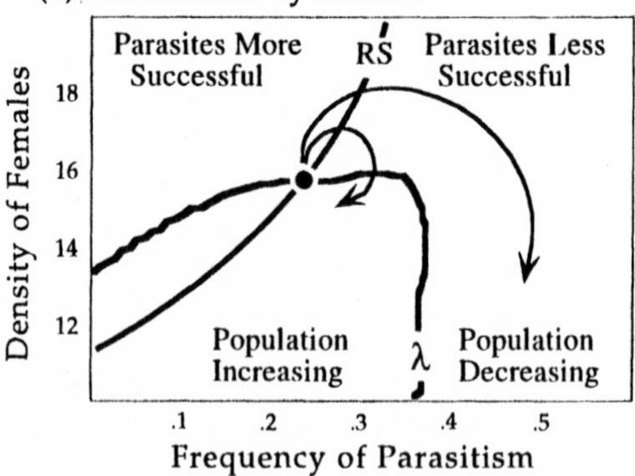

Figure 12-8 Results of a modeling exercise to examine the interaction between the density of breeding females and the frequency of conspecific brood parasitism on (a) the reproductive success of parasitic and nonparasitic females, (b) the population growth rate (λ), and (c) the combined behavioral and population dynamics. In (a) the dark line indicates where parasites and nonparasites have equal reproductive success. In (b) the dark line indicates the combinations of population density and frequency of parasitism that result in stability. In (c) the solid point is the intersection of the curves from (a) and (b).

ity because when nests are in short supply, many nonparasites are unable to breed. Parasitic females face no such limitation, provided that there are at least some hosts with nests. Hence the greater the population density, the greater the reproductive advantage of laying parasitically. This is the density-dependent component of fitness for females in the simulated population.

As the frequency of parasitism increases (moving along the abscissa in fig. 12-8a), the reproductive success of both parasites and nonparasites declines due to increasing clutch sizes and lower hatching success of eggs in parasitized nests (see fig. 12-7; also Semel et al., 1988, 1990 for similar data on wood ducks). However, the relative RS of a parasite declines more rapidly than that of a nonparasitic female because all parasitic eggs are, by definition, in enlarged (parasitized) clutches, whereas some nonparasitic eggs are in nests that have escaped parasitism). The expected hatching success of a parasitic egg is therefore lower on average, and this disadvantage increases as the frequency of parasitism increases. This defines the frequency-dependent components of fitness.

It is possible, then, to seek combinations of density and frequency at which the reproductive success of a parasite would equal that of a nonparasite, so there would be no net change in the frequency of parasitism in the population. These combinations are indicated by the solid line in fig. 12-8a. This isocline represents the evolutionarily stable strategy (ESS) from which any deviation in the frequency of the two behavioral tactics would return the population to that equilibrium.

What is somewhat surprising and has not been fully appreciated previously is that there is no single ESS because the ESS frequency of parasitism is contingent on population density. As density increases, the ESS frequency of parasitism also increases, and there are some combinations of parameters where an ESS is unattainable (e.g., below a density of 11 females in the population simulated in fig. 12-8a). This illustrates that any attempt to examine the evolutionary dynamics of CBP (or other alternative female tactics) must consider not only the frequency-dependent but also the density-dependent components of fitness (see also Nee and May, 1993).

Population Growth

In fig. 12-8b, isoclines of population growth rate are plotted as a function of female density and frequency of parasitism ($\lambda > 1$ indicates a growing population; $\lambda < 1$ indicates a declining population). When the density of females is high, absolute reproductive success of all females is reduced due to competition for limited nest sites and to large numbers of parasitic females laying eggs in a fixed number of host nests. Hence, the population declines ($\lambda < 1$). Likewise, when the frequency of parasitism is high, reproductive success is reduced because clutch sizes become very large, egg hatching success declines, and the population falls. The only situations where the population will grow ($\lambda > 1$) are when it is small and the frequency of parasitism is low (lower left portion of Figure 12-8b).

Under what conditions would the population be stable ($\lambda = 1$)? Combinations of density and frequency of parasitism leading to stability are shown by the darkest line in fig. 12-8b. Once again, there is no single equilibrium. Rather, the equilibrial population density varies with the frequency of parasitism. Initially, the stable density increases slightly as the frequency of parasitism increases. This is because hatching success is hardly affected by a few parasitic eggs, so more females are able to contribute offspring to the population via brood parasitism, even when nest sites are saturated. However, as the frequency of parasitism rises, this situation changes rapidly and population growth rates fall below 1. Above a threshold level of parasitism (> 0.40 in fig. 12-8b), there is no stable population density that can be obtained.

The Combined Dynamics

The most intriguing results of the model occur when the behavioral and population dynamics are combined (fig. 12-8c). Recall that there are two relevant equilibria: the equilibrial frequency of parasitism, dictated by the relative reproductive success of parasites and nonparasites, and the equilibrial population density, dictated by the effects of density and parasitism on population growth rates. Is there any combination of density and frequency at which both equilibria can be obtained? Indeed there is, at the intersection of the dark line that defines equal reproductive success for parasites and nonparasites and the dark line that defines a stable population ($\lambda = 1$; Figs. 12-8a,b). That intersection is shown as the solid point in Fig. 12-8c.

Consider why this is a joint equilibrium. Imagine that a hypothetical population is moved off the equilibrium point because of an increase in density (i.e., it moves upward from the solid point in fig. 12-8c, following the trajectory traced by the inner arrow). At this new point, parasitic females are more successful than nonparasites, and so they will increase in the population. Thus, the population moves to the right (i.e., the frequency of parasitism increase). Now the population growth rate is < 1. Here the population declines, so the trajectory moves down. Now parasites have lower relative RS than nonparasites, and the frequency of parasitism decreases, so the population moves left. Finally, as population growth rates become > 1, the population increases, and it ultimately moves upward again to return to the equilibrium point.

There are several interesting properties of this equilibrium. First, it is a joint equilibrium for both population density and parasite frequency. Any small deviation in either parameter will return the population to that point. Thus, demography sets the conditions under which parasitic and nonparasitic behavior can co-exist at an evolutionarily stable frequency. Conversely, the frequency of parasitism that is maintained by selection operating on the fitnesses of these alternative behaviors will ultimately influence the population size that is demographically stable. Clearly, there is an inextricable link between behavioral ecology and demography in this system.

The second interesting property of this equilibrium is that there are regions of "parameter space" in fig. 12-8c where instability occurs. Specifically, when the frequency of parasitism is high, the "inertia" of the population may prevent a return to the equilibrial point and the population may crash. In other words, the high frequency of parasitism results in such a low value of λ that the population cannot move back quickly enough into a region where $\lambda > 1$ before it goes extinct (this scenario is depicted by the outer arrow in fig. 12-8c).

To explore this further, we used simulations in which we systematically perturbed our model population from the joint equilibrium (fig. 12-9). We found that as the frequency of parasitism increases, there is an exponentially increasing risk of extinction. This risk is low when parasitism is moderate, but it skyrockets when the frequency of parasitism is > 80%. Surprisingly, the starting population size has only a minimal effect on this relationship. Populations of 12, 50, and 100 individuals became extinct when the frequency of parasitism was high, yet populations of 10, 12, and 42 individuals persisted when the frequency of parasitism was low. Clearly, CBP potentially impacts not only dynamics of local populations, but it can even cause extinction if its frequency becomes rapidly inflated.

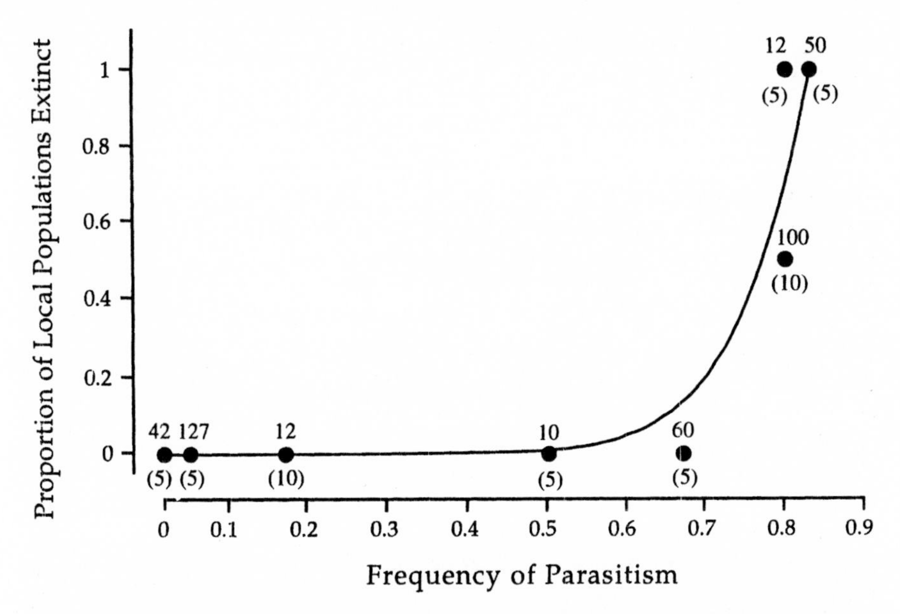

Figure 12-9 Results of a simulation model on the likelihood of local population extinction as a function of the frequency of conspecific brood parasitism. Values above each point indicate the population size for that set of simulations, and the values in parentheses below each point are the number of simulations conducted for those parameters. The logistic regression fit to the points is given by the equation $Y = 5.778 \times {}^{9.414}$.

Linking Behavioral Ecology and Resource Management: Effects of Nest-Boxes

The preceding section demonstrates, in theory, how CBP can influence demography. Of greatest relevance to conservation biologists, however, is the suggestion that high levels of CBP can cause populations to crash (Eadie and Fryxell, 1992; Nee and May, 1993). How could such high levels of parasitism arise? Paradoxically, CBP may be the result of providing artificial nest sites—the very programs designed to increase local population size and viability!

Our studies on wood ducks (Semel and Sherman, 1986, 1992, 1993, 1995; Semel et al., 1988, 1990; Sherman and Semel, 1989) demonstrated that high levels of CBP can lead to reduced reproductive success and productivity. At sites in Missouri and Illinois, Semel and Sherman (1986) and Semel et al. (1988) observed conflicts between females over access to boxes and up to eight different females laying eggs in each one; on average, more than four females contributed to each clutch (fig. 12-10a). Egg deposition rates did not differ significantly from a Poisson distribution, indicating that the number laid in a nest on any particular day was effectively random. In one case, five different females entered and left the same nest box in 21 min, and four of them laid eggs. Similar patterns of parasitism have been documented by others (reviewed by Bellrose and Holm, 1994). Bellrose and Holm (1994) reported many cases of aggression between females over boxes, some of which were so intense that eggs were crushed and females were wounded or killed. Similar strife was re-

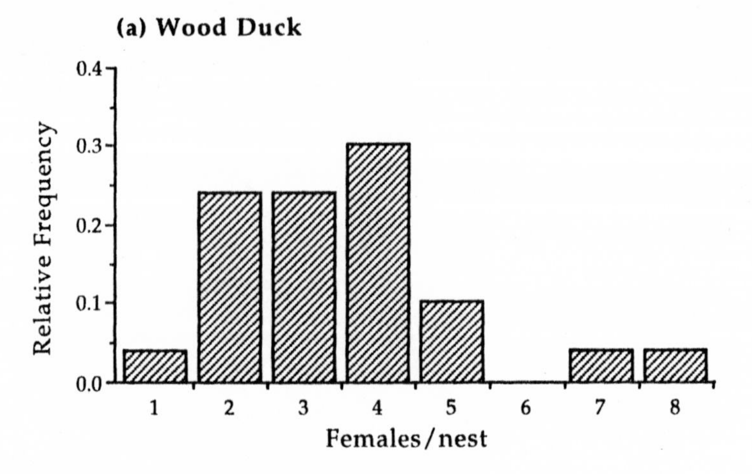

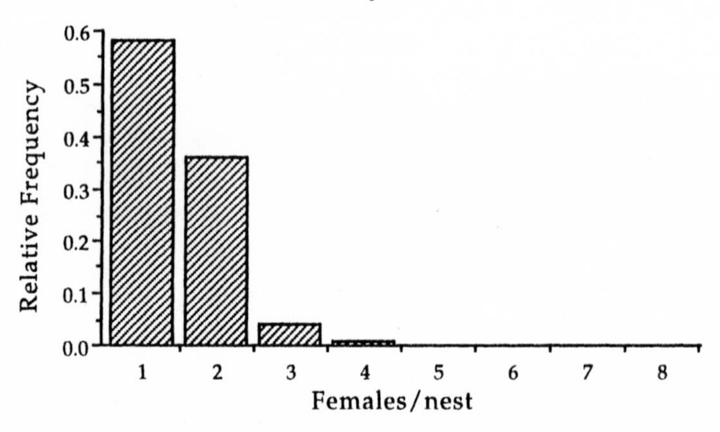

Figure 12-10 Frequency distributions of the number of females laying eggs in each nest in a population of (a) wood ducks (data from Semel and Sherman, 1986) and (b) Barrow's goldeneyes (data from Eadie, unpublished). Mean number of females per nest ($\pm$ 1 SE) for wood ducks was 4.10 $\pm$ 2.23; mean number of females per nest for goldeneyes was 1.49 $\pm$ 0.64.

ported in a population of mallards nesting at very high densities (Titman and Lowther, 1975).

What might cause such extreme levels of parasitism and aggressive behavior? Semel and Sherman (1986) and Semel et al. (1988) suggested that in wood ducks it was related to the placement of nest-boxes. Typically, boxes are erected at high densities, in highly visible locations (e.g., on poles over open water), and often they are tightly clustered (e.g., back-to-back "duplexes," or multiple-box "apartments"; McLaughlin and Grice, 1952; photos in Bellrose and Holm, 1994). The intent is to maximize nest-box occupancy, and this goal has

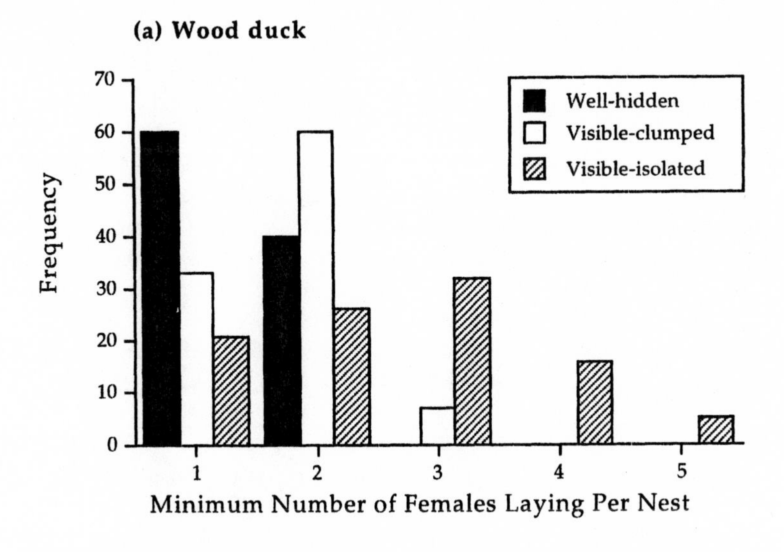

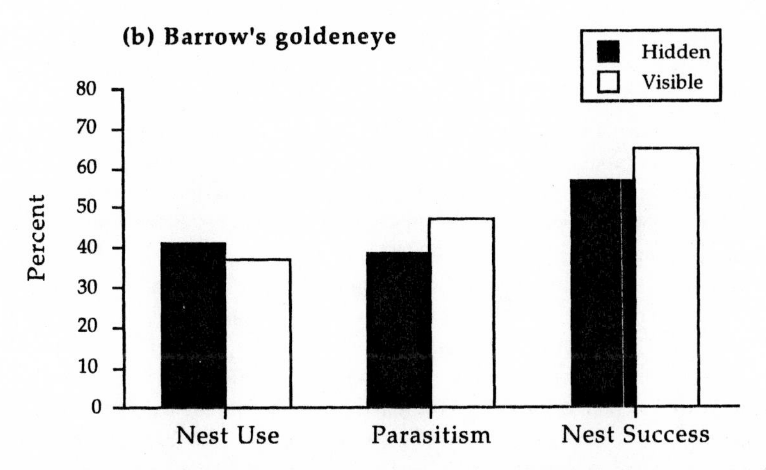

Figure 12-11 Effects of nest-box visibility on nest use and the frequency of brood parasitism. (a) Frequency distribution of the minimum number of wood duck females laying per nest as a function of the location of the nest site (well-hidden, visible and clumped, or visible and isolated from other boxes; data from Semel et al., 1988); (b) nest use, frequency of brood parasitism, and nest success of Barrow's goldeneyes in hidden and visible nest-boxes (nest use: G test with Williams' correction, $G_w = 1.40, p = .26, n = 416$ visible nest-boxes, 479 hidden boxes; brood parasitism: $G_w = 2.58, p = .11, n = 155$ visible boxes, 197 hidden boxes; nest success: $G_w = 2.47, p = .15, n = 136$ visible boxes, 161 hidden boxes). Data from Eadie (unpublished).

generally been met (e.g., Soulliere, 1990a). But consider how such a management practice might affect the birds' social behavior.

Wood ducks typically nest in tree cavities that are widely spaced and well-hidden. CBP appears to be facilitated by females observing conspecifics entering or leaving active nests (Semel and Sherman, 1986; Wilson, 1993). So, when boxes are grouped in visible locations, it becomes virtually impossible for females to visit their nest unobserved. Erecting boxes at high densities and in close proximity effectively creates a colonial nesting situation for a species that typically nests solitarily (Sherman and Semel, 1989). The results often are high box-occupancy rates, coupled with widespread CBP, intraspecific interference among females, and reduced egg hatching success.

These patterns are illustrated for an Illinois population of wood ducks in fig. 12-11a. Dispersed, hidden boxes were far less likely to be parasitized, or parasitized by multiple females, than boxes that were either dispersed or clustered on poles over open water. Recently Semel and Sherman (1995) manipulated the placement of nest-boxes and confirmed that boxes erected in the open suffered significantly higher rates of parasitism and significantly reduced egg hatching success compared to boxes erected on tree trunks in woodlands.

The management implications of these observations and our analyses are that high frequencies of parasitism, brought about by too many closely-spaced nest-boxes, can move a local population into a region of demographic instability, resulting in population decline (figs. 12-8, 12-9). Field studies of wood ducks provide empirical support for this scenario. Declines or sudden "crashes" in numbers characterize populations nesting in visible, clustered nest-boxes, and these declines often occur after a build up in population density and frequency of CBP (table 12-3, figs. 12-1–12-3).

It might be thought that even if egg hatching success per nest decreases as a result of frequent CBP, the "productivity" of a given population increases because more eggs are laid in total. If this were true, then so long as there is a net gain in the total number of offspring produced, the population-level consequences of CBP need not be of concern. The problem with this reasoning is that it ignores potential life history effects of increased female reproductive efforts (Lessells, 1992). In ducks generally, eggs are large relative to female body size (Bellrose and Holm, 1994); egg production is particularly energetically expensive in wood ducks (Drobney 1980). If visible, clustered nest boxes induce females to lay large numbers of eggs that have little chance of hatching, the net effect may be to reduce female survival (as well as population size). For example, Semel and Sherman (1995) found that females nesting in visible boxes returned 12% less frequently than females nesting in hidden boxes. Little consideration has been given to such hidden, long-term consequences of traditional nest box programs. The clear message is that if we ignore the behavioral ecology of wood ducks in our efforts to improve nesting habitats and thus reproduction, we run the risk of defeating our own purposes.

Brood Parasitism, Nest-Boxes, and Goldeneyes: An Alternative Perspective

We found surprisingly different effects of nest-box placement on CBP and population dynamics in Barrow's goldeneyes. Unlike wood ducks, goldeneyes are highly territorial (Savard, 1984, 1988; Savard and Smith, 1987). Males defend an exclusive site-specific territory during the egg-laying and early-incubation periods, apparently to provide food resources and an undisturbed place for their mate to feed during the energetically demanding

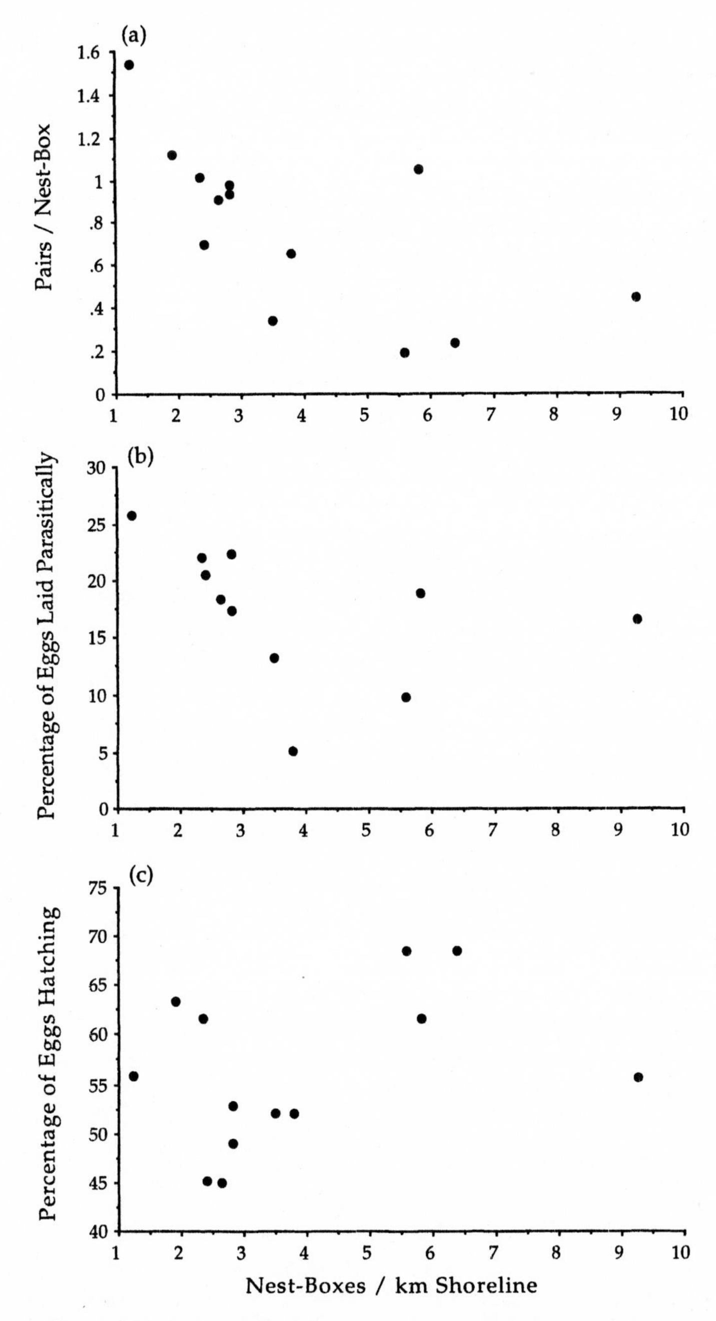

Figure 12-12 Relationships between nest-box density (per kilometer of lake shoreline) and (a) number of Barrow's goldeneye pairs per nest-box ($r_s = -.81$, $p = .02$), (b) frequency of conspecific brood parasitism ($r_s = -.71$, $p < .05$), and (c) percentage of eggs that hatched ($r_s = .46$, $p > .10$). Each point represents the average (over a 9-year study period) for one of 13 lakes in British Columbia. Data from Eadie (unpublished).

period of egg laying. Females also maintain exclusive, fixed territories after their ducklings hatch, apparently to ensure adequate food resources.

There are three consequences of goldeneye territoriality. First, it sets an upper limit on the number of pairs that can breed on a particular lake or pond (Savard, 1984, 1988; Savard and Smith, 1987). For this reason, addition of nest-boxes does not lead to rapid increases in local goldeneye populations. In fact, addition of boxes reduces the ratio of pairs per box (fig. 12-12a) because the same number of breeding pairs is distributed among a larger number of boxes. As a result, the frequency of parasitism decreases (fig. 12-12b), and egg hatching success does not diminish when boxes are added (fig. 12-12c).

Second, because goldeneye pairs aggressively exclude conspecifics, fewer females lay in the same nest-box than do wood ducks (fig. 12-10b). On average, 35% of nests were parasitized in the box-nesting goldeneye population studied by Eadie (1989). It should be noted that when high frequencies of parasitism do occur, they affect female reproductive success in the same frequency-dependent manner as in wood ducks (fig. 12-5). However, parasitism in box-nesting goldeneyes rarely reaches the extreme levels recorded for wood ducks.

Third, nest-box visibility does not seem to affect frequency of box use, CBP, or nest success in goldeneyes (fig. 12-11b). If anything, nest success (proportion of nests from which at least one duckling hatched) in visible boxes is slightly higher than in hidden boxes. In sum, the behavioral ecology of Barrow's goldeneyes is sufficiently different from that of wood ducks that factors facilitating widespread CBP in the wood ducks rarely affect goldeneyes.

May et al. (1991:1015) likened CBP to Hardin's "tragedy of the commons":

> A sole owner—a bird dealing with its own nest and fledglings in the absence of parasitism—can evolve behavior (such as an optimal clutch size) that maximizes the returns on its efforts. But in a "commons," exploiters will continue to enter the market—parasites will continue to lay eggs in others nests—until the point is reached where the marginal rate of return is zero (profits balances costs); everyone loses, locked into the perverse, and sometimes tragic, logic of the commons.

We think this metaphor is especially appropriate to the situation in wood ducks and goldeneyes.

The solution to the dilemma of the commons in a human economy is to construct regulatory mechanisms that effectively restore ownership (May et al., 1991). In a Darwinian economy, restricted access to nests might be effected through the evolution of vigilance and aggressive site defense (territoriality), as in goldeneyes, or surreptitious behavior near the (hidden) nest site, as in wood ducks. The problem is that our management practices, by effectively circumventing these proprietary mechanisms, can unwittingly cause the resource to become unregulated. Such is certainly the case with wood duck nest-box programs. By adding clusters of highly visible boxes, we have negated the evolved mechanisms by which the birds retain sole nest ownership. Elimination of such mechanisms results inevitably in overexploitation of resources to the point where the marginal rate of return is zero, and the tragedy of open access prevails. In goldeneyes, by contrast, proprietary use of a limited resource appears to be maintained through territoriality. Because access remains regulated, the tragedy is averted.

In sum, determining whether the commons is likely to become unregulated is important to the success of management and conservation protocols. Determining how the commons might become unregulated is an important insight that behavioral ecologists can provide.

Beyond Wood Ducks and Goldeneyes

Conspecific brood parasitism has been documented in more than 185 birds (Yom-Tov, 1980; Eadie et al., 1988; MacWhirter, 1989; Rohwer and Freeman, 1989; Sayler, 1992), of which 49 species nest in cavities (table 12-4). Among waterfowl, CBP occurs at least occasionally in all 38 hole-nesters (Sayler, 1992; see also Eadie et al., 1988; Rohwer and Freeman, 1989; Eadie, 1991). Frequencies of CBP within populations of cavity-nesting waterfowl (fig. 12-13; mean = 44.7% of nests parasitized) are significantly higher than in species that nest in emergent vegetation (25.2%) or in upland areas (12.4%; Kruskal-Wallis $H = 25.3$ $p < .0001$). Of particular interest are circumstances where CBP exceeds 60% of nests because the greatest potential for population destabilization occurs at these levels (figs. 12-8, 12-9). At least five studies in Sayler's (1992) sample of cavity-nesting species reported frequencies of CBP exceeding 60% (fig. 12-13).

Information on the frequency of parasitism within populations is available for only a few non-Anseriforme species. In some hole-nesters CBP occurs as often as in waterfowl. For example, the proportion of clutches parasitized among years was 12–46% in European starlings *Sturnus vulgaris* (Evans, 1988; Pinxten et al., 1993), 20% in eastern bluebirds *Sialia sialis* (Gowaty and Karlin, 1984; Gowaty and Bridges, 1991), 36% in purple martins *Progne subis* (Morton et al., 1990), 10–27% in white-fronted bee eaters *Merops bullockoides* (Emlen and Wrege, 1986), and 8% in house sparrows *Passer domesticus* (Kendra et al., 1988); the latter occasionally nest in holes. Conspecific brood parasitism can also occur at high frequencies in species that build enclosed mud or grass nests such as cliff swallows *Hirundo pyrrhonota* (22–43%; Brown and Brown, 1988), barn swallows *Hirundo rustica* (3–33%; Møller, 1987), and northern masked weavers *Ploceus taeniopterus* (25–35%; Jackson, 1992).

We do not yet know how CBP affects the population biology of these species. However, numerous studies have suggested links among nest-box density, population density, and frequency of CBP. Thus, Delnicki (1973) and McCamant and Bolen (1979) observed exceptionally high levels of CBP (74–91% of nests) in a population of black-bellied whistling ducks *Dendrocygna autumnalis* nesting in closely-spaced boxes. Evans (1988) reported that frequencies of CBP in starlings increased when the occupancy of nest-boxes was high. Gowaty and Bridges (1991) experimentally manipulated nest-box density in a population of eastern bluebirds and found that the proportion of nondescendent nestlings (i.e., a result of either CBP or extrapair fertilizations) could be as high as 60%, and was significantly higher in the high density treatment. Møller (1987) found that CBP was more frequent in barn swallows in years when populations were large. CBP also was more frequent in large colonies than in small groups of cliff swallows which do not nest in boxes (Brown and Brown, 1988, 1996).

Although CBP interferes with individual reproductive success in most species examined (Petrie and Møller, 1991), frequent CBP has not been linked to population declines in any of these species. However, evidence of a link between CBP and population dynamics may be elusive, even when it exists. Our empirical results and modeling efforts indicate that deleterious effects of CBP occur only when frequencies of parasitism are high. In natural populations, where nest sites are dispersed and densities are low, populations might rarely, if ever, reach levels of CBP where population instability or extinction is predicted (fig. 12-8). Hence, we might not expect to find evidence of such density dependence. Our concern, however, is that management practices may inadvertently move populations into that region

Table 12-4 Species of cavity and hole-nesting birds for which conspecific brood parasitism has been recorded (data from Yom-Tov, 1980; Eadie et al., 1988; Rohwer and Freeman, 1989; Sayler, 1992).

Common name	Species name
Spotted whistling duck	*Dendrocygna guttata*
Black-billed whistling duck	*Dendrocygna arborea*
Lesser whistling duck	*Dendrocygna javanica*
Black-bellied whistling duck	*Dendrocygna autumnalis*
Fulvous whistling duck	*Dendrocygna bicolor*
Comb duck	*Sarkidiornis melanotos*
Ruddy shelduck	*Tadorna ferruginea*
Cape shelduck	*Tadorna cana*
Australian shelduck	*Tadorna tadornoides*
Paradise shelduck	*Tadorna variegata*
Radjah shelduck	*Tadorna radjah*
Shelduck	*Tadorna tadorna*
Orinoco goose	*Neochen jubata*
Egyptian goose	*Alopochen aegyptiacus*
Hartlaub's duck	*Pteronetta hartlaubi*
Muscovy duck	*Cairina moschata*
White-winged wood duck	*Cairina scutulata*
Wood duck	*Aix sponsa*
Mandarin	*Aix galericulata*
Maned duck	*Chenonetta jubata*
Green pygmy goose	*Nettapus pulchellus*
Cotton pygmy goose	*Nettapus coromandelianus*
African pygmy goose	*Nettapus auritus*
Brazilian duck	*Amazonetta brasiliensis*
Ringed teal	*Callonetta leucophrys*
Gray teal	*Anas gibberifrons*
Chestnut teal	*Anas castanea*
Spot-billed duck	*Anas poecilorhyncha*
Speckled Yellow-billed teal	*Anas flavirostris*
Gray duck	*Anas superciliosa*
Bufflehead	*Bucephala albeola*
Barrow's goldeneye	*Bucephala islandica*
Common goldeneye	*Bucephala clangula*
Smew	*Mergellus albellus*
Hooded merganser	*Lophodytes cucullatus*
Red-breasted merganser	*Mergus serrator*
Chinese merganser	*Mergus squamatus*
Common merganser	*Mergus merganser*
Great tit	*Parus major*
Tree swallow	*Tachycineta bicolor*
Bank swallow	*Riparia riparia*
Cliff swallow	*Hirundo pyrrhonota*
Purple martin	*Progne subis*
Eastern bluebird	*Sialia sialis*
Western bluebird	*Sialia mexicana*
Mountain bluebird	*Sialia currucoides*
European starling	*Sturnus vulgaris*
House sparrow	*Passer domesticus*
White-fronted bee-eater	*Merops bullockoides*

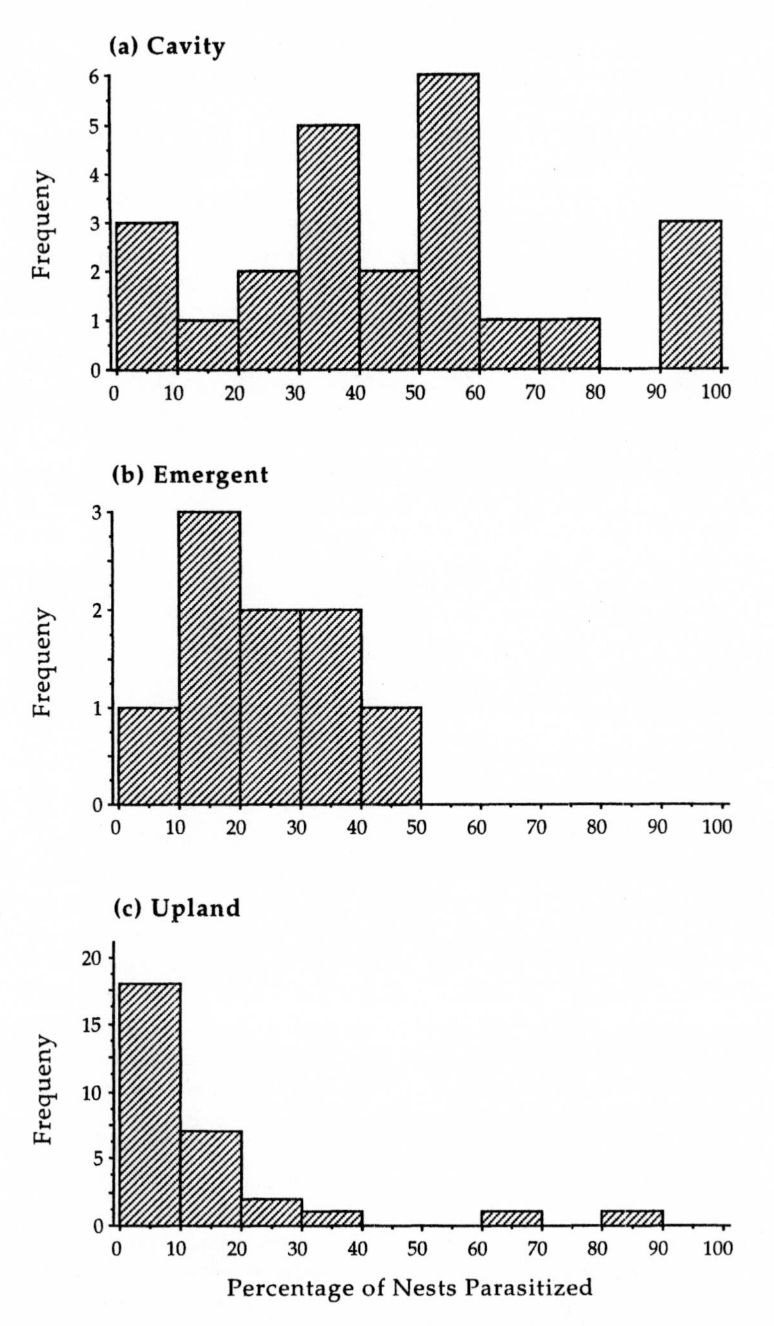

Figure 12-13 Histograms of the frequency of conspecific brood parasitism recorded in separate studies for (a) cavity-nesting waterfowl (24 studies on 10 species), (b) emergent nesting waterfowl (9 studies on 7 species), and (c) upland nesting waterfowl (30 studies on 14 species). Data from Sayler (1992).

(e.g., by the addition of highly visible boxes at high densities, such as some bluebird trails). It is then that we might be unpleasantly surprised.

Nest boxes are not the only habitat alterations that may concentrate breeding individuals to the point where social dynamics are altered. For example, habitat fragmentation may concentrate individuals into the few remaining suitable patches. If inflated local densities lead to high rates of CBP (e.g., Titman and Lowther, 1975), this could further destabilize a population. Habitat fragmentation might also increase interspecific brood parasitism by brown-headed cowbirds *Molothrus ater*, as well as competition between native cavity-nesting species and European starlings (Semel and Sherman 1995).

Provision of artificial nest structures other than boxes (e.g. nesting islands for waterfowl, nest platforms for geese, nest burrows for seabirds) could have similar effects. For example, Giroux (1985) found increased levels of CBP in American avocets (*Recurvirostra americana*) nesting on artificial islands, and Giroux (1981) and Lokemoen (1991) reported inflated levels of conspecific and interspecific brood parasitism among several species of waterfowl nesting on islands. Ultimately, any anthropogenic factor that increases the density of local populations could lead to elevated levels of CBP, density strife, and subsequent population instability. Lack of empirical information on these issues limits our abilities to employ a behavioral ecological approach more broadly.

Strengths and Weaknesses of the Behavioral Ecological Approach

Strengths

Our approach explicitly links behavioral dynamics with population dynamics. There have been surprisingly few efforts to examine the impact of intraspecific behavioral interactions on the dynamics of natural populations or to model the possible population trajectories that might result. Most conservation and management programs are undertaken with the assumption that population dynamics result primarily from processes extrinsic to the population, both biotic (predation, disease) and abiotic (weather, habitat). However, population processes also are affected by individuals' reproductive decisions; dispersal is an obvious example. Indeed, individual behaviors are sometimes more important to population dynamics than extrinsic factors. This chapter identified a specific behavioral interaction (CBP) that directly impacts population dynamics and explored how these factors are linked using both field studies and simulation modeling.

Models based on behavioral ecological data allow us make predictions for other populations or species. Behavioral ecological models explicitly incorporate costs and benefits of behavioral alternatives, measured in terms of individual reproductive success (Grafen, 1991). These models generate testable predictions about how changes in individuals' behaviors might influence populations dynamics. Such approaches offer the important opportunity of exploring regions of population instability and unpredictability before those regions are ever realized in the field.

Weaknesses

Most of our information on wood ducks and goldeneyes is from studies employing nest-boxes. Møller (1989, 1992) argued that nest-boxes represent a highly artificial breeding situation and that predation rates, age structure of the subpopulation using boxes, invertebrate parasite loads, nest-site preferences, and patterns of natal dispersal all may differ substantially between boxes and natural cavities. By contrast, Koenig et al. (1992) pointed out that habitat alteration by humans has resulted in such a marked decline in snags that adding nest-boxes may result in cavity densities that approximate conditions prior to human influence (i.e., the addition of nest-boxes restores natural conditions). Either way, paucity of information on CBP among birds breeding in natural cavities represents a significant limitation to our ability to generalize further from our results.

Current models may be too simplified and unrealistic to be of practical use to conservation managers. All models require numerous simplifying assumptions. For example, we assumed, as did Eadie and Fryxell (1992) and May et al. (1991), that females either parasitized or nested, but not both. However, we know that parasitism is not a "fixed" behavioral strategy (Bellrose and Holm, 1994); indeed, behavioral observations indicate that some females parasitize and nest in the same season (Eadie, Semel, and Sherman, unpublished data). Therefore we must be cautious about extrapolating too much from the results.

However, there are several reasons to believe that the relationships we discovered (figs. 12-8, 12-9) are reasonable. First, all the functions and parameters used in our model were based on empirically derived field data. Second, Eadie and Fryxell (1992) examined the sensitivity of the basic model to relaxation of several key assumptions, such as the form of the brood survival function (fig. 12-7) and the constraint that females act only as parasites or nonparasites, and the results were robust. Third, Nee and May (1993) independently developed an analytical model that, although similar in conceptual structure, was not based on the specific biology of any particular species. They arrived at remarkably convergent conclusions to ours. Even if one questions the predictive value of these modeling exercises, they have important heuristic value in demonstrating that social interactions among conspecifics can play a critical role in population stability and maintenance.

Recommendations

We do not wish to discourage use of nest-boxes in management and conservation. Rather, we encourage awareness of how nest-boxes can affect the intraspecific behavioral dynamics and reproduction of the target species. This points to a need for more information on the behavioral ecology of cavity-nesting birds in natural sites. Much of the available literature concentrates on a small number of species, predominantly populations nesting in boxes. If we are to understand the factors influencing population trajectories, we will need more and better comparative information on behavioral interactions and population ecology of birds using natural cavities versus nest-boxes.

If nest-boxes are to be erected, field experiments should be conducted to determine the optimum dispersion and density of boxes. In many cases, it may be advisable to mimic the natural dispersion and circumstances of cavities by separating boxes and hiding them. Semel and Sherman (1993, 1995) demonstrated that more ducklings were produced from the same number of wood duck nest-boxes when they were dispersed and hidden than when the boxes were clustered over open water. They also calculated that "nontraditional" nest-box placement could considerably reduce the number of boxes needed to meet the same production goal, thereby reducing material and personnel costs. The implication is that box placement strategies can enhance the effectiveness of a nest-box program. Such strategies should be developed in light of, rather than in spite of, the behavioral ecology of the target species.

Finally, we encourage management by experiment. Often we have little information on the basic biology of a species, let alone detailed insights into its behavioral ecology. Such information can potentially be obtained through our management actions. We will learn far more by experimentally creating a range of management alternatives than by pursuing a single management option. Accurate measurement requires sufficient variation with which to detect a response. If we provide only one management option, we have little scope for measuring the population's response; hence, we fail to learn and adapt. Adaptive management (Walters and Holling, 1990) has provided an important guideline for wildlife and fisheries biology and could prove even more valuable for conservation biology.

Summary

A substantial number of bird species nest in cavities. For many of these, nesting habitat had dwindled. As a result, a minimum of 100 species of cavity-nesting birds have declined sufficiently to warrant national and international concern. Traditional approaches to the conservation of cavity-nesting birds species include snag management, alteration of silviculture practices to maintain the heterogeneity of stand age structure, and lengthening of the harvest period to increase the abundance of mature trees. Another widely-used approach has been to supplement natural cavities with artificial nest sites.

Nest-boxes are a simple and cost-effective method to improve nesting habitat. However, providing high densities of highly visible nest-boxes has the potential to alter social dynamics and reproductive competition among individuals in a population. In some species, these changes may lead to precipitous declines in individual reproductive success, resulting in population instability and decline. Studies of wood ducks in eastern North America demonstrate these complications. Conspecific brood parasitism occurs frequently among wood ducks. In populations breeding in visible, clumped nest-boxes, up to 100% of nests may be parasitized and up to eight females may attempt to lay eggs in a single nest. Numerous studies have demonstrated a link between the density of nest-boxes, the frequency of CBP, and reduced egg hatching success. Typically, as nest-boxes are added to an area and breeding density increases, the frequency of CBP also increases and reproductive efficiency declines. In several populations, population crashes occurred when breeding densities and CBP were high.

A simple model illustrates how the behavioral dynamics of CBP can interact with population density to produce a variety of demographic outcomes. Notably, high frequencies of CBP can lead to unstable fluctuations and possibly to the extirpation of local populations. Extremely high levels of CBP and subsequent declines in population productivity result from nest-box placement. Boxes erected at high densities and in highly visible locations are more often parasitized and produce fewer ducklings than dispersed, hidden boxes (which mimic locations of natural tree cavities).

Conspecific brood parasitism also occurs in the Barrow's goldeneye, another cavity-nester. However, frequencies of CBP do not reach the same levels in goldeneyes as in wood ducks, apparently because goldeneyes are highly territorial, which limits local breeding densities. As a result, adding nest-boxes to lakes on which goldeneyes nest does not lead to an increase in CBP or reduced reproductive success. Differences between wood ducks and goldeneyes illustrate the importance of considering the behavioral ecology of the target species in devising conservation or management plans and the utility of behavioral ecological models that explicitly link population growth with the reproductive tactics and successes of individuals. Individuals that use nest-boxes in management (or research) should recognize that boxes may alter social structure, reproductive competition, and population dynamics of the target species. We therefore encourage a policy of management by experiment, whereby alternative density and dispersion patterns of artificial nest sites are evaluated simultaneously and systematically.

Acknowledgments We thank Tim Caro, Walt Koenig, Mike Lombardo, and an anonymous reviewer for many helpful comments on this chapter. Dirk Van Vuren offered sage advice and helpful consultation during the writing of this chapter. John Eadie's research was supported by grants from the Natural Science and Engineering Research Council of Canada, the Dennis Raveling Waterfowl Endowment, and a U.S. Department of Agriculture Hatch Grant administered by University of California. Paul Sherman and Brad Semel's research was supported by the Max McGraw Wildlife Foundation and a U.S. Department of Agriculture Hatch Grant administered by Cornell University.

References

Andersson M, 1984. Brood parasitism within species. In: Producers and scroungers: strategies of exploitation and parasitism (Barnard CJ, ed). London: Croom Helm, 195–227.

Andersson M, Eriksson MG, 1982. Nest parasitism in goldeneye *Bucephala clangula:* Some evolutionary aspects. Am Nat 120:1–16.

Bellrose FC, 1976. Ducks, geese and swans of North America. Harrisburg, Pennsylvania: Stackpole.

Bellrose FC, 1990. The history of wood duck management, In: Proceedings of the 1988 North American Wood Duck Symposium (Frederickson LH, Burger GV, Havera SP, Graber DA, Kirby RE, Taylor TS, eds). St. Louis, Missouri; 13–20.

Bellrose FC, Holm DJ, 1994. Ecology and management of the wood duck. Mechanicsburg, Pennsylvania: Stackpole.

Bock CE, Cruz A, Grant MC, Aid CS, Strong TR, 1992. Field experimental evidence for diffuse competition among southwestern riparian birds. Am Nat 140:815–828.

Brawn JD, Balda RP, 1988. Population biology of cavity nesters in Northern Arizona: do nest sites limit breeding densities? Condor 90:61–71.

Brown CR, Brown MB, 1988. Genetic evidence of multiple parentage in broods of cliff swallows. Behav Ecol Sociobiol 23:379–387.

Brown CR, Brown MB, 1996. Coloniality in the cliff swallow. Chicago: University of Chicago Press.

Brush T, 1983. Cavity use by secondary cavity-nesting birds and response to manipulations. Condor 85:461–466.

Caine LA, Marion WR, 1991. Artificial addition of snags and nest boxes to slash pine plantations. J Field Ornithol 62:97–106.

Clawson EL, Hartman GW, Fredrickson LH, 1979. Dump nesting in a Missouri wood duck population. H Wildl Manage 43:347–355.

Davis JW, 1983. Snags are for wildlife. In: Proceedings of a symposium on snag habitat management (Davis JW, Goodwin GA, Okenfels RA, eds). USDA Forest Service Technical Report RM-99. Fort Collins, Colorado: U.S. Department of Agriculture, 4–8.

Delnicki DE, 1973. Renesting, incubation behavior and compound clutches of the black-bellied tree duck in southern Texas. (MS thesis). Lubbock: Texas Tech University.

Drobney RD, 1980. Reproductive bioenergetics of wood ducks. Auk 97:480–490.

Eadie JM, 1989. Alternative reproductive tactics in a precocial bird: the ecology and evolution of brood parasitism in goldeneyes (PhD dissertation). Vancouver: University of British Columbia.

Eadie JM, 1991. Constraint and opportunity in the evolution of brood parasitism in waterfowl. Acta Congr Int Ornithol 20:1031–1040.

Eadie JM, Fryxell JM, 1992. Density dependence, frequency dependence, and alternative nesting strategies in goldeneyes. Am Nat 140:621–641.

Eadie JM, Gauthier G, 1985. Prospecting for nest sites by cavity-nesting ducks of the genus *Bucephala*. Condor 87:528–534.

Eadie JM, Kehoe FP, Nudds TD, 1988. Pre-hatch and post-hatch brood amalgamation in North American Anatidae: a review of the hypotheses. Can J Zool 66:1709–1721.

Eadie JM, Mallory ML, Lumsden HG, 1995. Common goldeneye (*Bucephala clangula*). In: The birds of North America, (Poole A, Gill F, eds). Washington, DC: American Ornithologists Union. 170:1–32.

Ehrlich PR, Dobkin DS, Wheye D, 1988. The birder's handbook: a field guide to the natural history of North American birds. New York: Simon and Schuster.

Ehrlich PR, Dobkin DS, Wheye D, Pimm SL, 1994. The birdwatcher's handbook: a field guide to the natural history of the birds of Britain and Europe. New York: Simon and Schuster.

Emlen ST, Wrege PH, 1986. Forced copulations and intraspecific parasitism: two costs of social living in the white-fronted bee-eater. Ethology 71:2–29.

Evans PGH, 1988. Intraspecific nest parasitism in the European starling *Sturnus vulgaris*. Anim Behav 36:1282–1294.

Giroux J-F, 1981. Interspecific nest parasitism by redheads on islands in southeastern Alberta. Can J Zool 59:2053–2057.

Giroux J-F, 1985. Nest sites and superclutches of American avocets on artificial islands. Can J Zool 63:1302–1305.

Gowaty PA, Bridges WC, 1991. Nestbox availability affects extra-pair fertilizations and conspecific nest parasitism in eastern bluebirds, *Sialia sialis*. Anim Behav 41:661–675.

Gowaty PA, Karlin AA, 1984. Multiple maternity and paternity in single broods of apparently monogamous eastern bluebirds (*Sialia sialis*). Behav Ecol Sociobiol 15:91–95.

Grafen A, 1991. Modelling in behavioural ecology. In Behavioural ecology, 3rd ed. (Krebs JR, Davies NB, eds). Oxford: Blackwell; 5–31.

Haramis GM, 1990. Breeding ecology of the wood duck: a review. In: Proceedings of the 1988 North American wood duck symposium (Frederickson LH, Burger GV, Havera SP, Graber DA, Kirby RE, Taylor TS, eds). St. Louis, Missouri; 45–60.

Haramis GM, Thompson DQ, 1985. Density-production characteristics of box-nesting wood ducks in a northern greentree impoundment. J Wildl Manage 49:429–436.

Hepp GR, Bellrose FC, 1995. Wood duck (*Aix sponsa*). In: The birds of North America, Poole A, Gill F, eds). Washington, DC: American Ornithologists Union, 169:1–24.

Jackson WM, 1992. Estimating conspecific nest parasitism in the northern masked weaver based on within-female variability in egg appearance. Auk 109:435–443.

Jones RE, Leopold AS, 1967. Nesting interference in a dense population of wood ducks. J Wildl Manage 31:221–228.

Kendra PE, Roth RR, Tallamy DW, 1988. Conspecific brood parasitism in the house sparrow. Wilson Bull 100:80–90.

King WB, !981. Endangered birds of the world. The ICBP bird red data book. Washington, DC: Smithsonian Institution.

Koenig WD, Gowaty PA, and Dickinson JL, 1992. Boxes, barns, and bridges: confounding factors or exceptional opportunities in ecological studies? Oikos 63:305–308.

Land D, Marion WR, O'Meara TE, 1989. Snag availability and cavity nesting birds in slash pine plantations. J Wildl Manage 53:1165–1171.

Lessells CM, 1992. The evolution of life histories. In Behavioral ecology: an evolutionary approach, 3rd ed (Krebs JR, Davies NB, eds). Oxford: Blackwell. 32–68.

Lokemoen JT, 1991. Brood parasitism among waterfowl nesting on islands and peninsulas in North Dakota. Condor 93:340–345.

McCamant RE, Bolen EG, 1979. A 12-year study of nest box utilization by black-bellied whistling ducks. J Wildl Manage 43:936–943.

McLaughlin CL, Grice D, 1952. The effectiveness of large-scale erection of wood duck boxes as a management procedure. Trans N Am Wildl Conf 17:242–59.

MacWhirter RB, 1989. On the rarity of intraspecific brood parasitism. Condor 91:485–492.

May RM, Nee S, Watts C, 1991. Could intraspecific brood parasitism cause population cycles? Acta Congr Int Ornithol 20:1012–1022.

Møller AP, 1987. Intraspecific nest parasitism and anti-parasite behavior in swallows, *Hirundo rustica.* Anim Behav 35:247–254.

Møller AP, 1989. Parasites, predators and nest boxes: facts and artefacts in nest box studies of birds? Oikos 56:421–423.

Møller AP, 1992. Nest boxes and the scientific rigour of experimental studies. Oikos 63:309–311.

Morse TE, Wight HM, 1969. Dump nesting and its effects on production in wood ducks. J Wildl Manage 33:284–293.

Morton ES, Forman L, Braun M, 1990. Extrapair fertilizations and the evolution of colonial breeding in purple martins. Auk 107:275–283.

Nee S, May RM, 1993. Population-level consequences of conspecific brood parasitism in birds and insects. J Theor Biol 161:95–109.

Newton I, 1994a. Experiments on the limitation of bird breeding densities. Ibis 136:397–411.

Newton I, 1994b. The role of nest sites in limiting the numbers of hole-nesting birds: a review. Biol Conserv 70:265–276.

Petrie M, Møller AP, 1991. Laying eggs in others' nest: intraspecific brood parasitism in birds. Trends Ecol Evol 6:315–320.

Pinxten R, Hanotte O, Eens M, Verheyen RF, Dhondt A, Burke T, 1993. Extra-pair paternity and intraspecific brood parasitism in the European starling *Sturnus vulgaris:* evidence from DNA fingerprinting. Anim Behav 45:795–809.

Raphael MG, White M, 1984. Use of snags by cavity-nesting birds in the Sierra Nevada. Wildlife Monographs no. 86:1–66.

Rohwer FC, Freeman S, 1989. The distribution of conspecific nest parasitism in birds. Can J Zool 67:239–253.

Sandström U, 1992. Cavities in trees: their occurrence, formation and importance for hole-nesting birds in relation to Silvicultural practise (dissertation). Uppsala: Swedish University of Agricultural Sciences.

Savard J-PL, 1984. Territorial behaviour of common goldeneye, Barrow's goldeneye, and bufflehead in areas of sympatry. Ornis Scand 15:211–216.

Savard J-PL, 1988. Winter, spring and summer territoriality in Barrow's goldeneye: characteristics and benefits. Ornis Scand 19:119–128.

Savard J-PL, Smith JNM, 1987. Interspecific aggression by Barrow's goldeneye: a descriptive and functional analysis. Behaviour 102:168–184.

Sayler RD, 1992. Ecology and evolution of brood parasitism in waterfowl. In: Ecology and management of breeding waterfowl (Batt BDJ, Afton AD, Anderson MG, Ankney CD, Johnson DH, Kadlec JA, Krapu GL, eds). Minneapolis: University of Minnesota Press; 290–322.

Semel B, Sherman PW, 1986. Dynamics of nest parasitism in wood ducks. Auk 103: 813–816.

Semel B, Sherman PW, 1992. Use of clutch size to infer brood parasitism in wood ducks. J Wildl Manage 56:495–499.

Semel B, Sherman PW, 1993. Answering basic questions to address management needs: case studies of wood duck nest box programs. Trans N Am Wildl Nat Res Conf 58:537–550.

Semel B, Sherman PW, 1995. Alternative placement strategies for wood duck nest boxes. Wildl Soc Bull 23:463–471.

Semel B, Sherman PW, Byers SM, 1988. Effects of brood parasitism and nest-box placement on wood duck breeding ecology. Condor 90:920–930.

Semel B, Sherman PW, Byers SM, 1990. Nest boxes and brood parasitism in wood ducks: A management dilemma. In: Proceedings of 1988 North American wood duck symposium (Frederickson LH, Burger GV, Havera SP, Graber DA, Kirby RE, Taylor TS, eds). St. Louis, Missouri; 163–170.

Sherman PW, Semel B, 1989. Behavioral ecology and the management of a natural resource. NY Food Life Sci Q 19:23–26.

Soulliere GJ, 1988. Density of suitable wood duck nest cavities in a northern hardwood forest. J Wildl Manage 52:86–89.

Soulliere GJ, 1990a. Regional and site-specific trends in wood duck use of nest boxes. In: Proceedings of the 1988 North American Wood Duck Symposium (Frederickson LH, Burger GV, Havera SP, Graber DA, Kirby RE, Taylor TS, eds), St. Louis, Missouri; 235–244.

Soulliere GJ, 1990b. Review of wood duck nest-cavity characteristics. In: Proceedings of the 1988 North American wood duck symposium (Frederickson LH, Burger GV, Havera SP, Graber DA, Kirby RE, Taylor TS, eds). St. Louis, Missouri; 153–162.

Titman RD, Lowther JK, 1975. The breeding behaviour of a crowded population of mallards. Can J Zool 53:1270–1283.

von Haartmann L, 1957. Adaptation in hole-nesting birds. Evolution 11:339–347.

Walters CJ, Holling CS, 1990. Large-scale management experiments and learning by doing. Ecology 71:2060–2068.

Weier RW, 1966. A survey of wood duck nest sites on Mingo National Wildlife Refuge in southeast Missouri. In: Wood duck management and research: a symposium (Trefethen JB, ed). Washington, DC: Wildlife Management Institute; 91–112.

Wilson SF, 1993. Use of wood duck decoys in a study of brood parasitism. J Field Ornithol 64:337–340.

Yom-Tov Y, 1980. Intraspecific nest parasitism in birds. Biol Rev 55:93–108.

13

The Importance of Mate Choice in Improving Viability in Captive Populations

Mats Grahn
Åsa Langefors
Torbjörn von Schantz

Captive Populations

There is a constant increase in the number of species threatened with extinction, and a number of measures are being used to save endangered species or populations, with varying success. One of these is captive breeding, and a growing number of species are today dependent on zoological parks for their existence. Indeed, the main enterprise for many conservation biologists is to maintain viable and reproductive populations of such species in captivity. Specifically, their goal is to create a self-sustaining population that is able to maintain 75–98% of its original heterozygosity for 20–200 years (Ebenhard, 1995; Wiese et al., 1996) in order to "maintain the population's evolutionary plasticity" (Vrijenhoek and Leberg, 1991). The methods for achieving this goal include minimization of inbreeding with the aid of pedigree records and reduction of genetic drift by planned matings and equalized family sizes (Haig et al., 1990). These measures will increase the ratio of effective population size to actual population size (N_e/N) and produce as much heterozygosity as possible in a limited population (Foose and Ballou, 1988).

A problem in captive populations, be it salmon in hatcheries, dairy cattle, or zoo animals, is the affliction of various diseases (table 13-1). The reasons for disease outbreaks are numerous. Populations are often much denser in captivity than in the wild, which facilitates transmission of pathogens. Exotic species introduced to a new environment for which they are not adapted are often more susceptible to the existing pathogens in that environment, for which native species have evolved resistance (Gjedrem and Aulstad, 1974; Zinn et al., 1977; Bakke et al., 1990; Brown et al., 1991). In many cases the artificial nature of multispecies facilities promotes cross-species transmission of disease (Landolt and Kocan, 1976; Partington et al., 1989; Snyder et al., 1996).

In this chapter, we argue from sexual selection theory that disease can be reduced in captive populations by allowing females to choose their mates. First, we discuss traditional management practices to maintain genetic variation; then we turn to the theory of sexual selection and relationship between mate choice and the immune system. Next, we introduce Atlantic salmon *Salmo salar*, a species in which there is a strong suggestion that absence of

Table 13-1 Examples of recent diseases in captive populations (most references are collected from Snyder et al., 1996, and Jacobson, 1993).

Species	Disease	Reference
Mammals		
Black-footed ferret (*Mustela nigripes*)	Distemper virus	Carpenter et al. (1976)
Birds		
Rothchild's mynah (*Leucospar rothchildii*)	Avian pox	Landolt and Kocan (1976)
Bali mynah (*Leucospar rothchildii*)	Atoxoplasmosis	Partington et al. (1989)
Puerto Rican plain pigeon (*Columbia inornata*)	Flukes of Tanaisia bragai (coccidiosis, capillaria)	Arnizaut et al. (1991)
Mauritius pink pigeon (*Nesoenas mayeri*)	Herpes virus	Snyder et al. (1985)
Whooping crane (*Grus americana*)	Equine encephalitis	Dein et al. (1986)
Hooded (*Grus monacha*) and Manchurian crane (*G. japonensis*)	Inclusion body disease	Docherty and Romaine (1983)
Reptiles		
Fer-de-lance (*Bothrops moojeni*)	Ophidian paramyxovirus	Foelsch and Leloup (1983)
Rock rattle snake (*Crotalus lepidus*)	Ophidian paramyxovirus	Jacobson et al. (1980) Jacobson (1993)
Argentine tortoise (*Geochelone chilensis*)	Herpes virus	Jacobson et al. (1985)
Green sea turtle (*Chelonia mydas*)	Chlamydiosis	Jacobson (1993)
Fish		
Arctic charr (*Salvelinus alpinus*)	Proliferative kidney disease	Brown et al. (1991)
Atlantic salmon (*Salmo salar*)	Furunculosis	Johnsen and Jensen (1994)

free choice in mating is associated with high levels of juvenile mortality. At the end of the chapter, we weigh the pros and cons of allowing mate choice and make new recommendations for captive-breeding programs.

The Maintenance of Genetic Variation in Captive Breeding

In natural populations many associations between overall genetic heterozygosity and fitness have been found (Vrijenhoek and Leberg, 1991; Mitton, 1993). Heterozygosity correlates with fitness characters such as growth rate, developmental rate and stability, metabolic efficiency, and measures of body condition and also with traits that directly reflect an individual's fitness such as disease resistance and survival (Ralls et al., 1979; Samollow and Soulé, 1983; Mitton and Grant, 1984; Danzmann et al., 1987, 1988; Sövenyi et al., 1988; Ferguson and Drahushchak, 1990; Clayton and Price, 1994; Jiménez et al., 1994). Moreover, reproductive success is often impaired by inbreeding because of increased juvenile mortality or reduced hatchability (Ralls et al., 1979; Brock and White, 1992; Bensch et al., 1994). Much of the genetic polymorphism in animal populations is in the form of rare recessive alleles at loci under stabilizing selection. In a large outbred population these rare deleterious alleles are maintained in a heterozygous form (Lande and Barrowclough, 1987; Mitton, 1993). In small populations, even with several dozen founders, some rare alleles will be lost, whereas others will become much more common. Subsequent mating within the confined population will expose recessive, deleterious alleles to selection because they now are more likely to occur in homozygous form (Lande and Barrowclough, 1987). This increase in genetic load among individuals will lead to an ever-increasing need for management of unhealthy animals that carry these alleles and an increased need for human intervention to get these animals to reproduce.

Given these findings, many captive-breeding programs aim at preserving and maximizing general heterozygosity by minimizing inbreeding, especially since captive individuals of a species at different zoos are often descendants of a single or a few ancestors. Current conservation programs now consider the minimum number of founder animals required when initiating a captive-breeding program, the relationship between the size of a captive population and maintenance of genetic variation, and optimal sex ratios (de Boer, 1992; Ebenhard, 1995). Because natal dispersal is reduced in a captive population, it was understood early on that special measures were needed to avoid deleterious effects of close kin mating (Ralls et al., 1979). Records of pedigrees for individual animals combined with studbooks and species survival plans (Foose and Ballou, 1988; Rabb, 1994; Ebenhard, 1995) facilitate selection of suitable mates (Foose and Ballou, 1988; Ballou and Cooper, 1992; de Boer, 1992; Holt, 1992).

Unfortunately, suitable sires and dams are not always kept in nearby zoos, and these programs often result in expensive transportation of individuals between distant zoos, which not only causes stress to the transported animals but also risks disease transmission. Modern technology has started to eliminate these problems and has made mating of captive animals more and more a biotechnological task. The technique of artificial insemination has been widely used in husbandry and has now also proved successful for zoo animals (Ballou and Cooper, 1992; Holt, 1992). From the breeders' point of view, one of the advantages with this technique is that it eliminates the risk of animals refusing to copulate. Forced matings can lead to spontaneous abortions, miscarriages, lower fertilization rates, or offspring with low viability (Klint and Enquist, 1981; Thomas et al., 1985; Hedrick, 1988) (table 13-2). Hence, when a pair, carefully pinpointed by consulting pedigree records, studbooks, and species survival plans, refuse to copulate, they can nevertheless be induced to sire offspring (Lindburg and Fitch-Snyder, 1994).

Mate incompatibility, poor breeding, or aggression are considered the main difficulties in achieving reliable reproduction for the majority of cases where reproductive disturbance (see table 13-2) has been observed in mammals managed under the North American Species Survival Plan program (see also Wielebnowski, chapter 6, this volume). Rarely are indifference or aggression considered as adaptions to minimize the risks of reproductive failure due to physiological or genetical incompatibility between the mates.

Mate Choice and Viability

In many animal species a female has behaviors and adaptations aimed at picking the best possible father for her young. By doing so a female will, in many cases, increase the qual-

Table 13-2 Selected examples of reproductive disturbances in captive populations.

Species	Nature of problem	Reference
Mammals		
Tiger (*Panthera tigris*)	Unsuccessful breeding, juvenile mortality	van Bemmel (1968)
Black rhinoceros (*Diceros bicornis*)	Low reproduction	Smith and Read (1992)
Kangaroo rat (*Dipodomys merriami*)	Unsuccessful breeding	Chew (1958)
Pygmy hippopotamus (*Choeropsis liberiensis*)	Juvenile mortality	Ralls et al. (1979)
Dorcas gazelle (*Gazella dorcas*)	Juvenile mortality	Ralls et al. (1979)
Birds		
Puerto Rican parrot (*Amazona vittata*)	Low fecundity, low egg hatchability	Brock and White (1992)
Whooping crane (*Grus americana*)	Infertility, incompatibility	Lewis (1990)

ity, survival, and future reproductive success of her offspring (Zahavi, 1975, 1977; Andersson, 1982, 1994; Hamilton and Zuk, 1982; Kodric-Brown and Brown, 1984). Interfering with the mechanisms that may allow females to choose healthy males can deprive the population of its ability to adapt to changes in the environment. In the wild, animals often have adaptations to avoid inbreeding. Females in natural populations often select mates on basis of traits that reflect male viability and health. To the extent that these traits reflect genetic differences, the same traits can easily be used by captive breeders to select sires of high quality. The ambition to maintain overall genetic variation, without taking sexual selection processes into account, is particularly accentuated in the conservation of anadromous salmonid fishes, for which many of the natural spawning sites are destroyed or made inaccessible. The Swedish program of culturing the Atlantic salmon in the Baltic Sea in hatcheries may be a key example where the preservation of genetic variation may obstruct the ultimate goal of the conservation program.

Mate Choice and Inbreeding Avoidance

In nature, inbreeding is often reduced by dispersal of individuals from their natal group or site before reaching sexual maturity (Pusey and Wolf, 1996). In most species the dispersal is sex biased, which decreases the likelihood that individuals encounter mates that are close kin. In species with no or limited dispersal where kin do encounter each other, animals may possess mechanisms for kin recognition to avoid the costs associated with inbreeding.

Recognition of close kin is a well-known phenomenon observed in animals ranging from insects to mammals (Olsén, 1989; Waldman et al., 1992; Winberg and Olsén, 1992; Smith, 1995; reviewed by Pfennig and Sherman, 1995). In black-tailed prairie dogs *Cynomys ludovicianus,* females do not come into estrus while their father is still present, and, given the opportunity to mate with close kin, they avoid it (Hoogland, 1982). Female guppies *Poecilia reticulata* prefer to mate with newly introduced males or with males with rare pigmentation patterns (Farr, 1977, 1980). Avoidance of mating with kin has also been observed in deer mice *Peromyscus maniculatus* and cactus mice *P. eremicus* (Hill, 1974; Dewsbury, 1982).

In chimpanzees *Pan troglodytes* females usually move to a new community when they become sexually mature. Before maturation, they live in close proximity to male relatives, but once they reach sexual maturity they avoid these males (Pusey, 1980).

Other mate-choice strategies that females may use to avoid inbreeding include two different sperm-competition mechanisms. Multiple matings, where females increase their reproductive success by mating with many males, have been observed in inbred populations of adders *Vipera berus* and sand lizards *Lacerta agilis* (Madsen et al., 1992; Olsson, 1992). Females that are involved in a social pair-bond may also perform multiple matings by seeking or accepting extrapair fertilizations. This widespread mating strategy is important for reducing inbreeding levels in, for example, the splendid fairy wren *Malurus splendens,* in which DNA fingerprinting revealed extrapair young in as many as 60% of the broods (Rowley et al., 1993).

Mate Choice and "Good Genes"

Some exaggerated ornamental traits carried by males of many species, such as bearing a tail twice as long as the rest of the body, require a special explanation, as they obviously have little survival value. To explain the evolution of such traits, Darwin (1859, 1871) proposed the mechanism of sexual selection. Traits promoting male competition for copulations, territories, or other resources related to attraction of mates would be selected for, thus explaining why males are often bigger than females and have ornaments such as antlers that can function as weapons in fights. The most striking extravagant male ornamental traits, like the peacock's tail, which apparently were of little or no value in fights, evolved, according to Darwin (1871), simply because females preferred such traits.

There are numerous suggestions as to what females can gain by being selective in their choice of mate (Fisher, 1915, cited in Harvey and Bradbury, 1991; Kirkpatrick and Ryan, 1991; Andersson, 1994). The "good genes" models assume and predict that (1) males differ in condition and viability and that this difference can be inherited by the offspring; (2) that females can assess this variation in male fitness by ornamental traits which are expressed in a condition-dependent mode such that more viable males express more exaggerated ornaments (Zahavi, 1975; Andersson, 1982, 1986; Hamilton and Zuk, 1982; Kodric-Brown and Brown, 1984).

The notion that genetic differences between males are responsible for the evolution of mate choice was first proposed by Fisher (1958), but another of Fisher's models has since acted as an effective road block to the search for genetic variation in traits related to mate choice. Any trait closely linked to fitness should rapidly reach genetic fixation, and heritable variation will become limited at the mutation equilibrium level (Fisher, 1958). A way out of this theoretical dilemma was proposed by Hamilton and Zuk (1982), who realized that traits affecting disease resistance are unlikely to become fixed because selection pressures constantly change due to a co-evolutionary "arms race" between parasites and their hosts. Moreover, condition is likely affected by many genes that are sensitive to epistatic effects (the effect of a gene may differ when combined with certain other genes) and environmental variation (Falconer, 1981), and thus it is unlikely to reach genetic fixation even under strong and consistent directional selection (Pomiankowski and Møller, 1995; von Schantz et al., 1995). Hamilton and Zuk suggested that females may choose to mate with males that have low parasite loads so as to produce healthy offspring that are genetically resistant to prevalent parasites and diseases. The idea is that healthy males can afford to have a conspicuous ornament that parasitized males cannot afford to acquire. To assure the condition-dependent expression of the trait, the ornament should hence impose a "handicap" (Zahavi, 1975) or cost to the bearer (e.g., Andersson, 1986). The same logic also applies to weapons and "badges" used in male–male competition, where the size may reflect condition and health (Clutton-Brock et al., 1982; Møller, 1992; Veiga, 1993). Because many condition-dependent traits can function both as weapons and ornaments, the reliability of the signal may be constantly verified in male–male competition (Kodric-Brown and Brown, 1984).

The condition dependence of male ornaments is supported by many studies showing that the expression of ornaments, such as tail ornaments in birds (Alatalo et al., 1988) and tarsial spurs in pheasants *Phasianus colchicus* (Wittzell and von Schantz, 1991), is highly variable both between and within males and that male ornamentation correlates with condition and survival (canine teeth in primates: Manning and Chamberlain, 1993; carotenoid pigmentation in fish and birds: Milinski and Bakker, 1990; Hill, 1990, 1991; Niccoletto, 1993; spurs in pheasants: von Schantz et al., 1989; combs in red jungle fowl *Gallus gallus:* Zuk et al., 1990; see also a general review by Andersson, 1994). Moreover, by controlled infection experiments, it has been shown that ornaments are sensitive to diseases (Houde and Torio, 1992), even more so than other morphological traits (Hillgarth, 1990; Zuk et al., 1990). In addition, laboratory studies on insects have shown that females increase the quality of their offspring when they have access to several males (Partridge, 1980; Crocker and Day, 1987; Simmons, 1987; Taylor et al., 1987; Moore, 1994; but see Schaeffer et al., 1984), and recent studies on birds in the wild indicate that females can increase offspring fitness by mating with more ornamented males (Norris, 1993; Møller, 1994; von Schantz et al., 1994; Hasselquist et al., 1996).

Sexual selection theory was proposed to explain female mating preferences for highly ornamented males in polygynous animals. However, as Darwin recognized, monogamy does not preclude the operation of sexual selection if females can benefit by mating with particular males. By seeking extrapair fertilizations with high-quality males, a female in a monogamous mating system can increase the quality of her young (Kempenaers et al., 1992). Thus, from the females' point of view, opportunities for female choice and access to several males is as important in monogamous as in polygynous breeding systems. The main difference is that the potential pay-off for ornamented males is lower in monogamous species than in polygynous, giving more subtle male ornaments among the former.

Folstad and Karter (1992) proposed that an immune-suppressive effect of testosterone may mediate a trade-off between body condition, health, and ornamentation. Elevated levels of testosterone during the mating season increase ornamentation but impair immune function. Healthy males may sustain higher levels of testosterone and are therefore also able to grow and sustain larger ornaments. Variation among males in genetic resistance to parasites and diseases will influence the likelihood of getting ill and thus also affect the expression of condition-dependent ornaments. More ornamented males show lower responses to pathogens, as measured by counts of white blood cells, in red jungle fowl (Zuk et al., 1995) and Arctic charr *Salvelinus alpinus* (Skarstein and Folstad, 1996). Genetic resistance may not always result in "parasite-free" individuals, but it can be manifested as a high tolerance to common parasites (Zinkernagel, 1996), and mate choice based on condition-dependent ornaments may promote matings with parasite-tolerant partners (Skarstein and Folstad, 1996). This provides a direct link between condition-dependent ornaments and the population's ability to track co-evolving diseases through mate choice.

Mate Choice and the Major Histocompatibility Complex

In vertebrates most of the immune system's genetic variation at the population level is located at the major histocompatibility complex (MHC), which contains several linked genes that code for cell-surface proteins. These genes are often inherited together on one chromosome as one haplotype and have a critical function in the vertebrate immune system. There are the two groups of MHC receptors, class I and class II, which present peptide antigens derived from pathogens to T-cells. MHC class I is expressed on all nucleated cells and presents intracellular peptides to cytotoxic T-cells. Class II is expressed on specialized phagocytic white blood cells (e.g., macrophages) but also on B-cells, and presents antigens of extracellular origin to T-helper cells. The MHC class II receptor is a dimeric protein formed by two subunits, α and β. The antigen binding part of the β chain is coded by the second exon of the MHC class II β (Roitt et al., 1993). Class II β genes have been found to be highly polymorphic in the majority of species studied, with most of the variation located in the second exon. Up to 20 different alleles have been found in local populations (Potts and Wakeland, 1990). Each variant of the MHC receptor is able to bind only a limited number of different peptides, and different MHC alleles thus allow the presentation of a different set of antigens. Unlike most of the variability of the t-cell and B-cell receptors, which is generated anew at the individual level by somatic mutations, the variation of MHC is inherited from the parents' gametes.

Several studies have found an association between specific MHC haplotypes and resistance to infectious diseases and autoimmune disorders in mammals and birds (Briles et al., 1983; Tiwari and Teraski, 1985; Hill et al., 1991; Nepom and Erlich, 1991; Han et al., 1992; McGuire et al., 1994; Rose, 1994). Therefore, it has been suggested that the high degree of within-population variability of MHC genes is the result of selection from pathogens. From the pathogens' point of view, different hosts are different "food patches," and even if a pathogen evolves the ability to avoid recognition by one specific MHC genotype, the most common "food patch," it may be less effective in infecting other hosts in a population with high MHC variation (Zinkernagel, 1996). In this system, uncommon MHC genotypes will be at an advantage compared to common ones, leading to frequency-dependent selection favoring rare MHC alleles (Clarke and Kirby, 1966; Howard, 1991; Hedrick, 1994). Furthermore, it has been suggested that heterozygotes may be able to detect a larger set of antigens and therefore be more resistant to diseases (Doherty and Zinkernagel, 1975; Black and Salzano, 1981; Wakeland et al., 1989; Kroemer et al., 1990). In mole rats *Spalax ehrenbergi,* some subspecies appear to have evolved a higher degree of MHC heterozygosity as a response to high parasite load, compared to less parasitized subspecies (Nevo and Beiles, 1992). Studies of MHC polymorphism in some mammal and bird populations have found a high allelic variation and an excess of heterozygotes, compared to Hardy-Weinberg expectations (reviewed by Potts and Wakeland, 1990, 1993; von Schantz et al., 1996; but see Ellegren et al., 1996; Edwards and Potts, in press).

In wild male pheasants, MHC genotype was found to correlate with spur length, which is a condition-dependent ornament in this species, and with survival (von Schantz et al., 1996). In other species MHC influences mating preferences directly. Female house mice preferred to mate with males that differed from themselves at the MHC genes (Potts et al., 1991). The cue they used to distinguish between males of different MHC-genotypes is the odor of the urine (Yamazaki et al., 1976). It is not known what the molecular basis of the odor cues is, but there are indications of imprinting on odors during the early juvenile stage

(Singh et al., 1987). Humans also prefer certain MHC-based odor types (Wedekind et al., 1995).

Hence, MHC variation may be important to mating preferences for at least two reasons: it can influence mate choice via inbreeding avoidance and may also affect condition-dependent ornaments used by females when assessing mate quality. The operation of female choice can, through mating preferences for males possessing large ornaments, select for MHC alleles that are both less sensitive to autoimmune disorders and more protective against common pathogens.

Mate Choice and Oxidative Stress

In fishes and birds, males often display sexual signals through reddish coloration in the plumage or skin. This reddish pigmentation consists of carotenoids, and females frequently prefer males with more reddish coloration (Endler, 1983; Kodric-Brown, 1989; Hill, 1990; Milinski and Bakker, 1990). In guppies and sticklebacks *Gasterosteus aculeatus* more reddish males are more vigorous and more resistant to pathogens (Milinski and Bakker, 1991; Frischknecht, 1993; Nicoletto, 1993). The main hypothesis to explain these correlations has been that males with more carotenoid pigmentation are better foragers and therefore more viable (Endler, 1983), but this ignores the carotenoids' function as important antioxidants (Lozano, 1994).

Oxidative metabolism generates free radicals. Respiratory processes in mitochondrial energy production, immune responses, detoxification of exogeneous toxic substances (such as persistant organochlorine pollutants), and metabolism of endogeneous sex steroids (e.g., testosterone) by cytochrome P450 enzymes all generate free radicals (Miller and Miller, 1966; Fougereau and Dausset, 1980; Gonzalez and Nebert, 1990; Klein, 1990; Wheeler and Guenther, 1991). Free radicals can damage a variety of critical molecules and physiological processes, including DNA (Weitberg et al., 1985), and can also affect expression of class II MHC molecules (Gruner et al., 1986) and the suppression of both T-cell– and B-cell–based immune reactions (Street and Sharma, 1975; Holsapple et al., 1991). Free radicals are especially harmful to cells with high energy turnover and to proliferating cells such as white and red blood cells, spermatozoa, and endotel and epitel cells (Anderson and Theron, 1990).

A series of naturally occurring antioxidant defense mechanisms normally prevent or limit free radical production and tissue damage. It is generally believed that the different types of antioxidants can complement each other so that the reaction of one antioxidant can spare others (Olanow, 1993). Some important antioxidants, such as carotenoids, vitamin E, and ascorbic acid, are often of dietary origin and are depleted during the course of their antioxidant action (Frei et al., 1992). Individuals under relatively high oxidative stress due to infection, diseases, or metabolism of persistent organochlorine pollutants, for example, are therefore likely to have lower deposits of these antioxidants. Hence all available data show that carotenoid pigmentation may function as a condition-dependent ornament (Hill, 1995).

In vertebrates the metabolism of persistent organochlorine pollutants is often processed by cytochrome P450 enzymes. One important function for cytochrome P450 enzymes is to convert these chemicals into highly water-soluble products that can be eliminated from the body. The cytochrome P450 enzymes are encoded by different loci in the genome. The allelic variation of many of the P450 genes is high (Gonzalez and Nebert, 1990). It is not unusual for a single chemical substrate to be metabolized to varying degrees by a handful of different P450 enzymes. Because different P450 enzymes differ in their capacity to detoxify a given chemical, heterozygotes may be better adapted to metabolize and excrete a broad assemblage of foreign toxic compounds (Gonzalez and Nebert, 1990). It is also clear that there are several distinct classes of P450 inducers that activate transcription of one or more P450 genes. One of these inducers is the Ah (aryl hydrocarbon) locus, which is triggered by exposure to dioxin and related polycyclic aromatic hydrocarbons. Studies on mammals and fish have shown that the Ah locus has an allelic variation (Burbach et al., 1992; Hahn and Karchner, 1995, 1996), and preliminary analyses indicate that the Ah locus may also be polymorphic in Atlantic salmon (T. von Schantz, unpublished data). In mice, some inbred strains with a given Ah allele are more sensitive to treatments with dioxin than are other strains (Thomas et al., 1972; Poland et al., 1987).

Hence, both the MHC and the detoxification mediated by the P450s affect the levels of free radicals and oxidative stress in the body, and they are encoded by genes or gene families characterized by a high degree of genetic polymorphism. Because the level of oxidative stress can be visualized by sexual signals, such as carotenoid-dependent pigmentation, these genetic systems may constitute an important justification for the "good genes" in the Hamilton and Zuk (1982) model for the evolution of female choice and male condition-dependent ornamentation.

The Atlantic Salmon Case

In spring and early summer, anadromous maturing male and female Atlantic salmon start to migrate from the sea into their native rivers. Once the fish have entered the river, they lose their silvery color and become brownish. The males also acquire reddish-colored spots on parts of their body, and the lower jaw becomes hook-shaped in form. The adipose fin is larger on males than on females, and the males' fin size increases throughout summer and reaches its maximum size at spawning (Naesje et al., 1988), which peaks in October and November. Spawning females deposit their eggs in sequences in up to eight spawning beds, and several males may fertilize the eggs (Jones, 1959; Hutchings and Myers, 1988). Male body size and the relative size of the hook of the lower jaw correlate with male dominance and mating success (Järvi, 1990). An experimental study of Atlantic salmon showed that although the size of the adipose fin did not correlate with male dominance, females repelled males with a small adipose fin from the spawning beds significantly more often than males with a larger adipose fin (Järvi, 1990). These results indicate that active female mate choice may have some influence on male reproductive success (Järvi, 1990).

Culturing of Atlantic salmon has been performed since the 1950s to compensate for the loss of natural spawning sites caused by damming for hydropower production (Eriksson and Eriksson, 1993). Maturing fish that are returning to their natal river from the sea in spring are trapped at the water outlet of the first hydroelectric power plant that the fish encounter on their upstream journey, often close to the river mouth. The maturing fish are kept in captivity until autumn when roe and milt are stripped from randomly paired spawners. Often the batch of roe from one female is fertilized by milt from one or two males. The roe are hatched in the cultured hatcheries, and the juvenile fish are released into the river as 1- or 2-year-old smolts. The cultured hatchery program explicitly states that all possible genetic variation shall be maintained in the different salmon stocks and no intentional selection for any traits (e.g., growth rate, size, color or disease resistance) is allowed either when collecting spawners or during the growth of the juveniles.

The Early Mortality Syndrome

Since 1974 a marked increase in mortality of salmon alevins has been observed in the Swedish hatcheries on the Baltic coast. A series of symptoms occur at the end of the yolk-sac reception stage, followed by a rapid and often total mortality of the whole clutch a few days after the first symptoms are observed. In 1992 the syndrome caused mortality of up to 95% of salmon alevins that were reared in Swedish hatcheries along the Baltic coast (Johansson et al., 1993). Because more than 90% of the recruits of Atlantic salmon to the Baltic sea consist of artificially reared smolts, the situation is alarming (Johansson et al., 1993). In spite of intense research, the ultimate causes of this maternally transmitted, non-infectious syndrome are unknown, but similar symptoms and mortality rates have been observed among salmonid alevins reared in hatcheries in the Great Lakes of North America (Skea et al., 1985; Fitzsimons et al., 1995) and in New York's Finger Lakes (Fisher et al., 1995a). Recent findings from these lake basins clearly show that thiamine (vitamin B_1) treatment of roe and alevins effectively eliminates the syndrome-related mortality (Fisher et al., 1995b); Fitzsimons, 1995). Studies in hatcheries around the Baltic Sea have revealed that there is a deficiency of thiamine in ova and yolk-sac fry that develop the syndrome (Amcoff et al., 1996), and although the reasons for this deficiency currently are unknown, subsequent experiments in Sweden and Finland have confirmed that thiamine treatment is an effective cure for the syndrome (Amcoff et al., 1996; Koski et al., 1996). Since 1995 virtually all alevins of Baltic salmon stripped in Swedish hatcheries undergo this treatment. In effect, the present mass breeding of Baltic salmon in compensation programs now face a situation where the majority of females from the Baltic salmon stocks are physiologically incapable of producing viable offspring outside hatcheries.

Good Genes and Conservation Actions

At spawning, male Atlantic salmon mobilize carotenoids from the flesh to the skin, where they are visualized as reddish color patterns. Females deposit their carotenoids in the roe, and carotenoids seem to be essential to the juveniles during the first feeding period. Fry and parr fed with diets supplemented with a carotenoid, astaxanthin, have higher growth and survival rates than fish fed without astaxanthin supplementation (Christiansen et al., 1994, 1995). An astaxanthin-enriched diet also gave Atlantic salmon parr improved disease resistance and higher levels of other antioxidants in the body, mainly vitamin E and ascorbic acid, than parr fed with normal diet (Christiansen et al., 1994, 1995).

Not surprisingly, Baltic salmon alevins hatched from clutches of more highly carotenoid-pigmented roe are less likely to suffer from the early mortality syndrome than clutches with less pigmentation (Lignell, 1994; Börjesson et al., 1996). Has the lack of mate choice and natural breeding contributed to the early mortality syndrome, and would a hatchery practice where sires and dams are selected on basis of their carotenoid pigmentation confer a sustainable cure for the syndrome?

We know for certain that the present thiamine treatment impedes natural selection for increased carotenoid pigmentation in females. However, we do not yet know if female salmonids base their mate choice on male coloration or even on male MHC genotype, but if they do there is certainly no room for such processes in today's cultured hatcheries. There are indications from Atlantic salmon that females are performing active mate choice for a male trait that does not necessarily correlate with male dominance (Järvi, 1990), and a recent experiment on arctic charr shows that juveniles can identify conspecifics on basis of their MHC genotype (H. Olsén, M. Grahn, J. Lohm, Å. Langefors, unpublished data). Moreover, controlled experiments on sib groups of Atlantic salmon show significant genetic effects on the variation in carotenoid pigmentation (Torrissen and Neavdal, 1988).

Except for those alleles that were randomly lost due to bottleneck effects at the initiation of the hatcheries, the Swedish culture of Atlantic salmon probably have contributed to the conservation of the allelic frequencies that were present in the 1950s in the different spawning stocks of salmon. Meanwhile, increased environmental pollution, which has been particularly prevalent in the Baltic Sea since the 1950s, may have changed selective regimes—for example, the salmon's metabolism of persistent organochlorine pollutants. Moreover, eutrophication and exploitation of fisheries have changed the composition of plant and animal communities in the Baltic Sea (Elmgren, 1989; Johansson et al., 1993), which may have led to changes in the food composition of the salmon.

Given sufficient genetic variation in the population, a shift in the genotype–environment interactions during growth may promote a distribution of genotypes at reproduction that differ from the original one regarding reproductive competence. Any captive-breeding program that continuously aims at preserving the original allelic frequencies may impair necessary evolutionary responses and eventually undermine the species' chances of survival in changing refuges.

Strengths and Weaknesses of Extended Mate Choice in Captive Breeding

The ambition to maintain genetic variation indiscriminately needs more thought. A breeding strategy aimed at maximizing heterozygosity by keeping an equal number of offspring from each adult while minimizing inbreeding may counteract the purging of deleterious alleles by natural selection and may eventually increase the genetic load of the population. This strategy may under some circumstances confer negative effects on the species' ability to survive when reintroduced into its original refuges.

We do not disagree with recommendations derived from population genetics theory that selection should be kept at a minimum to reduce the effect of genetic drift in very small populations. However, given that the effective population sizes are adequate (Lande and Barrowclough, 1987), we propose that mate choice and sexual selection be given more room in conservation breeding programs. In current conservation breeding practices, mate choice is considered at best irrelevant, or sometimes even deleterious because overall genetic diversity in a captive population can be reduced if variance in reproductive success is increased (Briton et al., 1994). Although the ambition of current practices in captive breeding to maximize genetic diversity does contribute to the preservation of alleles by increasing the effective population size, the current program does not distinguish between "beneficial" and deleterious alleles.

Major components of fitness such as health and fecundity are more sensitive to epistatic interactions between alleles at different loci than most other quantitative characters (Falconer, 1981). The operation of mate choice may reduce the break down of favorable allele combinations (Fisher, 1958; Pomiankowski and Møller, 1995) compared to a strict application of current breeding practices. We argue that if one allows for the operation of mate choice based on condition-dependent ornaments and a limited increase in variance of reproduction, at least among males, it may be possible to weed out deleterious alleles and still maintain genetic variation at important loci (Mitton, 1993). In the worst case, today's management strategies with no room for mate choice of healthy males may jeopardize the long-term survival of a population by not allowing it to track environmental changes in general and co-evolving diseases in particular.

Increasing mate choice may lead to faster adaptation to captivity. If this is true, it could have both beneficial and detrimental effects. It could be good if a species evolves resistance to new diseases, but fatal if they come to rely on a supplemented food nutrient. However, this can also be a problem without sexual selection and mate choice because selection and adaptation takes place in any captive population (Johnsson et al., 1996), and the only cure is careful evaluation of husbandry and a thorough knowledge of the animals' biology.

Although studies of ornaments such as tail feathers of swallows *Hirundo rustica,* (Møller, 1991), canine teeth of lowland gorillas *Gorilla gorilla* (Manning and Chamberlain, 1993), carotenoid pigmentation in fish and birds (Hill, 1990; Milinski and Bakker, 1990; Niccoletto, 1993) and tarsial spurs of pheasants (Grahn and von Schantz, 1994) have revealed that these ornaments are reliable indicators of condition and hence obvious traits for selecting sires of high quality, we stress that we do not always advocate extensive directed selection for sexual ornaments. In an experimental study on domestic chickens *Gallus domesticus* across nine generations, intense directional selection for male comb size, a condition-dependent male ornament promoted by female choice (Zuk et al., 1990), resulted not only in an increased ornamental size but also in an increased mortality among the male carrying the exaggerated ornament (von Schantz et al., 1995).

Birds and fishes fit the Hamilton and Zuk hypothesis well due to the brightness and conspicuousness of the ornaments of many of these species. However, many endangered, captive populations are mammalian species in which brightness is much less pronounced. Conspicuous male traits revealing male condition are therefore not frequently at hand in mammals. In addition, it appears that intrasexual selection (male–male competition) is of greater importance in mammals than intersexual selection (mate choice). Many mammalian species have a more or less hierarchical polygynous mating system, in which the most dominant male gets access to most of the females. It seems likely that males in good condition prevail among the dominants (Hausfater and Watson, 1976; Freeland, 1981). Unfortunately, allowing direct male–male competition is rarely practical in captive populations.

Recommendations for Captive Breeding

In spite of extensive efforts, we have found no comparative study that contrasts the consequences of traditional captive breeding with captive breeding mimicking a natural breeding system with extended mate choice. Hence, our foremost recommendation is that experimental studies should be initiated in captive-breeding populations to compare reproductive success, genetic variation, and offspring viability resulting from traditional "studbook" breeding and "free mate choice" breeding. For example, in a study of domestic pigeons *Columba livia,* Klint and Enquist (1981) found that spontaneously formed pairs had earlier egg-laying, laid more eggs, and had more fertile eggs than randomly assigned pairs. In the Mauritius kestrel *Falco punctatus,* eggs collected from free-breeding pairs had higher hatchability than eggs collected from pairs breeding in captivity (Jones et al., 1995).

It is, of course, not always practical or even possible to allow unrestricted mate choice in captive breeding. However, as a complement to pedigree and studbook records, there may be fairly simple ways to improve results from captive breeding when mate choice is not feasible. Even limited mate choice may be better than no choice at all.

It may be feasible at those zoos that harbor several mature males of a species to let them "show off" to females during the mating season. To minimize the risk of injuries, males should not be allowed physical contact; most fights in the wild are settled well before serious injuries happen. During these arena displays, female behavior would be monitored and the males' "popularity score" could be added to the other information in the studbook. This information could then be used to increase the number of offspring from popular males and decrease it from low-scoring males. Such "beauty contests" would greatly improve the collection of important data to benefit scientific study of mate choice, and they would, perhaps, even increase the appeal of zoos both to the animals and to the public. Such a study has been conducted in captive cheetahs *Acinonyx jubatus* (N. Wielebnowski, unpublished data).

One particularly successful conservation approach is the so-called hacking practice on wild breeding pairs of birds of prey (Jones et al., 1995). This procedure takes advantage of birds' ability to lay replacement clutches. After the first clutch has been completed by a pair living in the wild, the eggs are collected and artificially hatched. The young are raised in captivity for later release into the wild. This conservation practice does not interfere with mate choice but can very effectively increase the population's numbers. At least some elements of the practice of hacking can be extended to other species of birds and even to mammals. Collected eggs or young are the product of mate choice, and offspring could be nourished and protected in captivity to decrease juvenile mortality before being reintroduced into the wild. The hacking practice mimics the failure of a breeding attempt, a common event in the wild, which many species have adapted to by replacement clutches or, in case of mammals, an earlier start of a new estrous cycle. This practice, however, requires a thorough understanding of the species' specific needs to produce mature adults prepared for independent life in the wild.

Sometimes when the breeding pair is genetically similar the female may inhibit fertilization, implantation and development by several different mechanisms (Wedekind, 1994), resulting in low fertility and a high incidence of spontaneous abortions (Hedrick, 1988; Thomas et al., 1985). Thus, if such symptoms are observed, it seems to be a better idea to replace one of the mates instead of persistently trying to achieve offspring of that particular pair.

Some species have dispersal as the main mechanism for inbreeding avoidance, and it is likely that females of such species have poor kin-recognition mechanisms. They may therefore not refuse to mate with their fathers or brothers in confined populations. The risk for increased inbreeding may be avoided by keeping females separate from kin males at the time of mating. It may be important to allow young animals to learn and imprint on relevant cues in order to identify close kin by, for example, odors generated by their genetic parents and siblings (Aldhouse, 1989).

A variety of mammals may exhibit MHC-based mating preferences. A deficiency in frequency of MHC homozygotes among juveniles has been reported from seminatural or natural populations of both mammals (Potts et al., 1991) and birds (von Schantz et al., 1996). It may therefore turn out that MHC-genotyping of potential mates in captive-breeding programs, especially of endangered large mammals, may facilitate not only the pair formation and the likelihood for successful copulations but also improve the viability of the offspring (Thomas et al., 1985; Hedrick, 1988; Wedekind, 1994). MHC typing could therefore act as a complement to pedigree considerations (Hughes, 1991; Miller and Hedrick, 1991; Vrijenhoek and Leberg, 1991; Miller, 1995; Edwards and Pots, in press). Such MHC typing could be accomplished fairly easily and at low cost for a multitude of endangered captive populations of fish, birds, and mammals.

Knowledge about condition-dependent ornaments in the species, or in a related taxa, can confer economical and practical alternatives to finding parents of high genetic quality. Since condition-dependent ornaments are likely to integrate information from a large number of genes that affect body condition, these ornaments may be especially sensitive to genotype–environmental interactions (Pomiankowski and Møller, 1995). In fact, Hill (1995) argued that condition-dependent ornaments can be used as reliable and sensitive indicators of environmental quality.

There is clearly an urgent need to understand the processes of sexual selection in the Atlantic salmon. Given that female Atlantic salmon select mates on the basis of male carotenoid pigmentation, it would be logical to recommend that cultured hatcheries around the Baltic coast select for breeders on basis of this trait. By increasing the catch of potential spawners, but then discriminating against those individuals that have the least carotenoid pigmentation in the skin (males) and the roe, one could induce a certain amount of selection without reducing the effective population size. The current practice to treat all roe and alevins with thiamine cannot contribute in the long run to the conservation of wild Atlantic salmon in the Baltic sea. To the extent that differences in carotenoid pigmentation correlate with a heritable component of susceptibility to the early mortality syndrome, selection for this trait could at least theoretically lead to populations of Baltic salmon that are more resistant to the syndrome. We need more research on genetic variation and mate choice to direct wise conservation in this and other species.

Summary

Recent models and empirical studies of sexual selection support the notion that females have adaptations aimed at selecting among potential mates to increase fitness of their offspring. They choose mates on the basis of signals that reliably indicate health and condition. Mating preferences in many species decrease the likelihood of close inbreeding, and observations in some species have revealed a direct influence of genetic variation in the immune system on inbreeding avoidance. The link between mate choice, disease, and genetic variation is exemplified by the effect of genetic variation at the MHC on condition-dependent ornaments and kin-recognition. We argue that, contrary to the emerging practice in captive breeding of endangered species, female mate choice, to avoid close inbreeding and select for healthy males, may preserve genetic variation at important loci and allow the population to track environmental changes.

Acknowledgments We thank A. Kodric-Brown, M. Milinski, and P. A. Siri for comments on the chapter. The work was supported by grants from the Swedish Council for Forestry and Agricultural Research (to M.G. and T.v.S.), the Swedish Environmental Protection Agency (to T.v.S.), and the Swedish Natural Science Research Council (to T.v.S.).

References

Alatalo RV, Höglund J, Lundberg A. 1988. Patterns of variation in tail ornament size in birds. Biol J Linn Soc 34:363–374.

Aldhous P, 1989. The effects of individual cross-fostering on the development of intrasexual kin discrimination in male laboratory mice, *Mus musculus* L. Anim Behav 37: 741–750.

Amcoff P, Norrgren L, Börjesson H, Lindeberg J, 1996. Lowered concentrations of thiamine (vitamin B_1) in M74-affected feral Baltic salmon (*Salmo salar*). In: Report from the second workshop on reproduction disturbances in fish. Report 4534 (Bengtsson B-E, Hill C, Nellbring S, eds). Stockholm: Swedish Environmental Protection Agency; 38–39.

Anderson R, Theron AJ, 1990. Physiological potential of ascorbate, β-carotene and α-tocopherol individually and in combination in the prevention of tissue damage, carcinogenesis and immune dysfunction mediated by phagocyte-derived reactive oxidants. In: Aspects of some vitamins, minerals and enzymes in health and disease. World Review of Nutrition and Diet, vol 62 (Bourne GH, ed). Basel: Karger; 62:27–58.

Andersson M, 1982. Sexual selection, natural selection, and quality advertisement. Biol J Linn Soc 17:375–393.

Andersson M, 1986. Evolution of condition-dependent sex ornaments and mating preferences: sexual selection based on viability differences. Evolution 40:804–816.

Andersson M, 1994. Sexual selection. Princeton, New Jersey: Princeton University Press.

Arnizaut AB, Hayes L, Olsen GH, Torres JS, Ruiz C, Pérez-Rivera R, 1991. An epizootic of *Tanaisia bragai* in a captive population of Puerto Rican Plain Pigeon (*Columbia inornata wetmorei*). Ann NY Acad Sci 653:202–205.

Bakke TA, Jansen PA, Hansen LP, 1990. Differences in the host resistance of Atlantic salmon, *Salmo salar* L., stocks to the monogenean *Gyrodactulus salaris* Malmberg, 1957. J Fish Biol 37:577–587.

Ballou JD, Cooper KA, 1992. Genetic management strategies for endangered captive populations: the role of genetic and reproductive technology. Symp Zool Soc Lond 64:183–206.

van Bemmel ACV, 1968. Breeding tigers at Rotterdam Zoo. Int Zoo Yrbk 8:60–63.

Bensch S, Hasselquist D, von Schantz T, 1994. Genetic similarity between parents predicts hatching failure: nonincestous inbreeding in the great reed warbler? Evolution 48:317–326.

Black FL, Salzano FM, 1981. Evidence for heterosis in the HA system. Am J Hum Genet 33:894–899.

Börjesson H, Förlin L, Norrgren L, 1996. Investigation of antioxidants and prooxidants in salmon affected by the M74 syndrome. In: Report from the second workshop on reproduction disturbances in fish. Report 4534 (Bengtsson B-E, Hill C, Nellbring S, eds). Stockholm: Swedish Environmental Protection Agency; 95–96.

Briles WE, Briles RW, Taffs RE, Stone HA, 1983. Resistance to a malignant lymphoma in chickens is mapped to subregion of major histocompatibility (B) complex. Science 219:977–979.

Briton J, Nurthen RK, Briscoe DA, Frankham R, 1994. Modelling problems in conservation genetics using *Drosophila:* consequences of harems. Biol Conserv 69:267–275.

Brock MK, White BN, 1992. Application of DNA fingerprinting to the recovery program of the endangered Puerto Rican parrot. Proc Natl Acad Sci USA 89:11121–11125.

Brown JA, Thonney J-O, Holwell D, Wilson WR, 1991. A comparison of the susceptibility of *Salvelinus alpinus* and *Salmo salar ouananiche* to proliferative kidney disease. Aquaculture 96:1–6.

Burbach KM, Poland A, Bradfield CA, 1992. Cloning of the Ah-receptor cDNA reveals a distinctive ligand-activated transcription factor. Proc Natl Acad Sci USA 89: 8185–8189.

Carpenter JW, Appel MJG, Erickson RC, Novilla MN, 1976. Fatal vaccine-induced canine distemper virus infection in black-footed ferret. J Am Vet Med Assoc 169:961–964.

Chew RM, 1958. Reproduction of *Dipodomys merriami* in captivity. J Mammal 39:597–598.

Christiansen R, Glette J, Lie O, Torrissen O, Waagbo R, 1995. Antioxidant status and immunity in Atlantic salmon, *Salmo salar* L., fed semi-purified diets with and without astaxanthin supplementation. J Fish Dis 18:317–328.

Christiansen R, Lie O, Torrissen OJ, 1994. Effect of astaxanthin and vitamin A on growth and survival during first feeding of Atlantic salmon, *Salmo salar* L. Aquacult Fish Manage 25:903–914.

Clarke B, Kirby DRS, 1966. Maintenance of histocompatability polymorphisms. Nature 211:999–1000.

Clayton GM, Price J, 1994. Heterosis in resistance to *Ichthyophthirius multifiliis* infections in poeciliid fish. J Fish Biol 44:59–66.

Clutton-Brock TH, Guinness FE, Albon SD, 1982. Red deer: the behavior and ecology of two sexes. Chicago: University of Chicago Press.

Crocker G, Day T, 1987. An advantage to mate choice in the seaweed fly, *Coelopa frigida*. Behav Ecol Sociobiol 20:295–301.

Danzmann RG, Ferguson MM, Allendorf FW, 1987. Heterozygosity and oxygen consumption rate as predictors of growth and developmental rate in rainbow trout. Physiol Zool 60:211–220.

Danzmann RG, Ferguson MM, Allendorf FW, 1988. Heterozygosity and components of fitness in a strain of rainbow trout. Biol J Linn Soc 33:285–304.

Darwin C, 1859. On the origin of species. London: Murray.

Darwin C, 1871. The descent of man and selection in relation to sex, 1st ed. London: Murray.

de Boer LEM, 1992. Current status of captive breeding programmes. Symp Zool Soc Lond 64:5–16.

Dein FJ, Carpenter JW, Clark GG, Montali RJ, Crabbs CL, Tsai TF, Docherty DE, 1986. Mortality of captive whooping cranes caused by eastern equine encephalitis virus. J Am Vet Med Assoc 189:1006–1010.

Dewsbury DA, 1982. Avoidance of incestuous breeding between siblings in two species of *Peromyscus* mice. Biol Behav 7:157–169.

Docherty DE, Romaine RT, 1983. Inclusion body disease of cranes: a serological follow-up to the 1978 die-off. Avian Dis 27:830–835.

Doherty PC, Zinkernagel RM, 1975. Enhanced immunological surveillance in mice heterozygous at the H-2 gene complex. Nature 256:50–52.

Ebenhard T, 1995. Conservation breeding as a tool for saving animal species from extinction. Trends Ecol Evol 10:438–443.

Edwards S, Potts WK, in press. Polymorphism of genes in the major histocompatibility complex: implications for conservation genetics of vertebrates. In: Molecular approaches to conservation (Smith TB, Wayne RK, eds). AAAS Symposium.

Ellegren H, Mikko S, Wallin K, Andersson L, 1996. Limited polymorphism at major histocompatibility complex (MHC) loci in the Swedish moose. *A. alces*. Mole Ecol 5:3–9.

Elmgren R, 1989. Man's impact on the ecosystem of the Baltic Sea: energy flows today and at the turn of the century. Ambio 18:326–332.

Endler JA, 1983. Natural and sexual selection on color patterns in poeciliid fishes. Environ Biol Fishes 9:173–190.

Eriksson T, Eriksson L-O, 1993. The status of wild and hatchery propagated Swedish salmon stocks after 40 years of hatchery releases in the Baltic rivers. Fish Res 18:147–159.

Falconer DS, 1981. Introduction to quantitative genetics, 2nd ed. New York: Longmans.

Farr JA, 1977. Male rarity or novelty, female choice behavior, and sexual selection in the guppy. Evolution 31:162–168.

Farr JA, 1980. Social behavior patterns as determinants of reproductive success in the guppy, *Poecilia reticulata* Peters (Pisces: Poeciliidae): an experimental study of the effect of intermale competition, female choice, and sexual selection. Behaviour 74:38–91.

Ferguson MM, Drahushchak LR, 1990. Disease resistance and enzyme heterozygosity in rainbow trout. Heredity 64:413–417.

Fisher JP, Spitsbergen JM, Getchell R, Symula H, Skea J, Babenzein M, Chiotti T, 1995a. Reproductive failure in landlocked Atlantic salmon from New York's Finger Lakes: investigations into the etiology and epidemiology and of the "Cayuga Syndrome". J Aqua Anim Health 7:81–94.

Fisher JP, Spitsbergen JM, Iamonte T, Little EE, DeLonay A, 1995b. Pathological and behavioral manifestations of the "Cayuga Syndrome", a thiamine deficiency in larval landlocked Atlantic salmon. J Aqua Anim Health 7:269–283.

Fisher RA, 1958. The genetical theory of natural selection, 2nd ed. New York: Doyer.

Fitzsimons JD, 1995. The effect of B-vitamins on swim-up syndrome in Lake Ontario lake trout. J Great Lakes Res 21: suppl. T 286–289.

Fitzsimons JD, Huestis S, Williston B, 1995. Occurrence of a swim-up syndrome in Lake Ontario lake trout in relation to contaminants and cultural practices. J Great Lakes Res 21 Suppl. T: 277–285.

Foelsch DW, Leloup P, 1983. Infection enzootique grave dans un serpentarium. In: Proceedings of the International Colloquium on Pathology of reptiles and amphibians (Vago C, Matz G, eds). Angers: France; 25–31.

Folstad I, Karter AJ, 1992. Parasites, bright males and the immunocompetence handicap. Am Nat 139:603–622.

Foose TJ, Ballou JD, 1988. Management of small populations. Int Zool Yrbk 27:26–41.

Fougereau M, Dausset J, 1980. Progress in immunology I, vol 1. London: Academic Press.

Freeland WJ, 1981. Parasitism and behavioral dominance among male mice. Science 213:461–462.

Frei B, Stoker R, Ames BN, 1992. Small molecule antioxidant defenses in human extracellular fluids. Comm Cell & Mole Biol 5:23–45.

Frischknecht M, 1993. The breeding colouration of male three-spined sticklebacks (*Gasterosteus aculeatus*) as an indicator of energy investment in vigour. Evol Ecol 7:439–450.

Gjedrem T, Aulstad D, 1974. Selection experiments with salmon: I. Differences in resistance to vibrio disease of salmon parr (*Salmo salar*). Aquaculture 3:51–59.

Gonzalez FJ, Nebert DW, 1990. Evolution of the P450 gene superfamily: animal-plant "warfare" molecular drive, and human genetic differences in drug oxidation. Trends Genet 6:182–196.

Grahn M, von Schantz T, 1994. Fashion and age in pheasants: age difference in mate choice. Proc R Soc Lond B 255:237–241.

Gruner S, Volk H-D, Falck P, Von Baehr R, 1986. The influence of phagocytic stimuli on the expression of HLA-DR antigens; role of reactive oxygen intermediates. Eur J Immunol 16:212–215.

Haig SM, Ballou JD, Derrickson SR, 1990. Management options for preserving genetic diversity: reintroduction of Guam rails to the wild. Conserv Biol 4:290–300.

Hamilton WD, Zuk M, 1982. Heritable true fitness and bright birds: a role for parasites? Science 218:384–387.

Hahn ME, Karchner SI, 1995. Evolutionary conservation of the vertebrate Ah (dioxin) receptor: amplification and sequencing of the PAS domain of a teleost Ah receptor cDNA. Biochem J 310:383–387.

Hahn ME, Karchner SI, 1996. Identification of two Ah receptor (AhR) genes in the teleost *Fundulus heteroclitus*. Toxicologist 30:267.

Han R, Breitburd F, Marche PN, Orth G, 1992. Linkage regression and malignant conversion of rabbit viral papillomas to MHC class II genes. Nature 356:66–68.

Harvey PH, Bradbury JW, 1991. Sexual selection. In: Behavioural ecology: an evolutionary approach, 3rd ed (Krebs JR, Davies NB, eds). Oxford: Blackwell; 203–233.

Hasselquist D, Bensch S, von Schantz T, 1996. Correlation between male song repertoire, extra-pair paternity and offspring survival in the great reed warbler. Nature 381:229–232.

Hausfater G, Watson DF, 1976. Social and reproductive correlates of parasite ova emission by baboons. Nature 262:688–689.

Hedrick PW, 1988. HLA-sharing, recurrent spontaneous abortion, and the genetic hypothesis. Genetics 119:199–204.

Hedrick PW, 1994. Evolutionary genetics of the major histocompatibility complex. Am Nat 143:945–964.

Hill AVS, Allsopp CEM, Kwiatkowski D, Anstey NM, Twumasi P, Rowe PA, Bennett S, Brewster D, McMichael AJ, Greenwood BM, 1991. Common West African HLA antigen are associated with protection from severe malaria. Nature 352:595–600.

Hill GE, 1990. Female house finches prefer colorful males: sexual selection for a condition-dependent trait. Anim Behav 40:563–572.

Hill GE, 1991. Plumage coloration is a sexually selected indicator of male quality. Nature 350:337–339.

Hill GE, 1995. Ornamental traits as indicators of environmental health. Bioscience 45:25–28.

Hill JL, 1974. *Peromyscus:* effects of early pairing on reproduction. Science 186: 1042–1044.

Hillgarth N, 1990. Parasites and female choice in the ring-necked pheasant. Am Zool 30:227–233.

Holsapple MP, Snyder NK, Wood SC, Morris DL, 1991. A review of 2,3,7,8-tetrachloro-*p*-dioxin induced changes in immunocompetence: 1991 update. Toxicology 69:219–255.

Holt WV, 1992. Advances in artificial insemination and semen freezing in mammals. Symp Zool Soc Lond 64:19–35.

Hoogland JL, 1982. Prairie dogs avoid extreme inbreeding. Science 215:1639–1641.

Houde AE, Torio AJ, 1992. Effect of parasitic infection on male color pattern and female choice in guppies. Behav Ecol 3:346–351.

Howard JC, 1991. Disease and evolution. Nature 352:565–567.

Hughes AL, 1991. MHC polymorphism and the design of captive breeding programs. Conserv Biol 5:249–251.

Hutchings JA, Myers RA, 1988. Mating success of alternative maturation phenotypes in male Atlantic salmon, *Salmo salar.* Oecologia 75:169–174.

Jacobson ER, 1993. Implications of infectious diseases for captive propagation and introduction programs of threatened/endangered reptiles. J Zoo Wildl Med 24:245–255.

Jacobson ER, Clubb S, Gaskin JM, Gardiner C, 1985. Herpesvirus-like infection in Argentine tortoise. J Am Vet Med Assoc 187:1227–1229.

Jacobson ER, Gaskin JM, Simpson CF, Terell TG, 1980. Paramyxo-like virus infection in a rock rattlesnake. J Am Vet Med Assoc 177:796–799.

Järvi T, 1990. The effects of male dominance, secondary sexual characteristics and female mate choice on the mating success of male Atlantic salmon *Salmo salar.* Ethology 84:123–132.

Jiménez JA, Hughes KA, Alaks G, Graham L, Lacy RC, 1994. An experimental study of inbreeding depression in a natural habitat. Science 266:271–273.

Johansson N, Johnsson P, Svanberg O, Södergren A, Thulin J, 1993. Reproduction disturbances in Baltic fish. Report 4222. Stockholm: Swedish Environmental Protection Agency.

Johnson BO, Jensen AJ, 1994. The spread of furunculosis in salmonids in Norwegian rivers. J Fish Biol 45:47–55.

Johnsson JI, Petersson E, Jönsson E, Björnsson BT, Järvi T, 1996. Domestication and growth hormone alter antipredator behaviour and growth patterns in juvenile brown trout, Salmo trutta. Can. J. Fish. Aquat. Sci. 53:1546–1554.

Jones CG, Heck W, Lewis RE, Mungroo Y, Slade G, Cade T, 1995. The restoration of the Mauritius kestrel *Falco punctatus* population. Ibis 137 (supplement) 1:S173–S180.

Jones JW, 1959. The salmon. London: Collins.

Kempenaers B, Verheyen GR, van den Broeck M, Burke T, van Broeckhoven C, Dhondt AA, 1992. Extrapair paternity results from female preference for high-quality males in the blue tit. Nature 357:494–496.

Kirkpatrick M, Ryan MJ, 1991. The evolution of mating preferences and the paradox of the lek. Nature 350:33–38.

Klein J, 1990. Immunology. Oxford: Blackwell.

Klint T, Enquist M, 1981. Pair formation and reproductive output in domestic pigeons. Behav Process 6:57–62.

Kodric-Brown A, 1989. Dietary carotenoids and male mating success in the guppy: an environmental component to female choice. Behav Ecol Sociobiol 25:393–401.

Kodric-Brown A, Brown JH, 1984. Truth in advertising: the kinds of traits favored by sexual selection. Am Nat 124:309–323.

Koski P, Pakarinen M, Soivio A, 1996. A dose-response study of thiamine hydrochloride bathing for the prevention of yolk-sac mortality in Baltic salmon fry (M74 syndrome). In: Report from the second workshop on reproduction disturbances in fish. Report 4534 (Bengtsson B-E, Hill C, Nellbring S, eds). Stockholm: Swedish Environmental Protection Agency; 46.

Kroemer G, Bernot A, Béhar G, Chaussé A-M, Gastinel L-N, Guillemot F, Park I, Thoraval P, Zoorob R, Auffray C, 1990. Molecular genetics of the chicken MHC: current status and evolutionary aspects. Immunol Rev 113:119–145.

Lande R, Barrowclough GF, 1987. Effective population size, genetic variation, and their use in population management. In: Viable populations for conservation (Soulé ME, ed). Cambridge: Cambridge University Press; 87–123.

Landolt M, Kocan RM, 1976. Transmission of avian pox from starlings to Rothchild's mynahs. J Wildl Dis 12:353–256.

Lewis JC, 1990. Captive propagation in the recovery of the whooping crane. End Spec Update 8:46–48.

Lignell Å, 1994. Astaxanthin in yolk-sac fry from feral Baltic salmon. In: Report from the Uppsala workshop on reproductive disturbances in fish. Report 4346 (Norrgren L, ed). Stocholm: Swedish Environmental Protection Agency; 94–96.

Lindburg DG, Fitch-Snyder H, 1994. Use of behavior to evaluate reproductive problems in captive mammals. Zoo Biol. 13:433–455.

Lozano GA, 1994. Carotenoids, parasites, and sexual selection. Oikos 70:309–311.

Madsen T, Shine R, Loman J, Håkansson T, 1992. Why do female adders copulate so frequently? Nature 355:440–441.

McGuire W, Hill AVS, Allsopp CEM, Greenwood BM, Kwiatkowski D, 1994. Variation in the TNF-α promoter region associated with susceptibility to cerebral malaria. Nature 371:508–511.

Manning JT, Chamberlain AT, 1993. Fluctuating asymmetry, sexual selection and canine teeth in primates. Proc R Soc Lond B 251:83–87.

Milinski M, Bakker TCM, 1990. Female sticklebacks use male coloration in mate choice and hence avoid parasitized males. Nature 344:330–333.

Milinski M, Bakker TCM, 1991. Sexual selection: female sticklebacks recognize a male's parasitization by its breeding coloration. Verh Dtsch Zool Ges 1991:308.

Miller EC, Miller JA,. 1966. Mechanisms of chemical carcinogenesis: nature of proximate carcinogens and interactions with macromolecules. Pharmacol Rev 18:805–838.

Miller PS, 1995. Selective breeding programs for rare alleles: examples from Przewalski's horse and California condor pedigrees. Conserv Biol 9:1262–1273.

Miller PS, Hedrick PW, 1991. MHC polymorphism and the design of captive breeding programs: simple solutions are not the answer. Conserv Biol 5:556–558.

Mitton JB, 12993. Theory and data pertinent to the relationship between heterozygosity and fitness. In: The natural history of inbreeding and outbreeding. Theoretical and empirical perspectives (Thorhill NW, ed). Chicago: University of Chicago Press; 17–41.

Mitton JB, Grant MC, 1984. Associations among protein heterozygosity, growth rate and developmental stasis. Annu Rev Ecol Syst 145:479–499.

Møller AP, 1991. Sexual ornament size and the cost of fluctuating asymmetry. Proc R Soc Lond B 243:59–62.

Møller AP, 1992. Patterns of fluctuating asymmetry in weapons: evidence for reliable signaling of quality in beetle horns and bird spurs. Proc R Soc Lond B 248:199–206.

Møller AP, 1994. Sexual selection and the barn swallow. New York: Oxford University Press.

Moore AJ, 1994. Genetic evidence for the "good genes" process of sexual selection. Behav Ecol Sociobiol 35:235–241.

Naesje T, Hansen L-P, Järvi T, 1988. Sexual dimorphism in the adipose fin of Atlantic salmon (*Salmo salar*). J Fish Biol 33:955–956.

Nepom GT, Erlich H, 1991. MHC class-II molecules and autoimmunity. Annu Rev Immunol 9:493–525.

Nevo E, Beiles A, 1992. Selection for class II MHC heterozygosity by parasites in subterranean mole rats. Experimentia 48:512–515.

Nicoletto PF, 1993. Female sexual response to condition-dependent ornaments in the guppy, *Poecilia reticulata*. Anim Behav 46:441–450.

Norris K, 1993. Heritable variation in a plumage indicator of viability in male great tits *Parus major*. Nature 362:537–539.

Olanow CW, 1993. A radical hypothesis for neurodegeneration. Trends Neurosci 16: 439–444.

Olsén KH, 1989. Sibling recognition in juvenile Arctic charr, *Salvelinus alpinus* (L.). J. Fish Bill 34:571–581.

Olsén KH, Grahn M, Lohm J, Langefors Å. (in press). MHC and kin discrimination in juvenile Arctic charr, *Salvelinus alpinus* (L.). Anim Behav

Olsson M, 1992. Sexual selection and reproductive strategies in the sand lizard (*Lacerta agilis*) (PhD dissertation). Göteborg Sweden: University of Göteborg.

Partington CJ, Gardiner CH, Fritz D, Phillips LG, Montali RJ, 1989. Atoxoplasmosis in Bali mynahs (*Leucospar rothchildi*). J Zoo Wildl Med 20(3): 328–335.
Partridge L, 1980. Male choice increases a component of offspring fitness in fruit flies. Nature 283:290–291.
Pfennig DW, Sherman PW, 1995. Kin recognition. Sci Am 272:98–103.
Poland A, Glover E, Taylor BA, 1987. The murine *Ah* locus: a new allele and mapping to chromosome 12. Mol Pharmacol 32:471–478.
Pomiankowski A, Møller AP, 1995. A resolution of the lek paradox. Proc R Soc Lond B 260:21–29.
Potts WK, Manning CJ, Wakeland EK, 1991. Mating patterns in semi-natural populations of mice influence by MHC genotype. Nature 352:619–621.
Potts WK, Wakeland EK, 1990. Evolution of diversity at the major histocompatibility complex. Trends Ecol Evol 5:181–187.
Potts WK, Wakeland EK, 1993. Evolution of MHC genetic diversity: a tale of incest, pestilence, and sexual preference. Trends Gent 9:408–412.
Pusey AE, 1980. Inbreeding avoidance in chimpanzees. Anim Behav 28:543–552.
Pusey AE, Wolf M, 1996. Inbreeding avoidance in animals. Trends Ecol Evol 11:201–206.
Rabb GB, 1994. The changing roles of zoological parks in conserving biological diversity. Am Zool 34: 159–164.
Ralls K, Brugger K, Ballou J, 1979. Inbreeding and juvenile mortality in small populations of ungulates. Science 206:1101–1103.
Roitt I, Brostoff J, Male D. 1993. Immunology, 3rd ed. London: Mosby.
Rose NR, 1994. Avian models of autoimmune disease: lessons from the birds. Poultry Sci 73:984–990.
Rowley I, Russell E, Brooker M, 1993. Inbreeding in birds. In: The natural history of inbreeding and outbreeding. Theoretical and empirical perspectives (Thornhill NW, ed). Chicago: University of Chicago Press; 309–328.
Samollow PB, Soulé ME, 1983. A case of stress related heterozygote superiority in nature. Evolution 37:646–649.
Schaeffer SW, Brown CJ, Anderson WW, 1984. Does mate choice affect fitness? Genetics 107:94.
von Schantz T, Göransson G, Andersson G, Fröberg I, Grahn M, Helgée A, Wittzell H, 1989. Female choice selects for a viability-based male trait in pheasants. Nature 337: 166–169.
von Schantz T, Grahn M, Göransson G, 1994. Intersexual selection and reproductive success in the pheasant *Phasianus colchicus*. Am Nat 144:510–527.
von Schantz T, Tufvesson M, Grahn M, Göransson G, Wilhelmson M, Wittzell H, 1995. Artificial selection for increased comb size reduces other sexual characters and viability in domestic chicken (*Gallus domesticus*). Heredity 75:518–529.
von Schantz T, Wittzell H, Göransson G, Grahn M, Persson K, 1996. MHC genotype and male ornamentation: genetic evidence for the Hamilton-Zuk model. Proc R Soc Lond B 263:265–271.
Simmons LW, 1987. Female choice contributes to offspring fitness in the field cricket, *Gryllus bimaculatus* (de Geer). Behav Ecol Sociobiol 21:313–321.
Singh PB, Brown RE, Roser B, 1987. MHC antigens in urine as olfactory recognition cues. Nature 327:161–164.
Skarstein F, Folstad I, 1996. Sexual dichromatism and the immunocompetence handicap: an observational approach using Arctic charr. Oikos 76:359–367.
Skea JC, Symula J, Miccoli J, 1985. Separating starvation losses from other early feeding

fry mortality in steelhead trout (*Salmo gairdneri*), chinook salmon (*Oncorhynchus tshawytscha*) and lake trout (*Salvelinus namaycush*). Bull Environ Contam Toxicol 35:82–91.

Smith DG, 1995. Avoidance of close consanguineous inbreeding in captive groups of rhesus macaques. Am J Primatol 35:31–40.

Smith RL, Read B, 1992. Management parameters affecting the reproductive potential of captive, female black rhinoceros, *Diceros bicornis*. Zoo Biol 11:375–383.

Snyder B, Tahilsted J, Burgess B, Richard M, 1985. Pigeon herpesvirus mortalities in foster-reared Mauritius pink pigeons. Proc Am Assoc Zoo Vet 1985:69–70.

Snyder NFR, Derrickson SR, Beissinger SR, Wiley JW, Smith TB, Toone WD, Miller B, 1996. Limitations of captive breeding in endangered species recovery. Conserv Biol 10:338–348.

Sövenyi JF, Bercsényi M, Bakos J, 1988. Comparative examination of susceptibility of two genotypes of carp *Cyprinus carpio* L. to infection with *Aeromonas salmonicida*. Aquaculture 709:301–308.

Street A, Sharma RP, 1975. Alteration of induced cellular and humoral immune responce by pesticides and chemicals of environmental concern: quantitative studies of immunosuppression by DDT, Aroclor 1254, carbaryl, carbofuran, and methylparathion. Toxicol Appl Pharmacol 32:587–602.

Taylor CE, Pereda AD, Ferrari JA, 1987. On the correlation between mating success and offspring quality in *Drosophila melanogaster*. Am Nat 129:721–729.

Thomas ML, Harger JH, Wagener DK, Rabin BS, Gill III TJ, 1985. HLA sharing and spontaneous abortion in humans. Am J Obstet Gynecol 151:1053–1058.

Thomas PE, Kouri RE, and Hutton JJ, 1972. The genetics of aryl hydrocarbon hydroxylase induction in mice: a single gene difference between C57BL/6J and DBA/2J. Biochem Genet 6:157–168.

Tiwari JL, Teraski PI, 1985. HLA and disease associations. New York: Springer-Verlag.

Torrissen OJ, Naevdal G, 1988. Pigmentation of salmonids—variation in flesh carotenoids of Atlantic salmon. Aquaculture 68:305–310.

Veiga JP, 1993. Badge size, phenotypic quality, and reproductive success in the house sparrow: a study on honest advertisement. Evolution 47:1161–1170.

Vrijenhoek RC, Leberg PL, 1991. Let's not throw the baby out with the bathwater: a comment on management for MHC diversity in captive populations. Conserv Biol 5:252–254.

Wakeland EK, Boehme S, She JX, Lu CC, McIndoe RA, Cheng I, Ye Y, Potts WK, 1989. Ancestral polymorphisms of MHC class II genes: divergent allele advantage. Immun Res 9:123.

Waldman B, Rice JE, Honeycutt RL, 1992. Kin recognition and incest avoidance in toads. Am Zool 30:18–30.

Wedekind C, 1994. Mate choice and maternal selection for specific parasite resistance before, during and after fertilization. Phil Trans R Soc Lond B 346:303–311.

Wedekind C, Seebeck T, Bettens F, Paepke AJ, 1995. MHC-dependent mate preferences in humans. Proc R Soc Lond B 260:245–249.

Weitberg AB, Weitzman SA, Clark EP, Stossel TP, 1985. Effects of antioxidants on oxidant-induced sister chromatid exchange formation. J Clin Invest 75:1835–1841.

Wheeler CW, Guenther TM, 1991. Cytochrome P-450-dependent metabolism of xenobiotics in human lung. J Biochem Toxicol 6:163–169.

Wiese RJ, Willis K, Hutchins M, 1996. Conservation breeding in 1995: an update. Trends Ecol Evol 11:218–219.

Winberg S, Olsén KH, 1992. The influence of rearing conditions on the sibling odour preference of juvenile Arctic charr, *Salvelinus alpinus* L. Anim Behav 44:157–164.

Wittzell H, with von Schantz T, 1991. No repeatability of male ornaments in the pheasant *Phasianus colchicus*. In: Natural and sexual selection in the pheasant *Phasianus colchicus*. (PhD dissertation). Lund, Sweden: University of Lund; 59–71.

Yamazaki K, Boyse EA, Miké V, Thaler HT, Mathieson BJ, Abbot J, Boyse J, Zayas ZA, Thomas L, 1976. Control of mating preference in mice by genes in the major histocompatibility complex. J Exp Med 144:1324–1335.

Zahavi A, 1975. Mate selection—a selection for a handicap. J Theor Biol 53:205–214.

Zahavi A, 1977. The cost of honesty (further remarks on the handicap principle). J Theor Biol 67:603–605.

Zinkernagel RM, 1996. Immunology taught by viruses. Science 271:173–178.

Zinn JL, Johnson KA, Sanders JE, Fryer JL, 1977. Susceptibility of salmonid species and hatchery strains of chinook salmon (*Oncorhynchus tsawytscha*) to infections by *Ceratomyxa shasta*. J Fish Res Board Can 34:933–936.

Zuk M, Johnsen TS, Maclarty T, 1995. Endocrine-immune interactions, ornaments and mate choice in red jungle fowl. Proc R Soc Lond B 260:205–210.

Zuk M, Thornhill R, Ligon JD, Johnson K, 1990. Parasites and mate choice in red jungle fowl. Am Zool 30:235–244.

Part V

Dispersal and Inbreeding Avoidance

Patterns of dispersal, "the movement of an organism or propagule from its site or group of origin to its first or subsequent breeding site or group" (Shields, 1987:4; see also Lidicker, 1975), have ramifications for many aspects of conservation biology. First, the distance over which individuals disperse affects which habitat patches are included in a metapopulation (Wiens, 1985). Second, in any metapopulation, the rate of dispersal affects the probability with which vacant habitats are recolonized and hence affects population persistence (Hanski, 1989). On a larger scale, therefore, the judicial design of reserves, or more particularly, the proximity and juxtaposition of adjacent reserves, should take into account patterns of dispersal of organisms between reserves (Gutierrez and Harrison, 1996).

Third, dispersal affects the extent to which genetic variation is maintained in a population or metapopulation (Hedrick, 1996). Compared to a single large population, each partially isolated subpopulation or social group will have low N_e and will tend to become homozygous for neutral alles at a faster rate. Each subpopulation will also lose different alleles over time. Dispersal, whose primary function for the individual is thought to be inbreeding avoidance (see, e.g., Pusey and Packer, 1987), will, at a population level, introduce old alleles back into a subpopulation, thereby slowing loss of allelic diversity, and will set up new gene combinations, occasionally enabling adaptive gene complexes to spread through the whole population (Wright, 1931, 1932).

Fourth, dispersal can affect monitoring strategy. Frequent dispersal into vacant niches will mean that a single census of a population's size and distribution over a short time window may fail to identify the potential range and size to which the population can grow. In two thorough reviews in this section, Van Vuren and Waldman and Tocher explore aspects of dispersal as they relate to reserve design and genetic differentiation, respectively.

Van Vuren examines subtle but important conservation aspects of dispersal, mainly in North American mammals, by reviewing the literature on which sex disperses, how individuals locate a new home range, the direction and distance individuals move, and their survivorship during dispersal. Although these data are quantitative, information on dispersal is still difficult to collect despite radio tracking and intense field effort. Van Vuren's finding of an association between dispersal distance and body size is therefore helpful in estimating median dispersal distances in poorly studied species. Comparative data on dispersal are also useful for identifying the dietary groups, or taxa within a dietary niche, that have the shortest dispersal distances. These taxa can then be used as "umbrella dispersal species" in setting the minimum distance between suitable habitat patches. Similarly, the species with the most specific habitat requirements while dispersing sets the minimum habitat requirements necessary to connect reserves. Knowledge about dispersal patterns (e.g., the extent to which individuals cross open terrain), bears on the efficacy of corridors in connecting reserves. There is a rapidly growing literature on corridor use (Bentley and Catterall, 1997; Downes et al., 1997).

Although dispersal is common in mammals (but see Sherman et al., 1991), it seems to be less common in amphibians. In chapter 15, Waldman and Tocher assess the importance of kin recognition mechanisms in avoiding inbreeding in amphibian species in which gene flow between subpopulations is limited. Amphibians are of great conservation concern at present because parallel declines in populations have been observed in many parts of the world, and amphibia have long been regarded as an indicator taxon in toxicology because they occupy two habitats during their life cycle and have highly permeable skin (Blaustein and Wake, 1990). In an extensive review, Waldman and Tocher evaluate the principal hypotheses that have been advanced to account for amphibian declines. They then draw attention to the fact that although many frogs and toads live and return to isolated pools, genetic differentiation, as measured by mitochondrial DNA variability, is high, implying that inbreeding is avoided in ways other than through dispersal. There is abundant evidence of kin recognition in many frog and toad species at the tadpole stage (Hepper, 1986), but less is known about discrimination when choosing mates. Nevertheless, the authors present evidence that related male American toads *Bufo americanus* produce similar calls and that females prefer the calls of males that are less genetically similar to themselves. If such preferences are manifest in the wild, the prognosis is good: inbreeding will be relatively low even in isolated populations, and allelic variation at the major histocompatability complex may remain high, thereby maintaining immunological defenses.

In the last part of their chapter, Waldman and Tocher make suggestions for amphibian conservation, two of which independently reiterate the suggestions of other authors in this volume. First, they caution against the use of forced pairings of mates in captive-breeding programs because this may bypass mate-choice mechanisms that reduce disease risk in progeny (see Grahn, Langefors, and von Schantz, chapter 13). Second, they suggest that secretive frogs can be censused by counting calling males, reminiscent of the technique that McGregor and Peake (chapter 2) used to count corncrakes *Crex crex*. Third, they suggest that genetic variation in isolated amphibian populations should be surveyed regularly using noninvasive molecular techniques.

These two chapters advocate different methods to assess the extent of dispersal. Van Vuren points to the fact that molecular approaches reflect the outcome of the long-term behavior of populations, whereas records of individual dispersal indicate how a population is responding to current and changing environmental pressures, which are often anthropogenic in origin (see LaHaye et al., 1994). Moreover, although molecular approaches speak to successful dispersal, the records of individuals can address proximate factors influencing success, which may be more pertinent to conservation decisions. In contrast, Waldman and Tocher suggest that classical mark–recapture methods in herpetology overestimate gene flow; molecular analyses show that genetic differentiation in frogs and toads occurs over very short distances, less than 5 km. Indeed, it is the molecular analyses that suggest populations are potentially subject to considerable inbreeding and that individuals might employ kin recognition mechanisms to avoid it. Although different methods of estimating the extent of dispersal may be more appropriate for different taxa, these contrasting views suggest that using both observational and molecular techniques may be most prudent in addressing conservation problems at this stage.

References

Blaustein AR, Wake DB, 1990. Declining amphibian populations: a global phenomenon? Trends Ecol Evol 5:203–204.

Bentley JM, Catterall CP, 1997. The use of bushland, corridors, and linear remnants by birds in southeastern Queensland, Australia. Conser Biol 11:1173–1189.

Downes SJ, Handasyde KA, Elgar MA, 1997. The use of corridors by mammals in fragmented Australian eucalypt forests. Conser Biol 11:718–726.

Gutierrez RJ, Harrison S, 1996. Applying metapopulation theory to spotted owl management: a history and critique. In: Metapopulations and wildlife conservation (McCullough DR, ed). Covelo, California: Island Press; 167–185.

Hanski I, 1989. Metapopulation dynamics: does it help to have more of the same? Trends Ecol Evol 4:113–114.

Hedrick PW, 1996. Genetics of metapopulations: aspects of a comprehensive perspective. In: Metapopulations: aspects of a comprehensive perspective. In: Metapopulations and wildlife conservation (McCullough DR, ed). Covelo, California: Island Press; 29–51.

Hepper PG, 1986. Kin recognition: functions and mechanisms. A review. Biol Rev 61:63–93.

LaHaye WS, Gutierrez RJ, Akcakaya HR, 1994. Spotted owl metapopulation dynamics in southern Califonria. J Anim Ecol 63:775–785.

Lidicker WZ Jr, 1975. The role of dispersal in the demography of small mammals. In: Small mammals: their productivity and population dynamics (Golley FB, Petrusewicz K, Ryszkowski L, eds). London: Cambridge University Press; 103–128.

Pusey AE, Packer C, 1987. Dispersal and philopatry. In: Primate societies (Smuts BB, Cheney DL, Seyfarth RM, Wrangham RW, Struhsaker TT, eds). Chicago: University of Chicago Press; 250–266.

Sherman PW, Jarvis JUM, Alexander RD, 1991. The biology of the naked mole-rat. Princeton, New Jersey: Princeton University Press.

Shields WM, 1987. Dispersal and mating systems: investigating the causal connections. In: Mammalian dispersal patterns: the effects of social structure on population genetics (Chepko-Sade BD, Halpin ZT, eds). Chicago: University of Chicago Press; 3–24.

Wiens JA, 1985. Vertebrate responses to environmental patchiness in arid and semiarid ecosystems. In: The ecology

of natural disturbance and patch dynamics (Pickett STA, White PS, eds). New York: Academic Press; 169–193.
Wright S, 1931. Evolution in Mendelian populations. Genetics 16:97–159.
Wright S, 1932. The roles of mutation, inbreeding, cross-breeding and selection in evolution. Proc Sixth Int Cong Genet 356–366.

14

Mammalian Dispersal and Reserve Design

Dirk Van Vuren

Reserve Design

The proper design of nature reserves has been a matter of considerable debate (Diamond, 1975; Terborgh, 1975; Simberloff and Abele, 1976). In general, the best plan is large reserves and many of them (Soulé and Simberloff, 1986), but resources rarely are sufficient to achieve this goal. Thus, a major challenge for conservation biologists has been how to make do with reserves that are smaller than desired. Small reserves support small populations, and smaller populations are more likely to become extinct; however, extinction may be counteracted by immigration (Diamond, 1975). Research initially focused on effects of immigration at the community level. Extinction of some species may be balanced by colonization by others, resulting in an equilibrium; all else being equal, smaller reserves will support a small number of species (Diamond, 1975). Subsequently, Brown and Kodric-Brown (1977) noted that immigration of conspecifics could augment a declining species through the "rescue effect," thereby reducing extinction probabilities for individual species. Immigration results from dispersal of individuals, followed by settlement and reproduction; thus, conservation biologists have emphasized the importance of dispersal as a means of connecting separate reserves that may be inadequate by themselves. Dispersal, if sufficiently frequent, would reduce the chances of extinction in small reserves or, if extinction did occur, would ensure that the reserve would eventually be recolonized. This idea has been formalized as the metapopulation concept (Levins, 1969; Hanski and Gilpin, 1991). Corridors, linear strips of suitable habitat connecting reserves, have been advocated as a means of promoting dispersal between reserves (Simberloff and Cox, 1987; Noss, 1987). Translocation is considered an alternative when natural dispersal is inadequate (Griffith et al., 1989; Lubow, 1996; Hodder and Bullock, 1997).

Considering the potential importance of inter-reserve movement in reserve design, it is surprising how little attention conservation biologists have paid to research on dispersal (Simberloff, 1988). Behavioral ecologists have been studying dispersal for about two decades (Chepko-Sade and Halpin, 1987; Pusey and Packer, 1987), and their perspective may be helpful in designing effective reserves. In this chapter I outline the ways that dispersal has been considered in reserve design; review and evaluate pertinent contributions, primarily on mammals, from a behavioral ecological perspective; and make recommendations about reserve design and research needs.

Dispersal and Conservation Biology

Metapopulations

A metapopulation is a set of local populations connected by dispersing individuals (Hanski and Gilpin, 1991). Local populations that are small experience a relatively high probability of extinction, but if extinction were to be balanced by recolonization through dispersal, the metapopulation may persist. Application of the metapopulation concept to reserve design has become popular in conservation biology (Hanski and Gilpin, 1991; Doak and Mills, 1994; Harrison, 1994). A set of reserves, each too small to ensure long-term persistence of resident species, may yet prove effective if managed as a metapopulation; dispersal among reserves, however, must be adequate. Most metapopulation models have made simplifying assumptions that reduce dispersal to one (e.g., migration rate, *m*) or a few parameters. Few models have incorporated the behavior of dispersers, which may affect metapopulation persistence (Smith and Peacock, 1990; Ray et al., 1991).

Recently, the increasing reliance on metapopulation theory in reserve design has been called into question. Harrison (1994) pointed out that true metapopulations may be uncommon in nature. Further, most treatments of metapopulations have been theoretical, and the detailed data needed to parameterize models and apply them to rare species are extremely difficult to obtain (Doak and Mills, 1994; Wennergren et al., 1995). For example, the relatively few empirical applications of the metapopulation concept to vertebrates often have been constrained by inadequate data on dispersal (e.g., Lankester et al., 1991; Stacey and Taper, 1992; LaHaye et al., 1994; Lamberson et al., 1994; Lindenmayer and Lacy, 1995; Price and Gilpin, 1996). Harrison (1994) argued that the concept of the metapopulation would be more useful if it were broadened to include spatially distributed populations among which dispersal and turnover are possible but do not necessarily occur. Regardless of the uncertainty over the future role of metapopulations in conservation biology, it is certain that habitat fragmentation will continue, and effective conservation in fragmented habitats requires data on dispersal (Simberloff, 1988; Ebenhard, 1991; Doak and Mills, 1994; Fahrig and Merriam, 1994; Brawn and Robinson, 1996).

Corridors

Corridors have been proposed as a means of promoting dispersal between refuges (Diamond, 1975; Wilson and Willis, 1975). Corridors are inherently appealing, but application of the concept to reserve design remains controversial for several reasons (Noss, 1987; Simberloff and Cox, 1987; Hobbs, 1992; Simberloff et al., 1992; Mann and Plummer, 1995). In addition to the positive effect of conveying dispersers, corridors may convey negative effects such as diseases, fires, and other catastrophes (Simberloff and Cox, 1987; Simberloff et al., 1992; Hess, 1994). Corridors may be economically costly, to the extent of precluding other conservation options (Simberloff et al., 1992; Mann and Plummer, 1995). Furthermore, the proper design of a corridor, including features such as width, length, and habitat structure, is poorly understood (RL Harrison, 1992; Lindenmayer and Nix, 1993; Bennett et al., 1994; Ruefenacht and Knight, 1995; Andreassen et al., 1996b).

Perhaps most important, it is uncertain if animals will use corridors as they are designed to be used (Rosenberg et al., 1997). Research on mammals has shown that individuals do move *within* habitat corridors and that corridors may support resident populations (Wegner and Merriam, 1979; Henderson et al., 1985; Hansson, 1987; Bennett, 1990; Arnold et al., 1991; Prevett, 1991; Bennett et al., 1994; Cummings and Vessey, 1994; Ruefenacht and Knight, 1995; Kotzageorgis and Mason, 1996; Downes et al., 1997). Further, corridor width may influence movement rate within corridors (Andreassen et al., 1996a) or vulnerability to diseases (Stoner, 1996). Evidence of movement or residence within corridors, however, sheds little light on whether individuals will use corridors to move between patches or if corridors facilitate such movement. A few studies have established that dispersing mammals do use corridors; for example, some mountain lions *Felis concolor* (Beier, 1995), sugar gliders *Petaurus breviceps* (Suckling, 1984) and European red squirrels *Sciurus vulgaris* (Wauters et al., 1994) used identifiable corridors during dispersal. Additionally, Merriam and Lanoue (1990) showed that white-footed mice *Peromyscus leucopus* preferred corridors over surrounding agricultural lands as travel routes. Conversely, however, dispersing red foxes *Vulpes vulpes* did not use railway lines, believed to be important corridors, as dispersal routes (Trewhella and Harris, 1990).

The evidence that corridors facilitate movement is inconclusive, in part because goals and methods have varied among studies. Some studies of interpatch movement exploited existing patterns of fragmentation in agricultural landscapes. Bank voles *Clethrionomys glareolus* moved between two patches connected by a corridor but did not move between unconnected patches the same distance apart (Szacki, 1987). Wood mice *Apodemus sylvaticus* moving between patches, however, traveled either through corridors or across open fields (Zhang and Usher, 1991). Chipmunks *Tamias striatus* moving between patches may have used corridors, but this was not established (Bennett et al., 1994). Other studies employed experimental, artificial landscapes with replicates. Lorenz and Barrett (1990) found that confined house mice *Mus musculus* preferred one type of corridor over another when moving between patches, but they did not compare movement between patches with and without corridors. La Polla and Barrett (1993) found that movement between patches by meadow voles *Microtus pennsylvanicus* was more frequent if patches were connected by a corridor, but they acknowledged that patches were so close together that voles may have perceived corridors as extensions of patch habitat instead of as travel routes.

Corridors remain promising as components of reserve design, but a full evaluation requires information on how particular species use corridors, an assessment of negative effects of corridors, and controlled experiments that test whether corridors promote dispersal between reserves (Simberloff and Cox, 1987; Hobbs, 1992; Inglis and Underwood, 1992). Negative effects of corridors remain undocumented. Determining how corridors are used is inherently difficult and is confused by the fact that species may use corridors that are not perceived as such by humans (Kozakiewicz and Szacki, 1995). Further, experiments to assess corridor effect are difficult to conduct (Nicholls and Margules, 1991; Barrett et al., 1995).

Translocation

Despite our best efforts to design effective reserves, we may find in some cases that dispersal, even when facilitated by corridors, is insufficient for long-term persistence of species. Translocation is an option to restore or augment populations that are not sustained through natural dispersal. However, translocation is difficult, costly, and involves risks such as disease transmission (Nielsen, 1988; Cunningham, 1996). Further, translocation faces three problems concerning the behavior of animals after release. First, some animals will attempt to return home. Many species of mammals have homing ability, but the means by which animals find their way home is poorly understood (Joslin, 1977; Rogers, 1988). For some species, homing is accomplished by searching randomly until familiar landmarks are encountered (Wilson and Findley, 1972; Furrer, 1973; Van Vuren et al., 1997). Because success decreases rapidly with translocation distance, animals moved far enough are unlikely to return home. Others show evidence of true orientation and navigation (Fritts et al., 1984; Rogers 1987b), and these species may be able to home from long distances. Second, even if animals do not return home, many will abandon the release site and settle elsewhere (Fritts et al., 1984; Van Vuren et al., 1997). Third, survival of translocated animals may be low (O'Bryan and McCullough, 1985; Blanchard and Knight, 1995).

Survival and site fidelity may be affected by several factors. Poor survival may be related to the extensive movements often shown after release (Van Vuren et al., 1997). Among large mammals, omnivores and carnivores generally move more after release than do herbivores, for uncertain reasons (Rogers, 1988). The presence of conspecifics may play a role in translocation success. Some mammals live in cohesive social groups (Kleiman, 1989); additionally, presence of conspecifics may enhance site fidelity because of conspecific attraction (Smith and Peacock, 1990; Reed and Dobson, 1993). However, in many species releasing animals in various kinds of groups had little or no effect on translocation success (Hawkins and Montgomery, 1969; Fritts et al., 1984; Bright and Morris, 1994; Armstrong, 1995; Armstrong and Craig, 1995; Jones et al., 1997; but see Robinette et al., 1995). Finally, method of release influences site fidelity and may influence survival as well. Most translocations involve a hard release, in which animals are released abruptly after transport. A soft release involves retaining animals in a holding pen at the release site for a period of time before release (Davis, 1983; Fritts et al., 1984; Kleiman, 1989; Bright and Morris, 1994). Soft releases promote site fidelity and reduce postrelease movement (Davis, 1983, Bright and Morris, 1994; Bangs and Fritts, 1996), perhaps because hard-released animals are disoriented from sudden exposure to a novel environment (Bright and Morris, 1994).

Behavioral Ecology of Dispersal

Dispersal has been viewed and defined in various ways by different disciplines (Shields, 1987). Much of the work on dispersal has focused on its role in population processes, such as population regulation and population genetics (Barton, 1992; Krebs, 1992). An emphasis on population-level outcomes, however, has tended to obscure the individual nature of dispersal; dispersal fundamentally is a behavioral attribute of certain individuals, with important implications for individual fitness. Behavioral ecologists focus on the behavior of individuals, especially ecological influences and fitness consequences (Krebs and Davies, 1993). Dispersal usually is defined as the one-way movement of an animal away from its current home range to a new home range (Lidicker, 1975). Dispersal can also include movement away from a familiar social environment (Isbell and Van Vuren, 1996), but the locational aspect of dispersal is most important in reserve design.

Conservation biology often emphasizes broad-scale questions such as Will a reserve be recolonized sufficiently often? Will enough gene flow occur to prevent unwanted genetic consequences of fragmentation? Yet, this emphasis should not obscure the importance of research conducted at smaller scales. The individual-based perspective of behavioral ecologists is fundamental to answering the larger-scale questions of conservation biology (Lima and Zollner, 1996), a point which is underscored by the recent proliferation of individual-based models (Dunning et al., 1995). A surprisingly large and growing body of information about dispersal is available, much of it directly applicable to conservation questions, especially those about reserve design.

Who Disperses?

Dispersal occurs in most species of mammals and typically involves juveniles. But not all individuals disperse; some are philopatric, never leaving their natal home range (Waser and Jones, 1983). Among mammals, males are the predominantly dispersing sex. For example, female black bears *Ursus americanus* rarely disperse (Rogers, 1987a; Schwartz and Franzmann, 1992), and available evidence suggests the same for grizzly bears *Ursus arctors* (Blanchard and Knight, 1991). In a minority of mammals, dispersal is equal between the sexes, and in only a few is dispersal female biased (Greenwood, 1980; Dobson, 1982; Pusey, 1987; Wolff, 1994). Sex-biased dispersal has important implications for reserve design because dispersers that reach a reserve may be substantially or entirely one sex, usually males.

Finding a New Home

The enduring image of dispersing lemmings plunging blindly off cliffs is both inaccurate and misleading; dispersal should be viewed as a behavior that improves prospects for individual fitness, and dispersers make decisions accordingly (Murray, 1967). The process used by dispersers in finding a home is poorly understood, but three general patterns are evident in the literature that vary according to spatial scale (Van Vuren and Smallwood, 1996).

First, some dispersers detect a vacancy or opportunity adjacent to their home range and disperse through a local shift in home range. This pattern is well known in cooperatively breeding birds (e.g., Walters, 1990; Woolfenden and Fitzpatrick, 1990), but it also occurs in mammals. For example, dwarf mongooses *Helogale parvula* often disperse by moving to an adjacent home range when an opportunity becomes available (Rood, 1987), and red squirrels *Tamiasciurus hudsonicus* disperse to adjacent territories after the owner is experimentally removed (Boutin et al., 1993). Second, some dispersers discover opportunities beyond their home range through exploratory excursions, round-trip forays that allow assessment of prospects elsewhere before eventual dispersal (Johnson, 1989). For example, red squirrels (Larsen and Boutin, 1994), red foxes (Woollard and Harris, 1990), Merriam's kangaroo rats *Dipodomys merriami* (Jones, 1989), and Belding's ground squirrels *Spermophilus beldingi* (Holekamp, 1986) make exploratory excursions to gain familiarity beyond their home range before dispersing. Exploratory excursions may exceed several home range diameters in length. Third, many dispersers discover opportunities for settlement by chance encounter during one-way movements through previously unknown areas. These movements often are abrupt, rapid, and may cover long distances; examples are Columbian ground squirrels *Spermophilus columbianus* (Wiggett et al., 1989) and coyotes *Canis latrans* (DJ Harrison, 1992).

Factors that affect where a disperser eventually settles are incompletely understood and may be complex. A common theme in theoretical (e.g., Murray, 1967; Waser, 1985) and empirical studies (e.g., Jones, 1984; Wiggett et al., 1989) is that dispersers move until they find vacant, suitable habitat. Other factors, however, may also be important; Holekamp (1986) observed dispersing Belding's ground squirrels pass through but not settle in vacant habitat known to be suitable, and I have observed the same for yellow-bellied marmots *Marmota flaviventris*. Some species may use the presence of nearby conspecifics as an indicator of habitat suitability (Stamps, 1987). For polygynous mammals, males may disperse until they locate females that are undefended by another male (Wolff, 1994). Among gregarious carnivores, settlement may be influenced by the social environment in the new location (Waser, 1996).

Knowledge of how dispersers locate a new home range can affect reserve design. Dispersal to adjacent home ranges is local in scale and will be important primarily within a reserve. If, however, effective corridors are developed, movement among reserves might be encouraged with corridors that support resident populations, although such movement may be relatively slow. Dispersal after exploratory excursions occurs on a larger scale and can have effects both within and among reserves; exploratory excursions reported for some species exceed the median dispersal distance and may involve crossing unsuitable habitat (Van Ballenberghe, 1983; Wiggett et al., 1989; Van Vuren, 1990). Dispersal into unfamiliar areas occurs on the largest scale and may be most important for movements among reserves. An understanding of factors affecting choice of settlement site is also important. For example, dispersers that preferentially settle near conspecifics may be slower to recolonize vacant reserves (Smith and Peacock, 1990).

Dispersal Direction

The direction taken by dispersers is important in reserve design for two reasons. First, if direction is random, then for reserves at the edge of a reserve system, one-half or more of dispersers will be lost from the system. Second, if direction is nonrandom, knowledge of habitat features that affect dispersal direction may be useful in improving corridor design.

Studies on a variety of mammals show that dispersers emigrate in all compass directions (e.g., Storm et al., 1976; Bunnell and Harestad, 1983; Harris and Trewhella, 1988; Gese and Mech, 1991; Thomson et al., 1992), and for some species directionality is maintained during dispersal (Storm et al., 1976; DJ Harrison, 1992). Dispersal direction can be influenced by major topographic features such as sea coasts, rivers, and lakes (Bunnell and Harestad, 1983; DJ Harrison, 1992). Direction may also be influenced by habitat features. For example, white-tailed deer *Odocoileus virginianus* tend to disperse either up or down a river, keeping within riparian habitat (Dusek et al., 1989), and dispersing tigers *Panthera tigris* avoid cultivated land (Smith, 1993). Dispersing Columbian ground squirrels and black-tailed prairie dogs *Cynomys ludovicianus* often follow linear habitat features, including dirt roads (Knowles, 1985; Garrett and Franklin, 1988; Wiggett et al., 1989), and some yellow-bellied marmots disperse along naturally occurring, linear habitat features that provide suitable cover (Van Vuren, 1990).

We may view reserves as islands of suitable habitat surrounded by a sea of unsuitable habitat (Diamond, 1975), but we know too little about whether dispersers share the same perception (Stamps et al., 1987). Although habitat does influence dispersal direction for some species, the literature is also replete with accounts of dispersers crossing bleak terrain. For example, field voles *Microtus agrestis* commonly swim up to 620 m across open water when dispersing between islands (Pokki, 1981). To cross frozen lakes, common shrews *Sorex araneus* (Tegelström and Hansson, 1987) and mice *Peromyscus* spp.; Christianson, 1977) will traverse 1 km and 600 m of exposed ice, respectively. One common shrew traveled 4 km on the ice (Tegelström and Hansson, 1987). Further, there is increasing evidence that habitat unsuitable for residency may still be acceptable for dispersal. Most dispersing yellow-bellied marmots cross large areas entirely devoid of suitable habitat, and some of them swim rivers (Van Vuren, 1990). Tigers disperse through habitat that is inferior to that in which they eventually settle (Smith, 1993). Mountain sheep *Ovis canadensis* cross barren desert valleys between mountain ranges (Bleich et al., 1990). Clearly, the role of habitat in hindering dispersal or in funneling dispersers through corridors needs further study.

Dispersal Distance

The metapopulation model, as originally envisioned by Levins (1969), assumed that all patches are equally likely to be colonized, regardless of distance to other patches. The need for more realistic models in making management decisions has resulted in an increased emphasis on spatially explicit models that incorporate information on dispersal behavior, including spatial aspects of dispersal (Dunning et al., 1995; Wiens, 1996). Dispersal distance is a critical element in reserve design; if inter-reserve distance is too long for dispersers, the reserve will fail. Accurate estimates of dispersal distance, however, are difficult to obtain even for common species.

Dispersal distances vary greatly among mammals, ranging from a few meters for some voles (Lambin, 1994) and chipmunks (Elliott, 1978) to 886 km for a wolf (Fritts, 1983), a difference approaching five orders of magnitude. Dispersal involves a shift in home range location. Because spatial and temporal attributes of mammalian home ranges, such as home range size and rate of home range use, vary among species as a function of body size (McNab, 1963; Harestad and Bunnell, 1979; Lindstedt et al., 1986; Swihart et al., 1988), I hypothesize that dispersal distance does as well. I searched the literature on North American mammals (excluding marine mammals and bats) for measures of dispersal distance. In selecting suitable studies, I defined dispersal as a one-way move by an individual to a new, non-overlapping home range if individual home ranges were known; otherwise, I defined dispersal as a one-way move equal to or greater than the mean home range diameter for the population (or individual sexes, if available). Methods for determining dispersal distance may differ in accuracy. Trapping or observation, for example, may substantially underestimate dispersal distance in comparison with radio-telemetry (Koenig et al., 1996). I included studies that may have underestimated dispersal distance, but I excluded studies that likely overestimated distance—for example, reports of unusual long-distance moves (e.g., Fritts, 1983). In designing a reserve, it is best to be conservative.

Dispersal distances of mammals are usually right-skewed, with most animals moving shorter distances, so instead of using mean distance I calculated median distance for each study, representing the distance moved by 50% of dispersers. For distributions reported in intervals, I used the midpoint of the interval that included the median. Dispersal distances of mammals often differ between the sexes, so for most species ($n = 34$) I calculated medians separately for males and females. I calculated a weighted average among studies for medians for each sex, then an unweighted average of the sexes for each species. No dispersal distances have been reported for female grizzly bears, so I used a distance of one home range diameter. Sex was not reported for the remaining six species, so I calculated medians for males and females combined. For three of these (*Spermophilus beldingi, Tamiasciurus hudsonicus,* and *Scapanus townsendii*) the investigator reported that disper-

Table 14-1 Body mass, median dispersal distance, and total sample size of dispersers of North American mammals.

Species[a]	Mass (g)	Distance: Median (m)	Distance: *n*	Reference
Ursus arctos (O)	204,120	53,650	8	Berns et al. (1980), Blanchard and Knight (1991)
Odocoileus virginianus (H)	90,720	14,600	106	Tierson et al. (1985), Dusek et al. (1989), Nelson (1993)
Ursus americanus (O)	76,204	25,900	36	LeCount (1982), Young and Ruff (1982), Rogers (1987a), Elowe and Dodge (1989), Schwartz and Franzmann (1992)
Felis concolor (C)	67,000	65,800	40	Hemker et al. (1984), Logan et al. (1986), Maehr et al. (1991), Anderson et al. (1992), Lindzey et al. (1994), Beier (1995), Cunningham et al. (1995)
Odocoileus hemionus columbianus (H)	64,638	11,500	26	Zwickel et al. (1953), Bunnell and Harestad (1983), Harestad and Bunnell (1983)
Canis lupus (C)	37,422	57,700	118	Fritts and Mech (1981), Peterson et al. (1984), Ballard et al. (1987), Gese and Mech (1991)
Castor canadensis (H)	17,720	10,200	28	Beer (1955), Leege (1968)
Canis latrans (C)	15,890	41,200	58	Andelt and Gipson (1979) Bowen (1982), Woodruff and Keller (1982), Sumner et al. (1984), Roy and Dorrance (1985), D. J. Harrison (1992)
Taxidea taxus (C)	13,620	12,300	11	Messick and Hornocker (1981)
Lynx rufus (C)	9072	31,400	12	Berg (1979), Kitchings and Story (1979), Kitchings and Story (1984), Wassmer et al. (1988), Knick (1990)
Lutra canadensis (C)	8550	26,500	2	Melquist and Hornocker (1983)
Erethizon dorsatum (H)	7882	2800	4	Roze (1989)
Procyon lotor (O)	7264	10,600	25	Stuewer (1943), Ellis (1964), Urban (1970), Fritzell (1978), Lehman (1984), Clark et al. (1989)
Vulpes vulpes (C)	5448	12,300	117	Storm et al. (1976), Tullar and Berchielli (1982)
Marmota flaviventris (H)	3628	1088	68	Van Vuren (1990)
Urocyon cinereoargenteus (C)	3600	10,600	16	Tullar and Berchielli (1982), Nicholson et al. (1985)
Martes pennanti (C)	3459	10,400	13	Arthur et al. (1993)
Lepus californicus (H)	3039	1206	11	French et al. (1965)
Didelphis virginiana (O)	2724	1772	13	Gillette (1980)

(continued)

Table 14-1 (*Continued*)

Species[a]	Mass (g)	Distance		Reference
		Median (m)	*n*	
Mephitis mephitis (O)	2586	3280	16	Rosatte and Gunson (1984)
Vulpes macrotis (C)	2340	5535	19	L. Spiegel (unpublished data)
Lepus americanus (H)	1543	582	6	Adams (1959), O'Farrell (1965)
Ondatra zibethicus (H)	1275	225	19	Caley (1987)
Mustela nigripes (C)	883	2473	15	Biggins et al. (1985), Forrest et al. (1985, 1988)
Sciurus carolinensis (H)	500	535	25	Cordes and Barkalow (1972), Gull (1977)
Spermophilus columbianus (H)	466	345	86	Murie and Harris (1984), Hackett (1987)
Spermophilus beldingi (H)	282	243	70	Holekamp (1984)
Tamiasciurus hudsonicus (H)	254	180	37	Larsen and Boutin (1994)
Mustela erminea (C)	202	1370	14	Erlinge (1977)
Scapanus townsendii (C)	142	228	32	Giger (1973)
Dipodomys spectabilis (H)	123	100	69	Jones (1987), Jones et al. (1988)
Tamias striatus (H)	85	116	19	Burt (1940), Elliott (1978)
Dipodomys stephensi (H)	66	45	74	Price et al. (1994)
Microtus townsendii (H)	55	23	104	Lambin (1994)
Peromyscus californicus (O)	44	100	31	Ribble (1992)
Dipodomys merriami (H)	41	113	23	Jones (1989)
Microtus pennsylvanicus (H)	40	66	3	Madison (1980)
Peromyscus leucopus (O)	21	82	76	Keane (1990), Glass et al. (1991), Jacquot and Vessey (1995)
Peromyscus maniculatus (O)	19	133	100	Nicholson (1941), Dice and Howard (1951)
Blarina brevicauda (C)	18	162	2	Fitch (1958)

[a]Feeding category (C = carnivore, O = omnivore, H = herbivore) in parentheses.

sal distances did not differ between the sexes. I included the other three species (*Lepus californicus, L. americanus,* and *Blarina brevicauda*) because no other dispersal data were available for these species.

I obtained values for body mass from Harestad and Bunnell (1979), except for the following species: beaver *Castor canadensis* (Grasse and Putnam, 1950; Beer, 1955), river otter *Lutra canadensis* (Melquist and Hornocker, 1983), gray fox *Urocyon cinereoargenteus* (Lindstedt et al., 1986), kit fox *Vulpes macrotis* (L. Spiegel, unpublished data), muskrat *Ondatra zibethicus* (Errington, 1963), black-footed ferret *Mustela nigripes* (Forrest et al., 1988), ground squirrels (*Spermophilus* spp.; Armitage, 1981), ermine *Mustela erminea* (Erlinge, 1977), kangaroo rats (*Dipodomys* spp.; Jones, 1985), Townsend's vole *Microtus townsendii* (Lambin, 1994), white-footed mice (Silva and Downing, 1995), and short-tailed shrew *Blarina brevicauda* (Innes, 1994). Feeding categories (carnivore, omnivore, or herbivore) were assigned following Harestad and Bunnell (1979).

Median dispersal distances varied greatly among species, ranging from 23 m to 65.8 km

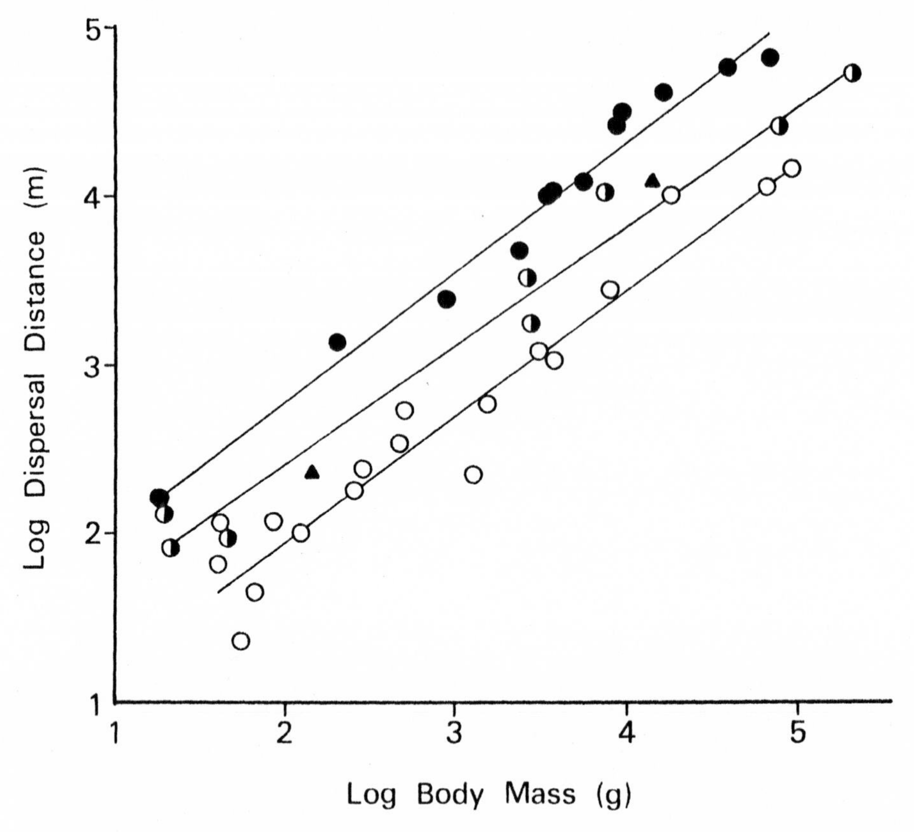

Figure 14-1 The relationship between median dispersal distance (m) and body mass (g) of North American mammals. Open circles are herbivores, half-shaded circles are omnivores, and fully shaded circles are carnivores. The regression line for carnivores does not include fossorial *(Scapanus townsendii)* and semifossorial *(Taxidea taxus)* species, which are indicated by fully shaded triangles. Axes are on the basis of $\log_{10}$.

(table 14-1). Body mass explained a significant portion ($p < .001$) of the variation in median dispersal distance for all species (fig. 14-1), and variation explained was increased when species were considered separately according to feeding category (table 14-2). Whether a species feeds above or below ground may also be important; fossorial (*Scapanus townsendii*) and semifossorial (*Taxidea taxus*) carnivores dispersed shorter distances for a given body size than did other carnivores (fig. 14-1). A similar relationship was reported for home range size of fossorial mammals (Harestad and Bunnell, 1979).

Analysis of covariance indicated no significant difference ($F_{2,32} = 0.501$, $p > .50$) among slopes for carnivores excluding fossorial and semifossorial species (0.77), omnivores (0.70), and herbivores (0.75), and none of these slopes differed significantly from 0.75 ($p > .10$). Thus, the relationship between dispersal distance and body size is consistent with that for size-dependent metabolic requirements (McNab, 1963). For a given body size, carnivores dispersed farthest, followed by omnivores, then herbivores. Harestad and Bunnell (1979) found the same relationship for home range size and suggested the cause was dif-

Table 14-2 Regression equations for double-logarithmic relationships between median dispersal distance (*D*, in meters) and body mass (*M*, in grams) of North American mammals, by feeding category.

Feeding category	Regression equation	r^2	n
All species	$D = 4.23M^{0.82}$	.85	40
Carnivores	$D = 11.73M^{0.81}$	.94	14
Carnivores except fossorial and semifossorial species	$D = 18.69M^{0.77}$	.98	12
Omnivores	$D = 10.84M^{0.70}$	.98	8
Herbivores	$D = 2.93M^{0.75}$	.93	18

Equations are presented both for all carnivores and for carnivores excluding fossorial (*Scapanus townsendii*) and semifossorial (*Taxidea taxus*) species.

fering densities of food resources; density is highest for herbivores and lowest for carnivores. Thus, the distance that a disperser moves may be related to size-dependent energetics and resource density.

Brown and Zeng (1989) found that the relation between dispersal distance and body size among desert rodents was nonlinear; distance was shortest for species weighing 40–140 g and increased for larger and smaller species. They proposed that the smallest species have such high energy requirements that they must disperse between the richest habitat patches, thus they move relatively long distances. In my analysis, there is an indication of such an increase in dispersal distance among the smallest species, below about 50 g (fig. 14-1). This pattern appears to result primarily from two differences among feeding categories: 1) for a given body size, dispersal distance of carnivores is greatest, for omnivores intermediate, and for herbivores least, presumably because of differing resource density; 2) the smallest carnivores and omnivores are smaller than the smallest herbivores, presumably because the energetic demands of small size mandate high-energy foods. Thus, my findings agree with the interpretation of Brown and Zeng (1989).

Dispersal tendency in mammals is male biased (Greenwood, 1980; Dobson, 1982), and for mammals in which both sexes disperse, dispersal distance may be male biased as well (Waser, 1985; Wolff, 1994). Among 33 species in table 14-1 for which data on each sex were available, the male/female ratio of median dispersal distances was >1.0 for 23 species and <1.0 for 8 species. These numbers should be interpreted with caution because of varying and sometimes small sample sizes, often extracted from several different studies, but they do suggest that males tend to disperse farther than females.

Allometric equations, like those reported here, only describe patterns; they are not predictive laws (Lindstedt et al., 1986). For many species of mammals, however, we currently have nothing but guesswork for determining inter-reserve distance when designing a refuge system that requires linkage through dispersal. The equations presented here (table 14-2) might provide a rough estimate of median dispersal distance if no empirical measures are available.

Survival of Dispersers

Dispersal seems dangerous. Dispersers often move through unfamiliar or even inhospitable terrain, where they may be vulnerable to predation, stress, or malnutrition. Thus, survival

Table 14-3 Survival estimates and causes of mortality (expressed as a percentage of total mortality) for dispersing mammals.

Species	Survival	Causes of mortality		Reference
		Predation	Humans[a]	
Carnivores				
Black bear (*Ursus americanus*)	0.50		Most	Schwartz and Franzmann (1992)
Mountain lion (*Felis concolor*)	0.57		100%	Lindzey et al. (1988)
	0.22		56%	Beier (1995)
Gray wolf (*Canis lupus*)	0.30		Most	Peterson et al. (1984)
Coyote (*Canis latrans*)	0.23		100%	Gese et al. (1989)
	0.47		86%	D. J. Harrison (1992)
Badger (*Taxidea taxus*)	0.38		100%	Messick and Hornocker (1981)
Red fox (*Vulpes vulpes*)	0.56			Woollard and Harris (1990)
Ungulates				
White-tailed deer (*Odocoileus virginianus*)	1.00			Nelson and Mech (1991)
Rodents				
Yellow-bellied marmot (*Marmota flaviventris*)	0.73	100%		Van Vuren and Armitage (1994)
Black-tailed prairie dog (*Cynomys ludovicianus*)	0.44	73%		Garrett and Franklin (1988)
Columbian ground squirrel (*Spermophilus columbianus*)	1.00			Hackett (1987)
	1.00			Wiggett et al. (1989)
Red squirrel (*Tamiasciurus hudsonicus*)	0.71	100%		Larson and Boutin (1994)
Common vole (*Microtus arvalis*)	0.98			Boyce and Boyce (1988)

[a]Human-caused mortality includes hunting, trapping, and collisions with vehicles.

of dispersers has long been believed to be low (Anderson, 1989). Experiments have shown that small mammals are more vulnerable to predation when in unfamiliar environments (Metzgar, 1967; Ambrose, 1972). Morality, however, may result more from greater movement than from lack of familiarity (Kaufman, 1974; Snyder et al., 1976). Gaines and McClenaghan (1980) suggested that survival of dispersers was inversely related to the amount of time spent in transit. Dispersers often move rapidly (Storm et al., 1976; Gillette, 1980; Wiggett et al., 1989), presumably to decrease time in transit. Boyce and Boyce (1988) concluded that the high survival (98%) of dispersing common voles *Microtus arvalis* resulted from the short duration of dispersal (mean = 39 min). Yellow-bellied marmots often move rapidly during dispersal, and dispersal distance, and presumably duration as well, is inversely related to survival (Van Vuren, 1990).

If survival of dispersers is as poor as has been believed, the prognosis for a fragmented reserve is bleak. A reserve system designed to be well within the dispersal capabilities of a species will still fail if most dispersers die while attempting to move between reserves. Until recently, almost nothing was known about the survival of dispersers (Johnson and Gaines, 1990). Enough estimates are now available to attempt some generalizations (table 14-3). Survival of dispersers varies among species and populations, ranging from 22% to 100%. For some species, especially rodents, survival is relatively high, and most mortality is caused by predation. Survival of carnivores is lower, and most mortality is caused by humans through hunting, trapping, and collisions with vehicles. Consequently, a reserve system that includes carnivores should incorporate measures to reduce human-caused mortality during dispersal between reserves.

The Behavioral Ecological Approach: An Assessment

The main strength of the behavioral ecological approach is that it provides the information on dispersal behavior needed to develop and parameterize realistic, spatially explicit models and design effective reserves. For a growing number of species, we have information such as who disperses, how they go about finding a new home, the direction they take and the factors that influence direction, the distance they travel, and their prospects for survival. Armed with this knowledge, we can estimate the consequences of dispersal and design reserves accordingly. We can estimate the distance between reserves that dispersers of a particular species are likely to traverse, habitats that they are likely or unlikely to use during dispersal, including potential corridors, and the factors that may affect settlement. We can identify mortality factors that might be manipulated to improve survival and predict the sex ratio of successful dispersers.

The principal weakness of the behavioral ecological approach is insufficient data. Although we know a surprising amount about dispersal of some species, for others we know too little. Further, some aspects of dispersal behavior, such as which factors influence choice of settlement site or influence the direction, distance, and route traveled to get there, are not sufficiently understood for many species. Developments in molecular genetic approaches may provide some of the needed data. Geographic variation in allozyme frequencies has been employed to infer the amount and extent of dispersal in mammals, but such comparisons between indirect and direct measures of gene flow have yielded equivocal results. In some cases these measures were reasonably concordant (Schwartz and Armitage, 1980; Daly and Patton, 1990), whereas in others direct measures of dispersal were substantially less than those inferred from patterns of genetic variation (Waser and Elliott, 1991; Dobson, 1994). Recently, DNA techniques have been developed that hold promise for providing more accurate estimates of certain aspects of dispersal, particularly dispersal distance and sex biases in dispersal tendency (Neigel et al., 1991; Neigel and Avise, 1993; Moritz, 1994; Avise, 1995; Webb et al., 1995). Application of DNA approaches to conservation, however, remains problematic, in part because of differing time scales (Moritz, 1994). DNA approaches reflect the long-term behavior of populations, yet it is often short-term questions that are of interest to conservation biology (Moritz, 1994). For example, current patterns of genetic variation in a rare species may represent the consequences of extensive dispersal during past abundance, yet what is needed for conservation is information about dispersal occurring today. Further, DNA approaches tell us about the genetic consequences of successful dispersal, but they tell us little about the factors influencing that success. Much of the information about dispersal we need for reserve design, such as factors affecting survival and settlement site choice and the influence of habitat on dispersal direction and distance, is currently outside the realm of molecular techniques. The behavioral ecological approach provides this information.

Recommendations

Reserves should be designed, in part, on the basis of the dispersal behavior of the species being protected. Dispersal in most mammals is biased toward males, with important implications for reserve effectiveness; translocation of females may be necessary. Dispersers that cross unfamiliar terrain are most likely to discover vacant reserves. In some species, however, such as Merriam's kangaroo rat (Jones, 1989), crossing large expanses of unfamiliar terrain is rare, and steps should be taken to promote movement of these species between reserves. Dispersal direction often appears random, but there is a growing body of evidence that direction may be influenced by landscape features that might be managed to funnel dispersers between reserves.

We need more information about the factors that affect the settlement decision and how reserve design and management might be manipulated to promote settlement. Knowing how far dispersers are likely to travel is essential. Rough estimates might be obtained from allometric equations (table 14-2), but these estimates should be used with caution and only in the absence of empirical measures. For species with sex-biased dispersal distances, inter-reserve distance might be restricted to that of the sex that disperses the shortest distance. We especially need information about the effects of human-caused habitat fragmentation on dispersal distance. Artificial fragmentation increased the distances moved by several species of small mammals, although it was uncertain if such movements constituted dispersal (Diffendorfer et al., 1995). Survival of dispersers might be increased by discovering and reducing factors likely to cause mortality during dispersal. For example, reserves for large carnivores should be established away from well-traveled roads, in areas where hunting and trapping are prohibited.

Corridors and translocation remain potential options if unfacilitated dispersal is inadequate, but each is problematic. Despite the inherent appeal of corridors, we do not know if they facilitate dispersal between reserves, and, even if corridors are shown to be effective, we know little about appropriate designs. Research is needed in both areas. Translocation is costly, difficult, and risky, and careful attention should be paid to factors influencing postrelease survival and site fidelity. Species lacking true navigational ability should be translocated far enough to preclude homing. A soft release, although more expensive and time-consuming, may improve site fidelity and survival.

Summary

Dispersal is an important component of reserve design, yet conservation biologists have paid little attention to the behavioral ecology of dispersal. Behavioral ecologists have been studying dispersal behavior for about two decades using an individual-based approach, and this perspective is fundamental to designing effective reserves.

Among most species of mammals, males are the predominantly dispersing sex. Dispersers find a new home by detecting a vacancy in an adjacent home range, by making round-trip exploratory excursions beyond their home range, or by chance encounter during abrupt and often long-distance movements through previously unknown areas. Factors affecting settlement site choice are poorly understood. Dispersers often emigrate in all compass directions. Direction may be influenced by topographic features, but dispersers sometimes cross large expanses of seemingly bleak terrain. Dispersal distance varies greatly among mammalian species, and body mass explains a substantial portion of this variation. The resulting allometric equations may provide a rough estimate of dispersal distance if no empirical data are available. Males of some species appear to disperse farther than females. Survival of dispersers is relatively high for rodents but is lower for carnivores, primarily because of human-caused mortality such as hunting, trapping, and collisions with vehicles.

Currently, metapopulation theory is popular as a model for reserve design; corridors are being advocated as a means of promoting dispersal between reserves; and translocation is considered an option if natural dispersal is inadequate. However, data to parameterize metapopulation models are inadequate; the effectiveness of corridors is unproven; and translocation is difficult, expensive, and risky. The behavioral ecological approach provides the information on dispersal behavior needed to develop and parameterize realistic models and design effective reserves. The principal shortcoming is insufficient data. Recent molecular approaches may provide a partial answer to this insufficiency, but these approaches face important limitations.

Acknowledgments I thank Tim Caro, Alice Clarke, Chris Ray, and Andrew Smith for insightful comments that substantially improved this chapter.

References

Adams L, 1959. An analysis of a population of snowshoe hares in northwestern Montana. Ecol Monogr 29:141–170.

Ambrose HW III, 1972. Effect of habitat familiarity and toe-clipping on rate of owl predation in *Microtus pennsylvanicus*. J Mammal 53:909–912.

Andelt WF, Gipson PS, 1979. Home range, activity and daily movements of coyotes. J Wildl Manage 43:944–951.

Anderson AE, Bowden DC, Kattner DM, 1992. The puma on Uncompahgre Plateau, Colorado. Colorado Div Wildl Tech Publ 40:1–116.

Anderson PK, 1989. Dispersal in rodents: a resident fitness hypothesis. Am Soc Mammal Spec Publ 9:1–141.

Andreassen HP, Halle S, Ims RA, 1996a. Optimal width of movement corridors for root voles: not too narrow and not too wide. J Appl Ecol 33:63–70.

Andreassen HP, Ims RA, Steinset OK, 1996b. Discontinuous habitat corridors: effects on male root vole movements. J Appl Ecol 33:555–560.

Armitage KB, 1981. Sociality as a life-history tactic of ground squirrels. Oecologia 48:36–49.

Armstrong DP, 1995. Effects of familiarity on the outcome of translocations, II. A test using New England robins. Biol Conserv 71:281–288.

Armstrong DP, Craig JL, 1995. Effects of familiarity on the outcome of translocations, I. A test using saddlebacks *Philesturnus carunculatus rufusater*. Biol Conserv 71:133–141.

Arnold GW, Weeldenburg JR, Steven DE, 1991. Distribution and abundance of two species of kangaroo in remnants of native vegetation in the central wheatbelt of Western Australia and the role of native vegetation along road verges and fencelines as linkages. In: Nature conservation 2: the role of corridors (Saunders DA, Hobbs RJ, eds). Chipping Norton, NSW, Australia: Surrey Beatty and Sons; 273–280.

Arthur SM, Paragi TF, Krohn WB, 1993. Dispersal of juvenile fishers in Maine. J Wildl Manage 57:868–874.

Avise JC, 1995. Mitochondrial DNA polymorphism and a connection between genetics and demography of relevance to conservation. Conserv Biol 9:686–690.

Ballard WB, Whitman JS, Gardner CL, 1987. Ecology of an exploited wolf population in south-central Alaska. Wildl Monogr 98:1–54.

Bangs EE, Fritts SH, 1996. Reintroducing the gray wolf to central Idaho and Yellowstone National Park. Wildl Soc Bull 24:402–413.

Barrett GW, Peles JD, Harper SJ, 1995. Reflections on the use of experimental landscapes in mammalian ecology. In: Landscape approaches in mammalian ecology and conservation (Lidicker WZ, Jr, ed). Minneapolis: University of Minnesota Press; 157–174.

Barton NH, 1992. The genetic consequences of dispersal. In: Animal dispersal: small mammals as a model (Stenseth NC, Lidicker WZ Jr., eds). London: Chapman and Hall; 37–59.

Beer JR, 1955. Movements of tagged beaver. J Wildl Manage 19:492–493.

Beier P, 1995. Dispersal of juvenile cougars in fragmented habitat. J Wildl Manage 59:228–237.

Bennett AF, 1990. Habitat corridors and the conservation of small mammals in a fragmented forest environment. Landscape Ecol 4:109–122.

Bennett AF, Heinen K, Merriam G, 1994. Corridor use and the elements of corridor quality: chipmunks and fencerows in a farmland mosaic. Biol Conserv 68:155–165.

Berg WE, 1979. Ecology of bobcats in northern Minnesota. In: Bobcat research conference proceedings (Escherich PC, Blum LG, eds). Scientific and Technical Series 6. National Wildlife Federation; 55–61.

Berns VD, Atwell GC, Boone DL, 1980. Brown bear movements and habitat use at Karluk Lake, Kodiak Island. Int Conf Bear Res Manage 4:293–296.

Biggins DE, Schroeder M, Forrest S, Richardson L, 1985. Movements and habitat relationships of radio-tagged black-footed ferrets. In: Black-footed ferret workshop proceedings (Anderson SH, Inkley DB, eds). Cheyenne, Wyoming: Wyoming Game and Fish Department; 11.1–11.17.

Blanchard BM, Knight RR, 1991. Movements of Yellowstone grizzly bears. Biol Conserv 58:41–67.

Blanchard BM, Knight RR, 1995. Biological consequences of relocating grizzly bears in the Yellowstone ecosystem. J Wildl Manage 59:560–565.

Bleich VC, Wehausen JD, Holl SA, 1990. Desert-dwelling mountain sheep: conservation implications of a naturally fragmented distribution. Conserv Biol 4:383–390.

Boutin S, Tooze Z, Price K, 1993. Post-breeding dispersal by female red squirrels (*Tamiasciurus hudsonicus*): the effect of local vacancies. Behav Ecol 4:151–155.

Bowen WD, 1982. Home range and spatial organization of coyotes in Jasper National Park, Alberta. J Wildl Manage 46:201–216.

Boyce CCK, Boyce JL III, 1988. Population biology of *Microtus arvalis*. II. Natal and breeding dispersal of females. J Anim Ecol 57:723–736.

Brawn JD, Robinson SK, 1996. Source-sink population dynamics may complicate the interpretation of long-term census data. Ecology 77:3–12.

Bright PW, Morris PA, 1994. Animal translocation for conservation: performance of dormice in relation to release methods, origin and season. J Appl Ecol 31:699–708.

Brown JH, Kodric-Brown A, 1977. Turnover rates in insular biogeography: effect of immigration on extinction. Ecology 58:445–449.

Brown JH, Zeng Z, 1989. Comparative population ecology of eleven species of rodents in the Chihuahuan Desert. Ecology 70:1507–1525.

Bunnell FL, Harestad AS, 1983. Dispersal and dispersion of black-tailed deer: models and observations. J Mammal 64:201–209.

Burt WH, 1940. Territorial behavior and populations of some small mammals in southern Michigan. Misc Publ Mus Zool Univ Mich 45:1–58.

Caley MJ, 1987. Dispersal and inbreeding avoidance in muskrats. Anim Behav 35:1225–1233.

Chepko-Sade BD, Halpin ZT, 1987. Mammalian dispersal patterns. Chicago: University of Chicago Press.

Christainson L, 1977. Winter movements of *Peromyscus* across a lake in northern Minnesota. J Mammal 58:244.

Clark WR, Hasbrouck JJ, Kienzler JM, Glueck TF, 1989. Vital statistics and harvest of an Iowa raccoon population. J Wildl Manage 53:982–990.

Cordes CL, Barkalow FS Jr, 1972. Home range and dispersal in a North Carolina gray squirrel population. Proc Southeast Assoc Game Fish Comm 26:124–135.

Cummings JR, Vessey SH, 1994. Agricultural influences on movement patterns of white-footed mice (*Peromyscus leucopus*). Am Midl Nat 132:209–218.

Cunningham AA, 1996. Disease risks of wildlife translocations. Conserv Biol 10:349–353.

Cunningham SC, Haynes LA, Gustavson C, Haywood DD, 1995. Evaluation of the interaction between mountain lions and cattle in the Aravaipa-Klondyke area of southeast Arizona. Arizona Game Fish Dept Tech Rep 17:1–64.

Daly JC, Patton JL, 1990. Dispersal, gene flow, and allelic diversity between local populations of *Thomomys bottae* pocket gophers in the coastal ranges of California. Evolution 44:1283–1294.

Davis MH, 1983. Post-release movements of introduced marten. J Wildl Manage 47:59–66.

Diamond JM, 1975. The island dilemma: lessons of modern biogeographic studies for the design of natural reserves. Biol Conserv 7:129–146.

Dice LR, Howard WE, 1951. Distance of dispersal by prairie deermice from birthplaces to breeding sites. Contrib Lab Vertebr Biol Univ Mich 50:1–15.

Diffendorfer JE, Gaines MS, Holt RD, 1995. Habitat fragmentation and movements of three small mammals (*Sigmodon, Microtus,* and *Peromyscus*). Ecology 76:827–839.

Doak DF, Mills LS, 1994. A useful role for theory in conservation. Ecology 76:615–626.

Dobson FS, 1982. Competition for mates and predominant juvenile male dispersal in mammals. Anim Behav 30:1183–1192.

Dobson FS, 1994. Measures of gene flow in the Columbian ground squirrel. Oecologia 100:190–195.

Downes SJ, Handasyde KA, Elgar MA, 1997. Variation in the use of corridors by introduced and native rodents in south-eastern Australia. Biol Conserv 82:379–383.

Dunning JB Jr, Stewart DJ, Danielson BJ, Noon BR, Root TL, Lamberson RH, Stevens EE, 1995. Spatially explicit population models: current forms and future uses. Ecol Appl 5:3–11.

Dusek GL, Mackie RJ, Herriges JD Jr, Compton BB, 1989. Population ecology of white-tailed deer along the lower Yellowstone River. Wildl Monogr 104:1–68.

Ebenhard T, 1991. Colonization in metapopulations: a review of theory and observations. Biol J Linn Soc 42:105–121.

Elliott L, 1978. Social behavior and foraging ecology of the eastern chipmunk (*Tamais striatus*) in the Adirondack Mountains. Smithsonian Contrib Zool 265:1–107.

Ellis RJ, 1964. Tracking raccoons by radio. J Wildl Manage 28:363–368.

Elowe KD, Dodge WE, 1989. Factors affecting black bear reproductive success and cub survival. J. Wildl Manage 53:962–968.

Erlinge S, 1977. Spacing strategy in stoat *Mustela erminea.* Oikos 28:32–42.

Errington PL, 1963. Muskrat populations. Ames: Iowa State University Press.

Fahrig L, Merriam G, 1994. Conservation of fragmented populations. Conserv Biol 8:50–59.

Fitch HS, 1958. Home ranges, territories, and seasonal movements of vertebrates of the Natural History Reservation. Univ Kansas Publ Mus Nat Hist 11:63–326.

Forrest SC, Biggins DE, Richardson L, Clark TW, Campbell TM III, Fagerstone KA, Thorne ET, 1988. Population attributes for the black-footed ferret (*Mustela nigripes*) at Meeteetse, Wyoming, 1981–1985. J Mammal 69:261–273.

Forrest SC, Clark TW, Richardson L, Campbell TM III, 1985. Black-footed ferret habitat: some management and reintroduction considerations. Wyoming Bur Land Manage Tech Bull 2:1–49.

French NR, McBride R, Detmer J, 1965. Fertility and population density of the black-tailed jackrabbit. J Wildl Manage 29:14–26.

Fritts SH, 1983. Record dispersal by a wolf from Minnesota. J Mammal 64:166–167.

Fritts SH, Mech LD, 1981. Dynamics, movements, and feeding ecology of a newly protected wolf population in northwestern Minnesota. Wildl Monogr 80:1–79.

Fritts SH, Paul WJ, Mech LD, 1984. Movements of translocated wolves in Minnesota. J Wildl Manage 48:709–721.

Fritzell EK, 1978. Aspects of raccoon (*Procyon lotor*) social organization. Can J Zool 56:260–271.

Furrer RK, 1973. Homing of *Peromyscus maniculatus* in the channelled scablands of east-central Washington. J Mammal 54:466–482.

Gaines MS, McClenaghan LR Jr, 1980. Dispersal in small mammals. Annu Rev Ecol Syst 11:163–196.

Garrett MG, Franklin WL, 1988. Behavioral ecology of dispersal in the black-tailed prairie dog. J Mammal 69:236–250.

Gese EM, Mech LD, 1991. Dispersal of wolves (*Canis lupus*) in northeastern Minnesota, 1969–1989. Can J Zool 69:2946–2955.

Gese EM, Rongstad OJ, Mytton WR, 1989. Population dynamics of coyotes in southeastern Colorado. J Wildl Manage 53:174–181.

Giger RD, 1973. Movements and homing in Townsend's mole near Tillamook, Oregon. J Mammal 54:648–659.

Gillette LN, 19980. Movement patterns of radio-tagged opossums in Wisconsin. Am Midl Nat 104:1–12.

Glass GE, Korch GW, Gomez JE, Childs JE, 1991. Use of exotic antigens to measure reproduction and dispersal in *Peromyscus leucopus*. Can J Zool 69:528–530.

Grasse JE, Putnam EF, 1950. Beaver management and ecology in Wyoming. Wy Game Fish Comm Bull 6:1–52.

Greenwood PJ, 1980. Mating systems, philopatry and dispersal in birds and mammals. Anim Behav 28:1140–1162.

Griffith B, Scott JM, Carpenter JW, Reed C, 1989. Translocation as a species conservation tool: status and strategy. Science 245:477–480.

Gull J, 1977. Movement and dispersal patterns of immature gray squirrels (*Sciurus carolinensis*) in east-central Minnesota (MS thesis). Minneapolis: University of Minnesota.

Hackett DF, 1987. Dispersal of yearling Columbian ground squirrels (PhD dissertation). Edmonton: University of Alberta.

Hanski I, Gilpin M, 1991. Metapopulation dynamics: brief history and conceptual domain. Biol J Linn Soc 42:3–16.

Hansson L, 1987. Dispersal routes of small mammals at an abandoned field in central Sweden. Holoarctic Ecol 10:154–159.

Harestad AS, Bunnell FL, 1979. Home range and body weight—a reevaluation. Ecology 60:389–402.

Harestad AS, Bunnell FL, 1983. Dispersal of a yearling male black-tailed deer. Northwest Sci 57:45–48.

Harris S, Trewhella WJ, 1988. An analysis of some of the factors affecting dispersal in an urban fox (*Vulpes vulpes*) population. J Appl Ecol 25:409–422.

Harrison DJ, 1992. Dispersal characteristics of juvenile coyotes in Maine. J Wildl Manage 56:128–138.

Harrison RL, 1992. Toward a theory of inter-refuge corridor design. Conserv Biol 6:293–295.

Harrison S, 1994. Metapopulations and conservation. In: Large-scale ecology and conservation biology (Edwards PJ, May RM, Webb NR, eds). London: Blackwell Scientific; 111–128.

Hawkins RE, Montgomery GG, 1969. Movements of translocated deer as determined by telemetry. J Wildl Manage 33:196–203.

Hemker TP, Lindzey FG, Ackerman GG, 1984. Population characteristics and movement patterns of cougars in southern Utah. J Wildl Manage 48:1275–1284.

Henderson MT, Merriam G, Wegner J, 1985. Patchy environments and species survival: chipmunks in an agricultural mosaic. Biol Conserv 31:95–105.

Hess GR, 1994. Conservation corridors and contagious disease: a cautionary note. Conserv Biol 8:256–262.

Hobbs RJ, 1992. The role of corridors in conservation: solution or bandwagon? Trends Ecol Evol 7:389–392.

Hodder KH, Bullock JM, 1997. Translocations of native species in the UK: implications for biodiversity. J Appl Ecol 34:547–565.

Holekamp KE, 1984. Natal dispersal in Belding's ground squirrels (*Spermophilus beldingi*). Behav Ecol Sociobiol 16:21–30.

Holekamp KE, 1986. Proximal causes of natal dispersal in Belding's ground squirrels (*Spermophilus beldingi*). Ecol Monogr 56:365–391.

Inglis G, Underwood AJ, 1992. Comments on some experimental designs proposed for experiments on the biological importance of corridors. Conserv Biol 6:581–583.

Innes DGL, 1994. Life histories of the Soricidae: a review. In: Advances in the biology of shrews (Merritt JM, Kirkland GL Jr, Rose RK, eds). Carnegie Museum of Natural History Special Publication 18. Pittsburgh, Pennsylvania: CMNH 111–136.

Isbell LA, Van Vuren D, 1996. Differential costs of locational and social dispersal and their consequences for female group-living primates. Behaviour 133:1–36.

Jacquot JJ, Vessey SH, 1995. Influence of the natal environment on dispersal of white-footed mice. Behav Ecol Sociobiol 37:407–412.

Johnson ML, 1989. Exploratory behavior and dispersal: a graphical model. Can J Zool 67:2325–2328.

Johnson ML, Gaines MS, 1990. Evolution of dispersal: theoretical models and empirical tests using birds and mammals. Annu Rev Ecol Syst 21:449–480.

Jones ML, Mathews NE, Porter WF, 1997. Influence of social organization on dispersal and survival of translocated female white-tailed deer. Wildl Soc Bull 25:272–278.

Jones WT, 1984. Natal philopatry in bannertailed kangaroo rats. Behav Ecol Sociobiol 15:151–155.

Jones WT, 1985. Body size and life-history variables in heteromyids. J Mammal 66:128–132.

Jones WT, 1987. Dispersal patterns in kangaroo rats. In: Mammalian dispersal patterns (Chepko-Sade BD, Halpin ZT, eds). Chicago: University of Chicago Press; 119–127.

Jones WT, 1989. Dispersal distance and the range of nightly movements in Merriam's kangaroo rats. J Mammal 70:27–34.

Jones WT, Waser PM, Elliot LF, Link NE, Bush BB, 1988. Philopatry, dispersal, and habitat saturation in the banner-tailed kangaroo rat, *Dipodomys spectabilis*. Ecology 69:1466–1473.

Joslin JK, 1977. Rodent long distance orientation ("homing"). Adv Ecol Res 10:63–89.

Kaufman DW, 1974. Differential predation on active and inactive prey by owls. Auk 91:172–173.

Keane B, 1990. Dispersal and inbreeding avoidance in the white-footed mouse, *Peromyscus leucopus*. Anim Behav 40:143–152.

Kitchings JT, Story JD, 1979. Home range and diet of bobcats in eastern Tennessee. In: Bobcat research conference proceedings (Escherich PC, Blum LG, eds). Scientific and Technical Series 6. National Wildlife Federation; 47–52.

Kitchings JT, Story JD, 1984. Movements and dispersal of bobcats in east Tennessee. J Wildl Manage 48:957–961.

Kleiman DG, 1989. Reintroduction of captive mammals for conservation. Bioscience 39:152–161.

Knick ST, 1990. Ecology of bobcats relative to exploitation and a prey decline in southeastern Idaho. Wildl Monogr 108:1–42.

Knowles CJ, 1985. Observations on prairie dog dispersal in Montana. Prairie Nat 17:33–40.

Koenig WD, Van Vuren D, Hooge PN, 1996. Detectability, philopatry, and the distribution of dispersal distances in vertebrates. Trends Ecol Evol 11:514–517.

Kotzageorgis GC, Mason CF, 1996. Range use, determined by telemetry, of yellow-necked mice (*Apodemus flavicollis*) in hedgerows. J Zool 240:773–777.

Kozakiewicz M, Szacki J, 1995. Movements of small mammals in a landscape: patch restriction or nomadism? In: Landscape approaches in mammalian ecology and conservation (Lidicker WZ Jr, ed). Minneapolis: University of Minnesota Press; 78–94.

Krebs CJ, 1992. The role of dispersal in cyclic rodent populations. In: Animal dispersal: small mammals as a model (Stenseth NC, Lidicker WZ Jr, eds). London: Chapman and Hall; 160–175.

Krebs JR, Davies NB, 1993. An introduction to behavioural ecology, 3rd ed. Oxford: Blackwell Scientific.

LaHaye WS, Gutiérrez RJ, Akçakaya HR, 1994. Spotted owl metapopulation dynamics in Southern California. J Anim Ecol 63:775–785.

Lamberson RH, Noon BR, Voss C, McKelvey KS, 1994. Reserve design for territorial species: the effects of patch size and spacing on the viability of the northern spotted owl. Conserv Biol 8:185–195.

Lambin X, 1994. Natal philopatry, competition for resources, and inbreeding avoidance in Townsend's voles (*Microtus townsendii*). Ecology 75:224–235.

Lankester K, Van Apeldoorn R, Meelis E, Verbloom J, 1991. Management perspectives for populations of the Eurasian badger (*Meles meles*) in a fragmented landscape. J Appl Ecol 28:561–573.

La Polla VN, Barrett GW, 1993. Effects of corridor width and presence on the population dynamics of the meadow vole (*Microtus pennsylvanicus*). Landscape Ecol 8:25–37.

Larsen KW, Boutin S, 1994. Movements, survival, and settlement of red squirrel (*Tamiasciurus hudsonicus*) offspring. Ecology 75:214–223.

LeCount AL, 1982. Characteristics of a central Arizona black bear population. J Wildl Manage 46:861–868.

Leege TA, 1968. Natural movements of beavers in southeastern Idaho. J Wildl Manage 32:973–976.

Lehman LE, 1984. Raccoon density, home range, and habitat use on south-central Indiana farmland. Ind Div Fish Wildl Pittman-Robertson Bull 15:1–66.

Levins R, 1969. Some demographic and genetic consequences of environmental heterogeneity for biological control. Bull Entomol Soc Am 15:237–240.

Lidicker WZ Jr, 1975. The role of dispersal in the demography of small mammals. In: Small mammals: their productivity and population dynamics (Golley FB, Petrusewicz K, Ryszkowski L, eds). London: Cambridge University Press; 103–128.

Lima SL, Zollner PA, 1996. Towards a behavioral ecology of ecological landscapes. Trends Ecol Evol 11:131–135.

Lindenmayer DB, Lacy RC, 1995. Metapopulation viability of arboreal marsupials in fragmented old-growth forests: comparison among species. Ecol Appl 5:183–199.

Lindenmayer DB, Nix HA, 1993. Ecological principles for the design of wildlife corridors. Conserv Biol 7:627–630.

Lindstedt SL, Miller BJ, Buskirk SW, 1986. Home range, time, and body size in mammals. Ecology 67:413–418.

Lindzey FG, Ackerman BB, Barnhurst D, Hemker TP, 1988. Survival rates of mountain lions in southern Utah. J Wildl Manage 52:664–667.

Lindzey FG, Van Sickle WD, Ackerman BB, Barnhurst D, Hemker TP, Laing SP, 1994. Cougar population dynamics in southern Utah. J Wildl Manage 58:619–624.

Logan KA, Irwin LL, Skinner R, 1986. Characteristics of a hunted mountain lion population in Wyoming. J Wildl Manage 50:648–654.

Lorenz GC, Barrett GW, 1990. Influence of simulated landscape corridors on house mouse (*Mus musculus*) dispersal. Am Midl Nat 123:348–356.

Lubow BC, 1996. Optimal translocation strategies for enhancing stochastic metapopulation viability. Ecol Appl 6:1268–1280.

McNab BK, 1963. Bioenergetics and the determination of home range size. Am Nat 97:133–140.

Madison DM, 1980. Movement types and weather correlates in free-ranging meadow voles. Proc East Pine Meadow Vole Symp 4:34–42.

Maehr DS, Land ED, Roof JC, 1991. Social ecology of Florida panthers. Nat Geogr Res Explor 7:414–431.

Mann CC, Plummer ML, 1995. Are wildlife corridors the right path? Science 270:1428–1430.

Melquist WE, Hornocker MG, 1983. Ecology of river otters in west central Idaho. Wildl Monogr 83:1–60.

Merriam G, Lanoue A, 1990. Corridor use by small mammals: field measurement for three experimental types of *Peromyscus leucopus*. Landscape Ecol 4:123–131.

Messick JP, Hornocker MG, 1981. Ecology of the badger in southwestern Idaho. Wildl Monogr 76:1–53.

Metzgar LH, 1967. An experimental comparison of screech owl predation on resident and transient white-footed mice (*Peromyscus leucopus*). J Mammal 48:387–391.

Moritz C, 1994. Applications of mitochondrial DNA analysis in conservation: a critical review. Mol Ecol 3:401–411.

Murie JO, Harris MA, 1984. The history of individuals in a population of Columbian ground squirrels: source, settlement, and site attachment. In: The biology of ground-dwelling squirrels (Murie JO, Michener GR, eds). Lincoln: University of Nebraska Press; 353–373.

Murray BG Jr, 1967. Dispersal in vertebrates. Ecology 48:975–978.

Neigel JE, Avise JC, 1993. Application of a random walk model to geographic distributions of animal mitochondrial DNA variation. Genetics 135:1209–1220.

Neigel JE, Ball RM Jr, Avise JC, 1991. Estimation of single generation migration distances from geographic variation in animal mitochondrial DNA. Evolution 45:423–432.

Nelson ME, 1993. Natal dispersal and gene flow in white-tailed deer in northeastern Minnesota. J Mammal 74:316–322.

Nelson ME, Mech LD, 1991. Wolf predation risk associated with white-tailed deer movements. Can J Zool 69:2696–2699.

Nicholls AO, Margules CR, 1991. The design of studies to demonstrate the biological importance of corridors. In: Nature conservation 2: the role of corridors (Saunders DA, Hobbs RJ, eds). Chipping Norton, NSW, Australia: Surrey Beatty and Sons; 49–61.

Nicholson AJ, 1941. The homes and social habits of the wood-mouse (*Peromyscus leucopus noveboracensis*) in southern Michigan. Am Midl Nat 25:196–223.

Nicholson WS, Hill EP, Briggs D, 1985. Denning, pup-rearing, and dispersal in the gray fox in east-central Alabama. J Wildl Manage 49:33–37.

Nielsen L, 1988. Definitions, considerations, and guidelines for translocation of wild animals. In: Translocation of wild animals (Nielsen L, Brown RD, eds). Milwaukee: Wisconsin Human Society and Kingsville, Texas: Caesar Kleberg Wildlife Research Institute; 12–51.

Noss RF, 1987. Corridors in real landscapes: a reply to Simberloff and Cox. Conserv Biol 1:159–164.

O'Bryan MK, McCullough DR, 1985. Survival of black-tailed deer following relocation in California. J Wildl Manage 49:115–119.

O'Farrell TP, 1965. Home range and ecology of snowshoe hares in interior Alaska. J Mammal 46:406–418.

Peterson RO, Woolington JD, Bailey TN, 1984. Wolves of the Kenai Peninsula, Alaska. Wildl Monogr 88:1–52.

Pokki J, 1981. Distribution, demography and dispersal of the field vole, *Microtus agrestis* (L.), in the Tvarminne archipelago, Finland. Acta Zool Fenn 164:1–48.

Prevett PT, 1991. Movement paths of koalas in the urban-rural fringes of Ballarat, Victoria: implications for management. In: Nature conservation 2: the role of corridors (Saunders DA, Hobbs RJ, eds). Chipping Norton, NSW, Australia: Surrey Beatty and Sons; 259–271.

Price MV, Gilpin M, 1996. Modelers, mammalogists, and metapopulations: designing Stephens' kangaroo rat reserves. In: Metapopulations and wildlife conservation (McCullough DR, ed). Washington, DC: Island Press; 217–240.

Price MV, Kelly PA, Goldingay RL, 1994. Distances moved by Stephens' kangaroo rat (*Dipodomys stephensi* Merriam) and implications for conservation. J Mammal 75:929–939.

Pusey AE, 1987. Sex-biased dispersal and inbreeding avoidance in birds and mammals. Trends Ecol Evol 2:295–299.

Pusey AE, Packer C, 1987. Dispersal and philopatry. In: Primate societies (Smuts BB, Cheney DL, Seyfarth RM, Struhsaker TT, Wrangham RW, eds). Chicago: University of Chicago Press; 250–266.

Ray C, Gilpin M, Smith AT, 1991. The effect of conspecific attraction on metapopulation dynamics. Biol J Linn Soc 42:123–134.

Reed JM, Dobson AP, 1993. Behavioural constraints and conservation biology: conspecific attraction and recruitment. Trends Ecol Evol 8:253–256.

Ribble DO, 1992. Dispersal in a monogamous rodent, *Peromyscus californicus*. Ecology 73:859–866.

Robinette KW, Andelt WF, Burnham KP, 1995. Effect of group size on survival of relocated prairie dogs. J Wildl Manage 59:867–874.

Rogers LL, 1987a. Effects of food supply and kinship on social behavior, movements, and population growth of black bears in northwestern Minnesota. Wildl Monogr 97:1–72.

Rogers LL, 1987b. Navigation by adult black bears. J Mammal 68:185–188.

Rogers LL, 1988. Homing tendencies of large mammals: a review. In: Translocation of wild animals (Nielsen L, Brown RD, eds). Milwaukee: Wisconsin Humane Society and Kingsville, Texas: Caesar Kleberg Wildlife Research Institute: 76–91.

Rood JP, 1987. Dispersal and intergroup transfer in the dwarf mongoose. In: Mammalian dispersal patterns (Chepko-Sade BD, Halpin ZT, eds). Chicago: University of Chicago Press; 85–103.

Rosatte RC, Gunson JR, 1984. Dispersal and home range of striped skunks, *Mephitis mephitis,* in an area of population reduction in southern Alberta. Can Field Nat 98:315–319.

Rosenberg DK, Noon BR, Meslow EC, 1997. Biological corridors: form, function, and efficacy. Bioscience 47:677–687.

Roy LD, Dorrance MJ, 1985. Coyote movements, habitat use, and vulnerability in central Alberta. J Wildl Manage 49:307–313.

Roze U, 1989. The North American porcupine. Washington, DC: Smithsonian Institution Press.

Ruefenacht B, Knight RL, 1995. Influences of corridor continuity and width on survival and movement of deermice. Biol Conserv 71:269–274.

Schwartz CC, Franzmann AW, 1992. Dispersal and survival of subadult black bears from the Kenai Peninsula, Alaska. J Wildl Manage 56:426–431.

Schwartz OA, Armitage KB, 1980. Genetic variation in social mammals: the marmot model. Science 207:665–667.

Shields WM, 1987. Dispersal and mating systems: investigating their causal connections. In: Mammalian dispersal patterns (Chepko-Sade BD, Halpin ZT, eds). Chicago: University of Chicago Press; 3–24.

Silva M, Downing JA, 1995. Handbook of mammalian body masses. Boca Raton, Florida: CRC Press.

Simberloff D, 1988. The contribution of population and community biology to conservation science. Annu Rev Ecol Syst 19:473–511.

Simberloff DS, Abele LG, 1976. Island biogeography theory and conservation practice. Science 191:285–286.

Simberloff D, Cox J, 1987. Consequences and costs of conservation corridors. Conserv Biol 1:63–71.

Simberloff D, Farr JA, Cox J, Mehlman DW, 1992. Movement corridors: conservation bargains or poor investments? Conserv Biol 6:493–504.

Smith AT, Peacock MM, 1990. Conspecific attraction and the determination of metapopulation colonization rates. Conserv Biol 4:320–323.

Smith JLD, 1993. The role of dispersal in structuring the Chitwan tiger population. Behaviour 124:165–195.

Snyder RL, Jenson W, Cheney CD, 1976. Environmental familiarity and activity: aspects of prey selection for a ferruginous hawk. Condor 78:138–139.

Soulé ME, Simberloff D, 1986. What do genetics and ecology tell us about the design of nature reserves? Biol Conserv 35:19–40.

Stacey PB, Taper M, 1992. Environmental variation and the persistence of small populations. Ecol Appl 2:18–29.

Stamps JA, 1987. Conspecifics as cues to territory quality: a preference of juvenile lizards (*Anolis aeneus*) for previously used territories. Am Nat 129:629–642.

Stamps JA, Buechner M, Krishnan VV, 1987. The effects of edge permeability and habitat geometry on emigration from patches of habitat. Am Nat 129:533–552.

Stoner KE, 1996. Prevalence and intensity of intestinal parasites in mantled howling monkeys (*Alouatta palliata*) in northeastern Costa Rica: implications for conservation biology. Conserv Biol 10:539–546.

Storm GL, Andrews RD, Phillips RL, Bishop RA, Siniff DB, Tester JR, 1976. Morphology, reproduction, dispersal, and mortality of midwestern red fox populations. Wildl Monogr 49:1–82.

Stuewer FW, 1943. Raccoons: their habits and management in Michigan. Ecol Monogr 13:203–257.

Suckling GC, 1984. Population ecology of the sugar glider, *Petaurus breviceps*, in a system of fragmented habitats. Aust Wildl Res 11:49–75.

Sumner PW, Hill EP, Wooding JB, 1984. Activity and movements of coyotes in Mississippi and Alabama. Proc Conf Southeast Assoc Fish Wildl Agencies 38:174–181.

Swihart RK, Slade NA, Bergstrom BJ, 1988. Relating body size to rate of home range use in mammals. Ecology 69:393–399.

Szacki J, 1987. Ecological corridor is a factor determining the structure and organization of a bank vole population. Acta Theriol 32:31–44.

Tegelström H, Hansson L, 1987. Evidence of long distance dispersal in the common shrew (*Sorex araneus*). Z Säugetierkunde 52:52–54.

Terborgh J, 1975. Faunal equilibria and the design of wildlife preserves. In: Tropical ecological systems (Golley FB, Medina E, eds). New York: Springer-Verlag; 369–380.

Tierson WC, Mattfeld GF, Sage RW Jr, Behrend DF, 1985. Seasonal movements and home ranges of white-tailed deer in the Adirondacks. J Wildl Manage 49:760–769.

Trewhella WJ, Harris S, 1990. The effect of railway lines on urban fox (*Vulpes vulpes*) numbers and dispersal movements. J Zool 221:321–326.

Tullar BF Jr, Berchielli LT Jr, 1982. Comparison of red foxes and gray foxes in central New York with respect to certain features of behavior, movement and mortality. NY Fish Game J 29:127–133.

Urban D, 1970. Raccoon populations, movement patterns, and predation on a managed waterfowl marsh. J Wildl Manage 34:377–382.

Van Ballenberghe V, 1983. Extraterritorial movements and dispersal of wolves in southcentral Alaska. J Mammal 64:168–171.

Van Vuren D, 1990. Dispersal of yellow-bellied marmots (PhD dissertation). Lawrence: University of Kansas.

Van Vuren D, Armitage KB, 1994. Survival of dispersing and philopatric yellow-bellied marmots: what is the cost of dispersal? Oikos 69:179–181.

Van Vuren D, Kuenzi AJ, Loredo I, Leider AL, Morrison ML, 1997. Translocation as a nonlethal alternative for managing California ground squirrels. J Wildl Manage 61:351–359.

Van Vuren D, Smallwood KS, 1996. Ecological management of vertebrate pests in agricultural systems. Biol Agric Hortic 13:39–62.

Walters JR, 1990. Red-cockaded woodpeckers: a 'primitive' cooperative breeder. In: Cooperative breeding in birds (Stacey PB, Koenig WD, eds). Cambridge: Cambridge University Press; 69–101.

Waser PM, 1985. Does competition drive dispersal? Ecology 66:1170–1175.

Waser PM, 1996. Patterns and consequences of dispersal in gregarious carnivores. In: Carnivore behavior, ecology, and evolution, vol 2 (Gittleman JL, ed). Ithaca, New York: Cornell University Press; 267–295.

Waser PM, Elliot LF, 1991. Dispersal and genetic structure in kangaroo rats. Evolution 45:935–943.

Waser PM, Jones WT, 1983. Natal philopatry among solitary mammals. Q Rev Biol 58:355–390.

Wassmer DA, Guenther DD, Layne JN, 1988. Ecology of the bobcat in south-central Florida. Bull Florida State Mus Biol Sci 33:159–228.

Wauters L, Casale P, Dhondt AA, 1994. Space use and dispersal of red squirrels in fragmented habitats. Oikos 69:140–146.

Webb NJ, Ibrahim KM, Bell DJ, Hewitt GM, 1995. Natal dispersal and genetic structure in a population of the European wild rabbit (*Oryctolagus cuniculus*). Mol Ecol 4:239–247.

Wegner JF, Merriam G, 1979. Movements by birds and small mammals between a wood and adjoining farmland habitats. J Appl Ecol 16:349–357.

Wennergren U, Ruckelshaus M, Kareiva P, 1995. The promise and limitations of spatial models in conservation biology. Oikos 74:349–356.

Wiens JA, 1996. Wildlife in patchy environments: metapopulations, mosaics, and management. In: Metapopulations and wildlife conservation (McCullough DR, ed). Washington, DC: Island Press; 53–84.

Wiggett DR, Boag DA, Wiggett ADR, 1989. Movements of intercolony natal dispersers in the Columbian ground squirrel. Can J Zool 67:1447–1452.

Wilson DE, Findley JS, 1972. Randomness in bat homing. Am Nat 106:418–424.

Wilson EO, Willis EO, 1975. Applied biogeography. In: Ecology and evolution of communities (Cody ML, Diamond JM, eds). Cambridge: Harvard University Press; 522–534.

Wolff JO, 1994. More on juvenile dispersal in mammals. Oikos 71:349–352.

Woodruff RA, Keller BL, 1982. Dispersal, daily activity, and home range of coyotes in southeastern Idaho. Northwest Sci 56:199–207.

Woolfenden GE, Fitzpatrick JW, 1990. Florida scrub jays: a synopsis after 18 years of study. In: Cooperative breeding in birds (Stacey PB, Koenig WD, eds). Cambridge: Cambridge University Press; 241–266.

Woollard T, Harris S, 1990. A behavioural comparison of dispersing and non-dispersing foxes (*Vulpes vulpes*) and evaluation of some dispersal hypotheses. J Anim Ecol 59:709–722.

Young BF, Ruff RL, 1982. Population dynamics and movements of black bears in east central Alberta. J Wildl Manage 46:845–860.

Zhang Z, Usher MB, 1991. Dispersal of wood mice and bank voles in an agricultural landscape. Acta Theriol 36:239–245.

Zwickel F, Jones G, Brent H, 1953. Movement of Columbian black-tailed deer in the Willapa Hills area, Washington. Murrelet 34:41–46.

15

Behavioral Ecology, Genetic Diversity, and Declining Amphibian Populations

Bruce Waldman
Mandy Tocher

Amphibian Decline

All over the world, populations of frogs and toads have declined precipitously in recent years. Declines have been noted in diverse habitats including areas far removed from known sources of human interference (e.g., Crump et al., 1992). The pattern is puzzling. Populations of certain species are dwindling even as sympatric species are thriving (e.g., Richards et al., 1993; Drost and Fellers, 1996). In some areas virtually all frogs have been wiped out; yet elsewhere few if any species appear to have been affected (Pechmann et al., 1991). The specter of mass extinctions within an entire vertebrate class is ominous not only because amphibians are key components of many ecosystems (Stebbins and Cohen, 1995), but because whatever factors are responsible might also be affecting other vertebrates including ourselves (Colborn and Clement, 1992).

The scale of the problem has been realized only recently because few long-term studies exist, fluctuations in amphibian populations are not unusual, and effects vary tremendously among species. Anecdotal accounts are much more numerous than long-term quantitative data, but available data support the hypothesis that species worldwide are declining in synchrony (see table 15-1). Amphibian populations regularly vary in size from year to year. These oscillations can be dramatic both in magnitude and frequency (Pechmann and Wilbur, 1994; cf. Blaustein, 1994), or more subtle and difficult to detect (Reed and Blaustein 1995; Hayes and Steidl, 1997). During the past 40 years, major crashes of frog populations periodically have been reported, but often they have been followed by remarkable recoveries (e.g., Bragg, 1954, 1960; Reed, 1957; Gibbs et al., 1971; Ash, 1997). The large number of concurrent reports now emerging suggests, however, that these events are occurring with increasing frequency to produce an observable overall trend.

No single factor or group of factors can account for the global population declines that we are witnessing. Causation needs to be analyzed on a case-by-case basis. Some previously common amphibians have dropped in numbers simply because they have been overcollected (Gibbs et al., 1971; Hairston and Wiley, 1993), and others have disappeared as their habitat has been destroyed (deMaynadier and Hunter, 1995), but most declines remain

Table 15-1 Long term surveys of amphibian populations.

Taxon	Locality	Maximum years of study	Nature of data	Population status	References
Australasia					
Bufo marinus	Australia (introduced)	60	Historical records and population monitoring	Increasing	Tyler (1994)
Rheobatrachus silus	Australia	20	Regular searches	Possibly extinct	Czechura and Ingram (1990), Tyler (1991), Ingram and McDonald (1993), Richards et al. (1993), Laurance (1996)
Taudactylus diurnus	Australia	20	Regular searches	Possibly extinct	Czechura and Ingram (1990), Tyler (1991), Ingram and McDonald (1993), Richards et al. (1993), Laurance (1996)
Litoria aurea	Australia	57	Historical records	Declining	Osborne et al. (1996), White and Pyke (1996)
		28	Regular searches and population monitoring		
Philoria frosti	Australia	10	Historical records and recent surveys	Declining	Hollis (1995)
Leiopelma archeyi	New Zealand	12	Population monitoring	Declining	Bell (1994, 1996)
Leiopelma pakeka	New Zealand	16	Population monitoring	Stable	Bell (1994)
Various species[a]	Borneo	22	Periodic transect sampling (3 times over 22 years)	Fluctuating	Voris and Inger (1995)
Central/South America					
Eleutherodactylus coqui	Puerto Rico	15	Population monitoring	Fluctuating	Woolbright (1991), Stewart (1995)
Eleutherodactylus karlschmidti	Puerto Rico	30	Regular searches	Possibly extinct	Hedges (1993)
Atelopus varius	Costa Rica	20	Regular surveys	Declining	Pounds and Crump (1994)
Bufo perigenes	Costa Rica	17	Anecdotal, not systemic	Possibly extinct	Crump et al. (1992), Pounds and Crump (1994)

(continued)

Table 15-1 *(Continued)*

Taxon	Locality	Maximum years of study	Nature of data	Population status	References
Various species[b]	Costa Rica	5	Population monitoring	Declining	Lips (1998)
Cycloamphus fuliginosus	Brazil	14	Population monitoring	Declining	Weygoldt (1989)
Hylodine frogs[c]	Brazil	14	Population monitoring	Declining	Weygoldt (1989)
Various species[d]	Brazil	35	Field observations and museum specimens	Declining	Heyer et al. (1988)
North America					
Bufo boreas	Western USA	8	Population monitoring	Declining	Carey (1993), Corn and Vertucci (1992), Bradford et al. (1992)
		77	Historical records and recent surveys		Drost and Fellers (1996), Fisher and Shaffer (1996)
Bufo canorus	Western USA	20	Population monitoring	Declining	Kagarise Sherman and Morton (1993)
Hyla regilla	Western USA	77	Historical records and recent surveys	Declining	Drost and Fellers (1996)
Rana aurora	Western USA	77	Historical records and recent surveys	Declining	Moyle (1973), Hayes and Jennings (1986), Drost and Fellers (1996), Fisher and Shaffer (1996)
Rana boylii	Western USA	77	Historical records and recent surveys	Declining	Drost and Fellers (1996)
Rana cascadae	Western USA	15	Historical records and recent surveys	Declining	Fellers and Drost (1993)
Rana muscosa	Western USA	77	Historical records and recent surveys	Declining	Zweifel (1955), Bradford (1989), Bradford et al. (1992, 1993), Drost and Fellers (1996)
Scaphiopus hammandii	Western USA	>12	Historical records and recent surveys	Declining	Fisher and Shaffer (1996)
Scaphiopus intermontanus	Western USA	77	Historical records and recent surveys	Declining	Orchard (1992), Drost and Fellers (1996)

Pseudacris regilla	Western USA	84	Historical records and population monitoring	Fluctuating	Weitzel and Panik (1993), Fisher and Shaffer (1996)
Ambystoma californiense	Western USA	>12	Historical records and recent surveys	Declining	Fisher and Shaffer (1996)
Taricha rivularis	Western USA	13	Population monitoring	Stable	Twitty (1966)
Taricha granulosa torosa	Western USA	>12	Historical records and recent surveys	Declining	Fisher and Shaffer (1996)
Rana pipiens	Western and central USA, Canada	69	Regular searches, historical records and recent surveys	Declining	Hammerson (1982), Corn and Fogleman (1984), Clarkson and Rorabaugh (1989), Corn and Vertucci (1992), Koonz (1992), Roberts (1992) Corn (1994), Lanoo et al. (1994)
Rana catesbeiana	Western and central USA, (introduced)	77	Historical records and recent surveys	Increasing	Lanoo et al. (1994), Drost and Fellers (1996)
Bufo cognatus	Central USA	69	Historical records and recent surveys	Increasing	Lanoo et al. (1994)
Pseudacris triserata	Central USA	20	Population monitoring	Declining	D. Smith (unpublished data)
Hyla versicolor/ chrysoscelis	Central USA	69	Historical records and recent surveys	Declining	Lanoo et al. (1994)
Acris crepitans	Central USA, Canada	69	Historical records and recent surveys	Declining	Oldham (1992), Lanoo et al. (1994)
Necturus maculosus	Central USA	69	Historical records and recent surveys	Declining	Lanoo et al. (1994)
Ambystoma tigrinum	Central USA	69	Historical records and recent surveys	Declining	Lanoo et al. (1994)
Bufo americanus	Eastern USA	12	Population monitoring	Fluctuating	Pechmann et al. (1991)
	Central USA	69	Historical records and recent surveys	Possibly increasing	Lanoo et al. (1994)
Various species[e]	Eastern USA	15	Population monitoring	Fluctuating	B. Waldman (unpublished data)
	Central USA	>63	Historical records and recent surveys	Stable	Busby and Parmelee (1996)

(continued)

Table 15-1 (*Continued*)

Taxon	Locality	Maximum years of study	Nature of data	Population status	References
Pseudacris ornata	Eastern USA	12	Population monitoring	Stable	Pechmann et al. (1991)
Rana sylcatica	Eastern USA	7	Population monitoring	Fluctuating	Berven (1990)
Ambystoma maculatum	Eastern USA	5	Population monitoring	Fluctuating	Husting (1965)
Ambystoma opacum	Eastern USA	12	Population monitoring	Fluctuating	Pechmann et al. (1991)
Ambystoma talpoideum	Eastern USA	12	Population monitoring	Fluctuating	Pechmann et al. (1991)
Plethodon cinereus	Eastern USA	14	Population monitoring	Fluctuating	Jaeger. (1980)
Plethodon glutinosus	Eastern USA	15	Transect sampling	Fluctuating	Hairston (1983, 1987), Hairston and Wiley (1993)
Plethodon jordani	Eastern USA	15	Transect sampling	Fluctuating	Hairston (1983, 1987), Hairston and Wiley (1993)
Plethodon shenandoah	Eastern USA	14	Population monitoring	Declining	Jaeger (1980)
Desmognathus salamanders[f]	Eastern USA	20	Transect sampling	Stable	Hairston and Wiley (1993)
Various species[g]	Eastern USA	14	Historical records and recent surveys	Stable	Delis et al. (1996)
Various species[h]	Eastern USA	16	Population monitoring	Fluctuating	Semlitsch et al. (1996)
Europe					
Bombina bombina	Denmark	10	Annual surveys	Declining	Fog (1993)
Hyla arborea	Denmark	40	Historical records, recent records	Declining	Fog (1993)

Bufo bufo	Norway, Denmark	23	Population monitoring	Declining	Semb-Johannson (1992), Fog (1988)
	Britain	75	Interviews of residents	Declining	Cooke and Ferguson (1976)
Rana temporaria	Britain	75	Interviews of residents	Declining	Cooke and Ferguson (1976)
Bufo calamita	Britain	20	Range surveys	Declining	Bcebee (1976, 1977), Beebee et al. (1990), Banks et al. (1994), Banks and Bebee (1987)
Africa					
Arthroleptis poecilonotus	Ivory Coast	4	Population monitoring	Fluctuating	Barbault (1984)

[a] *Amolops phaeomerus, Amolops poecilus, Rana blythi, Rana ibanorium, Rana ingeri, Rana kuhli, Rana hosei, Rana chalconota, Rana signata, Rhacophorus pardalis, Bufo asper, Bufo divergens, Ansonia leptopus, Pedostibes hosei, Leptobrachium montanum.*

[b] *Atelopus chiriquiensis, Rana vibicaria, Hyla calypsa, Hyla rivularis, Hyla picadoi, Bufo fastidiosus, Eleutherodactylus punctariolus, Eleutherodactylus melanostictus, Oedipina grandis, Bolituglossa minutula.*

[c] *Hylodes lateris trigatus, Hylodes babax, Crossodactylus gaudichaudii, Crossodactylus dispar, Centrolenlella eurygnatha, Phyllomedus exilis, Colostethus offersoides.*

[d] *Crossodactylus dispar, Crossodactylus gaudichaudii, Cycloramphus boraciencis, Cycloramphus semipalmatus, Hylodes asper, Thoropa miliaris.*

[e] Of 46 species found before 1930, 37 were found in 1993. Two species found in 1993 were not found before 1930. Only 4 species are thought to be locally extinct.

[f] *Desmognathus quadramaculatus, Desmognathus monticola, Desmognathus ochrophaeus, Desmognathus oeneus.*

[g] *Eleutherodactylus planirostris, Acris gryllus, Pseudacris ocularis, Pseudacris nigrita, Hyla gratiosa, Hyla femoralis, Hyla squirella, Hyla cinerea, Rana capito, Rana catesbeiana, Rana grylio, Rana utricularia, Scaphiopus holbrooki, Bufo quercicus, Bufo terrestris, Gastrophryne carolinensis.*

[h] *Ambystoma opacum, Ambystoma talpoideum, Ambystoma tigrinum, Eurcycea quadridigitata, Notophthalmus viridescens, Bufo terrestris, Gastrophryne carolinenses, Pseudacris crucifer, Pseudacris nigrita, Pseudacris ornata, Rana clamitans, Rana utricularia, Scaphiopus holbrooki.*

mysterious. Often, perhaps no individual factor is responsible, but factors act synergistically to increase mortality and ultimately produce observable declines (Wake, 1991; Carey, 1993; Carey and Bryant, 1995; Kiesecker and Blaustein, 1995; Long et al., 1995). We believe that amphibian declines can be understood only by examining first how individuals respond to environmental stresses, and next how they interact with one another in populations facing these stresses. Behavioral ecology can provide insight into the dynamics of declining populations and can provide methodology to identify and monitor populations most at risk.

Causes of Population Declines

Numerous factors have been suggested as potential contributors to amphibian declines (table 15-2; see also Kuzmin, 1994, 1996). Primary suspects include both physical and biotic environmental changes. Susceptible individuals may be directly affected, or environmental changes may induce shifts in population structure or community dynamics that ultimately lead to declines (Barbault, 1991). As populations decrease in size, their vulnerability to stochastic events increases (Lande, 1993), and they may be drawn into "extinction vortices" in which environmental, demographic, and genetic factors reinforce one another to hasten declines (Gilpin and Soulé, 1986). In this synopsis, we highlight current controversies and areas under active investigation.

Changes in the Physical Environment

Ultraviolet Radiation

Increased transmission of ultraviolet radiation through the atmosphere due to depletion of stratospheric ozone may threaten amphibian populations. Significant increases in levels of UV radiation have been detected in many areas (Kerr and McElroy, 1993), but populations at extreme southern latitudes and those at higher altitudes may be subjected to especially high doses (La Marca and Reinthaler, 1991; Basher et al., 1994; Madronich et al., 1995; McKenzie et al., 1996). UV-B radiation can cause potentially lethal damage to developing amphibian embryos, larvae, and metamorphosing individuals (Worrest and Kimeldorf, 1975, 1976). The idea that amphibian declines may be linked to the thinning ozone layer (e.g., Hayes and Jennings, 1990; Wyman, 1990; Wake, 1991) has recently led to experimental tests of effects of UV-B exposure or survivorship of eggs and larvae. Embryos of at least one species of frog (*Rana cascadae*), toad (*Bufo boreas*), and salamander (*Ambystoma gracile*) show increased mortality when exposed to natural levels of UV-B as compared with embryos covered with UV-B filters; but other frogs (e.g., *Pseudacris regilla, Rana aurora*) appear unaffected (Blaustein et al., 1994c, 1995a, 1996a; Ovaska et al., 1997). Blaustein et al. (1994c, 1996a) hypothesised that species that typically spawn in sites exposed to sunlight have evolved efficient DNA repair mechanisms and found, consistent with this, higher levels of key repair enzymes in *P. regilla* and *R. aurora* than in the other amphibians tested (also see Hays et al., 1996).

The UV-B findings have been widely reported in the popular media (see Phillips, 1994) as "the first to directly link a specific environmental cause, other than habitat loss, to reports on scattered worldwide declines and even local extinctions of some frogs, toads, and salamanders" (Cohn, 1995, p. 12). Many details are lacking from the published experimental

Table 15-2 Suggested causes of declines in natural populations.

Factors	Species	Locality	References
Ultraviolet radiation	*Bufo boreas*	Western USA	Blaustein et al. (1994c)
	Rana cascadae	Western USA	Blaustein et al. (1994c)
	Ambystoma gracile	Western USA	Blaustein et al. (1995a)
	Triturus alpestris	Austria	Nagl and Hofer (1997)
Climate and weather patterns	*Pseudophyrne corroboree*	Australia	Osborne (1989)
	Hylodine frogs[a]	Brazil	Weygoldt (1989)
	Cycloamphus fuliginosus	Brazil	Weygoldt (1989)
	Various species[b]	Brazil	Heyer et al. (1988)
	Bufo perigenes	Còsta Rica	Crump et al. (1992), Pounds and Crump (1994)
	Eleutherodactylus coqui	Puerto Rico	Woolbright (1991, 1996), Stewart (1995)
	Rana pipiens	Western USA	Corn and Fogleman (1984)
	Rana muscosa	Western USA	Corn and Fogleman (1984), Bradford (1991)
	Bufo canorus	Western USA	Kagarise Sherman and Morton (1993)
	Bufo boreas	Western USA	Carey (1993)
	Rana cascadae	Western USA	Fellers and Drost (1993)
	Ambystoma tigrinum	Western USA	Wissinger and Whiteman (1992)
	Notophthalmus perstriatus	Eastern USA	Dodd (1993)
	Bufo quercicus	Eastern USA	Dodd (1994)
	Bufo terrestris	Eastern USA	Dodd (1994)
Acidification	Hylodine frogs[a]	Brazil	Weygoldt (1989)
	Cycloramphus fuliginosus	Brazil	Weygoldt (1989)
	Ambystoma tigrinum	Western USA	Harte and Hoffman (1989)
	Ambystoma maculatum	Eastern USA	Pough (1976), Pough and Wilson (1977), Albers and Prouty (1987)
	Ambystoma jeffersonianum	Eastern USA	Freda and Dunson (1986), Horne and Dunson (1994, 1995b)
	Rana temporaria	Britain	Beattie and Tyler-Jones (1992)
	Bufo calamita	Britain	Beebee et al. (1990)
	Bufo bufo	Scandinavia	Semb-Johansson (1992)
Pesticides, herbicides, toxic chemicals, eutrophication	*Bufo periglenes*	Costa Rica	Pounds and Crump (1994)
	Rana pretiosa	Western USA	Kirk (1988)
	Pseudacris crucifer	Eastern Canada	Russell et al. (1995)
	Pseudacris triseriata	Eastern Canada	Hecnar (1995)
	Rana temporaria	Britain	Cooke (1973)
	Triturus vulgaris	Britain	Watt and Oldham (1995)
Predation by exotic species			
Fish	*Litoria aurea*	Australia	Morgan and Buttemer (1996), Pyke and White (1996), White and Pyke (1996)

(*continued*)

Table 15-2 (*Continued*)

Factors	Species	Locality	References
	Rana muscosa	Western USA	Hayes and Jennings (1986), Bradford (1989), Bradford et al. (1993), Drost and Fellers (1996)
	Rana aurora	Western USA	Hayes and Jennings (1986), Fisher and Shaffer (1996)
	Rana cascadae	Western USA	Fellers and Drost (1993)
	Rana pretiosa	Western USA/ Canada	Orchard (1992)
	Taricha torosa	Western USA	Gamradt and Kats (1996), Fisher and Shaffer (1996)
	Various species[c]	Eastern Canada	Hecnar and McCloskey (1997b)
	Hyla arborea	Europe	Fog (1993), Brönmark and Edenhamm (1994)
Bullfrogs	*Atelopus mucubajiensis, Atelopus pinangoi*	Venezuela	La Marca and Reinthaler (1991)
	Rana aurora	Western USA	Moyle (1973), Fisher and Shaffer (1996)
	Rana muscosa	Western USA	Moyle (1973)
	Rana pipiens	Western USA	Schwalbe and Rosen (1988), Clarkson and Rorabough (1989), Hammerson (1982)
	Rana pretiosa	Western USA/ Canada	Dumas (1966), Orchard (1992)
	Rana onca	Westen USA	Cowles and Bogert (1936)
	Rana clamitans	Eastern Canada	Hecnar and McClosky (1997a)
Crayfish	*Taricha torosa*	Western USA	Gamradt and Kats (1996), Gamradt et al. (1997)
Competition with other frogs	*Bufo calamita*	Britain	Beebee (1977)
Degradation, destruction, fragmentation of habitat	Various species[d]	Australia	Richards et al. (1993)
	Rana onca	Western USA	Jennings (1988)
	Rana aurora	Western USA	Moyle (1973)
	Rana muscosa	Western USA	Moyle (1973)
	Plethodon vehiculum	Western Canada	Dupuis et al. (1995)
	Plethodontid salamanders[e]	Eastern USA	Ash (1997)
	Ambystoma cingulatum	Eastern USA	Means et al. (1996)
	Ambystoma talpoideum	Eastern USA	Raymond and Hardy (1991)
	Various species[f]	Eastern USA	Petranka et al. (1993, 1994)
	Various species[g]	Eastern USA	Delis et al. (1996)
	Various species[h]	Eastern USA	Hecnar and McClosky (1996)
	Bufo calamita	Britain	Beebee (1977), Banks and Beebee (1987), Banks et al. (1994)
	Bufo bufo	Britain	Cooke (1972b) Cooke and Ferguson (1976)

(*continued*)

Table 15-2 (*Continued*)

Factors	Species	Locality	References
	Rana temporaria	Britain	Cooke (1972b) Cooke and Ferguson (1976)
Succession and natural changes	*Ambystoma tigrinum*	Western USA	Harte and Hoffman (1989),
	Rana pipiens	Western USA	Corn and Fogleman (1984)
	Rana sylvatica	Eastern USA	Berven (1990)
Disease and pathogens	*Taudactylus acutirostris*	Australia	Mahony and Dennis (1994), Laurance et al. (1996)
	Various species[i]	Australia	Laurance (1996)
	Hylodine frogs[a]	Brazil	Weygoldt (1989)
	Cycloramphus fuliginosus	Brazil	Weygoldt (1989)
	Various species[j]	Costa Rica	Lips (1998)
	Ambystoma tigrinum	Western USA	Collins et al. (1988), Worthylake and Hovingh (1989)
	Rana muscosa	Western USA	Bradford (1991)
	Bufo canorus	Western USA	Kagarise Sherman and Morton (1993)
	Bufo boreas	Western USA	Carey (1993), Blaustein et al. (1994b)
	Bufo hemiophrys	Western USA	Taylor et al (1995)
	Rana sylvatica	Eastern USA	Nyman (1986)
	Rana temporaria	Britain	Cunningham et al. (1993), Drury et al. (1995)
	Alytes obstericans	Spain	Marquez et al. (1995)
Human study	*Bufo canorus*	Western USA	Kagarise Sherman and Morton (1993)
Overharvesting	*Rana tigrina*	India	Abdulali (1986)
	Rana aurora	Western USA	Jennings and Hayes (1985)
	Rana pipiens	USA/Mexico/ Canada	Gibbs et al. (1971)

[a]*Hylodes lateristrigatus, Hylodes babax, Crossodactylus gaudichaudii, Crossodactylus dispar, Centrolenlella eurygnatha, Phyllomedusa exilis, Colostethus olfersoides.*

[b]*Crossodactylus dispar, Crossodactylus gaudichaudii, Cycloramphus boraciencis, Cycloramphus semipalmatus, Hylodes asper, Thoropa miliaris.*

[c]*Rana pipiens, Pseudacris crucufer, Pseudacris triseriata, Hyla versicolor, Notophthalmus viridescens, Ambystoma laterale, Ambystoma maculatum.*

[d]*Litoria nyakalensus, Taudactylus rheophilus, Taudactylus acutirostris, Litoria nannotis, Litoria rheocola, Nyctimystes dayi.*

[e]*Plethodon jordani, Plethodon oconaluftee, Eurycea wilderae, Desmognathus ochrophaeus, Plethodon serratus.*

[f]*Plethodon jordani, Plethodon glutinosus, Plethodon yonahlossee, Desmognathus ochrophaeus, Desmognathus quadramaculatus, Desmognathus monticola, Desmognathus fuscus, Eurycea wilderae, Gyrinophilus porphyriticus, Pseudotritin ruber, Notophthalmus viridescens.*

[g]*Eleutherodactylus planirostris, Acris gryllus, Pseudacris ocalaris, Pseudacris nigrita, Hyla gratiosa, Hyla femoralis, Hyla squirella, Hyla cinerea, Rana capito, Rana catesbeiana, Rana grylio, Rana utricularia, Scaphiopus holbrooki, Bufo quericus, Bufo terrestris, Gastrophryne carolinensis.*

[h]*Pseudacris crucifer, Notophthalmus viridescens, Hyla versicolor, Rana sylcatica, Ambystoma maculatum.*

[i]*Taudactylus diurnus, Rheobatrachus silus, Litoria pearsoniana, Mixophyes iteratus, Mixophyes fleayi, Taudactylus eungellensis, Rheobatrachus vitellinus, Litoria lorica, Litoria nyakalensis, Litoria rheocola, Litoria nannotis, Taudactylus rheophilus, Taudactylus acutirostris, Nyctimystes dayi.*

[j]*Atelopus chiriquiensis, Rana vibicaria, Hyla calypsa, Hyla rivularis, Hyla picadoi, Bufo fastidiosus, Eleutherodactylus punctariolus, Eleutherodactylus melanostictus, Oedipina grandis, Bolitoglossa minutula.*

methods, however, and the results and interpretations are controversial (Licht, 1995, 1996; Roush, 1995; Licht and Grant, 1997; cf. Blaustein, 1995, and subsequent letters; Blaustein et al., 1995b, 1996b). The possibility that UV-B affected embryos indirectly, for example by leaching toxic solutes from experimental enclosures, was not considered. Moreover, the jelly coats surrounding amphibian eggs serve as natural UV-B filters; by removing some layers, the researchers may have inadvertently biased their results (Grant and Licht, 1995). Egg jelly rapidly swells from the time of oviposition, and this may provide considerable protection after a few hours. Because many amphibians breed at night, newly laid eggs might be safe from UV-B damage during their most vulnerable period. UV-B radiation fails to penetrate even 1 cm beneath the surface of many ponds because dissolved organic matter, algae, and waterborne particles effectively block it Licht and Grant, 1997; Nagl and Hofer, 1997). Most amphibians thus would be protected, but those that breed in shallow, high-altitude mountain ponds with clear water, such as the newt *Triturus alpestris,* may be vulnerable (Nagl and Hofer, 1997).

Many species declines cannot be explained by UV-B effects (e.g., Blaustein et al., 1996a). For example, the Australian hylid frogs *Litoria aurea* and *L. raniformis* have experienced massive range contractions during recent years, while populations farther south in New Zealand, which have experienced larger increases in ambient UV-B radiation (McKenzie et al., 1995), continue to thrive (Osborne et al., 1996; Waldman, 1996; White and Pyke, 1996). Indeed, survival to the larval stage appears to be unaffected by UV-B radiation both in *Litoria aurea* and its sympatric congeners whose populations are stable (van de Mortel and Buttemer, 1996). Certainly we need to continue to evaluate UV-B radiation as a potential cause of amphibian declines, especially in the Southern Hemisphere where UV-B radiation flux is highest in the spring and summer (McKenzie et al., 1995), commensurate with the breeding season of many amphibians. In contrast, UV-B exposure in the North Hemisphere has increased most in the autumn and winter (Blumthaler, 1993; Kerr and McElroy, 1993) when few amphibians are breeding. UV-B is unlikely to explain declines in recent years of frogs and toads in tropical and subtropical regions (e.g., montane populations of Australian rainforest frogs; Richards et al., 1993; Trenerry et al., 1994; Laurance et al., 1996) because UV-B levels at these latitudes should not appreciably differ now from 15 years ago (Madronich and de Gruiji, 1993).

Climate Change

On a global scale, climate change might conceivably explain simultaneous changes in distant populations. Even short-term variation in weather patterns can dramatically affect amphibian populations. Amphibians need moist habitats to live and reproduce, so desiccation is an incessant threat. The abundance and composition of the invertebrate prey in their diet is likely to shift with changing environmental conditions. Reproductive behaviors, spatial distributions, and vulnerability to predation and disease are likely to change as conditions become warmer or drier (Beebee, 1995; Donnelly and Crump, 1997).

Some recent population declines coincide with changing weather patterns. Persistent droughts may have resulted in extinction of populations of leopard frogs *Rana pipiens* in North America (Corn and Fogleman, 1984) and corroboree frogs *Pseudophryne corroboree* in Australia (Osborne, 1989). Altered weather patterns may be responsible for wide-ranging declines of frog populations in Brazil (Heyer et al., 1988; Weygoldt, 1989) and the possible extinction of the golden toad *Bufo periglenes* in Costa Rica (Crump et al., 1992). Berven (1990) found that the survivorship of adult wood frogs *Rana sylvatica* declined in months with reduced rainfall, and Semlitsch et al. (1996) documented a close relationship between rainfall and breeding success in numerous species over a 16-year period. Prolonged dry spells led to massive die-offs of adult and juvenile *Eleutherodactylus coqui* (Stewart, 1995). Then as populations began to recover, they disappeared following severe damage inflected by a hurricane (Stewart, 1995), only to reappear as the forest recovered (Woolbright, 1996). But many amphibian declines cannot be adequately explained by changing weather patterns (e.g., Beebee, 1977; Czechura and Ingram, 1990; Drost and Fellers, 1996; Laurance, 1996).

Acid Rain

Amphibians have complex life cycles, typically with aquatic larvae that metamorphose into adults that live on land or in water. Degradation of either larval or adult habitat can have serious effects on populations. Acid rain was suggested early on as a serious threat to amphibian populations because of its devastating effects on larval survivorship (Pough, 1976). Acidic pulses resulting from snowmelt dramatically alter the chemistry of temporary ponds in which many amphibians breed and, when these pulses coincide with highly vulnerable early embryonic stages, significant mortality occurs (Pierce, 1985). The natural acidification of soils also may be accelerated, causing mortality or reproductive failure in terrestrial amphibians (Wyman and Jancola, 1992). Developmental aberrations and decreased larval growth rates have been noted in acidified environments (Beattie and Tyler-Jones, 1992; Horne and Dunson, 1995a). In recent times, many breeding habitats in eastern North America (Pough, 1976; Portnoy, 1990; Glooschenko et al., 1992), Britain (Beebee et al., 1990), and Scandinavia (Hagström, 1977) have become sufficiently acidified to cause breeding failures.

Yet selection favors acid tolerance, and some species persist in acidified conditions, although surviving individuals may be impaired (e.g., Pierce and Harvey, 1987; Andrén et al., 1989; Rowe et al., 1992; Grant and Licht, 1993; Kiesecker, 1996). In other regions, such as in western North America where numerous species are in severe decline, acid rain normally does not appear to be a problem (e.g., Corn and Vertucci, 1992; Bradford et al., 1994; Corn, 1994; but see Harte and Hoffman, 1989, 1994; Vertucci and Corn, 1994, 1996). Dunson et al. (1992) caution that the loss of even a single population as a result of acid rain has yet to be rigorously documented. Effects of acidification on amphibian populations vary depending on many factors, including the presence or absence of toxic metals, dissolved organic matter, and the algal communities present (Horne and Dunson, 1995a). Some amphibians may detect acidified conditions prior to spawning, and then search for more suitable localities in which to breed, but individuals inhabiting acidified ponds apparently lack these behavioral discrimination abilities (Whiteman et al., 1995).

Pesticides, Herbicides and Fertilizers

Their complex life cycle also makes amphibians especially vulnerable to toxic chemicals such as fertilizers (Hecnar, 1995; Watt and Oldham, 1995), pesticides (Vardia et al., 1984; Kirk, 1988; Hall and Henry, 1992; Berrill et al., 1994; Russell et al., 1995), and herbicides (Bidwell and Gorrie, 1995) used in agriculture (reviewed in Power et al., 1989; Devillers and Exbrayat, 1992; Tyler, 1994). Early embryonic stages are extremely sensitive to environmental insults (Rugh, 1962). Mortality may result from exposure to agricultural chemicals and industrial pollutants (Porter and Hakanson, 1976; Cooke, 1981; Mahaney, 1994;

Boyer and Grue, 1995; Lefcort et al., 1997). Sublethal concentrations can retard growth and disrupt the capability of amphibians to survive and reproduce (Carey and Bryant, 1995), for example by changing feeding behaviors (Hecnar, 1995) or altering behavioral responses to predators (Cooke, 1971, 1981). These effect are likely to be exacerbated in acidified environments (Jung and Jagoe, 1995).

Many of these chemicals mimic reproductive hormones, and more insidious perhaps than their direct toxicological effects is the disruption they can cause the endocrine system. Abnormal sexual behaviors, retarded gonadal development, reduced gamete production, and immunosuppression have been noted in a variety of exposed vertebrates including humans (Colborn and Clement, 1992). Effects potentially are far reaching, as chemicals can spread vast distances through the upper atmosphere in mist or fog to contaminate remote, otherwise pristine, regions (Pounds and Crump, 1994). Even at low concentrations, possibly below detectable limits, toxicants may be able to interfere with normal reproduction and contribute to population declines (Carey and Bryant, 1995).

Changes in the Biotic Environment

Deforestation and destruction of wetlands destroys the habitat necessary for many frogs, toads, and salamanders to live and breed (e.g., Beebee, 1977; Raymond and Hardy, 1991; Gibbs, 1993; Petranka et al., 1994; Dupuis et al., 1995; Means et al., 1996; Ash, 1997; reviewed in deMaynadier and Hunter, 1995). Some species do quite well at exploiting disturbed habitat, but many succumb to intensive grazing and other agricultural practices (Anderson, 1993). Introduced predators on—or competitors to—native fauna can wreak havoc on ecosystems and cause population cashes (e.g., Hayes and Jennings, 1986; Sexton and Phillips, 1986; Bradford et al., 1993; Bronmark and Edenhamn, 1994). For example, mosquitofish currently being released into streams to control mosquito larvae are decimating populations of California newts (Gamradt and Kats, 1996) and Australian hylid frogs (Morgan and Buttemer, 1996; Pyke and White, 1996). Crayfish represent another threat to amphibians (Gamradt and Kats, 1996; Axelsson et al., 1997). Widespread stocking of game fish known to prey on larvae of threatened frogs appears to be a principal cause of amphibian declines in the western United States (Hayes and Jennings, 1986; Bradford, 1989, 1991; Bradford et al., 1993; Fisher and Shaffer, 1996). Yet some wildlife managers advocate expansion of these practices under the apparent misimpression that abiotic factors such as UV-B radiation have been firmly established as causal agents of population declines (e.g., Stienstra, 1995).

Fragmentation

Fragmentation, the disruption of continuous habitat into smaller isolated patches, appears to be directly associated with declines of some amphibians (e.g., Laan and Verboom, 1990; Mann et al., 1991; Kattan, 1993; Sjögren Gulve, 1994; Edenhamn, 1996). Extinctions and population declines also have been noted in species inhabiting apparently pristine montane habitat, where populations historically have been restricted to narrow geographical ranges (e.g., Crump et al., 1992; Richards et al., 1993; Stewart, 1995; Lips 1998). Fragmentation of populations has increased in recent times due to urban expansion, agricultural development, and logging, and because new dispersal barriers have been created such as the introduction of predatory fish into waterways. In other regions fragmentation may be less of a problem; for example, few species appear to be declining in the southeastern United States, where populations are dense and habitat is relatively continuous (Wake, 1991). Fragmentation contributes to declines in two ways: by altering the demography and by modifying the genetic structure off populations.

Demographic Effects

Local populations are normally interconnected by the migration of individuals among them. The network comprises a metapopulation (e.g., Gill, 1978; Hanski and Gilpin, 1991; Sjögren, 1991a; Sjögren Gulve, 1994; Hanski et al., 1995a). Migration into peripheral populations may be essential to prevent their extinction (Brown and Kodric-Brown, 1977; Sjögren, 1991b; Burkey, 1995). Moreover, when one population goes extinct within a metapopulation, another should eventually be reestablished at the same locality through recolonization. This may happen rapidly if potential source populations are nearby and if substantial dispersal occurs from them, but may take many generations, or fail to occur at all, when populations are distantly scattered and dispersal rates are low (Travis, 1994). Recent empirical studies show that frog populations respond as theory predicts: the probability that local populations go extinct increases in proportion to the distance between populations (Sjögren, 1991a; Sjögren Gulve, 1994; Edenhamn, 1996).

Barriers further occlude migration and can lead to a progressive breakdown of the metapopulation, thereby preventing recolonization and accelerating species declines (Hanski and Gilpin, 1991; Sarre, 1995). Such efforts have been documented in a variety of frogs (e.g., Laan and Verboom, 1990). What constitutes a barrier varies among species; although forest-dwelling species may find themselves being constricted into islands by deforestation, for example, other species successfully exploit such disturbed habitat (e.g., Inger et al., 1974; Edenhamn, 1996; Poynton, 1996; Tocher, 1996). Metapopulations can collapse to extinction without warning even in habitats that are only gradually degrading (Hanski et al., 1995b). Reports of rapid simultaneous declines of numerous species within a region, even as some species appear to flourish (e.g., Richards et al., 1993; Trenerry et al., 1994; Drost and Fellers, 1996; Lips, 1998), are thus consistent with the predictions of metapopulation theory.

Genetic Effects

Dispersal among local populations not only ensures their persistence by maintaining stable population sizes but enlarges genetically effective population sizes (N_e). Gene flow builds up a reservoir of genetic variation that may be necessary to generate evolutionary responses to environmental changes (Allendorf and Leary, 1986). Moreover, both the likelihood and potential costs of inbreeding decrease as population sizes increase.

As barriers arise to gene flow and metapopulations break down, populations become more genetically homogeneous. Less-common alleles are increasingly lost from populations by genetic drift and individuals become more homozygous (Falconer, 1989). Mildly deleterious mutations may become fixed within populations, increasing the risk of extinction (Lynch and Gabriel, 1990; van Noordwijk, 1994; Lande, 1995). After localized extinction events, those populations that recover are likely to be initiated by fewer founders, which in turn further reduces levels of genetic variation within populations and increases differentiation among populations (McCauley, 1993; also see Heyer et al., 1988). Consistent with this model, Reh and Seitz (1990) found little genetic polymorphism in common frog *Rana temporaria* populations isolated by motorways or railroad lines, and Sjögren (1991b) found especially low genetic variation in peripheral populations of pool frogs *Rana lessonae*.

The incidence of breeding among close relatives is likely to increase. Deleterious genetic effects then result from the expression of recessive deleterious alleles or overdominance (Wright, 1977; Charlesworth and Charlesworth, 1987). In ectothermic vertebrates, hatching success, growth rates, developmental stability, size at maturity, and courtship behaviors all show evidence of inbreeding depression (Waldman and McKinnon, 1993; Vrijenhoek, 1994; Madsen et al., 1996). Conversely, outbreeding can benefit progeny by conferring on them fitness benefits associated with increased heterozygosity (Naylor, 1962; Falconer, 1989). Heterozygous tiger salamanders *Ambystoma tigrinum* grow faster and have greater metabolic efficiency than homozygous individuals (Pierce and Mitton, 1982; Mitton et al., 1986; cf. Chazal et al., 1996). Survival correlates positively with the level of multilocus heterozygosity in overwintering juvenile western toads *Bufo boreas* (Samollow and Soulé, 1983; cf. McAlpine and Smith, 1995), and reproductive success of green treefrogs *Hyla cinerea* correlates positively with their heterozygosity (McAlpine, 1993).

Genetic variation confers important benefits both on individuals and on populations, especially in stressful environments (Hoffmann and Parsons, 1991; Keller et al., 1994). Some amphibian populations lack much genetic variation but nonetheless persist over many years with little sign of inbreeding depression in stable habitats (Sjögren, 1991a,b; Hitchings and Beebee, 1996; also see Edenhamn, 1996). Theory and laboratory studies predict, however, that once a threshold of inbreeding is reached, populations are certain to go extinct, often without any warning (Frankham, 1995b; Lande, 1995; Lynch et al., 1995).

Disease

Population declines sometimes are associated with identifiable bacterial, fungal, or viral pathogens. Epidemics of the bacterium *Aeromonas hydrophila,* known to cause symptoms of "red-leg" (a highly contagious disease especially common in captive populations) may have been responsible for massive die-offs of natural populations of adult American toads *Bufo americanus* (Dusi, 1949), larval wood frogs *Rana sylvatica* (Nyman, 1986), larval and adult mountain yellow-legged frogs *Rana mucosa* (Bradford, 1991), and larval and adult midwife toads *Alytes obstetricans* (Marquez et al., 1995). *Aeromonas hydrophila* and other bacteria are commonly isolated from populations of dying frogs, toads, and salamanders (e.g., Dusi, 1949; Hunsaker and Potter, 1960; Hird et al., 1981; Worthylake and Hovingh, 1989; Kagarise Sherman and Morton, 1993; Taylor et al., 1995).

Pathogenic fungal infestations similarly have been linked to mortality of embryos and larvae in populations of various frogs and toads (*Scaphiopus bombifrons*: Bragg and Bragg, 1958; *Rana pipiens, Bufo terrestris*: Bragg 1962; *Rana calamita, Rana temporaria*: Beattie et al., 1991; *Bufo boreas*: Blaustein et al., 1994b). Fungi may be the primary agents of decline in some species, and they in turn increase the vulnerability of their hosts to secondary bacterial infections (Taylor et al., 1995). The geometry of egg-laying can create conditions that promote fungal infestation. Species that deposit their eggs communally appear more at risk than those that scatter their eggs, presumably because fungi are more readily transmitted over close distances (Kiesecker and Blaustein, 1997). Moreover, aberrant oviposition behavior, which may be induced by environmental changes, can exacerbate these effects—for example, rather than wrapping their eggs in strings around vegetation as toads typically do, *Bufo bufo* in some localities have been observed to lay their eggs in compact clumps from which no embryos survived (B. Waldman, unpublished data). Although rapid fungal infestation is commonly associated with the mortality of clumped embryos, asphyxiation due to insufficient diffusion of oxygen through the gelatinous egg mass (Strathmann and Chaffee, 1989; Seymour, 1994) needs to be considered as a possible cause of death. Oxygen depletion of the water due to eutrophication is likely to compound this problem.

Patterns of population declines ("extinction waves") in Australia, Central America, and the western United States are typical of epidemics involving highly virulent infectious agents (Scott, 1993; Laurance et al., 1996; Lips, 1998). Iridoviruses isolated from dying frogs in Australia (Speare and Smith, 1992; Speare, 1995) and elsewhere (Crawshaw, 1992; Cunningham et al., 1993, 1996; Drury et al., 1995, and studies cited therein) have been targeted as a likely cause, and Laurance et al. (1996) speculate that these viruses have been spread worldwide by exotic aquarium fishes (but see Alford and Richards, 1997; Hero and Gillespie, 1997; Laurance et al., 1997). Parasites also are known to cause epidemics in captive amphibian populations (Ippen and Zwart, 1996) and apparently can cause rapid declines in wild populations. Unidentified protozoan parasites (*Perkinsus*-like protists; Winstead and Couch, 1988) have been found in the skin of frogs succumbing to epidemics in remote regions of Panama and Australia (Blakeslee, 1997). Skin lesions caused by these parasites ultimately may kill frogs by interfering with their respiration and osmoregulation (Nichols et al., 1996). Parasites also may be responsible for the large number of deformities recently observed in anuran tadpoles in central and eastern North America (Sessions and Ruth, 1990; Rebuffoni, 1995), although pesticides or other agricultural chemicals may be implicated in some cases (Ouellet et al., 1997).

Healthy individuals may remain healthy even in the presence of pathogens because of their immunological competence. Immune systems, when weakened due to environmental stress or viral action, potentially leave individuals vulnerable to bacteria, fungi, or parasites from which they are normally shielded (Baldwin and Cohen, 1975), and with which under ordinary circumstances they may live commensally (e.g., Rigney et al., 1978; Olson et al., 1992).

Synergistic Interactions

The causes of amphibian declines defy simple explanations probably because complex interactions among environmental stresses produce effects of greater magnitude than are revealed in investigations of single factors alone. Environmental stresses break down defenses against pathogens and parasites (Harbuz and Lightman, 1992). UV-B, for example, can increase susceptibility to viral, bacterial, fungal, and parasitic infections (Longstreth et al., 1995). Long et al. (1995) found that, although neither acidic pH nor UV-B alone influenced the survival of leopard frog *Rana pipiens* embryos, embryonic survival was reduced under conditions of simultaneous exposure to both acidic pH and UV-B. Similarly, pathogenic fungi more readily attack embryos already weakened by exposure to cold temperature, acidic pH (Gascon and Planas, 1986; Banks and Beebee, 1988; Beattie et al., 1991), or UV-B radiation (Kiesecker and Blaustein, 1995). Many environmental pollutants increase the susceptibility of amphibians to disease (Carey and Bryant, 1995). Climate change can produce behavioral and ecological changes that increase disease risk (Donnelly and Crump, 1997).

One possible pathway for these synergistic effects is through the induction of immunosuppression (Carey, 1993). Environmental changes that appear harmless nonetheless may cause sublethal stress, which in turn increases the secretion of adrenocortical hormones that can reduce the animal's ability to fight disease (Saad, 1988; Saad and Płytycz, 1994). Even handling frogs momentarily causes elevated corticosterone levels over extended durations (Licht et al., 1983; Paolucci et al. 1990; Zerani et al., 1991; Coddington and Cree, 1995), so just by studying declining populations we may be stressing them and hastening their demise (Carey, 1993; Kagarise Sherman and Morton, 1993). Corticosterone also effectively reduces circulating levels of sex hormones, thereby inhibiting courtship behaviors (Licht et al., 1983; Moore and Miller, 1984; Moore and Zoeller, 1985; Paolucci et al., 1990). The cumulative effects of environmental stresses on endocrine and immune system function can result in increased mortality, decreased reproductive rates, and ultimately population declines (Harbuz and Lightman, 1992; Carey, 1993; Carey and Bryant, 1995).

Whether declining amphibian populations suffer immunosuppression—attributable to low pH, cold temperature, or other factors (e.g., Carey et al. 1996a,b)—needs to be addressed with standard immunological methods (reviewed in Zuk, 1996). Comparisons of corticosterone levels between declining and nondeclining populations have yet to be made, but results would be difficult to interpret. Corticosterone causes a decrease in circulating lymphocytes but also a transient increase in circulating heterophils, which are primary defenses against bacteria (Horton, 1994). Moreover, corticosteroids mediate many physiological processes, and high levels do not always indicate impaired immune function (Moynihan et al., 1994).

The most puzzling extinctions worldwide have occurred in montane populations. Because amphibians are ectothermic, their body temperatures correspond to the cold ambient temperatures typical there. Low body temperatures can induce immunosuppression (Saad and Ali, 1992; Carey, 1993). In this state, amphibians may become susceptible to pathogens or parasites to which they are normally immune; indeed, *Aeromonas hydrophila* and other bacteria associated with die-offs can be found in and on healthy individuals (Hazen et al., 1978; Hird et al., 1981; Palumbo, 1993). Available evidence suggests that many environmental changes (shifts in temperature, water balance, UV radiation, and water chemistry) represent stresses that make animals vulnerable to disease (reviewed in Carey, 1993). When pathogens increase in number and virulence, they can reach a critical level at which even healthy individuals become susceptible and epidemics may ensue (Ewald, 1994).

Disease risk also rises as genetic homogeneity increases within populations (Waldman, 1988). To the extent that species inhabiting restricted montane habitats consists of genetically isolated populations (Inger et al., 1974), they may be more vulnerable to disease. Genetically similar individuals share similar immunological defense systems. When a pathogen overcomes one individual, it likely will overcome others as well. Thus the rate at which pathogens spread, and whether they can build up to a critical level in the population, is likely to depend on the degree of uniformity among the host genotypes present (Schmitt and Antonovics, 1986; Hamilton, 1987; Schmid-Hempel, 1994; Lively and Apanius, 1995). In a similar way, individuals within genetically homogeneous populations are likely to show correlated responses to environmental peturbations, increasing the risk of cataclysmic declines.

Effects of inbreeding can accumulate over time within populations (Lande and Barrowclough, 1987; Lande, 1994, 1995), gradually decreasing reproductive rates and increasing susceptibility to disease, parasites, predators, competitors, and climatic changes (Frankham, 1995a). Extinctions of inbred populations may be triggered by demographic or environmental causes, yet ultimately result from the cumulative effects of inbreeding (Frankham, 1995b), especially in changing environments (Hoffmann and Parsons, 1991; Keller et al., 1994).

Dispersal, Natal Philopatry, and Genetic Population Structure

Studies of the behavioral ecology of amphibians suggest that genetic homogeneity poses special problems for them. Amphibians show remarkable site fidelity (reviewed in Sinsch, 1990; Waldman and McKinnon, 1993). Many frogs (e.g., Jameson, 1957), toads (e.g., Bogert, 1947), and salamanders (Twitty et al., 1964; Holomuzki, 1982; Kleeberger and Werner, 1982) migrate year after year to particular sites to breed—even if the ponds are paved over into parking lots (Heusser, 1969). This site consistency suggests that amphibians may not readily recolonize extirpated populations (Blaustein et al., 1994a), and efforts of environmental planners to move breeding populations even short distances often fail (see Dodd and Seigel, 1991).

Inferences drawn from adult behavior may be deceptive, however, as significant proportions of juveniles have been found to disperse in those frog and toad populations that have been surveyed in longitudinal studies. Breden (1987) marked Fowler's toads *Bufo woodhousei fowleri* as larvae or at metamorphosis and found that most returned to breed in their natal ponds, but some dispersed to nearby ponds, and a few migrated over long distances. Most wood frogs *Rana sylvatica* also demonstrate natal philopatry, but a minority regularly disperse, usually nearby but sometimes to ponds up to 2.5 km away (Berven and Grudzien, 1990). Although adult pool frogs *Rana lessonae* show extreme site fidelity, more than a third of each juvenile cohort emigrates from natal ponds (Sjögren Gulve, 1994). Similarly, most common toads *Bufo bufo* return to their natal pond to breed, but one out of six disperses to other ponds (Reading et al., 1991).

Although juvenile migration normally ensures some gene flow between local populations (Breden, 1987), the large proportion of philopatric individuals is likely to give rise to genetic differentiation among populations, especially when populations are small (Endler, 1973; Nurnberger and Harrison, 1995). Genetic substructuring increases the vulnerability of amphibians to declines in two ways. First, because relatively few individuals disperse even in undisturbed habitat, metapopulation structure can be easily disrupted by fragmentation. Second, high levels of genetic homogeneity within local populations, even in undisturbed habitat, can lead to inbreeding depression and increased risk of disease. As possibilities for dispersal are reduced, fragmentation leads ultimately to even further depletion of genetic variation and higher levels of inbreeding (see Reh and Seitz, 1990). To the extent that inbreeding decreases fitness or genetic homogeneity increases disease risk, fragmentation creates a positive feedback loop that escalates the probability of population extinction with each generation.

Estimates of dispersal typically have been determined with mark–recapture studies. These present numerous logistical challenges and they tend to overestimate levels of gene flow (Endler, 1979). Genetic population structure can be determined more readily by molecular analyses. Because protein polymorphisms usually vary only over larger geographic ranges (e.g., Inger et al., 1974; Gartside, 1982; Larson et al., 1984), few studies have examined genetic differentiation among nearby amphibian populations. Driscoll et al. (1994) found significant genetic subdivision among populations of the threatened Australian frog *Geocrinia alba,* both within and among creeks, over distances of 5 km. Genetic subdivision of local populations of the common frog *Rana temporaria* was found only where localities were surrounded by substantial barriers such as motorways and railroad lines (Reh and Seitz, 1990). Analyses of nuclear DNA variation among declining populations of the leopard frog, *Rana pipiens,* reveal some differentiation over 3.5 km (Kimberling et al., 1996). Similar analyses that we currently are conducting on threatened populations of the New

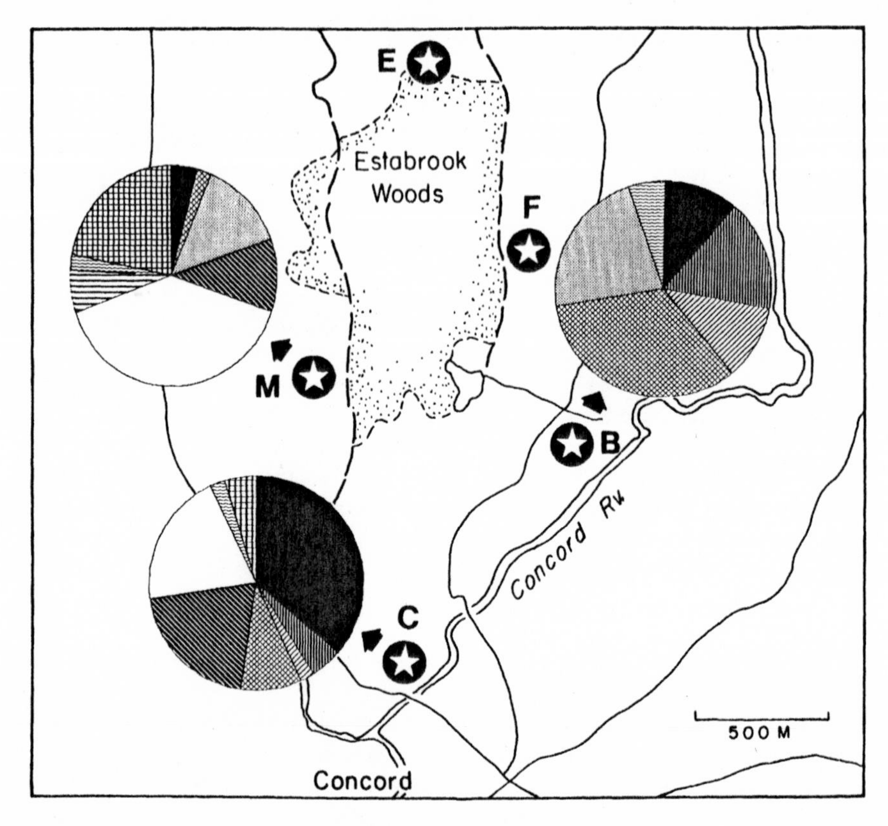

Figure 15-1 Distribution of common mtDNA haplotypes among toads mating at three primary breeding sites surrounding the Estabrook Woods (B, Beecher Pond; C, Concord Center; M, Mink Pond). Advertisement vocalizations of males were recorded at Beecher and Mink Ponds and at two additional localities (E, Evans Pond; F, Freeman Pond). Frequencies of haplotypes (each denoted by a unique shading pattern) are shown in pie charts. Haplotypes were determined as composite restriction fragment length polymorphisms, based on digests with four restriction enzymes (from Waldman et al., 1992).

Zealand frog *Leiopelma archeyi* suggest that populations are genetically subdivided over distances as small as 20 m.

By examining variation in mitochondrial DNA (mtDNA), Waldman et al. (1992) found striking levels of genetic differentiation among local breeding populations of American toads (*Bufo americanus*). During most of the year, toads in these study populations inhabit a wooded reserve. During the spring, however, individuals migrate, some over distances of several kilometers, to ponds at the edge of the reserve. There they typically remain for several days before breeding and then disperse. Although the ponds are in close proximity to one another (distances range from 0.8 to 2.2 km), mitochondrial DNA haplotypes of individuals mating in these localities tend to differ (fig. 15-1). Genetic similarities within ponds, and differences among ponds, persist year after year, suggesting strong natal philopatry of females (Waldman et al., 1992). Nuclear DNA fingerprinting also points to large numbers of male progeny returning to natal sites to breed (B. Waldman et al., unpublished data). Although *Bufo americanus* populations fluctuate in size from year to year (table 15-1), the species readily exploits disturbed habitat and appears to be thriving. Even had we not studied these toad populations over many years, the substantial genetic variation apparent within populations would have suggested that the species was not in decline.

Kin Recognition and Inbreeding Avoidance

The typical amphibian population structure is conducive to high levels of inbreeding. Some individuals disperse and are unlikely to inbreed. But most individuals—of both sexes—return to natal localities to breed (e.g., Berven and Grudzien, 1990). Once there, they encounter siblings and other close relatives as potential mates. Dispersal does not provide a reliable means of inbreeding avoidance for amphibians as it does for many other vertebrates (Pusey and Wolf, 1996; also see Kiester, 1985). In declining populations, the frequency of mating between close relatives is likely to increase.

If amphibians can recognize their close kin, they might avoid mating with them. Over many years, we have studied natural populations of *Bufo americanus* (Waldman et al., 1992; Waldman and McKinnon, 1993; Waldman, 1997) for evidence of inbreeding and possible mate choice. Toads are easy to observe, perhaps because they are poisonous and not easily intimidated by potential predators, and they appear to behave normally even when approached or handled. Thus we were able to employ both observational and experimental approaches to estimate the frequency of inbreeding and to analyze mechanisms of mate choice.

Upon arrival at breeding ponds each spring, male toads produce trilled calls which can be easily heard up to 1 km away. These probably serve as advertisement signals for females and may attract conspecific males. At any given time, more males than females are typically visible in a breeding aggregation. Males scramble for females, and indiscriminately clasp males and females, as well as other similarly-sized frogs, and even inanimate objects. Clasped males utter a release call which effects their release. Females usually arrive at ponds unamplexed, and upon entering the water they tend to remain submerged and repeatedly approach males. Females often act as if they are evaluating particular males. Females swim slowly toward males, lifting their heads slightly out of the water, and finally approach one and initiate amplexus (Licht, 1976; Howard, 1988; Sullivan, 1992; B. Waldman, personal observation). Females sometimes are clasped while swimming, however, and in this case they have little opportunity to exercise mate choice (Howard, 1988). Pairs occasionally are harassed by unpaired males; the unpaired male struggles with the amplexed male for access to the female. Attempts to dislodge amplexed males are rarely successful (Howard, 1988; B. Waldman, personal observation). Most matings occur within a period of 2–5 days at each site, and pairs may remain in amplexus for 24 h or more before oviposition occurs. Polygyny thus is infrequent (Gatz, 1981; Howard, 1988).

Do Toads Mate with Close Relatives?

We analyzed variation in both mitochondrial and nuclear DNA to assess the frequency of matings between close relatives in natural conditions and the mechanisms by which mate choice occurs. Mitochondrial DNA is maternally inherited, so siblings, as well as more distant matrilineal relatives, share identical fragment patterns. If individuals differ in mtDNA haplotypes (i.e., mitochondrial genotypes), then barring mutations, they cannot be siblings. The application of mtDNA to studies of population structure and sexual selection would be limited if particular haplotypes predominated, but often this is not the case. In many vertebrates, mtDNA undergoes more rapid evolution than nuclear DNA. Consequently, mitochondrial markers can vary extensively among individuals both within and among populations (reviewed in Avise, 1994), sometimes approaching hypervariable minisatellite regions of the nuclear genome in their diversity (used for genetic fingerprinting; Jeffreys, 1987).

We captured toad pairs as they were laying eggs, and we later genetically typed them. We inferred that individuals were not siblings if they differed in their mitochondrial haplotypes (generated with a series of restriction enzymes). Mates with identical haplotypes might be siblings, or they might be related through a more distant female ancestor. Of 86 mated pairs collected, members of only two pairs had identical haplotypes. Randomly generated pairings of males and females present at each pond during each season, however, led to a null expectation of 12 matings between individuals bearing identical haplotypes. Thus, significantly fewer individuals mated with close relatives than would be expected if pairing were random, but the expected frequencies are low (Waldman et al., 1992). Brother–sister matings appear to be exceedingly rare; even the two pairs whose haplotypes matched may have been only distant relatives.

Similarities in Vocalizations of Related Males

Behavioral observations suggest that most female toads have an opportunity to choose their mates and that mate choice is based in part on their assessment of advertisement calls broadcast by males (Sullivan, 1992). Frogs often choose larger or older males by detecting features of their calls that correlate with these characters (reviewed in Ryan, 1991). We were curious whether females might avoid mating with their close kin based on information that they could discern from males' calls. In the field, we recorded calls of 15 males in each of four breeding populations. Additionally, we noted information on their size and environmental factors including temperature. Later we genetically typed each male by obtaining nuclear DNA fingerprints of each using multilocus minisatellite probes (Jeffreys, 1987). We estimated the genetic similarity of calling males within ponds by the proportion of bands they shared (Wetton et al., 1987), and determined proportions of bands shared by siblings and half-siblings by controlled crosses in the laboratory. Next we analyzed similarities in both temporal and frequency components of males' calls as a function of the callers' genetic similarity and relatedness.

Genetically similar males, including brothers and more distant relatives, do indeed produce similar advertisement vocalizations. In each breeding population, calls of close relatives were similar in their temporal components (e.g., pulse duration, interpulse interval, rise time, and call duration), whereas calls of genetically dissimilar individuals were much more variable (fig. 15-2) (Waldman et al., 1992). The effects of genetic similarity on call structure were additional to those of body size and temperature. Females thus might discriminate between close and distant relatives on the basis of hearing their calls and also discriminate between smaller and larger (or older) males based on the same call parameters. The pitch of the call, or dominant frequency, appears to be a better trait for females to use to assess male size, however, and our study revealed no correlation between dominant frequency and genetic similarity (Waldman et al., 1992).

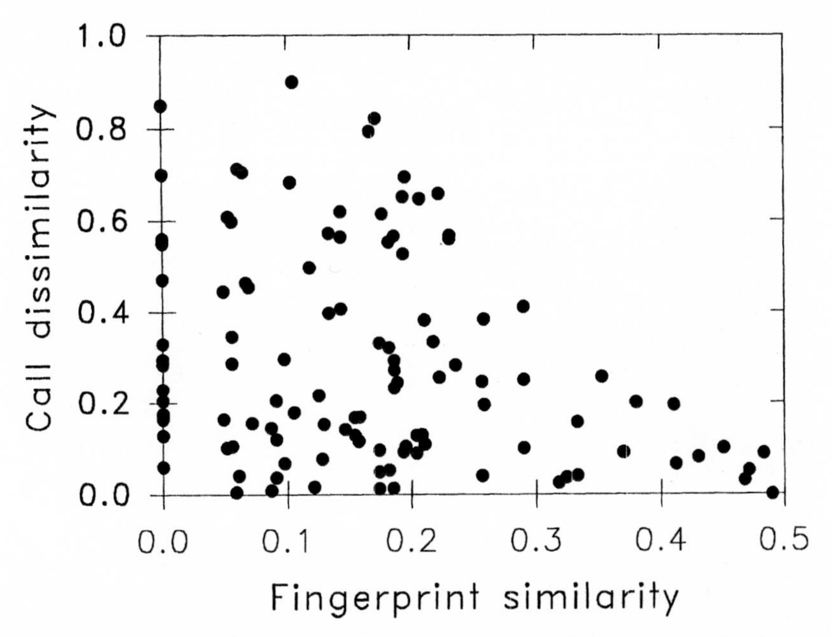

Figure 15-2 Call dissimilarity as a function of genetic similarity among calling males. Shown here are analyses of one component of the advertisement call, the interpulse interval, from males collected at Freeman Pond. Similar results are obtained with other call parameters and at each pond. Identical calls have a dissimilarity value of 0, and increasing dissimilar calls have higher values. Fingerprint similarity values increase with relatedness *(r)* and inbreeding *(F)* coefficients (from Waldman et al., 1992).

Female Mate Choice

Kinship information encoded in males' calls might enable females to recognize their close relatives and to avoid mating with them or even to choose an "optimally" related mate (Bateson, 1983). But can females decode this information and use it? To answer this question, we individually tested 29 females in a laboratory arena, observing their responses to recorded calls of two males, alternately broadcast from speakers on either side of the arena (Waldman, 1997; B. Waldman et al., unpublished data). In each test, both males were from the same pond as the female subject, but they were probably unfamiliar to her because we had recorded the males in the field during a previous year. Additionally, we matched stimulus calls for call duration and sound intensity. We genetically typed the females, comparing their multilocus DNA fingerprints (Jeffreys, 1987) with those of the males only after the behavioral tests had been completed. Thus the behavioral tests were conducted blindly.

Females can discriminate between relatives and nonrelatives on the basis of their calls. Subjects showed a strong preference to approach the speaker broadcasting the call of the male genetically less similar to themselves (fig. 15-3). Only 2 of 29 females preferred the call of the genetically more similar male. Seven females showed no consistent preference for either call, but their DNA fingerprints reveal that they were no more closely related to one male than to the other (Waldman, 1997). Given our observations of pair formation in

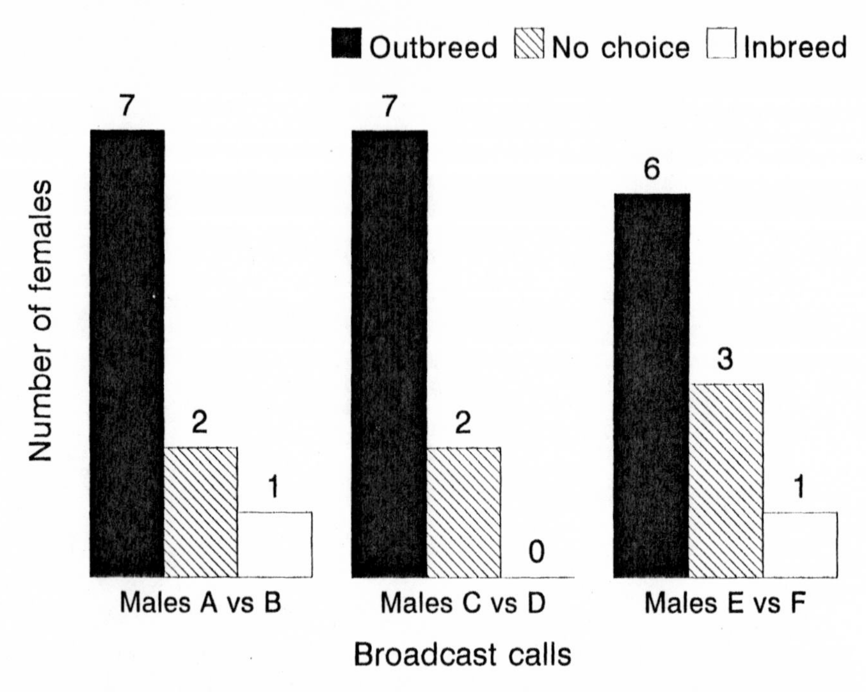

Figure 15-3 Responses of females to calls of males broadcast alternately from speakers on either side of an indoor test arena. Calls of three pairs of males were broadcast, and different females were tested with each pair. Females were judged to have made a choice if they moved within 20 cm of a speaker. DNA fingerprints of males and females were compared only after behavioral tests had been completed. Twenty females approached the speaker broadcasting the call of the male to which they were less closely related (i.e., they "outbred"). Only two females approached the speaker broadcasting the call of the male to which they were more closely related (i.e., they "inbred"). Five females failed to make a choice, but they were equally closely related to the two males whose calls were broadcast (Waldman, 1997).

the field, we interpret these results as evidence of mate selection. Female toads appear able to preferentially choose nonrelatives as mates.

We still do not know how females identify call features of their close relatives and nonrelatives. Kin-recognition mechanisms of larval amphibians permit them to discriminate not only kin from nonkin, but also between different classes of relatives (e.g., siblings from half-siblings). Specific preferences appear to be acquired through the learning of one's own traits or those of close kin during development (reviewed in Waldman, 1991; Blaustein and Waldman, 1992; Balustein and Walls, 1995). Advertisement vocalizations may be genetically determined, but females do not produce these calls, and they probably never have an opportunity to learn their brothers' calls. Mate choice may involve a kin-recognition system in which females evaluate males' calls based on a genetically encoded recognition template (Waldman, 1987).

Kin Recognition and Immune Function

Mate choice affects not only levels of inbreeding, but it may also affect immune function and disease resistance. Allelic variation at loci of the major histocompatibility complex (MHC) is thought to play a key role in immunological defense and effectively permits hosts to respond to much more rapidly evolving pathogens (Haldane, 1949; Hamilton, 1982). Individuals with heterozygous MHC alleles, if they can recognize and respond to more foreign antigenic epitopes, may be conferred enhanced disease resistance (e.g., Doherty and Zinkernagel, 1975). Individuals that mate disassortatively at MHC loci avoid deleterious consequences of inbreeding and accrue general benefits of heterozygosity, but especially reduced susceptibility to infectious diseases (Potts and Wakeland, 1993; Edwards and Potts, 1996; but see Hill et al., 1991; Potts et al., 1994). Disassortative mating also generates variation among individuals in their MHC types (Hedrick, 1992) and thus variation in their susceptibility to particular pathogens. This should further boost disease resistance within populations by reducing opportunities for pathogen transmission.

Models predict that populations with little MHC variation should suffer higher incidences of disease than would genetically heterogeneous populations. O'Brien and Evermann (1988) hypothesized that population declines in cheetah *Acinonyx jubatus,* black-footed ferret *Mustela nigripes,* bighorn sheep *Ovis canadensis,* and other species, are attributable to decreased immunological resistance resulting from extreme MHC monomorphism. Yet many mammals appear perfectly healthy despite having few polymorphic MHC alleles (Reimann and Miller, 1983; Edwards and Potts, 1996), and cheetah populations seem to be constrained more by predation than by lack of genetic diversity (Caro and Laurenson, 1994; also see Nunney and Campbell, 1993; Caughley, 1994).

The biochemical and functional similarities between amphibian MHC loci and those of other vertebrates (Flajnik and Du Pasquier, 1990; Horton, 1994) suggest the possibility that mate choice in frogs may result from selection to reduce susceptibility to disease. Natural outbred populations of leopard frogs *Rana pipiens* and bullfrogs *Rana catesbeiana* are sufficiently variable that no two individuals are likely to share exactly the same MHC alleles (Roux and Volpe, 1975). Comparative studies suggest that while MHC alleles regulate immunity in all amphibians, levels of polymorphism vary among species (Plytycz, 1984) with frogs and toads typically being more variable than salamanders (Kaufman et al., 1995). We suggest that comparisons of MHC polymorphisms between declining and nondeclining species, and between healthy and diseased conspecific populations, might establish possible genetic correlates that would aid us in understanding why populations decline. More generally, such comparisons might begin to resolve the role of MHC variation in conferring disease resistance.

The evolution of genetically based kin-recognition systems at first appears enigmatic in that extreme polymorphism is necessary at marker loci and kin selection should deplete this variation (Alexander, 1979; Crozier, 1986). Frequency-dependent selection for mate choice or host–parasite interactions appears necessary to sustain variation in the cues used by this type of recognition system (Crozier, 1987; Grosberg, 1988; Klein and O'Huigin, 1994). Only MHC loci are sufficiently polymorphic to produce unambiguous markers of overall kinship identity (Waldman et al., 1988), but the polymorphism presumably is driven by selection for effective immune function (e.g., in *Xenopus* frogs; Sato et al., 1993). Conversely, genetically based outbreeding mechanisms may have evolved not to reduce homozygosity throughout the genome but particularly to maintain polymorphisms at MHC loci responsible for immune function (Potts and Wakeland, 1993; Potts and Slev, 1995). Evidence is accumulating that odors associated with histocompatibility genes influence both altruism and mate choice in organisms as diverse as sea squirts, rodents, and humans (reviewed by Brown and Eklund, 1994; Wedekind et al., 1995). Work is currently underway in our laboratory to examine in amphibians the extent to which kin recognition is attributable to MHC variation. If kin recognition is indeed regulated by MHC alleles, depletion of MHC genetic diversity that is likely to accompany population declines would deprive threatened species of an evolved mechanism to avoid inbreeding (Edwards and Potts, 1996).

Behavioral Ecology and Amphibian Conservation

Environmental perturbations do not necessarily directly cause population declines. Rather, in some cases they may precipitate ecological and behavioral changes that in turn increase mortality or decrease recruitment. For example, breeding habitat may be altered because of changes in weather or anthropogenic activities. To utilize available habitat, amphibians may need to modify their reproductive timing, perhaps breeding later in the year or during shorter periods. Intra- and interspecific competition for calling and oviposition sites may subsequently intensify, with males altering their calling behavior in response to acoustical interference (Donnelly and Crump, 1997). Patterns of intra- and interspecific competition among larvae, and their interactions with predators and prey, will change in unpredictable ways (e.g., Warner et al., 1993; Kiesecker, 1996). When suitable habitat becomes restricted, amphibians might be found in denser aggregations, which might be more readily exploited by pathogens and predators. Simultaneously, food resources may decrease in availability due to the altered social structure. Changes in population dynamics and community structure can have profound effects on demography and reproductive strategies that lead to population declines (Donnelly and Crump, 1997).

Subtle behavioral and ecological effects may offer the first indication that a population is under threat long before declining numbers are noted. The life history of many amphibians is characterized by high larval mortality due to predation and desiccation (Wilbur, 1980). Pesticides, herbicides, fertilizers, and other toxicants alter the swimming behavior of amphibian larvae, making them more vulnerable to predators (reviewed in Power et al., 1989; Devillers and Exbrayat, 1992; Carey and Bryant, 1995; Hecnar, 1995). UV-B radiation similarly can induce erratic swimming behavior (Nagl and Hofer, 1997). Behavioral changes can make larvae easier to capture. For example, dragonfly larvae prey more efficiently on *Hyla cinerea* tadpoles exposed to toxic metals than on controls (Jung and Jagoe, 1995). *Rana temporaria* tadpoles exposed to pesticides become hyperactive and are selectively preyed on by newts (Cooke, 1971). Changes in habitat selection, feeding behavior, and pigmentation patterns also can put exposed larvae at increased risk (e.g., Cooke 1972a; Watt and Oldham, 1995). Even if a potential toxicant is shown to be harmless to amphibians in laboratory studies, through its indirect effects the same agent may prove lethal in nature.

Perspectives gained from behavioral ecology are important in interpreting how genetic stochasticity affects declining populations. As populations become smaller, they are more likely to go extinct due to purely random events, and inbreeding increasingly contributes to this risk (Lande, 1994). Conservation biologists believed for most of the past two decades that populations with an effective size of at least 500 can maintain sufficient genetic variation to survive (Franklin, 1980; Soulé, 1980). Lande (1995), however, recently argued that sustainable populations must be larger by least an order of magnitude (i.e., $N_e > 5000$) because of the accumulation of mildly deleterious alleles within small populations. N_e represents the number of individuals in an idealized panmictic population, so censused population sizes usually need to be considerably larger to achieve the requisite genetic variation. Most amphibian populations have N_e values significantly smaller than either of these threshold values (Waldman and McKinnon, 1993), suggesting that all may be destined for extinction. Yet, our finding that amphibians can use behavioral mechanisms to selectively outbreed points to the need to carefully reevaluate such models based on realistic assumptions (Barton and Turelli, 1989). The minimum population sizes necessary to counter the effects of genetic drift might be reduced, and the expected longevity of populations lengthened, as a consequence of mechanisms that enforce outbreeding.

When populations diminish substantially in size and number, captive rearing may be attempted to minimize risks of environmental stochasticity. In these conditions, allowing amphibians to choose their mates may be important in reducing deleterious effects of inbreeding and minimizing the risk of disease (see Grahn et al., chapter 13, this volume). Consider the case of the endangered Wyoming toad *Bufo hemiophrys baxteri.* Until the mid-1970s, this toad was fairly common in the Laramie Basin of Wyoming, but it then declined rapidly (Baxter et al., 1982), and within a decade it was presumed extinct (Lewis et al., 1985). A single population was discovered in 1987, but the number of remaining individuals decreased from 394 in 1991 to 155 in 1992 due to adult mortality resulting apparently from bacterial and fungal infections (Corn, 1993; Taylor et al., 1995). As the species seemed destined for extinction, in 1994 the Wyoming Game and Fish Department removed all known remaining toads from the wild (Corn, 1994). Toads are now being bred in zoos so that tadpoles can be released back into the wild. Our findings suggest that females choose males with which they are genetically compatible as mates. Pairing males and females haphazardly or randomly for breeding, as is currently being done, thus may not be the safest strategy. Rather, natural conditions should be simulated to allow individuals to choose their mates. Should this fail, genetically dissimilar individuals can be identified and paired with the aid of DNA fingerprinting techniques. With numbers of dwindling, planning matings to maximize genetic variation may seem esoteric. Yet to ignore potential genetic consequences of mate choice could be disastrous for species recovery programs.

By considering amphibian conservation from the perspective of behavioral ecology, we can gain new insights on the dynamics of population declines and possible solutions to the problem. Yet behavioral ecology offers no panacea, and its incorporation into conservation programs presents disadvantages as well. Research into the behavioral ecology of declining populations can be painstaking and time consuming, but the rapid declines that we are witnessing may require quick decisions. Behavioral studies are likely to reveal complex effects that vary under different ecological regimes, so just as with conventional approaches, general answers may be elusive. Most troubling, our studies of threatened populations may subject them to new stresses that hasten their decline (Kagarise Sherman and Morton, 1993; Lips, 1998).

Recommendations

Behavioral Tools for Population Surveys

Male frogs of most but not all species produce calls to advertise to females and often to communicate to other males. These vocalizations are species specific and presumably function to facilitate species recognition (Gerhardt, 1994). Calls of many species are easily detected from some distance by human listeners. Counting calling males, or numbers of aggregations of calling males, thus can serve as an efficient method to detect species within areas and to estimate relative population densities (Zimmerman, 1994; McGregor and Peake, chapter 2, this volume). Many frogs are naturally secretive and others live in habitat that is difficult to survey by visual methods. Not only are visual surveys slow and laborious, but in many cases microhabitats may be seriously disrupted or even destroyed in the process of searching for individuals. Tocher's (1996) recent use of audio surveys to examine effects of habitat fragmentation on species diversity of central Amazonia frogs illustrates the power of such methodology to expedite research into the causes of amphibian declines.

The advantages of using calling behavior as a measure of population density may often outweigh its disadvantages. Still, caution needs to be exercised in interpreting results. Zimmerman (1994) discusses numerous assumptions underlying this methodology and suggests possible controls (e.g., for observer reliability). The number of frogs calling in a chorus does not necessarily correspond to the number of frogs present; calling rates tend to decrease as chorus size increases (e.g., Höglund and Robertson, 1988). Two points especially need to be emphasized. First, the use of vocalizations to monitor either species diversity or population numbers is only valid if one can establish that calling individuals are successfully reproducing. Second, many frogs are relatively long-lived, living 20 years or more (e.g., Bell, 1994). Because of their site fidelity, frogs may return to call from, and even mate at, particular breeding localities long after the sites have become unsuitable for supporting larvae. Visual surveys of adults can suffer from the same weaknesses. Population declines of long-lived species may have been underway for years, even decades, before surveys hint at any problems. By then, declines may be irreversible.

While the species specificity of frog calls makes them suitable for determining the presence or absence of species, and to some extent the numbers of frogs present, vocalizations encode more specific genetic information as well. Geographic variation in mating calls of conspecific frogs has been noted for some time (e.g., Snyder and Jameson, 1965; Capranica et al., 1973). Only recently, however, has such variation been investigated on a more fine grained scale. Substantial differences in call characters between populations of cricket frogs *Acris crepitans* in different ecological settings within 65 km of one another were noted by Ryan and Wilczynski (1988). More detailed studies reveal call variation among populations, some considerably closer (Ryan and Wilczynski, 1991), although females appeared not to discriminate among calls from different populations (Ryan et al., 1992). Our work reveals that *Bufo americanus* calls may vary substantially on a microgeographic scale, over distances of 1 km or less, even in extremely similar environments (Waldman et al., 1992). As we described earlier, males' calls resemble one another as a function of their genetic similarity (Waldman et al., 1992).

Small, genetically homogeneous populations are most at risk of extinction due to a variety of causes. We suggest that conservation biologists might initially identify threatened populations by assaying variation in heritable behavioral traits. For each target species, baseline data are first needed to establish that behavioral differences are correlated with genetic differences, which may be determined either though pedigree analysis or genetic fingerprinting. With this information available, audio surveys might be used, not just for censusing, but also for monitoring genetic variation within populations. Assessment of genetic variation by analysis of calls or other behavioral cues can potentially provide a quick means of identifying populations in possible danger without disrupting habitat or manipulating subjects. Where problems are suspected, genetic variation then can be assayed more precisely by molecular techniques to establish appropriate management plans.

Many traits, both morphological and behavioral, show evidence of fluctuating asymmetry when individuals are stressed (Hoffman and Parsons, 1991). Environmental stresses, including pesticides, pollutants, acidification, and abnormal temperature regimes, appear to interfere with the developmental pathways that normally produce bilaterally symmetric traits (reviewed in Leary and Allendorf, 1989). Populations with reduced genetic variation, highly inbred individuals, and individuals with low levels of heterozygosity are most likely to show evidence of these asymmetries (Palmer and Strobeck, 1986; Leary and Allendorf, 1989). As we have argued here, fragmentation depletes genetic variation and decreases the ability of individuals to cope with environmental perturbations. Not surprisingly, individuals in fragmented populations show higher levels of fluctuating asymmetry than do those in undisturbed habitat (Sarre, 1996). The measurement of fluctuating asymmetries thus constitutes another simple means of identifying populations that may be declining (Leary and Allendorf, 1989; Sarre et al., 1994; Clarke, 1996). Recently, Alford et al. (1997) found that museum specimens of two species of tropical wetland frogs, *Litoria genimaculata* and *Litoria nannotis,* showed increasing evidence of fluctuating asymmetry in a variety of morphological characters during the months preceding the onset of their population declines. We suggest that the monitoring of behaviors that are likely to show evidence of fluctuating asymmetry, such as locomotory abilities of tadpoles or adults, may even more readily identify populations that are being adversely affected by environmental and genetic stresses.

Assessing Genetic Variation

Recent advances in genetic fingerprinting technology make direct molecular assessment of population variation feasible, inexpensive, and use relatively noninvasive procedures. DNA can be extracted from blood or from toe clips, which typically are obtained when marking amphibians in field studies (Donnelly et al., 1994). Should concern arise about possible effects of toe-clipping (Kagarise Sherman and Morton, 1993), dead skin shed by individuals can be used for genetic typing (B. Waldman, unpublished data). By use of the polymerase chain reaction (PCR), microgram quantities of DNA extracted from such tissues can be amplified repeatedly to generate sufficient material for genetic analyses. Thus, individuals no longer need to be sacrificed to genetically characterize populations. PCR methodology facilitates ongoing genetic monitoring of threatened species. Tissue samples can be stored in the field in ethanol or in other preservatives (Reiss et al., 1995), thus making genetic sampling possible even in remote locations.

As seen in our work on *Bufo,* a variety of approaches can be taken to assess levels of genetic variation (reviewed in Avise, 1994; Avise et al., 1995; Lambert and Millar, 1995; Ferraris and Palumbi, 1996; Hillis et al., 1996). Analyses of variation in mitochondrial or nuclear genes may be more appropriate depending on the question of interest. Mitochondrial (mtDNA) genomes are small and thus more tractable, but they are also matrilineally inherited; thus they fail to reveal information on male dispersal and reproductive success. Analyses of mtDNA traditionally have focused on geographic population structure rather than on localized population variation (e.g., Norman et al., 1994). Our work suggests, however, that amphibian mitochondrial genomes are sufficiently variable to generate maternal fingerprints that may be useful for studying genetic relationships within populations and fine-scaled population substructuring. Using PCR, primers can be designed to amplify the control region (D-loop segment), which typically shows the greatest variation. Amplified products can be digested and characterized by fragment size to examine polymorphisms (RFLPs; restriction fragment length polymorphisms), sequenced to determine nucleotide substitutions (e.g., Yang et al., 1994), or screened for variants by examining fragment conformation or stability (heteroduplexes, SSCP, DGGE, TGGE) (reviewed in Lessa and Applebaum, 1993; Hillis et al., 1996; see also Norman et al., 1994). Screening techniques can be used to design primers for specific mitochondrial and nuclear genes, such as those of the MHC, which may serve as especially useful markers in captive breeding (see Grahn et al., chapter 13, this volume).

Nuclear DNA is more commonly fingerprinted for population studies. Currently several methods are available for analyzing nuclear variation, each with advantages and disadvantages. Until recently, variation was usually assayed by using probes that hybridize with particular sequences (VNTRs; sequences that consist of a variable number of tandem repeat nucleotides). Multilocus probes, such as those isolated by Jeffreys (1987), nonspecifically detect minisatellite loci containing these VNTR regions, which usually are homologous across taxa. This methodology can be used even to characterize individual tadpoles from pieces of their tails (D'Orgeix and Turner, 1995), but larger quantities of tissue typically are required to type individuals confidently. Single-locus probers provide less ambiguous genetic data, which facilitate the computation of population parameters (e.g., N_e; see Scribner et al., 1997).

The identification of even shorter repeat sequences, termed "microsatellite" loci, makes possible precise and meaningful estimates of genetic variation for conservation studies (e.g., Scribner et al., 1994). Although inbreeding, population substructuring, and population size may affect interpretations (Pena, 1995; Nauta and Weissing, 1996), the utility of microsatellites for population studies is undisputed (Jarne and Lagoda, 1996). Microsatellite sequences can be amplified by PCR, so only minute tissue samples are required. Unfortunately, primers must be designed for each species (or species group) and for each locus, and the process is laborious and time consuming (Queller et al., 1993).

The use of random primers (RAPDs; random amplified polymorphic DNA) overcomes these difficulties, but suffers from problems of interpretation and repeatability (Hadrys et al., 1992; Bielawski et al., 1995; Masters, 1995). Nonetheless, RAPD primers quickly and inexpensively generate numerous genetic markers (Micheli and Bova, 1997), which can provide a useful measure of genetic variation that should allow conservation biologists to readily target populations of special concern. Kimberling et al.'s (1996) study of gene flow among *Rana pipiens* population serves as a good example of the versatility of RAPDs. With sufficient sample sizes, population parameters (degree of inbreeding and population subdivision, levels of heterozygosity, effective population sizes, and gene flow) can be estimated using these markers (Lynch and Milligan, 1994). Differentiation can be partitioned within and among populations by using AMOVA (analysis of molecular variance) procedures (Excoffier et al., 1992), originally derived for analyses of mtDNA haplotypes, but equally applicable to RAPD data (e.g., Haig et al., 1994). Recent work raises the possibility that the ease of use of RAPDs may be combined with the precision of microsatellites by using anchored primers to amplify short, interspersed repeats common to the genomes of most species (Zietkiewicz et al., 1994; Weising et al., 1995).

Conservation Strategies

We suggest that surveys of amphibians routinely should include genetic monitoring programs using noninvasive methods such as those we described here. Levels of genetic differentiation among populations need to be determined to identify appropriate management units (Moritz, 1994) and to determine source populations toward which conservation efforts should be directed (Templeton et al., 1990; Dias, 1996). Populations characterized by low levels of genetic variation merit special attention because of their increased vulnerability to environmental perturbations and pathogens. Behavioral bioassays and analyses of fluctuating asymmetry may facilitate the identification of populations contaminated by toxicants or exposed to other stresses. Searches for particular pathogens (e.g., Laurance et al., 1996), in contrast, may be less fruitful because amphibians in declining populations appear to succumb to organisms normally present in their environment.

Managers potentially can boost levels of genetic variation within populations by facilitating gene flow between populations. This can be done either by providing corridors or by conducting periodic translocations among separated populations (Nunney and Campbell, 1993). As few as 0.5 migrants per generation among populations should reduce levels of inbreeding and eventually restore genetic heterogeneity within populations (Lacy, 1987). In practice, attempts to translocate amphibian populations often have not been successful (Dodd and Seigel, 1991; but see Burke, 1991; Reinert, 1991), and translocated populations rapidly lose allelic diversity (Stockwell et al., 1996). The translocation of individuals to maximize genetic diversity within populations may represent a better solution. To prevent the transmission of pathogens, individuals must be thoroughly screened before relocation (Dodd and Seigel, 1991; Cunningham, 1996; Hess, 1996). Outbreeding depression represents another risk, especially if transferred animals are adapted to different environments (Waldman and McKinnon, 1993). Clearly, translocations should not be conducted between populations characterized by genetic differences that may qualify them as separate management units (e.g., Green, 1994; Kimberling et al., 1996). Despite the potential pitfalls, carefully conducted translocations offer the promise of reversing the loss of genetic diversity that contributes to amphibian declines.

Summary

Reports of population declines of amphibians, especially frogs and toads, have increased in recent years. Population extinctions may be now occurring with increasing frequency. Amphibian populations are adversely affected by the destruction or fragmentation of habitat, ultraviolet radiation, changing weather patterns, acidification, herbicides and pesticides, pollutants, and the introduction of exotic predators or competitors. Yet these factors do not explain all population declines, and many cannot be accounted for by these factors.

Environmental changes that appear harmless nonetheless may cause sublethal stress. Multiple factors may act synergistically to impair immune function and increase susceptibility to disease. Disease is associated with declines of many amphibian populations. Because genetically similar individuals share immunological vulnerabilities, pathogens can more readily build up to critical epidemic levels when populations are genetically homogeneous.

Many declines have occurred in regions where amphibian populations are fragmented. Fragmentation disrupts metapopulations, increases the probability that local populations will go extinct, and decreases the probability of recolonization. Recolonized populations will suffer reduced genetic variation. Amphibians are highly philopatric, which further reduces gene flow between populations. Levels of genetic homogeneity within populations may be high, leading to inbreeding depression, especially in stressful environments. Despite the potential for inbreeding, our field studies on toads demonstrate significant inbreeding avoidance in natural populations. Closely related mates have similar advertisement calls, and call similarity between males is correlated with their genetic similarity. By means of playback experiments, we demonstrated that females can discriminate between calls of males more and less closely related to themselves, preferentially choosing the more distantly related male. Kin recognition in some taxa is mediated by cues encoded by histocompatibility alleles. Variation in these cues may result from selection for disease resistance. Behavioral mechanisms that promote outbreeding may offset, at least partially, the depletion of genetic variation expected in small populations and can boost disease resistance.

Behavioral traits, such as larval swimming behavior, may serve as early indicators of environmental stresses that threaten amphibian populations. Mating calls can be used to establish species ranges and population densities. We advocate routine genetic monitoring as a means to identify populations of threatened and endangered species that are at most risk of extinction. Recent analytical innovations based on PCR permit genetic fingerprinting of individuals without causing undue harm to members of surveyed populations. Our work suggests that genetic variation within populations can be rapidly estimated from nonintrusive acoustical analyses of call variability.

Acknowledgments We thank Tim Caro, Bob Lacy, Bill Laurance, Murray Littlejohn, and three anonymous referees for their thoughtful comments on the chapter, Ross Alford, Andrew Balmford, Lee Berger, Phil Bishop, Terry Burke, Cindy Carey, Margaret Carpenter, Gary Fellers, Carl Gerhardt, Anna Goebel, Bill Hamilton, Tyrone Hayes, Richard McKenzie, Alexander Nagl, Dale Roberts, David Smith, Sharon Taylor, Randy Thornhill, and Penny Watt for discussion on points we have raised; M. Donnelly, Per Edenhamn, Scott Edwards, Bill Laurance, Karen Lips, Richard McKenzie, Rob Oldham, and Will Osborne for making available manuscripts prior to publication; Curt Lively for providing a copy of Haldane's elusive paper; and Peggy Lundgren for translations. BW's research on frog population genetics has been supported by grants from the National Science Foundation (USA), the Foundation for Research, Science and Technology, (NZ), the New Zealand Lottery Grants Board, Harvard University, and the University of Canterbury.

References

Abdulali H, 1986. On the export of frog legs from India. J Bombay Nat Hist Soc 82: 347–375.

Albers PH, Prouty RM, 1987. Survival of spotted salamander eggs in temporary woodland ponds of coastal Maryland. Environ Pollut 46:45–61.

Alexander RD, 1979. Darwinism and human affairs. Seattle: University of Washington Press.

Alford RA, Bradfield KS, Richards SJ, 1997. Predicting declines in rainforest frog populations. Abstracts, Annual Research Meeting of the Cooperative Research Centre for Tropical Rainforest Ecology and Management. Townsville: James Cook University of North Queensland.

Alford RA, Richards SJ, 1997. Lack of evidence for epidemic disease as an agent in the catastrophic decline of Australian rain forest frogs. Conserv Biol 11:1026–1029.

Allendorf FW, Leary RF, 1986. Heterozygosity and fitness in natural populations of animals. In: Conservation biology: the science of scarcity and diversity (Soulé ME, ed). Sunderland, Massachusetts: Sinauer Associates; 57–76.

Anderson S, 1993. Livestock management effects on wildlife, fisheries and riparian areas: a selected literature review. Elko, Nevada: Humboldt National Forest.

Andrén C, Mårdén M, Nilson G, 1989. Tolerance to low pH in a population of moor frogs, *Rana arvalis,* from an acid and a neutral environment: a possible case of rapid evolutionary response to acidification. Oikos 56:215–223.

Avise JC, 1994. Molecular markers, natural history and evolution. New York: Chapman and Hall.

Avise JC, Haig SM, Ryder O, Lynch M, Geyer CJ, 1995. Descriptive genetic studies: applications in population management and conservation biology. In: Population management for survival and recovery: analytical methods and strategies in small population conservation (Ballou JD, Gilpin M, Foose TJ, ed). New York: Columbia University Press; 183–244.

Ash AN, 1997. Disappearance and return of plethodontid salamanders to clearcut plots in the southern Blue Ridge mountains. Conserv Biol 11:983–989.

Axelsson E, Nyström P, Sidenmark J, Brönmark C, 1997. Crayfish predation on amphibian eggs and larvae. Amphibia-Reptilia 18:217–228.

Baldwin WM III, Cohen N, 1975. Alloimplant extrusion: a link between invertebrate and vertebrate defense systems? Immunogenetics 2:73–79.

Banks B, Beebee TJC, 1987. Factors influencing breeding site choice by the pioneering amphibian *Bufo calamita.* Hol Ecol 10:14–21.

Banks B, Beebee TJC, 1988. Reproductive success of natterjack toads *Bufo calamita* in two contrasting habitats. J Anim Ecol 57:475–492.

Banks B, Beebee TJC, Cooke AS, 1994. Conservation of the natterjack toad *Bufo calamita* in Britain over the period 1970–1990 in relation to site protection and other factors. Biol Conserv 67:111–118.

Barbault R, 1984. Stratégies de reproduction et démographie de qualques amphibians anoures tropicaux. Oikos 43:77–87.

Barbault R, 1991. Ecological constraints and community dynamics: linking community patterns to organismal ecology. The case of tropical herpetofaunas. Acta Oecol 12:139–163.

Barton NH, Turelli M, 1989. Evolutionary quantitative genetics: how little do we know? Annu Rev Genet 23:337–370.

Basher RE, Zhen X, Nichol S, 1994. Ozone-related trends in solar UV-B series. Geophys Res Lett 24:2713–2716.

Bateson P, 1983. Optimal outbreeding. In: Mate choice (Bateson P, ed). Cambridge: Cambridge University Press; 257–277.

Baxter GT, Stromberg MR, Dodd CK Jr, 1992. The status of the Wyoming toad, *Bufo hemiophrys baxteri.* Environ Conserv 9:348,338.

Beattie RC, Aston RJ, Milner AGP, 1991. A field study of fertilization and embryonic development in the common frog (*Rana temporaria*) with particular reference to acidity and temperature. J Appl Ecol 28:346–357.

Beattie RC, Tyler-Jones R, 1992. The effects of low pH and aluminum on breeding success in the frog *Rana temporaria.* J Herpetol 26:353–360.

Beebee TJC, 1976. The natterjack toad in the British Isles, a study of past and present status. Br J Herpetol 5:515–521.

Beebee TJC, 1977. Environmental change as a cause of natterjack toad (*Bufo calamita*) declines in Britain. Biol Conserv 11:87–102.

Beebee TJC, 1995. Amphibian breeding and climate. Nature 374:219–220.

Beebee TJC, Flower RJ, Stevenson AC, Patrick ST, Appleby PG, Fletcher C, Marsh C, Natkanski J, Rippey B, Battarbee RW, 1990. Decline of the natterjack toad *Bufo calamita* in Britain: palaeoecological, documentary and experimental evidence for breeding site acidification. Biol Conserv 53:1–20.

Bell BD, 1994. A review of the status of New Zealand *Leiopelma* species (Anura: Leiopelmatidae), including a summary of demographic studies in Coromandel and on Maud Island. NZ J Zool 21:341–349.

Bell BD, 1996. Aspects of the ecological management of New Zealand frogs: conservation status, location, identification, examination and survey techniques. Ecological Management (Department of Conservation, Wellington) 4:91–111.

Berrill M, Bertram S, McGillivray L, Kolohon M, Pauli B, 1994. Effects of low concentrations of forest-use pesticides on frog embryos and tadpoles. Environ Toxicol Chem 13:657–664.

Berven KA, 1990. Factors affecting population fluctuations in larval and adult stages of the wood frog (*Rana sylvatica*). Ecology 71:1599–1608.

Berven KA, Grudzien TA, 1990. Dispersal in the wood frog (*Rana sylvatica*): implications for genetic population structure. Evolution 44:2047–2056.

Bidwell JR, Gorrie JR, 1995. Acute toxicity of a herbicide to selected frog species. Final report. Perth: Western Australia Department of Environmental Protection.

Bielawski JP, Noack K, Pumo DE, 1995. Reproducible amplifications of RAPD markers from vertebrate DNA. Biotechniques 18:856,858–860.

Blakeslee S, 1997. New culprit in widespread deaths of frogs. New York Times, 16 September.

Blaustein AR, Hoffman PD, Kiesecker JM, Hays JB, 1996a. DNA repair activity and resistance to solar UV-B radiation in eggs of the red-legged frog. Conserv Biol 10:1398–1402.

Blaustein AR, 1994. Chicken little or Nero's fiddle? A perspective on declining amphibian populations. Herpetologica 50:85–97.

Blaustein AR, 1995. Ecological research. Science 269:1201–1202.

Blaustein AR, Edmond B, Kiesecker JM, Beatty JJ, Hokit DG, 1995a. Ambient ultraviolet radiation causes mortality in salamander eggs. Ecol Appl 5:740–743.

Blaustein AR, Hoffman PD, Hokit DG, Kiesecker JM, Walls SC, Hays JB, 1994c. UV repair and resistance to solar UV-B in amphibian eggs: a link to population declines? Proc Natl Acad Sci USA 91:1791–1795.

Blaustein AR, Hokit DG, O'Hara RK, Holt RA, 1994b. Pathogenic fungus contributes to amphibian losses in the Pacific Northwest. Biol Conserv 67:251–254.

Blaustein AR, Kiesecker JM, Hokit DG, Walls SC, 1995b. Amphibian declines and UV radiation. Bioscience 45:514–515.

Blaustein AR, Kiesecker JM, Walls SC, Hokit DG, 1996b. Field experiments, amphibian mortality, and UV radiation. Bioscience 46:386–388.

Blaustein AR, Wake DB, Sousa WP, 1994a. Amphibian declines: judging stability, persistence, and susceptibility of populations to local and global extinctions. Conserv Biol 8:60–71.

Blaustein AR, Waldman B, 1992. Kin recognition in anuran amphibians. Anim Behav 44:207–221.

Blaustein AR, Walls SC, 1995. Aggregation and kin recognition. In: Amphibian biology, vol 2. Social behaviour (Heatwole H, ed). Chipping Norton, New South Wales: Surrey Beatty; 568–602.

Blumthaler M, 1993. Solar UV measurements. In: UV-B radiation and ozone depletion (Tevini M, ed). Boca Raton, Florida: Lewis Publishers, 71–94.

Bogert CM, 1947. A field study of homing in the Carolina toad. Am Mus Nov 1355:1–24.

Boyer R, Grue CE, 1995. The need for water quality criteria for frogs. Environ Health Perspect 103:352–357.

Bradford DF, 1989. Allotopic distribution of native frogs and introduced fishes in high Sierra Nevada lakes of California: implications of the negative effect of fish introductions. Copeia 1989:775–778.

Bradford DF, 1991. Mass mortality and extinction in a high-elevation population of *Rana muscosa.* J Herpetol 25:174–177.

Bradford DF, Gordon MS, Johnson DF, Andrews RD, Jennings WB, 1994. Acidic deposition as an unlikely cause for amphibian population declines in the Sierra Nevada, California. Biol Conserv 69:155–161.

Bradford DF, Swanson C, Gordon MS, 1992. Effects of low pH and aluminum on two declining species of amphibians in Sierra Nevada, California. J Herpetol 26:369–377.

Bradford DF, Tabatabai F, Graber DM, 1993. Isolation of remaining populations of the native frog, *Rana muscosa,* by introduced fishes in Sequoia and Kings Canyon National Parks, California. Conserv Biol 7:882–888.

Bragg AN, 1954. Decline in toad populations in central Oklahoma. Proc Ok Acad Sci 33:70.

Bragg AN, 1960. Population fluctuation in the amphibian fauna of Cleveland County, Oklahoma during the past twenty-five years. Southwest Nat 5:165–169.

Bragg AN, 1962. *Saprolegnia* on tadpoles again in Oklahoma. Southwest Nat 7:79–80.

Bragg AN, Bragg WN, 1958. Parasitism of spadefoot tadpoles by *Saprolegnia.* Herpetologica 14:34.

Breden F, 1987. The effect of post-metamorphic dispersal on the population genetic structure of Fowler's toad, *Bufo woodhousei fowleri.* Copeia 1987:386–395.

Brönmark C, Edenhamn P. 1994. Does the presence of fish affect the distribution of tree frogs (*Hyla Arborea*)? Conserv Biol 8:841–845.

Brown JH, Kodric-Brown A, 1977. Turnover rates in insular biogeography: effect of immigration on extinction. Ecology 58:445–449.

Brown JL, Eklund A, 1994. Kin recognition and the major histocompatibility complex: an integrative review. Am Nat 143:435–461.

Burke RL, 1991. Relocations, repatriations, and translocations of amphibians and reptiles: taking a broader view. Herpetologica 47:350–357.

Burkey TV, 1995. Extinction rates in archipelagoes: implications for populations in fragmented habitats. Conserv Biol 9:527–541.

Busby WH, Parmelee JR, 1995. Historical changes in a herpetofaunal assemblage in the Flint Hills of Kansas. Am Midl Nat 135:81–91.

Capranica RR, Frishkopf LS, Nevo E, 1973. Encoding of geographic dialects in the auditory system of the cricket frog. Science 182:1272–1275.

Carey C, 1993. Hypothesis concerning the causes of the disappearance of boreal toads from the mountains of Colorado. Conserv Biol 7:355–362.

Carey C, Bryant CJ, 1995. Possible interrelations among environmental toxicants, amphibian development, and decline of amphibian populations. Environ Health Perspect 103(suppl 4):13–17.

Carey C, Maniero G, Harper CW, Snyder G, 1996a. Measurements of several aspects of immune function in toads (*Bufo marinus*) after exposure to low pH. In: Modulators of immune responses: the evolutionary trail (Stolen JS, Fletcher TC, Bayne CJ, Secombes

CJ, Zelikoff JT, Twerdok LE, Anderson DP, eds). Fair Haven, New Jersey: SOS Publications; 565–577.
Carey C, Maniero GD, Stinn JF, 1996b. Effect of cold on immune function and susceptibility to bacterial infection in toads (*Bufo marinus*). In: Adaptations to the cold: tenth international hibernation symposium (Geiser F, Hulbert AJ, Nicol SC, eds). Armidale, New South Wales: University of New England Press; 123–129.
Caro TM, Laurenson MK, 1994. Ecological and genetic factors in conservation: a cautionary tale. Science 263:485–486.
Caughley G, 1994. Directions in conservation biology. J Anim Ecol 63:215–244.
Charlesworth D, Charlesworth B, 1987. Inbreeding depression and its evolutionary consequences. Annu Rev Ecol Syst 18:237–268.
Chazal AC, Krenz JD, Scott DE, 1996. Relationship of larval density and heterozygosity to growth and survival of juvenile marbled salamanders (*Ambystoma opacum*). Can J Zool 74:1122–1129.
Clarke GM, 1996. Relationships between fluctuating asymmetry and fitness: how good is the evidence? Pacific Conserv Biol 2:146–149.
Clarkson RW, Rorabaugh JC, 1989. Status of leopard frogs (*Rana pipiens* complex: Ranidae) in Arizona and southeastern California. Southwest Nat 34:531–538.
Coddington EJ, Cree A, 1995. Effect of acute captivity stress on plasma concentrations of corticosterone and sex steroids in female whistling frogs, *Litoria ewingi.* Gen Comp Endocrinol 100:33–38.
Cohn JP, 1995. Enzymes mediate UV damage. Bioscience 45:11–12.
Colborn T, Clement C (eds), 1992. Chemically-induced alterations in sexual and functional development: the wildlife/humans connection. Princeton, New Jersey: Princeton Scientific.
Collins JP, Jones TR, Berna HJ, 1988. Conserving genetically-distinctive populations: the case of the Huachuca tiger salamander (*Ambystoma tigrinum stebbinsi* Lowe). General Technical Report RM 166. For Collins, Colorado: US Forest Service; 45–53.
Cooke AS, 1971. Selective predation by newts on frog tadpoles treated with DDT. Nature 229:275–276.
Cooke AS, 1972a. The effects of DDT, dieldrin and 2,4-D on amphibian spawn and tadpoles. Environ Pollut 3:51–68.
Cooke AS, 1972b. Indications of recent changes in status in the British Isles of the frog (*Rana temporaria*) and the toad (*Bufo bufo*). J Zool 167:161–178.
Cooke AS, 1973. The effects of DDT, when used as a mosquito larvicide, on tadpoles of the frog *Rana temporaria.* Environ Pollut 5:259–273.
Cooke AS, 1981. Tadpoles as indicators of harmful levels of pollution in the field. Environ Pollut A 25:123–133.
Cooke AS, Ferguson PF, 1976. Changes in status of the frog (*Rana temporaria*) and the toad (*Bufo bufo*) on part of the East Anglian Fenland in Britain. Biol Conserv 9:191–198.
Corn PS, 1993. Recent trends in the population of Wyoming toads (*Bufo hemiophrys baxteri*). Paper presented at the meeting of the Society for the Study of Amphibians and Reptiles, Indiana University, Bloomington, 7–12 August.
Corn PS, 1994. What we know and don't know about amphibian declines in the west. General technical report RM 247. Fort Collins, Colorado: US Forest Service. 59–67.
Corn PS, Fogelman JC, 1984. Extinction of montane populations of northern leopard frog (*Rana pipiens*) in Colorado. J Herpetol 18:147–152.
Corn PS, Vertucci FA, 1992. Descriptive risk assessment of the effects of acid deposition on Rocky Mountain amphibians. J Herpetol 26:361–369.

Cowles RB, Bogert CM, 1936. The herpetology of the Boulder Dam region (Nev., Ariz., Utah). Herpetologica 1:33–42.

Crawshaw GJ, 1992. The role of disease in amphibian decline. In: Declines in Canadian amphibian populations: designing a national monitoring strategy (Bishop CA, Pettit KE, eds). Occasional Paper 76. Ottawa: Canadian Wildlife Service; 60–62.

Crozier RH, 1986. Genetic clonal recognition abilities in marine invertebrates must be maintained by selection for something else. Evolution 40:1100–1101.

Crozier RH, 1987. Genetic aspects of kin recognition: concepts, models, and synthesis. In: Kin recognition in animals (Fletcher DJC, Michener CD, eds). Chichester, UK: John Wiley; 55–73.

Crump ML, Hensley FR, Clark KL, 1992. Apparent decline of the golden toad: underground or extinct? Copeia 1992:413–420.

Cunningham AA, 1996. Disease risks of wildlife translocations. Conserv Biol 10:349–353.

Cunningham AA, Langton TES, Bennett PM, Drury SEN, Gough RE, Kirkwood JK, 1993. Unusual mortality associated with poxvirus-like particles in frogs (*Rana temporaria*). Vet Rec 133:141–142.

Cunningham AA, Langton TES, Bennett PM, Lewin JF, Drury SEN, Gough RE, MacGregor SK, 1996. Pathological and microbiological findings from incidents of unusual mortality of the common frog (*Rana temporaria*). Phil Trans R Soc Lond B 351:1539–1557.

Czechura GV, Ingram GJ, 1990. *Taudactylus diurnus* and the case of the disappearing frogs. Mem Queensland Mus 29:361–365.

Delis PR, Mushinsky HR, McCoy ED, 1996. Decline of some west-central Florida anuran populations in response to habitat degradation. Biodiv Conserv 5:1579–1595.

deMaynadier PG, Hunter ML Jr, 1995. The relationship between forest management and amphibian ecology: a review of the North American literature. Environ Rev 3:230–261.

Devillers J, Exbrayat JM, 1992. Ecotoxicity of chemicals to amphibians. Philadelphia: Gordon and Breach.

Dias PC, 1996. Sources and sinks in population biology. Trends Ecol Evol 11:326–330.

Dodd CK Jr, 1993. Cost of living in an unpredictable environment: the ecology of striped newts *Notophthalmus perstriatus* during a prolonged drought. Copeia 1993:605–614.

Dodd CK Jr, 1994. The effects of drought on population structure, activity, and orientation of toads (*Bufo quercicus* and *B. terrestris*) at a temporary pond. Ethol Ecol Evol 6:331–349.

Dodd CK Jr, Seigel RA, 1991. Relocations, repatriation, and translocation of amphibians and reptiles: are they conservation strategies that work? Herpetologica 47:336–350.

Doherty PC, Zinkernagel RM, 1975. Enhanced immunological surveillance in mice heterozygous at the H-2 gene complex. Nature 256:50–52.

Donnelly MA, Guyer C, Juterbock JE, Alford RA, 1994. Techniques for marking amphibians. In: Measuring and monitoring biological diversity. Standard methods for amphibians (Heyer WR, Donnelly MA, McDiarmid RW, Hayek L-A C, Foster MS, eds). Washington, DC: Smithsonian Institution Press; 277–284.

Donnelly MA, Crump ML, 1994. Potential effects of climate change on two neotropical amphibian assemblages. Climatic Change, in press.

D'Orgeix CA, Turner BJ, 1995. Multiple paternity in the red-eyed treefrog *Agalychnis callidryas* (Cope). Mol Ecol 4:505–508.

Driscoll D, Wardell-Johnson G, Roberts JD, 1994. Genetic structuring and distribution patterns in rare southwestern Australian frogs: implications for translocation programmes. In: Reintroduction biology of Australian and New Zealand fauna (Serena M, ed). Chipping Norton, New South Wales: Surrey Beatty; 85–90.

Drost CA, Fellers GM, 1996. Collapse of a regional frog fauna in the Yosemite area of the California Sierra Nevada, USA. Conserv Biol 10:414–425.

Drury SEN, Gough RE, Cunningham AA, 1995. Isolation of an iridovirus-like agent from common frogs (*Rana temporaria*). Vet Rec 137:72–73.

Dumas PC, 1966. Studies of the *Rana* species complex in the Pacific Northwest. Copeia 1966:60–74.

Dunson WA, Wyman RL, Corbett ES, 1992. A symposium on amphibian declines and habitat acidification. J Herpetology 26:349–352.

Dupuis LA, Smith JNM, Bunnell F, 1995. Relation of terrestrial-breeding amphibian abundance to tree-stand age. Conserv Biol 9:645–653.

Dusi JL, 1949. The natural occurrence of "redleg," *Pseudomonas hydrophila,* in a population of American toads, *Bufo americanus.* Ohio J Sci 49:70–71.

Edenhamn P, 1996. Spatial dynamics of the European tree frog (*Hyla arborea* L.) in a heterogeneous landscape (doctoral thesis). Uppsala: Swedish University of Agricultural Sciences.

Edwards SV, Potts WK, 1996. Polymorphism of genes in the major histocompatibility complex (*Mhc*): implications for conservation genetics of vertebrates. In: Molecular approaches to conservation (Smith TB, Wayne RK, eds). Oxford: Oxford University Press; 214–237.

Endler JA, 1973. Gene flow and population differentiation. Science 179:243–250.

Endler JA, 1979. Gene flow and life history patterns. Genetics 93:263–284.

Ewald PW, 1994. Evolution of infectious disease. Oxford: Oxford University Press.

Excoffier L, Smouse PE, Quattro JM, 1992. Analysis of molecular variance inferred from metric distances among DNA haplotypes: application to human mitochondrial DNA restriction data. Genetics 131:479–491.

Falconer DS, 1989. Introduction to quantitative genetics, 3rd ed. Burnt Mill, Harlow, UK: Longman.

Fellers GM, Drost CA, 1993. Disappearance of the Cascades frog *Rana cascadae* at the southern end of its range, California, U.S.A. Biol Conserv 65:177–181.

Ferraris JD, Palumbi SR (eds), 1996. Molecular zoology: advances, strategies, and protocols. New York: Wiley-Liss.

Fisher RN, Shaffer HB, 1996. The decline of amphibians in California's Great Central Valley. Conserv Biol 10:1387–1397.

Flajnik MF, Du Pasquier L, 1990. The major histocompatibility complex of frogs. Immunol Rev 113:47–63.

Fog K, 1988. Reinvestigation of 1300 amphibian localities recorded in the 1940s. Mem Soc Fauna Flora Fenn 64:134–135.

Fog K, 1993. Oplaeg til forvaltningsplan for Danmarks padder og krybdyr [Management plan for Danish amphibians and reptiles]. Copenhagen: Skov-og Naturstyrelsen.

Frankham R, 1995a. Conservation genetics. Annu Rev Genet 29:305–327.

Frankham R, 1995b. Inbreeding and extinction: a threshold effect. Conserv Biol 9:792–799.

Franklin IR, 1980. Evolutionary changes in small populations. In: Conservation biology: an evolutionary-ecological perspective (Soulé ME, Wilcox BA, eds). Sunderland, Massachusetts: Sinauer Associates; 135–149.

Freda J, Dunson WA, 1986. Effects of low pH and other chemical variables on the local distribution of amphibians. Copeia 1986:454–466.

Gamradt SC, Kats LB, 1996. Effect of introduced crayfish and mosquitofish on California newts. Conserv Biol 10:1155–1162.

Gartside DF, 1982. The *Litoria ewingi* complex (Anura: Hylidae) in south-eastern Australia. VI. Geographic variation in transferrins of four taxa. Aust J Zool 30:103–113.

Gascon C, Planas D, 1986. Spring pond water chemistry and the reproduction of the wood frog *Rana sylvatica.* Can J Zool 64:543–550.

Gatz AJ Jr, 1981. Non-random mating by size in American toads, *Bufo americanus.* Anim Behav 29:1004–1012.

Gerhardt HC, 1994. The evolution of vocalizations in frogs and toads. Annu Rev Ecol Sys 25:293–324.

Gibbs EL, Nace GW, Emmons MB. 1971. The live frog is almost dead. Bioscience 21:1027–1034.

Gibbs JP, 1993. Importance of small wetlands for the persistence of local populations of wetland-associated animals. Wetlands 13:25–31.

Gill DE, 1978. Effective population size and interdemic migration rates in a metapopulation of the red-spotted newt, *Notophthalmus viridescens* (Rafinesque). Evolution 32:839–849.

Gilpin ME, Soulé ME, 1986. Minimum viable populations: processes of species extinction. In: Conservation biology: the science of scarcity and diversity (Soulé ME, ed). Sunderland, Massachusetts: Sinauer Associates; 19–34.

Glooschenko V, Weller WF, Smith PGR, Alvo R, Archbold JHG. 1992. Amphibian distribution with respect to pond water chemistry near Sudbury, Ontario. Can J Fish Aquat Sci 49(suppl)1:114–121.

Grant KP, Licht LE, 1993. Acid tolerance of anuran embryos and larvae from central Ontario. J Herpetol 27:1–6.

Grant KP, Licht LE, 1995. Effects of ultraviolet radiation on life-history stages of anurans from Ontario, Canada. Can J Zool 73:2292–2301.

Green DM, 1994. Genetic and cytogenetic diversity in Hochstetter's frog, *Leioplema hochstetteri,* and its importance for conservation management: NZ J Zool 21: 417–424.

Grosberg RK, 1988. The evolution of allorecognition specificity in clonal invertebrates. Q Rev Biol 63:377–412.

Hadrys H, Balick M, Schierwater B, 1992. Applications of random amplified polymorphic DNA (RAPD) in molecular ecology. Mol Ecol 1:55–63.

Hagström T, 1977. Grodornas försvinnande i en försurad sjö [The extinction of frogs in a lake acidified by atmospheric pollution]. Sveriges Nat 11:367–369.

Haig SM, Rhymer JM, Heckel DG, 1994. Population differentiation in randomly amplified polymorphic DNA of red-cockaded woodpeckers. *Picoides borealis.* Mol Ecol 3:581–595.

Hairston NG, 1983. Growth, survival and reproduction of *Plethodon jordani:* trade-offs between selective pressures. Copeia 1983:1024–1035.

Hairston NG Sr, 1987. Community ecology and salamander guilds. Cambridge: Cambridge University Press.

Hairston NG Sr, Wiley RH, 1993. No decline in salamander (Amphibia: Caudata) populations: a twenty-year study in the Southern Appalachians. Brimleyana 18:59–64.

Haldane JBC, 1949. Disease and evolution. Ricerc Sci (suppl)19:68–76.

Hall RJ, Henry PFP, 1992. Assessing effects of pesticides on amphibians and reptiles: status and needs. Herpetol J 2:65–71.

Hamilton WD, 1982. Pathogens as causes of genetic diversity in their host populations. In: Population biology of infectious disease agents (Anderson RM, May RM, eds). Berlin: Springer-Verlag; 269–296.

Hamilton WD, 1987. Kinship, recognition, disease, and intelligence: constraints of social evolution. In: Animal societies: theories and facts (Itô Y, Brown JL, Kikkawa J, eds). Tokyo: Japan Scientific Societies Press; 81–102.

Hammerson GA, 1982. Bullfrog eliminating leopard frogs in Colorado? Herpetol Rev 13:115–116.

Hanski I, Gilpin M, 1991. Metapopulation dynamics—brief history and conceptual domain. Biol J Linn Soc 42:3–16.

Hanski I, Pakkala T, Kuussaari M, Lei G, 1995a. Metapopulation persistence of an endangered butterfly in a fragmented landscape. Oikos 72:21–28.

Hanski I, Pöyry J, Pakkala T, Kuussaari M, 1995b. Multiple equilibria in metapopulation dynamics. Nature 377:618–621.

Harbuz MS, Lightman SL, 1992. Stress and the hypothalamo-pituitary-adrenal axis: acute, chronic and immunological activation. J Endocrinol 134:327–339.

Harte J, Hoffman E, 1989. Possible effects of acidic deposition on a Rocky Mountain population of the tiger salamander *Ambystoma tigrinum*. Conserv Biol 3:149–158.

Harte J, Hoffman E, 1994. Acidification and salamander recruitment. Bioscience 44:125–126.

Hayes JP, Steidl RJ, 1997. Statistical power analysis and amphibian population trends. Conserv Biol 11:273–275.

Hayes MP, Jennings MR, 1986. Decline of ranid frog species in western North America: are bullfrogs (*Rana catesbeiana*) responsible? J Herpetol 20:490–509.

Hayes MP, Jennings MR, 1990. Vanishing new mystery. Mainstream 21:20–23.

Hays JB, Blaustein AR, Kiesecker JM, Hoffman PD, Pandelova I, Coyle D, Richardson T, 1996. Developmental responses of amphibians to solar and artificial UV-B sources: a comparative study. Photochem Photobiol 64:449–456.

Hazen TC, Fliermans CB, Hirsch RP, Esch GW, 1978. Prevalence and distribution of *Aeromonas hydrophila* in the United States. Appl Environ Microbiol 36:731–738.

Hecnar SJ, 1995. Acute and chronic toxicity of ammonium nitrate fertilizer to amphibians from southern Ontario. Environ Toxicol Chem 14:2131–2137.

Hecnar SJ, M'Closkey RT, 1996. Regional dynamics and the status of amphibians. Ecology 77:2091–2097.

Hecnar SJ, M'Closkey RT, 1997a. Changes in the composition of a ranid frog community following bullfrog extinction. Am Midl Nat 137:145–150.

Hecnar SJ, M'Closkey RT, 1997b. The effects of predatory fish on amphibian species richness and distribution. Biol Conserv 79:123–131.

Hedges SB, 1993. Global amphibian declines: a perspective from the Caribbean. Biodivers Conserv 2:290–303.

Hedrick PW, 1992. Female choice and variation in the major histocompatibility complex. Genetics 132:575–581.

Hero J-M, Gillespie GR, 1997. Epidemic disease and amphibian declines in Australia. Conserv Biol 11:1023–1025.

Hess G, 1996. Disease in metapopulation models: implications for conservation. Ecology 77:1617–1632.

Heusser H, 1969. Die Lebensweise der Erdlkröte, *Bufo bufo* (L.); Das Orientierungsproblem. Rev Suisse Zool 76:443–518.

Heyer WR, Rand AS, da Cruz CAG, Peixoto OL, 1988. Decimations, extinctions, and colonizations of frog populations in southeast Brazil and their evolutionary implications. Biotropica 20:230–235.

Hill AVS, Allsopp CEM, Kwiatkowski D, Anstey NM, Twumasi P, Rowe PA, Bennett S, Brewster D, McMichael AJ, Greenwood BM, 1991. Common west African HLA antigens are associated with protection from severe malaria. Nature 352:595–600.

Hillis DM, Moritz C, Mable BK (eds), 1996. Molecular systematics, 2nd ed. Sunderland, Massachusetts: Sinauer Associates.

Hird DW, Diesch SL, McKinnell RG, Gorham E, Martin FB, Kurtz SW, Dubrovolny C, 1981. *Aeromonas hydrophila* in wild-caught frogs and tadpoles (*Rana pipiens*) in Minnesota. Lab Anim Sci 31:166–169.

Hitchings SP, Beebee TJC, 1996. Persistence of British natterjack toad *Bufo calamita* Laurenti (Anura: Bufonidae) populations despite low genetic diversity. Biol J Linn Soc 57:69–80.

Hoffmann AA, Parsons PA, 1991. Evolutionary genetics and environmental stress. Oxford: Oxford University Press.

Höglund J, Robertson JGM, 1988. Chorusing behaviour, a density-dependent alternative mating strategy in male common toads (*Bufo bufo*). Ethology 79:324–332.

Hollis GJ, 1995. Reassessment of the distribution, abundance and habitat of the Baw Baw frog *Philoria frosti*. Victorian Nat 112:190–201.

Holomuzki JR, 1982. Homing behavior of *Desmognathus ochrophaeus* along a stream. J Herpetol 16:307–309.

Horne MT, Dunson WA, 1994. Exclusion of the Jefferson salamander, *Ambystoma jeffersonianum*, from some potential breeding ponds in Pennsylvania: effects of pH, temperature, and metals on embryonic development. Arch Environ Contam Toxicol 27:323–330.

Horne MT, Dunson WA, 1995a. The interactive effects of low pH, toxic metals, and DOC on a simulated temporary pond community. Environ Pollut 89:155–161.

Horne MT, Dunson WA, 1995b. Toxicity of metals and low pH to embryos and larvae of the Jefferson salamander, *Ambystoma jeffersonianum*. Arch Environ Contam Toxicol 29:110–114.

Horton JD, 1994. Amphibians. In: Immunology: a comparative approach (Turner RJ, ed). Chichester, UK: John Wiley: 101–136.

Howard RD, 1988. Sexual selection on male body size and mating behaviour in American toads, *Bufo americanus*. Anim Behav 36:1796–1808.

Hunsaker D II, Potter FE Jr, 1960. "Red leg" in a natural population of amphibians. Herpetologica 16:285–286.

Husting EL, 1965. Survival and breeding structure in a population of *Ambystoma maculatum*. Copeia 1965:352–362.

Inger RF, Voris HK, Voris HH, 1974. Genetic variation and population ecology of some southeast Asian frogs of the genera *Bufo* and *Rana*. Biochem Genet 12:121–145.

Ingram GJ, McDonald KR, 1993. An update on the decline of Queensland's frogs. In: Herpetology in Australia: a diverse discipline (Lunney D, Ayers D, eds). Mosman: Royal Zoological Society of New South Wales; 297–303.

Ippen R, Zwart P, 1996. Infectious and parasitic diseases of captive reptiles and amphibians, with special emphasis on husbandry practices which prevent or promote diseases. Rev Sci Tech Office Int Epizoo 15:43–54.

Jaeger RG, 1980. Density-dependent and density-independent causes of extinction of a salamander population. Evolution 34:617–621.

Jameson DL, 1957. Population structure and homing responses in the Pacific tree frog. Copeia 1957:221–228.

Jarne P, Lagoda PJL, 1996. Microsatellites, from molecules to populations and back. Trends Ecol Evol 11:424–429.

Jeffreys AJ, 1987. Highly variable minisatellites and DNA fingerprints. Biochem Soc Transact 15:309–317.

Jennings MR, 1988. *Rana onca*. Cat Am Amphib Rept 417:1–2.

Jennings MR, Hayes MP, 1985. Pre-1900 overharvest of California red-legged frogs (*Rana aurora draytonii*): the inducement for bullfrog (*Rana catesbiana*) introduction. Herpetologica 41:94–103.

Jung RE, Jagoe CH, 1995. Effects of low pH and aluminum on body size, swimming performance, and susceptibility to predation of green tree frog (*Hyla cinerea*) tadpoles. Can J Zool 73:2171–2183.

Kagarise Sherman C, Morton ML, 1993. Population declines of Yosemite toads in the eastern Sierra Nevada of Califonria. J Herpetol 27:186–198.

Kattan G, 1993. The effects of forest fragmentation on frogs and birds in the Andes of Columbia: implications for watershed management. In: Forest remnants in the tropical landscape: benefits and policy implications (Doyle JK, Schelhas J, eds). Washington, DC: Smithsonian Institution Press; 11–13.

Kaufman J, Völk H, Wallny H-J, 1995. A "minimal essential Mhc" and an "unrecognized Mhc": two extremes in selection for polymorphism. Immunol Rev 143:63–88.

Keller LF, Arcese P, Smith JNM, Hochachka WM, Stearns SC, 1994. Selection against inbred song sparrows during a natural population bottleneck. Nature 372:356–357.

Kerr JB, McElroy CT, 1993. Evidence for large upward trends of ultraviolet-B radiation linked to ozone depletion. Science 262:1032–1034.

Kiesecker J, 1996. pH-mediated predator-prey interactions between *Ambystoma tigrinum* and *Pseudacris triseriata*. Ecol Appl 6:1325–1331.

Kiesecker JM, Blaustein AR, 1995. Synergism between UV-B radiation and a pathogen magnifies amphibian embryo mortality in nature. Proc Nat Acad Sci USA 92:11049–11052.

Kiesecker JM, Blaustein AR, 1997. Influences of egg laying behavior on pathogenic infection of amphibian eggs. Conserv Biol 11:214–220.

Kiester AR, 1985. Sex-specific dynamics of aggregation and dispersal in reptiles and amphibians. Contrib Mar Sci 27(suppl):425–434.

Kimberling DN, Ferreira AR, Shuster SM, Keim P, 1996. RAPD marker estimation of genetic structure among isolated northern leopard frog populations in the south-western USA. Mol Ecol 5:521–529.

Kirk JJ, 1988. Western spotted frog (*Rana pretiosa*) morality following forest spraying of DDT. Herpetol Rev 19:51–53.

Kleeberger SR, Werner JK, 1982. Home range and homing behavior of *Plethodon cinereus* in northern Michigan. Copeia 1982:409–415.

Klein J, O'Huigin C, 1994. MHC polymorphism and parasites. Phil Trans R Soc Lond B 346:351–358.

Koonz W, 1992. Amphibians in Manitoba. In: Declines in Canadian amphibian populations: designing a national monitoring strategy (Bishop CA, Pettit KE, eds). Occasional Paper 76, Ottawa: Canadian Wildlife Service; 19–20.

Kuzmin SL, 1994. The problem of declining amphibian populations in the Commonwealth of Independent States and adjacent territories. Alytes 12:123–134.

Kuzmin SL, 1996. Threatened amphibians in the former Soviet Union: the current situation and the main threats. Oryx 30:24–30.

Laan R, Verboom B, 1990. Effects of pool size and isolation on amphibian communities. Biol Conserv 54:251–262.

Lacy RC, 1987. Loss of genetic diversity from managed populations: interacting effects of drift, mutation, immigration, selection, and population subdivision. Conserv Biol 1:143–159.

La Marca E, Reinthaler HP, 1991. Population changes in *Atelopus* species of the Cordillera de Mérida, Venezuela. Herpetol Rev 22:125–128.

Lambert DM, Millar CD, 1995. DNA science and conservation. Pacific Conserv Biol 2:21–38.

Lande R, 1993. Risks of population extinction from demographic and environmental stochasticity and random catastrophes. Am Nat 142:911–927.

Lande R, 1994. Risk of population extinction from fixation of new deleterious mutations. Evolution 48:1460–1469.

Lande R, 1995. Mutation and conservation. Conserv Biol 9:782–791.

Lande R, Barrowclough GF, 1987. Effective population size, genetic variation, and their use in population management. In: Viable populations for conservation (Soulé ME, ed). Cambridge: Cambridge University Press; 87–123.

Lanoo MJ, Lang K, Waltz T, Phillips GS, 1994. An altered amphibian assemblage: Dickinson County, Iowa, 70 years after Frank Blanchard's survey. Am Mid Nat 131:311–319.

Larson A, Wake DB, Yanev KP, 1984. Measuring gene flow among populations having high levels of genetic fragmentation. Genetics 106:293–308.

Laurance WF, 1996. Catastrophic declines of Australian rainforest frogs: is unusual weather responsible? Biol Conserv 77:203–212.

Laurance WF, McDonald KR, Speare R, 1996. Epidemic disease and the catastrophic decline of Australian rain forest frogs. Conserv Biol 10:406–413.

Laurance WF, McDonald KR, Speare R, 1977. In defense of the epidemic disease hypothesis. Conserv Biol 11:1030–1034.

Leary RF, Allendorf FW, 1989. Fluctuating asymmetry as an indicator of stress: implications for conservation biology. Trends Ecol Evol 4:214–217.

Lefcort H, Hancock KA, Maur KM, Rostal DC, 1997. The effects of used motor oil, silt, and the water mold *Saprolegnia parasitica* on the growth and survival of mole salamanders (genus *Ambystoma*). Arch Environ Contam Toxicol 32:383–388.

Lessa EP, Applebaum G, 1993. Screening techniques for detecting allelic variation in DNA sequences. Mol Ecol 2:119–129.

Lewis DL, Baxter GT, Johnson KM, Stone MD, 1985. Possible extinction of the Wyoming toad, *Bufo hemiophrys baxteri.* J Herpetol 199:166–168.

Licht LE, 1976. Sexual selection in toads (*Bufo americanus*). Can J Zool 54:1277–1284.

Licht LE, 1995. Disappearing amphibians? Bioscience 45:307.

Licht LE, 1996. Amphibian decline still a puzzle. Bioscience 46:172–173.

Licht LE, Grant KP, 1997. The effects of ultraviolet radiation on the biology of amphibians. Amer Zool 37:137–145.

Licht P, McCreery BR, Barnes R, Pang R, 1983. Seasonal and stress related changes in plasma gonadotropins, sex steroids, and corticosterone in the bullfrog, *Rana catesbeiana.* Gen Comp Endocrinol 50:124–145.

Lips K, 1998. Decline of a tropical montane amphibian fauna. Conserv Biol, 12:106–117.

Lively CM, Apanius V, 1995. Genetic diversity in host-parasite interactions. In: Ecology of infectious diseases in natural populations (Grenfell BT, Dobson AP, eds). Cambridge: Cambridge University Press; 421–449.

Long LE, Saylor LS, Soulé ME, 1995. A pH/UV-B synergism in amphibians. Conserv Biol 9:1301–1303.

Longstreth JD, de Gruijl FR, Kripke ML, Takizawa Y, van der Leun JC, 1995. Effects of increased solar ultraviolet radiation on human health. Ambio 24:153–165.

Lynch M, Conery J, Bürger R, 1995. Mutational accumulation and the extinction of small populations. Am Nat 146:489–518.

Lynch M, Gabriel W, 1990. Mutation load and the survival of small populations. Evolution 44:1725–1737.

Lynch M, Milligan BG, 1994. Analysis of population genetic structure with RAPD markers. Mol Ecol 3:91–99.

Madronich S, de Gruij FR, 1993. Skin cancer and UV radiation. Nature 366:23.

Madronich S, McKenzie RL, Caldwell MM, Björn LO, 1995. Changes in ultraviolet radiation reaching the earth's surface. Ambio 24:143–152.

Madsen T, Stille B, Shine R, 1996. Inbreeding depression in an isolated population of adders *Vipera berus*. Biol Conserv 75:113–118.

Mahaney PA, 1994. Effects of freshwater petroleum contamination on amphibian hatching and metamorphosis. Environ Toxicol Chem 13:259–265.

Mahony M, Dennis A, 1994. The cause of local extinction and population decline in day frogs (genus *Taudactylus*) of the wet tropics area. Reproductive biology and recruitment in the sharp-snouted day frog (*T. acutirostris*). Newcastle, New South Wales: Department of Biological Sciences, University of Newcastle.

Mann W, Dorn P, Brandl R, 1991. Local distribution of amphibians: the importance of habitat fragmentation. Global Ecol Biogeogr Lett 1:36–41.

Marquez R, Olmo JL, Bosch J. 1995. Recurrent mass mortality of larval midwife toads *Alytes obstetricans* in a lake in the Pyrenean mountains. Herpetol J 5:287–289.

Masters BS, 1995. The use of RAPD markers for species identification in desmognathine salamanders. Herpetol Rev 26:92–95.

McAlpine S, 1993. Genetic heterozygosity and reproductive success in the green treefrog, *Hyla cinerea*. Heredity 70:553–558.

McAlpine S, Smith MH, 1995. Genetic correlates of fitness in the green treefrog, *Hyla cinerea*. Herpetologica 51:393–400.

McCauley DE, 1993. Genetic consequences of extinction and recolonization in fragmented habitats. In: Biotic interactions and global change (Kareiva PM, Kingsolver JG, Huey RB, eds). Sunderland, Massachusetts: Sinauer Associates; 217–233.

McKenzie RL, et al. 1995. Surface ultraviolet radiation. In: Scientific assessment of ozone depletion: 1994. Global ozone research and monitoring project, report 37. Geneva: World Meterological Organization; 9.1–9.22.

McKenzie RL, Bodeker GE, Keep DJ, Kotkamp M, Evans J. 1996. UV radiation in New Zealand: north-to-south differences between two sites, and relationship to other latitudes. Weather Climate 16:17–27.

Means DB, Palis JG, Bagget M, 1996. Effects of slash pine silviculture on a Florida population of flatwoods salamander. Conserv Biol 10:426–437.

Micheli MR, Bova R (eds), 1997. Fingerprinting methods based on arbitrarily primed PCR. Berlin: Springer-Verlag.

Mitton JB, Carey C, Kocher TD, 1986. The relation of enzyme heterozygosity to standard and active oxygen consumption and body size of tiger salamanders, *Ambystoma tigrinum*. Physiol Zool 59:574–582.

Moore FL, Miller LJ, 1984. Stress-induced inhibition of sexual behavior: corticosterone inhibits courtship behaviors of a male amphibian (*Taricha granulosa*). Horm Behav 18:400–410.

Moore FL, Zoeller RT, 1985. Stress-induced inhibition of reproduction: evidence of suppressed secretion of LH-RH in an amphibian. Gen Comp Endocrinol 60:252–258.

Moritz C, 1994. Defining 'evolutionary significant units' for conservation. Trends Ecol Evol 9:373–375.

Morgan LA, Buttemer WA, 1996. Predation by the non-native fish *Gambusia holbrooki* on small *Litoria aurea* and *L. dentata* tadpoles. Aust Zool 30:143–149.

Moyle PB, 1973. Effects of introduced bullfrogs, *Rana catesbeiana*, on the native frogs of the San Joaquin Valley, California. Copeia 1973:18–22.

Moynihan JA, Cohen N, Ader R, 1994. Stress and immunity. In: Neuropeptides and immunoregulation (Scharrer B, Smith EM, Stefano GB, eds). Berlin: Springer-Verlag; 120–138.

Nagl AM, Hofer R, 1997. Effects of ultraviolet radiation on early larval stages of the Alpine newt, *Triturus alpestris,* under natural and laboratory conditions. Oecologia. 110: 514–519.

Nauta MJ, Weissing FJ, 1996. Constraints on allele size at microsatellite loci: implications for genetic differentiation. Genetics 143:1021–1032.

Naylor AF, 1962. Mating systems which could increase heterozygosity for a pair of alleles. Am Nat 96:51–60.

Nichols DK, Smith AJ, Gardiner CH, 1996. Dermatitis of anurans caused by fungal-like protists. Proc Am Assoc Zoo Vet 1996:220–221.

van Noordwijk AJ, 1994. The interaction of inbreeding depression and environmental stochasticity in the risk of extinction of small populations. In: Conservation genetics (Loeschcke V, Tomiuk J, Jain SK, eds). Basel: Birkhauser Verlag; 132–146.

Norman JA, Moritz C, Limpus CJ, 1994. Mitochondrial DNA control region polymorphisms: genetic markers for ecological studies of marine turtles. Mol Ecol 3:363–373.

Nunney L, Campbell KA, 1993. Assessing minimum viable population size: demography meets population genetics. Trends Ecol Evol 8:234–239.

Nurnberger B, Harrison RG, 1995. Spatial population structure in the whirligig beetle *Dineutus assimilis:* evolutionary inferences based on mitochondrial DNA and field data. Evolution 49:266–275.

Nyman S, 1986. Mass mortality in larval *Rana sylvatica* attributable to the bacterium, *Aeromonas hydrophila.* J Herpetol 20:196–201.

O'Brien SJ, Evermann JF, 1988. Interactive influence of infectious disease and genetic diversity in natural populations. Trends Ecol Evol 3:254–259.

Oldham MJ, 1992. Declines in Blanchard's cricket frog in Ontario. In: Declines in Canadian amphibian populations:designing a national monitoring strategy (Bishop CA, Pettit KE, eds). Occasional Paper 76. Ottawa: Canadian Wildlife Service; 30–31.

Oldham RS, Latham DM, Hilton-Brown D, Towns M, Cooke AS, Burn A, 1997. The effect of ammonium nitrate fertiliser on frog (*Rana temporaria*) survival. Agric Ecosyst Environ 61:69–74.

Olson ME, Gard S, Brown M, Hampton R, Morck DW, 1992. *Flavobacterium indologenes* infection in leopard frogs. J Am Vet Med Assoc 201:1766–1770.

Orchard SA, 1992. Amphibian population declines in British Columbia. In: Declines in Canadian amphibian populations: designing a national monitoring strategy (Bishop CA, Pettit KE, eds). Occasional Paper 76. Ottawa: Canadian Wildlife Service; 10–13.

Osborne WS, 1989. Distribution, relative abundance and conservation status of Corroboree frogs, *Pseudophryne corroboree* Moore (Anura: Myobatrachidae). Aust Wildl Res 16:537–547.

Osborne WS, Littlejohn MJ, Thomson SA, 1996. Former distribution and apparent disappearance of the *Litoria aurea* complex from the Southern Tablelands of New South Wales and the Australian Capital Territory. Aus Zool 30:190–198.

Ouellet M, Bonin J, Rodrigue J, Desgranges JL, Lair S, 1997. Hindlimb deformities (Ectromelia, Ectrodactyly) in free-living anurans from agricultural habitats. J Wildl Dis 33:95–104.

Ovaska K, Davis TM, Novales Flamarique I, 1997. Hatching success and larval survival of the frogs *Hyla regilla* and *Rana aurora* under ambient and artificially enhanced solar ultraviolet radiation. Can J Zool 75:1081–1088.

Palmer AR, Strobeck C, 1986. Fluctuating asymmetry: measurement, analysis, patterns. Ann Rev Ecol Syst 17:391–421.

Palumbo SA, 1993. The occurrence and significance of organisms of the *Aeromonas hydrophila* group in food and water. Med Microbiol Lett 2:339–346.

Paolucci M, Esposito V, Di Fiore MM, Botte V, 1990. Effects of short postcapture confinement on plasma reproductive hormone and corticosterone profiles in *Rana esculenta* during the sexual cycle. Boll Zool 57:253–259.

Pechmann JHK, Wilbur HM, 1994. Putting declining amphibian populations in perspective: natural fluctuations and human impacts. Herpetologica 50:65–84.

Pechmann JHK, Scott DE, Semlitsch RD, Caldwell JP, Vitt LJ, Gibbons JW, 1991. Declining amphibian populations: the problem of separating human impacts from natural fluctuations. Science 253:892–895.

Pena SDJ, 1995. Pitfalls of paternity testing based solely on PCR typing of minisatellites and microsatellites. Am J Hum Genet 56:1503–1504.

Petranka JW, Brannon MP, Hopey ME, Smith CK, 1994. Effects of timber harvesting on low elevation populations of southern Appalachian salamanders. Forest Ecol Manag 67:135–147.

Petranka JW, Eldridge ME, Haley KE, 1993. Effects of timber harvesting on southern Appalachian salamanders. Conserv Biol 7:363–370.

Phillips K, 1994. Tracking the vanishing frogs: an ecological mystery. New York: St. Martin's Press.

Pierce BA, 1985. Acid tolerance in amphibians. Bioscience 35:239–243.

Pierce BA, Harvey JM, 1987. Geographic variation in acid tolerance of Connecticut wood frogs. Copeia 1987:94–103.

Plytycz B, 1984. Differential polymorphism of the amphibian MHC. Dev Comp Immunol 8:727–732.

Porter KR, Hakanson DE, 1976. Toxicity of mine drainage to embryonic and larval boreal toads (Bufonidae: *Bufo boreas*). Copeia 1976:327–331.

Portnoy JW, 1990. Breeding biology of the spotted salamander *Ambystoma maculatum* (Shaw) in acidic temporary ponds at Cape Cod, USA. Biol Conserv 53:61–75.

Potts WK, Manning CJ, Wakeland EK, 1994. The role of infectious disease, inbreeding and mating preferences in maintaining MHC genetic diversity: an experimental test. Phil Trans R Soc Lond B 346:369–378.

Potts WL, Slev PR, 1995. Pathogen-based models favoring MHC genetic diversity. Immunol Rev 143:181–197.

Potts WK, Wakeland EK, 1993. Evolution of MHC genetic diversity: a tale of incest, pestilence and sexual preference. Trends Genet 9:408–412.

Pough FH, 1976. Acid precipitation and embryonic mortality of spotted salamanders, *Ambystoma maculatum*. Science 192:68–70.

Pough FH, Wilson RE, 1977. Acid precipitation and reproductive success of *Ambystoma* salamanders. Water Air Soil Pollut 7:531–544.

Pounds JA, Crump ML, 1994. Amphibian declines and climate disturbance: the case of the golden toad and the harlequin frog. Conserv Biol 8:72–85.

Power T, Clark KL, Harfenist A, Peakall DB, 1989. A review and evaluation of the amphibian toxicological literature. Technical report 61. Ottawa: Canadian Wildlife Service.

Poynton JC, 1996. Diversity and conservation of African bufonids (Anura): some preliminary findings. African J Herpetol 45:1–7.

Pusey A, Wolf M, 1996. Inbreeding avoidance in animals. Trends Ecol Evol 11:201–206.

Pyke GH, White AW, 1996. Habitat requirements for the green and golden bell frog *Litoria aurea* (Anura: Hylidae) Aust Zool 30:224–232.

Queller DC, Strassman JE, Hughes CR, 1983. Microsatellites and kinship. Trends Ecol Evol 8:285–288.

Raymond LR, Hardy LM, 1991. Effects of a clearcut on a population of the mole salamander *Ambystoma talpoideum,* in an adjacent unaltered forest. J Herpetol 25:509–512.

Reading CJ, Loman J, Madsen T, 1991. Breeding pond fidelity in the common toad, *Bufo bufo.* J Zool 225:201–211.

Rebuffoni D, 1995. Those strange frogs: mutants—or what? Minneapolis Star Tribune, 25 November, B3.

Reed CF, 1957. Contributions to the herpetofauna of Virginia, 2: the reptiles and amphibians of Northern Neck. J Wash Acad Sci 47:21–23.

Reed JM, Blaustein AR, 1995. Assessment of "nondeclining" amphibian populations using power analysis. Conserv Biol 9:1299–1300.

Reh W, Seitz A, 1990. The influence of land use on the genetic structure of populations of the common frog *Rana temporaria.* Biol Conserv 54:239–249.

Reimann J, Miller RG, 1983. Polymorphism and MHC gene function. Dev Comp Immunol 7:403–412.

Reinert HK, 1991. Translocation as a conservation strategy for amphibians and reptiles: some comments, concerns, and observations. Herpetologica 47:357–363.

Reiss RA, Schwert DP, Ashworth AC, 1995. Field preservation of Coleoptera for molecular genetic analyses. Environ Entomol 24:716–719.

Richards SJ, McDonald KR, Alford RA, 1993. Declines in populations of Australia's endemic tropical rainforest frogs. Pacific Conserv Biol 1:66–77.

Rigney MM, Zilinksy JW, Rouf MA, 1978. Pathogenicity of *Aeromonas hydrophila* in red leg disease in frogs. Curr Microbiol 1:175–179.

Roberts W, 1992. Declines in amphibian populations in Alberta. In: Declines in Canadian amphibian populations: designing a national monitoring strategy (Bishop CA, Pettit KE, eds). Occasional Paper 76. Ottawa: Canadian Wildlife Service, 14–16.

Roush W, 1995. When rigor meets reality. Science 269:313–315.

Roux KH, Volpe EP, 1975. Evidence for a major histocompatibility complex in the leopard frog. Immunogenetics 2:577–589.

Rowe CL, Sadinksi WJ, Dunson WA, 1992. Effects of acute and chronic acidification on three larval amphibians that breed in temporary pools. Arch Environ Contam Toxicol 23:399–350.

Rugh R, 1962. Experimental embryology, 3rd ed. Minneapolis, Minnesota: Burgess.

Russell RW, Hecnar SJ, Haffner GD, 1995. Organochlorine pesticide residues in southern Ontario spring peepers. Environ Toxicol Chem 14:815–817.

Ryan MJ, 1991. Sexual selection and communication in frogs. Trends Ecol Evol 6:351–355.

Ryan MJ, Perrill SA, Wilczynski W, 1992. Auditory tuning and call frequency predict population-based mating preferences in the cricket frog, *Acris crepitans.* Am Nat 139:1370–1383.

Ryan MJ, Wilczynski W, 1988. Coevolution of sender and receiver: effect on local mate preference in cricket frogs. Science 240:1786–1788.

Ryan MJ, Wilczynski W, 1991. Evolution of intraspecific variation in the advertisement call of a cricket frog (*Acris crepitans,* Hylidae). Biol J Linn Soc 44:249–271.

Saad AH, 1988. Corticosteroids and immune systems of non-mammalian vertebrates: a review. Dev Comp Immunol 12:481–494.

Saad AH, Ali W, 1992. Effects of season, sex and temperature on the immune responses of the Reuss's toad, *Bufo regularis*. J Egypt-German Soc Zool 7A:197–208.

Saad AH, Flytycz B, 1994. Hormonal and nervous regulation of amphibian and reptilian immunity. Folia Biol 42:63–78.

Samallow BP, Soulé ME, 1983. A case of stress related heterozygote superiority in nature. Evolution 37:646–649.

Sarre S, 1995. Mitochondrial DNA variation among populations of *Oedura reticulata* (Gekkonidae) in remnant vegetation: implications for metapopulation structure and population decline. Mol Ecol 4:395–405.

Sarre S, 1996. Habitat fragmentation promotes fluctuating asymmetry but not morphological divergence in two geckos. Res Popul Ecol (Kyoto) 38:57–64.

Sarre S, Dearn JM, Georges A, 1994. The application of fluctuating asymmetry in the monitoring of animal populations. Pacific Conserv Biol 1:118–122.

Sato K, Flajnik MF, Du Pasquier L, Katagiri M, Kasahara M, 1993. Evolution of the MHC: isolation of class II β-chain cDNA clones from the amphibian *Xenopus laevis*. J Immunol 150:2831–2843.

Schmid-Hempel P, 1994. Infection and colony variability in social insects. Phil Trans R Soc Lond B 346:313–321.

Schmitt J, Antonovics J, 1986. Experimental studies of the evolutionary significance of sexual reproduction. IV. Effect of neighbor relatedness and aphid infestation on seedling performance. Evolution 40:830–836.

Schwalbe CR, Rosen PC, 1988. Preliminary report on effect of bullfrogs on wetland herpetofaunas in southeastern Arizona. General Technical Report RM 166. Fort Collins, Colorado: US Forest Service; 166–173.

Scott NJ, 1993. Postmetamorphic death syndrome. Froglog 7:1–2.

Scribner KT, Arntzen JW, Burke T, 1994. Comparative analysis of intra- and interpopulation genetic diversity in *Bufo bufo*, using allozyme, single-locus microsatellite, minisatellite, and multilocus minisatellite data. Mol Biol Evol 11:737–748.

Scribner KT, Arntzen JW, Burke T, 1997. Effective number of breeding adults in *Bufo bufo* estimated from age-specific variation at minisatellite loci. Mol Ecol 6:701–712.

Semb-Johansson A, 1992. Declining populations of the common toad (*Bufo bufo* L.) on two islands in Oslofjord, Norway. Amphibia-Reptilia 13:409–412.

Semlitsch RD, Scott DE, Pechmann JHK, Gibbons JW, 1996. Structure and dynamics of an amphibian community. Evidence from a 16-year study of a natural pond. In: Long-term studies of verterbrate communities (Cody ML, Smallwood JA, eds). San Diego: Academic Press; 217–248.

Sessions SK, Ruth SB, 1990. Explanation for naturally occurring supernumerary limbs in amphibians. J Exp Zool 254:38–47.

Sexton OJ, Phillips C, 1986. A qualitative study of fish-amphibian interactions in 3 Missouri ponds. Trans Ms Acad Sci 20:25–35.

Seymour RS, 1994. Oxygen diffusion through the jelly capsules of amphibian eggs. Israel J Zool 40:493–506.

Sinsch U, 1990. Migration and orientation in anuran amphibians. Ethol Ecol Evol 2: 65–79.

Sjögren P, 1991a. Extinction and isolation gradients in metapopulations: the case of the pool frog (*Rana lessonae*). Biol J Linn Soc 42:135–147.

Sjögren P, 1991b. Genetic variation in relation to demography of peripheral pool frog populations (*Rana lessonae*). Evol Ecol 5:248–271.

Sjögren Gulve P, 1994. Distribution and extinction patterns within a northern metapopulation of the pool frog, *Rana lessonae*. Ecology 75:1357–1367.

Snyder WF, Jameson DL, 1965. Multivariate geographic variation of mating call in populations of the Pacific tree frog (*Hyla regilla*). Copeia 1965:129–142.

Soulé ME, 1980. Thresholds for survival: maintaining fitness and evolutionary potential. In: Conservation biology: an evolutionary-ecological perspective (Soulé ME, Wilcox BA, eds). Sunderland, Massachusetts: Sinauer Associates; 151–170.

Speare R, 1995. Preliminary study on diseases in Australian wet tropics amphibians: deaths of rainforest frogs at O'Keefe Creek, Big Tableland. Final report. Brisbane: Queensland Department of Environment and Heritage.

Speare R, Smith RJ, 1992. An iridovirus-like agent isolated from the ornate burrowing frog *Limnodynastes ornatus* in northern Australia. Dis Aquat Org 14:51–57.

Stebbins RC, Cohen NW, 1995. A natural history of amphibians. Princeton, New Jersey: Princeton University Press.

Stewart MM, 1995. Climate driven population fluctuations in rain forest frogs. J Herpetol 29:437–446.

Stienstra T, 1995. Are trout devouring polliwogs? San Francisco Sunday Examiner and Chronicle, 2 July; D-12.

Stockwell CA, Mulvey M, Vinyard GL, 1996. Translocations and the preservation of allelic diversity. Conserv Biol 10:1133–1141.

Strathmann RR, Chaffee C, 1984. Constraints on egg masses. II. Effect of spacing, size, and number of eggs on ventilation of masses of embryos in jelly, adherent groups, or thin-walled capsules. J Exp Mar Biol Ecol 84:85–93.

Sullivan BK, 1992. Sexual selection and calling behavior in the American toad (*Bufo americanus*). Copeia 1992:1–7.

Taylor SK, Williams ES, Mills KW, Boerger-Fields AM, Lynn CJ, Hearne CE, Thome ET, Pistono SL, 1995. A review of causes of mortality and the diagnostic investigation for pathogens of the Wyoming toad (*Bufo hemiophrys baxteri*). Cheyenne, Wyoming: US Fish and Wildlife Service.

Templeton AR, Shaw K, Routman E, Davis SK, 1990. The genetic consequences of habitat fragmentation. Ann M Bot Gardens 77:13–27.

Tocher M, 1996. The effect of deforestation and forest fragmentation on a central Amazonian frog community (PhD thesis). Christchurch: University of Canterbury.

Travis J, 1994. Calibrating our expectations in studying amphibian populations. Herpetologica 50:104–108.

Trenerry MP, Laurance WF, McDonald KR, 1994. Further evidence for the precipitous decline of endemic rainforest frogs in tropical Australia. Pacific Conserv Biol 1:150–153.

Twitty VC, 1966. Of scientists and salamanders. San Francisco: WH Freeman.

Twitty VC, Grant D, Anderson O, 1964. Long distance homing in the next *Taricha rivularis*. Proc Nat Acad Sci USA 51:51–58.

Tyler MJ, 1991. Declining amphibian populations—a global phenomena? An Australian perspective. Alytes 99:43–50.

Tyler MJ, 1994. Australian frogs: a natural history, revised edition. Chatswood, New South Wales: Reed.

van de Mortel TF, Buttemer WA, 1996. Are *Litoria aurea* eggs more sensitive to ultraviolet-B radiation than eggs of sympatric *L. peronii* or *L. dentata*? Aust Zool 30:150–157.

Vardia HK, Sambasiva Rao P, Durve VS, 1984. Sensitivity of toad larvae to 2,4-D and endosulfan pesticides. Arch Hydrobiol 100:395–400.

Vertucci F, Corn S, 1994. Acidification and salamander recruitment (reply). Bioscience 44:126–127.

Vertucci FA, Corn PS, 1996. Evaluation of episodic acidification and amphibian declines in the Rocky Mountains. Ecol Appl 6:449–457.

Voris HK, Inger RF, 1995. Frog abundance along streams in Bornean forests. Conserv Biol 9:679–683.

Vrijenhoek RC, 1994. Genetic diversity and fitness in small populations. In: Conservation genetics (Loeschcke V, Tomiuk J, Jain SK, eds). Basel: Birkhauser Verlag; 37–53.

Wake DB, 1991. Declining amphibian populations. Science 253:860.

Waldman B, 1987. Mechanisms of kin recognition. J Theor Biol 128:159–185.

Waldman B, 1988. The ecology of kin recognition. Ann Rev Ecol Syst 19:543–571.

Waldman B, 1991. Kin recognition in amphibians. In: Kin recognition (Hepper PG, ed). Cambridge: Cambridge University Press; 162–219.

Waldman B, 1996. Frogs fight on. NZ Sci Month 7(5):2.

Waldman B, 1997. Kinship, sexual selection, and female choice in toads. Adv Ethol 32:200.

Waldman B, Frumhoff PC, Sherman PW, 1988. Problems of kin recognition. Trends Ecol Evol 3:8–13.

Waldman B, McKinnon JS, 1993. Inbreeding and outbreeding in fishes, amphibians, and reptiles. In: The natural history of inbreeding and outbreeding: theoretical and empirical perspectives (Thornhill NW, ed). Chicago: University of Chicago Press; 250–282.

Waldman B, Rice JE, Honeycutt RL, 1992. Kin recognition and incest avoidance in toads. Am Zool 32:18–30.

Warner SC, Travis J, Dunson WA, 1993. Effect of pH variation on interspecific competition between two species of hylid tadpoles. Ecology 74:183–194.

Watt PJ, Oldham RS, 1995. The effect of ammonium nitrate on the feeding and development of larvae of the smooth newt, *Triturus vulgaris* (L.), and on the behaviour of its food source, *Daphnia*. Freshwat Biol 33:319–324.

Wedekind C, Seebeck T, Bettens F, Paepke AJ, 1995. MHC-dependent mate preferences in humans. Proc R Soc Lond B 260:245–249.

Weising K, Atkinson RG, Gardner RC, 1995. Genomic fingerprinting by microsatellite-primed PCR: a critical evaluation. PCR Meth Appl 4:249–255.

Weitzel NH, Panik HR, 1993. Long-term fluctuations of an isolated population of the Pacific chorus frog (*Pseudacris regilla*) in northwestern Nevada. Great Basin Nat 53: 379–384.

Wetton JH, Carter RE, Parkin DT, Walters D, 1987. Demographic study of a wild house sparrow population by DNA fingerprinting. Nature 327:147–149.

Weygoldt P, 1989. Changes in the composition of mountain stream frog communities in the Atlantic mountains of Brazil: frogs as indicators of environmental deteriorations? Studies Neotrop Fauna Environ 243:249–255.

White AW, Pyke GH, 1996. Distribution and conservation status of the green and golden bell frog *Litoria aurea* in New South Wales. Aust Zool 30:177–189.

Whiteman HH, Howard RD, Whitten KA, 1995. Effects of pH on embryo tolerance and adult behavior in the tiger salamander, *Ambystoma tigrinum tigrinum*. Can J Zool 73:1529–1537.

Wilbur HM, 1980. Complex life cycles. Ann Rev Ecol Sys 11:67–93.

Winstead JT, Couch JA, 1988. Enhancement of protozoan pathogen *Perkinsus marinus* infections in American oysters *Crassostrea virginica* exposed to the chemical carcinogen n-nitrosodiethylamine (DENA). Dis Aquat Org 5:205–213.

Wissinger SA, Whiteman HH, 1992. Fluctuation in a Rocky Mountain population of salamanders: anthropogenic acidification or natural variation? J Herpetol 26:377–391.

Woolbright LL, 1991. The impact of hurricane Hugo on forest frogs in Puerto Rico. Biotropica 23:462–467.

Woolbright LL, 1996. Disturbance influences long-term population patterns in the Puerto

Rican frog, *Eleutherodactylus coqui* (Anura: Leptodactylidae). Biotropica 28: 493–501.

Worrest RC, Kimeldorf DJ, 1975. Photoreactivation of potentially lethal, UV-induced damage to boreal toad (*Bufo boreas boreas*) tadpoles. Life Sci 17:1545–1550.

Worrest RC, Kimeldorf DJ, 1976. Distortions in amphibian development induced by ultraviolet-B enhancement (290–315 nm) of a simulated solar spectrum. Photochem Photobiol 24:377–382.

Worthylake KM, Hovingh P, 1989. Mass mortality of salamanders (*Ambystoma tigrinum*) by bacteria (*Acinetobacter*) in an oligotrophic seepage mountain lake. Great Basin Nat 49:364–372.

Wright S, 1977. Evolution and the genetics of populations, vol 3. Experimental results and evolutionary deductions. Chicago: University of Chicago Press.

Wyman RL, 1990. What's happening to the amphibians? Conserv Biol 4:350–352.

Wyman RL, Jancola J, 1992. Degree and scale of terrestrial acidification and amphibian community structure. J Herpetol 4:392–401.

Yang Y-J, Lin Y-S, Wu J-L, Hui C-F, 1994. Variation in mitochondrial DNA and population structure of the Taipei treefrog *Rhacophorus taipeianus* in Taiwan. *Mol Ecol* 3:219–228.

Zerani M, Amabili F, Mosconi G, Gobbetti A, 1991. Effects of captivity stress on plasma steroid levels in the green frog, *Rana esculenta,* during the annual reproductive cycle. Comparative Biochem Physiol 98A:491–496.

Zietkiewicz, Rafalski A, Labuda D, 1994. Genome fingerprinting by simple sequence repeat (SSR)-anchored polymerase chain reaction amplification. Genomics 20:176–183.

Zimmerman BL, 1994. Audio strip transects. In: Measuring and monitoring biological diversity. Standard methods for amphibians (Heyer WR, Donnelly MA, McDiarmid RW, Hayek L-A C, Foster MS, eds). Washington; DC: Smithsonian Institution Press; 92–97.

Zuk M, 1996. Disease, endocrine-immune interactions, and sexual selection. Ecology 77:1037–1042.

Zweifel RG, 1955. Ecology, distribution, and systematics of frogs of the *Rana boylei* group. Univ Cal Pub Zool 54:207–292.

Part VI

Human Behavioral Ecology

Animal utilization is viewed as an important conservation strategy in many parts of the world (IUCN/UNEP/WWF, 1991). The premise is that forms of utilization such as subsistence hunting, tourist hunting, and commercial harvest provide an economic incentive to conserve wild habitats in which target species live. Animal exploitation will only be a viable conservation tool if it is sustainable in the longer term, but analyses of sustainability usually focus on economic returns (Clark, 1985). Analyses of the ability of populations to withstand exploitation are less common and often rely on crude information. For example, Robinson and Redford (1991) had to calculate species' densities in neotropical rainforests using body weights and intrinsic rates of natural increase because precise information on exploited species' densities were unavailable. In addition, analyses of population growth rates under different offtake regimes rarely consider the behavior of individuals except under circumstances when populations fall to critically low levels (e.g., the Allee effect) (Goodman, 1987). Nor do analyses consider possible changes in the behavior of individuals that may result from the process of exploitation itself—for example, alterations in habitat use.

In the first chapter in this section, FitzGibbon reviews a series of individual behavioral attributes of prey populations that could affect a population's response to hunting pressure. These include classic antipredator strategies such as grouping and flight behavior, as well as ranging and habitat choice. As might be expected, many of these behavior patterns change in response to human hunting pressure. The significance of considering these factors is that they influence the relative vulnerability of different age and sex classes and consequently a population's ability to maintain a positive growth rate in the face of exploitation.

Human behavior has been studied from an evolutionary perspective for 20 years (Chagnon and Irons, 1979; Betzig, 1997), and many sophisticated models and empirical data exist on foraging behavior in premodern societies (Kaplan and Hill, 1992). Because multiple-use areas incorporating local hunters are such an important component of conservation strategy in developing countries nowadays (Robinson and Redford, 1994), the importance of understanding the nature of subsistence hunting has become pressing. In the second half of chapter 16, FitzGibbon reviews a suite of strategies used by subsistence hunters that follow from optimal foraging theory. These include preferences for various size, age, and sex classes of prey, the area in which they choose to hunt, the time they hunt, and the effects of employing weapons of varying sophistication. Optimal foraging models help identify the decision rules that subsistence hunters seem to follow and thus the likelihood with which specific management strategies may be accepted by local people.

In chapter 17, Alvard further develops the theme of optimal foraging in traditional societies by exploring the question; Are indigenous neotropical hunters conservationists? Alvard pits five predictions from optimal foraging theory against those derived from a conservationist strategy, tests them using observational data derived from a study of the Piro in eastern Peru, and then broadens the database to include six other hunting societies. Results show unequivocally that Piro hunters do not avoid vulnerable species or sex and age classes that would minimize their impact on prey populations. Moreover, they continue to hunt in depleted areas against conservation predictions. These findings are interesting insofar as they provide high-quality data refuting the idea that hunter-gatherers are prudent conservationists. This has been an important controversy to resolve given the new and increasing reliance on local people as long-term protectors of wildlife (Western et al., 1994). Second, as in FitzGibbon's review, the findings suggest which management solutions have a chance of being accepted locally.

Chapter 18 uses another branch of human behavioral study, evolutionary psychology, to predict patterns of resource exploitation, this time in modern societies. This is a relatively new departure in this subdiscipline: to date, research has been concerned with predicting general patterns of human behavior such as cognition (Wang, 1996), social exchange (Cosmides and Tooby, 1992), and social relationships (Daly and Wilson, 1988; Buss and Schmitt, 1993). Using evolutionary arguments about sex differences in risk taking derived from other studies, Wilson, Daly, and Gordon make predictions about how males and females would view hypothetical choices about taking health risks and about degrading the environment. They find that males would accept greater costs to personal health than women and would make decisions that were more environmentally detrimental. Evolutionary predictions can also be made as to how factors such as life stage, social status, and parenthood affect attitudes and actions toward the environment. Such knowledge is useful in suggesting which groups will be most sympathetic to conservation initiatives and in attempts to change public opinion through lobbying and advertising.

Although conservation biology acknowledges the central role of human action in conserving wild organisms and wild places, we still know relatively little about the social and environmental factors that underlie decisions to exploit resources, what to capture and cut, which to harvest, and how much to mine. Nevertheless, the constraints on decision making ultimately rest on economic factors, such as market forces, and political factors, such as national policies concerning land-use legislation (McCay and Acheson, 1987; Bromley, 1991), that are currently outside the realm of behavioral ecology. Evolutionary theory has the potential to shed light on the rationale behind human decision making about the environment while operating within these constraints.

References

Betzig L (ed), 1997. Human nature: a critical reader. New York: Oxford University Press.

Bromley DW, 1991. Environment and economy: property rights and public policy. Oxford: Blackwell Scientific Publications.

Buss DM, Schmitt D, 1993. Sexual strategies theory: an evolutionary perspective on human mating. Psychol Rev 100:204–232.

Chagnon NA, Irons WG (eds), 1979. Evolutionary biology and human social behavior: an anthropological perspective. North Scituate, Massachusetts: Duxbury Press.

Clark CW, 1985. Bioeconomic modelling and fisheries management. New York: Wiley Interscience.

Cosmides L, Tooby J, 1992. Cognitive adaptations to social exchange. In: The adapted mind (Barkow J, ed). New York: Oxford University Press; 163–228.

Daly M, Wilson M, 1988. Homicide. New York: Aldine de Gruyter.

Goodman D, 1987. The demography of chance extinction. In: Viable populations for conservation (Soule ME, ed). Cambridge: Cambridge University Press; 11–34.

IUCN/UNEP/WWF, 1991. Caring for the earth: a strategy for sustainable living. Gland: International Union for the Conservation of Nature and Natural Resources.

Kaplan H, Hill K, 1992. The evolutionary ecology of food acquisition. In: Evolutionary ecology and human behavior (Smith EA, Winterhalder B, eds). New York: Aldine de Gruyter; 167–201.

McCay BJ, Acheson JM (eds), 1987. The question of the commons. Tuscon: University of Arizona Press.

Robinson JG, Redford KH, 1991. Sustainable harvest of neotropical forest mammals. In: Neotropical wildlife use and conservation (Robinson JG, Redford KH, eds). Chicago: University of Chicago Press; 415–429.

Robinson JG, Redford KH, 1994. Community-based approaches to wildlife conservation in neotropical forests. In: Natural connections: perspectives in community-based conservation (Western D, Wright RM, Strum SC, eds). Washington, DC: Island Press; 300–319.

Wang XT, 1996. Evolutionary hypotheses of risk-sensitive choice: age differences and perspective change. Ethol Sociobiol 17:1–16.

Western D, Wright RM, Sturm SC, 1994. Natural connections: perspectives in community-based conservation. Washington, DC: Island Press.

16

The Management of Subsistence Harvesting: Behavioral Ecology of Hunters and Their Mammalian Prey

Clare FitzGibbon

Subsistence Hunting: The Problem

Wildlife populations have always been a traditional source of meat, skins, and other essential items for local people, and hunting remains an important subsistence activity in many parts of the world. Wild animals contribute a minimum of 20% of the animal protein in people's diets in at least 62 countries (Prescott-Allen and Prescott-Allen, 1982). In the Serengeti ecosystem, in Tanzania alone, the total annual offtake of harvested ungulates exceeds 160,000 animals and benefits 1 million people (Campbell and Hofer, 1995). Subsistence harvesting can be defined as harvesting by an individual when the direct products of hunting are consumed or used by the hunters and their dependents (Caughley and Gunn, 1996). Commercial hunting (where products are sold before their benefits accrue to the hunter) is generally on a much larger scale than subsistence hunting, and its potential impact on prey populations is therefore much greater (Lavigne et al., 1996).

The ecological impact of hunting has only recently started to receive attention (Redford and Robinson, 1985). Although the idea of hunter-gatherers living in harmony with nature was popular among anthropologists in the 1960s and 1970s, this is no longer the prevailing view, and it is now clear that traditional societies often overharvest their prey (Diamond, 1988; Alvard, chapter 17, this volume). Increases in human population density and habitat loss have exacerbated the problem. A number of studies in a variety of habitats and countries have now demonstrated that subsistence hunting can have deleterious effects on wildlife populations and that this has severe consequences for the ecological functioning of the forest (Redford, 1992). In some areas, the effects of harvesting on mammal populations may be even more extreme than the relatively well-publicized effects of deforestation or international trade (Redford, 1992; Bodmer, 1995b). Although overharvesting is considered to be mainly a problem in developing countries, in North America, overhunting (usually illegally) is the leading cause of endangerment and local extinction among mammals (Hayes, 1991).

Overwhelmed by the difficulties of monitoring and regulating wildlife harvesting, many countries have responded to the threat posed by harvesting by banning it. Banning is often ineffective because it is difficult to enforce (e.g., Ojasti, 1984), particularly given the poor resources available to most wildlife departments in developing countries. In addition, although subsistence harvesting poses a major threat to some prey species, the continued freedom to hunt is essential for maintaining the traditional lifestyles of many people. Hunting is not only important as a source of food, but also to control crop predators (FitzGibbon et al., 1995; Jorgenson, 1995) and as a source of income (Clay, 1988). It may also have conservation benefits. It is one of the few ways in which local communities can derive benefits from wildlife, and by offsetting some of the direct and indirect costs of forest conservation, communities thus have an interest in the continued existence of natural habitats (Balmford et al., 1992; Bodmer et al., 1994). The banning of hunting removes this key incentive for forest conservation. Generally, the available evidence suggests that the sale of wildlife products to satisfy external demands almost invariably leads to overexploitation and increases the "extinction potential" of target species, whereas, with adequate regulation, hunting of wildlife for personal consumption or to supply local communities on a limited commercial basis can be a sustainable activity (Ehrenfeld, 1970; Geist, 1988, 1994; Lavigne et al., 1996). Authorities must therefore confront the daunting task of managing game hunting by local people.

An effective harvesting management program requires information on the sustainability of harvesting different prey species as well as their relative importance in the diet of hunters and their commercial value, so that management recommendations will take into account the requirements of the local people and therefore have greater chances of success. The key questions for conservation biologists are what determines the sustainability of harvesting and can harvesting be managed to ensure sustainability. To decide what species are most suitable for harvesting, it is necessary to examine the susceptibility of different species to overhunting and the biological factors, including ecology and behavior, that underlie this susceptibility (Bodmer et al., 1988; Bodmer, 1994). Educational programs are then required to promote harvesting of species that are able to support high levels of offtake, combined with harvesting controls to reduce hunting of seriously affected species (Silva and Strahl, 1991).

Sustainable Harvesting: The Theory

In general, the aim of a wildlife harvesting management program is to conserve biodiversity by ensuring that harvests are sustainable. However, the conservation goals of wildlife managers may be very different from those of the indigenous hunters (Redford and Stearman, 1993). Biodiversity for conservationists usually implies maintaining the full range of genetic, species, and ecosystem diversity in a particular area at its natural abundance (Redford and Stearman, 1993). To indigenous people, preserving biodiversity often means preventing the large-scale destruction of habitats. While subsistence harvesting may be compatible with the biodiversity conservation aims of indigenous hunters, many studies have demonstrated that even low levels of subsistence harvesting alter biodiversity as defined by conservationists. A compromise is one in which traditional lifestyles are maintained that conserve "acceptable" levels of biodiversity.

Sustainability requires harvesting a population at the same rate or lower than its natural rate of growth (Clark, 1990). However, most unharvested populations are not increasing, and consequently the population must be stimulated to increase before sustained harvesting can take place. In natural habitats, the only way to achieve this is by reducing the number of animals competing for essential resources, thus increasing fecundity and reducing mortality. The level of sustained offtake will depend on the population density. If the density is reduced a little, the rate of increase induced will be small, and the sustained yield will be a small proportion of a relatively large population. Alternatively, if the density is reduced drastically, the induced rate of increase will be large, and the sustained yield will be a large proportion of what is now a relatively small population. The maximum sustainable yield (MSY) is taken from an intermediate density at which the induced rate of increase multiplied by the density is at a maximum. Clearly, MSYs cannot be achieved without loss of biodiversity as described above (i.e., without reduction in natural abundances), but can be compatible with the long-term maintenance of populations.

Two main factors influence the number of animals that can be harvested from a population: intrinsic rate of increase and density. The intrinsic rate of increase, r_{max}, the highest rate of increase that can be attained by a population not limited by food, space, resource competition, or predation, is related to the age of reproduction, *a,* the age of last reproduction, *w,* and the annual birth rate of female offspring, *b,* in the following way (Cole, 1954):

$$1 = e^{-r_{max}} + be^{-r_{max}(a)} - be^{-r_{max}(w+1)}$$

Thus species that breed earlier and more often can withstand higher levels of harvesting. The life span of a species and thus its age of last reproduction has relatively little effect on r_{max}. In mammal populations, the intrinsic rate of natural increase is inversely related to adult body size (Fenchel, 1974; Western, 1979; Hennemann, 1983). Although smaller bodied species can increase more rapidly than can populations of larger-bodied species, there is considerable variation among species of comparable size, and a strong effect of phylogeny. For example, in an analysis of neotropical forest mammals, Robinson and Redford (1986b) found that primates and tapirs tended to have lower rates of increase than expected for their body size; rodents, lagomorphs, or ungulates were higher than expected; and carnivores were close to expectation.

The total production of the population depends not only on the natural rate of increase of the population but on the number of adult females reproducing and therefore will be greater in higher density populations. Among mammals, density is closely related to the average adult body mass of a species, its diet, habitat, and biogeographic area. Larger species tend to occur at lower densities, while herbivores tend to live at higher densities and carnivores at lower densities than other mammals (Clutton-Brock and Harvey, 1977; Eisenberg, 1980; Peters and Wassenberg, 1983; Robinson and Redford, 1986a).

Variability in Susceptibility to Harvesting: The Evidence

Few studies have been able to determine whether current harvesting levels are sustainable, and little effort has been invested in determining why, in practice, some harvests are sustainable and others not. However, in Amazonia, Bodmer (1994) used differences in the density of species between slightly and persistently hunted sites to measure the susceptibility of mammals to overhunting. He demonstrated that density differences correlated with the intrinsic rate of natural increase of species (r_{max}). Thus rodents and artiodactyls, which have relatively high values of r_{max}, showed little difference in their densities, while primates and lowland tapirs, which have comparatively low values of r_{max}, showed considerable differences in their densities. By using estimates of population densities and rates of population increase to obtain first estimates of MSYs with which to compare with current offtake levels (Robinson and Redford, 1991), FitzGibbon et al. (1995) were able to show that current harvest levels of elephant shrews, duikers, and squirrels in Arabuko-Sokoke Forest were probably sustainable, while harvests of primates and larger ungulates were not. High reproductive rates of elephant shrews and smaller prey species compared with the low rates of primates and larger ungulates were partly responsible for the differences, but not entirely. Fa et al. (1995) found that current harvesting levels of five primate species and one ungulate in Bioko, Equatorial Guinea, were unlikely to be sustainable; in some cases harvests exceeded 20 times the predicted levels of potential sustainable harvests. Correlating the ratio of actual harvests to sustainable harvests (calculated from data presented in Fa et al., 1995) as a measure of overharvesting with r_{max} values suggests that approximately 25% of the variation in the extent of overharvesting can be explained by r_{max} (correlation coefficients are $r = -.500$, $n = 12$ for Bioko; $r = -.563$, $n = 17$ for Rio Muni). For neither site did prey density have a significant influence on vulnerability to overharvesting (for Bioko, $r = -.241$, $n = 12$ and for Rio Muni, $r = -.136$, $n = 17$).

What Contribution Can Behavioral Ecology Make to the Development of Sustainable Subsistence Hunting?

Although studies have demonstrated that a species' intrinsic rate of increase, resulting from differences in body size and phylogeny, is a key factor influencing susceptibility to overharvesting, it is clear that other factors are also involved. The rate at which a species is harvested will depend on a range of behavioral factors both on the part of the prey and the hunter. Behavioral ecology may therefore help in the development of sustainable harvesting management plans by determining which behavior patterns contribute to inter- and intraspecific variation in vulnerability to harvesting. It may also help us determine the consequences of harvesting, in terms of its effects on the social organization and behavior of prey, and thus predict how particular prey species will respond to different levels of harvesting (Greene et al., Chapter 11, this volume). Appreciating the factors influencing human hunting behavior may enable managers to understand what determines prey selectivity and how variations in habitat, prey availability, and hunting methods influence the impact of hunting on prey populations. In the next section, some behavior patterns that are likely to influence either inter- or intraspecific vulnerability to harvesting are described. In the following section, the determinants of human hunting behavior are explored, and the consequences for prey populations are outlined.

Impact of Prey Behavior on Vulnerability to Harvesting

Social Behavior

In many prey species, grouping has evolved as a form of predator defense (reviewed in Krebs and Davies, 1993). The antipredator advantages of grouping include improved predator detection and defense as well as a reduced risk of predation as a result of the dilution effect. The dilution benefits of grouping rely on the fact that the number of attacks and the number of individuals killed in a predator–prey encounter does not increase in proportion to group size. Overall, the benefits of grouping primarily accrue to species inhabiting open habits, as the difficulties of maintaining groups and monitoring other group members are considerable in forested habitats. In addition, individuals living in forested habitats often rely on concealment, which is easier when alone or in small groups.

Few data are available on the effects of grouping on vulnerability to human predation, but a number of social species make use of group defenses to reduce vulnerability to human hunters (e.g., musk oxen *Ovibos moschatus*), and some authors report hunters being reluctant to harvest grouped prey because of the risk of injury (e.g., the Achuara avoid large groups of white-lipped peccary *Tayassu pecari;* Ross, 1978). The increased alertness of large groups of prey is likely to result in hunters favoring individuals on their own or in smaller groups, just as natural predators frequently do (e.g., cheetahs *Acinonyx jubatus* hunting Thomson's gazelles *Gazella thomsoni;* FitzGibbon, 1990).

There is some evidence that group living species may be more vulnerable to human hunters under certain conditions. Groups are easier to locate than solitary individuals, and once the group has been found, more than one individual can be killed (Peres, 1990). For example, Sinclair (1977) describes how the herding behavior of buffalo *Syncerus caffer* is exploited by hunters to eradicate local herds in cooperative drives into snares lines. Similar techniques were used by Chipewyan to catch large numbers of caribou *Rangifer tarandus* in Alaska (Heffley, 1981). Hunters may also make use of the social behavior of prey species to attract them. For example, M. Alvard (personal communication) reports that if Piro hunters in Peru catch a social primate that is still alive, they will try to force it to make distress calls that attract other members of the troop. Once these other individuals have been lured into range, they can then be killed. A similar technique is used to catch capybaras *Hydrochaeris hydrochaeris;* one hunter carries a young capybara as a lure while another walks ahead with a torch and a club to intercept any capybaras that approach.

Differences in grouping behavior between age and sex classes may result in differential predation rates. For example, the preference of hunters to stalk solitary individuals is likely to result in increased predation of males, which are more likely to be on their own or in small groups (FitzGibbon, 1990). Adult blue duikers *Cephalophus monticola* usually live in family groups consisting of one male, one female, and their offspring younger than a year and a half, while subadults live singly and tend to be found away from preferred habitats (Dubost, 1980). Consequently, Hudson (1991) argues that because net hunters tend to set nets in areas with high duiker densities, subadults are less likely to be caught. Nevertheless, a repeat netting session in a recently harvested area is likely to catch a high proportion of subadults which have moved in to occupy the vacant territories.

Flight Behavior

Variation in vigilance levels and flight distances is likely to predispose different species and different age or sex classes to selection by hunters. Females of sexually dimorphic or polygynous species are generally more vigilant and likely to flee from predators than are males (Berger and Cunningham, 1988; Prins and Iason, 1989; FitzGibbon, 1990; Berger, 1991), probably as a consequence of the presence of young, relatively smaller body size, and/or group size differences. Such differences are likely to be manifest when humans are the predators. For example, male Thomson's gazelles flee at shorter distances from humans than females (C. FitzGibbon, personal observation), whereas female black rhinoceros *Diceros bicornis* flee farther from approaching humans than males and consequently may be more prone to human predation (Berger and Cunningham, 1995). Males may also be more vulnerable during rutting because of reduced vigilance toward predators and increased focus on defending access to females (e.g., in moose *Alces alces;* Winterhalder, 1981). Similarly, archaeological research on Steller's sea lions *Eumetopias jubatus* suggests that prehistoric hunters primarily killed males (75% of the harvest), probably because they were hunted during the breeding season when males maintain and vigorously defend territories (Hildebrandt, 1984; Hildebrandt and Jones, 1992; Jones and Hildebrandt, 1995). Females and juveniles flee into the ocean when attacked, but males stay and defend the beach. The inability to flee as fast as other individuals may predispose pregnant females to harvesting, particularly when hunting techniques rely on active pursuit (e.g., capybaras; Moreira, 1995; and some primates; Alvard and Kaplan, 1991).

Variation in flight distances may also influence the vulnerability of different age classes. Behavioral observations by Dubost (1980) suggested that young blue duikers less than 10 months old were more likely to remain quiet and stay hidden rather than to jump and run when startled. This makes them less vulnerable to net hunters such as the Aka (Hudson, 1991). Bodmer (1994) suggests that older peccaries may be more wary of hunters and therefore less likely to be caught, although his data did not confirm the suggestion. Archaeological research suggests that nursery groups of newborn harbor seals *Phoca vitulina* were exploited to the near-exclusion of adults of this species because the adults tended to be more vigilant than the naive young (Hildebrandt, 1984; Hildebrandt and Jones, 1992; Jones and Hildebrandt, 1995).

Temporal Activity Patterns

Although it is predicted that nocturnal species are less likely to be caught by hunters than diurnal ones (Ross, 1978), simply as a result of reduced opportunities for capture, few quantitative data are available to confirm this. Nevertheless, nocturnal species, particularly primates, make up a small proportion of captures in most locations; for example, Britton (1941) reports that sloths are rarely harvested because of their nocturnal behavior patterns. Alternatively, hunters may make use of the fact that animals can be blinded at night and use spot lights to facilitate capture. For example, hunters in Arabuko-Sokoke use strong flashlights to blind duikers temporarily (C. FitzGibbon, personal observation). In addition, species with regular behavioral routines, such as visiting waterholes at particular times, may be more susceptible to hunters.

Trail and Burrow Use

Species that make trails through vegetation or use burrows, particularly simple, single-exit ones, are more vulnerable to snaring than those species that have less-defined movement patterns. For example, one of the reasons four-toed elephant shrews *Petrodomus tetradactylus* are caught much more frequently than the golden-rumped species *Rhynchocyon chrysopygus* in Arabuko-Sokoke Forest, Kenya, is that the former make trails through the leaf litter along which snares can be set (FitzGibbon et al., 1995). Pigs are also vulnerable because they travel along trails (M. Alvard, personal communication). The sedentary and localized beaver *Castor canadensis* is susceptible to overtrapping because it can be easily snared at the entrance of its lodge (Nelson, 1982).

Attraction to Baits

Species that can be attracted into traps using baits, calls, or decoys are particularly vulnerable to harvesting. Baiting of primates and rodents is widespread, but it also applies to other groups. For example, hunters in Korup, Cameroon, attract bay duikers *Cephalophus dorsalis* by calling (Infield, 1988). Many American peoples focus on species that are attracted to their gardens—for example, the Mayas (Jorgenson, 1995), Afroecuadorians, and Cachi

(Suarez et al., 1995). Ungulates are easily attracted to burnt grasslands, even a few hours after burning, and many indigenous people make use of this fact. North American Indians burn reeds along lakeside margins to increase root growth, attracting muskrats *Ondata zibethicus* (Lewis, 1982). Hunters may also make use of animals' attraction to natural features such as water holes and salt licks. For example, Amazonian Indians wait in ambush for tapirs *Tapirus terrestris* attracted to such features (Ross, 1978).

Ranging Behavior and Habitat Choice

Dispersal and ranging behavior will also affect a species' vulnerability to harvesting, influencing the extent to which individuals move from protected areas into hunting zones and their ability to recolonize depleted areas (Novaro, 1995; Vincent, 1995). In predator–prey theory, dispersal of prey from predator-free patches (sources) into harvested areas (sinks) is considered a key factor in prey persistence (Roff, 1974; Hillborn, 1979). Harvested populations frequently survive only as a result of immigration from other areas (e.g., culpeo foxes *Dusicyon culpaeus:* Novaro, 1995; bobcats *Felis rufus:* Knick, 1990). Species may depend on the preservation of refuges large enough to maintain a sufficient flow of recruits (e.g., bobcats in Idaho; Knick 1990). Obviously, the dispersal ability of the prey species will influence the scale of patchiness required to maintain a population.

Migratory species may be particularly vulnerable to overharvesting. This is partly because species that have regular movement patterns may be favored as prey, given the ease with which hunters can locate and intercept them. For example, the spring and fall migrations of caribou make them vulnerable to capture as they move through mountain passes (Burch, 1972). In addition, it is difficult to assess declines in migratory species, particularly when such species form large aggregations during migration. As a result there is some suggestion that even those societies that do practice some conservation ethic will overharvest migratory species, tending to take as many prey as possible when they do appear, knowing they may vanish tomorrow and not reappear for many months (Nelson, 1982). The "Pleistocene overkill" sites in North America involved herding animals that were probably migratory.

Even local, predictable movements can increase a species' vulnerability to harvesting. Large herds of caribou tend to stay close to lake shores and avoid areas without thick bush or woodland (Burch, 1972). In the summer, the caribou are plagued by warble flies and select higher areas that catch the wind. Any predictable movement increases the chances of hunters successfully locating and stalking them. In contrast, in winter, moose are more difficult to locate because they become confined to small feeding patches where the snow is thin, and they do not leave long trails (Bishop and Rausch, 1974).

Differences in ranging behavior between males and females may result in intraspecific variation in vulnerability. For example, in Serengeti National Park, the majority of wildebeest *Connochaetes taurinus* caught in snares are males (88–98%; Georgiadis, 1988) because males tend to be at the front of herds as they move into new areas and consequently meet snares first. Males may in general be more vulnerable than females as a result of increased activity and movement associated with searching for mates.

Response to Harvesting

The way in which species alter their behavior in response to harvesting can have a major impact on their vulnerability. Many species become more wary of humans, increasing vigilance levels and flight distances and altering movement patterns, which reduces harvesting rates. For example, geese quickly learn to avoid hunting areas in preference for protected sites (Ebbinge, 1991; Giroux, 1991), capybaras become more nocturnal and increase their vigilance and flight distance (Azcarate, 1981), and large Neotropical primates become extremely shy of humans whenever they have been hunted in the past (Hernandez-Camacho and Cooper, 1976; Peres, 1990). In addition, where hunting pressure is intense, the capybara, a savannah-living animal, is found more frequently in forest (Cordero and Ojasti, 1981). However, changes in behavior may also have damaging effects on food intake and reproductive rates, with individuals trading off the benefits of reduced predation risk against the costs of reduced foraging time or access to high-quality food resources (Lima and Dill, 1990). For example, Paraguayan caimans *Caiman yacare* respond to harvesting by reduced egg guarding, which increases predation rates on eggs and reduces reproductive success (Crawshaw, 1991).

Hunting may also break up family groups, resulting in reduced breeding success. For example, orphaned juvenile geese survive less well, and breeding is frequently disrupted by the loss of a mate as a result of shooting (Ebbinge, 1991). In addition, successful breeding in social species may be group-size dependent, and thus removal of a few individuals from a group may prevent the group from successfully breeding again (e.g., social capybaras; Herrera and Macdonald, 1987; Moreira and Macdonald, 1996). Also, solitary living, monogamously mated species that live at low density may be vulnerable to harvesting because of the time required to make new pair bonds if one member of the pair is killed (e.g., seahorses; Vincent, 1995; see Greene et al., chapter 11, this volume). Yet, in other species, missing partners are replaced very quickly (e.g., dik-dik *Madoqua kirkii;* Brotherton, 1994), and the potential for disruption of birth rates is low.

Impact of Human Hunting Behavior

Understanding and predicting the impact of human hunting behavior on prey has centered around the use of optimal foraging models. Although these models were originally developed for understanding the factors influencing nonhuman foraging behavior, their utility to researchers of human foraging behavior rapidly became apparent. The underlying assumption is that either the nutritional status of an individual or the time needed to acquire necessary nutritional requirements determines fitness. Models have been primarily used first to predict prey choice (i.e., predict the food items the forager will attempt to exploit and those it will ignore in favor of continued search for more preferred foods), and second, when resources are clumped together in patches, to determine whether a forager will enter a patch and how long it will stay in that patch before moving on to another.

These optimal foraging models have been combined with population biology models to simulate the dynamic relationships between hunter-gatherers and their prey resources (e.g., Belovsky, 1988; Winterhalder et al., 1988). These models can be used to predict how the abundance of hunters influences prey abundance and the conditions under which overexploitation is likely to occur. The model developed by Winterhalder et al. (1988), however, suggests that neither the dynamic properties nor the equilibrium density of the foraging population can be predicted from only the biomass of prey and the requirements of the foragers. Steady growth to equilibrium, damped cycles, stable limit cycles, and extinction of either the predator or its prey are all possible outcomes, but small changes in factors such as foraging effort and prey density can influence the dynamic properties of the system. Although the predictions of these models have yet to be thoroughly tested, it is clear that optimal foraging models provide a means to predict the impact of hunter-gatherers on prey populations under a variety of conditions and constraints.

Extent of Hunting versus Plant Gathering/Farming

A number of studies have noted that hunter-gatherers devote large amounts of time and energy to hunting, even when gathering plant food or farming would have maximized energy-return rates. For example, virtually all South American horticulturalists obtain much higher caloric return rates from farming than they do from hunting or fishing, yet most spend considerably more time hunting and fishing than farming (Hames, 1992). One obvious reason is that meat contains high-quality protein, the nine essential amino acids that the human body cannot synthesize, and fats that are important for the absorption, transportation, metabolism, and storage of fat-soluble vitamins (Kelly, 1995). Linear programming models, a method of solving for behavioral optima under specified contraints, have been used to predict the combination of food types that best satisfies the nutrient-maximizing or time-minimizing goals that best suit a particular society and have been used to understand why so much time and effort are put into hunting as compared to gathering (e.g., the !Kung, Belovsky, 1987). Gathered foods increase in the diet as hunted foods are reduced in abundance (Belovsky, 1987), but hunter-gatherers will continue to hunt prey populations even when they become scarce.

As a result of the high demand for meat, humans are capable of having severe impacts on prey populations even when prey are scarce. The use of a population model built around optimal foraging models suggests that hunter-gatherers are likely to have a greater impact on prey populations when gathered foods are more abundant, allowing larger human populations to be maintained and the exploitation of hunted foods to remain high even when hunted foods are rare (Belovsky, 1988). Consequently, the model predicts that overharvesting is more likely in areas of high primary productivity (Belovsky, 1988). This prediction is counter to that of traditional overexploitation models, where prey extinction is more likely at low primary productivities. In addition, the model suggests that the key to the demise of hunted foods is not their productivity or abundance, but the productivity of gathered foods.

Selecting Prey Species

Prey-choice models predict that the prey type that yields the highest rate upon encounter should always be pursued. Other lower-ranked resources should be included sequentially until the next-most-profitable resource yields a lower rate of return upon encounter than that which could be obtained by continuing to search for and pursue the more profitable items. Most of the tests of the models have been qualitative rather than quantitative, predicting directional tendencies in prey choice or diet breadth in relation to directional changes in parameters such as return rate or abundance. Predictions include the fact that large prey are generally more profitable than small prey (as long as handling costs do not increase in proportion to body weight), that low-ranked resources should drop out of the diet when search costs decrease and hence overall return rate increases (e.g., Winterhalder, 1977, 1981), and that diet breadth should increase with a decrease in the density of highly profitable prey items (e.g., Hames and Vickers, 1982).

A number of studies have reported hunters' preference for large or medium-sized prey over smaller prey (Mittermeier, 1987; Infield, 1988; Hawkes et al., 1991; Alvard, 1993;

Bodmer, 1994, 1995a; Colell et al., 1994; Raez-Luna, 1995; see table 16-1). The preference for medium-sized prey is particularly marked in the case of hunters harvesting for commercial markets because carcasses often need to be transported over long distances and in addition are easier to sell and are economically more viable (Fa et al., 1995). This increasing preference for large prey as travel time increases is predicted by central-place foraging models, which are characterized by a pattern in which the forager makes an outward trip, forages, and then returns home with the captured prey (Orians and Pearson, 1979). As radial travel time increases, these models predict that the optimal diet breadth is progressively restricted to prey species of value, and the optimal load for the return trip increases in size.

The body size of harvested prey is important in determining susceptibility to overharvesting because larger species tend to have lower reproductive rates and lower densities and therefore are more vulnerable to overharvesting (see above). The only factor that may favor larger species is the fact that they tend to be distributed over a wider area (Arita et al., 1990). Consequently, local hunting may be less likely to exterminate a whole population than intensive harvesting of a smaller species.

The preference for larger prey depends partly on prey abundance, with smaller prey becoming more profitable as large prey decrease in abundance (Hames and Vickers, 1982). Consequently, Hames and Vickers (1982) predicted that in zones of high hunting pressure, hunters should shoot large animals when they are found, but also prey on the more abundant smaller game, and in zones of low hunting pressure where large game are more abundant, hunters should be more likely to ignore small game. Testing these predictions on the hunting strategies of three Amazonian societies showed that in all three cases, the proportion of large game to small game hunted decreased with increased hunting pressure. Among the Yekwana and Yanomano, significantly fewer kills of small game (pacas *Agouti paca* and armadillos *Dasypus novemcinctus*) occurred in distant zones than would be expected given the probabilities derived from binomial distribution theory.

Maximizing mean daily meat returns over the long-term (long-term rate maximization) may not be the most suitable measure for determining an optimum foraging strategy because the regularity with which meat is obtained (i.e., the short-term temporal variation) may be equally important. Although specializing on larger game may maximize mean daily return rates, success rates for big game hunting, measured as the chance of acquiring a carcass on any particular day, are often lower than for smaller prey (e.g., for the Hadza in East Africa; Hawkes et al., 1991). Hunters can reduce the variance in meat supply by including smaller game in the suite of prey they exploit, or when large food packages are acquired by different hunters, by sharing carcasses between them (Cashdan, 1985; Kaplan and Hill, 1985; Winterhalder, 1986). Thus, sharing increases the emphasis on large prey items, whereas reducing the risk of failure on any day puts emphasis on smaller prey. This may be important when small children need to be fed or when the risk of starvation is high (Hawkes et al., 1991). The need to provide some meat reliably each day, if only a little, is one reason elephant shrews are such popular prey in Arabuko-Sokoke (FitzGibbon et al., 1995).

Despite individual preferences for larger bodied species, hunters often harvest more smaller species (table 16-1), particularly those that reproduce rapidly, because these species tend to maintain high densities and therefore are encountered more often and so are more easily caught. For example, Bodmer (1994), studying hunters in Tahuayo, Peru, found that in terms of the actual numbers of animals taken, harvests were positively correlated with reproductive productivity. An examination of a number of studies demonstrates that this may also apply in other areas (e.g., Bodmer, 1995a; FitzGibbon et al., in press). Another

Table 16-1 A comparison of prey choice in seven subsistence hunting tropical communities.

Place	Relative actual harvests[a]	Reasons for actual harvests and preferred prey	Reference
Tamshiyacu-Tahuayo Peru	Paca > 2 peccary spp. > agouti > 3 primate spp. > red brocket deer > white-fronted capuchin > tapir > squirrel	Actual harvests correlate with r_{max}, not body size, economic value, or density. Hunters prefer larger species with greater economic value.	Bodmer (1994)
Arabuko-Sokoke Kenya	Elephant shrews > Sykes monkeys > aardvarks > bushpigs > baboons > mongooses > squirrels > duikers*	Ease of capture. Positive correlation between r_{max} and actual harvests. Duikers preferred for their quality meat.	FitzGibbon et al. (1995)
Bioko, Equatorial Guinea[b]	Ogilby's duiker* > Emin's rat > brush-tailed porcupine > blue duiker* > 6 primate spp.	Actual harvest not significantly correlated with r_{max}, density, or body size. Hunters prefer species that can easily be sold at market.	Fa et al. (1995)
Bioko, Equatorial Guinea[c]	Blue duiker* > Emin's rat brush-tailed porcupine > Ogilby's duiker > 1 squirrel spp. > 3 primate spp. > 2 squirrel spp. and 4 primates	Actual harvests significantly correlated with r_{max}, but not density or body size. Larger species preferred for selling in markets.	Colell et al. (1994)
Amazon	Peccary* > capuchin monkey > deer howler and spider monkeys,* agouti > tapir,* capybara, paca, and ocelot	Maximizing return rates. Larger species preferred.	Alvard (1993, 1994)
Korup, Cameroon	Blue and bay duiker* > porcupine > red colobus > drill > white-nosed monley > Ogilby's duiker > mona monkey > water chevrotain > bush pig > red colobus > russet-eared guenon > 2 primates and 1 duiker spp.	Ease of capture. Hunters prefer larger species and those important as crop pests.	Infield (1988)
Serengeti, Tanzania	Wildebeest > impala > zebra > topi > buffalo * > giraffe* > waterbuck* > Thomson's gazelle	Abundance and ease of capture. Relative to abundance, bush species are taken more than mixed habitat or plain species due to the setting of snares in thickets.	Arcese et al. (1995)

[a]The species preferred by hunters are indicated with an asterisk (usually those that are always pursued once located), and where possible the reason for this preference is given. Preferred prey species are not always those caught most frequently, usually because they are large-bodied, low-density species.

[b]Data taken from market surveys, so smaller species, particularly rodents not suitable for selling, are underrepresented. Correlations of harvest with r_{max}, density, and body size carried out using data from Fa et al. (1995). Similar analyses from another site (Rio Muni) also using data from Fa et al. (1995) revealed the same lack of significant correlations.

[c]Data from hunters, so results reflect true harvests. Analysis of correlation between r_{max}, density, and body size carried out using a subsample of 12 species for which r_{max}, density, and body size data were available from Fa et al. (1995). Correlation between r_{max} and actual harvests was 0.593, $n = 12$.

reason hunters may be constrained from taking as many large prey as they would wish is that the harvesting of large prey often depends on cooperation between hunters. For example, while individual Aka hunters from central Africa regularly trap small rodents, setting nets for larger prey, such as duikers, requires the cooperation of more than 30 adults (Hudson, 1991).

Selection of Specific Age Classes

Assuming that handling costs are equal across different age categories, optimal foraging theory predicts that adults are more profitable prey than juveniles because they are larger. In most subsistence harvesting situations, however, there are few opportunities for age-biased selection. Bodmer (1994), for example, states that low visibility in dense Amazonian forest results in animals being harvested randomly with respect to age. Alvard (1995a) found that harvests of peccaries (*Tayassu* spp.) and red brocket deer *Mazama americana* by Amazonian hunters were approximately random with respect to age (although the age classes were understandably broad). Marks (1973) found that Valley Bisa hunters concentrated on adult animals, as they provided better returns for their hunting effort. In contrast, capybara harvesting, which primarily takes place in open savannah habitat, mainly removes larger adults, which results in the age structure of harvesting populations being heavily biased toward younger individuals (Herrera and Macdonald, 1987). Differences in age-class susceptibility (see above) may also affect selectivity.

Theoretical considerations suggest that the impact of harvesting on a population's reproduction rate can be minimized by harvesting animals with a high risk of mortality or low reproductive potential (Caughley, 1977). Studies of the reproductive success of known aged individuals (e.g., red deer *Cervus elaphus;* Clutton-Brock et al., 1982) suggest that, in general, harvesting prime-aged individuals will have a greater effect on the population than harvesting either young or old animals. In social capybaras, younger females have relatively low rates of pregnancy and smaller litter sizes than older females (Moreira and Macdonald, 1993), so the selective harvest of larger, older females is likely to have a detrimental effect on reproductive rates (Moreira and Macdonald, 1996). However, behavioral ecological studies suggest that the effects of removing large proportions of subadult animals may have more severe consequences than theoretical studies suggest, particularly in social species. Subadults often contribute significantly to the reproductive success of the breeding adults by looking out for predators, finding food, and defending the territory, and in some cases they may be essential for successful reproduction (e.g., in social mongooses, a prey species in East Africa).

Selection of Different Sexes

Assuming that handling costs are equal for male and female prey, optimal foraging theory predicts that the larger sex (usually the male) is more profitable and therefore should be hunted most often. As with age selection, in most subsistence harvesting situations, there are few opportunities for sex-biased selection. Trapping or netting is rarely sex specific, and even when shooting the prey, limited visibility often makes identification of males and females difficult. There are few data available on the sex ratio of prey harvests from subsistence hunters. Marks (1973) found that Valley Bisa hunters in Zambia killed mainly male ungulates; Alvard (1995a) found little evidence for preferential selection of male prey by the Piro in Amazonia; whereas Peres (1990) reported that hunters in western Brazilian Amazonia preferentially targeted breeding female primates because the infants they carry may be captured alive and sold as pets. Irrespective of hunters' preferences for males and females, it is possible that one sex may be more vulnerable because of differences in behavior or size. For example, there is some evidence that female primates are more vulnerable than males because of the mobility costs of pregnancy and carrying infants (Alvard and Kaplan, 1991).

Male-biased harvests have been shown to increase the MSYs of populations both in theory (Fowler, 1981; Harris and Kochel, 1981) and in practice (Catto, 1976; Nelson and Peek, 1982; Fairall, 1985), under the assumption that many species have a polygynous social structure in which each male mates with several females, and consequently that a female-biased population sex ratio will not have a detrimental effect on female reproductive rates (but see Ginsberg and Milner-Gulland, 1994; Greene et al., chapter 11, this volume). Some of the effects of female-biased populations may well be advantageous—for example, reduced aggression and mortality among males and reduced harassment of females by bachelor males (Ginsberg and Milner-Gulland, 1994).

Behavioral ecological data suggest that the effects of male-biased harvests may be more detrimental than originally believed, particularly for certain species (Ginsberg and Milner-Gulland, 1994). Potential deleterious effects include disruption of territorial and group structure, increased mortality of young born out of season, artificial selection for inferior males, or an inadequate number of males to inseminate females. Moreira (1995), modeling the effects of harvesting on capybara populations, found that hunting had greater impacts on populations when it was sex biased (either way) than when it was not because of the different roles of males and females in this social species. In addition, although Fairall (1985) argued that in impala highly male-biased hunting is unlikely to limit female fecundity because there are "excess males in bachelor herds," Ginsberg and Milner-Gulland (1994) point out that detailed behavioral data on known individuals (Jarman, 1979) suggest that an excess of males in impala cannot be assumed from observations of bachelor groups. In a single breeding season, a territory may be held by several males. Hence excess males may actually be breeding males who have lost or not yet gained a territory.

There is some evidence that low proportions of males in ungulate populations can result in reduced female fecundity (e.g., elk *Cervus elaphus*: Hines and Lemos, 1979; Prothero et al., 1979; Smith, 1980; caribou: Bergerud, 1974). Removing a high proportion of adult males might result in breeding by partially incompetent yearlings (e.g., elk: Prothero et al., 1979) or in an inadequate number of dominant males being available to fertilize receptive females. The consequences of these changes are generally unknown. However, red deer hinds mated by young, incompetent males are more likely to be damaged physically than those mated by dominant males (Clutton-Brock et al., 1982). Ginsberg and Milner-Gulland (1994) have modeled the effects of female-biased sex ratios on the ability of male ungulates to inseminate females and suggest that it can be a problem at extreme levels (e.g., 12 females to 1 male).

Deciding Where to Live and Hunt

The distribution of settlements in relation to prey distribution and density has been modeled by Horn (1968) and Wilmsen (1973). Their models predict that if resources are predictably and evenly dispersed, then the more efficient pattern of settlement is regular dispersion of small forager social units. In contrast, if the available resources are clumped together and move unpredictably throughout a large range, then the optimal strategy is aggregation of the foraging population at the center of that range. The ability of humans to directly exchange information about the location of prey resources enhances the value of central-place aggregation (Winterhalder, 1982). In areas where prey are predictably and evenly dispersed, the concentration of people into fewer larger settlements is likely to reduce harvesting efficiency and result in the depletion of prey populations close to villages. As prey abundances decrease, the range of prey species harvested will increase. By reducing travel costs, hunters can be encouraged to travel farther afield in search of prey.

Thus, if even harvest of resources is the preferred pattern of utilization, effective management would encourage the proliferation of small settlements or the introduction of more efficient means of transport (cars, snowmobiles, motorized canoes, etc.) to reduce search and travel costs. Alternatively, it might be preferable to concentrate harvest in a few areas, around settlements, leaving other areas as prey refuges, depending on the distribution of prey, prey selection, and harvesting patterns. Many hunting societies have become concentrated in a few permanent settlements that are larger than the traditional settlement units as a result of economic and in some cases government inducements (e.g., the Cree; Winterhalder, 1982).

Natural environments are patchy at some scale or another, with the result that prey are not distributed evenly across the available habitat and hunters must decide where to hunt and how long to stay in each area. Deciding where to hunt in a patchy or heterogeneous habitat has been the subject of optimal foraging models using the marginal value theorem (Charnov, 1976; Charnov et al., 1976). These models predict that hunters should remain in a patch as long as the expected returns from the next unit of foraging time in the patch are higher than the expected returns from searching for and exploiting other patches. In practice, this means that it is not normally worthwhile for hunters to harvest a patch completely before moving on to a new area, with the result that some animals are left (a potential breeding population) in each patch. For example, Cree only shoot the most readily available muskrats before moving on to the next population concentration (Winterhalder, 1982). Similarly, although beaver lodges may have three to five underwater openings, snares are set at only two or three, suggesting that the return is greater on two snares at the next beaver lodge than on additional snares at the same lodge (Winterhalder, 1981).

Differences in hunting success in different habitats can dramatically influence hunters' decisions about where to hunt and the prey species caught. Snaring and netting, for example, tend to take place in wooded habitats, where animals often move along defined trails where traps can be set (Arcese et al., 1995). As a result, in the Serengeti, Tanzania, ungulates living in woodland are far more vulnerable to overharvesting than those living in mixed habitat or open areas (Arcese et al., 1995).

Deciding how far to travel in search of prey is another problem faced by hunters and one that is strongly dependent on the available information about the distribution of prey resources and the costs of travel. Central-place foraging models predict that, as prey resources become depleted close to a settlement, the optimal forager will travel farther whenever the cost is offset by the harvesting benefits of more distant hunting sites. The lower the travel costs, the greater the evenness in resource use around a settlement.

Many traditional hunting societies were reported to maintain unhunted areas as sources of future prey or to rotate harvesting areas to provide opportunities for prey populations to recover. For example, the Mbuti divided their hunting territories so as to maintain a central "no mass land" which acted as a wildlife sanctuary where no hunting took place (Turnbull, 1983), while the Montagnais-Naskapi randomized hunting excursions and avoided repeated hunts in one area (Kelly, 1995).

Differences in harvesting patterns can substantially affect the impact that different people have on prey populations. For example, while Latin American *campesinos* tend to be sedentary, indigenous people harvest over a wider area, are less sedentary, and harvest on a rotational system, allowing populations time to recover (Ojasti, 1984; Redford and Robinson, 1987). Thus, for a variety of reasons, harvesting is often patchily distributed, being more intense close to villages and markets, and resulting in source and sink areas. Many hunted species may be able to tolerate high levels of exploitation as a result of such heterogeneous spatial distribution of the hunting pressure (Novaro, 1995), and this may therefore favor prey persistence.

Limiting Harvesting to Specific Times

In general, subsistence harvesting tends to take place throughout the year, although it may be more intensive at some times than others (for example, when other food is in short supply or when less time is required for tending crops; Infield, 1988; Colell et al., 1994). However, harvesting specific species may only be profitable at particular times of year. For example, although Ache men always pursue armadillos when they are encountered above ground, they only dig them from their burrows, a very labor-intensive process, during the late warm-wet and early dry-cold season when their prey are fat (Hill et al., 1987).

The effect of harvesting on prey species is likely to vary through the year. Compared with species that breed over longer time periods, species that have a very peaked birth or mating season may be more severely affected by harvesting than would be expected from the number of animals removed. Harvesting during the rutting season can result in the disruption of territorial structure, increased male–male conflict, and, as a result, reduced rates of conception (Gruver et al., 1984; Ginsberg and Milner-Gulland, 1994). Females not inseminated during their first cycle may either not conceive in that year or continue cycling, which could lead to a decrease in birth synchrony. For example, hunting of saiga *Saiga tatarica* during the mating season resulted in a reduction in conception rates from 85% in 1 year olds and 96% in adults to 55% and 86%, respectively (Bannikov et al., 1961).

Birthing is often carefully timed to fit in with seasonal change, for example, the onset of spring or the rainy season, or is highly synchronized to reduce predation. Consequently, delays due to disruption of normal mating patterns can result in increased juvenile mortality and a consequent decline in population size. For example, wildebeest calves born outside the main birthing peak experience higher mortality rates as a result of predation (Estes, 1976). In red deer, a delay in breeding for a single cycle of 18 days can result in a 36% decline in a female's reproductive success as a result of increased calf mortality and a mother's reduced fertility the following year (Clutton-Brock et al., 1983, 1987).

The timing of harvesting with respect to the main period of adult mortality can also influence the effect of harvesting on a prey population. For example, in the Serengeti, Tanzania, the main period of poaching is early in the dry season, preceding the period when intraspecific competition for scarce high-quality food would lead to increased wildebeest mortality. Illegal hunting therefore reduces the intensity of intraspecific competition (Campbell and Hofer, 1995).

Defense of Hunting Territories

Some hunter-gatherer societies defend hunting territories (e.g., the Mbuti; Turnbull, 1983), while others do not (e.g., the Shoshoni Indians of North America, Dyson-Hudson and Smith, 1978). Behavioral ecological models predict that territoriality should be found only where resources are economically defendable (Brown, 1964)—that is, when the benefits of defense outweigh the costs. Moderately dense, predictable, and evenly distributed resources create conditions that lead to the evolution of territoriality (Brown, 1964; Dyson-Hudson and Smith, 1978). Once territories are established, they create conditions more favorable to long-term conservation of resources because the forager is assured of the delayed benefit of short-term restraint (Winterhalder, 1982). Consequently, conservation of overdepleted prey species may be more likely when hunting territories are defended.

Selecting Prey Based on Religious, Tribal, and Personal Preferences

Religion not only influences whether wild animals are considered suitable food, it also dictates the type of species that can be harvested. For example, many Pemón indians are Adventists, which permits the hunting of only birds and deer among the higher vertebrates. Bushbabies are avoided in Bioko as being evil (Colell et al., 1994), and a variety of taboos may prevent the harvesting of particular species, such as chimpanzees in Korup (Infield, 1988). Individual preferences have a considerable effect on prey-species composition. For example, the Mayas avoid tapirs because the meat tastes unpleasant (Jorgenson, 1995). Some food taboos may have their origins in long-term information transfer about the low profitability of rarely encountered prey (Kaplan and Hill, 1992) or in the fact that some species are more vulnerable to overharvesting than others (Ross, 1978).

Commercialization

Commercialization not only results in the intensification of hunting efforts (Lavigne et al., 1996) but also in the concentration on medium-sized species, which can easily be transported to and sold at market; for example, 80% of antelopes but only 10% of rodents go to market in Bioko (Colell et al., 1994). The extent to which meat is shipped to market rather than being sold locally or consumed by the family depends on the buying power of the market (Juste et al., 1995).

Use of More Intensive Hunting Methods

Although the use of shotguns is often mentioned as one of the major factors contributing to overhunting (Mittermeir, 1987), the use of modern weapons does not necessarily result in increased offtake levels (Hames, 1979; Alvard, 1995b), but it may increase success rates of hunters. However, some taxa may be more vulnerable than others to overharvesting by shooting. Primates tend to be very visible, they are highly mobile and difficult to catch with traditional weapons, but they are within range of modern shotguns (Mittermeir, 1987). Winterhalder (1981) reports how access to nets, steel traps, and firearms substantially raised the speed and reliability with which the Cree could catch fish and game once they were encountered, lowering their handling times. As a result, a larger range of species could be taken, as prey that had previously been considered too small or too difficult to capture were now suitable foraging targets. Subsequently, access to boats and snowmobiles decreased the search time associated with foraging and reversed the earlier trend toward increasing diet breadth. As search time decreased, foragers grew more selective in their choice of prey.

Subsistence Harvesting and Behavioral Ecology: Conclusion

While in theory the rate at which a prey species can be harvested sustainably depends on its intrinsic rate of increase, the level at which it is harvested is strongly dependent on the behavior of both prey and predator. Understanding the factors that influence the susceptibility of different prey species to harvesting and the extent and organization of human hunting behavior is therefore essential for the development of an effective harvesting management program. Behavioral ecology has emphasized the different roles that individuals play in a population and how reproductive success varies according to age, sex, and individual status. Consequently, behavioral ecology can also help determine how the impact of harvesting depends on both the number and type of individuals removed. At the moment, the management of subsistence harvesting is usually restricted to broad recommendations such as setting up nonextractive areas, completely banning harvesting, or preferentially selecting males (e.g., Bodmer, 1995a). Information from behavioral ecologists will enable wildlife managers to refine offtake models, make more accurate predictions as to the impact of harvesting on prey populations, and refine management recommendations with respect to individual species.

The problem, of course, is that behavioral data are often difficult to obtain, particularly in tropical forests where much subsistence harvesting takes place. Tropical forest species are often solitary and shy, making behavioral research extremely time consuming and often requiring the use of expensive radio-tracking equipment to maintain contact with the animals. In addition, difficulties are exacerbated in tropical situations where resources are lacking and species are often close to extinction before action is taken. Immediate decisions are often required without recourse to information from detailed field studies. However, in some cases, detailed knowledge of one species may enable harvesting recommendations to be generalized to other related species, based on only limited behavioral information.

Realistically, it is in the understanding of human hunting behavior that behavioral ecology probably has most to offer conservation managers in the near future. Optimal foraging models can help identify decision rules that hunter-gatherers appear to follow in their efficient use of resources and can suggest specific and effective management practices (Winterhalder, 1982). The success of management programs will depend on their acceptance by hunters, and it is easier to persuade hunters to accept management recommendations that minimize the disruption of their optimal hunting behavior (Bodmer, 1994). Thus, for example, banning the hunting of a species that is not highly ranked is unlikely to cause great contention, but the conservation of a more favored species may require a more sensitive approach. In some cases, optimal foraging models have identified circumstances in which game conservation is consistent with maximally effective exploitation. Consequently, little or no intervention is required in the interests of conservation. For example, in patchy habitats, the efficient forager leaves some animals, a potential breeding population, in each patch. An optimal forager from a central place does not necessarily devastate resources close to home and ignore those farther away.

Alternatively, diet-breadth models suggest that certain highly ranked prey species will be pursued whenever encountered, and therefore conservation management may be necessary for these. In addition, optimal foraging models can often help predict how a change in hunting behavior—for example, the use of more effective weapons or more effective transport—is likely to influence the pattern and intensity of hunting. They may also help explain how demographic processes such as increases or decreases in population density are likely to influence animal communities and human foraging patterns and under what conditions hunters might deplete particular prey resources to extinction.

Recommendations

Although it is clear that behavioral ecology can contribute much to understanding and predicting the effects of harvesting on prey species, it is less clear to what extent wildlife managers will make use of this information. Controlling the extent and nature of subsistence harvesting is notoriously difficult, particularly in the tropical forests of the developing world where many threatened species exist and infrastructure is limited. Not only will it be difficult to enforce recommendations but, in addition, we are often dealing with situations where people are hunting not to improve their standard of living but for the food necessary to stay alive.

Consequently, we need realistic recommendations that provide sensible options for local people and that are likely to be adopted. Only if behavioral ecology can help conservation biologists make such recommendations will it have contributed anything to the conservation of threatened species.

Summary

Wildlife harvesting is an important subsistence activity in many parts of the world. Although subsistence hunting poses a major threat to some prey species, the continued freedom to hunt is essential for maintaining the traditional lifestyles of many people as a source of food and income and to control crop predators. It is also one of the few ways in which local communities can derive benefits from wildlife, offsetting some of the direct and indirect costs of forest conservation.

Although in theory the rate at which a prey species can be harvested sustainably depends on its intrinsic rate of increase, the level at which it is actually harvested strongly depends on the behavior of both prey and predator. The response of a species to harvesting is influenced by behavior, the number and type of individuals removed, and the role that these individuals play in the population. Understanding the factors influencing human hunting behavior may enable managers to understand what determines prey selectivity and how variations in habitat, prey availability, and hunting methods influence the impact of hunting on prey populations.

Behavioral factors that may influence inter- and intraspecific vulnerability to harvesting include social behavior, vigilance and flight, activity patterns, use of trails, ranging behavior, habitat choice, and attraction to baits. Many species are reported to respond to harvesting by changes in behavior such as increased shyness, which reduces their susceptibility to harvesting. However, these changes may have detrimental effects on reproductive rates as a result of individuals trading off the costs of reduced foraging efficiency or parental care against the benefits of reduced predation risk, the indirect effects of human predation.

The determinants of human hunting behavior have been explored and the consequences for prey populations outlined. Optimal foraging models have been used to explain the extent of subsistence harvesting versus plant gathering and farming and the type of prey that should be harvested, including its age and sex. Predictions include the fact that large prey are generally more profitable than small prey, and low-ranked resources should drop out of the diet when search costs decrease and hence overall return rate increases. The body size of harvested prey is important in determining susceptibility to overharvesting because larger species tend to have lower reproductive rates and therefore are more vulnerable. Behavioral ecological data suggest that the consequences of harvesting particular age and sex classes may not be as clear as originally thought.

Optimal foraging models have also been used to explain the spatial pattern of hunting, how long to stay in a particular resource patch, how far to travel in search of prey, and the role of travel costs in these decisions. Such decisions are important because many hunted species may be able to tolerate high levels of exploitation only as a result of heterogeneous spatial distribution of the hunting pressure. Other aspects of human hunting include the timing of harvesting, the defense of hunting territories, commercialization of hunting, the use of intensive hunting methods, and the role of religious, tribal, and personal preferences in prey choice. Optimal foraging models have also been combined with population biology models to simulate the dynamic relationships between hunter-gatherers and their prey resources.

It is clear that behavioral ecology can contribute much to understanding and predicting the effects of harvesting prey species, but it is less clear to what extent wildlife managers will be able to make use of this information. Consequently, there is a need for realistic recommendations that provide sensible options for local people as these are likely to be the ones that are adopted.

Acknowledgments I thank Michael Alvard, Rachel Brock, Tim Caro, and an anonymous reviewer for many helpful comments on the original draft of this chapter.

References

Alvard M, 1993. Testing the "ecologically noble savage" hypothesis: interspecific prey choice by Piro hunters of Amazonian Peru. Hum Ecol 21:355–387.

Alvard M, 1995a. Intraspecific prey choice by Amazonian hunters. Curr Anthropol 36:789–818.

Alvard M, 1995b. Shotguns and sustainable hunting in the Neotropics. Oryx 29:58–66.

Alvard M, Kaplan H, 1991. Procurement technology and prey mortality among indigenous neotropical hunters. In: Human predators and prey mortality (Stiner M, ed). Boulder, Colorado: Westview Press; 79–104.

Arcese P, Hando J, Campbell K, 1995. Historical and present-day anti-poaching efforts in Serengeti. In: Serengeti II: Dynamics, management and conservation of an ecosystem (Sinclair ARE, Arcese P, eds). Chicago: University of Chicago Press; 506–533.

Arita HT, Robinson JG, Redford KH, 1990. Rarity in neotropical forest mammals and its ecological correlates. Conserv Biol 4:181–192.

Azcarate BT, 1981. Sociobiologia y manejo del capibara. Do-ana Acta Vert 6-7:1–228.

Balmford A, Leader-Williams N, Green MJB 1992. Protected areas of Afrotropical forest: history, status and propects. In Tropical rain forests—an atlas for conservation, vol 2. Africa (Collins M, Sayer JA, eds). London MacMillan; 69–80.

Bannikov AG, Zhirnov LV, Lebedeva LS, Fandeev AA, 1961. The biology of the Saiga. Moscow: Moskovskoi Veterinarnoi Akademii.

Belovsky GE, 1987. Hunter-gatherer foraging: a linear programming approach. J Anthropol Archaeol 3:29–76.

Belovsky GE, 1988. An optimal foraging-based model of hunter-gatherer population dynamics. J Anthropol Archaeol 7:329–372.

Berger J, 1991. Pregnancy incentives, predation constraints and habitat shifts: experimental and field evidence for wild bighorn sheep. Anim Behav 41:61–77.

Berger J, Cunningham C, 1988. Size-related effects on search times in North American ungulates females. Ecology 69:177–183.

Berger J, Cunningham C, 1995. Predation, sensitivity, and sex: why female black rhinoceroses outlive males. Behav Ecol 6:65–72.

Bergerud AT, 1974. Rutting behaviour of the Newfoundland caribou. In: The behaviour of ungulates and its relation to management (Geist V, Walther F, eds). Gland: International Union for the Conservation of Nature; 395–435.

Bishop RH, Rausch RA, 1974. Moose population fluctuations in Alaska. Nat Can 101:559–593.

Bodmer RE, 1994. Managing Amazonian wildlife: biological correlates of game choice of detribalized hunters. Ecol Appl 5:872–877.

Bodmer RE, 1995a. Comments on Alvard M, 1995. Intraspecific prey choice by Amazonian hunters. Curr Anthropol 36:804.

Bodmer RE, 1995b. Priorities for the conservation of mammals in the Peruvian Amazon. Oryx 29:23–28.

Bodmer RE, Fang TG, Moya LI, 1988. Primates and ungulates: a comparison of susceptibility to hunting. Primate Conserv 9:79–83.

Bodmer RE, Fang TG, Moya LI, Gill R, 1994. Managing wildlife to conserve Amazonian forests: population biology and economic considerations of game hunting. Biol Conserv 67:29–35.

Britton SW, 1941. Form and function in the sloth. Rev Biol 16:13–34.

Brotherton PNM, 1994. The evolution of monogamy in mammals (PhD thesis). Cambridge: University of Cambridge.

Brown JL, 1964. The evolution of diversity in avian territorial systems. Wilson Bull 76:160–169.

Burch ES, 1972. The caribou/wild reindeer as a human resource. Am Antiq 37:339–368.

Campbell K, Hofer H, 1995. People and wildlife: spatial dynamics and zones of interaction. In: Serengeti II: Dynamics, management and conservation of an ecosystem (Sinclair ARE, Arcese P, eds). Chicago: University of Chicago Press; 534–570.

Cashdan E, 1985. Coping with risk: reciprocity among the Basarwa of Northern Botswana. Man 20:454–474.

Catto G, 1976. Optimal production from a blesbok herd. J Environ Manage 4:105–121.

Caughley G, 1977. Analysis of vertebrate populations. London: John Wiley & Sons.

Caughley G, Gunn A, 1996. Conservation biology in theory and practice. Cambridge, Massachusetts: Blackwell Scientific.

Charnov EL, 1976. Optimal foraging, the marginal value theorem. Theor Popul Biol 9:129–136.

Charnov EL, Orians G, Hyatt K, 1976. Ecological implications of resource depression. Am Nat 110:247–259.

Clark C, 1990. Mathematical bioeconomics: the optimal management of renewable resources. New York: John-Wiley and Sons.

Clay JW, 1988. Indigenous peoples and tropical forest: models of land use and management from Latin America. Cambridge, Massachusetts: Cultural Survival Inc.

Clutton-Brock TH, Guiness FE, Albon SD, 1982. Red deer: behavioural ecology of two sexes. Edinburgh: Edinburgh University Press.

Clutton-Brock TH, Guiness FE, Albon SD, 1983. The costs of reproduction to red deer hinds. J Anim Ecol 52:367–383.

Clutton-Brock TH, Harvey PH, 1977. Species differences in feeding and ranging behaviour of primates. In: Primate ecology (Clutton-Brock TH, ed). New York: Academic Press; 557–584.

Clutton-Brock TH, Major M, Albon SD, Guiness FE, 1987. Early development and population dynamics in red deer. I. Density-dependent effects on juvenile survival. J Anim Ecol 56:53–67.

Cole LC, 1954. The population consequences of life history phenomena. Q Rev Biol 29:103–137.

Colell M, Maté C, Fa JE, 1994. Hunting among Moka Bubis in Bioko: dynamics of faunal exploitation at the village level. Biodiver Conserv 3:939–950.

Cordero RGA, Ojasti J, 1981. Comparison of the capybara populations of open and forested habitats. J Wildl Manage 45:267–271.

Crawshaw PG, 1991. Effects of hunting on the reproduction of the Paraguayan Caiman (*Caiman yacare*) in the Pantanal of Mato Grosso, Brazil. In: Neotropical wildlife use and conservation (Robinson JG, Redford KH, eds). Chicago: University of Chicago Press; 145–154.

Diamond J, 1988. The golden age that never was. Discover 9:70–79.

Dubost G, 1980. Ecology and social behaviour of the blue duiker, a small African forest ruminant. Z Tierpsychol 54:205–266.

Dyson-Hudson R, Alden Smith E, 1978. Human territoriality: an ecological reassessment. Am Anthropol 80:21–41.

Ebbinge BS, 1991. The impact of hunting on mortality rates and spatial distribution of geese wintering in the Western Palearctic. Ardea 79:143–157.

Ehrenfeld DW, 1970. Biological conservation. Toronto: Holt, Rinehart and Winston.

Eisenberg JF, 1980. The density and biomass of tropical mammals. In: Conservation biology, an evolutionary-ecological perspective (Soulé ME, Wilcox BA, eds). Sunderland, Massachusetts: Sinauer Associates; 35–55.

Estes RD, 1976. The significance of breeding synchrony in the wildebeest. E Afr Wildl J 14:135–152.

Fa JE, Juste J, Perez del Val J, Castroviejo J, 1995. Impact of market hunting on mammal species in Equatorial Guinea. Conserv Biol 9:1107–1115.

Fairall N, 1985. Manipulation of age and sex ratios to optimize production from impala (*Aepyceros melampus*) populations. S Afr J Wildl Res 15:85–88.

Fenchel T, 1974. Intrinsic rate of natural increase: the relationship with body size. Oecologia 14:317–326.

FitzGibbon CD, 1990. Why do hunting cheetahs prefer male gazelles? Anim Behav 40:837–845.

FitzGibbon CD, Mogaka H, Fanshawe JH, 1995. Subsistence hunting in Arabuko-Sokoke Forest, Kenya and its effects on mammal populations. Conserv Biol 9:1116–1126.

FitzGibbon CD, Mogaka H, Fanshawe JH, in press. Threatened mammals, subsistence harvesting and high human population densities: a recipe for disaster? In: Evaluating the sustainability of hunting in tropical forests. (Robinson J, Bennett E, eds). Columbia University Press.

Fowler C, 1981. Comparative population dynamics in large mammals. In: Dynamics of large mammal populations (Fowler C, Smith T, eds). New York: John Wiley & Sons; 437–455.

Geist V, 1988. How markets in wildlife meat and parts, and the sale of hunting privileges, jeopardize wildlife conservation. Conserv Biol 2:1–12.

Geist V, 1994. Wildlife conservation as wealth. Nature 368:491–492.

Georgiadis N, 1988. Efficiency of snaring the Serengeti migratory wildebeest. Serengeti Ecological Monitoring Project Serengeti Wildlife Research Institute, Tanzania.

Ginsberg JR, Milner-Gulland EJ, 1994. Sex-biased harvesting and population dynamics in ungulates: implications for conservation and sustainable use. Conserv Biol 8:157–166.

Giroux JF, 1991. Roost fidelity of pink-footed geese *Anser brachyrhyncus* in north east Scotland. Bird Study 38:112–117.

Gruver BJ, Guynn DC, Jacobsen HA, 1984. Simulated effects of harvest strategy on reproduction in white-tailed deer. J Wildl Manage 48:535–541.

Hames RB, 1979. Comparison of the efficiencies of the shotgun and the bow in neotropical forest hunting. Hum Ecol 7:219–252.

Hames R, 1992. Time allocation. In Evolutionary ecology and human behaviour (Alden Smith E, Winterhalder B, eds). New York: Aldine de Gruyter; 203–236.

Hames R, Vickers W, 1982. Optimal foraging theory as a model to explain variability in Amazonian hunting. Am Ethnol 9:358–378.

Harris L, Kochel I, 1981. A decision-making framework for population management. In: Dynamics of large mammal populations (Fowler C, Smith T, eds). New York: John Wiley & Sons; 221–239.

Hawkes K, O'Connell J, Blurton Jones N, 1991. Hunting income patterns among the Hadza: big game, common goods, foraging goals, and the evolution of the human diet. Phil Trans R Soc Lond B 334:243–251.

Hayes JP, 1991. How do mammals become engandered? J Wildl Manage 19:210–215.

Heffley S, 1981. Northern Athabaskan settlement patterns and resource distributions: an application of Horn's model. In: Hunter-gatherer foraging strategies (Winterhalder B, Alden Smith E, eds). Chicago: University of Chicago Press; 127–147.

Hennemann WW, 1983. Relationship among body mass, metabolic rate and the intrinsic rate of natural increase in mammals. Oecologia 56:104–108.

Hernandez-Camacho J, Cooper RW, 1976. The non-human primates of Columbia. In: Neotropical primates: field studies and conservation (Thorington RW, Heltne PG, eds). Washington, DC: National Academy of Sciences; 35–69.

Herrera EA, Macdonald DW, 1987. Group stability and the structure of a capybara population. Symp Zool Soc Lond 58:115–130.

Hildebrandt W, 1984. Late-period hunting adaptations on the north coast of California. J Cal Great Basin Anthropol 6:189–206.

Hildebrandt W, Jones TL, 1992. Evolution of marine mammal hunting: a view from the California and Oregon coasts. J Anthropol Archaeol 11:360–401.

Hill K, Kaplan H, Hawkes K, Hurtado A, 1987. Foraging decisions among Aché hunter-gatherers: new data and implications for optimal foraging models. Ethol Sociobiol 8:1–36.

Hillborn R, 1979. Some long-term dynamics of predator-prey models with diffusion. Ecol Model 6:23–39.

Hines WW, Lemos JC, 1979. Reproductive performance by two age-classes of male Roosevelt elk in south-western Oregon. Research Report no. 8. Oregon Department of Fish and Wildlife.

Horn HS, 1968. The adaptive significance of colonial nesting in the Brewers blackbird (*Euphagus cyanocephalus*). Ecology 49:682–694.

Hudson J, 1991. Nonselective small game hunting strategies: an ethnoarchaeological study of Aka Pygmy sites. In: Human predators and prey mortality (Stiner M, ed). Boulder, Colorado: Westview Press; 105–120.

Infield M, 1988. Hunting, trapping and fishing in villages within and on the periphery of the Korup National Park. Final report. Godalming, UK: World Wildlife Fund.

Jarman MV, 1979. Impala social behaviour: territory, hierarchy, mating and the use of space. Berlin: Verlag Paul Parey.

Jones TL, Hildebrandt WR, 1995. Reasserting a prehistoric tragedy of the commons: reply to Lyman. J Anthropol Archaeol 14:78–98.

Jorgenson JP, 1995. Maya subsistence hunters in Quintana Roo, Mexico. Oryx 29 :49–57.

Juste J, Fa JE, Perez del Val J, Castroviejo J, 1995. Market dynamics of bushmeat species in Equatorial Guinea. J Appl Ecol 32:454–467.

Kaplan H, Hill K, 1985. Food sharing among Aché foragers: tests of explanatory hypotheses. Curr Anthropol 26:223–245.

Kaplan H, Hill K, 1992. The evolutionary ecology of food acquisition. In: Evolutionary ecology and human behaviour (Alden Smith E, Winterhalder B, eds). New York: Aldine de Gruyter; 167–202.

Kelly RL, 1995. The foraging spectrum: diversity in hunter-gatherer lifeways. Washington, DC: Smithsonian Institution Press.

Knick ST, 1990. Ecology of bobcats relative to exploitation and a prey decline in southeastern Idaho. Wildl Monogr 108:1–42.

Krebs J, Davies NB, 1993. Introduction to behavioural ecology. Oxford: Blackwell Scientific.

Lavigne DM, Callaghan CJ, Smith RJ, 1996. Sustainable utilization: the lessons of history. In: The exploitation of mammal populations (Taylor VJ, Dunstone N, eds). London: Chapman and Hall; 250–265.

Lewis HT, 1982. Fire technology and resource management in Aboriginal North America and Australia. In: Resource managers (Williams NM, Hunn ES, eds). Melbourne: Australian Institute of Aboriginal Studies; 45–68.

Lima SL, Dill LM, 1990. Behavioral decisions made under the risk of predation: a review and prospectus. Can J Zool 68:619–640.

Marks SA, 1973. Prey selection and annual harvest of game in a rural Zambian community. E Afr Wildl J 11:113–128.

Mittermeier R, 1987. Effects of hunting on rain forest primates. In: Primate conservation in the tropical rain forest (Marsh C, Mittermeier R, eds). New York: Alan Liss; 109–148.

Moreira JR, 1995. The reproduction, demography and management of capybaras (*Hydrochaeris hydrochaeris*) on Marajó Island, Brazil (DPhil thesis). Oxford: University of Oxford.

Moreira JR, Macdonald DW, 1993. The population ecology of capybaras (*Hydrochaeris hydrochaeris*) and their management for conservation in Brazilian Amazonia. In: Biodiversity and environment: Brazilian themes for the future (Mayo SJ, Zappi DC, eds). London: Linnean Society of London; 26–27.

Moreira JR, Macdonald DW, 1996. Capybara use and conservation in South America. In: The exploitation of mammal populations (Taylor VJ, Dunstone N, eds). London: Chapman and Hall; 88–101.

Nelson RK, 1982. A conservation ethic and environment: the Koyukon of Alaska. In:

Resource managers (Williams NM, Hunn ES, eds). Melbourne: Australian Institute of Aboriginal Studies; 211–228.

Nelson L, Peek J, 1982. Effect of survival and fecundity on rate of increase of elk. J Wildl Manage 46:535–540.

Novaro AJ, 1995. Sustainability of harvest of culpeo foxes in Patagonia. Oryx 29:18–22.

Ojasti J, 1984. Hunting and conservation of mammals in Latin America. Acta Zool Feen 172:177–181.

Orians GH, Pearson NP, 1979. On the theory of central place foraging. In: Analysis of ecological systems (Horn DJ, Stairs GR, Mitchell RD, eds). Columbus: University of Ohio Press; 155–177.

Peres CA, 1990. Effects of hunting on western Amazonian primate communities. Biol Conserv 54:47–59.

Peters RH, Wassenberg K, 1983. The effect of body size on animal abundance. Oecologia 60:89–96.

Prescott-Allen R, Prescott-Allen C, 1982. What's wildlife worth? Washington, DC: International Institute for Environment and Development.

Prins HHT, Iason GR, 1989. Dangerous lions and nonchalant buffalo. Behaviour 108: 262–296.

Prothero WL, Spillett JJ, Ralph DF, 1979. Rutting behaviour of yearling and mature elk: some implications for open bull hunting. In: North American elk: ecology, behaviour and management (Boyce MS, Hayden-Wing LD, eds). Laramie: University of Wyoming; 160–165.

Ráez-Luna EF, 1995. Hunting large primates and conservation of the Neotropical rain forests. Oryx 29:43–48.

Redford KH, 1992. The empty forest. Bioscience 42:412–422.

Redford KH, Robinson JG, 1985. Hunting by indigenous peoples and conservation of game species. Cult Surv Q 9:41–43.

Redford KH, Robinson JG, 1987. The game of choice: patterns of indian and colonist hunting in the neotropics. Am Anthropol 89:650–667.

Redford KH, Stearman AM, 1993. Forest-dwelling native Amazonians and the conservation of biodiversity: interests in common or in collision? Conserv Biol 7:248–255.

Robinson JG, Redford KH, 1986a. Body size, diet and population density of neotropical forest mammals. Am Nat 128:665–680.

Robinson JG, Redford KH, 1986b. Intrinsic rate of natural increase in Neotropical forest mammals: relationship to phylogeny and diet. Oecologia 68:516–520.

Robinson JG, Redford KH, 1991. Sustainable harvest of neotropical forest mammals. In: Neotropical wildlife use and conservation (Robinson JG, Redford KH, eds). Chicago: University of Chicago Press; 415–429.

Roff DA, 1974. The analysis of a population model demonstrating the importance of dispersal on a heterogeneous environment. Oecologia 15:259–275.

Ross EB, 1978. Food taboos, diet and hunting strategy: the adaptation to animals in Amazon cultural ecology. Curr Anthropol 19:1–36.

Silva JL, Strahl SD, 1991. Human impact on populations of Chachalacas, Guans, and Curassows (Galliformes: Cracidae) in Venezuela. In: Neotropical wildlife use and conservation (Robinson JG, Redford KH, eds). Chicago: University of Chicago Press; 37–51.

Sinclair ARE, 1977. The African buffalo: a study of resource limitation of populations. Chicago: University of Chicago Press.

Smith J, 1980. Managing elk in the Olympic Mountains. In: Proceedings of the western states elk workshop, Cranbrook, British Columbia (Macgregor W, ed); 67–111.

Suarez E, Stallings J, Suárez L, 1995. Small-mammal hunting by two ethnic groups in north-western Ecuador. Oryx 29:35–42.

Turnbull CM, 1983. The Mbuti pygmies, change and adaptations. New York: Holt, Reinhart and Winston.

Vincent ACJ, 1995. Trade in seahorses for traditional chinese medicines, aquarium fishes and curios. TRAFFIC Bull 15:125–128.

Western D, 1979. Size, life-history and ecology in mammals. Afr J Ecol 17:185–204.

Wilmsen EN, 1973. Interaction, spacing behaviour and the organization of hunting bands. J Anthropol Res 29:1–31.

Winterhalder B, 1977. Foraging strategy adaptations of the boreal forest Cree: an evaluation of theory and models from evolutionary ecology (PhD dissertation) Ithaca, New York: Cornell University.

Winterhalder B, 1981. Foraging strategies in the boreal forest: an analysis of Cree hunting and gathering. In: Hunter-gatherer foraging strategies (Winterhalder B, Alden Smith A, eds). Chicago: University of Chicago Press; 66–98.

Winterhalder B, 1982. The boreal forest, Cree-Ojibwa foraging and adaptive management. In: Resources and dynamics of the boreal zone (Wein R, Riewe R, Methven I, eds). Ontario: Association of Canadian Universities for Northern Studies; 331–345.

Winterhalder B, 1986. Diet choice, risk, and food sharing in a stochastic environment. J Anthropol Archaeol 5:369–392.

Winterhalder B, Baillargeon W, Cappelletto F, Randolph Daniel JI, Prescott C, 1988. The population ecology of hunter-gatherers and their prey. J Anthropol Archeaol 7:289–328.

17

Indigenous Hunting in the Neotropics: Conservation or Optimal Foraging?

Michael Alvard

Subsistence Hunters and Conservation of Their Faunal Resources

Much work of conservationists arises from the desire to balance individuals' needs with the long-term goal of conserving biological diversity. One result is that solutions often include persuading people to behave in ways that are contrary to their own short-term self-interest. This conflict is apparent in the context of subsistence or traditional peoples and their use of species and habitats that conservation biologists consider threatened or endangered. Some conservationists view native peoples as allies whose goals are essentially isomorphic with their own (e.g., Alcorn, 1993), while others consider native people to be at least part of the problem (Redford and Stearman, 1993). There is no question that people and their use of natural resources are the ultimate cause of the conservation dilemma. It is human activity that leads to the destruction of ecosystems, the extinction of species, and the loss of biodiversity. A corollary to this truth is the implication that "natural" (not influenced by human action) processes that lead to extinction, habitat loss, or loss of biodiversity are an acceptable part of the way nature works (e.g., predation by nonhuman predators). The standard view of humans as despoilers of nature, however, is often reserved for industrial and postindustrial societies (Oelschlaeger, 1991; Budianski, 1995).

Indigenous people are frequently afforded a separate status. They are often considered closer to nature, less intrusive, even more like nonhuman animals with regard to the resources they use. Following this line of thinking, traditional people cannot be any more harmful to their environment than can the rest of nature's creatures. The conventional opinion of anthropology, the current belief of many conservationists, and much of the lay public, is that subsistence people are benign stewards of nature (Gorsline and House, 1974; Dasmann, 1976; Reichel-Domatoff, 1974; Nelson, 1982; Posey, 1985; Hughes, 1983; Todd, 1986; Clay, 1988; Bunyard, 1989; Oelschlaeger, 1991; Pearce, 1992). The perception of natives as people yet not-people has a long history in European thought and is still current in many subtle ways. For example, witness the belief that the New World was unspoiled wilderness only inhabited by Indians before 1492 (see Simms, 1992; Denevan, 1992). In conservation circles, native people are viewed as a fix to many conservation problems (Cox and Elmquist, 1991; Gadgil et al., 1993; Peres, 1994). An oft-cited example is the country of Columbia. Its leaders granted 28 million ha of lowland tropical rainforest to indigenous people with the assumption that since they live in harmony with nature, their stewardship will ensure the forests will be conserved (Bunyard, 1989). Indeed, many extractive reserves rely on the assumed "benign stewardship" of native people to protect plants and animals (Fearnside, 1989; Anderson, 1990; Poffenberger, 1990).

There is a growing consensus that this view is in error and that it needs to be acknowledged that exploitative behavior was and is a part of the behavioral repertoire of traditional peoples (Clad, 1985; Hames, 1987, 1991; Johnson, 1989; Redford, 1991; Diamond, 1992; Alvard, 1993a; Headland, 1994; Low, 1996). Some of the strongest evidence in support of this view is that prehistoric groups, entirely devoid of any possible "contamination" by western culture, had significant disruptive effects on their resource base.

For example, the extinction of many mammal species in the late Pleistocene has been attributed by a number of researchers to overkill by human hunters (Martin, 1984; Owen-Smith, 1989; Burney, 1993). Humans arrived on Madagascar around 1500 B.P. and hunted gigantic lemurs to extinction (Tattersall, 1982; Dewar, 1984). The arrival of the Maoris, the "native" people of New Zealand, approximately 1000 years ago was quickly followed by the extinction of 34 species of birds, the most well known being the moa (Trotter and McCulloch, 1984). A large number of bird extinctions followed the first migrations of humans into and across the Pacific Ocean (Steadman and Olson, 1985; Olson, 1989). Archaeological evidence from Paleoindian kill sites indicate excessive wastage of meat (Wheat, 1972). Evidence for a negative impact on flora includes a direct correlation with the appearance and intensification of agriculture and the establishment of grasslands in previously forested tropical regions, including the highland of New Guinea (Golson, 1977) and upland regions of Sulawesi, Indonesia (M. Alvard, personal observation).

Now, as conservationists realize that solutions to many problems need to be addressed at the grass-roots level, it is imperative to understand how local users of resources affect their environment. Indigenous people deserve title to their homelands, but there is much to lose if notions of natives as "ecologically noble" are not accurate and no other measures of environmental protection are taken (Peres, 1994).

Previous Perspectives

The view of native peoples as normatively "ecologically noble" has many origins and will not be reviewed here (see Alvard, 1993b). It is useful, however, to examine briefly two common errors—one methodological, the other conceptual—that have contributed to the confusion. Although much has been written about indigenous peoples and their relationship to the resources they use, researchers have directed little empirical work at the question of native conservation. Traditional methods in anthropology are based on the naive assumption that people behave as informants say they do. Because many native belief systems apparently emphasize a reverence for nature, the conclusion is that people act accordingly (see Callicott, 1989; Vecsey, 1990; Hames, 1991; Budianski, 1995). An example is the long-held perception that Eastern religions are different from the Western religions in that nature and the individual are not conceptually dichotomized. This has led to the conclusion that developing Eastern cultures are in a better position to avoid the environmental disasters beset on the West. Interestingly, it is along the slopes of the Himalayas in Tibet where some of the worst cases of deforestation and erosion are occurring—with local, devout Buddhists the culprits (see Tuan, 1968, 1970; Low and Heinen, 1993). Another example comes from the work of Reichel-Domatoff (1974). His work with the Tukano of Eastern Colombia painted a picture of a society living with a world view dominated by the spiritual reverence of nature and a Shamanistic tradition steeped with a conservation ethic. Unfortunately, the data were collected from interviews with one man living in the urban area of Bogata. The informant had not lived with his people for 10 years. Nonetheless, his second-hand description of how the people thought about nature was interpreted as how they actually behaved (see Vickers, 1995).

A serious error made by researchers who *have* looked at native resource-use more systematically arises from the failure to appreciate the difference between the effects of behavior and behavioral design. Hames (1991) argues that it is critical that the suspected conservation behavior be shown to be designed to conserve resources and is not a collateral effect of other behaviors. It is well known that a number of behaviors exhibited by native people result in conservationlike outcomes. Many traditional food taboos, for example, result in smaller harvests of endangered prey species. Ross (1978) argues that Achuara Indians of Peru taboo prey species susceptible to overexploitation. McDonald (1977) referred to such taboos as "a primitive environmental protection agency." One problem with this conclusion is that it is not clear that prey conservation is the reason that the taboo behaviors originated or are maintained. Thus, it is ambiguous whether such behavior should be considered genuine conservation or should be better identified by the term epiphenomonal conservation. In his discussion of the subsistence strategy of Sahaptin foragers of the North American Colombia Plateau, Hunn (1982) reasoned that the apparent balance these people had with their environment resulted from low population densities, limited technology, and a highly mobile foraging pattern. In such a context, the Sahaptin simply did not have the capability to overexploit resources. Hunn referred to this as epiphenomonal conservation, which can be defined as behavior that produces sustained harvests and/or other conservationlike results, yet is the result of noncostly behavior or is the by-product of behaviors whose purpose is other than conservation.

Recent work with the Wana of Sulawesi (M. Alvard, unpublished data) shows that the labor requirements of the agricultural cycle draw men away form hunting and trapping and may allow time for prey populations to rebound. This is not why the Wana harvest their rice, however. Conversely, some animals, notably wild pigs, are attracted to the secondary forests created by the disturbances of Wana agriculture—again, this is not why the Wana create the disturbance. Simply observing a native group living at equilibrium with its environment is not sufficient evidence to label them conservationists (Hames, 1991; Alvard, 1993a; Vickers, 1995).

Native Conservation and Behavioral Ecology

In response to the inadequacy of previous approaches, researchers have turned to behavioral ecology as a theoretical and methodological framework to address the question of native conservation. Optimal foraging theory (OFT), in particular, has been useful for examining issues of resource use by humans with subsistence economies. Early theoretical work argued that animals altruistically restrained their reproduction to conserve resources and prevent habitat degradation (Wynne-Edwards, 1962). Subsequent arguments of Williams

(1966) and others have ruled out such selfless behavior or, as I refer to it here, truly altruistic conservation. Altruistic conservation would consist of costly restraint on the part of the actor, with the benefits accruing to the group. The actor in this case may receive some benefits, but they are shared by the group and are small compared to the initial costs. It should be noted that although the arguments against group selection rule out evolution producing altruistic conservation, they do not exclude the possibility of selfish conservation as an evolved strategy. Selfish conservation behavior could conceivably evolve if the long-term benefits that accrue to the actor (appropriately discounted because of time delay) are greater than the initial costs (Hill, 1993). Whether or not this behavior is common and in what circumstances it is most likely to occur remain both empirical and theoretical questions that behavioral ecology may help answer.

A classic example that has demonstrated the advantages of an optimal foraging perspective on a conservation problem is Smith's (1983) critique of Feit's (1973) interpretation of Waswanipi Cree hunting practices. The Cree are a subsistence hunting population living in the Subarctic region of Quebec, Canada. Feit observed that the Cree rotated their hunting areas on a semiannual basis and interpreted this behavior as a means of game conservation. As game in one area was depleted, rather than hunt the game to local extinction, he argued, Cree hunters moved their hunting efforts to less depleted areas. Smith noted that the movement of hunting zones or patches as described by Feit is also predicted by the marginal value theory (MVT) of optimal foraging theory. Rather than change patches to conserve game, the MVT predicts that hunters will change patches so as to maximize hunting returns (Charnov, 1976). When returns in the current patch drop to a point equal to that of the entire habitat, hunters switch to a less depleted patch (see also Winterhalder, 1981). Smith argues it is short-term returns, not long-term sustainability, that probably motivate hunting decisions in this case. While theoretically more satisfying, the OFT interpretation of the Cree hunting decisions predicts the same behavior as the conservation interpretation, and Smith offered no way to test between the alternatives.

Conservation Defined in a Behavioral Ecological Framework

Clearly, behavioral effects or outcomes are important for understanding conservation. Simply observing outcomes, however, can only provide evidence to reject the hypothesis of conservation. If a group of hunters depletes their prey, most would agree not to label them conservationists. But, because of the possibility of epiphenomonal conservation, outcomes such as sustainable harvests do not necessarily indicate genuine proactive conservation. Additional criteria are needed to discern epiphenomonal from genuine conservation.

A working definition of conservation must have two components. First, it must include the notion of restraint that is inherent in the idea of conservation. Actual conservers check their level of resource use; they limit their consumption to some point below what would be optimal in the short term (Alvard, 1993a,b, 1994, 1995a). Second, the criterion of design must be satisfied. The costly behavior must be designed to result in long-term benefits via sustained future harvests of the resource in question (Hames, 1991; Alvard, 1995a; Smith, 1995). As Smith (1995:810) states, "If natural selection (of genetic or cultural variants) favors something because it prevents or mitigates resource depletion, this would be an example of true conservation by this definition." These two criteria provide a means of distinguishing epiphenomonal conservation from genuine, as well as placing conservation squarely within the reach of cost–benefit analyses common to behavioral ecological research.

Conservation and Optimal Foraging

Optimal foraging theory provides an empirical and theoretically sound method of measuring the costs and benefits of resource acquisition (see Stephens and Krebs, 1986). It has two advantages as a tool for testing for the conservation proclivities of native hunters. First, it is grounded in a firm body of evolutionary theory. Second, OFT can make precise predictions about resource acquisition behaviors that contrast with the predictions made by the conservation hypothesis.

Optimal foraging theory is linked to evolutionary theory by two basic assumptions. The first is that more food, up to a point, increases individual survivorship and fertility. The second is that minimizing the amount of time spent acquiring food allows a forager to engage in other fitness-enhancing activities (Kaplan and Hill, 1992). In such a context, foragers are expected to maximize their short-term harvesting rate (Stephens and Krebs, 1986).

As I will show, the simple OFT models provide a strong test in contrast to the conservation hypotheses and provide a basis on which to build and test what might be called behavioral conservation theory. Using the parlance of foraging theory, conservation is a strategy where current rate maximization is sacrificed in return for sustainable use of the resource population into the future. *If there is no short-term cost in terms of rate maximization, the presumed conservationlike behavior cannot be distinguished from optimal foraging.* The assumption of selfish, short-term rate maximization made by the basic prey-choice models and the assumption of long-term conservation goals implicit in a conservation strategy are the major contrasts between the two hypotheses. These two contrasting assumptions produce mutually exclusive predictions about hunters' behaviors that can be tested using field data.

The Piro Hunting Project

I conducted a study to test contrasting predictions generated from OFT and from the assumptions of native conservation. I reasoned that if indigenous people possess the intimate knowledge of their environment that is often attributed to them, it is not unreasonable to assume that native hunters are aware of the reproductive parameters and limitations of their prey species. Native hunters following a conservation strategy would be expected to use their knowledge to minimize their impact on their prey. Hunters could accomplish this goal in a number of ways. Conservation tactics that I tested for among the Diamante Piro include selective interspecific and intraspecific prey choice and selective patch-choice decisions. These analyses are the focus of this paper. I also performed tests to examine evidence of game depletion and harvest sustainability. The conservation predictions as well as the alternative OFT predictions are presented in table 17-1 and briefly reviewed below.

The Diamante Piro inhabit the Neotropical rainforests of Southeastern Peru. I collected data during two field sessions in the study community of Diamante. The village is located on the Alo Madre de Dios River, on the border of Manu National Park in the department of Madre de Dios, Peru. I was present in Diamante from August 1988 from May 1989 and again from October 1990 to May 1991. Although the Piro practice a small of amount of wage labor and market activities, their economy is essentially subsistence based. Hunting, fishing, gathering, and horticulture produce 95% of the calories in their diet. The Piro are proficient and frequently hunt and fish with bows, but they obtain the majority of their game (85% by weight) with shotguns.

Table 17-1 Conservation and alternative optimal foraging theory (OFT) predictions.

Type	OFT prediction	Conservation prediction
Interspecific prey choice	Hunters choose prey types that maximize return rate and ignore those that do not.	Hunter choose species that are less vulnerable to local extinction.
Intraspecific prey choice	Hunters take each sex in proportion to the sex ratio in the prey population, assuming both sexes are in the optimal diet.	For polygynous species, hunters choose males in greater proportion than their abundance.
Intraspecific prey choice	Hunters take adult individuals and ignore immature individuals if they fall out of the body-size range of the optimal diet.	Hunter choose younger and older rather than prime-aged individuals.
Patch choice	Hunters choose the most profitable patches.	Hunters choose patches in proportion to prey abundance.
Depletion	Prey may or may not be locally depleted.	Prey are not locally depleted.

The important prey for the Piro shotgun hunters include many types common to the diets of native groups across Amazonia (Vickers, 1984; Redford and Robinson, 1987). The Brazilian tapir *Tapirus terrestris,* and collared peccary *Tayassu tajacu* are the two species that provide the most meat in the Piro diet (Alvard, 1993b). Red brocket deer *Mazama americana,* capybaras *Hydrochaeris hydrochaeris,* and agoutis *Dasyprocta variegata* are taken occasionally. Black spider monkeys *Ateles paniscus* and red howler monkeys *Alouatta seniculus* are the two important primates in the diet. Other game include capuchin monkeys *Cebus apella* and cracid (Cracidae) game birds.

The data I present here are from 79 directly observed hunts. I collected additional data on 120 randomly selected unobserved hunts through systematic interviews with the hunters. Records of the offtake for a subsample of the village were kept (~140 individuals; 37,003 consumer days). I also collected the mandibles of all game killed when possible (see Alvard, 1993a, 1994, 1995a,b).

Hypotheses and Predictions

Interspecific Prey Choice

The ability of different prey to withstand various levels of harvest without depletion varies with the population dynamics of the species (Caughley, 1977; Winterhalder and Lu, 1997). Although every species is able to withstand some level of harvest, some are particularly susceptible to overexploitation and local extinction because of slow reproductive rates and/or low population densities (Robinson and Ramirez, 1982). Hunters interested in conserving prey would identify those species most susceptible to uncontrolled harvesting and refrain from killing more than would be sustainable (e.g., Ross, 1978).

For the purposes of the study, I used the maximum intrinsic rate of increase (r_{max}) as a measure of a prey species' vulnerability to overhunting and local extinction. Although density is also an important factor, r_{max} is positively correlated with density (Robinson and

Table 17-2 Maximum intrinsic rate of increase for Piro prey species (from Robinson and Redford, 1986).

Species	r_{max}
Spider monkey	0.08
Capuchin monkey	0.14
Cracid birds	0.15
Howler monkey	0.16
Tapir	0.20
Red brocket deer	0.40
Capybara	0.69
Collared peccary	0.84
Agouti	1.10

A rough estimate for cracid birds was calculated from data in Silva and Strahl (1991).

Redford, 1986a; Thompson, 1987). In addition, density will vary with among habitats, but r_{max} does not. Table 17-2 presents estimates of r_{max} for a number of the Piro prey species. These estimates are obtained from Robinson and Redford (1986b) who used Cole's (1954) equation to calculate r_{max} from age of first and last reproduction and the annual birth rate. If the hunters engage in any restraint to sustain the fauna around Diamante, the conservation prediction is that this practice should be most apparent in the species identified as most vulnerable to overhunting. For the Piro these are the large primate species (howler and spider monkeys), tapir, and the cracid game birds.

The basic prey-choice model of OFT and its predictions are well known. Details of the model can be found in Stephens and Krebs (1986) and its application to the Piro case in Alvard (1993a). According to the model, hunters are assumed to make decisions that maximize the foraging return rate, measured in terms of resource acquired per unit time spent foraging (Stephens and Krebs, 1986). From this point of view, selective harvests are the result of how profitably different prey can be killed, rather than how their removal will affect the sustainability of the harvest (Pyke et al., 1977). To determine which types a hunter should pursue, prey types are ranked according to their profitability, calculated as the amount of energy harvested per time handling the prey type. The "optimal diet" refers to those prey types that, if pursued upon encounter, will maximize the hunter's return rate. A prey type is included in the optimal diet if the average expected return rate for pursuing that type when encountered is higher than the average expected return rate for continued search for all higher ranked items (Stephens and Krebs, 1986). The predictions of OFT are unambiguous: in order to maximize the return rate, hunters should pursue those prey types in the optimal diet and ignore those that are not.

Figure 17-1 plots two types of data for each ranked prey item. The first is the item's profitability (i.e., the mean return rate from pursuit). This is the average return in calories divided by the time required for handling (pursue, kill, and field process) the item. Handling begins upon encounter and is mutually exclusive from search. This value includes unsuccessful pursuits. The second data are the mean return rates from continued search for higher ranked prey. This is the average expected return if the hunter ignores the encounter and continues to search for higher ranked items. This value is the average return in calories for higher ranked prey divided by both handling and search time.

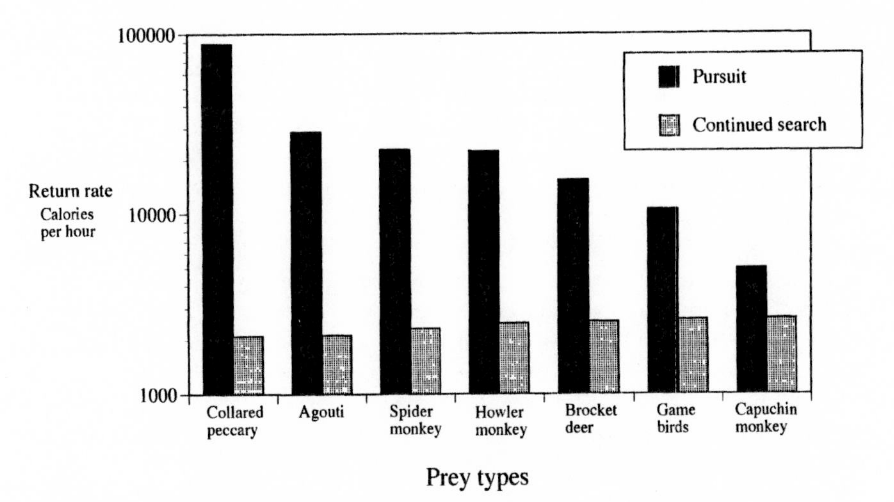

Figure 17-1 Results of the prey-choice analysis. At every encounter, hunters have the option of pursuit or continued search for higher ranked prey. The bars represent the average return rate for each decision for each species. The rates are calculated from observed pursuits. For each prey type in the graph, the rate-maximizing decision is for hunters to pursue upon encounter (see text for details). Reprinted from Alvard 1993a with permission.

Collared peccary is the most profitable prey type, followed by agouti, spider and howler monkeys, brocket deer, game birds, and finally the capuchin monkey. These prey are either large packages of calories and/or types that can be taken with relatively short pursuits. Insufficient numbers of capybara and tapir pursuits were observed to include these species in the analysis. This is also true for numerous small species such as small primates (tamarins, titi monkeys, squirrel monkeys), squirrels, and most birds. Conservative estimates of handling times places both capybara and tapir well within the optimal diet and places the small species outside the diet of the shotgun hunters (see Alvard, 1993a). The OFT prediction is that the types in the optimal diet should be pursued by hunters at every opportunity regardless of their vulnerability to overhunting. The predicted prey choice will be compared to the actual prey choice of Piro hunters. Restrained harvesting of prey vulnerable to extinction, which are nonetheless in the optimal diet, would support the conservation model.

Intraspecific Prey Choice by Age and Sex

Selective intraspecific prey choice is also a tactic available to hunters whose goal is to conserve prey. Selective harvest of individual prey that have low reproductive value can significantly mitigate the impact of hunting compared to nonselective harvesting (Caughley, 1977). If Piro hunters choose age categories to limit their impact on prey population growth, hunters will bias their kill toward immatures and older adult animals, while avoiding prime-aged animals. This is equivalent to the tactics employed by Slobodkin's prudent predators (1961, 1968; see also MacArthur, 1960).

With respect to sex, male-biased harvests have been shown to mitigate the impact of hunting on game populations (Hayne and Gwynn, 1977; Fowler 1981; Harris and Kochel, 1981; Nelson and Peek, 1982). This is because males generally have greater variance in reproductive success than females. It is possible, particularly in species with polygynous or promiscuous mating systems, for a few males to monopolize many females and sire the majority of the offspring produced. Such a pattern essentially renders a significant proportion of the males in the population superfluous with regard to population growth (see Greene et al., chapter 11, this volume).

In theory, the OFT prey-choice model can be refined to predict hunter intraspecific prey choice as well. In this case, rather than have species define the prey types, the age and sex classes of each species define the types. The methodological limitation to this approach is that sufficient numbers of pursuits for each age and sex category for each species need to be observed to calculate the profitability for each type. This type of analysis requires a much larger sample of observed hunts than is available for the Piro.

As an alternative, I assumed that handling times were equal across age and sex categories within species. With this simplifying assumption, body size (number of calories) determines intraspecific prey choice. Adults are larger and more profitable than younger animals, and in sexually dimorphic species males are usually larger and more profitable than female. Piro hunters, whose goal is to maximize returns, are expected to take each sex in proportion to the sex ratio in the population, assuming both sexes are in the optimal diet. This will be true unless one sex is significantly smaller than the other and falls out of the diet. Hunters will prefer adult individuals and ignore immature individuals if they fall out of the body-size range of the optimal diet (see below).

Patch Choice

As implied by Feit's (1973) work discussed earlier, judicious patch choice is also a way subsistence hunters could conserve prey. Hames (1987, 1991) examined this question and predicted a number of behaviors from conserving hunters. Using empirical data collected among Amazonian Indians, he investigated the allocation of hunting time to zones with different prey abundances. Foraging theory predicts that hunters will allocate more time to areas where prey densities are higher and return rates are greater than to areas where return rates are lower. Using data from Yanomamo and Ye'kwana Indians in Venezuela, Hames (1987, 1991) showed that return rates increased significantly with distance from the villages, indicating local depletion, and that hunters spent significantly more time in those areas where return rates where higher. Hames also recognized the problem originally identified by Smith (1983): these observations are consistent with a conservation strategy that predicts hunters will avoid depleted areas.

The central-place foraging context of village-dwelling hunters provides a way to test for conservation. Hames (1991) noted that hunters traveling to more distant and productive hunting zones must first move through depleted areas near the village. Hunters whose goal is to avoid local depletion are predicted to avoid taking game they opportunistically encounter while traveling through the depleted areas. They pay the cost required of genuine conservation by ignoring animals they could otherwise kill. Rate-maximizing foragers are predicted to pursue game in their optimal diet wherever they are encountered, including depleted areas. While he had no quantitative data to evaluate the hypothesis, Hames noted that during his follows of hunters, they always pursued game in depleted areas during travel to undepleted areas.

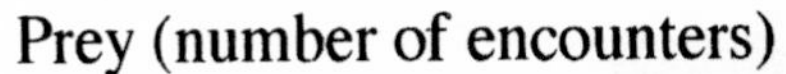

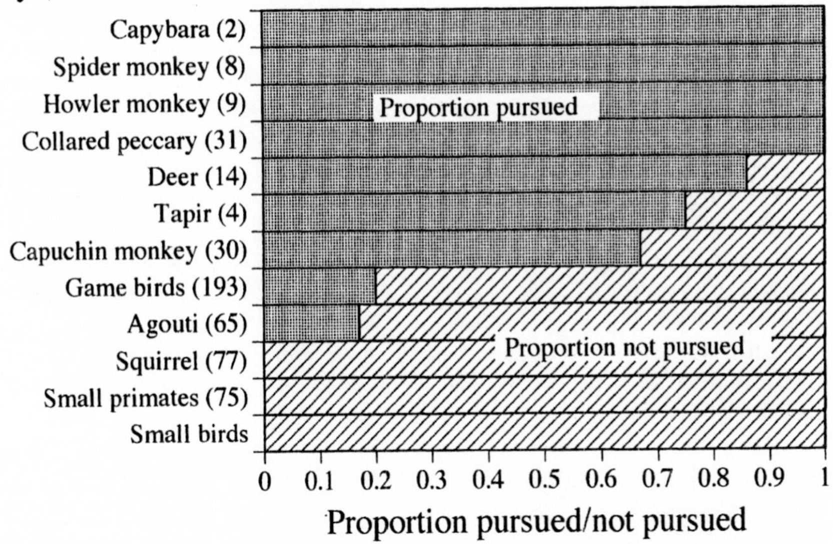

Figure 17-2 Proportion of encounters pursued by shotguns hunters. Number of hunts sampled = 64. Small bird encounters were too numerous to count. (Figure reprinted from Alvard, 1995a, with permission.)

Results

Interspecific Prey Choice

Figure 17-2 presents pursuit data from the observed hunt sample for selected prey species. The prey items that are always pursued by Piro shotgun hunters closely match those predicted by foraging theory. Collared peccaries were pursued at every encounter except for two cases during unobserved hunts when ammunition was not available. Spider monkeys, howler monkeys, capybara, deer, and tapir were pursued at nearly every encounter. Contrary to the conservation hypothesis, hunters displayed no restraint in killing the large primate species and tapirs. Consistent with the foraging hypothesis, the shotgun hunters ignored the less profitable small primates, squirrels, and small birds.

Among some of the middle-ranked prey, however, there is considerable deviation from the all-or-nothing pursuit decision predicted by the simple prey-choice model (Krebs and McCleery, 1984). The Piro show a partial preference by pursuing 14% of the agoutis, 24% of the spix's guans *Penelope jacquacu*, and 66% of the capuchin monkeys encountered. Although agoutis are not among the species identified as more vulnerable to over hunting, the cracids and capuchins are candidates for local depletion. This presents the possibility that the Piro hunters ignore harvesting opportunities to conserve these species.

There are two reasons not to invoke the partial preferences as evidence of conservation. First, the Piro partial preferences are not consistent with the prediction of the conservation strategy. Some of the species for which the Piro have partial preferences are not particularly vulnerable to local extinction. Agoutis, for example, are exceedingly fast breeders. Also, as noted above, the hunters do not display partial preferences for the species where it is most expected, the vulnerable large primates.

Second, there are a number of more complex models within the scope of foraging theory that can explain these preferences. The first is related to the fact that hunters are limited in the number of shotgun shells and hence the number of attacks they can make during the course of a hunt. A significant number of the ignored game bird encounters, for example, occurred early in the hunt, with the hunters' stated intent to use the shell for something larger, but if unsuccessful, return to kill the bird on the way back (for more details see Alvard, 1993a).

Hill et al. (1987) and Smith (1991) suggest that context-specific variation in encounters also plays a role in patterning preferences. Agoutis are a good example. Hunters often encounter agoutis as they give alarm barks and flee. Such encounters are never pursued by shotgun hunters. Most of the agoutis were killed in situations when the agouti was observed by the hunter but the agouti was unaware the hunter was near. The latter type of encounter provides high returns because of low handling times. Variability in agouti encounters explains why they are the second most profitable prey item, yet hunters did not always pursue them.

Intraspecific Prey Choice

Before comparing the age–sex profile of kills with wild-living prey populations, note that the age–sex structures of the actual prey populations in the immediate Diamante area were not known. When available, prey census data collected at Cocha Cashu, a biological research station located approximately 90 km from Diamante inside Manu Park, are used for comparisons (Terborgh, 1983). Otherwise, comparisons are made with other censused wild populations.

The results presented in table 17-3 are from analyses using both field sessions' data and two age categories of resolution. The table shows that in almost every case the ratio of adults killed to immatures killed is either statistically indistinguishable from the censused populations or is biased toward adults. Piro hunters are either taking peccaries randomly with respect to age or are focusing on adult individuals; the young-dominated tactic predicted by the conservation hypothesis is not apparent. The same is true for deer: 82% of the harvested deer were adults, as were 54% of the capybara. The primate kills were significantly biased toward adults compared to the censused populations. Spider and howler monkey harvests ranged from 75% to 90% adult. Adults account for between 51% and 54% of the howler populations in the Venezuelan llanos (Neville, 1972, 1976), and 54% of the spider monkey study groups at Cocha Cashu in Manu Park (Symington, 1988).

Analysis of the dentition on the prey mandibles collected in 1990–91 allows the additional category of "old adult" to be discerned. These data indicate old adults, identified by extreme tooth wear and tooth loss, make up a small proportion of the Piro kills. The analysis of dental data shows no bias toward old individuals for any species, and most of the adults are prime aged. Old adults accounted for between 8% and 14% of the kill. Comparative data from censused populations fine grained enough to distinguish the proportion of old adults do not exist for any of the species, but it can be assumed that old adults are a small proportion of any animal population that is stable or especially one that is growing. The results suggest that Piro hunters are not selectively targeting old adults.

The distribution of points on a triangle graph (fig. 17-3) shows that the Piro primate kills are prime-adult biased, with very few immatures or old among the kill. The peccary and

Table 17-3 Chi-square test results comparing age profiles of harvested prey with censused populations. Table reprinted from Alward (1995a) with permission.

	Kill sample			Tested against				
Species	Immature	Adults	n	Immature	Adults	References	χ^2adj	p
Collared peccary	0.27	0.73	141	0.31	0.69	Kiltie and Terborgh (1983)	0.37	.543
				0.24	0.76	Bissonette (1982)	0.19	.659
				0.45	0.55	Sowls (1984)	9.15	.003
				0.44	0.56	Castellanos (1983)	8.17	.004
				0.26	0.74	Sowls (1984)	0.01	.956
Red brocket deer	0.18	0.82	27	0.39	0.61	McCullough (1984)	1.85	.173
Capybara	0.46	0.54	13	0.30	0.70	Ojasti (1973)	0.19	.626
				0.42	0.58	Herrera and MacDonald (1987)	0.04	.850
Spider monkey	0.14	0.86	29	0.47	0.54	Symington (1988)	5.84	.016
				0.34	0.66	Klein (1972)	2.18	.140
Howler monkey	0.14	0.86	44	0.51	0.49	Neville (1972)	12.09	.001
				0.54	0.46	Rudran (1979)	13.95	.0002

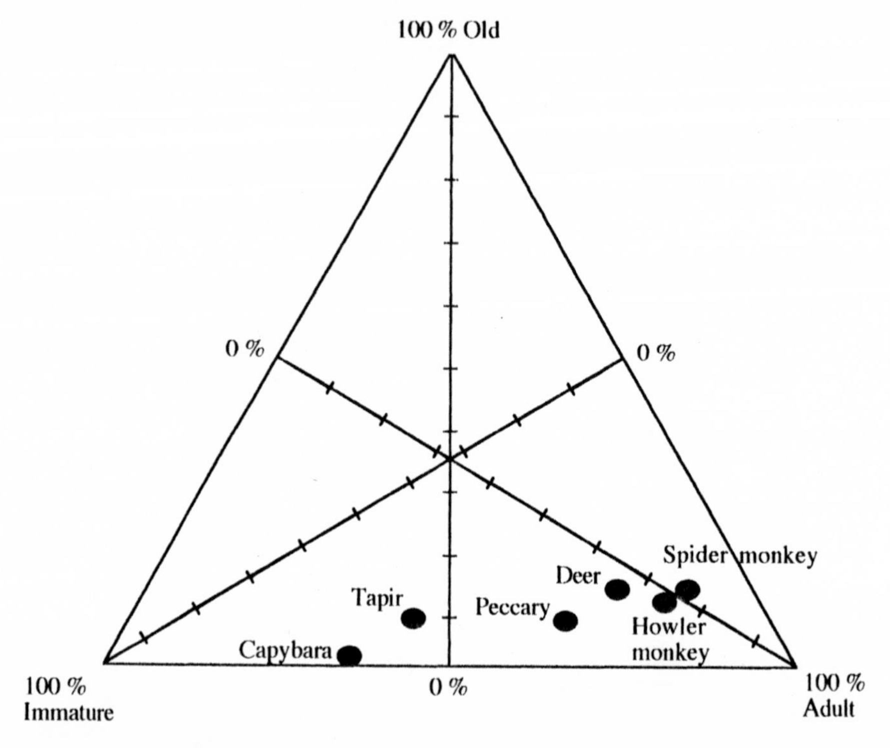

Figure 17-3 Harvest age profiles of important Piro prey species. Figure reprinted from Alvard 1995a with permission.

deer harvests fall into a range of the diagram that indicates a pattern of nonselective prey choice with respect to age. That is, prey choice for these species seems to mirror the living structures of their populations (see Stiner, 1991). The tapir and capybara kills have a slight immature bias.

Examining these results with respect to the alternative hypotheses provides support for the optimal foraging hypothesis. The hunters are targeting the large, prime-aged primate adults, in spite of the fact that these species are among the most vulnerable to overhunting (Alvard, 1993a). Interestingly, tapir are also vulnerable to overhunting, and there is a bias toward young tapir, as would be predicted by the conservation hypothesis. There is also a bias toward young capybara, but these animals are much less vulnerable to depletion. This interspecific variability in intraspecific prey choice cannot be understood with reference to hunters' hypothesized conservation goals; it can be explained with reference to OFT, however (Alvard, 1995a).

Assuming that body size is the primary factor affecting prey type profitability, smaller bodied immatures are less profitable than adults of the same species. The immatures of some species, however, are larger than the adults of others. For example, young tapir and capybara rapidly grow to the range of body sizes (>5 kg) of prey pursued by the Piro. Other species, the primates in particular, grow slowly and do not obtain a weight of 5 kg for a number of years after birth.

If return-rate maximization, rather than conservation, is the motivating factor for

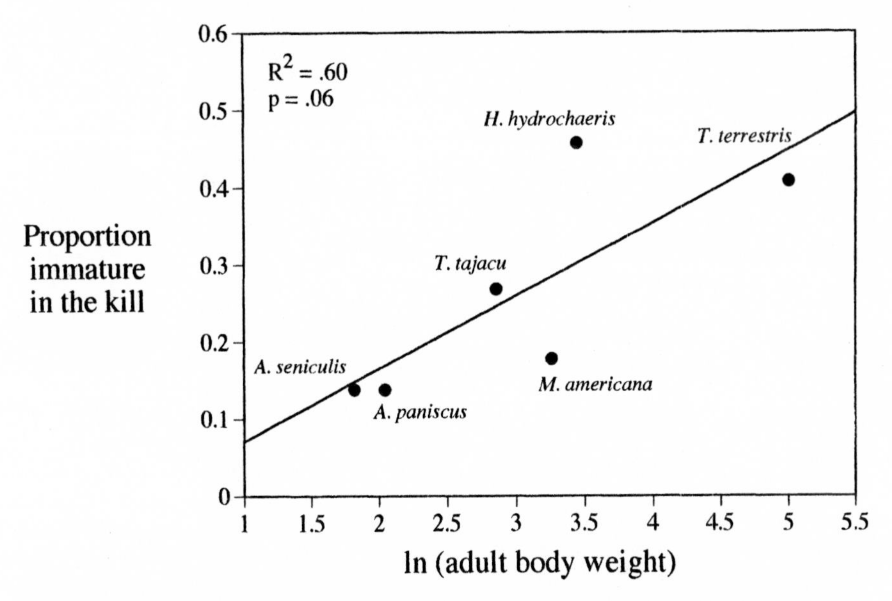

Figure 17-4 The proportion of the kill that is immature as a function of the log adult prey weight. (Figure reprinted from Alvard, 1995a, with permission.)

hunters, immature body size should predict their proportion among the kill. Hunters should kill immatures that weigh more than 4–5 kg in higher proportion than immatures that weigh less. Figure 17-4 shows a significant positive relationship between the proportion of the kill that is immature and the adult body weight of the species. Hunters are more likely to pursue immatures of species such as tapir and capybara whose young are large bodied. For smaller species, such as the primates, hunters focus on adults. This explains the distribution of points found in the triangle graph of fig. 17-3. It also provides support for the hypothesis that return-rate maximization motivates age selection by Piro hunters rather than a desire to conserve prey populations.

Table 17-4 compares the adult sex ratios of species for which there are adequate estimates from wild-living populations to the sex ratio of adults harvested by the Piro. There is no significant difference between the two samples (Alvard, 1995a). The sex ratios of most of the kills at Diamante are biased away from 100:100 toward females. This result is contrary the prediction generated from the assumption of conservation-oriented hunting behavior. Even after controlling for any skew from 100:100 likely to occur in the prey populations, Diamante shotgun hunters do not selectively avoid females and do not kill a disproportionate number of males for any species. Hunters stated the desire to kill larger individuals, for example the males of howler and capuchin monkeys, but this not apparent in the data. Both the desire for larger prey items and the hunters' occasional acceptance of smaller individuals within the range of profitable body size are choices consistent with OFT.

Patch Choice

The Piro are classic central-place foragers (Orians and Pearson, 1979). Hunting occurs along trails that radiate into the forest, and men bring game back to the village where they

Table 17-4 Chi-square test results comparing sex ratios of harvested prey with censused populations.

Species	Sex ratio of adult Diamante kill	Sex ratio of censused population	χ^2	p	References
Red brocket deer	50:100 (n = 21)	96:100	—	0.11	Branan and Marchinton (1987); hunter's kill in Venezuela.
Spider monkey	33:100 (n = 20)	34:100	—	0.62	Symington (1988), Van Roosmalen (1980) for adults
Howler monkey	64:100 (n = 36)	60:100	0.015	0.90	Mean of ratios for adults cited in Crockett and Eisenberg (1987)
Brown capuchin monkey	100:100 (n = 12)	101:100			Izawa (1975), Robinson and Janson (1987)
Collared peccary	104:100 (n = 106)	much variability 120:100	0.146	0.70	Kiltie and Terborgh (1983), Bissonette (1982), Sowls (1984), Castellanos (1983)
Tapir	120:100 (n = 11)	—	—	—	No information available
Capybara	0:100 (n = 4)	77:100	—	—	Herrerra and MacDonald (1987)

Numbers for censused populations are means of values from references. For samples less than 25, exact probabilities were calculated. Tests were not done for species with a sample of less than 5 for either sex. Table reprinted from Alvard 1995a with permission.

provision women and children. Evidence for depletion around the village is substantial. Figure 17-5 shows that hunters' return rates increased significantly as they moved away from the village. The return rate for portions of hunts from 0 to 4 km away from the village was 0.98 kg/h, whereas from 4 to 8 km the return rate was 3.18 kg/h. This is the same pattern observed by Hames (1987, 1991) for both Yanomamo and Ye'kwana villages. Encounters with spider monkeys, capuchin monkeys, and cracid game birds occur significantly less often near the village (<5 km) than farther away (>5 km) (Alvard, 1993b; Alvard et al., 1997). It is difficult to test for the relationship between return rates and hunting time allocation using these data, however. Hunters indicated they preferred to hunt in areas farther away, but hunting time is confounded with travel time because hunters must spend time traveling through near zones to hunt in far zones.

Return rates not only varied with distance from the village, but also among different hunting areas surrounding the village. The question of time allocation and return rates can be examined with this data set without the confounding problem of search versus travel time. Hunters traveled to each zone either by canoe or by walking through villages or gardens. Figure 17-6 plots the relationship between the return rate in hunting zones recognized by Piro hunters (each about 45 km^2) and the number of hours spent hunting in that patch for the sample of 122 unobserved hunts. As can be seen, there is a significant positive relationship between the two variables (R^2 = .95, p = .001). Hunters chose to hunt in areas with higher return rate and avoided areas where returns were low and animals were apparently depleted (see Alvard, 1994).

These results match those found for the Yanamomo and Ye'kwana by Hames: (1) the areas around Diamante are depleted, in part due to central-place hunting, and (2) hunters tend to focus on areas where returns are greater. These results are consistent with both a conser-

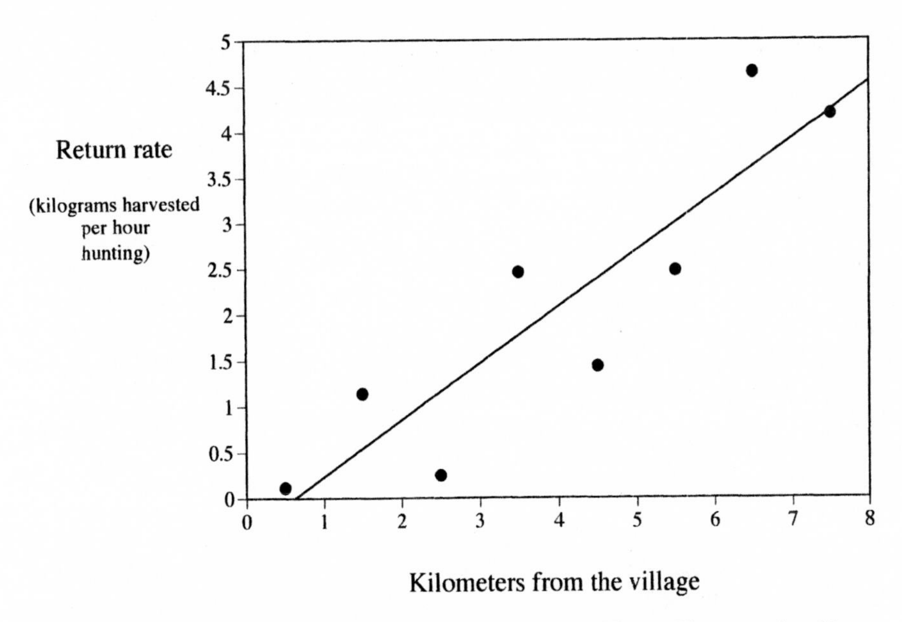

Figure 17-5 Return rate as a function of distance from the village. (Figure reprinted from Alvard 1994, with permission.)

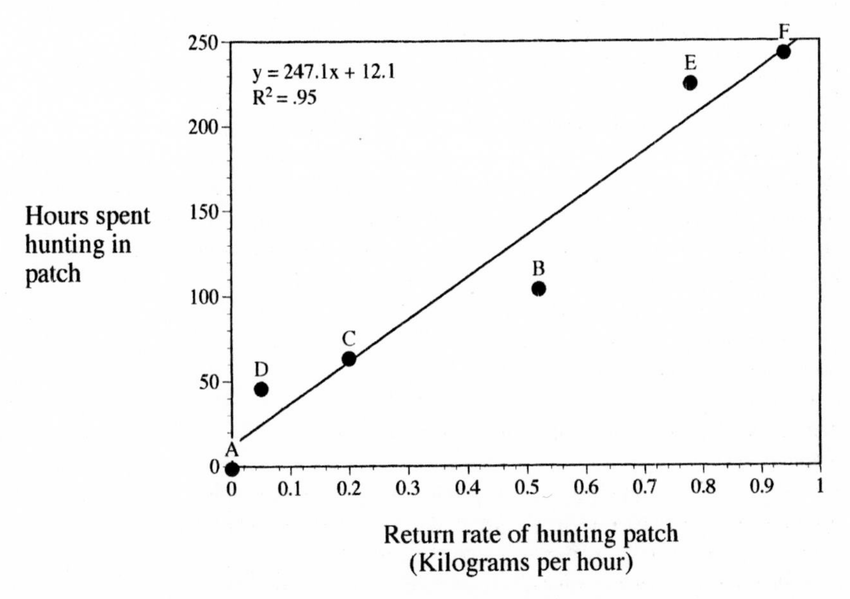

Figure 17-6 Hours spent by hunters in hunting patches as a function of the average return rate of the patches. Each letter represents one of six hunting patches surrounding Diamante. (Figure reprinted from Alvard, 1994, with permission.)

vation and an optimal foraging hypothesis. Hames suggested that examining prey-choice decisions in the depleted areas can discern between conservationists and hunters set to maximize their return rates. Conservationists are predicted to ignore pursuit opportunities in the nearby depleted areas on their travels to the more distant productive areas. Rate maximizers are predicted to pursue all encounters regardless of their location.

Records indicate that the hunters pursued the five large prey species (spider and howler monkey, collared peccary, deer, and tapir) at almost every opportunity, even though species such as large primates and tapir are vulnerable to local extinction (Alvard, 1993a). The data show that even in the areas < 4 km from the village, hunters readily pursued vulnerable prey types (Alvard, 1994).

For the species ignored on occasion—three cracid game bird species (*Penelope jacquacu, Aburria pipile,* and *Mitu mitu*), agoutis, and capuchin monkeys—hunters were not more likely to overlook these species when encountered close to the village (<5 km) compared to away from the village (>5 km) (χ^2 1.853, $p = .173$, $N = 180$).

These results indicate there was no systematic restraint by hunters that could be attributed to the fact they were hunting in depleted areas. Hunters spent less time in areas where returns were low and animals less common, but they were not less likely to pursue the animals they encountered in those areas. The hypothesis that hunters ignore pursuit opportunities in depleted areas to conserve prey resources can be rejected for the Piro case.

Other Field Work with Humans and Implications for Conservation

Table 17-5 presents studies that have used optimal foraging and whose results have a bearing on the questions discussed in this paper. Most of the prey-choice studies show concordance between OFT predictions and foragers' actual prey and patch choices. There is little support for indigenous game management or conservation. Not included in this table is work that uses strictly theoretical approaches and modeling. Rogers (1991) uses a population genetics approach and suggests conservation is unlikely to evolve even in a context where inheritance of the conserved resource by decedents is likely. Winterhalder (Winterhalder et al., 1988; Winterhalder and Lu, 1997) uses OFT to model the long-term effects of human hunting on prey and to model the development of conservationlike effects and prey extinctions. Both these approaches have the advantage of introducing time depth into the analysis.

Strategies and Weaknesses

The results presented in this paper do not support the hypothesis that the Piro hunt in a way designed to conserve their faunal resources. This study, and the others cited above, suggest a cautious reappraisal by conservationists concerning how indigenous people use resources. Knowing that native peoples are not going to act as altruists is one step toward incorporating them into realistic conservation solutions (Alvard, 1993a).

In addition to this general insight, an operational benefit to users of an OFT–behavioral ecology approach is that predictions can be made concerning areas of greatest conflict between resource users and management schemes created by conservationists. Naive tinkering with the suite of prey that hunters are allowed to kill may produce unexpected results

Table 17-5 Field studies using optimal foraging theory (OFT), with results that bear on the issue of conservation.

People	Primary hunting technology	Prey	Location	Results	References
Ache	Shotgun and bow	Large/medium-sized ungulates, rodents, primates	Paraguay; rainforest	OFT predictions generally upheld, "complete absence of any practices obviously designed to check overharvesting of resources" (1985:236)	Kaplan and Hill (1985)
Cree	Rifle and trap	Large ungulates, small mammals, birds, fish	Northern Ontario, Canada; boreal forest	OFT predictions generally upheld, "conservation may be the incidental effect of efficient foraging in a heterogeneous environment" (1981:97).	Winterhalder (1981, 1982)
Wana	Trap, blowgun, spear, and dog	Large ungulates, primates, rodents, birds, bats, fish	Sulawesi, Indonesia; rainforest	No obvious conservation. Some vulnerable species depleted.	Alvard (in press)
Piro	Shotgun and bow	Large/medium-sized ungulates, rodents, primates, birds	Amazonian Peru; rainforest	OFT predictions generally upheld, "Piro hunters' decisions are more accurately predicted by optimal foraging theory than by the conservation hypothesis" (1995:800).	Alvard (1993a,b, 1994, 1995a,b)
Inujjuamiut	Rifle and trap	Large marine and terrestrial mammals, birds, fish	Canadian Arctic; tundra	OFT predictions generally upheld, "little support to the idea of indigenous game management among the Inuit" (1991:256).	Smith (1991)
Yanamomo, Ye'kwana	Shotgun and bow	Large/medium-sized ungulates, rodents, primates, birds, fish	Venezuela; rainforest	"Hunters and fishers who have depleted game resources hunt and fish more intensively." (1987:96) "The correlation between hunting zone intensity and hunting zone efficiency is positive and efficient. . . . [supporting] both the conservation and efficiency hypothesis" (1987:99).	Saffirio and Hames (1983) Hames (1987, 1991)
Hadza	Bow	Large terrestrial mammals	Tanzania; savannah woodland	"A Hadza hunter does indeed maximize their average rate of meat aquisition by generally ignoring [small prey taxon]" (1991:86).	Hawkes et al. (1991)

(B. Winterhalder and F. Lu, unpublished data). For example, one prediction of OFT not discussed in detail here is that, as the most profitable prey deplete or disappear, hunters are expected to increase their diet breadth by including previously unhunted, lower ranked game (Stephens and Krebs, 1986). This means that if high-ranked game are excluded from the diet through the threat of sanctions, for example, hunters may begin to include other species previously ignored but equally endangered.

Winterhalder (1982) uses examples from his work with the Ojibwa-Cree to show how optimal foraging models could inform and improve management schemes involving indigenous people. Following the work of Orians and Pearson (1979), Winterhalder notes that the diet breadth of a group of central-place hunters is predicted to change depending on the distance they must travel to a hunting patch. Hunters will focus on the highest ranked species and drop the lower ranked types the farther they must travel. Such information is critical as many management plans contemplate resettlement of communities away from their primary sources of prey.

There are, however, a number of methodological weaknesses with the prey-choice model as it was applied in this case. For example, the time span over which foraging theory assumes the return rate is maximized is referred to as "long-term" (Stephens and Krebs, 1984). This is rather ambiguous, yet the time scale over which the return rate is maximized is very important when dealing with issues of conservation. Generations are often required for exploitive foraging decisions to deplete prey populations. From this perspective, the span over which most OFT analyses assume maximization is relatively short term (over the course of one or several foraging bouts; Stephens and Krebs, 1986). The ambiguity stems from what Hill (1995) has pointed out is an assumption of the basic OFT models—foragers do not affect the density of their prey. Although this factor could be incorporated into future models, the simple model is useful because its predictions contrast with the conservation predictions.

The lack of sex and age profiles for the hunted prey populations also weakens the results. The difficulties involved in censusing hunted populations around Diamante required comparing the Piro kill to censused prey populations at other sites. To mitigate this problem, comparisons of the Piro harvest were made, in most cases, with multiple sites and/or across multiple years.

Another weakness is the assumption that all age and sex types of a particular species are equally vulnerable to hunters. This is not always true (see Fitzgibbon, chapter 16, this volume). Female primates may be more vulnerable to bow pursuits than males because of the mobility costs of pregnancy and carrying infants (Alvard and Kaplan, 1991). Male Steller's sea lions *Eumetopias jubatus* are much easier to kill than females during the breeding season (Hildebrandt and Jones, 1992). Assuming equal handling times for each sex might lead to the erroneous conclusion that a male-biased sea lion harvest was conservation when it was not (Lyman, 1995). Concluding that hunters are not acting conservatively when they are is a more difficult error to make, however. This is because conserving hunters are predicted to avoid killing individuals of high reproductive value regardless of how easy they are to kill (Alvard, 1995a).

In anticipation of less-than-ideal data for any one test, the overall research design included a number of tests, each made with numerous species to test the general hypotheses. Conclusions based on any one test for any one species may be questioned; conclusions based on many tests that all point to the same answer are less likely to be spurious.

Although there is general awareness of its importance in producing behavioral variability in humans, the mechanism of cultural transmission has yet to be integrated into behavior ecological models. These processes may be able to explain hunting decisions that prove difficult to understand otherwise. For example, these mechanisms may explain some of the food taboos of apparently profitable game reported by many anthropologists (Ross, 1978). While it might be argued that this is what cultural anthropologists have been advancing for years, the difference is that the cultural transmission models such as those developed by Boyd and Richerson (1985) are devised to be testable—a notable deficiency of previous attempts to understand seemingly nonadaptive cultural behavior. They also rest on a firm foundation of evolutionary theory.

One last weakness, not limited to behavioral ecology, is that models are not designed to produce value-laden decisions. For example, the following question cannot be answered using the models above: "What is more important, the health, well being, and cultural integrity of this human community, or the natural (and perhaps endangered) resources they exploit?" Some have argued that this is a fatal shortcoming (Scheper-Hughes, 1995). Others have argued more convincingly that value-laden decisions can only be made if informed with objectively produced knowledge (D'Andrade, 1995). A realistic understanding of how and why native peoples make subsistence decisions, be they conservation oriented or not, is critical for designing collaborative management schemes advocated by many professional conservationists.

Recommendations

The first recommendation is that land-tenure policy decisions should not be based on the conservation performance of native peoples. This does not relegate the results presented in this chapter to academic exercise, however, but rather emphasizes the fact that native people should not have to endure the additional burden of living according to the expectations of good-hearted but misinformed conservationists.

The second recommendation stems from this chapter's conclusion: it is unlikely that normative descriptions of native people as natural conservationists are accurate and that genuine conservation by native people is probably uncommon. The impact that indigenous people have on their environments can quickly change from benign to damaging as development leads to economic opportunities (Peres, 1994). Conservation schemes, such as extractive reserves (Salafsky et al., 1993), should assume that individuals will act according to their own self-interest even if it is contrary to what is best for the long-term viability of the natural resources they are using.

Many workers do agree that conservation, or at least conservationlike effects, can develop depending on how constraints change. Territoriality and rule enforcement, common to societies with centralized authority, are probably minimal requirements for the development of conservationlike behavior and sustainable use (Hames, 1987, 1991; Rogers, 1991). Private or common property ownership is useful to prevent poaching by nongroup members, and enforcement keeps within-group cheaters in check (Berkes et al., 1989; Rasker et al., 1992). The resources exploited in many extractive reserves are open access, providing little incentive to individuals to conserve for the long term (Ciroacy-Wantrup and Bishop, 1985). Note that to be theoretically correct, these are "conservationlike" behaviors; coerced or enticed restraint is epiphenomonal conservation. Individuals who are coerced into restraint are not genuinely conserving. They are not paying a short-term cost but are rather avoiding one. The issue is important because it emphasizes that people are unlikely to be altruists and must be assured of some benefit in return for their present-day restraint.

The sustainability analysis of the Piro harvest suggests that the Piro could completely remove primates from their diet and make up the difference in increased, yet sustainable, peccary harvests (Alvard et al., 1997). Such a scheme, however, asks hunters to ignore encountered primates in their optimal diet and suffer lower return rates as a result. Hunters would either return from hunts with less meat or have to spend more time hunting to obtain the same quantity. One solution to assure local people's cooperation is to provide incentives that are equal or greater than the costs they are asked to endure. Restraint may be encouraged through enticement, making the short-term benefits of conservation to individual resource users sufficiently high, or through enforcement, by making the costs of not conserving prohibitively high. These ideas are not new in the field of conservation (McNeely, 1988; Schultz, 1990; Rasker et al., 1992; EGEAB, 1994), but they are given additional support by evolutionary theory. One specific advantage of the OFT method is that the actual costs due to hunting restrictions can be determined in terms of lost calories or time. This could help set the level of incentive required to modify hunters' behavior.

Summary

Debate has arisen concerning the commonly held view that native people are normatively conservationist. Multiple cases of traditional native groups exploiting resources to extinction are evidence that "native" is not synonymous with "natural conservationist." Both methodical and conceptual problems have contributed to an inaccurate characterization of natives peoples. A preference for nonempirical methodology in anthropology has produced a database that is essentially unreliable.

The Piro hunting project was designed to test for conservation behaviors among a group of subsistence hunters living in the rainforests of eastern Peru. Possible conservation tactics tested for were interspecific prey choice, intraspecific prey choice, and patch-choice decisions. Additional tests were performed to examine for evidence of game depletion and harvest sustainability. Results indicate that hunters did not avoid vulnerable species such as primates and tapir, but rather chose prey consistent with the predictions of OFT. Hunters did not avoid sex and age types that would minimize their impact on the prey populations. The area near the Piro village was locally depleted, and yet the hunters did not refrain from killing vulnerable animals in this area. Hunters focus their efforts in the areas where they expect the highest returns. Studies of other subsistence hunters that incorporate OFT models have generally found no evidence for game management or conservation behaviors. At the same time, the OFT models have seen general success in predicting hunter's decisions.

A behavioral ecological approach illustrates the cost-and-benefit calculus guiding the decision-making process of human subsistence resource users. Conservation professionals can take advantage of the substantial predictive power that results from this knowledge, for example, by anticipating conflicts of interests between resource users and management schemes. While this research suggests that voluntary conservation may be unlikely in simple societies, restraint may be encouraged through enticement or enforcement.

Acknowledgments This research was funded by the Charles Lindbergh Foundation, the LSB Leakey Foundation, the Tinker Inter-American Research Foundation, the University of New Mexico, and the National Science Foundation (grant BNS-8717886 to H. Kaplan).

References

Anderson A, 1990. Alternatives to deforestation: steps toward sustainable use of the Amazon rain forest. New York: Columbia University Press.

Alcorn J, 1993. Indigenous people and conservation. Conserv Biol 7:424–426.

Alvard M, 1993a. A test of the ecologically noble savage hypothesis: interspecific prey choice by neotropical hunters. Hum Ecol 21:355–387.

Alvard M, 1993b. Testing the ecologically noble savage hypothesis: conservation and subsistence hunting in Amazonian Peru (PhD dissertation). Albuquerque: University of New Mexico.

Alvard M, 1994. Prey choice in a depleted area. Hum Nature 5:127–154.

Alvard M, 1995a. Intraspecific prey choice by Amazonian hunters. Curr Anthropol 36:789–818.

Alvard M, 1995b. Shotguns and sustainable hunting in the Neotropics. Oryx 29:58–66.

Alvard M, in press. The impact of traditional subsistence hunting and trapping on prey populations: data from the Wana of upland Central Sulawesi, Indonesia. In: Sustainability of Hunting in tropical forests. (Robinson J, Bennett L, eds).

Alvard M, Kaplan H, 1991. Procurement technology and prey mortality among indigenous neotropical hunters. In: Human predators and prey mortality (Stiner M, ed). Boulder, Colorado: Westview Press; 79–104.

Alvard M, Robinson J, Redford K, Kaplan H, 1997. The sustainability of subsistence hunting in the Neotropics. Conserv Biol 11:977–982.

Berkes F, Feeny D, McCay B, Acheson J, 1989. The benefits of the common. Nature 340:91–93.

Bissonette J, 1982. Ecology and social behavior of the collared peccary in Big Bend National Park, Texas. Scientific Monograph Series no. 16. Washington, DC: National Park Service.

Boyd R, Richerson P, 1985. Culture and the evolutionary process. Chicago: University of Chicago Press.

Branan W, Marchinton RL, 1987. Reproductive ecology of white-tailed and red brocket deer in Surinam. In: Biology and management of the cervidae (Wemmer C, ed). Washington DC: Smithsonian Institute Press; 344–351.

Budianski S, 1995. Nature's keepers. New York: Free Press.

Bunyard P, 1989. The Columbian Amazon: policies for the protection of its indigenous peoples and their environment. Cornwall, UK: Ecological Press.

Burney D, 1993. Recent animal extinctions: recipes for disaster. Am Sci 81:530–541.

Callicott J, 1989. In defense of the land ethic: essays in environmental philosophy. Albany, State University of New York Press.

Castellanos H, 1983. Aspectos de la organizacion social del baquiro de collar, Tayassu tajacu, en el estado Guarico—Venezuela. Acta Biol Venez 11:127–143.

Caughley G, 1977. Analysis of vertebrate populations. London: John Wiley and Sons.

Charnov E, 1976. Optimal foraging: the marginal value theorem. Theor Popul Biol 9:129–136.

Ciroacy-Wantrup S, Bishop R, 1985. Common property as a concept in natural resource policy. Nat Resources J 4:127–137.

Clad J, 1985. Conservation and indigenous peoples: a study of convergent interests. In: The human dimension in environmental planning (McNeely J, Pitt D, eds). Gland: International Union for the Conservation of Nature; 45–62.

Clay J, 1988. Indigenous peoples and tropical forests: models of land use and management from Latin America. Cambridge: Cultural Survival, Inc.

Cole L, 1954. The population consequences of life history phenomena. Q Rev Biol 29:103–137.

Cox P, Elmquist T, 1991. Indigenous control of tropical rain forest reserves: an alternative strategy for conservation. Ambio 20:317–321.

Crockett C, Eisenberg J, 1987. Howlers: variations in group size and demography. In: Primate societies (Smuts B, Cheney D, Seyfarth R, Wrangham R, Struhsaker T, eds). Chicago: University of Chicago Press; 54–68.

D'Andrade R, 1995. Objectivity and militancy: a debate. Curr Anthropol 36:399–408.

Dasmann R, 1976. Life-styles and nature conservation. Oryx 13:281–286.

Denevan W, 1992. The pristine myth: the landscape of the Americas in 1492. Ann Assoc Am Geogr 82:369–385.

Dewar R, 1984. Extinctions in Madagascar: the loss of the subfossil fauna. In: Quaternary extinctions (Martin P, Klein R, eds). Tucson: University of Arizona Press; 574–593.

Diamond J, 1992. The third chimpanzee. New York: HarperCollins.

EGEAB (Expert Group on Economic Aspects of Biodiversity), 1994. Economic incentive measures for the conservation and sustainable use of biological diversity: conceptual framework and guidelines for case studies. Paris: Organization for Economic Cooperation and Development.

Fearnside P, 1989. Extractive reserves in Brazilian Amazonia. Bioscience 39:387–393.

Feit H, 1973. The ethno-ecology of the Waswanipi Cree; or how hunters can handle their resources. In: Cultural ecology (Cox B, ed). Toronto: McClelland and Stewart; 115–125.

Fowler C, 1981. Comparative population dynamics in large mammals. In: Dynamics of large mammal populations (Fowler C, Smith T, eds). New York: John Wiley and Sons; 437–455.

Gadgil M, Berkes F, Folke C, 1993. Indigenous knowledge for biodiversity conservation. Ambio 22:151–156.

Golson, J. 1977. No room at the top: agricultural intensification in the New Guinea Highlands. In: Sunda and Sahul: prehistoric studies in Southeast Asia, Melanesia and Australia Allen (Golson J, Jones R, eds). London: Academic Press; 601–638.

Gorsline J, House L, 1974. Future primitive. Planet Forum 3:1–13.

Hames R, 1987. Game conservation or efficient hunting? In: The question of the commons (McCay B, Acheson J, eds). Tucson: University of Arizona Press; 97–102.

Hames R, 1991. Wildlife conservation in tribal societies. In: Culture, conservation and ecodevelopment (Oldfield M, Alcorn J, eds). Boulder, Colorado: Westview Press; 172–199.

Harris L, Kochel I, 1981. A decision-making framework for population management. In: Dynamics of large mammal populations (Fowler C, Smith T, eds). New York: John Wiley and Sons; 221–239.

Hayne D, Gwynn J, 1977. Percentage does in total kill as a harvest strategy. In: Proceedings of the Joint Northwest-Southwest Deer Study Group Meeting, Fort Pickett, Virginia; 117–127.

Hawkes K, O'Connell J, Blurton Jones N, 1991. Hunting income patterns among the Hadza: big game, common goods, foraging goals and the evolution of the human diet. In: Foraging strategies and natural diets of monkeys, apes and humans (Whitten A, Widdowson E, eds). Oxford: Oxford University Press; 83–91.

Headland T, 1994. Ecological revisionism: recent attacks against "myths" in anthropology, and the role of historical ecology in searching out the truth. Paper presented at the Conference on Historical Ecology, Tulane University, New Orleans, Louisiana, 99–11 June.

Herrera E, MacDonald D, 1987. Group stability and the structure of a capybara population. Symposia of the Zoological Society of London 58:115–130.

Hildebrandt W, Jones T, 1992. Evolution of marine mammal hunting: a view from the California and Oregon coasts. J Anthropol Archeol 11:360–401.

Hill K, 1993. Life history theory and evolutionary anthropology. Evol Anthropol 2:78–88.

Hill K, 1995. Comment on, "Intra specific prey choice by Amazonian hunters." Curr Anthropol 36:805–807.

Hill K, Kaplan H, Hawkes K, Hurtado A, 1987. Foraging decisions among Ache hunter-gatherers: new data and implications for optimal foraging models. Ethol Sociobiol 8:1–36.

Hughes J, 1983. American indian ecology. El Paso: Texas Western Press.

Hunn E, 1982. Mobility as a factor limiting resource use in the Columbia Plateau of North America. In: Resource managers: North American and Australian hunter-gatherers (Williams N, Hunn E, eds). Boulder, Colorado: Westview Press; 17–43.

Izawa K, 1975. Group sizes and compositions of monkeys in the upper Amazon Basin. Primates 17:367–399.

Johnson A, 1989. How the Machiguenga manage resources: conservation or exploitation of nature? Adv Econ Bot 72:13–222.

Kaplan H, Hill K, 1985. Food sharing among Ache foragers: tests of explanatory hypotheses. Curr Anthropol 26:223–245.

Kaplan H, Hill K, 1992. The evolutionary ecology of food acquisition. In: Evolutionary ecology and human behavior (Smith E, Winterhalder B, eds). New York: Aldine de Gruyter; 167–202.

Kiltie R, Terborgh J, 1983. Observations on the behavior of rain forest peccaries in Peru: why do white-lipped peccaries form herds? Z Tierpsychol 62:241–255.

Klein L, 1972. The ecology and social organization of the spider monkey, *Ateles belzebuth* (PhD dissertation). Berkeley: University of California.

Krebs J, McCleery R, 1984. Optimization in behavioral ecology. In: Behavioral ecology: an evolutionary approach (Krebs J, Davies N, eds). Sunderland, Massachusetts: Sinauer Associates; 91–121.

Low B, 1996. Behavioral ecology of conservation in traditional societies. Human Nature 7:353–379.

Low B, Heinen J, 1993. Population, resources and environment. Implications of human behavioral ecology for conservation. Popul Environ 15:7–41.

Lyman R, 1995. Comment on, "Intra Specific prey choice by Amazonian hunters." Curr Anthropol 36:808–809.

MacArthur R, 1960. On the relation between reproductive value and optimal predation. Proc Nat Acad Sci 46:143–145.

Martin P, 1984. Prehistoric overkill. In: Quaternary extinctions (Martin P, Klein R, eds). Tucson: University of Arizona Press; 354–403.

McCullough D, 1984. Lessons from the George Reserve, Michigan. In: White tailed deer: ecology and management (Halls L, ed). Harrisburg, Pennsylvania: Stackpole Books; 211–242.

McDonald D, 1977. Food taboos: a primitive environmental protection agency. Anthropos 72:734–748.

McNeely J, 1988. Economics and biological diversity: using economic incentives to conserve biological resources. Gland: International Union for the Conservation of Nature.

Nelson L, Peek J, 1982. Effect of survival and fecundity on rate of increase of elk. J Wildl Manage 46:535–540.

Nelson R, 1982. A conservation ethic and environment: the Koyukon of Alaska. In:

Resource managers: North American and Australian hunter-gatherers (Williams N, Hunn E, eds). Boulder, Colorado: Westview Press; 211–238.

Neville M, 1972. The population structure of red howler monkeys (*Alouatta seniculus*) in Trinidad and Venezuela. Folia Primatol 17:56–86.

Neville M, 1976. The population and conservation of howler monkeys in Venezuela and Trinidad. In: Neotropical primates: field studies and conservation (Thorington R, Heltne P, eds). Washington, DC: National Academy Press; 101–121.

Oelschlaeger M, 1991. The idea of wilderness: prehistory to the age of ecology. New Haven, Connecticut: Yale University Press.

Olson S, 1989. Extinction on islands: man as a catastrophe. In: Conservation for the twenty-first century (Western D, Pearl M, eds). New York: Oxford University Press; 50–53.

Ojasti J, 1973. Estudio Biológico del Chigüire o Capibara. Caracas, Venezuela: Fondo Nacional de Investigaciones Agropecuarias.

Orians G, Pearson N, 1979. On the theory of central place foraging. In: Analysis of ecological systems (Mitchell R, ed). Columbus: Ohio State University Press; 155–177.

Owen-Smith N, 1987. Pleistocene extinctions: the pivotal role of megaherbivores. Paleobiology 13:351–362.

Pearce F, 1992. First aid for the Amazon. New Scientist 3:42–46.

Peres C, 1994. Indigenous reserves and nature conservation in Amazonian forests. Conserv Biol 8:586–588.

Poffenberger M, 1990. Keepers of the forest. West Hartford, Connecticut: Kumarian Press.

Posey D, 1985. Native and indigenous guidelines for new Amazonian development strategies: understanding biodiversity through ethnoecology. In: Change in the Amazon (Hemming J, ed). Manchester: Manchester University Press; 156–181.

Pyke G, Pulliam H, Charnov E, 1977. Optimal foraging: a selective review of theory and tests. Q Rev Biol 52:137–154.

Rambo A, 1978. Bows, blowpipes, and blunderbusses: ecological implications of weapons change among Malyasian Negritos. Malay Nat J 32:209–216.

Rasker R, Martin M, Johnson R, 1992. Economics: theory versus practice in wildlife management. Conserv Biol 6:338–349.

Redford K, 1991. The ecologically noble savage. Orion 9:24–29.

Redford K, Robinson J, 1987. The game of choice: patterns of Indian and colonist hunting in the neotropics. Am Anthr 89:650–667.

Redford K, Stearman A, 1993. Forest-dwelling native Amazonians and the conservation of biodiversity: interests in common or in collision? Conserv Biol 7:248–255.

Reichel-Dolmatoff G, 1974. Amazonian cosmos: the sexual and religious symbolism of the Tukano Indians. Chicago: University of Chicago Press.

Robinson J, Janson C, 1987. Capuchins, squirrel monkeys, and Atelines: Sociological Convergence with old world primates. In: Primate societies (Smuts B, Cheney D, Seyfarth R, Wrangham R, Struhsaker T, eds). Chicago: University of Chicago Press; 69–82.

Robinson J, Ramirez J, 1982. Conservation biology of Neotropical primates. Special publication 6. Pymatuning Laboratory of Ecology, University of Pittsburgh; 329–344.

Robinson J, Redford K, 1986a. Body size, diet, and population density of neotropical forest mammals. American Naturalist 128:665–680.

Robinson J, Redford K, 1986b. Intrinsic rate of natural increase in Neotropical forest mammals: relationship to phylogeny and diet. Oecologia 68:516–520.

Rogers A, 1991. Conserving resources for children. Hum Nature 1:73–82.

Ross E, 1978. Food taboos, diet, and hunting strategy: the adaptation to animals in Amazon cultural ecology. Curr Anthropol 19:1–36.

Rudran R, 1979. The demography and social mobility of a red howler (*Alouatta seniculus*) population in Venezuela. In: Vertebrate ecology in the Northern Neotropics (Eisenberg J, ed). Washington, DC: Smithsonian Institution Press; 107–126.

Saffirio J, Hames R, 1983. The forest and the highway. In: Working papers on South American Indians, vol 6 (Hames R, ed). Bennington, Vermont: Bennington College Press; 7–29.

Salafsky N, Dugelby B, Terborgh J, 1993. Can extractive reserves save the rain forest? An ecological and socioeconomic comparison of nontimber forest product extraction systems in Petén Guatemala, and West Kalimantan, Indonesia. Conserv Biol 7: 39–52.

Scheper-Hughes N, 1995. The primacy of the ethical: propositions for a militant anthropology. Curr Anthropol 36:409–440.

Schultz D, 1990. Designing shared-savings incentive programs for energy efficiency: balancing carrots and sticks. Berkeley, California: Lawrence Berkeley Laboratory.

Silva J, Strahl S, 1991. Human impact on populations of Chachalaca, Guans and Currassows (Galliformes: Cracidae) in Venezuela. In: Neotropical wildlife use and conservation (Robinson J, Redford K, eds). Chicago: University of Chicago Press; 37–52.

Simms S, 1992. Wilderness as human landscape. In: Wilderness tapestry: an eclectic approach to preservation (Zeveloff S, Vause L, McVaugh W, eds). Reno: University of Nevada Press; 183–201.

Slobodkin L, 1961. Growth and regulation of animal populations. New York: Holt, Rinehart and Winston.

Slobodkin L, 1968. How to be a predator. Am Zool 8:43–51.

Smith E, 1983. Anthropological applications of optimal foraging theory: a critical theory. Curr Anthropol 24:625–651.

Smith E, 1991. Inujjuamiut foraging strategies: evolutionary ecology of an Arctic hunting economy. New York: Aldine.

Smith E, 1995. Comment on, "Intra specific prey choice by Amazonian hunters". Curr Anthropol 36:810–812.

Sokal R, Rohlf F, 1981. Biometry, 2nd ed. New York: WH Freeman.

Sowls L, 1984. The peccaries. Tucson: University of Arizona Press.

Steadman D, Olson S, 1985. Bird remains from an archaeological site on Herders Island, South Pacific: man-caused extinctions on an "uninhabited" island. Proc Natl Acad Sci 82:6191–6195.

Stephens D, Krebs J, 1986. Foraging theory. Princeton, New Jersey: Princeton University Press.

Stiner M, 1991. The ecology of choice: procurement and transport of animal resources by upper Pleistocene hominids in West-Central Italy (PhD dissertation). Albuquerque: University of New Mexico.

Symington M, 1988. Demography, ranging patterns and activity budgets of black spider monkeys (*Ateles paniscus chamek*) in the Manu National Park, Peru. Am J Primatol 15:45–67.

Tattersall I, 1982. The primates of Madagascar. New York: Columbia University Press.

Terborgh J, 1983. Five new world primates. Princeton: Princeton University Press.

Thompson S, 1987. Body size, duration of parental care and the intrinsic rate of natural increase in eutherian and metatherian mammals. Oecologia 71:201–209.

Todd J, 1986. Earth dwelling: the Hopi environmental ethos and its architectural symbolism—a model for the deep ecology movement (PhD dissertation), Santa Cruz: University of California.

Trotter M, McCulloch B, 1984. Moas, men and middens. In: Quaternary extinctions (Martin P, Klein R, eds). Tucson: University of Arizona Press; 708–727.

Tuan Y, 1968. Discrepancies between environmental attitude and behaviour: examples from Europe and China. Can Geogr 12:176–191.

Tuan Y, 1970. Our treatment of the environment in ideal and actuality. Am Sci 58:244–249.

van Roosmalen M, 1980. Habitat preferences, diet, feeding strategy and social organization of the black spider monkey (*Ateles p. paniscus* Linnaeus 1758) in Surinam. Arnhem, The Netherlands: Rijksinstituut voor Natuubeheer.

Vecsey C, 1990. Religion in native North America. Moscow: University of Idaho Press.

Vickers W, 1995. From opportunism to nascent conservation. Hum Nature 5:307–338.

Wheat J, 1972. The Olsen-Chubbock site: a Paleo-Indian bison kill. Society for American Archeology Memoir 26. Washington DC.

Williams G. 1966. Adaptation and natural selection. Princeton, New Jersey: Princeton University Press.

Winterhalder B, 1981. Foraging strategies in the boreal forest: an analysis of Cree hunting and gathering. In: Hunter-gatherer foraging strategies (Winterhalder B, Smith E, eds). Chicago: University of Chicago Press; 66–98.

Winterhalder B, 1982. The boreal forest, Cree-Ojibwa foraging and adaptive management. In: Resources and dynamics of the boreal zone (Wein R, Riewe R, Methven I, eds). Ontario: Association of Canadian Universities for Northern Studies; 331–345.

Winterhalder B, Baillargeon W, Cappelletto F, Daniel I, Prescott C, 1988. The population ecology of hunter-gatherers and their prey. J Anthropol Archaeol 7:289–328.

Winterhalder B, Lu F, 1997. A forager-resource population ecology model and implications for indigenous conservation. Conserv Biol 11:1–12.

Wynne-Edwards V, 1962. Animal dispersion in relation to social behavior. New York: Hafner.

18

The Evolved Psychological Apparatus of Human Decision-Making Is One Source of Environmental Problems

Margo Wilson
Martin Daly
Stephen Gordon

Resource Exploitation by *Homo Sapiens*

It has become increasingly difficult to ignore or deny the fact that the Earth's biota are in crisis. The abundance and diversity of flora and fauna have been and are being diminished at an accelerating pace, both as a direct result of human exploitation and as an indirect result of habitat loss and environmental degradation (Wilson, 1992). Despite the efforts of parties with economic interests antagonistic to conservation, it is no longer possible for informed citizens to doubt the reality of these trends, nor is there reason to doubt that the diversity and abundance of species will continue to decline for some time as a result of human numbers and activities. What is controversial is what to do about it (Clark, 1991).

The accumulation and dissemination of information about the crisis and its roots in human action are clearly not all that is required to bring about an effective remedial response. Yet, according to Ridley and Low (1994), many conservationists have assumed, at least implicitly, that if people were fully informed of the problems and their causes, they would change their priorities and activities in order to conserve resources for the future, and by relying on that assumption conservationists have implicitly embraced an unrealistic model of human beings as rational collectivists. Education is not sufficient, Ridley and Low argue, because natural selection has not designed human psychology to give priority to either the common good or the distant future, but to relatively short-term gains and positional advantages in a zero-sum intraspecific competition. According to this argument, the forces that have shaped human nature over evolutionary time have been forces that favor rapid, thorough exploitation of our resource base rather than stewardship. The human animal is not exceptional in this regard: because selection is predominantly a matter of within-species differentials in reproductive success, the phenotypes that proliferate are precisely those that enable organisms to exploit resources sooner and more effectively than their competitors, especially conspecifics, and to externalize or pass on to future others the costs of that resource exploitation.

The popular notion that aboriginal people who are uncorrupted by "western" values are reverent conservationists appears to be a romantic myth. The evidence from present-day hunting and foraging societies (Hames, 1987, 1991; Alvard and Kaplan, 1991; Alvard, 1993, 1994, Chapter 17, this volume), from ethnographic accounts of nonstate societies (Low, 1996), and from studies of human history and prehistory (Diamond, 1992; Kay, 1994) lends scant support to the idea that nonindustrialized foragers abide by a conservation ethic, nor to the proposition that greedy modern westerners are exceptional in their reluctance to subordinate their present wants to the future or to the common good. Moreover, although the conflict between human wants, on the one hand, and conservation goals, on the other, is often discussed in terms of human survivorship and comfort, human resource exploitation goes beyond these "essentials": nonindustrialized peoples, like westerners, deplete resources in ostentatious displays of resource-accruing potential and success in social competition (for industrialized societies, see Kaplan and Hill, 1985; Hawkes, 1993; and for western societies, see Frank, 1985; Ng and Wang, 1993; Howarth, 1996).

In our view, the Ridley and Low argument is overstated in the extent to which they suggest that current understanding of the natural selective process implies a "selfish" as opposed to a more collectivist evolved social psychology (Daly and Wilson, 1994). *Homo sapiens* is, after all, a social species with many psychological adaptations for social actions (e.g., Daly and Wilson, 1988; Cosmides, 1989; Simpson and Kenrick, 1997). Nevertheless, Ridley and Low's general point seems to be well taken: both theory and the available data on human behavior support the thesis that *Homo sapiens* is not by nature a conservationist, and hence that recognizing environmental problems, deploring them, and gaining a sophisticated understanding of their sources in our actions, may still not be enough to motivate the behavioral changes required to rectify them.

In this chapter, we argue for a more evolutionarily and psychologically informed model of *Homo economicus,* since economics is possibly the most relevant discipline to guide the development of incentive structures which will alleviate the current conservation crisis. This more realistic economic model will necessarily have to consider variations in human preferences and decision-making in relation to variables such as sex, age, and parental status that behavioral ecologists and other evolutionists consider fundamental. To illustrate our argument, we focus primarily on sex differences and age. The possibility that men and women "value" environmental goods somewhat differently is a topic that has hitherto received surprisingly little attention (Low and Heinen, 1993), despite an obvious selectionist rationale for predicting evolved sex differences in such domains as the subjective acceptability of various sorts of risks in the pursuit of status and resources. We also briefly discuss how a selectionist perspective on life history suggests that preferences and decision-making are also likely to have evolved to vary systematically with age.

Toward an Evolutionarily Informed Model of *Homo Economics*

The social science with an obvious role to play in remediating the current global crisis is economics. It is economic forces that drive technological innovations with their associated risks of contamination, despoliation, and expropriation. The developing field of ecological economics (e.g., Costanza, 1991) has much to say about common pool resource use and conservation incentives, consumer practices, monetary valuation of environmental goods, and the processes and consequences of externalizing costs, including pricing costs of foregone future resource use.

Economic ways of thinking make sense to evolutionary ecologists, who for decades have borrowed concepts like cost–benefit analysis, marginal values, investment, and profitability. Recently, the flow of ideas between these disciplines has become bidirectional. Several economists are now considering how past selection pressures have designed psychological processes underlying preferences, cooperation, and other aspects of economic transactions (e.g., Becker, 1976; Rubin and Paul, 1979; Frank, 1985, 1988; Bergstrom and Bagnoli, 1993; Samuelson, 1993; Simon, 1993; Bergstrom, 1995; Binmore et al., 1995; Mulligan, 1997; Sethi and Somanathan, 1996; Ben-Ner and Putterman, 1998; Romer, 1995), and some have begun to attend to variations in preferences and utility functions in relation to variables that evolutionists would consider central, such as sex, age, and parental and other kinship statuses (e.g., Rubin and Paul, 1979; Becker, 1981; Bergstrom, 1995; Eckel and Grossman, 1996; Mulligan, 1997). However, the dominant model of *Homo economicus* continues to be a folk psychological one in which preferences are translated into action by "rational" processes of deliberative decision-making that do not necessarily correspond to the psychological machinery that has actually evolved (Daly and Wilson, 1997).

Traditional economic analysis has assumed not only that actors are rational utility maximizers, but also that there is a unitary currency of utility in which all "goods" can be valued. (See Sunstein, 1994, for a critique of the assumption of a unitary currency of utility.) These assumptions make the application of economic decision theory to the behavior of nonhuman animals seem metaphorical. But the application of cost–benefit models to *Homo sapiens* is really no less metaphorical. All complex animals confront the problem of how to value seemingly incommensurate goods in a common "currency." How many prospective calories will cover the predation risk cost of foraging activity *X*? Is mating opportunity *Y* sufficiently valuable to warrant accepting prospective injury risk *Z* by competing for it? From this comparative perspective, the real innovation in the invention of money was not that it reduces disparate utilities to one, but that it facilitates otherwise difficult reciprocal exchange. Money permits the elaboration of economic transactions by eliminating the necessity that one party trust the other to reciprocate in future, as well as by enabling exchanges in which the "buyer" does not otherwise have a commodity presently desired by the "seller." Unfortunately, this fungibility of assets in modern economic systems increases the appeal of destructive resource exploitation because exploiters can take their profits and invest elsewhere.

Some readers may protest that the costs, benefits, and trade-offs that we invoke in explaining risky decision-making by animals are only statistical characterizations of the natural selective past, whereas for human actors prospective costs and benefits are actually calculated and considered and hence are proximate determinants of behavioral choices. Perhaps so, but the model of decision makers as conscious and rational deliberators is, in fact, just as problematic when applied to people as when applied to kangaroo rats or starlings. Experimental psychologists have shown that people do not have the sort of privileged insight into the determinants of their own decisions that rational actor models presume and that the sense of having engaged in conscious deliberation and reasoned choice is largely illusory and after the fact (e.g., Nisbett and Wilson, 1977; Nisbett and Ross, 1980; Kahneman et al., 1982; Marcus, 1986). Although there are controversies about how best to characterize the psychological processes that produce human choice behavior, the evidence is unequivocally contrary to the assumption that people engage in the sort of simple rational calculus of utility maximization customarily attributed to *Homo economicus* (e.g., Kahneman and Tversky, 1979, 1984; Nisbett and Ross, 1980; Loewenstein and Thaler, 1989; Shafir, 1993; Gigerenzer and Hoffrage, 1995; Cosmides and Tooby, 1996; Hoffman et al., 1996).

Consider, for example, the classic demonstration by Kahneman and Tversky (1979) that people weigh alternatives very differently when exactly the same end states are framed as gains versus losses. Most people prefer a sure $1500 gain over letting a coin toss determine whether they would get $1000 or $2000, and this "risk aversion" is not hard to rationalize: it apparently reflects the diminishing marginal utility of money, presumably because each successive dollar's incremental effect on our expected well-being really is smaller than the last. (The difference between being penniless or a millionaire is much greater than the additional impact of a second million.) However, if people are presented with exactly the same alternative outcomes framed as an initial award of $2000 followed by a choice between relinquishing $500 or taking a 50% chance on being obliged to relinquish $1000, most switch to "risk acceptance" (preferring the gamble). This is very much harder to rationalize in terms of the curvilinear utility of money. Losing any ground whatever from a state already attained apparently has a strong negative emotional valence.

How are the mental processes that produce such apparent inconsistencies of preference to be understood? Adaptationist thinking suggests several testable hypotheses. One is that voluntarily relinquishing prior gains has evolved to be aversive in the specific context of social bargaining because in ancestral environments, to relinquish prior gains was to advertise weakness, inviting future demands for additional concessions. Another hypothesis is that people may be averse to alternatives that take more time or require more steps, ultimately because delay and complexity have entailed risk of defection or duplicity. Even those decision theorists who have been critical of the assumption that people are rational utilitarians with full conscious knowledge of their own preferences (e.g., Kahneman and Tversky, 1984; Loewenstein and Thaler, 1989; Shafir, 1993; Knetsch, 1995) and who have thus attempted to model the psychological processes that produce these "irrational" effects have yet to consider such possibilities. In addition to its value as a cautionary tale against simple rational-actor models, Kahneman and Tversky's gain–loss framing effect is potentially interesting with respect to decisions about how to pitch conservation efforts to the public: the emotional appeal of a campaign to avoid the loss of what we already possess may be more powerful than the appeal of promised gains through remediation.

Another area in which economic analysis might benefit from considering how the evolved human psyche works is in efforts to attach prices to nonmarket resources. Certain "goods," such as air, have not ordinarily been monopolizable, exchangeable, or partible, and have not traditionally been treated as property, nor even thought of as resources. Other "goods," such as the tranquility or beauty of a setting, are clearly threatened by various sorts of economic exploitation and must somehow be valued in decisions about whether the gains from that exploitation are sufficient to offset the losses in these nonmarket resources. Armed with a unitary currency (money) and the conception of human decision-makers as capable of articulating veridical, rational preferences, economists interested in placing values on nonmarket goods have invented the "contingent valuation method" (CVM; e.g., Carson and Mitchell, 1993; Goodwin et al., 1993; Willis and Garrod, 1993; Cummings and Harrison, 1994; Smith, 1994; Heyde, 1995).

In a CVM study, a sample of people are asked how much they would be willing to pay to retain or attain some benefit. Ideally, respondents in a CVM study are given sufficient relevant information to permit a meaningful answer to some question such as how much would you be willing to pay in order to engage in a recreational activity *X* at place *Y* under conditions *Z* on a total of *N* days in the next year, or what is the maximum additional amount that you would pay before deciding that *X* is too expensive (e.g., Cummings et al., 1986; Carson and Mitchell, 1993). Critics of this method have been alarmed by the growing use of CVM studies in policy making and in legal decisions concerning compensation and have decried the presumption that it is appropriate or even possible to place dollar values on such goods as human health, aesthetic worth, or species survival (e.g., Sunstein, 1994; Heyde, 1995). Moreover, when CVM survey data are used to determine the damages to be paid by environmental despoilers, as they have been and are being used, then the incentive structures for decision makers planning environmentally hazardous endeavors may become such that damaging even the recreational resources of the wealthy will be more costly (and hence more to be avoided) than damaging resources that are crucial to the lives and health of much larger numbers of people of lesser means (see also Boyce, 1994).

But the problems with the CVM are not limited to the questionable justness of its policy applications. There are good reasons to doubt that people are capable of giving meaningful, valid answers to CVM questions (e.g., Fischoff, 1991; Kahneman and Knetsch, 1992; Kahneman et al., 1993; Cummings and Harrison, 1994; Guagnano et al., 1994; Binger et al., 1995; Gregory et al., 1995; Loewenstein and Adler, 1995). Answers to CVM questions regularly violate the expectation that increments in the quantity of a good will increase its subjective value, for example, as may be illustrated by Kahneman's (1986) demonstration that different groups of people attached almost the same average dollar value in extra taxes to preserving the fish stocks of lakes in a small area of the province of Ontario as they were willing to pay for all the lakes in Ontario. Professed willingness to pay is also apt to be greatly exaggerated until respondents are reminded of the many possible demands on their limited means. For example, Hamilton, Ontario, residents who were asked how much they would be willing to pay to improve boating conditions in the local harbor gave a mean answer that was 30-fold higher if this was the first such CVM question in the interview than if it was the second (Dupont, 1996).

Being asked to put a price on certain environmental goods may be so out of the normal context in which a preference would be elicited that it is impossible to give a meaningful response. Indeed, it is questionable whether the sorts of preferences that the CVM obliges interviewees to articulate even exist prior to the questioning or are instead constructed in ways affected not only by the stable attributes of the respondent (as the CVM assumes), but also by the circumstances of the interview and the contextual framing of the task (Fischoff, 1991; Boyce et al., 1992; Kahneman and Knetsch, 1992; Irwin et al., 1993; Baron and Greene, 1996). Ajzen et al. (1996), for example, showed that respondents who had been "primed" by the inclusion of do-gooder bromides (e.g., "It's better to give than to receive") in an ostensibly unrelated word-unscrambling task committed almost twice as much to a public good from which they would derive no personal benefit as did respondents who had unscrambled only neutral control sentences.

It is also questionable whether even cooperative respondents are able to predict what they would really do or pay if the situation ceased to be hypothetical (Bohm, 1994; Loewenstein and Adler, 1995), and it is even more questionable whether they have conscious access tc the determinants of their choices. Nevertheless, CVM researchers ask people to articulate just these things and accept the answers at face value. When Kahneman and Knetsch (1992) proposed, for example, that professions of willingness to pay for environmental protection or remediation might represent "the purchase of moral satisfaction" rather than the specific environmental benefit's value to the respondent, several CVM researchers announced that they had disconfirmed this hypothesis by showing that respondents who were instructed to choose "the reason" for their choice of dollar values from a menu mainly picked something else (e.g., Loomis et al., 1993; MacDonald and McKenney, 1996).

If we are going to price nonmarket goods in making tough decisions among alternatives that all have negative aspects, as it seems we must, then we need to move beyond these simplistic conceptions of decision makers as rational and decision criteria as consciously accessible. Recent efforts (e.g., Gigerenzer et al., 1988, 1991; Cosmides, 1989; Gigerenzer and Hoffrage, 1995; Cosmides and Tooby, 1996; Wang, 1996) have begun to incorporate evolutionary psychological models into explanations for the seemingly irrational aspects of the ways in which people process information and order their priorities. Success in this endeavor partly depends on correctly hypothesizing the nature of the adaptive problems that emotional reactions and other psychological processes were designed to solve in order to clarify the functional organization of complex psychological phenomena involved in decision-making under uncertainty, risk-taking, discounting the future, collective action, cooperating in use of common pool resources, and many other aspects of decision-making relevant to conservation of resources, species, and habitats.

Risk as Variance of Expected Payoffs

An adaptationist perspective on human psychology and action could contribute to understanding of several aspects of the contemporary ecological crisis. The need to elucidate the psychological adaptations of most direct and remediable relevance to the continuing population explosion is one obvious example. Another area in which evolutionary theorizing has already contributed is in identifying the circumstances under which the restraint of selfish consumption in cooperative ventures is realizable and those under which opportunities for "cheating" make cooperation unstable (Axelrod, 1984; Cosmides and Tooby, 1989; Boone, 1992; Hawkes, 1992). But in addition to the much-discussed problems entailed by the natural selective advantages enjoyed by the most prolific and selfish phenotypes, the ways in which selection has shaped such subtle specifics as time preferences, social comparison processes, and sex differences may also have important implications for conservation and environmental remediation efforts. If we are to mitigate the ills caused by human reluctance to reduce resource accumulation and consumption, for example, it seems important to elucidate the precise ways in which human decision-making discounts the future and how this discounting responds to uncertainty, both in ontogeny and in facultative responsiveness to variable aspects of one's immediate situation. The perceived costs of giving up present consumption depend on one's material circumstances, but little is known about subjective valuations and perceptions of uncertainties as a function of material and social circumstances.

Experimental studies of nonhuman animal foraging decisions have established the ecological validity of a risk-preference model based on variance of expected payoffs. Rather than simply maximizing the expected (mean) return in some desired commodity such as food, animals should be, and demonstrably are, sensitive to variance as well (Real and Caraco, 1986). Whereas seed-eating birds generally prefer to forage in low variance microhabitats as compared to ones with a similar expected yield but greater variability, for example, they switch to preferring the high variance option when their body weight or blood sugar is so low as to predict that they will starve unless they can find food at a higher than average rate (Caraco et al., 1980). Although the high variance option increases the bird's chances of getting exceptionally little, a merely average yield is really no better, and the starving birds accept the risk of finding even less in exchange for at least some chance of finding enough to survive. Such experiments have produced essentially similar results in several species of seed-eating birds (Caraco and Lima, 1985; Barkan, 1990), as well as in rats (Kagel et al., 1986; Hastjarjo et al., 1990).

It may be possible to understand risk acceptance by human explorers, adventurers, and warriors in analogous terms. Even taking dangerous risks to unlawfully acquire the resources of others might be perceived as a more attractive option when safer, lawful means of acquiring material wealth yield a pittance, although the expected mean return from a life of robbery may be no higher and the expected life span shorter. Interestingly, variations in robbery and homicide rates between places are better explained by variance in income than by absolute values of poverty (e.g., Hsieh and Pugh, 1993).

There is also experimental evidence that human decision-making is sensitive to variance as well as to expected returns. Psychologists and economists, using various hypothetical lottery or decision-making dilemmas, have documented that people's choices among bets of similar expected value are affected by the distribution of rewards and probabilities (e.g., Lopes, 1987, 1993). They are also influenced by whether numerically equivalent outcomes are portrayed as gains or losses as discussed above (Kahneman and Tversky, 1979). The underlying psychological dimension governing these choices among alternative, uncertain outcomes has been conceptualized as one ranging from "risk-averse" to "risk-seeking" (or "risk-prone" or "risk-accepting"). In the experimental nonhuman studies described above, the starving below-weight animal preferring the high variance option would be deemed risk-seeking. Diversity in risk aversion or risk seeking could be mediated psychologically by either variation in the subjective utilities of the outcomes or variation in perceptions of the probabilities associated with each outcome or both (Real, 1987).

Sex Differences in Risk Acceptance and Resource Use?

Consideration of the ways in which sexual selection differentially affects the sexes suggests that women and men confronted by uncertainty might have different subjective utilities or subjective probabilities and that these psychological determinants of risk acceptance or aversion might also vary in relation to life-history variables and cues indicative of expected success in intrasexual competition. Psychologists studying risk acceptance have documented sex differences and age effects but have focused mainly on stable individual differences (e.g., Trimpop, 1994; Zuckerman, 1994) and have scarcely addressed how risk preferences may be affected by social and material cues of one's life prospects and by one's relative social and material success.

The rationale for anticipating sex differences in the way people value and exploit the environment, as well as differences in willingness to risk damaging one's health, is an argument that has been applied to other aspects of risk taking and to sexually differentiated adaptations for intrasexual competition (e.g., Wilson and Daly, 1985, 1993). Its premise is that ancestral males were subject to more intense sexual selection (the component of selection due to differential access to mates) than were ancestral females, with resultant effects on various sexually differentiated attributes.

Successful reproduction, in *Homo* as in most mammals, has always required a long-term commitment on the part of a female, but not necessarily on the part of a male. Female fitness has been limited mainly by access to material resources and by the time and energy demands of each offspring, whereas the fitness of males, the sex with lesser parental investment, is much more affected by the number of mates (Trivers, 1972; Clutton-Brock, 1991). It follows that the expected fitness payoffs of increments in "mating effort" (by which term we encompass both courtship and intrasexual competition over potential mates) diminish much more rapidly for females than for males, and it is presumably for this reason that such effort constitutes a larger proportion of total reproductive effort for men than for women. One hypothesis inspired by these considerations is that men may find rapid resource accrual, resource display, and immediate resource use somewhat more appealing than women and that men may be more inclined to disparage risks and discount the future in their decisions about acquiring and expending resources (Low and Heinen, 1993).

Following Bateman (1948), Williams (1966), and Trivers (1972), sex differences in the variance in reproductive success are widely considered indicative of sex differences in intrasexual competition. Relatively high variance generally entails both a bigger prize for winning and a greater likelihood of failure, both of which may exacerbate competitive effort and risk acceptance. Bigger prizes warrant bigger bets, and a high probability of total reproductive failure means an absence of selection against even life-threatening escalations of competitive effort on the part of those who perceive their present and probable future standing to be relatively low. Although it is worth cautioning that fitness variance represents only the potential for selection and that variations in fitness could in principle be nonselective (Sutherland, 1985), intrasexual fitness variance appears to be a good proxy of the intensity of sexual selection because it is a good predictor of the elaboration of otherwise costly sexually selected adaptations. In comparative studies, sex differences in such attributes as weaponry for intraspecific combat are apparently highly correlated with the degree of effective polygamy of the breeding system—that is, with sex differences in fitness variance (e.g., Clutton-Brock et al., 1980). It is also worth cautioning that there can be other evolutionary explanations for sex differences in risk acceptance besides the Bateman–Williams–Trivers theory of sexual selection (see, for example, Regelmann and Curio, 1986), but this theory currently appears to be the one of greatest relevance to mammals in general and humans in particular.

All evidence suggests that the human animal is and long has been an effectively polygynous species, albeit to a lesser degree than many other mammalian species. Successful men can sire more children than any one woman could bear, consigning other men to childlessness, and this conversion of success into reproductive advantage is ubiquitous across cultures (Betzig, 1985). Of course, great disparities in status and power are likely to be evolutionary novelties, no older than agriculture, but even among relatively egalitarian foraging peoples, who make their living much as most of our human ancestors did, male fitness variance consistently exceeds female fitness variance (Howell, 1979; Hewlett, 1988; Hill and Hurtado, 1995). Moreover, in addition to the evidence of sex differences in the variance of marital and reproductive success in contemporary and historically recent societies, human morphology and physiology manifest a suite of sex differences consistent with the proposition that our history of sexual selection has been mildly polygynous: size dimorphism with males the larger sex, sexual bimaturism with males later maturing, and sex differential senescence with males senescing faster (Harcourt et al., 1981; Møller, 1988).

If the fitness of our male ancestors was more strongly status dependent than that of our female ancestors, as seems likely, then from the perspective of sexual selection theory, men may be expected to be more sensitive than women to cues of their status relative to their rivals. If intrasexual competition among men has largely depended on acquisition of resources (both material and social), which were converted into reproductive opportunities, and if there has been a history of high variance in the distribution of resources and reproductive opportunities, then the masculine psyche is likely to have evolved to accept greater risk in its efforts to acquire, display, and consume resources, especially when accepting a small payoff has little or no more value than no payoff, as, for example, when a small payoff leaves a poor man still unmarriageable. This argument treats risk as variance in the magnitude of payoffs for a given course of action. In life-threatening circumstances people often take the riskier (higher variance) course of action. But people also take great risks when present circumstances are perceived as "dead ends." For example, history reveals that successful explorers, warriors, and adventurers have often been men who had few alternative prospects for attaining material and social success. Later-born sons of aristocratic families were the explorers and conquerors of Portuguese colonial expansion, for example, while inheritance of the estate and noble status went to first-born stay-at-home sons (Boone, 1988). Similarly, later-born sons and other men with poor prospects have been the ones who risked emigration among more humble folk, too (e.g., Clarke, 1993), a choice which sometimes paid off handsomely, as in European colonial expansion, but must surely have more often led to an early death.

Sex Difference in Disdain for Health Risks?

One of the many domains within which men manifest greater risk acceptance than women is in health monitoring and preventive health care. Apparently, the average number of physician contacts per year is greater for males than females before puberty, but between 15 and 45 years of age, women visit physicians almost twice as often as men (Woodwell, 1997), even after one has accounted for birth-related visits and sex differences in rates of accident and illness. We hypothesize that men will also disregard the health hazards of various environmental contaminants more than women. And if men are relatively insensitive to the risks that they themselves incur, it seems likely that they will also be relatively insensitive to the risks that their activities entail for other people and for other fauna and flora.

One way to test these ideas is to ask people how they would behave in hypothetical dilemmas. As an example of this approach, we asked 173 introductory psychology students (90 women and 83 men) at McMaster University in Hamilton, Ontario, to consider the following hypothetical situation and then answer questions as if the situation applied to them.

> Imagine that you presently live in a mid-sized southern Ontario city of 300,000 people, where you were born and where most of your family and friends still reside. You have been looking for work and you suddenly find yourself with two job offers to choose between.
>
> If you accept Baylor & Wilson's offer of employment at $30,000 per annum, you can continue to live and work in your home town. If you accept Smithers and Company's offer of $35,000 [$50,000] instead, you will be relocated to a city of 600,000 people in another province. From what you've heard, this city sounds like an interesting and beautiful place to live, but air pollution levels and respiratory disease rates are twice [ten times] what they are in the city where you now live.
>
> Which offer do you accept?
>
> Baylor & Wilson ______ Smithers & Company ______

The alternatives in square brackets were presented to distinct sets of subjects, making a 2x2x2 between-groups experimental design: male versus female subjects x the magnitude of the incentive to move ($5000 versus $20,000 higher salary) x the magnitude of the deterrent costs in air quality and attendant health hazard (2-fold versus 10-fold).

Although all subjects were university students, at the same life stage and almost unanimously unmarried and childless, women and men responded somewhat differently to the experimental variables (Fig. 18-1). As we predicted on the basis of the arguments above, men were attracted by an extra financial incentive more than were women, although not significantly so. More striking, and statistically significant, was the differential response to en-

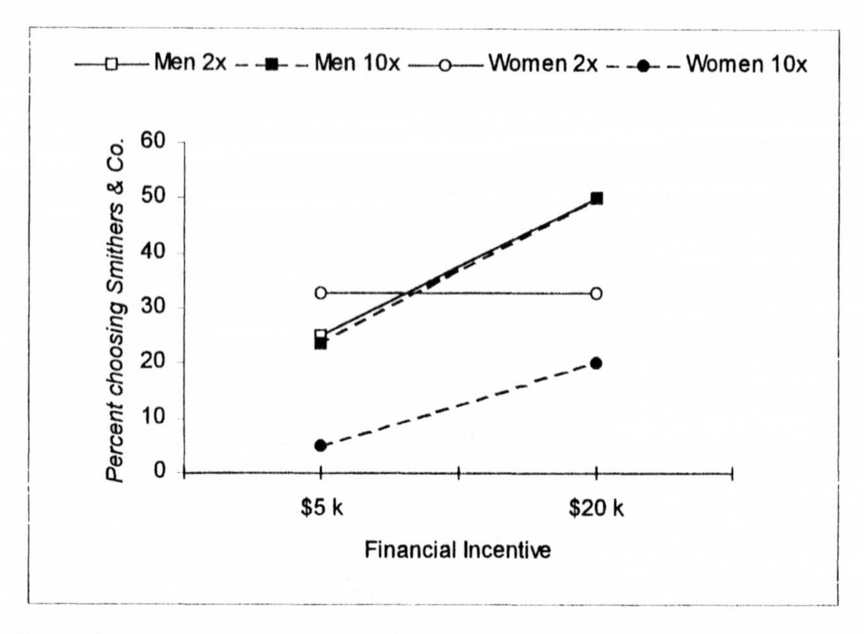

Figure 18-1 Percentage of men and women choosing the job at Smithers & Company, which would entail moving far away, either a $5,000 or $20,000 incentive above the hometown job, and either a 2 times or 10 times greater risk of respiratory problems than that of the hometown. Women were significantly more deterred by the health risk than males (p = .03 by logit loglinear analysis). The tendency for men to be more attracted by a financial incentive was not significant (p = .14).

vironmental risk: women were substantially deterred by higher costs in air quality and health hazards, but men were completely unaffected by this variable, choosing identically regardless of whether the stated costs were 2-fold or 10-fold. Other evidence also indicates that women may be more concerned about environmental health hazards than men (e.g., Flynn et al., 1994; Sachs, 1996, 1997). In a previous study involving a similar dilemma (but no variation of financial incentives and health risks), Wilson et al. (1996) found that men were significantly more likely than women to say they would accept a promotion "which would significantly boost your career" but would require moving to a city where the respiratory health risk was 10% higher than that of the hometown. In this earlier version, there were many parents among the subjects, and 41% of those who were parents said they would accept the promotion, compared to 81% of those without children, a difference that remained significant when the age of respondents was controlled.

Earlier in this chapter, we criticized "contingent valuation" studies for asking people how much they would be willing to pay for a particular benefit and taking their answers at face value, and we must acknowledge that the results we report here may have similar validity problems. Unlike CVM studies of nonmarket goods, however, we have asked people to consider a situation that is likely to be a common experience of most people: deciding to take one job rather than another, with benefits and costs associated with both. In principle, data from people's actual decisions between different employment opportunities can be compared with our results (as sometimes can be done and sometimes has been done in validating CVM results with "revealed preference" analyses of what people have actually paid for different goods or benefits; Smith, 1994; Carson et al., 1996). We attach no significance to the specific percentages of men and women choosing the Smith & Co. employment opportunity, but only to the sex difference in the impacts of an imagined financial incentive and an imagined health hazard. The apparently greater willingness of men than of women to treat health hazards as acceptable costs of opportunities for financial benefit should be further tested with real-world data on choices among different job opportunities.

Sex Difference in Disregard of Environmental Degradation?

In addition to the expectation that men are more likely than women to disdain personal health risks in their pursuit of economic and status advantages, we hypothesize that men are more likely than women to disregard or downplay environmental degradation. Support for this proposition already exists (e.g., Mohai, 1992; Sachs, 1997), but the possibility that it is a reflection of the male's psyche's greater prioritizing of present profits as a result of differential histories of sexual selection has not been articulated or explored.

The rate at which one "discounts the future" is the rate at which the subjective value of future consumption diminishes relative to the alternative of present consumption (or, the "interest rate" required to motivate foregoing consumption). If *A* discounts the future more steeply than *B*, then *A* will value a given present reward relative to expected future rewards more highly than *B* and will be less tolerant of what psychologists call "delay of gratification." Hence, variable willingness to engage in nonsustainable modes of resource exploitation such as clearcutting or otherwise expending one's capital may be construed, at least in part, as variation in the rates at which decision makers discount the future.

Do men discount the future more steeply than women in the specific realm of conservation decisions? Wilson et al. (1996, p. 154) asked another set of 104 McMaster University people (36 men and 68 women ranging in age from 17 to 24) to consider the following dilemma:

> Imagine you are farming a tract of land. Your father, like his father before him, lived off the profits from the farm without taking additional wage work elsewhere. You were fortunate to earn a scholarship to university to study agriculture, and now that you have inherited the farm you are considering changing the techniques of farming to be more specific and business-like. Prior to inheriting the farm you had a successful career as a broker specializing in agricultural commodities. [After your wife died suddenly, you've decided to leave that job to return to the farm. Your two children are delighted about the prospect of living on the farm.] Presently, you are pondering whether to follow one course of action (Plan A) or another (Plan B).
> Plan A: Convert the farm entirely to hybrid corn production for livestock feed. Corn is extremely profitable to grow, but it requires heavy chemical fertilization which over time will percolate into the water table with a very high probability that the land will not be usable in 60 years without heavy chemical supplements.
> Plan B: Convert the farm entirely to hay for livestock feed. Hay in good years can bring a good market price, but generally hay yields a modest profit. On the other hand, hay production does not diminish the quality of the soil and chemical supplements are not needed.
> Which plan did you choose? A or B? ______

Men were significantly more likely to choose the soil-degrading option (39% of men and 16% of women, fig. 18-2). In order to determine whether these "decision makers" were uti-

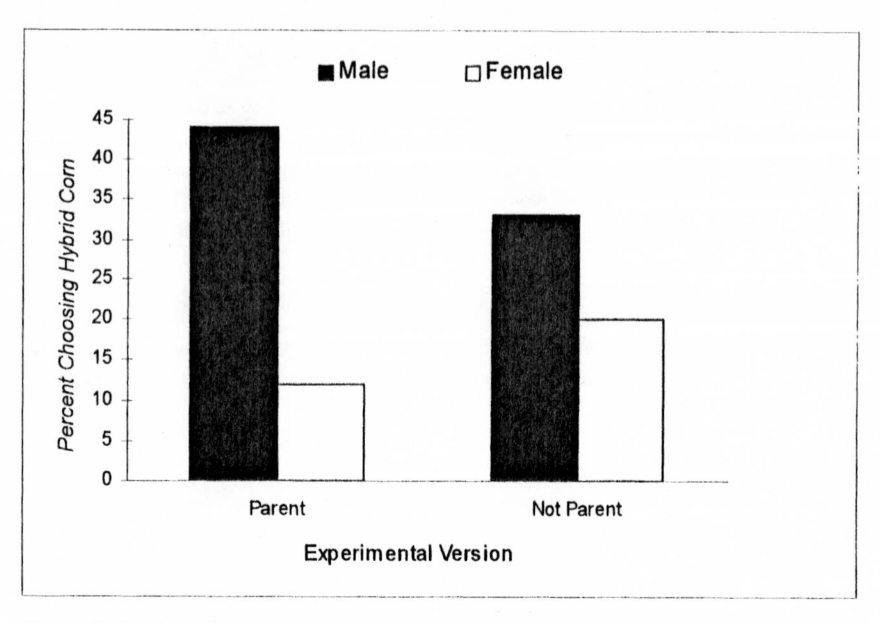

Figure 18-2 Percentage of men and women choosing the soil-degrading option (plan A: hybrid corn) according to the experimental version of the hypothetical dilemma they considered (being a widowed parent of two children, versus parental status unmentioned). Men were significantly more likely to choose the soil-degrading option ($\chi^2 = 6.6, p < .01$), but neither parental status ($p = .91$) nor the interaction of sex of the subject by parental status ($p = .27$) was significant by logit loglinear analysis.

lizing sound economic logic, we asked them to rate their agreement (on seven-point Likert scales) with propositions that might reflect the reasoning behind their choices. As expected, the proposition that "because you can always invest the profits from farming in other economic ventures including other farmland, you should weight profit over damage to the land" was endorsed significantly more strongly ($p < .0001$) by those who chose corn than by those who chose hay, but there was no significant effect of sex of subject.

In this scenario, one factor that might be expected to influence decisions that may have long-term negative effects on the quality of your farm is whether your children are likely to continue farming. This was the rationale for adding the two bracketed sentences ["After your wife died . . ."] for half the subjects. We had anticipated that parental status would increase the likelihood that subjects of both sexes would be deterred from planting corn due to the possible long-term costs, but inclusion of this sentence did not result in any detectable difference in the choice of crops (fig. 18-2). Perhaps imagining that one has children cannot evoke the mindset of actual parenthood. (In this sample, only four people were married and only two had children.) Another possibility is that some subjects interpreted the existence of children as a source of increased demand for imminent cash flow. (And it may be relevant that the experiment was conducted in a region where it has become the norm that farmland is retained only until suburban real estate developers are prepared to pay the farmer's asking price.)

We also anticipate that the percentage of people choosing corn versus hay might vary with expertise or other characteristics of the sample, so, for example, economics majors rather than psychology majors may be more likely to choose corn, and conservation biology majors may be more likely to choose hay than our sample of psychology majors. However, we predict that, in general, a sex difference in choice will remain; departures from this expectation may reveal interesting insights into the determinants of decision-making relevant to conservation efforts.

What other factors might we expect to influence the steepness of discounting functions? If we assume that there is an evolved, facultative decision process behind such discounting, then obvious candidates are life expectancy and other sources of variable, subjective probability that one will retain control of the resources in question in the future. Wilson and Herrnstein (1985) have argued, on the basis of diverse evidence, that men who engage in predatory violence and other risky criminal activity have different "time horizons" than law-abiding men, weighing the near future relatively heavily against the long term. What these authors failed to note is that facultative adjustment of one's personal time horizons could be an adaptive response to predictive information about one's prospects for longevity (Daly and Wilson, 1990; Rogers, 1991, 1994; Hawkes, 1992; Gardner, 1993; Wilson and Daly, 1997) and the stability of one's social order and ownership rights.

Much of the social science literature on discounting and time horizons treats tolerance of delayed gratification as a proxy for intelligence. We see this as an anthropomorphic stance predicated on the claim that the capacity to plan far ahead and adjust present behavior to long-term future expectancies is a hallmark of complex cognitive capacity in which the human animal is unmatched. From an evolutionary adaptationist perspective, however, discounting and delay of gratification represent essentially the same issue as that addressed by Fisher (1930) and all subsequent life-history theorists: how is the future optimally weighted in deciding present allocations of effort (e.g., Roitberg et al., 1992; Clinton and LeBoeuf, 1993). The answers depend on the expected present and future reproductive payoffs associated with each alternative, expectations that may vary facultatively in response to available cues, and these issues are as germane to nonhuman animals (and plants) as to sophisticated cognizers. From this perspective, what selects for willingness to delay gratification is a high likelihood that present somatic effort can be converted to future reproduction. Thus, rather than reflecting stupidity, short time horizons are likely to characterize those with short life expectancies, those whose sources of mortality are not strongly or predictably dependent on their actions, and those for whom the expected fitness returns of present striving are positively accelerated rather than exhibiting diminishing marginal returns.

How human beings and other animals discount the future has been described in considerable detail by experimental psychologists, but a fuller understanding of these processes awaits the infusion of evolutionary adaptationist insights (Bateson and Kacelnik, 1996; Benson and Stephens, 1996; Kacelnik and Bateson, 1996; McNamara, 1996). The most noteworthy conundrum concerns the shape of discount functions, which are often, perhaps typically, hyperbolic rather than "rationally" exponential (Kirby and Herrnstein, 1995; Green and Myerson, 1996). The puzzling thing about hyperbolic discount functions is that they engender predictable reversals of preference between alternative futures with different time depths and hence predictable regret of what will become bad decisions in retrospect (e.g., Hoch and Loewenstein, 1991; Roelofsma, 1996). Suppose, for example, that a large reward two weeks hence is preferable to a smaller reward one week hence. If future discounting is hyperbolic, then as time passes the appeal of the more imminent reward rises more steeply than that of the more distant, until it may come to be preferred when almost at hand. One consequence is that people and other animals may even invest effort in erecting impediments to their own anticipated future lack of "self-control" or capacity to delay gratification (Kirby and Herrnstein, 1995). Understanding why the psychological underpinnings of time preference have evolved to produce such seemingly maladaptive internal struggles and why the evolved human psyche defies normative economic theory by discounting different utility domains at different rates (Chapman, 1996) may provide important clues for understanding why "waste" and inefficiency are so hard to eradicate. (See Kacelnik, 1997, for a possible adaptationist explanation for hyperbolic discounting).

Life-Stage Patterns of Risk Preference: Young Men as the Most Risk-Accepting Demographic Group

One may also hypothesize that sexually differentiated valuations of natural resources may be especially conspicuous in those life stages in which males have been selected to compete for reproductive opportunities most intensely. By this reasoning, the life-stage in which laying claim to resources and expending rather than conserving them should be most attractive is that in which such behavior would have had the greatest expected fitness payoff for our ancestors. There is reason to believe that that lifestage for men is and has long been young adulthood (Daly and Wilson, 1990). Once men are husbands, they have something to lose in intrasexual competition, and once they are fathers, concern for their offspring's well-being may result in alterations of their valuations of the environment, especially if the resources would be those of recurring value from one generation to the next, such as land or water rights. Remarkably, however, effects of parenthood on environmental attitudes and behavior are virtually unstudied.

Several lines of evidence about life-span development support the idea that young men constitute a demographic class specialized by a history of selection for maximal competitive effort and risk-taking. Young men appear to be psychologically specialized to embrace danger and confrontational competition (e.g., Gove, 1985; Jonah, 1986, Lyng, 1990, 1993; Bell and Bell, 1993).

Risk of death as a result of external causes (accidents, homicides, and suicides) is greater in men than in women and is maximally sexually differentiated in young adulthood, both in the modern west (Wilson and Daly, 1985, 1993; Holinger, 1987; Daly and Wilson, 1990), and in nonstate, foraging societies more like those in which we evolved (Hewlett, 1988; Hill and Hurtado, 1995). The fact that men senesce faster and die younger than women even when they are protected from external sources of mortality suggests that these sex differences in mortality have prevailed long enough and persistently enough that male physiology has evolved to discount the future more steeply than female physiology. In the case of homicides, young men are not only the principal victims but also the principal perpetrators; indeed, men's likelihood of killing is much more peaked in young adulthood than is the risk of being killed (Daly and Wilson, 1988, 1990). All of these facts can be interpreted as reflections of an evolved life span schedule of risk proneness.

An alternative to this hypothesis, however, is that age patterns reflect responses to changes in relevant circumstances that happen to be correlated with age. Mated status, for example, would be expected to inspire a reduction in dangerous risk-taking because access to mates is a principal issue inspiring competition, and married men have more to lose than their single counterparts. Marital status is indeed related to the probability of committing a lethal act of competitive violence, but age effects remain conspicuous when married and unmarried men are examined separately (Daly and Wilson, 1990). Similarly, men are most likely to be economically disadvantaged in young adulthood, and poverty, too, is a risk factor in intrasexual competitive homicide, but young adulthood and unemployment status are again separable risk factors for homicide (Daly and Wilson, 1990).

Dangerous acts are adaptive choices if the positive fitness consequences are large enough and probable enough to offset the costs (Daly and Wilson, 1988). Disdain of danger to oneself is especially to be expected where available risk-averse alternatives are likely to produce a fitness of zero: if opting out of dangerous competition maximizes longevity but never permits the accrual of sufficient resources to reproduce, then selection will favor opting in (Rubin and Paul, 1979; Enquist and Leimar, 1990).

From a psychological point of view, it is interesting to inquire how age- and sex-specific variations in effective risk-proneness are instantiated in perceptual and/or decision processes. As we noted above, one possible form of psychological mediation entails flexible time horizons or discount rates. Other psychological processes with the effect of promoting risk-taking can also be envisaged. One could become more risk prone as a result of one or more of the following: intensified desire for the fruits of success, intensified fear of the stigma of nonparticipation, finding the adrenalin rush of danger pleasurable in itself, underestimating objective dangers, overestimating one's competence, or ceasing to care whether one lives or dies. As drivers, for example, young men both underestimate objective risks and overestimate their own skills in comparison to older drivers (Finn and Bragg, 1986; Matthews and Moran, 1986; Brown and Groeger, 1988; Trimpop, 1994). There is also some evidence that the pleasure derived from skilled encounters with danger diminishes with age (Gove, 1985; Lyng, 1990, 1993). In general, sensation-seeking inclinations, as measured by preferences for thrilling, dangerous activities, are higher in men than in women and decrease with age in a pattern quite like that of violent crime perpetration (Zuckerman, 1994).

Youths are especially unlikely to seek medical assistance or other health-enhancing preventive measures (Millstein, 1989; Adams et al., 1995), and young men are the demographic group most willing to take risks with drugs and intoxicants and to risk contracting sexually transmitted diseases (Irwin, 1993; Millstein, 1993). Relative disdain for their own lives can also be inferred from the fact that men's suicide rates maximally surpass women's in young adulthood (Holinger, 1987; Gardner, 1993).

In this context, it may be worth noting that the data in fig. 18-1 and 18-2 were collected almost entirely from young adults, in whom risk acceptance and sex differences therein may be most pronounced. However, the subjects were also people with good economic prospects and life expectancies, and these factors should have diminished risk acceptance. Because of their demographic uniformity, these samples were unsuitable for assessing the possibility of differential responses according to age, marital, and parental status. Whether this artificial technique is suitable for exploring life-span developmental changes and differences between economic classes and other life circumstances remains to be seen.

It is clear that the most risk-prone demographic classes accept risk in diverse domains, and it seems likely that the same association would hold in comparing individuals within demographic categories. But the degree to which risk proneness is domain general is still largely an open question. Zuckerman (1994) has argued that sensation-seeking is a stable personality characteristic: a domain-general mindset which is highly correlated with individual differences in neuron membrane physiology, and he has developed a "sensation-seeking scale," on which men score significantly higher than women, and both sexes (but especially men) score highest in young adulthood. We asked subjects who participated in the hypothetical job choice dilemma (fig. 18-1) to complete Zuckerman's "thrill and adventure seeking" scale, and we, too, found a significant sex difference (average score for males was 7.6 and for females 6.1; $t = 3.54$, df $= 119$, $p < .001$). However, sensation-seeking scores were not associated with subjects' choice responses to the dilemma, and we are currently conducting research aimed at assessing the degree to which risk acceptance is consistent within individuals across different contexts and alternative operationalizations of risk.

Rogers (1994, 1997) has brought evolutionary reasoning to bear on the issue of optimal age-specific rates of future discounting, given the age-specific mortality and fertility schedules of human populations. His analysis suggests that people of both sexes should have evolved to have the shortest time horizons and to be maximally risk accepting in young adulthood. More specifically, his theoretical curve of age-specific optimal discount rates looks much like the human life-span trajectory of reckless risk proneness that may be inferred from data on accidental death rates and homicide perpetration. The claim that optimal discount rates decline as one ages may seem paradoxical, given the argument that indicators of a short or uncertain expected future life span should be cues favoring risk acceptance. The factors responsible for Rogers' counterintuitive result are certain peculiarities of human life history and sociality, namely, gradually diminishing fertility long before death and a shifting allocation of familially controlled resources between personal reproductive efforts and descendants' reproductive efforts.

Economists such as Norgaard and Howarth (1991) and Common (1995) consider it a conceptual error to extend the concept of future discounting beyond the individual actor's reasonably expected life span and argue that conserving resources for future generations is an issue of resource allocation and equity, instead. But to behavioral ecologists, one's descendants are an extension of one's self, and organisms may be expected to have evolved to act in ways that will promote their fitness both before and after their deaths. Thus, appropriate modeling of the factors affecting optimal discount rates requires consideration of the psychology of human kinship and lineage investment (Rogers, 1991; Kaplan, 1996).

Conclusions and Recommendations

We believe that effective solutions to environmental and conservation problems require a sophisticated understanding of their sources in human desires and actions. Answering this challenge will surely require an integration of conceptual and empirical contributions from several disciplines. Our thesis has been that the use of the Darwinian/Hamiltonian selectionist paradigm of behavioral ecology as metatheory for psychology and economics may constitute one particularly promising route toward productive interdisciplinary synthesis.

A strength of bringing this behavioral ecological perspective to bear on the study of human decision-making that impacts conservation and environmental degradation is that it has drawn attention to the likelihood of variations with respect to sex, life stage, parenthood, social status, inequity, and life expectancy cues, and unites these variables in a theoretical framework capable of generating predictions. This perspective has also contributed to the growing realization that research and education are insufficient to stem the tide of environmental degradation without sophisticated attention to modifying incentive structures, as argued by Ridley and Low (1994). And as we argued in criticizing some CVM studies, thinking evolutionarily draws attention to the fact that the functional organization of the human mind is not designed to produce accurate introspections, but rather to produce effectively reproductive action in ancestral environments, an understanding that sensitizes the researcher to the potential pitfalls of opinion polling.

A weakness is that the individualistic focus of evolutionary psychologists and behavioral ecologists has yet to shed much light on political processes, especially in state-level societies with complex governmental and other institutions, with the result that the implications drawn from evolutionists' insights are still likely to be rather far removed from practical policy recommendations.

The suggestion that our evolved "human nature" is a source of environmental exploitation and degradation is not a claim that nothing can be done, but a warning that effective conservation and remediation strategies will have to incorporate an understanding of relevant evolved psychological processes in order to modify human action.

Summary

The serious reduction in abundance and diversity of the earth's flora and fauna is a fact, but what can be done about it remains controversial. We argue that the use of the Darwinian/Hamiltonian selectionist paradigm of behavioral ecology as metatheory for psychology and economics offers a promising route to a sophisticated understanding of human desires and actions which are the sources of and solutions to conservation problems.

Our critique of the "contingent valuation method" widely used by economists centers on the point that the functional organization of the human mind is not designed to produce accurate introspections but rather to produce effectively reproductive action in ancestral environments.

A strength of the behavioral ecological perspective in developing hypotheses relevant to exploitation and despoliation is that it has drawn attention to the likelihood of variations in human decision-making with respect to sex, life stage, parenthood, social status, inequity, and life expectancy cues. In two experimental studies concerning sex differences in hypothetical decisions, men were significantly more likely than women to prefer a crop with higher profit but higher risk of soil degradation, and the men were more willing to treat personal health hazards as acceptable costs of opportunities for financial benefits.

Acknowledgments Our studies of environmental attitudes, perceptions, and decision-making have been supported by grants from the Great Lakes University Research Fund (GLURF grant 92-102), the Tri-Council Eco-Research Programme of Canada (TriCERP grant 922-93-0005), and by a John D. and Catherine T. MacArthur Foundation grant to the Preferences Network. We thank Tim Caro, Monique Borgerhoff Mulder, and two anonymous readers for comments on a previous draft.

References

Adams PF, Schoenborn CA, Moss AJ, Warren CW, Kann L, 1995. Health-risk behaviors among our nation's youth: United States, 1992. Bethesda, Maryland: Vital Health Statistics series 10, no. 192. National Center for Health Statistics.

Ajzen I, Brown TC, Rosenthal LH, 1996. Information bias in contingent valuation: effects of personal relevance, quality of information, and motivational orientation. J Environ Econ Manage 30:43–57.

Alvard M, 1993. Testing the noble savage hypothesis: interspecific prey choice by Piro hunters of Amazonian Peru. Hum Ecol 21:355–387.

Alvard M, 1994. Conservation by native peoples: prey choice in depleted habitats. Hum Nat 5:127–154.

Alvard M, Kaplan H, 1991. Procurement technology and prey mortality among indigenous neotropical hunters. In: Human predators and prey mortality (Stiner MC, ed). Boulder, Colorado: Westview Press, 79–104.

Axelrod R, 1984. The evolution of cooperation. New York: Basic Books.

Barkan CPL, 1990. A field test of risk-sensitive foraging in black-capped chickadees (*Parus atricapillus*). Ecology 71:391–400.

Baron J, Greene J, 1996. Determinants of insensitivity to quantity in valuation of public goods: contribution, warm glow, budget constraints, availability, and prominence. J Exp Psychol Appl 2:107–125.

Bateman AJ, 1948. Intra-sexual selection in Drosophila. Heredity 2:349–368.

Bateson M, Kacelnik A, 1996. Rate currencies and the foraging starling: the fallacy of the averages revisited. Behav Ecol 7:341–352.

Becker G, 1976. Altruism, egoism, and genetic fitness: economics and sociobiology. J Econ Lit 14:817–826.

Becker GS, 1981. A treatise on the family. Cambridge, Massachusetts: Harvard University Press.

Bell NJ, Bell RW, 1993. Adolescent risk taking. Newbury Park, California: Sage.

Ben-Ner A, Putterman L, 1998. Values and institutions in economic analysis. In: Economics, values and organizations (Ben-Ner A, Putterman L, eds). New York: Cambridge University Press, 3–69.

Benson M, Stephens DW, 1996. Interruptions, trade-offs, and temporal discounting. Am Zool 36:506–517.

Bergstrom TC, 1995. On the evolution of altruistic ethical rules for siblings. Am Econ Rev 85:58–81.

Bergstrom TC, Bagnoli M, 1993. Courtship as a waiting game. J Polit Econ 101:185–202.

Betzig L, 1985. Despotism and differential reproduction: a Darwinian view of history. New York: Aldine de Gruyter.

Binger BR, Copple R, Hoffman E, 1995. Contingent valuation methodology in the natural resource damage regulatory process: choice theory and the embedding phenomenon. Nat Res J 35:443–459.

Binmore KG, Samuelson L, Vaughn R, 1995. Musical chairs: modeling noisy evolution. Games Econ Behav 11:1–35.

Bohm P, 1994. CVM spells responses to *hypothetical* questions. Nat Res J 34:37–50.

Boone JL, 1988. Parental investment, social subordination and population processes among the 15th and 16th century Portuguese nobility. In: Human reproductive behaviour: a Darwinian perspective (Betzig LL, Borgerhoff Mulder M, Turke P, eds). Cambridge: Cambridge University Press; 201–219.

Boone JL, 1992. Competition, conflict, and the development of social hierarchies, In: Evolutionary ecology and human behavior (Smith EA, Winterhalder B, eds). New York: Aldine de Gruyter; 301–337.

Boyce J, 1994. Inequality as a cause of environmental degradation. Ecol Econ 11:169–178.

Boyce RR, Brown TC, McClelland GH, Peterson GL, Schulze WD, 1992. An experimental examination of intrinsic values as a source of the WTA-WTP disparity. Am Econ Rev 82:1366–1373.

Brown ID, Groeger JA, 1988. Risk perception and decision taking during the transition between novice and experienced driver status. Ergonomics 31:585–597.

Caraco T, Lima SL, 1985. Foraging juncos: interaction of reward mean and variability. Anim Behav 33:216–224.

Caraco T, Martindale S, Whittam TS, 1980. An empirical demonstration of risk-sensitive foraging preferences. Anim Behav 28:820–830.

Carson RT, Flores NE, Martin KM, Wright JL, 1996. Contingent valuation and revealed preference methodologies: comparing the estimates for quasi-public goods. Land Econ 72:80–99.

Carson RT, Mitchell RC, 1993. The value of clean water: the public's willingness to pay for boatable, fishable, and swimmable quality water. Water Resources Res 29:2445–2454.

Chapman GB, 1996. Temporal discounting and utility for health and money. J Exp Psychol Learn Memory Cognit 22:771–791.

Clark CW, 1991. Economic biases against sustainable development. In: Ecological economics: the science and management of sustainability (Costanza R, ed). New York: Columbia University Press; 319–343.

Clarke AL, 1993. Behavioral ecology of human dispersal in 19th century Sweden (PhD dissertation). Ann Arbor: University of Michigan.

Clinton WL, LeBoeuf BJ, 1993. Sexual selection's effects on male life history and the pattern of male mortality. Ecology 74:1884–1892.

Clutton-Brock TH, 1991. The evolution of parental care. Princeton, New Jersey: Princeton University Press.

Clutton-Brock TH, Albon SD, Harvey PH, 1980. Antlers, body size and breeding group size in the Cervidae. Nature 285:565–567.

Common M, 1995. Sustainability and policy: limits to economics. Cambridge: Cambridge University Press.

Cosmides L, 1989. The logic of social exchange: has natural selection shaped how humans reason? Studies with the Wason selection task. Cognition 31:187–276.

Cosmides L, Tooby J, 1989. Evolutionary psychology and the generation of culture, Part II. Case study: a computational theory of social exchange. Ethol Sociobiol 10:51–97.

Cosmides L, Tooby J, 1996. Are humans good intuitive statisticians after all? Rethinking some conclusions from the literature on judgment under uncertainty. Cognition 58:1–73.

Costanza R, 1991. Ecological economics: the science and management of sustainability. New York: Columbia University Press.

Cummings RG, Brookshire DS, Schulze WD, 1986. Valuing public goods: the contingent valuation method. Totowa, New Jersey: Rowman & Allenheld.

Cummings RG, Harrison GW, 1994. Was the Ohio court well informed in its assessment of the accuracy of the contingent valuation method? Nat Res J 34:1–36.

Daly M, Wilson M, 1988. Homicide. Hawthorne, New York: Aldine de Gruyter.

Daly M, Wilson M, 1990. Killing the competition: female/female and male/male homicide. Hum Nat 1:81–107.

Daly M, Wilson M, 1994. Comment on "Can selfishness save the environment?" Human Ecol Rev 1:42–45.

Daly M, Wilson M, 1997. Crime and conflict: homicide in evolutionary perspective. Crime and Justice 22:251–300.

Diamond JM, 1992. The third chimpanzee. New York: Harper Collins.

Dupont D, 1996. Contingent valuation study of recreational opportunities in the Hamilton Harbour ecosystem. In: Proceedings of the 3rd Annual EcoWise Workshop, McMaster University.

Eckel CC, Grossman PJ, 1996. The relative price of fairness: gender differences in a punishment game. J Econ Behav Org 30:143–158.

Enquist M, Leimar O, 1990. The evolution of fatal fighting. Anim Behav 39:1–9.

Finn P, Bragg BWE, 1986. Perception of the risk of an accident by young and older drivers. Accident Anal Prev 18:289–298.

Fischoff B, 1991. Value elicitation. Is there anything in there? Am Psychol 46:835–847.

Fisher RA, 1930. The genetical theory of natural selection. Oxford: Clarendon Press.
Flynn J, Slovic P, Mertz CK, 1994. Gender, race, and perception of environmental health risks. Risk Analysis 14:1101–1108.
Frank RH, 1985. Choosing the right pond: human behavior and the quest for status. New York: Oxford University Press.
Frank R, 1988. Passions within reason. New York: Norton.
Gardner W, 1993. A life-span rational-choice theory of risk taking. In: Adolescent risk taking (Bell NJ, Bell RW, eds). Newbury Park, California: Sage; 66–83.
Gigerenzer G, Hell W, Blank H, 1988. Presentation and content: the use of base rates as a continuous variable. J Exp Psychol Hum Percept Perform 14:513–525.
Gigerenzer G, Hoffrage U, 1995. How to improve Bayesian reasoning without instruction: frequency formats. Psychol Rev 102:684–704.
Gigerenzer G, Hoffrage U, Kleinbölting H, 1991. Probabilistic mental models: a Brunswikian theory of confidence. Psychol Rev 98:506–528.
Goodwin BK, Offenbach LA, Cable TT, Cook PS, 1993. Discrete/continuous contingent valuation of private hunting access in Kansas. J Environ Manage 39:1–12.
Gove WR, 1985. The effect of age and gender on deviant behavior: a biopsychological perspective. In: Gender and the life course (Rossi AS, ed). New York: Aldine; 115–144.
Green L, Myerson J, 1996. Exponential versus hyperbolic discounting of delayed outcomes: risk and waiting time. Am Zool 36:496–505.
Gregory R, Lichstenstein S, Brown TC, Peterson CL, Slovic P, 1995. How precise are monetary representations of environmental improvements? Land Econ 71:462–473.
Guagnano GA, Dietz T, Stern PC, 1994. Willingness to pay for public goods: a test of the contribution model. Psychol Sci 5:411–415.
Hames R, 1987. Game conservation or efficient hunting? In: The question of the commons: the culture and ecology of communal resources (McCay BJ, Acheson JM, eds). Tucson, Arizona: University of Arizona Press; 92–107.
Hames R, 1991. Wildlife conservation in tribal societies. In: Biodiversity: culture, conservation, and ecodevelopment (Oldfield ML, Alcorn JB, eds). Boulder, Colorado: Westview Press.
Harcourt AH, Harvey PH, Larson SG, Short RV, 1981. Testis weight, body weight, and breeding system in primates. Nature 293:55–57.
Hastjarjo T, Silberberg A, Hursh SR, 1990. Risky choice as a function of amount and variance in food supply. J Exp Anal Behav 53:155–161.
Hawkes K, 1992. Sharing and collective area. In: Evolutionary ecology and human behavior (Smith EA, Winterhalder B, eds). Hawthorne, New York: Aldine de Gruyter; 269–300.
Hawkes K, 1993. Why hunter-gatherers work: an ancient version of the problem of public goods. Curr Anthropol 34:341–361.
Hewlett BS, 1988. Sexual selection and paternal investment among Aka pygmies. In: Human reproductive behaviour (Betzig L, Borgerhoff Mulder M, Turke P, eds). Cambridge, Massachusetts: Cambridge University Press.
Heyde JM, 1995. Is contingent valuation worth the trouble? Univ Chicago Law Rev 62:331–362.
Hill K, Hurtado AM, 1995. Ache life history. Hawthorne, New York: Aldine de Gruyter.
Hoch SJ, Loewenstein GF, 1991. Time-inconsistent preferences and consumer self-control. J Consum Res 17:492–507.
Hoffman E, McCabe K, Smith VL, 1996. Social distance and other-regarding behavior in dictator games. Am Econ Rev 86:653–660.
Holinger PC, 1987. Violent deaths in the United States. New York: Guilford Press.

Howarth RB, 1996. Status effects and environmental externalities. Ecol Econ 16:25–34.
Howell N, 1979. The demography of the Dobe !Kung. New York: Academic Press.
Hsieh CC, Pugh MD, 1993. Poverty, income inequality, and violent crime: a meta-analysis of recent aggregate data studies. Crim Justice Rev 18:182–202.
Irwin CE, 1993. Adolescence and risk taking. In: Adolescent risk taking (Bell NJ, Bell RW, eds). Newbury Park, California: Sage, 17–28.
Irwin JR, Slovic P, Lichtenstein S, McClelland GH, 1993. Preference reversals and the measurement of environmental values. J Risk Uncert 6:5–18.
Jonah BA, 1986. Accident risk and risk-taking behaviour among young drivers. Accident Anal Prev 18:255–271.
Kacelnik A, 1997. Normative and descriptive models of decision making: time discounting and risk sensitivity. In: Characterizing human psychological adaptations (Bock G and Cardew G, eds). Ciba Foundation Symposium 208. London: Wiley; 51–70.
Kacelnik A, Bateson M, 1996. Risky theories—the effects of variance on foraging decisions. Am Zool 36:402–434.
Kagel JH, Green L, Caraco T, 1986. When foragers discount the future: constraint or adaptation? Anim Behav 34:271–283.
Kahneman D, 1986. Comments on the contingent valuation method. In: Valuing environmental goods: an assessment of the contingent valuation method (Cummings RG, Brookshire DS, Schulze WD, eds). Totowa, New Jersey: Rowman & Allanheld; 185–193.
Kahneman D, Knetsch JL, 1992. Valuing public goods: the purchase of moral satisfaction. J Environ Econ Manage 22:57–70.
Kahneman D, Ritov I, Jacowitz KE, Grant P, 1993. Stated willingness to pay for public goods: a psychological perspective. Psychol Sci 4:310–315.
Kahneman D, Slovic P, Tversky A, 1982. Judgment under uncertainty. New York: Cambridge University Press.
Kahneman D, Tversky A, 1979. Prospect theory: an analysis of decision under risk. Econometrika 47:263–291.
Kahneman D, Tversky A, 1984. Choices, values, and frames. Am Psychol 39:341–350.
Kaplan H, 1996. A theory of fertility and parental investment in traditional and modern societies. Yrbk Phys Anthropol 39:91–135.
Kaplan H, Hill K, 1985. Hunting ability and reproductive success among male Ache foragers: preliminary results. Curr Anthropol 26:131–133.
Kay CE, 1994. Aboriginal overkill. Hum Nat 5:359–398.
Kirby KN, Herrnstein RJ, 1995. Preference reversals due to myopic discounting of delayed reward. Psychol Sci 6:83–89.
Knetsch JL, 1995. Asymmetric valuation of gains and losses and preference order assumptions. Econ Inq 33:134–141.
Loewenstein G, Adler D, 1995. A bias in the prediction of tastes. Econ J 105:929–937.
Loewenstein G, Thaler RH, 1989. Intertemporal choice. J Econ Perspect 3:181–193.
Loomis J, Lockwood M, DeLacy T, 1993. Some empirical evidence on embedding effects in contingent valuation of forest protection. J Environ Econ Manage 24:45–55.
Lopes LL, 1987. Between hope and fear: the psychology of risk. Adv Exp Soc Psychol 20:255–295.
Lopes LL, 1993. Reasons and resources: the human side of risk taking. In: Adolescent risk taking (Bell NJ, Bell RW, eds). Newbury Park, California: Sage; 29–54.
Low BS, 1996. Behavioral ecology of conservation in traditional societies. Human Nat 7:353–379.
Low BS, Heinen J, 1993. Population, resources, and environment. Popul Environ 15:7–41.

Lyng S, 1990. Edgework: a social psychological analysis of voluntary risk taking. Am J Sociol 95:851–856.

Lyng S, 1993. Dysfunctional risk taking: criminal behavior as edgework. In: Adolescent risk taking (Bell NJ, Bell RW, eds). Newbury Park, California: Sage; 107–130.

Marcus GB, 1986. Stability and change in political attitudes: observe, recall, and "explain." Polit Behav 8:21–44.

Matthews ML, Moran AR, 1986. Age differences in male drivers' perception of accident risk: the role of perceived driving ability. Accident Anal Prev 18:299–313.

MacDonald H, McKenney DW, 1996. Varying levels of information and the embedding problem in contingent valuation: the case of Canadian wilderness. Can J Forest Res 26:1295–1303.

McNamara JM, 1996. Risk-prone behaviour under rules which have evolved in a changing environment. Am Zool 36:484–495.

Millstein SG, 1989. Adolescent health. Challenges for behavioral scientists. Am Psychol 44:837–842.

Millstein SG, 1993. Perceptual, attributional, and affective processes in perceptions of vulnerability through the life span. In: Adolescent risk taking (Bell NJ, Bell RW, eds). Newbury Park, California: Sage; 55–65.

Mohai P, 1992. Men, women, and the environment: an examination of the gender gap in environmental concern and activism. Society Nat Res 5:1–19.

Møller AP, 1988. Ejaculate quality, testes size and sperm competition in primates. J Hum Evol 17:479–488.

Mulligan CB, 1997. Parental priorities and economic inequality. Chicago: University of Chicago.

Ng YK, Wang J, 1993. Relative income, aspiration, environmental quality, individual and political myopia: why may the rat-race for material growth be welfare reducing? Math Soc Sci 26:3–23.

Nisbett RE, Ross L, 1980. Human inference: strategies and shortcomings of social judgment. Englewood Cliffs, New Jersey: Prentice Hall.

Nisbett RE, Wilson T, 1977. Telling more than we can know: verbal reports on mental processes. Psychol Rev 84:231–259.

Norgaard RB, Howarth RB, 1991. Sustainability and discounting the future. In: Ecological economics: the science and management of sustainability (Costanza R, ed). New York: Columbia University Press; 88–101.

Real L, 1987. Objective benefit versus subjective perception in the theory of risk-sensitive foraging. Am Nat 130:399–411.

Real L, Caraco T, 1986. Risk and foraging in stochastic environments. Annu Rev Ecol Syst 17:371–390.

Regelmann K, Curio E, 1986. Why do great tit (*Parus major*) males defend their brood more than females do? Anim Behav 34:1206–1214.

Ridley M, Low B, 1994. Can selfishness save the environment? Hum Ecol Rev 1:1–13.

Roelofsma PHMP, 1996. Modelling intertemporal choices: an anomaly approach. Acta Psychol 93:5–22.

Rogers AR, 1991. Conserving resources for children. Hum Nat 2:73–82.

Rogers AR, 1994. Evolution of time preference by natural selection. Am Econ Rev 84:460–481.

Rogers AR, 1997. The evolutionary theory of time preference. In: Characterizing human psychological adaptations (Bock G and Cardew G, eds). (Ciba Foundation Symposium 208). London: Wiley; 231–252.

Romer PM, 1995. Preferences, promises, and the politics of entitlement. In: Individual and social responsibility (Fuchs VR, ed). Chicago: University of Chicago Press; 195–220.

Roitberg BD, Mangel M, Lalonde RG, Roitberg CA, van Alphen JJM, Vet L, 1992. Seasonal dynamic shifts in patch exploitation by parasitic wasps. Behav Ecol 3:156–165.

Rubin PH, Paul CW, 1979. An evolutionary model of taste for risk. Econ Inq 17:585–596.

Sachs C, 1996. Gendered fields: rural women, agriculture and environment. Boulder, Colorado: Westview Press.

Sachs C, 1997. Resourceful natures, women, and environment. Washington, DC: Francis & Taylor.

Samuelson PA, 1993. Altruism as a problem involving group versus individual selection in economics and biology. Am Econ Rev 83:143–148.

Sethi R, Somanathan E, 1996. The evolution of social norms in common property resource use. Am Econ Rev 86:766–788.

Shafir E, 1993. Choosing versus rejecting: why some options are both better and worse than others. Memory Cognit 21:546–556.

Simon HA, 1993. Altruism and economics. Am Econ Rev 83:156–161.

Simpson J, Kenrick D, 1997. Evolutionary social psychology. Englewood Cliffs, New Jersey: Lawrence Erlbaum Associates.

Smith VK, 1994. Lightning rods, dart boards, and contingent valuation. Nat Res J 34:121–152.

Sunstein CR, 1994. Incommensurability and valuation in law. Mich Law Rev 92:779–861.

Sutherland WJ, 1985. Chance can produce a sex difference in variance in mating success and explain Bateman's data. Anim Behav 33:1349–1352.

Trimpop RM, 1994. The psychology of risk taking behavior. Amsterdam: North-Holland.

Trivers RL, 1972. Parental investment and sexual selection. In: Sexual selection and the descent of man, 1871–1971 (Campbell B, ed). Chicago: Aldine; 136–179.

Wang XT, 1996. Domain-specific rationality in human choices: violations of utility axioms and social contexts. Cognition 60:31–63.

Williams GC, 1966. Adaptation and natural selection. Princeton, New Jersey: Princeton University Press.

Willis KG, Garrod GD, 1993. Valuing landscape: a contingent valuation approach. J Environ Manage 37:1–22.

Wilson EO, 1992. The diversity of life. Cambridge, Massachusetts: Belknap Press.

Wilson JQ, Herrnstein RJ, 1985. Crime and human nature. New York: Simon & Schuster.

Wilson M, Daly M, 1985. Competitiveness, risk-taking and violence: the young male syndrome. Ethol Sociobiol 6:59–73.

Wilson M, Daly M, 1993. Lethal confrontational violence among young men. In: Adolescent risk taking (Bell NJ, Bell RW, eds). Newbury Park, California: Sage; 84–106.

Wilson M, Daly M, 1997. Life expectancy, economic inequality, homicide, and reproductive timing in Chicago neighbourhoods. British Medical J 314:1271–1274.

Wilson M, Daly M, Gordon S, Pratt A, 1996. Sex differences in valuations of the environment? Popul Environ 18:143–160.

Woodwell DA, 1997. National ambulatory medical care survey: 1995 Summary. Advance Data, Number 286. Bethesda, Maryland: National Center for Health Statistics.

Zuckerman M, 1994. Behavioral expressions and biosocial bases of sensation seeking. Cambridge: Cambridge University Press.

Afterword

19

Behavioral Ecology and Conservation Policy: On Balancing Science, Applications, and Advocacy

Daniel Rubenstein

A Dilemma

Planet Earth is at risk. Rapid population growth, degradation of landscapes, extinction of species, and the pollution of air and water is creating a crisis for the biosphere whose unprecedented proportions are only just being defined, let alone understood. With respect to biodiversity alone, our efforts at eliminating species far outstrip our abilities for estimating the magnitude of the carnage we are inflicting. If ecology is the science concerned with the relationship of organisms to each other and their environment and conservation biology is the study of how to conserve and manage biological diversity, then behavioral ecologists have much to offer. Because behavioral ecology is about how ecology shapes behavior, it can help define conservation problems more precisely and show how understanding the function, or survival value, of behavior can help eliminate some of these environmental problems.

Accomplishing these two tasks, however, is not straightforward. Because two of the aims of conservation biology are the protection and sustainable use of Earth's biological diversity, human interventions—along with their conflicting interests and values—are inevitable. Consequently, conservation biology has a divided "personality," being composed of both basic and applied elements. While it is important to gain a detailed understanding of how species and their ecosystems function, it is essential that this knowledge be used to manage species and manipulate their ecosystems for desired ends. Thus behavioral ecologists addressing conservation issues must in part be research scientists learning about basic behavioral, life history, and ecological attributes of individuals so that they can monitor, or at times even predict, changes that will occur when environments are altered. But behavioral ecologists must also be environmental physicians, diagnosing the health of species and their ecosystems and offering treatments that can cure problems. Thus environmentally concerned behavioral ecologists are conservation biologists because they are in the knowledge business, unraveling the workings of the living world while using their skills and insights to conserve biodiversity and ensure the proper workings of pristine and perturbed ecosystems. As scientists they use experimental manipulations, systematic observations, and theoretical modeling with appropriate controls to guard against possible latent biases as they generate and apply new knowledge.

Behavioral ecologists and conservation biologists alike, however, face a special problem: they are also people. And as people, strongly held values color perspective and help shape personal goals and actions. Many behavioral ecologists are imbued with strong ideals that guide them to live in harmony with nature and encourage them to participate actively in the political arena. As such they are biological conservationists who are activists that use insights from ecology to guide their activism whose agenda is in part being shaped by beliefs, values, and intuition.

Consequently, behavioral ecologists feel a tension that clearly exists between the need to ensure, on the one hand, the objectivity that underscores the impartiality of research science, or the practice of clinical medicine with its deductive diagnoses and humane treatments, and on the other hand, the need to act, impelled by values based on environmental concern. Unfortunately, intellectually based activism can become confused with popular environmentalism and its zealous faith in absolute truths. Blurring the distinction between ecologists as research scientists and environmental physicians might not be all bad because it could both focus "pure" research that could too easily become esoteric and limit the implementation of mitigation strategies that are often superficial, if not simplistic. But if science is to influence the shaping of effective policy, then maintaining the distinction between the science of scientists and their political actions is critical.

It is the aim of this chapter to chart a course that can guide behavioral ecologists through the Scylla of rational but time-consuming research programs that provide basic insights that can shape effective management plans, and the Charybdis of forceful environmental action that sometimes compromises the biological foundations of accurate ecology. Being able to play these different roles without losing scientific credibility and without sacrificing the ability to shape conservation policy requires that behavioral ecologists know the players, scripts, and structural constraints associated with each of the theaters. By examining a series of case studies, we will see how the tenets of behavioral ecology and the results of long-term studies have already provided basic scientific insights that have helped formulate realistic management strategies. At the same time we will explore how such studies can be made more useful and help shape effective public policy and management practices, hence fostering the cause of environmentalism without sacrificing scientific credibility.

The Importance of Behavioral Ecology

Concepts from population and community ecology have had a dramatic impact on conserving species. Keystone predation (Paine, 1966), island biogeography (MacArthur and Wilson, 1967), harvesting theory (Beddington and May, 1980) and demography (Caughley, 1977) have all been used to shape strategies for increasing biodiversity, siting and sizing nature reserves, assessing maximal sustainable yields for fisheries and whaling industries, and determining minimum viable population sizes. Vigorous debate concerning the utility of using such theories for designing particular management plans has ensued (Mills et al., 1993), but without doubt such principles have provided valuable starting points for organizing thinking about how to solve real-world problems.

Can the first principles of behavioral ecology similarly serve to crystalize thinking about conservation strategies? The answer should be an unequivocal "yes." Although concepts from traditional population and community ecology have had tremendous impact, their utility is often limited because they essentially treat all individuals in populations as equals. Behavioral ecology, however, is about diversity and understanding why individuals respond differently to similar environmental circumstances. Ultimately such an understanding derives from the notion that differences among individuals are real, having been shaped by natural selection. Typically, selection maintains behavioral polymorphisms and favors individuals that facultatively respond to their environment, one which is often dominated by the actions of others. Thus, if conservation models, be they qualitative or quantitative, are to be used in predicting the dynamics of animal populations and the consequences of various conservation strategies, they must be made realistic. And this can best be accomplished by incorporating a detailed understanding of not only the behavior of the species but also the environmental selective forces responsible for shaping the behavior of the individuals that compose them.

Incorporating a focus on individuals and their variation into realistic models should not be difficult for two reasons. First, behavioral ecology is rich in "first principles." Optimal foraging and life-history theory, decision-making rules in competitive and predator–prey situations, as well as models of social, mating, and breeding strategy illuminate how individuals move, aggregate, acquire resources, and reproduce. In turn, these actions by individuals affect how their populations migrate, impact their landscape or habitats, and grow. Because these outcomes influence the genetic structure, demography, and dynamics of populations, concepts from behavioral ecology should be able to provide powerful insights into the functioning of populations. Moreover, behavioral ecology specializes in long-term field studies, so details on variation in birth and death rates as well as on lifetime reproductive success that are necessary to calibrate, or even generate, realistic models are often available. Thus incorporating a behavioral ecological approach into the study of conservation should not only help uncover subtle but important aspects of a species' or population's biology but also help in diagnosing and healing ailing populations or the ecosystems they inhabit.

Behavioral Ecology and the Science of Conservation Biology

Much basic behavioral ecological research has been focused on understanding the patterns and processes of individual species and the communities they inhabit. The empirical results and theoretical insights derived from these studies shed light on how systems work and often provide the framework upon which management and conservation decisions derive. Underwood (1995) describes this as "available and directed research"; level 1 (table 19-1). Observations on the ranging patterns of species to establish the boundaries of national parks is perhaps the most basic form in which this type of research is employed. Only after the seasonal migratory patterns of wildebeest were known were the boundaries of the Serengeti National Park and the Ngorongoro Conservation Area established (Grzimek and Grzimek, 1959). In this way a sustainable park was created and conflicts with humans were mitigated at the outset.

Although use of "off-the-shelf" knowledge has value, it often leaves the researcher in a defensive position; decision makers are asking the questions and hence setting the agenda. Because the data are often based on descriptions from different systems that are at best not too dissimilar from those needing assistance, it also puts researchers in the position of making difficult predictions about processes. When ordinary scientific uncertainty is added to the mix, ecologists often find themselves in untenable situations. Prescribed actions are

Table 19-1 Hierarchy of research types.

Level	Type	Purpose	Examples
1	Available and directed	Assess impacts of current action based on present observations. Use basic models tuned to existing patterns.	Viability analyses and impact statements
2	Applied and environmental	Assess impact of managerial decisions. Treat management actions as experiments in progress and test their predictions.	Reintroduction operations, environmental remediation schemes and alternative harvesting strategies
3	Basic and strategic	Design new experiments and develop new models based upon limitations or failures of implementing previous ones. Shift focus from understanding patterns to processes and mechanisms.	Consequences of individual decision-making; in particular dynamics of sex ratio adjustments, sex differences in behavior, Allee affects
4	Managerial and policy making	Understand how policy makers and managers make decisions and choose courses of action. Apply sociobiological reasoning to understanding the behavior of institutions as societies and actions of their members and other stakeholders.	Analyses of organizational structures, legal frameworks, legislative procedures, economic systems, and human motivations

rarely effective, and the effectiveness of science is doubted. As a result, Underwood (1995) argues that ecologists should become more proactive and expand their research to include three additional domains (table 19-1). First, even when providing existing "off-the-shelf" (level 1) research to managers, scientists should inform users about the generality and applicability of applying the findings to a particular problem because they were most likely obtained for a different population, in a different system, and at a different scale. Second, ecologists should begin researching the consequences of management and conservation decisions ("applied and environmental research"; level 2). In effect, decisions about whether to intervene, and in what ways, represent large-scale experiments. Because they are derived from hypotheses that make strong predictions, their outcomes can be measured, and the fit to the predictions can and should be evaluated.

Third, new research programs should be established when previous attempts at conservation or management have failed ("basic and strategic research"; level 3). Interventions are likely to fail for many reasons. To know whether they did so because the off-the-shelf research was applied at the wrong scale or to a system with novel processes, or to species or populations with different behavioral repertoires, it is necessay to perform postmortems if future attempts at solving real-world problems are to succeed. Generating fundamentally new understandings of ecological problems is perhaps the best way to ensure that basic research ultimately has ecological applications.

Last, ecological research must examine the dynamics of management ("managerial and policy making research"; level 4). Understanding why scientific knowledge is misapplied or often ignored when decisions are made will require an understanding of what motivations and rewards shape the behavior of managers. Underwood (1995) argues that leaving such inquiry to the domain of social scientists and humanists leaves scientists as servants of managers. If scientific thought is truly to inform policy, these roles must become more balanced. As many of the case studies described below will show, had behavioral research been performed at a level higher in the hierarchy, applications of the findings would have been more useful in shaping policy.

Certain tenets of behavioral ecology are shaping conservation strategy by offering insights into how to prevent the demise of species. Although most species will be lost due to habitat degradation, fragmentation, or loss, some have already succumbed to excess harvesting by humans. To stop these trends it is important to understand how animals behave and how their strategic responses have evolved so that their behavior can be exploited to develop strategies that enhance survival. Only in this way will it be possible to move beyond the most basic off-the-shelf research to the higher levels proposed by Underwood, where both effectiveness and the chances of adoption will be enhanced. Here I explore a number of case studies that show how the demise of species can be prevented by improving our understanding of how the tenets of behavioral ecology shape viable population sizes, strategies of sustainable resource use, and patterns of biodiversity by drawing upon theories of optimal foraging, life-history evolution, mating systems, and the dynamic relationships that exist between predators and prey.

Optimal Foraging and Applications of Bet Hedging

Because natural selection favors behavior that maximizes an individual's fitness, it often pays individuals living in temporally changing environments to hedge their bets. To do this they often diversify their behavior and reduce variance in offspring number, a major component of lifetime reproductive success (Seger and Brockmann, 1987). Sometimes such behavior coincides with maximizing a population's growth rate, but this need not be the case (Eadie et al., chapter 12, this volume). In fact, natural selection of the group selectionist kind is necessary if maximizing a population's success is to be favored directly. Much theory has been developed to explore the role of diversifying behavior with respect to life-history evolution (Schaffer and Gadgil, 1975; Stearns, 1976; Gillespie, 1977; Real, 1980; Rubenstein, 1982; Bulmer, 1984), but perhaps the best empirical examples showing that animals behave in accordance with predictions of the models emerges from studies of optimal foraging. Caraco and co-workers (1980), for example, showed that foraging birds did best if they were sensitive to the variability of food rewards in addition to the mean rate of return. When energy requirements were less than the expected reward of either foraging option, individuals avoided taking risks and chose the less variable one. Only when energetic needs exceeded either option's expected rate of return was the more variable option chosen.

Human exploitation of marine fisheries has often resulted in overfishing and in bringing fisheries to the brink of extinction. Reliance on standard principles of maximal sustainable yield (MSY) or even optimal sustainable use (OSU) has not worked. Most recently the highly productive cod fishery off the Canadian Atlantic coast has collapsed and has highlighted the problem of managing fisheries. An indefinite moratorium is in place for the Great Banks, and temporary moratoriums have been imposed on neighboring fishing groups with the hope that populations will grow and the fishery will recover.

Fishery biologists are well aware of the role that environmental fluctuations play in introducing uncertainty into estimating levels of sustainable catches. Typically managers use conservative catch criteria that attempt to maintain catch levels below MSY values or try to maintain stocks above the MSY level. Lauck et al (in press) argue that these recommendations are seriously flawed because they assume that current stocks are accurately known, which is never the case. Estimation errors of 50% are not atypical, but reliance on this method of determining a sustainable catch implicitly assumes that with better methods stock assessment can be improved.

The alternative view is that many aspects of the natural world are never knowable with sufficient certainty, and this leads Lauck and co-workers (in press) to favor a completely different management strategy. Rather than base quotas on a best guess that ignores uncertainty, they suggest that managers take a page from the behavioral ecological literature: use bet hedging to diversify their own behavior by managing fisheries to reduce variation in catch. An effective way of doing this would require exploiting only part of the resource while protecting the rest in Marine Protected Areas, or "no-take" zones (Shackell and Willison, 1995). Obviously, for any given level of harvesting, the average catch for the combined areas would be lower than if the entire area were open to fishing at this same level, and this would necessarily afford the fishery some protection.

But viewing the problem as only affecting the mean misses the point. If the entire stock were open to fishing, then any inadvertent overfishing generated by unpredictable appearances of extremely harsh environmental conditions would drive the entire population dangerously close to zero. Even though there will be intervening good years, damage associated with severe declines could make it difficult for populations to recover. With the entire fishery open to harvesting, variation in yield will be high. By setting aside a portion of the habitat to protect a fraction of the fishery, the likelihood of excessive depletions is reduced, which in turn reduces the variance of the harvest. As Lauck and his co-workers show, the larger the reserve, the better the policing, or the more fecund the species, the better such reserves will be at hedging against environmental uncertainty. Because yield increases with harvesting intensity but with diminishing returns, the percent decreases in the long-term yield for the population is likely be smaller than the percentage of the range that is set aside as the refuge. As a result, for whatever the size of the protected area, it is likely that the exploitable area can be harvested more intensively than would otherwise be the case.

Whether these ideas are adopted by managers and find their way into policy depends upon a number of factors. As Ludwig et al. (1993) argue, fishery management has been a spectacular failure despite much scientific analysis. Disagreements among scientists are common, and the prospects of resolving these disagreements are virtually nil because controlled experiments, even on a small scale, would involve short-term losses for the industry. As a result, Ludwig and co-workers suggest that effective management can only result when human motivations, usually greed, are incorporated into the system and when science is limited to identifying rather than remedying the problem. But a tenet of behavioral ecology as well as economics, bet hedging, may go a long way to mitigating the problem of over fishing precisely because it confronts the issue of uncertainty head-on (i.e., level 3 research) and provides a conservative means of managing when information about the state of the world is poorly known and when human greed and human error are likely to prevail.

Life-History Theory and Demography

Natural selection favors individuals who transmit the most genes to future generations. How this is best accomplished varies depending on environmental circumstances. In some situations, individuals that produce many small young early in life are favored, whereas under other circumstances individuals delaying maturity and investing lavishly in only a few young have the advantage (Horn and Rubenstein, 1984; Lessells, 1991). The allocation of limited resources to balance the conflicting demands of survival and reproduction defines a life history. From a conservation perspective, understanding why certain evolutionary patterns evolve and knowing when they are malleable is essential if intervention and management are to be successful. Knowing why a particular type of life history is adaptive under one set of environmental conditions and not another, or why some life histories are plastic while others are not, will make some interventions more effective than others in particular circumstances. Being armed with this understanding before a management problem is implemented or even designed is the best antidote to costly and possibly harmful practices.

Atlantic Salmon

One of the most important life-history stages is age of first reproduction because it tends to be correlated with many other features (e.g., longevity, fecundity) of a life history (Rubenstein, 1993). The benefits of breeding early in life are many. At least in expanding populations (hopefully the case for endangered, but now protected, or recovering species), maturing early tends to accelerate the spread of genes into future generations because offspring have their offspring quickly and they disproportionately contribute to the growth of the population. Also, breeding early reduces the period of juvenile vulnerability, which can be high and is typically greater than that of an adult. Nevertheless, there are costs associated with breeding early. Perhaps the two most important are small size and inexperience, both factors that could limit subsequent longevity or fecundity. For example, smaller size often means smaller ovaries and fewer eggs for females of many taxa, especially among invertebrates, whereas for males it typically means limited intrasexual competitiveness and hence lowered access to mates.

That human behavior can have an impact on changing this important transition is nowhere more apparent than in Atlantic salmon (*Salmo salar*) populations. In Atlantic salmon, increased fishing has changed not only the age of first reproduction but also the entire pattern of sexual development (Montgomery, 1983). Typically salmon develop in freshwater streams for 1 or 2 years and then smolt by migrating to the sea. There they forage on zooplankton and continue to grow. Once they attain a certain size, they become sexually mature and return to rivers, traveling upstream until they reach spawning grounds. Some parr, however, never reach critical smolting size. They remain in the stream and become sexually mature at small sizes and at early ages. Under pristine conditions, the fraction of the male population adopting this alternative route to maturity is small. With intensive fishing reducing the number of adults maturing at sea and thus reducing the number able to return to the rivers, the direct maturing parr that spend their entire lives in the streams are now no longer at such a competitive disadvantage. Their survival prospects are high and the number of large, superior competitors they are likely to encounter is reduced. Consequently, their life-history strategy is selectively advantageous and is increasing in frequency. The consequences on the fishery are likely to be profound. Fewer and fewer fish will migrate to the sea, and even with increased harvesting effort, catches will continue to decline. Fortunately, the species will survive but with morphologies and behavior quite different from what we are accustomed to seeing. By providing an explanation for why such profound populationwide changes are occurring, life-history theory can reveal where intervention is likely to be most effective (i.e., level 1 research).

As awareness of the likelihood of global change grows, ecologists have begun studying the responses of organisms to climate change (Peters and Lovejoy, 1992), especially with respect to life-history events (Rubenstein, 1992). Here too, anthropogenic actions are likely to impact salmon life-histories and the viability of the fishery. Because the salmon fishery is highly profitable and is dependent on species that spend part of their lives in fresh water and part out to sea, understanding how climate change will affect the growth of stocks, which in turn will determine optimal catch size necessary for sustainable harvesting. Mangel (1994a) modeled growth, development, and behavior of salmon in both streams and oceanic environments and found that, once at sea, increased water temperatures would lower growth and adult survivorship and induce earlier maturity. The combined consequence of these changes is an early return to natal streams in all but the fastest growing individuals. These fast growers actually delay maturing for an additional year. Such populationwide changes would be the indirect effect of temperature-induced increased winds that would disrupt zooplankton patchiness and lower feeding rates. For parr developing in streams, Mangel's models suggest that increases in water temperature will enhance growth and survival and induce smolting after 1, rather than 2, years. Changing the parr–smolt transformation point as well as the age at sexual maturity should have major implications for the fishery. As females accelerate development, they will mature earlier and at smaller sizes, thus lowering their fecundity. In addition, their survival will be reduced. Such changes do not augur well for the fishery and dictate that harvesting levels must be significantly reduced.

The implications of these hypothetical predictions are profound. Because many of the models' formulations are based on best-guess assumptions, their predictions should only be viewed as possibilities. Nonetheless, the models identify not only areas where further behavioral and developmental research is needed, but they also underscore the importance of "knowing your organism," something behavioral ecologists routinely do. Without understanding the intricate details about what behaviors salmon exhibit, it would be difficult to meld the physical dynamics of oceans with those driven by behavior and physiology. By appreciating that natural selection shapes life histories, more realistic understandings of how environmental changes will affect the survivorship and fecundity schedules that are so crucial for a species' survival (level 3 research) are possible.

Asiatic Wild Ass

Reintroduction of the Asiatic wild ass (onager) *Equus hemionus* into areas of Israel and Palestine, where it flourished until the turn of the century, provides another example of where attention to details of the dynamics of life-history evolution mean success or failure. One of the goals of the Israeli government is to reintroduce biblical animals to Judea and Samaria. In 1982 the first onagers were moved from a breeding reserve, Hai-Bar Yotvata, to Makhtesh Ramon, a large erosional crater in the center of the Negev Desert. The first release contained only males, and they quickly dispersed and many were shot when they moved near the Israeli-Jordanian border. A second attempt involving two males and six females took place in 1983 and was followed with additional releases in 1984 (two males and five females) and 1987 (five males and three females). All but two individuals were between 2 and 5 years of age, and the two older ones (aged 6 and 17) died shortly after release.

The population has been continually monitored since 1983, and it became apparent that the population was growing very slowly (Saltz and Rubenstein, 1995). By 1993 the population only contained 16 breeding females, up from the original 14. But recruitment has suddenly improved since there now are three 2-year old females, four female yearlings and nine female foals. With the ranks of reproductive females swelling, the population is growing. But why did this demographic transition take so long? And why is the population growing quickly now? These were questions that the government asked us to answer.

Much attention has been paid to the logistical features of reintroductions that are crucial for assuring success. That only males were initially introduced underscores the need to pay attention to detail. In accordance with International Union for the Conservation of Nature guidelines, a feasibility study was completed before the project began. Release sites were prepared so that the transplanted individuals could habituate to the habitat, and postrelease monitoring was performed. Nevertheless, the population size remained almost constant for nearly a decade. The problem underlying this stasis was that basic facts about the behavioral and evolutionary ecology of the species were not known. In particular, it was not appreciated that wild asses could facultatively adjust sex ratio nor that captivity could dramatically limit fertility.

Trivers and Willard (1973) were the first to suggest that differences in the ability of individual females to invest in the rearing of their young should lead to individual differences in primary sex ratios. They argued that mothers with sufficient resources should invest in the sex with the higher variance in reproductive success as long as this investment could increase the chances that such offspring would be those producing the most offspring. In polygynous species of ungulates, males exhibit higher variances than females, and in many species (Clutton-Brock et al., 1984) levels of parental investment affect the subsequent reproductive success of offspring. For species in which competition for critical resources is of the "contest" variety in which winners exclude losers from acquiring critical resources, females of high rank are often in above-average bodily condition and they produce more sons than daughters. When competition is of the "scramble" type in which success is shaped by utilization efficiency, dominance has little influence on who acquires the most resources. Instead, age seems to be the determinant of sex ratio bias; at least for Asiatic wild asses, middle-aged females give birth to sons, whereas both young and old females give birth to daughters (Saltz and Rubenstein, 1995) (fig. 19-1). Because all the females released into the crater were between ages 2 and 5—the male-producing years—very few females were recruited into the population, and the population did not grow. Over time, however, we predicted that the age structure of the population would change, and as more females begin to enter the female-producing years, the population should begin to grow. In fact this appears to be happening. Since 1993, 9 of the 13 foals born were female.

The population also initially failed to grow because the fecundity of females transferred to the crater was extremely low. Fewer than 30% of all reintroduced females had given birth within 2 years of the translocation, and for females aged 5 or less, fewer than 50% bore young (Saltz and Rubenstein, 1995). For females born in the crater, however, between 80% and 100% of females ≤ 5 years of age have given birth (fig. 19-2). Many hypotheses have been proposed to explain this abrupt change. Perhaps mating opportunities were reduced because of excessively small population size (an Allee effect; Allee, 1931). Although it is true that the population contained only one reproductively active male, he regularly pa-

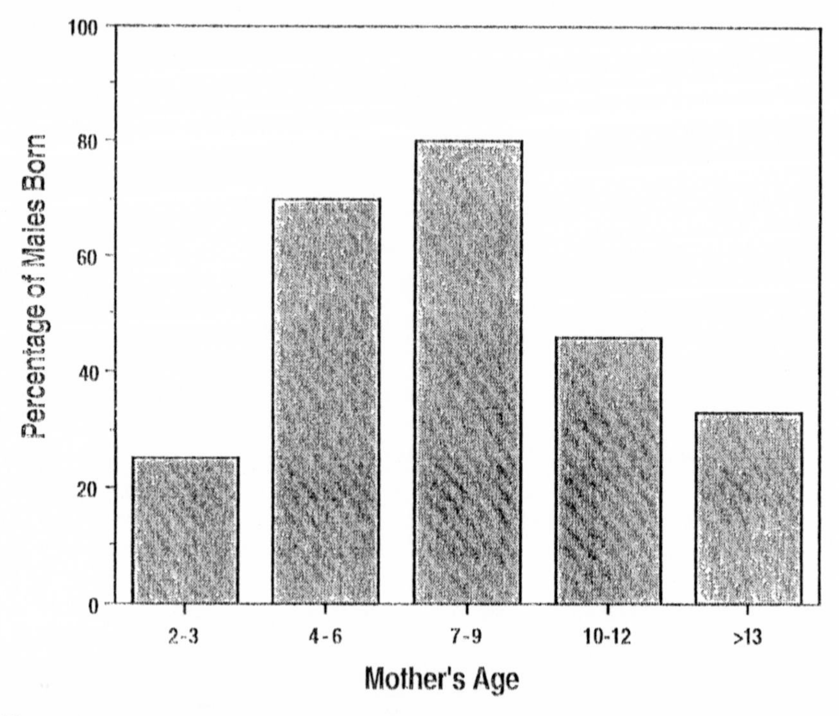

Figure 19-1 Percentage of females in the Ramon onager herd giving birth to sons as a function of mother's age in years.

trolled the entire crater and made contact with females regularly. Thus sperm limitation is unlikely. Alternatively, it is possible that excessive vigilance or inefficient foraging by females, both consequences of small group sizes, might have lowered bodily condition enough to limit conception success. Given that female fat levels as judged by rump scores have always been high and have not changed over time, this additional Allee effect seems unlikely. Perhaps inbreeding depression could have been responsible for diminished reproductive success, although this also seems unlikely because per capita breeding success has improved over time. Similarly, youthful inexperience is unlikely to be the primary causative agent because young wild-reared females breed prolifically. Rather, it seems most likely that either stress associated with handling or some other residual, long-lasting effect of prolonged captivity prior to release is responsible for lowering the fecundity of the reintroduced females; evidence of handling reducing the reproductive capacity of wild dogs *Lycaon pictus* has just emerged (Creel, 1996).

Thus, had the Nature Reserves Authority been apprised of some basic tenets of behavioral ecology, the reintroductions might have succeeded more quickly and might have been more economic. By reintroducing older females, who were most likely to have given birth to a daughter, who in turn would most likely have given birth to a daughter, rather than middle-aged females, the population would have grown more quickly. But with the results of new studies to be built upon these initial insights, even this strategy could be improved upon. If we knew for sure that the lowered fecundity does not result from transport or the

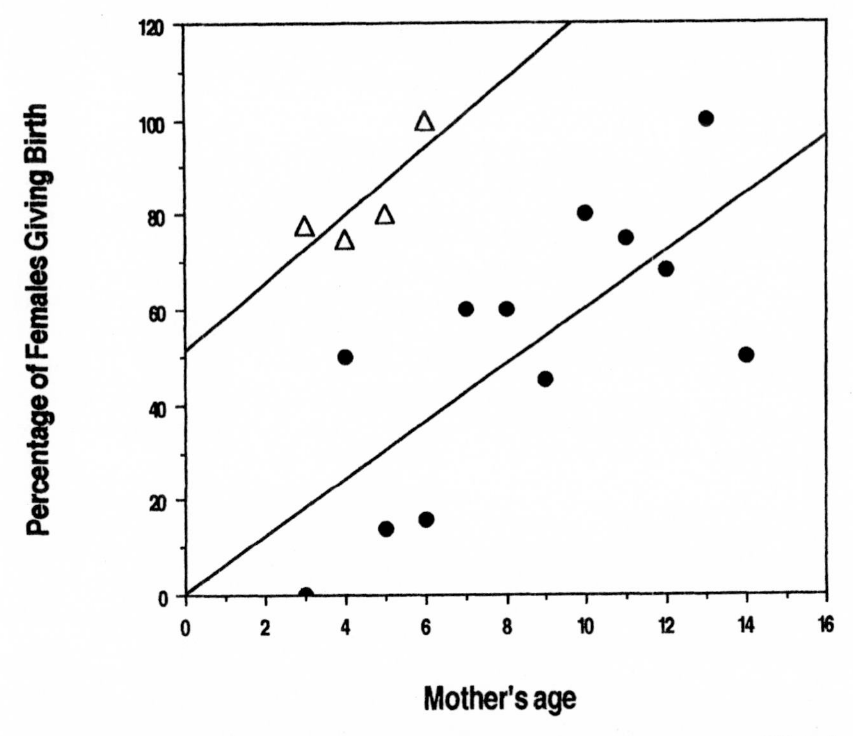

Figure 19-2 Percentage of adult female onagers giving birth at known ages. Filled circles denote translocated females ($y = 0.4 + 6.0x$; $F_{1,10} = 11.89, p = .006; R^2 = .54$). Open triangles indicate females born in the wild ($y = 51.3 + 7.1x$; $F_{1,2} = 3.74, p < .2; R^2 = .65$).

residual affects of captivity and if we assume that fecundity is enhanced after females acquire familiarity with the habitat and the all-important distribution of critical resources during the relatively peaceful prepubescent period, then introducing older females with their yearling daughters might be an even better strategy. Such a double addition of females producing mostly daughters would accelerate the recruitment process even more.

Clearly, the need to understand how environments shape behavior—the essential ingredient of behavioral ecology—would have been, and can still be, instrumental in devising effective and economical conservation strategy. Because the Nature Reserves Authority asked us to determine the problem and because they are planning to use our findings when designing further reintroductions, it is possible for scientific first-principles to make their way into effective management policy. That the Nature Reserves Authority has placed a behavioral ecologist as head of the reintroduction program shows that conservation planning is moving up the hierarchy to Underwood's (1995) basic and strategic research.

Mating Systems, Recruitment, and Population Viability

Perhaps the most powerful determinant of a population's ability to sustain itself is its ability to recruit new individuals. This can be accomplished either by increasing the survival prospects of females or by increasing their fecundity, and effective management should attempt to enhance both. By knowing something about mating systems and the processes of sexual selection that shape them, managers will gain insights into how to augment survival and fecundity by adjusting natural processes in meaningful ways. Unavoidable evolutionary trade-offs might limit any plan's effectiveness, but at least some undesirable and unintended consequences might be avoided.

Sexual selection often results in sexual dimorphism in morphology and behavior. In many polygyous species, only males exhibit weapons used in combat, and females invest more time and energy in rearing and protecting young. Appreciating that such differences are common can make a difference in ensuring that endangered species can increase their reproductive potential. This is likely to be especially true for those charismatic megafauna, where their extreme body size ordinarily limits their recruitment potential. Certainly, increasing either adult or juvenile survival is essential, and actions increasing both adult and juvenile mortality should be avoided. But doing so may not always be straightforward. Appreciating how prevalent and important sex differences are could complicate management by necessitating different action plans for males and females.

Black Rhinoceros

One species for which ignorance is particularly problematic is the black rhinoceros *Diceros bicornis.* Populations of black rhinos have been reduced by 97%, from 65,000 to less than 2500, in the last 25 years. Successful poaching for horns has led some nations to implement policies of dehorning in an attempt to reduce the slaughter by making rhinos less desirable to poachers (Berger, 1994). If effective, such a strategy would allow rhinos to continue roaming freely in their natural habitat. An alternative policy is to translocate rhinos to fenced areas where they can be monitored closely. Both strategies can foster ecotourism, but the former, by limiting the ability to monitor individual animals, still leaves rhinos at risk if dehorning fails.

Berger and co-worker's study (1994) of the behavior of horned and dehorned rhinos suggests that the dehorning strategy is likely to fail because horns regrow quickly and poachers appear to kill rhinos irrespective of horn size. Given that horns regrow at a rapid rate (anterior horns, 5.3 cm/year; posterior horns, 2.3 cm/year), rhinos regain value in just a few years if not dehorned again. But even if these assessments are overly pessimistic (Loutit and Montgomery, 1994), it is essential to know if dehorning disrupts normal behavior and puts individual rhinos at risk. According to Berger (1994) and Cunningham (Berger and Cunningham, 1995), the answer is unequivocally yes. Although dehorned mothers were no more likely to flee from predators than their intact counterparts, the disappearance of offspring being reared by dehorned mothers in areas of abundant spotted hyenas *Crocuta crocuta* and lions *Panthera leo,* but not those reared by normal horned females, suggests that horns play an important role in defense. Thus the population's ability to recruit is likely to be limited severely by dehorning; offspring appear to suffer immediate risks, and although dehorning might enhance a female's chances of survival in the short run, her medium- to long-term reproductive prospects are limited.

Despite morphological similarities among male and female rhinos, there are profound sexual differences in parental behavior. In response to predators, be they hyenas, lions, or humans, female rhinos with young are both more vigilant and more likely to respond actively than nonparous females or especially males (Berger, 1994). When offspring are older, mothers are likely to charge predators, but when offspring are young they tend to flee. Because repeated flight is likely to force all females to move long distances, chronic human disturbance is likely to put males at greater risk with respect to poaching because they stay put. In turn, as males become relatively rare, especially in small populations, sperm limitation of the kind generated by an Allee effect as described above and further elaborated by Dobson and Lyles (1989) and by Dobson and Poole (Chapter 8, this volume) could further limit the population's ability to grow.

Dunnocks

Knowledge of how changes in mating systems alter the per capita reproductive success of individual males and females should caution managers intent on altering a population's resource base or sex ratio. Davies's (1989) study on dunnocks *Prunella modularis* showed that changes in the ease at which females could acquire food dictated the size of their feeding territory. When food was made easy to acquire, territories were reduced in area. As a result, male territories became larger than those of individual females, and the mating system shifted from monogamy to polygyny. Conversely, when females defended large territories, especially if they were vegetatively and structurally complex, polyandry developed. Davies measured the reproductive success of individual males and females under these different mating regimes, so he was able to show convincingly that changes occurred and that their magnitudes could affect a population's recruitment. As fig. 19-3 shows, polygynous males have the highest reproductive success, but they contribute little apart from genes (unless an Allee affect limits sperm availability overall) to the growth of the population. And because females mated polygynously do much worse than if they were polyandrous or even monogamous, managers should do everything possible to avoid altering the landscape, or resource base, in ways that would encourage polygyny if it were important to promote population sizes of dunnocks or other species.

Sperm Whales

Direct application of these rules governing mating system and social evolution to conservation policy is rare. Ignorance of the dynamics of mating behavior has also led to a false sense of security when designing or implementing conservation plans for other large mammal populations. Models by Dobson and Poole (1997, Chapter 8, this volume) suggest that poaching the largest must males will prevent a large number of female elephants *Loxondata*

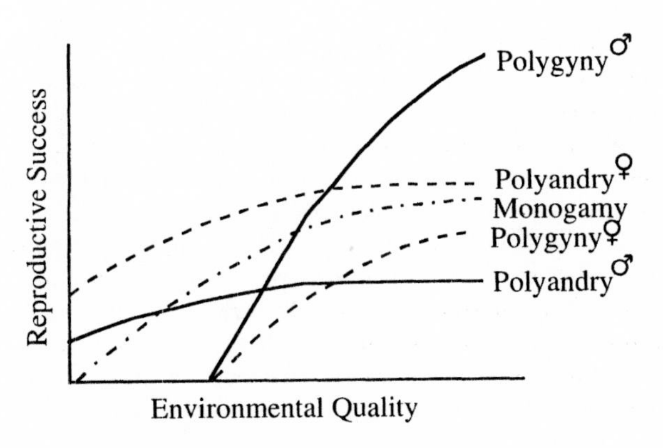

Figure 19-3 Environmentally induced sex differences in reproductive success (RS) for male and female dunnocks. Solid lines denote male RS, dashed lines denote female RS for polygamous mating systems. The dotted-dashed line depicts the RS for monogamous mates (adapted from Davies and Houston, 1984).

africana from mating. As a result, animals that exhibit low recruitment, in part because of large size and elephantlike "roving male" mating systems with strong male differentiation, will have a difficult time reversing long-term population declines once they begin.

In a similar way, ignorance about the mating system of sperm whales *Physeter macrocephalus*, the aquatic equivalent of elephants, has led to the implementation of inappropriate harvesting patterns that have almost resulted in their extinction. Because of their large size and extreme K-selected life-history characteristics, sperm whales were thought for years to exhibit harem-defense polygynous breeding systems. The reason for this mistaken classification was the result of using data from harvests in which the number of breeding females vastly exceeded the number of breeding males. Inferring that such ratios were the result of closed membership social groupings was uncalled for. Yet based on this assessment, it was assumed that most males would be superfluous. This inference encouraged the International Whaling Commission to allow the disproportionate hunting of males because most would be reproductively unnecessary and would be unable to fertilize females. Unfortunately for the sperm whale, the similarities with elephant societies are all too real (Weilgart et al., 1995). In both species, females are the core of the society and form permanent groups. In sperm whales these groups periodically merge when diving deep for squid. The young, which cannot follow their mothers for the entire dive, associate with adult female kin remaining on the surface. Males, however, leave their natal groups upon attaining sexual maturity, but they are excluded from associating with estrous females by older and more senior males. Consequently, the younger males remain at the higher latitudes in the cooler more productive waters when older males migrate with breeding females to the tropical breeding grounds. Once on the breeding grounds, mature males roam between female groups seeking reproductively active females.

Thus the mistaken notions that males are haremic and that subordinate males on the breeding grounds are superfluous has resulted in a severe reduction in the population of breeding males. And with the younger males thousands of kilometers away, it is not surprising that pregnancy rates among females have declined. Models developed by the scientific committee of the Whaling Commission predict that even a complete ban on sperm whaling will not result in an increase in the population for nearly 20 years. Such a complete ban would enable enough young males time to mature. Only then would all females be mated every year. Clearly, knowledge about the mating system of sperm whales could have prevented the overharvesting of mature males. Unraveling the mating system showed that all males found associating with females near the tropics were reproductively active and important for maximizing population recruitment. Their demise ensured the demise of the entire sperm whale population.

Understanding the dynamics of mating systems, especially the way in which environmental forces shape male–female relationships, can even have genetic implications for conservation. Although a severely reduced population must have its ecological population size boosted before its genetic effective population size increases, changes in mating system will influence both. Any move from polygyny to monogamy will increase the number of different males siring offspring. Because effective population size is essentially the harmonic mean of the reciprocals of the number of breeding males and females, any increase in the egalitarian nature of breeding will increase effective population size. So attention to mating system will influence both important demographic and genetic characteristics of a population (see Creel, Chapter 10, this volume). Unless behavioral ecologists work closely with managers and policy makers, this knowledge will be ignored because it cannot be gathered during simple surveys or short-term studies. Mating systems tend to be species specific with subtle details revealing if and how they can change facultatively as environmental circumstances change. It is only by arming managers with this detailed knowledge that a management plan will be effective. But before employing the power of these evolutionary rules becomes a regular part of any plan, managers must learn to appreciate their importance.

Predators, Prey, Arms Races, and Ecotourism

Predators and their prey are locked in struggles in which each is selected to counter the actions of the other. As predators move through a sequence of stages from detecting to approaching, to capturing and then consuming, prey attempt to disrupt this progression by either increasing vigilance, hiding, fleeing, or aggregating. Because the effectiveness of any particular tactic exhibited by the predator or the type of disruption selected by the prey depends on the costs and benefits of each and when in the sequence it operates (Endler, 1991), no one action will be best in every circumstance. Because the approach of even the most benign human, the ecotourist, is not unlike the approach of a predator, conservation biologists should be able to learn much about how to structure wildlife viewing so as not to disturb activities as important as foraging and mating.

It is true that wild animals often habituate more easily to tourists than to natural predators, but tourists often force wildlife to override this adaptive response by "pushing their luck" to get closer and closer to their subject. Any visitor to an African game park has seen a tourist van go off road to get close enough for that full-frame photo of an otherwise resting cheetah *Acinonyx jubatus*. Repeated disturbances of this kind can have profound consequences on altering the activity budgets of these creatures who are supposedly being protected in the reserve. It seems somewhat ironic that the increasing pressures by society and governments for wildlife protection to pay for itself, often via ecotourism, might end up hurting precisely those species that need the most help. Because the alternative of eliminating ecotourism is unacceptable, understanding how to control it requires research on how different species respond to different types and levels of human interference. Because behavioral ecologists are adept at identifying how environmental forces shape antipredator, foraging, and reproductive behavior, they are in a unique position to assess when and how different types of interference will have the greatest impact.

The finding that rhino females, but not males, run long distances from approaching humans highlights this problem and suggests that unless females can be habituated, their reproductive success could be hindered by frequent tourist visits. Although males will stay put, presumably because they are defending resources females desire, they may only do so as long as females regularly frequent their ranges. If the disturbance is too great, then it is possible that even males will leave the area, and the economic gains associated with tourism could be eliminated by the unintended consequences of its own popularity.

Such a dramatic effect in such an exotic site underscores the problem, but the consequences of visitations to nature reserves near urban centers in developed nations could be even more pronounced. Fortunately, studies on the impact of human visitation are emerging, but some of the results will challenge the ability of managers to balance the needs of wildlife with those of tourists. Klein and co-workers (1995) studied the impact of visitation on the activities of 49 species of waterbirds in a Florida wildlife refuge. Their results were striking. First, bird watchers, whether traveling on foot or in vehicles, not only displaced most species from preferred feeding sites, but they did so in ways that created nonlinear, or threshold, effects; only a few disturbances were enough to cause massive departures. Displaced birds took refuge with birds already foraging farther away, intensifying competition and reducing foraging success for all. Second, vehicular disturbance had more of an impact than did visitors traveling along footpaths. In part this resulted from the fact that there were more vehicles on the roadways than there were walkers on the paths. But in addition, 96% of the vehicles stopped and disgorged their passengers at least once, thus intensifying this type of disturbance. Different species were affected differently: some such as egrets *Egretta thula,* willets *Catotrophorus semipalmatus,* and sanderlings *Calidris alba* were more sensitive, but in general, migrants were more disturbed than were residents.

In one sense, the management implications of the study are obvious: human movements must be limited. But how should this be accomplished? Should quotas be imposed, or should more sophisticated rationing be instituted by restricting viewing times or access to routes? Deciding among alternatives will require better scientific data and will necessitate that refuge managers work with behavioral ecologists to decide what the next questions should be and what sorts of data should be collected. For example, determining which species need the most protection is essential and will only be possible from further studies on how diminished short-term foraging success influences longer-term reproductive success. Where species go when disturbed and whether the impact of disturbance is the same at different points in the tidal cycle or at different times of the year also must be ascertained. Answering these questions, however, will require more detailed studies on individually recognizable birds, a trademark of virtually all behavioral ecological studies (McGregor and Peake, Chapter 2, this volume). In addition, the detailed study of the behavior of individuals must be begun on the birdwatchers themselves. Any attempt at changing the behavior of birdwatchers will require an assessment of what they do in the reserve as well as knowledge about what they are least likely to give up when pursuing an activity they truly enjoy. Because their use of the reserve helps ensure that local, regional, and national governments continue to establish and maintain them, controlling human behavior will require delicate balancing of competing needs.

Understanding what shapes the behavior of prey in response to predators is only half the problem. Understanding what makes predators effective and how their actions change as environments change could provide insights into how better to control human predation, especially in relation to fisheries. Mangel (1994b) has developed spatially explicit models that predict the search patterns and movements of predators based on two factors: 1) the likelihood of additional prey being in the immediate vicinity if prey are already present; and 2) the likelihood that prey, once disturbed by the predator's presence, will reaggregate at given distances from the predator. The product of these two factors gives the probability that there will be undisturbed prey at various distances from the predator, so a clever predator should go where this probability is highest. From this simple formulation it is easy to see that predator search will be less area restricted when the initial distribution is more even or when the likelihood of predators dispersing prey is large. Thus, as fish shoals get smaller and more widely dispersed, it is clear why operators of fishing vessels have a tendency to concentrate their efforts into smaller and smaller areas. By increasing the pressure on the fishery in this way, the fishery is potentially further destabilized. In essence, the interaction of these two factors provides a mechanism to account for the searching movements of predators that goes beyond the mere descriptive patterns that emerge from recording the locations of sitings or attacks. By developing such models, behavioral ecologists can better understand the rules underlying searching patterns and thus forecast likely areas where impacts will be high. In turn, this should help in developing effective harvesting limits, deploying enforcement effort, and determining the appropriate size of protected areas, if such a strategy is adopted.

Keystones, Behavior, and Biodiversity

Not all species in ecological communities have the same impact in determining their structure. All play roles in recycling nutrients and in directing the flows of energy. But some organize and shape their communities out of proportion to their abundance (Paine, 1966). Such species are viewed as "keystones," and many conservation biologists appreciate that identifying the existence of a keystone species in a community will make preserving the functioning of the entire ecosystem that much easier. Moreover, the process of identifying the role every species plays in contributing to the fundamental organization of the assemblage has the added benefit of identifying the degree to which guild members are ecological, or functional, equivalents (Paine, 1995). This usually requires an understanding of the details of a species' foraging behavior, and this is where behavioral ecologists, armed with the tenets of foraging theory and the ability to follow the fates of individuals, become invaluable.

At the turn of the century, fur trappers had virtually exterminated the sea otter *Enhydra lutris* over most of its range. As conservation efforts led to the protection of sea otters, the species has begun recovering, and it is not uncommon to see areas along the Pacific coast where they are abundant. Sea otters are a keystone species (Estes and Palmisano, 1974), and where they are abundant the communities they inhabit are very different from those where they are absent. Because sea otters prefer to feed on sea urchins which consume aquatic vegetation, the removal of sea urchins has a dramatic affect on increasing the abundance and diversity of aquatic vegetation. In the presence of sea otters, kelp forests flourish, and their fronds provide refuges for juvenile and adult fish.

Although the effects of keystone species are typically felt across trophic levels, they can manifest themselves within tropic levels as well. The community of mammals grazing on North America's short-grass prairie represents a dramatic example. Prairie dogs (*Cynomys* spp.) live in reproductive groups called coteries, but these coteries aggregate into colonies that can cover hundreds of hectares. At the turn of the century these colonies occupied between 40 and 100 million ha of mixed short-grass prairie (Miller et al., 1994). Today their range has been limited to around 600,000 ha, largely as a result of government-sponsored control programs designed to help the livestock industry. Original estimates of prairie dogs reducing livestock range productivity by 50–75% (Merriam, 1902) appear to be exaggerated given that the level of competition between prairie dogs and livestock is only about 4–7% (Miller et al., 1994). In fact, detailed studies of prairie dogs' foraging and social behavior show that prairie dogs tend to facilitate the grazing of bison *Bison bison* and pronghorn antelope *Antilocapra americana,* two supposed competitors (Kreuger, 1986). Experiments with exclosures, coupled with detailed observations of bite and step rates along with nearest neighbor distances, showed that bison and prairie dogs feed better in the presence of each other on the edge of extant colonies and that the nitrogen content of the shoots growing there was also elevated when compared to controls. Pronghorn antelope also generally feed more efficiently at the center of prairie dog colonies, although there was no difference between existing and poisoned colonies. Apparently the behavior of prairie dogs, by altering the soil and increasing the abundance of dicots, have long-term effects on the structure of the community of grazing mammals.

Clearly, prairie dogs increase the diversity of the vegetation that ensures the existence of an abundant population of grazers. But their impact is felt even more widely; the poisoning of prairie dogs has been cited as a cause in the decline of the specialized predators such as the black-footed ferret *Mustela nigripes,* the swift fox *Vulpes velox,* and ferruginous hawk *Buteo regalis*, as well as the mountain plover *Charadrius montanus*, which needs open, short-grass habitats for nesting (Miller et al., 1994). It may prove that monies spent on prairie dog eradication will necessitate spending additional monies to protect a host of endangered species that would otherwise be thriving.

In communities of both prairie dogs and sea otters, attention to the behavior of certain species has illuminated the critical role that each plays in organizing the species assemblage in which it lives. Despite the clarity of the implications of these studies, segments of human society that want to exploit some of the other species intertwined in the community discount, ignore, or even attempt to discredit these scientific findings. Despite the economic boon that the charismatic sea otter provides to the Californian economy and the benefit it provides by creating a nursery and refuge for fish, local fisherman blame the sea otter for the demise of the local fishery and want otter populations reduced. Similarly, special interests that thrive on livestock grazing want prairie dog eradication programs to continue despite the fact that livestock could probably benefit from prairie dogs much as bison do. Moreover, it would appear to be more cost effective to protect prairie dogs rather than to protect the many individual species whose fates are tied to that of this keystone species. This antiscience sentiment frustrates conservationist biologists and will be difficult to overcome because it rises from deeply held attitudes and values that derive from myths, history, and greed. But is activism, often with its roots in environmentalist movements, an effective way of injecting science into the process of creating and then implementing effective policy? Much depends on the tactics that activist behavioral ecologists are likely to adopt.

Conservation Biology and Activist Behavioral Ecologists

Many behavioral ecologists want to apply their science to the conservation of biodiversity. As the previous examples have shown, understanding how the tenets of behavioral ecology and the methodology of following the actions and fates of individuals can make a difference in designing effective management programs and policies. Not wanting to sit passively and wait for their insights to be discovered and then applied, behavioral ecologists want to design relevant studies from the outset in order to accelerate the process. Such activism would move behavioral ecological research up Underwood's hierarchy to "applied and environmental research," and it would bring behavioral ecologists directly into the policy arena.

Problems and Pitfalls

Although behavioral ecologists conduct comparative or experimental studies in unbiased ways, the questions asked are often shaped by personal concerns, philosophies, and values. This injection of subjectivity into the selection of a research program should not matter as long as researchers ensure that conclusions and inferences are value-free. But activist scientists have to remain on their guard and be aware of what potential biases they can bring to their work. Because environmentalists typically pursue protectionist strategies based on a system of values, not on inferences derived from scientific study, tension can be created if activist behavioral ecologists align themselves with environmentalists. If behavioral ecologists are not careful and act upon their values instead of upon scientific knowledge, they can change from being "conservation biologists" into "biological conservationists," and as environmentalists, potentially jeopardize their scientific credibility. The dilemma is then how to balance actively influencing policy while remaining a respected and credible scientist.

The problem, and its ultimate solution, stems from the nature of how science is done and what it can tell the public, managers, policy makers, and leaders about how the natural world works. Good science rests upon assumptions that can be easily evaluated. If some are wrong, or cannot be substantiated, possibly because they rest upon hidden beliefs, then the conclusions, predictions, and applications of an empirical study or a theoretical model can be discounted or even rejected. If assumptions are sound and the study is well conceived and carried out expertly, then its results are typically accepted as valid. Unfortunately, neither the results nor the process of science is easily understood or readily accepted outside the scientific community. Not even environmentalists, let alone their adversaries, automatically accept the results of good scientific research; unless conclusions support preconceived notions, they are often discredited or ignored. What then is an activist conservation biologist to do?

Because conservation biologists are at times viewed as diagnosticians and healers, improving the health of species and habitats, examining the often-cited example of physicians and their proactive attack on the tobacco industry might provide some clues. It is often argued that medical researchers and clinical physicians rightly advocate banning cigarettes on behalf of their patients' health. The facts are abundantly clear that smoking kills, now that the haze of obfuscatory "science" generated by the tobacco companies has finally been blown away. Consequently, doctors would be remiss if they did not advocate a ban on the production and sale of cigarettes. Because no uncertainty remains about cause and effect, such activism is justified and called for. But whether such bans will become policy will depend as much on the desires of a supermajority of the populace as on their acceptance of the scientific evidence. Vested economic interests and the cry that millions of jobs will be lost are powerful forces for maintaining the status quo. Even economic arguments that medical costs will rise as a result of smoke-induced illness are weakened when viewed from the alternative perspective that more people living longer lives will in the long-run generate even higher medical costs. Despite all forms of advocacy, to ensure that the results of effective science can be applied to saving lives, individual physicians will continue to have to persuade individual patients to either stop smoking or never to start. If this analogy holds and offers any insights, then activist behavioral ecologists should be free to advocate on behalf of endangered species and habitats either from the bottom-up at the local level, or from the top-down in the larger policy arena.

Strategies: Purists and Pragmatists

Although activists come in many guises, they usually divide into two camps: "front-line purists" and "behind-the-lines pragmatists." Front-line purists do what it takes to protect an endangered species or a threatened habitat and its imperilled biodiversity. Typically they believe that only fundamental change in the way the system operates will lead to effective conservation. Behind-the-line pragmatists are more likely to accommodate existing structures and support actions that are more politically and economically palatable. They accept the notion that scientists are one of many stakeholders, each with a claim to "knowing what is right." Consequently, these pragmatists believe that science is only part of the solution, that scientists do not have all the answers. Science produces knowledge about how the natural world works and proposes ways to gain new knowledge when gaps in knowledge are identified. It can predict likely outcomes of various decisions, and it can identify acceptable boundaries within which human activity will not permanently harm the survival of species or the functioning of ecosystems. But science cannot specify where on this landscape of possibilities a local community, nation, or even an international community should be. Behind-the-line pragmatists realize that other stakeholders bring to the debate different perspectives, often with an antiscience bias, whereas front-line purists have a hard time accepting the distinction between the need to do good and the notion that what might be right may involve more than scientific fact.

Behind-the-line pragmatists will join with others to solve a problem mutually. But they assume that what other stakeholders expect of them—open mindedness—must also be accorded to them, and as a result, they believe that other stakeholders will allow science to play a critical role in shaping the solution. If this does not happen and front-line purists believe that it cannot, then collaborative and cooperative activities of stakeholders will disintegrate into arguments about beliefs, with those of the most powerful triumphing. Thus it is no wonder that pessimistic behavioral ecologists believing that science will always be superseded by powerful and entrenched political and economic pressures join the ranks of front-line purists. When they do so they will be free to advocate for one course of action or another. But to retain their credibility, they must clearly identify the facts and the existing limits to knowledge. Only by doing so can they highlight the origin of the uncertainties that lead to their best guesses about what is most likely to happen and what plan of action they champion pursuing. If the distinction is not made, then best guesses might be construed to be little more than value judgments, and both the public and policy makers could become confused and lose faith in the ability of science to suggest alternative courses of action and assess their biological and economic consequences.

For behind-the-line pragmatists, belief in the efficacy of discussion and negotiation is predicated on at least a partial leveling of the playing field by gaining acceptance of the value of science. But for behind-the-line pragmatic behavioral ecologists to succeed in shaping policy, their activism must take a variety of forms and operate on many fronts. Some of these efforts will produce results that percolate up from the bottom, whereas others will flow from the top.

Tactics: Activist Ways and Means

One way activist behavioral ecologists can make a difference is to use specific knowledge about particular endangered species or the degraded environments they inhabit for their protection and that of the additional biodiversity they harbor. Typically, behavioral ecologists study species in natural areas, such as reserves, parks, or ranches where access by the local populace is limited. Because these species and their lands represent resources with economic value denied to nearby residents, it is essential that local communities be made aware of what has been learned about the behavior of the animals and the ecology of their environments. In this way they may be able to derive some economic gain by acting in nonharmful ways. If this knowledge can be used to foster low-impact ecotourism or even to design resource harvesting schemes in accordance with Clark's proviso (1976) that harvesting be limited to species whose increase in value when alive is greater than the value of money, then by sharing the profits of these enterprises, local and hence broad based bottom-up support for conservation can be created.

Pragmatic behavioral ecologists must also be advocates for the utility of science and demonstrate the important role that science can play in informing conservationist policies. Educating other stakeholders about how science operates and what it can and cannot say about how species and their ecosystems work is an essential bottom-up strategy. To do this, science must be forced out of the ivory tower and scientific journals. Popular accounts of interesting studies, whitepapers and their shortened executive summaries, and editorials are all effective. Above all, these accessible publications should not shy away from stressing the fact that scientists are skeptics and that they make assumptions that often do not agree with those of others (but are usually transparent and open for debate) and that is why scientific disagreements about predictions are common. They should also underscore the notion that uncertainty is a pervasive part of the natural world and that not coming to grips with it has led to misguided and dangerous management strategies or no management at all (Ludwig et al., 1993).

Despite the ubiquity of uncertainty, there is no need to let the call for yet another study co-opt scientists and prevent them from drawing conclusions on the basis of data in hand. In fact, the widespread appreciation of uncertainty should serve as a clarion for more research into understanding complexity per se and how to manage in the face of it (level 3, basic and strategic research). The American public already cries out for just such research when it comes to understanding the weather, perhaps the most uncertain and complex process that individuals experience. Vast sums of money are spent on satellite data gathering and computer modeling so that individuals can organize their lives despite the inaccuracy of many of the model's predictions. Rather than ignoring spectacular failures, new research projects requiring ever more sophisticated and expensive science and technology are called for. It should be every behind-the-line pragmatist's hope that the need to understand how complexity and environmental uncertainty has put so many species in peril becomes part of every human's everyday consciousness.

But education should also be directed at the powerful interests residing at the top of the decision-making pyramid. Leaders in government and industry must be made aware that investing in science and then ignoring its findings is expensive and wasteful. The utility of behavioral ecology in identifying and then rectifying environmental problems must be highlighted, and examples of failed policy must be identified. Moreover, they must be shown, as do many of the case studies described above, how better science in the form of behavioral ecology could have done a better job if only its basic tenets had been included in the research program and decision-making process from the beginning. It must be the goal of all behind-the-line pragmatists to convince policy makers that science must be used to define the scope and nature of the problem. Activist scientists must help set the research agenda, not the managers, because the types of questions they can ask are limited by predetermined goals and the severe structural constraints under which they operate. It is also essential that science be used to monitor the effectiveness, as well as the unintended consequences, of whatever policy is implemented. The success of the Montreal Protocols in accelerating the reduction of Chlorofluorocarbon emissions should serve as a model and underscores the need to include science throughout the decision-making loop.

Stakeholders, Partnerships, and Valuing Science

To effect such a seismic shift will require a major change in the value that high-level managers and decision makers accord science. Only behind-the-lines pragmatists have a chance of effecting such a shift; continuing to shout all-or-nothing polemics simply polarizes the debate and hardens already entrenched positions. Using subtle powers of persuasion and keeping the faith that change is on the way is not enough, however. Activist behavioral ecologists must continue to use their novel approach to fight shoddy science that often ignores both the assumptions that underlie their studies of the environmental forces that shape behavior as well as the insights that are derived from detailed long-term studies on individuals. In this way the effectiveness and utility of good science is put continually on display. Then the challenge becomes making other stakeholders appreciate the value of good science. Otherwise there is no point in gaining important new knowledge if it is ultimately ignored. While lip service is often paid by some stakeholders to how useful science has been in cleaning up polluted air and water, they often claim either that the task is complete and that science has served its purpose or that the cost of regulation is too high. Changing the values that control the behavior of these powerful stakeholders will not be easy and will require that behavioral ecologists learn how managerial and policy decisions are made. In this way even if the topography of the so-called playing field is not leveled, at least the location of the hills and valleys will be understood.

Entering the fray at this level is the ultimate top-down gambit. Behavioral ecologists must learn how the legislative process operates, how judicial judgments are arrived at, where both political and economic power really resides, and how values and beliefs shape the perspectives of different stakeholders and motivate their actions (level 4 research). At the moment, however, there is no recipe for success that behavioral ecologists can use, but a perfect case—the controversy of managing livestock grazing on the grasslands of western Northern America—highlights what elements must be considered when joining with other stakeholders to solve collectively thorny real-world problems.

Grasslands and Grazers: Activist Science Pursuing Effective Policy

As discussed above, prairie dogs play a major role in structuring the community of grazers as well as many other species that benefit from their activities. Detailed community and behavioral ecological studies have revealed some surprising and unanticipated findings, in particular that some presumed competitive relationships are actually mutualistic. Other studies attempting to evaluate the extent to which native bison and cattle are competitors shows that the interactions are subtle. Bison show preferences for grasses and cattle show a limited tendency to specialize on forbs, but much depends on the patchiness of the landscape, whether the grazers are completely free-roaming or confined to fenced pastures, and at what time of year the measurements are taken (Plumb and Dodd, 1993). In fact, foraging theory is proving helpful in assessing the degree to which the decision rules employed by each species make them analogous herbivores. Still other studies (Dyer et al., 1993; McNaughton, 1993; Painter and Belsky, 1993) examine the extent to which grazing at different intensities changes the quality, abundance, and species composition of grasses on grasslands.

Clearly there is much scientific data emerging on the impact that different grazers have on each other and on the plants they eat. But as behind-the-line pragmatists realize, scientists are only one of many stakeholders involved in this controversy. Native Americans and early settlers have historical roots to the land and have harvested wildlife or raised livestock for generations. Residents of small rural communities and larger urban centers earn livelihoods from supplying livestock herders. Government employees facilitate (rangeland researchers) and regulate (local managers and more distant bureaucrats) these activities, groups with special interests profit from livestock grazing, lawmakers and judges make or in.erpret the rules governing the management of grasslands, and environmentalists and activist scientists care about the wild living resources that have to cope with the livestock. All have strong opinions because each is concerned about how changing the status quo in terms of altering what species are allowed to graze where and how extensively will impact either their pocketbook or the native species that inhabit the landscape.

There can be no doubt that grazing can and does have dramatic consequences for the structure and functioning of native grasslands. Not all studies purporting to document the harmful effects of livestock grazing are of high quality or apply across the landscape (Brown and McDonald, 1995), but enough are to warrant the involvement of scientists. And as the above case studies demonstrate, behavioral ecologists have the tools to enrich the scientific quality of the debate. But what should activists do? The editor of *Conservation Biology*, Reed Noss (1994) asked whether conservation biologists should "link arms with activists in efforts to reform grazing practices." Although he did not answer his own question, the debate raged on in the pages of the journal for over a year (e.g., Brussard et al., 1994; Fleischner, 1994). As we have seen, scientists can become activists, but the tactics they adopt will depend on how much faith they have in the system's ability to use science in deciding on the best course of action.

Front-line purists will take to the streets as citizen scientists and write, lobby, speak out, and generally agitate, but they will have to ensure that they identify where factual knowledge ends and where value judgments begin when making their case if they are to remain honest scientists and maintain their credibility. If behind-the-lines pragmatists have done their job by ensuring that scientific evidence will be used in deciding among alternative plans of action, then pragmatists can illuminate the debate by showing how different ecological factors (drought, fire, soil and nutrient heterogeneity, and species-specific behavior of wildlife and livestock) affect grasslands and the species they support.

Doing this, however, places the activist behavioral ecologist on the horns of a scientific dilemma. In its present form, most off-the-shelf research (level 1) is at best of marginal use because it was produced to solve a particular problem that typically has only limited similarities to the issue at hand. Yet real-world problems as thorny as the issue of grazing and multiple land-use schemes have an urgency that precludes the luxury of tailoring research to assessing and then solving the particular problem (level 2 and 3 research). To resolve the dilemma, activist behavioral ecologists must begin creating better, more generic off-the-shelf research. One way of doing this is to modify any level 3 study and extend its usefulness by making the models or findings more general. One such example comes from a study originally designed to evaluate alternative tactics for regulating a feral horse population by assessing the consequences of each tactic on the long-term demographic and genetic structures of the population. With minimal effort it has been possible to recast the model by parameterizing both ecological and life-history features in terms of size-scaling rules so that the model can be applied to the management of virtually any population ranging from mice to elephants (Rubenstein and Dobson, in preparation).

But making it easy to employ scientific thinking and the "what-if" theorizing in stakeholder discussions is only part of the solution. Activist behavioral ecologists will also have to change the values of the other stakeholders by demonstrating that the merits of managing the grasslands to ensure their ecological integrity is at least as valuable as managing them to maximize their primary and secondary productivity. To succeed at causing such a major shift in perspective will require understanding what values motivate the different stakeholders. Perhaps the interaction of values, beliefs, and scientific knowledge will convince all stakeholders that it will be possible to alter stocking levels, adjust patterns of herd movement, and modify the use of fire and a reliance on supplemental water in ways that minimize grazing impact without causing economic hardship for those seeking to enhance economic profit. Much of the rangeland is already degraded, and economic benefits will accrue simply by reclaiming them. Clearly, money will be needed to subsidize these changes, and healthy debate about the value of using such subsidies for these purposes rather than for other worthy causes (environmental or not) will necessarily ensue. But trade-offs and compromises will only emerge if desires, beliefs, and good science are all out in the open. By understanding both the cultural perspectives of the various stakeholders and the structures and operations of various legal, legislative, and regulatory institutions, laws and policies can be amended. New agendas for research will be necessary to predict and then validate the behavioral responses that will result from managing interactions between ecological forces and processes in novel ways. Partnerships among stakeholders that respect science and the scientists that uncover useful and fundamental knowledge offer the only hope for successful collaborations that will bring about these changes. Behavioral ecologists are one group of scientists that when active will have much to say when crafting these new policies and management practices.

Summary

Behavioral ecology has much to offer in solving real-world environmental problems. First, behavioral ecology focuses on the individuals and incorporating an understanding of how the environment shapes their behavior will make models more realistic. Second, behavioral ecology is rich in "first principles" that can provide insights into how recruitment can be enhanced in endangered populations.

Applying tenets of behavioral ecology will be difficult. Not only does the science have a "split personality," being part basic and part applied, but scientists' values color their perception and action. Scientists can improve the effectiveness of their science by moving from using the off-the-shelf research for gaining insight into analogous problems to performing postmortems on management interventions as if they were experiments or even to working with managers to design new management plans as experiments. Explorations of case studies highlighting how such large-scale experiments can be analyzed in retrospect, or how they can be designed a priori to be made more effective, illustrate the importance of drawing upon principles of optimal foraging theory, life-history evolution, individual decision-making, and mating strategy.

Activist scientists have to choose a strategy. "Purists" do what it takes to have their values or scientific understanding put into practice, while "pragmatists" will work with other stakeholders to come to some mutual understanding as to what is the best outcome given a series of constraints. Both have to ensure scientific objectivity by stressing that personal values only enter the scientific process at the point where questions are posed. Pragmatists must also stress if open mindedness is expected of them, then others must be open to the dictates of scientific study.

Activist tactics are varied, but common to all are needs to educate stakeholders as to the importance and utility of science and persuasive presence of uncertainty. Educational initiatives from top-down must stress the role that science can play in defining problems and monitoring their solutions at regular intervals, while those from the bottom-up must ensure that local stakeholders benefit from the eventual action plan.

References

Allee WC, 1931. Animal aggregations: a study in general sociology. Chicago: University of Chicago Press.

Beddington JR, May RM, 1980. Maximum sustainable yields in systems subject to harvesting at more than one trophic level. Math Biosci 51:261–281.

Berger J, 1994. Science, conservation and black rhinos. J Mammal 75:298–308.

Berger J, Cunningham C, 1995. Predation, sensitivity, and sex: why female black rhinoceroses outlive males. Behav Ecol 6:57–64.

Berger J, Cunningham C, Gawuseb AA, 1994. The uncertainty of data and dehorning black rhino. Conserv Biol 8:1149–1152.

Brown JH, McDonald W, 1995. Livestock grazing and conservation on southwestern rangelands. Conserv Biol 9:1664–1647.

Brussard PF, Murphy DD, Tracy CR, 1994. Cattle and conservation—another view. Conserv Biol 8:919–921.

Bulmer M, 1984. Delayed germination of seeds: Cohen's model revisited. Theor Popul Biol 26:367–377.

Caraco T, Martindale S, Pulliam HR, 1980. An empirical demonstration of risk-sensitive foraging preferences. Anim Behav 28:820–830.

Caughley G, 1977. Analysis of vertebrate populations. New York: John Wiley and Sons.

Clark CW, 1976. Mathematical bioeconomics: the optimum management of renewable resources. New York: Wiley Interscience.

Clutton-Brock TH, Albon SD, Guinness FE, 1984. Maternal dominance, breeding success and birth sex ratios in red deer. Nature 308:358–360.

Creel S, 1996. Conserving wild dogs. Trends Ecol Evol 11:337.

Davies NB, 1989. Sexual conflict and the polygamy threshold. Anim Behav 38:226–234.

Davies NB, Houston AI, 1984. Territory Economics. In: Behavioural Ecology, 2nd Ed. (Krebs JR, Davies NB, eds). Oxford: Blackwell Scientific; 148–169.

Dobson AP, Lyles AM, 1989. The population dynamics and conservation of primate populations. Conserv Biol 3:362–380.

Dobson AP, Poole, JH, 1997. Ivory poaching and viability of African elephant populations. Conservation Biology (in press).

Dyer MI, Turner CL, Seastedt TR, 1993. Herbivory and its consequences. Ecol Appl 3:10–16.

Endler JA, 1991. Interactions between predators and prey. In: Behavioural Ecology, 3rd Ed (Krebs JR, Davies NB, eds). Oxford: Blackwell Scientific; 169–202.

Estes JA, Palmisano JF, 1974. Sea otters: their role in structuring nearshore communities. Science 185:1058–1060.

Fleischner TL, 1994. Ecological costs of livestock grazing in western North America. Conserv Biol 8:629–644.

Gillespie J, 1977. Natural selection for variances in offspring numbers: a new principle. Am Nat 111:1010–1014.

Grzimek B, Grzimek M, 1959. Serengeti shall not die. Berlin: Ullstein.

Horn HS, Rubenstein DI, 1984. In: Behavioural Ecology, 2nd Ed: (Krebs JR, Davies NB, eds). Oxford: Blackwell Scientific; 169–202.

Klein ML, Humphrey SR, Percival HF, 1995. Effects of ecotourism on distribution of waterbirds in a wildlife refuge. Conserv Biol 9:1454–1465.

Krueger K, 1986. Feeding relationships among bison, pronghorn, and prairie dogs: an experimental analysis. Ecology 67:760–770.

Lauck T, Clark CW, Mangel M, Munro GR (in press). Implementing the precautionary principle in fisheries management through marine reserves. Ecol Appl.

Lessells CM, 1991. The evolution of life histories. In: Behavioural Ecology, 3rd Ed: (Krebs JR, Davies NB, eds). Oxford: Blackwell Scientific; 32–68.

Loutit B, Montgomery S, 1994. The efficacy of Rhino dehorning: too early to tell! Conserv Biol 8:923–924.

Ludwig D, Hilborn R, Walters C, 1993. Uncertainty, resource exploitation, and conservation: lessons from history. Science: 260:17, 36.

MacArthur R, Wison EO, 1967. The theory of island biogeography. Princeton, New Jersey: Princeton University Press.

McNaughton SJ, 1993. Grasses and grazers, science and management. Ecol Appl 3:17–20.

Mangel M, 1994a. Climate change and salmonid life history variation. Deep Sea Res 41:75–106.

Mangel M, 1994b. Spatial patterning in resource exploitation and conservation. Phil Trans R Soc Lond B 343:93–98.

Merriam CH, 1902. The prairie dog of the Great Plains. In: Yearbook of the U.S. Department of Agriculture 1901. Washington, DC: U.S. Government Printing Office; 257–270.

Miller B, Ceballos G, Reading R, 1994. The prairie dog and biotic diversity. Conserv Biol 8:677–681.

Mills LS, Soule ME, Doak DF, 1993. The keystone-species concept in ecology and conservation. Bioscience 43:219–224.

Montgomery WL, 1983. Parr excellence. Nat Hist 83(6):59–64.

Noss RF, 1994. Cows and conservation biology. Conserv Biol 8:613–616.

Paine RT, 1966. Food web complexity and species diversity. Am Nat 100:65–75.

Paine RT, 1995. A conversation on refining the concept of keystone species. Conserv Biol 9:962–964.

Painter EL, Belsky AJ, 1993. Application of herbivore optimization theory to rangelands of the western United States. Ecol Appl 3:2–9.

Peters RL, Lovejoy TE, 1992. Global climate change and species diversity. New Haven, Connecticut: Yale University Press.

Plumb GE, Dodd JL, 1993. Foraging ecology of bison and cattle on a mixed prairie: implication for natural area management. Ecol Appl 3:631–643.

Real L, 1980. Fitness, uncertainty and the role of diversification in evolution and behavior. Am Nat 115:623–638.

Rubenstein DI, 1982. Risk, uncertainty and evolutionary strategies. In: Current problems in sociobiology (King's College Sociobiology Group ed.), Cambridge: Cambridge University; 91–112.

Rubenstein DI, 1992. The greenhouse effect and changes in animal behavior: effects on social structure and life-history. In: Global warming and biological diversity (Peters RL, Lovejoy TE, eds). New Haven, Connecticut: Yale University Press; 180–190.

Rubenstein DI, 1993. On the evolution of juvenile life-styles in mammals. In: Juvenile primates: life history, development and behavior (Pereira ME, Fairbanks LA, eds). New York: Oxford University Press; 38–56.

Saltz D, Rubenstein DI, 1995. Population dynamics of a reintroduced Asiatic wild ass (*Equus hemionus*) herd. Ecol Appl 5:327–335.

Schaffer WM, Gadgil MD, 1975. Selection of optimal life histories in plants. In: Ecology and evolution of communities (Cody ML, Diamond JM, eds). Cambridge, Massachusetts: Harvard University Press; 142–157.

Seger J, Brockmann J, 1987. What is bet-hedging? Oxf Surv Evol Biol 4:182–311.

Shackell NL, Willison JHM, 1995. Marine protected areas and sustainable fisheries.

Wolfville, Nova Scotia: Centre for Wildlife and Conservation Biology, Acadia University.

Stearns SC, 1976. Life-history tactics: a review of the ideas. Q Rev Biol 51:3–47.

Trivers RL, Willard DE, 1973. Natural selection of parental ability to vary sex ratio of offspring. Science 179:90–92.

Underwood AJ, 1995. Ecological research and (and research into) environmental management. Ecol Appl 5:232–247.

Weilgart L, Whitehead H, Payne K, 1996. A colossal convergence. Am Sci 84:278–287.

Epilogue

20

How Do We Refocus Behavioral Ecology to Address Conservation Issues More Directly?

Tim Caro

I began this book by asking whether principles and methods used in behavioral ecology were of conservation significance. How should a new graduate student with an interest in behavioral ecological concepts and conservation biology focus his or her research? The chapters that followed highlighted the ways in which knowledge of the adaptive significance of behavior *can* change predictions about population persistence and the outcome of management schemes. Only in a few instances, however, *did* they actually demonstrate that individuals' adaptive solutions to environmental problems altered the fate of populations or the success of conservation interventions. Now that the models and verbal arguments linking these disciplines have been laid out, the next phase of research must be to show empirically that variation in individual behavior does affect population dynamics and recovery programs. To achieve this, I see five main ways to proceed: streamline the methodology, alter the focus toward questions of greater conservation significance, study factors that affect reproductive parameters, investigate the effects of human activities on animal populations, and examine ways in which humans exploit and relate to biological resources. First, however, I briefly address the main difficulties in linking these disciplines since this acts as a springboard for making behavioral ecological work more relevant in future.

Difficulties in Linking Behavioral Ecology to Conservation Biology

Three problems were repeatedly raised by the contributors to this book. The principal difficulty in applying behavioral ecological issues to conservation biology is the length of time involved. Behavioral ecological research is labor intensive and time consuming because it involves recognizing individuals, extended periods of observation, collecting sufficient sample sizes, and often following individuals over several years to gain information on reproductive parameters. Conservation problems, on the other hand, often demand rapid solutions, either because a habitat is under imminent threat or a species is dwindling fast. Second, relatively little behavioral ecological work has been carried out on endangered species or in disturbed habitats, both of which are of interest to conservation biologists. Finally, there are still too few behavioral ecological studies of any species to make meaningful generalizations from viable to endangered populations or species; moreover, scientists are often reluctant about making extrapolations. How can we remedy these difficulties?

Streamline the Methodology

Some of the most sought-after data used in constructing minimum viable populations (MVPs) come from long-term demographic information that takes years to collect. In many situations, however, time is limited because changing circumstances or politics force quick management decisions. Often, proposals have to be put forward based on incomplete data or else they will not be considered at all. Therefore, there are strong arguments for conducting field studies over a short time frame or for quickly deriving best estimates of life history variables and reproductive parameters in the course of collecting long-term data.

Because conservation problems are increasingly viewed as being solved at a habitat rather than at a species level, especially in regard to legislative protection (Noss and Cooperrider, 1994; Meffe and Carroll, 1997), it may be more useful to conduct simultaneous studies on several species living in the same habitat rather than concentrating on one. For instance, the length and width of a corridor between protected areas would best be designed with a knowledge of dispersal rates and habitat preferences of dispersers from several terrestrial species, large and small, rather than from one flagship species (see Downes et al., 1997). Comparative information has additional usefulness if measures are standardized across species; this is relatively easy to achieve under the umbrella of a single research program.

Research Areas That Force Consideration of Conservation Problems

In advance, it is sometimes difficult to see how an investigation of individual survival and reproductive strategies could be relevant to management policy. This may be especially true for biologists beginning their first research project. One way to overcome this impasse is to work on rare taxa, as conservation questions are inevitably raised by studying endangered or threatened species. For example, Komdeur's (1992) choice of an endangered warbler on which to investigate the ecological constraints and benefits of cooperative breeding led to a greater understanding of the constraints on population growth and other aspects of small population demography (Komdeur and Deerenberg, 1997). Similarly, Laurenson's study of reproductive strategies of endangered female cheetahs *Acinonyx jubatus* (Laurenson, 1995) led to a reassessment of the importance of genetic factors in conservation (Caro and Laurenson, 1994). Second, other things being equal, information on a rare species is likely to have greater significance for conservation than the same amount of research effort directed at a common species; often the utility of this information becomes apparent only later. A good example is the biological data that Forsman and his colleagues (1984) collected on the spotted owl *Strix occidentalis*. Third, the mere presence of someone studying an endangered species or population in the field, however esoteric their research, often draws beneficial local, regional, or national attention to the population. Disadvantages of working on rare species are that they usually exist at low densities, and there may be restrictions and ethical considerations in carrying out manipulations.

Another way to make behavioral ecological research more relevant to conservation biology is to work outside protected areas where species are in some way affected by human influence. At present, most behavioral ecologists elect to study animals in relatively undisturbed habitats in order to restrict sources of variation to those of evolutionary significance. Yet it is increasingly recognized that apparently pristine ecosystems are subject to human disturbance. For example, Hofer et al. (1993) noted that up to 10% of spotted hyena *Crocuta crocuta* mortality in the Serengeti National Park resulted from hyenas being caught in poachers' snares set for other species. By devising sampling that incorporates several sites that include multiple-use areas or agricultural land, behavioral ecologists will be forced to consider the influence of factors such as competition from domestic stock, restricted foraging opportunities, pollution, or avoidance of human predators when they try to understand behavioral decision making and its consequences.

Whether these pieces of advice are heeded or not, behavioral ecologists need to be more circumspect in presenting baseline demographic data. For example, Greene et al. (chapter 11, this volume) had great difficulty in finding published life tables for African ungulates, despite numerous studies on these species. As a matter of course, information on the mean and variance in group size, reproductive success for both sexes, the size and overlap of home ranges, rates of dispersal, and frequencies of intra- and interspecific interactions should be presented in publications as soon as they become available because these data are vital in predicting time to extinction, in predicting loss of genetic diversity, and in devising monitoring strategies (Caro and Durant, 1995).

Behavioral Factors That Affect Population Growth Rate and N_e

Behavioral ecological findings are particularly relevant to conservation of species when the behavior of individuals affects a population's growth rate. Thus, if social behavior affects age at first reproduction (e.g., reproductive suppression) or aspects of reproductive rate such as interbirth interval (e.g., dominance), litter size (e.g., life history trade-offs), or offspring sex ratio (e.g., age or dominance), or juvenile or adult survivorship (e.g., intraspecific competition), then the nature of such behavior and the circumstances under which it appears are directly relevant to population persistence. Indeed, the impact of adaptive behavior on a population's growth rate is a central theme that emerges from this book. For example, Vincent and Sadovy (Chapter 9) pointed out that sex ratio and age at maturity depend on social structure in some fish species, Rubenstein (Chapter 19) documented adaptive sex ratio manipulation in onagers *Equus hemionus,* and Greene et al. (Chapter 11) modeled the importance of infanticide in reducing offspring production.

Behavioral factors are also relevant to conservation when social behavior affects N_e by influencing the ratio of breeding males to females and variance in family size (Creel, Chapter 10, this volume). Thus the factors that affect the operational sex ratio, variation in reproductive success in males and females, and phonemena such as siblicide and infanticide can all affect N_e. Ecological factors causing variation in effective breeding population size over time also have potential impact on N_e. For an incoming student, then, the investigation of the circumstances under which social behavior affects population growth parameters or N_e will be a profitable avenue for research.

Effects of Human Disturbance

Missing from many analyses presented in this book are direct observations of individuals' responses to anthropogenic disturbance (but see Eadie et al., Chapter 12, and Harcourt, Chapter 3). For conservation purposes, we need to examine behavioral responses to rapidly changing environments (level 2, Underwood, 1992; see Rubenstein, Chapter 19, this volume). For example, planners can assess the resilience of species to habitat and population alterations with a knowledge of the flexibility of foraging behavior, demography, and dispersal (Weaver et al., 1996; Wolff et al., 1997). Why should behavioral ecologists be interested in responses to environmental change? Certainly, the speed with which a species adapts to new evolutionary pressures such as predators or parasites determines who wins an evolutionary arms race. As examples, extensive observational and experimental work has investigated the match between antipredator adaptations and predator distributions (Goldthwaite et al., 1990) and the rate at which host bird species come to reject nest parasites over large time scales (Rothstein, 1990). Another, more general answer is that, because most natural populations are now being perturbed by anthropogenic forces, behavioral ecologists can no longer afford to ignore the effects of such changes either conceptually or methodologically (see Stamps and Buechner, 1985). For example, in cichlid fish species in Lake Victoria, increased turbidity caused by people has interfered with mate choice and relaxed sexual selection, and has blocked reproductive isolation, all of which depend on fish being able to view conspecific coloration (Seehausen et al., 1997). Arguments over the rate of adaptive change have beleagured evolutionary studies of human behavior to the extent of causing the field to split into evolutionary psychologists and behavioral ecological anthropologists (Borgerhoff Mulder et al., 1997; Sherman and Reeve, 1997). To avoid such arguments, behavioral ecologists studying nonhumans need to marshall empirical evidence to document adaptive rates of change to ecological perturbations.

Anthropogenic disturbance comes in many forms ranging from encroachment to development and may even include disturbance due to conservation interventions. Some studies have started to examine effects of land clearance on life history variables and reproduction. For example, Martin and Clobert (1996) have argued that widespread and sustained reduction of forest cover in Europe over 5000 years has given birds sufficient time to adapt to living in cleared areas where nest predation is low; in contrast, North American birds still largely nest in forests where nest predation is high. In accordance with life history theory (Charlesworth, 1980), fecundity of European songbirds is higher than North American avifauna after controlling for body size and phylogeny. Some studies, just underway, are investigating the effects of hunting on animal social structure and reproductive rates. Preliminary data suggest that severe disruption may occur at even moderate levels of offtake (see Haber, 1996). Pronghorn antelope *Antilocapra americana* shift from resource-defense to harem-defense polygyny in response to density changes due to hunting pressure (Byers and Kitchen, 1988). More dramatically, tusklessness in elephants *Loxondata africana* is increasing in heavily poached populations, with unknown consequences for mate choice and the mating system (Jackmann et al., 1995).

Studies of behavioral changes in response to human activity are well documented in wildlife biology literature. Grizzly bears *Ursus arctos* avoid open roads (McLellan and Shackleton, 1988), and wolves *Canis lupus* recolonizing Wisconsin select areas with low road density (Mladenoff et al., 1995). Nevertheless, the adaptive consequences of such behavior, in terms of, say, reproduction, are normally undocumented, as are the ramifications of such behavior on other species (but see Isbell and Young, 1993).

Regarding the effects of conservation strategies, some studies have examined the impact of photographic tourism on animal behavior and distributions, especially in birds (see Burger and Gochfield, 1991; Klein et al., 1995; Rodgers and Smith, 1995; Johns, 1996), but, again, consequences of changes in behavior for individual reproduction and survival are usually unknown (but see Anderson and Keith, 1980).

A special instance of human disturbance is when researchers handle animals to monitor them. Aside from ethical concerns involved in handling (Cuthill, 1991), additional questions are raised when endangered populations are involved. Considerable controversy has arisen over the issue of whether wild dogs *Lycaon pictus* suffer increased mortality as a result of being radio collared and vaccinated (Burrows et al., 1994; Ginsberg et al., 1994; De Villiers et al., 1995). Unfortunately, few attempts have been made to assess the effects of different types of monitoring and invasive methodology on endangered species (but see Harrison et al., 1991; Laurenson and Caro, 1994); I expect this will become a prerequisite of endangered species studies in the future.

Behavioral Ecology of Human Exploitation

Human exploitation of animals and plants has been studied in many traditional and premodern societies (Redford and Padoch, 1992; Western et al., 1994; Redford and Mansour, 1997), and some generalizations are beginning to emerge. First, animal exploitation in Neotropical habitats may be sustainable for savannah-dwelling, large-bodied mammals with high rates of increase, but species living in forest that have lower population sizes are unlikely to support sustained hunting pressure (Robinson and Redford, 1991; Bodmer et al., 1997). Second, in Neotropical forests, extraction of nontimber products has been found to be less profitable in the short term than several other forms of land use (Browder, 1992). For example, in areas of high biodiversity where certain fruiting trees are uncommon, exploitation becomes difficult (Salafsky et al., 1993). Third, the economics and type of extraction are subject to market forces beyond local control (Schwartzman, 1989). When commercial markets for fruits in Neotropical forests first appear, prices increase and methods of harvesting become increasingly destructive (Vasquez and Gentry, 1989). Later, if widespread commercial production takes off elsewhere, the value of forest products may fall rapidly. As a result of these considerations, there is now an emerging consensus that plant extractive reserves can only serve as one of several remedial efforts directed toward the conservation of Neotropical forests.

Differences in types of multiple-use areas and modes of extraction nevertheless call for additional studies of resource use in the tropics to understand the biological factors and economic circumstances under which patterns of exploitation can be sustainable. The former include exploited populations' intrinsic rates of increase, reproductive responses to offtake, and behavioral avoidance of hunters, whereas the latter encompass use of more efficient methods of extraction and regulating increased demand following establishment of new markets.

Turning to human behavior in another conservation context, studies of nonconsumptive tourism have been initiated that examine people's effects on landscapes, culture, and biological communities (e.g., Harrison, 1992). How such consequences might influence decisions that ecotourists make are poorly known, however. We know even less about the behavior of consumptive tourists such as hunters and fishermen.

We also need to know the trade-offs that underlie people's willingness to overexploit or conserve resources. Can they be influenced by education (Caro et al., 1994) or by ameliorating their economic situation, and how quickly can this occur? Our proclivities to exploit or conserve must have been shaped by selection (Wilson et al., chapter 18, this volume), and it is sad that we understand so little about the biological origins and maintenance of our destructive predispositions as we watch the outcome of these selective processes wreak havoc across the globe.

Conclusion

The chapters in this book suggest that behavioral ecologists in academic institutions have a central role to play in advancing conservation theory and solving conservation problems (see also Arcese et al., 1997; Beissinger, 1997). The chapters additionally suggest that conservation biologists need to take greater note of the way in which animals find adaptive solutions to social and environmental problems in constructing their models and management plans. To some extent this will demand altering priorities. Behavioral ecologists will need to put more time into practical concerns rather than producing publications, and conservation biologists will have to scour the literature more thoroughly. If this is to occur, however, the consequent sacrifices must be recognized by facilitators in academia and conservation, by deans, managers, granting agencies, and politicians. They too must allow more time and energy for productive interactions because ultimately output, and recognition of that productivity, should be measured in terms of rescuing species rather than papers and management plans produced.

Summary

The chapters in this book demonstrate how behavioral ecology can be relevant to conservation biology; the next phase of research must be to show that it is. There is a discrepancy between the time it takes to carry out a field study and the speed at which conservation problems must be solved.

Links between the disciplines can be strengthened by conducting field studies over shorter frames and on several species simultaneously. Research should be carried out on rare taxa or in disturbed habitats. Ecological or social factors are important topics of study because they affect population growth rates and N_e. Observations of individual animals' reproductive responses to anthropogenic disturbance need systematic investigation. Finally, studies of resource use by people can inform us whether multiple-use areas are effective.

Politicians and granting organizations need to alter their criteria for judging career success in order to steer effort toward solving conservation problems.

Acknowledgments I thank Steve Albon and Joel Berger for comments.

References

Anderson DW, Keith JO, 1980. The human influence on seabird nesting success: conservation implications. Biol Conserv 18:65–80.

Arcese P, Keller LF, Cary JR, 1997. Why hire a behaviorist into a conservation or manage-

LIBRARY, UNIVERSITY OF CHESTER

ment team? In: Behavioral approaches to conservation in the wild. (Clemmons JR, Buchholz R, eds). Cambridge: Cambridge University Press; 48–71.

Beissinger SR, 1997. Integrating behavior into conservation biology: potentials and limitations. In: Behavioral approaches to conservation in the wild. (Clemmons JR, Buchholz R, eds). Cambridge: Cambridge University Press; 23–47.

Bodmer RE, Eisenberg JF, Redford KH, 1997. Hunting and the likelihood of extinction of Amazonian mammals. Conser Biol 11:460–466.

Borgerhoff Mulder M, Richerson PJ, Thornhill NW, Voland E, 1997. The place of behavioral ecological anthropology in evolutionary social science. In: Human by nature: between biology and the social sciences (Weingart P, Mitchell SD, Richerson PJ, Maasen S, eds). Mahwah, New Jersey: Lawrence Erlbaum Associates; 253–282.

Browder JO, 1992. The limits of extractivism. Bioscience 42:174–182.

Burger J, Gochfeld M, 1991. Human distance and birds: tolerance and response distances of resident and migrant species in India. Environ Conserv 18:158–165.

Burrows R, Hofer H, East ML, 1994. Demography, extinction and intervention in a small population: the case of the Serengeti wild dogs. Proc R Soc Lond B 256:281–292.

Byers JA, Kitchen DW, 1988. Mating system shift in a pronghorn population. Behav Ecol Sociobiol 22:355–360.

Caro TM, Durant SM, 1995. The importance of behavioral ecology for conservation biology: examples of Serengeti carnivores. In: Serengeti II: Dynamics, management, and conservation of an ecosystem (Sinclair ARE, Arcese P, eds). Chicago: University of Chicago Press; 451–472.

Caro TM, Laurenson MK, 1994. Ecological and genetic factors in conservation: a cautionary tale. Science 263:485–486.

Caro TM, Pelkey N, Grigione M, 1994. Effects of conservation biology education on attitudes toward nature. Conserv Biol 8:846–852.

Charlesworth B, 1980. Evolution in age-structured populations. Cambridge: Cambridge University Press.

Cuthill I, 1991. Field experiments in animal behavior: methods and ethics. Anim Behav 42:1007–1014.

De Villiers MS, Meltzer DGA, van Heerden J, Mills MGL, Richardson PRK, van Jaarsveld AS, 1995. Handling-induced stress and mortalities in African wild dogs (*Lycaon pictus*). Proc R Soc Lond B 262:215–220.

Downes SJ, Handasyde KA, Elgar MA, 1997. The use of corridors by mammals in fragmented Australian eucalypt forests. Conser Biol 11:718–726.

Forsman ED, Meslow EC, Wight HM, 1984. Distribution and biology of the spotted owl in Oregon. Wildl Monogr 87:1–64.

Ginsberg JR, Alexander KA, Creel S, Kat PW, McNutt JW, Mills MGL, 1994. Handling and survivorship of African wild dog (*Lycaon pictus*) in five ecosystems. Conserv Biol 9:665–674.

Goldthwaite RO, Coss RG, Owings DH, 1990. Evolutionary dissipation of an antisnake system: differential behavior by California and arctic ground squirrels in above- and below-ground contexts. Behaviour 112:246–269.

Haber GC, 1996. Biological, conservation, and ethical implications of exploiting and controlling wolves. Conserv Biol 10:1068–1081.

Harrison D (ed), 1992. Tourism and the less developed countries. London: Bellhaven Press.

Harrison S, Quinn JF, Baughman JF, Murphy DD, Ehrlich PR, 1991. Estimating the effects of scientific study on two butterfly populations. Am Nat 137:227–243.

Hofer H, East M, Campbell KLI, 1993. Snares, commuting hyaenas, and migratory herbivores: humans as predators in the Serengeti. Symp Zool Soc Lond 65:347–366.

Isbell LA, Young TP, 1993. Human presence reduces predation in a free-ranging vervet monkey population in Kenya. Anim Behav 45:1233–1255.

Jackmann H, Berry PSM, Imae H, 1995. Tusklessness in African elephants: a future trend. Afr J Ecol 33:230–235.

Johns BG, 1996. Responses of chimpanzees to habituation and tourism in the Kibale forest, Uganda. Biol Conserv 78:257–262.

Klein ML, Humphrey SR, Percival HF, 1995. Effects of ecotourism on distribution of waterbirds in a wildlife refuge. Conserv Biol 9:1454–1465.

Komdeur J, 1992. Importance of habitat saturation and territory quality for evolution of cooperative breeding in the Seychelles warbler. Nature 358:493–495.

Komdeur J, Deerenberg C, 1997. The importance of social behavior studies for conservation. In: Behavioral approaches to conservation in the wild (Clemmons JR, Buchholz R, eds). Cambridge: Cambridge University Press; 262–276.

Laurenson MK, 1995. Behavioral costs and constraints of lactation in free-living cheetahs (*Acinonyx jubatus*). Anim Behav 50:815–826.

Laurenson MK, Caro TM, 1994. Monitoring the effects of non-trivial handling in free-living cheetahs. Anim Behav 47:547–557.

McLellan BN, Shackleton DM, 1988. Grizzly bears and resource extraction industries: effects of roads on behavior, habitat use and demography. J Appl Ecol 25:451–460.

Martin TE, Clobert J, 1996. Nest predation and avian life-history evolution in Europe versus North America: a possible role for humans? Am Nat 147:1028–1046.

Meffe GK, Carrol CR, 1997. Principles of conservation biology, 2nd ed. Sunderland, Massachusetts: Sinauer Associates.

Mladenoff DJ, Sickley TA, Haight RG, Wydeven AP, 1995. A regional landscape analysis and prediction of favorable gray wolf habitat in the northern Great Lakes region. Conserv Biol 9:279–294.

Noss RF, Cooperrider AY, 1994. Saving nature's legacy: protecting and restoring biodiversity. Washington, DC: Island Press.

Redford KH, Mansour J, 1997. Traditional peoples and biodiversity conservation in large tropical landscapes. Covelo, California: Island Press.

Redford KH, Padoch C (eds), 1992. Conservation of neotropical forests: working from traditional resource use. New York: Columbia University Press.

Robinson JG, Redford KH (eds), 1991. Neotropical wildlife use and conservation. Chicago: University of Chicago Press.

Rodgers JA Jr, Smith TH, 1995. Set-back distances to protect nesting bird colonies from human disturbance in Florida. Conserv Biol 9:89–99.

Rothstein SI, 1990. A model system for coevolution: avian brood parasitism. Annu Rev Ecol Syst 21:481–508.

Salafsky N, Dugelby BL, Terborgh JW, 1993. Can extractive reserves save the rain forest? An ecological and socioeconomic comparison of nontimber product extraction systems in Peten, Guatemala and West Kalimatan, Indonesia. Conserv Biol 7:39–52.

Schwartzman S, 1989. Extractive reserves: the rubber tappers' strategy for sustainable use of the Amazon rain forest. In: Fragile lands in Latin America: the search for sustainable uses (Browder J, ed). Boulder, Colorado: Westview Press; 150–165.

Seehausen O, van Alphen JJM, Witte F, 1997. Cichlid fish diversity threatened by eutrophication that curbs sexual selection. Science 277:1808–1811.

Sherman PW, Reeve HK, 1997. Forward and backward: alternative approaches to studying human social evolution. In: Human nature: a critical reader (Betzig L, ed). New York: Oxford University Press; 147–158.

Stamps JA, Buechner M, 1985. The territorial defense hypothesis and the ecology of insular vertebrates. Q Rev Biol 60:155–181.

Underwood AJ, 1992. Ecological research and (and research into) environmental management. Ecol Appl 5:232–247.

Vasquez R, Gentry AW, 1989. Use and misuse of forest-harvested fruits in the Iquitos area. Conserv Biol 3:350–361.

Weaver JL, Paquet PC, Ruggiero LF, 1996. Resilience and conservation of large carnivores in the rocky mountains. Conserv Biol 10:964–976.

Western D, Wright RM, Strum SC (eds), 1994. Natural connections: perspectives in community-based conservation. Washington, DC: Island Press.

Wolff JO, Schauber EM, Edge WD, 1997. Effects of habitat loss and fragmentation on the behavior and demography of gray-tailed voles. Conserv Biol 11:945–956.

Taxonomic Index

Subject Index

Printed in the United Kingdom
by Lightning Source UK Ltd.
105548UKS00001B/118